Biomechanics of the Musculo-skeletal System

Third Edition

Biomechanics of the Musculo-skeletal System

Third Edition

Editors

BENNO M. NIGG

WALTER HERZOG

University of Calgary
Calgary, Alberta, CANADA

Telephone (+44) 1243 779777

Email (for orders and customer service enquiries): cs-books@wiley.co.uk
Visit our Home Page on www.wileyeurope.com or www.wiley.com

Other Wiley Editorial Offices

John Wiley & Sons Inc., 111 River Street, Hoboken, NJ 07030, USA

Jossey-Bass, 989 Market Street, San Francisco, CA 94103-1741, USA

Wiley-VCH Verlag GmbH, Boschstr. 12, D-69469 Weinheim, Germany

John Wiley & Sons Australia Ltd, 33 Park Road, Milton, Queensland 4064, Australia

John Wiley & Sons (Asia) Pte Ltd, 2 Clementi Loop #02-01, Jin Xing Distripark, Singapore 129809

John Wiley & Sons Canada Ltd, 6045 Freemont Blvd., Mississauga, Ontario, Canada L5R 4J3
Wiley also publishes its books in a variety of electronic formats. Some content that appears in print may not be available in electronic books.

Library of Congress Cataloging-in-Publication Data

(Applied for)

British Library Cataloguing in Publication Data

A catalogue record for this book is available from the British Library

ISBN - 13 978-0- 470- 01767-8 (HB)

Produced from pdf files supplied by the authors
Printed and bound in Great Britain by Antony Rowe Ltd, Chippenham, Wiltshire
This book is printed on acid-free paper responsibly manufactured from sustainable forestry in which at least two trees are planted for each one used for paper production.

TABLE OF CONTENTS

CONTRIBUTORS

Authors

Steven K. Boyd
Department of Mechanical and Manufacturing Engineering
University of Calgary
2500 University Drive, NW
Calgary, Alberta, CANADA, T2N 1N4

Walter Herzog
Human Performance Laboratory
Faculty of Kinesiology
The University of Calgary
2500 University Drive NW
Calgary, Alberta, CANADA, T2N 1N4

Benno M. Nigg
Human Performance Laboratory
Faculty of Kinesiology
The University of Calgary
2500 University Drive NW
Calgary, Alberta, CANADA, T2N 1N4

Nigel G. Shrive
Department of Civil Engineering
The University of Calgary
2500 University Drive NW
Calgary, Alberta, CANADA, T2N 1N4

Gail M. Thornton
Departments of Surgery and Civil Eng.
The University of Calgary
3330 Hospital Drive NW
Calgary, Alberta, CANADA, T2N 4N1

Ton J. van den Bogert
Department of Biomechanical Engineering
Cleveland Clinic Foundation
9500 Euclid Avenue
Cleveland, Ohio, USA, 44195

Vinzenz von Tscharner
Human Performance Laboratory
Faculty of Kinesiology
The University of Calgary
2500 University Drive NW
Calgary, Alberta, CANADA, T2N 1N4

Ronald F. Zernicke
Human Performance Laboratory
Faculty of Kinesiology
The University of Calgary
2500 University Drive NW
Calgary, Alberta, CANADA, T2N 1N4

Co-authors

Gerald K. Cole
Human Performance Laboratory
Faculty of Kinesiology
The University of Calgary
2500 University Drive NW
Calgary, Alberta, CANADA, T2N 1N4

Salvatore Federico
Human Performance Laboratory
Faculty of Kinesiology
The University of Calgary
2500 University Dr. N.W.
Calgary, Alberta, CANADA, T2N 1N4

Cy B. Frank
Department of Surgery
The University of Calgary
3330 Hospital Drive NW
Calgary, Alberta, CANADA, T2N 1N4

Stefan Judex
Department of Mechanical Engineering
The University of Calgary
2500 University Drive NW
Calgary, Alberta, CANADA, T2N 1N4

Caeley Lorincz
Human Performance Laboratory
Faculty of Kinesiology
The University of Calgary
2500 University Drive NW
Calgary, Alberta, CANADA, T2N 1N4

Ian C. Wright
Human Performance Laboratory
Faculty of Kinesiology
The University of Calgary
2500 University Drive NW
Calgary, Alberta, CANADA, T2N 1N4

For further information contact the Human Performance Laboratory,
University of Calgary, Alberta, CANADA

Tel: (403) 220-3436
Fax: (403) 284-3553

PREFACE

This is the third edition of Biomechanics of the Musculo-skeletal System. Developments and changes in the knowledge base of biomechanics made this new edition necessary. Major changes have been made in chapter 2 (Biological Materials) and chapter 3 (Measuring Techniques). The development of biomechanics in these areas was substantial in the last 10 years and changes in the book recognize this progress. Content of other chapters was minimally changed.

Biomechanics of the Musculo-skeletal System is intended for students and researchers in biomechanics. In writing this book we have assumed that the reader has a basic background in mathematics and mechanics. Through this book, the authors hope to stimulate growth and development in the exciting field of biomechanics.

At present, thousands of researchers work in areas that are included under the umbrella of the term biomechanics. The research interests among biomechanists differ greatly. This book focuses on the biomechanics of the musculo-skeletal system and it contains chapters on specifically selected areas:

Definition and Historical Highlights
Biological Materials
Measuring Techniques
Modelling

Chapter 1 contains a definition of biomechanics followed by a section on selected historical highlights. Ideas, thoughts, and concepts about the human body and its movement existed in ancient Greek and Roman times, and were at the centre of interest for science and art through the centuries. Important contributions and historical developments are discussed within the socio-cultural environment, since historical developments took place within, and were determined by, the framework of the cultural conditions of that time.

Chapter 1 also includes a summary of basic principles in mechanics as they relate to the musculo-skeletal system. This section begins with the general equations of three-dimensional motion. The variables in these equations are explained, and a summary on how these variables can be determined experimentally or theoretically is given. This chapter does not, however, provide the information typical in an advanced textbook in mechanics.

Chapter 2, Biological Materials, discusses the morphology, histology, and mechanics of bone, cartilage, ligament, tendon, and muscle as they relate to biomechanics. Selected injury and injury-repair mechanisms of the different biological materials are described and compared. Additionally, biomechanical properties are discussed as a function of age, gender, and activity. Chapter 2 emphasizes the mechanical aspects of the biological materials.

Chapter 3, Measuring Techniques, describes selected methods for quantifying biomechanical data experimentally. This chapter includes sections on force measurements, accelerometry, measurement of motion with optical methods, electromyography, strain measurements, and discusses methods for determining inertial parameters of the human or animal body. The primary goals of this chapter are to explain the principles involved in the experimental techniques, and to compare the different techniques. Little emphasis is given to technical details since they are subject to rapid change.

Chapter 4, Modelling, starts with a philosophical discussion of the importance and the place of mathematical modelling in biomechanics. Then the concept of force system analysis is discussed. The sections in this chapter are extensive because of the difficulty of finding publications which introduce modelling in biomechanics. The section on energy considerations is an introduction to thermomechanical systems, with some considerations given to the energetics of contraction of skeletal muscle. The last section of chapter 4, discusses simulation as a tool of biomechanical research.

All authors worked at one time at the University of Calgary, Canada. In addition to collaborating on this publication, the authors were involved in many joint research projects, providing their particular expertise in specific areas of biomechanics. Many examples in the book are taken from the authors' own work. The attempt was made, however, to distinguish between the general principles and the specific examples.

The various drafts of the book chapters or sections have been given to many colleagues for review in the process of writing. We gratefully acknowledge the helpful and valuable contributions of the following principal reviewers.

J. Andrews (The University of Iowa, USA); M.F. Bobbert (Free University Amsterdam, Netherlands); J. Denoth (ETH Zürich, Switzerland); J. Falck (The University of Calgary, Canada); G. Gautschi (KISTLER Instruments, Winterthur, Switzerland); J.G. Hay (The University of Iowa, USA); R. Lieber (The University of California, USA); K. Nicol (University of Münster, Germany); R. Norman (The University of Waterloo, Canada); J.P. Paul (The University of Strathclyde, Scotland); M.R. Yeadon (Loughborough University, UK).

Many co-workers of the different biomechanics research groups at the University of Calgary have contributed intellectually to the development of the final product. They provided critical comments, discussions, and were involved in the internal review process. We are indebted to them for their help in shaping the current form of the book. They include:

Allinger, Todd
Anton, Michael
Baroud, Gamal
Bergman, Joseph
Coza, Aurel
Damson, Erich
de Koning, Jos
Donoghue, Orna
Forcinito, Mario
Hamilton, Gordon
Leonard, Timothy
Liu, Wen
Mattar, Johnny
Miller, Janice
Morlock, Michael
Naumann, Andre
Neptune, Rick
Nurse, Matthew
Pitzel, Barb
Prilutzky, Boris
Reinschmidt, Christoph
Ronsky, Janet
Stefanyshyn, Darren
Stergiou, Pro
Vaz, Marco
Wilson, Jackie

Additionally, we are indebted to many co-workers at the University of Calgary for their help in preparing different aspects of the book: Dale Oldham, Doug Woeppel and Colin Roeke for the preparation of the figures; Karen Turnbull and Micki Kosman for the scientific editing of the text; Glenda McNeil and Byron Tory for their help in computer-related aspects; Ursula Heinz and Veronica Fisher for their help in many administrative capacities; Aurel Coza for technical support.

The third edition uses the same cover as the second edition, but a different background color. The graphical background and the runner in various biomechanical appearances was created by Adidas Research and Innovation. They (Dr. S. Luthi and his group) have allowed us to use their design for the book cover, which is greatly appreciated.

Textbooks usually contain (often stupid) errors and inconsistencies. The authors welcome communications from students and colleagues about concerns or errors found in this third edition of Biomechanics of the Musculo-skeletal System or any other information that may make possible future editions a better product.

Calgary, September, 2006 Benno M. Nigg
Walter Herzog

1 INTRODUCTION

Biomechanics is a discipline of science that is newly developed and becoming established. Two developments are often seen in such a situation: first, scientists attempt to define the new discipline, and second, scientists attempt to understand the roots and the historic development of their new discipline. This introductory chapter provides information on these two aspects. A proposed definition of biomechanics is introduced and selected historical facts are discussed, describing their possible relevance to the development of modern biomechanics. Since biomechanics is a young scientific discipline, historical highlights are not well defined, opening many avenues for discussion and speculation. The reader may want to digest this chapter with care. Finally, a short comment about mechanics is presented, emphasizing how the different variables in the equations of motion can be determined.

1.1 DEFINITION OF BIOMECHANICS

NIGG, B.M.

Our undying curiosity concerning our own and other species' biological functioning dates back as far as any other scientific attempt to probe specific aspects of life. Initial studies included attempts at understanding the function of the lung, heart, peripheral and central nervous systems, muscles, and joints. One discipline born of this early exploration was the study of the mechanics of living systems; a science we today call *biomechanics.* The definition of biomechanics has not yet achieved a consensus. Some authors propose broad definitions, others propose rather limited ones. This book uses an adaptation of Hay's (1973) definition of biomechanics:

Biomechanics is the science that examines forces acting upon and within a biological structure and effects produced by such forces.

External forces acting upon a system are quantified using sophisticated measuring devices. Internal forces, which result from muscle activity or external forces, or both, are assessed using implanted measuring devices or estimations from model calculations. Possible results of external and internal forces are:

- Movements of segments of interest,
- Deformation of biological material, and
- Biological changes in the tissue(s) on which they act.

Consequently, biomechanical research studies or quantifies, or both:

- Movement of different body segments and factors that influence movement, e.g., body alignment, weight distribution, equipment,
- Deformation of biological structures and factors that influence their deformity, and
- Biological effects of locally acting forces on living tissue; effects such as growth and development, or overload and injuries.

The definition suggested by Hay is analogous to the one proposed by Hatze (1971). Hatze stated: "Biomechanics is the science which studies structures and functions of biological systems using the knowledge and methods of mechanics" (my rather liberal translation from German).

Biomechanical research addresses several different areas of human and animal movement. It includes studies on (a) the functioning of muscles, tendons, ligaments, cartilage, and bone, (b) load and overload of specific structures of living systems, and (c) factors influencing performance. Subjects of study for human biomechanics include the disabled and the non-disabled, athletes, and non-athletes. Subjects range in age from children through to the elderly.

1.2 SELECTED HISTORICAL HIGHLIGHTS

NIGG, B.M.

This chapter discusses factors that are considered important in the development of biomechanics over the centuries. The facts of mechanical or biological scientific development are discussed within the socio-cultural framework in which they emerged. The facts have been researched carefully. The interpretation, however, is a personal view, and some readers may not agree with some of the speculations and conclusions that are made.

The historical highlights have been arbitrarily divided into periods which, for the development of biomechanics, have specific importance in the view of the author. These periods are:

Antiquity	650 B.C. to 200 A.D.
Middle Ages	200 B.C. to 1450 A.D.
Italian Renaissance	1450 A.D. to 1600 A.D.
Scientific Revolution	1600 A.D. to 1730 A.D.
Enlightenment	1730 A.D. to 1800 A.D.
The Gait Century	1800 A.D. to 1900 A.D.
The Twentieth Century	1900 A.D. to . . . .

These periods do not have a well defined beginning and end. The years indicated, therefore, are just an approximate guideline, allowing the reader to locate the different periods.

1.2.1 THE SCIENTIFIC LEGACY OF ANTIQUITY

Ancient cultures such as the Maya, Egyptians, Mesopotamians, and Phoenicians had well-developed knowledge in many scientific disciplines, including astronomy and mathematics. Their attempts to understand nature combined knowledge and myth. The ancient Greeks were the first to attempt to separate knowledge from myth, developing what we would call today true scientific inquiry. The emergence and development of Greek science from the sixth to the fourth century B.C. was influenced by the Greeks' relative freedom from political and religious restrictions. Their coastal harbors and inland caravan routes, which provided access to trade with Egyptians, Phoenicians, and the whole of Asia, also provided the freedom necessary to inspire their ideas. The politics, commerce, and climate of Greece created leisure time, which the Greeks used to pursue mental and physical excellence, enhancing the opportunity for scientific discovery. Harmony of mind and body required athletic activity to complement the pursuit of knowledge. Kinematic representations of Greek athletics dominated artistic media, demonstrating an interest in sport and human movement. Greek artists developed increasingly sophisticated methods and media for depicting motion. Greek sculpture illustrated the dynamics of movement and a working knowledge of surface anatomy.

Scientific development initiated by the Greeks can be characterized by four stages:

- The emergence of natural philosophy,
- The golden age of Greek science,
- The Hellenistic age, and
- The conquest by the Roman Empire.

THE EMERGENCE OF NATURAL PHILOSOPHY

The *natural philosophy* movement was founded by Thales (624-545 B.C.). It was based on the study of natural science detached from religion. During this period, Greek science developed the basic elements of mathematics, astronomy, mechanics, physics, geography, and medicine. Two personalities of particular interest in the context of this book are Pythagoras and Hippocrates.

Pythagoras (about 582 B.C.), born in a time of awakening, formed a brotherhood based on his philosophical and mathematical ideas, which included both men and women living together in a communal atmosphere. His disciples were divided into two hierarchical groups: the scientists and the listeners. The scientists were permitted to ask the master questions and to postulate their own ideas, while the listeners were confined primarily to study.

According to Pythagoras "...all things have form, all things are form, and all forms can be defined by numbers" (Koestler, 1968). His arithmetic was based on dots that could be drawn in the sand or with pebbles to depict geometrical shapes. Numbers like 3, 7, and 10 formed triangles, while 4, 9, and 16 formed squares. Some numbers, such as 10, were given mythical qualities. Experimentation with these shapes and numbers may have resulted in Pythagoras' famous theorem for rectangular triangles: $a^2 + b^2 = c^2$. His definitions of the universe and the human body were based on his mathematical analysis of music. He regarded the universe and the human body as musical instruments, whose strings required balance and tension to produce harmony. Pythagoras' geocentric analysis of the universe was based on spherical planets, whose path around the earth created a musical hum. The hum for each planet was unique, producing an eternal music box.

Pythagoras attempted to synthesize Egyptians and Babylonian concepts to create a foundation for future investigations. He believed that mathematical relations held the secrets of the universe, and his lifelong devotion to mathematics set the tone and direction of science in the Western world. He continues to shape our lives 2500 years after his death. His mathematics are used by schoolchildren, and his philosophies permeate our language. The word philosophy itself, along with harmony, tone, and tonic, have a Pythagorean heritage.

Hippocrates (460-370 B.C.) travelled extensively throughout Greece and taught in Cos. The exceptional fame he enjoyed during his lifetime continued long after his death. Hippocrates' rational scientific approach to diagnosis freed medicine from supernatural constraints. For him, observation was based entirely on sense perceptions, and diagnostic errors were admitted and analyzed. Hippocrates was well aware of the limitations of his era, expecting future generations to question and improve upon his work. His belief in the principle of causality "...that chance does not exist, for everything that occurs will be found to do so for a reason" (Sarton, 1953) confirms his commitment to forming a rational science. Hippocrates' emphasis on observation and experience of the senses pioneered the use of rational scientific thought in the practice of medicine.

THE GOLDEN AGE OF GREEK SCIENCE

The fourth century B.C. represented the *golden age* of Greek science, with Plato and Aristotle the main exponents of science in this age. In this time, politics and ethics displaced the natural sciences as a means of attaining truth and knowledge. The disintegration of Greek civilization during the fourth century B.C. set the tone for the teachings of Plato and Aristotle. A century of constant war and political corruption demoralized the population. Materialism twisted the philosophy of personal excellence in mind and body, as a population of spectators developed to watch gladiators duel to the bitter end. In this environment of instability and violence, a great emphasis was placed on the higher morality of contemplation.

Plato (427-347 B.C.) believed the world of the senses to be an illusory shadow of reality. Ideas were the only reality, and true knowledge could not be obtained through the study of nature. The pursuit of truth required contemplation, not action. Plato discouraged artisans, engineers, and scientists from mechanical trades. His universe was subject to the powers of its divine creator, whereby the stars and planets were spiritual entities whose movements and nature were derived from these powers.

Aristotle's father, Nicomachus, served as physician to Amyntas II of Macedon, the grandfather of Alexander the Great. As a boy, Aristotle (384-322 B.C.) was influenced both by his life at court and medicine. For 20 years, Aristotle studied in Athens, spending several years at the Academy as Plato's pupil. His intellectual curiosity and independent mind eventually led him to question Platonic philosophy. Aristotle's garden, or Lyceum, provided a forum for his teachings, eventually becoming the site of his famous Peripatetic school. One of Aristotle's famous pupils was the young Alexander the Great, who was greatly influenced by Aristotle's ethics and politics. Aristotle was a universal research scientist as well as a social commentator. His interests lay in mechanics, physics, mathematics, zoology, physiology, chemistry, botany, and psychology. He believed the aim of science was to explain nature, and that mathematics provided a good model for a well-organized science.

Aristotle believed the senses revealed reality, and ideas were merely abstractions towards mental concepts (Keele, 1983). True knowledge of nature could only be gained through careful observation. Aristotle believed that contemplation was superior to mechanical labor. As a result, his methods for observation did not include verification or experimentation. Aristotle's universe consisted of the four observable elements: fire, air, water, and earth. The elements arranged themselves in concentric spheres, with the earth being the central sphere. Spheres of water, air, and fire extended outward from the centre. Beyond these were the spheres of the planets, stars, and then nothingness. The moon divided this universe into terrestrial and celestial zones, where movements were differentiated. Natural movements, such as a falling stone, were always in straight lines in the terrestrial zone, while movements in the celestial zone were always circular, eternal, and beyond terrestrial laws of nature.

Aristotle believed life was capable of mechanical expression. According to Aristotle, every motion presupposed a mover. His peculiar definition of motion required that all that is moved, be moved by something else. The motor must either be present within the mobile or be in direct contact with it. Action at a distance was inconceivable. The motions of falling bodies and projectiles fascinated Aristotle. He assumed the average velocity of a falling body over a given distance was proportional to the weight of the falling body, and inversely proportional to the density of the medium. He believed that the nature of objects, not externalities such as distance, determined the speed at which they fell.

Aristotle's theory of the four elements (fire, air, water, and earth) and their four characteristic qualities (heat, cold, humidity, and dryness) combined to produce the four main humors of the body (blood, phlegm, yellow bile, and black bile). Aristotle believed that the heart was the source of human intelligence. Muscular movements were the result of pneuma, or breath, which passed through the heart to the rest of the body. The concepts of movement and change dominated Aristotle's study of nature. His treatise *About the Movements of Animals*, which was based on observation, described movement and locomotion for the first time. The text provided the first scientific analysis of gait, and the first geometrical analysis of muscular action. Mechanical comparisons illustrated a deeper understanding of the functions of bones and muscles, and Aristotle explained ground reaction forces: "...for just as the pusher pushes, so the pusher is pushed" (Cavanagh, 1990).

THE HELLENISTIC AGE

The conquests of Alexander the Great gave rise to the Hellenistic age. In the third century B.C., Alexandria became the centre for scientific specialization. The Museum of Alexandria, which contained an observatory and laboratories for physiology and anatomy, provided a safe haven for the research scientist. The seclusion of the Museum and the political freedoms of Alexandria, enhanced the development of both anatomy and physiology.

Herophilos (about 300 B.C.) initiated the foundation of modern anatomy by creating a systematic approach to dissection, identifying numerous organs for the first time. Herophilos was the first to distinguish between tendon and nerve, and attributed sensibility to nerve. He also made a distinction between arteries and veins, stating that arteries were six times as thick as veins, and that arteries contained blood, not air (Sarton, 1953). Rejecting Aristotle's assertion that the heart was the seat of intelligence, Herophilos revived Alcmaeon's (about 500 B.C.) belief that the brain was the locus of intelligence.

Erasistratis (about 280 B.C.) studied at the Museum of Alexandria as a young contemporary of Herophilos, possibly as his assistant. Erasistratis distinguished himself as more of a theorist than Herophilos by applying physical concepts to the understanding of anatomy. He was the first to describe the muscles as organs of contraction. Like Hippocrates, Erasistratis rejected supernatural causes in explaining nature.

Archimedes (287-212 B.C.) was the son of Phedias, an astronomer who supported his early interest in mathematics and astronomy. Archimedes' fame resulted from his inventions, which included mechanical weapons such as catapults, ingenious hooks, and concave mirrors used to attack and burn marauding Roman ships. Using his compound pulley and lever, Archimedes claimed that he would be able to move the earth – if he only had a place to stand to do so! Archimedes used geometrical methods to measure curves, and the area and volume of solids. Without the use of calculus, Archimedes used a close approximation for π to measure volumes and areas.

In his treatise *Equilibrium of Planes*, Archimedes proved that "...two magnitudes, whether commensurable or not, balance at distances reciprocally proportional to them" (Sarton, 1953). The distances considered were the respective distances of their centre of gravity from the fulcrum. He also demonstrated how to find the centre of gravity of a parallelogram, triangle, parallel trapezium, and parabolic segment. Archimedes' treatise *Floating Bodies* described the principle of water displacement, which he is said to have discovered while bathing! Movement occupied Archimedes' investigations as he solved the problem of how to move a given weight by a given force.

Archimedes' application of Euclidian methodology to mechanics formed the basis of rational mechanics, and his mathematical investigation of Aristotle's mechanical theories established statics and hydrostatics. His analyses, proofs, and methodology dominated statics and hydrostatics until the times of Simon Stevin (1548-1620) and Galileo Galilei (1564-1642). Archimedes died during the sack of Syracuse in 212 B.C. He is said to have been absorbed in the contemplation of geometrical figures drawn in the sand and to have shouted "keep off!" to an approaching Roman soldier. Not appreciating Archimedes' particular mathematical concern, the soldier killed him on the spot.

THE ROMAN EMPIRE'S CONQUEST

The emergence of the Roman Empire during the first century B.C. eroded the foundations of Greek science. During this period, emphasis was placed on ethics, and scientific activity was reduced to the development of fortification, military weapons, and sources of amusement.

Hero of Alexandria (about 62 A.D.) taught physics at the University of Alexandria. Hero produced treatises on mechanics, optics, and pneumatics, which contained the principles used in his inventions. He is credited with inventing the first rudimentary steam engine, in which steam propelled a globe in circular revolutions. By using compressed gases and saturated water vapors, Hero devised machines that opened and closed curtains for puppeteers, operated doors, and created a miniature carousel with mechanical birds. He invented a magic jug that used the principle of the siphon to control the flow of water. Hero's mechanical inventions amused and entertained, but were never used to provide practical improvements.

Galen (129-201 A.D.) began medical training in Pergamos. Inspired by a dream, his father ensured that Galen studied with the best physicians of the time. Galen's voracious appetite for learning, exceptional brilliance, overriding pride and ambitions, and energetic and sometimes violent nature manifested themselves in his passion to impress the world. Appointed physician to the College of Gladiators at the age of 28, Galen probably became the first sports physician and team doctor in history. For four years, he practiced surgery and dietetics among the gladiators, gaining substantial knowledge of the human body and human motion. Recognizing Galen's immense talents, Marcus Aurelius appointed him physician to the Emperor, where he remained for 20 years. Galen devoted much of this time to research, publishing over 500 medical treatises. He also wrote numerous philosophical books, such as *On the Passions of the Soul and its Errors*, which was largely autobiographical. In fact, our knowledge of the work of many of his predecessors came from Galen's writings. After fire destroyed the temple of Pax and the majority of Galen's manuscripts, he returned to Pergamos, where he died at the age of 70.

Galen believed medicine was a comprehensive science that included both anatomy and physiology. Like Aristotle, Galen was a teleologist, believing that nature arranged everything according to a distinct plan, with the soul ruling the body, and the senses and organs subordinate and subservient to the soul. *De Usu Partium* (On the Use of the Parts) gave Galen his transcendent influence on the medical sciences. The first text on physiology, it quickly became the uncontested authority on medicine, and remained so for 1300 years. It was first printed in 1473, and over 700 copies were in existence by 1600, emphasizing the importance of the understanding of structure and function in diagnosis and therapy. Anatomy did not experience comparable growth until Vesalius (1514-1564) and William Harvey (1578-1657).

De Motu Musculorum (On the Movements of Muscles) embodied Galen's lifelong passion for the mechanism of movement. He made enormous advances in the understanding of muscles, establishing the science of myology. Galen's emphasis on structure revealed the "...innervation that makes muscle substance into a muscle proper" (Bastholm, 1950). The flesh of the muscle contained webs of nerve endings that Galen believed transmitted *spiritus animalius* from the brain into the muscle, stimulating movement. Galen's Pneuma theory included the beliefs that the arteries contained blood, and that nerve endings pervaded muscle flesh and contained pneuma, Aristotle's teleological cause of movement.

Galen looked to the function of muscles to distinguish between skeletal muscles and muscle parts such as the heart and the stomach. He believed each muscle had only one possible movement, and muscular movement that required contraction and the ability to contract resulted from a property existing throughout the muscle. Galen described tonus and distinguished between agonistic and antagonistic muscles, and motor and sensory nerves.

Obsessed with the elegance of mathematics and trained by the best physicians of his time, Galen sought to raise medicine to the level of an exact science. As an anatomist, his descriptions of muscles and nerves far surpassed those of his predecessors. His correlation of injuries to various parts of the spinal cord and the cranial and cervical nerves, which provided paralysis of certain organs, were and are considered among the most brilliant medical observations of all times. Galen's era discouraged human dissection, and his observations were based almost entirely on the dissections of dogs, pigs, and apes. However, Galen was aware of this deficiency, encouraging students to travel to Alexandria to view the only human skeletons available to science. Unlike Hippocrates, Galen uncritically applied his principles, even in light of experience that contradicted them. Galen was the indisputable authority in his own time, and the Middle Ages further entrenched his authority, discouraging any investigations that revealed contradictory findings.

RELEVANCE TO BIOMECHANICS

The relevance of Antiquity to biomechanics lies in four major aspects:

- Separating knowledge and myth,
- Developing mechanical and mathematical paradigms,
- Developing anatomical paradigms, and
- Performing the first biomechanical analysis of the human body.

Thales was the first to attempt to separate knowledge and myth. Additionally, some of the ancient Greeks attempted to include observations to develop theories. Pythagoras, Aristotle, and Archimedes laid the foundations for mechanical and mathematical analysis in general, and for mechanical and mathematical analysis of movement, specifically. Hippocrates, Herophilos, Erasistratis, and Galen developed the first anatomical and neuro-physiological concepts of the human body. Aristotle wrote the first book about human movement *(About the Movements of Animals),* a first scientific analysis of human and animal movement based on observation and describing muscular action and movement.

1.2.2 THE MIDDLE AGES

During the Middle Ages (200 B.C.-1450 A.D.) scientific development decreased, as religious and spiritual development increased. Arab scholars saved the scientific investigations of antiquity from disappearing completely by translating the works from Greek to Arabic. Neoplatonic doctrines, initially established during the third century B.C., represented the main link between antiquity and medieval Europe. Plato's *Timaeus* provided the doctrine of macrocosm and microcosm, which compared the nature and structure of the universe to the nature and structure of man. Analogies were drawn combining the four elements (humors, organs, seasons, compass points and astronomical events) to describe nature, motion, and change. St. Augustine (354-430 A.D.) incorporated Neoplatonic doctrines into Christian beliefs. Discouraging scientific inquiry, St. Augustine explained that "...the only type of knowledge to be desired was the knowledge of God and the Soul, and that no profit was to be had from investigating the realm of nature" (Koestler, 1968). During the thirteenth century, Aristotle's writings were revived from Arabic translations and St. Thomas Aquinas (1227-1274) integrated Aristotelian philosophy into Christian beliefs. universities run by the Dominicans and Franciscans embraced Aristotle's doctrines.

The Middle Ages contributed little to the development of the field of biomechanics. The developments of the scientific age of antiquity remained relatively dormant throughout the period. Even the revival of Aristotle's writings did not create a renewed interest in investigating and understanding nature. Medicine existed, but any interest in the human body and human locomotion was discouraged. universities became strongholds of Aristotelian thought, but did not conduct critical research.

RELEVANCE TO BIOMECHANICS

The relevance of the Middle Ages to the development of biomechanics is minimal. Scientific development, in general, was discouraged in this period. Consequently, the earlier interest in human and animal anatomy, physiology, and locomotion was put to sleep for more than 1200 years. However, movement depiction flourished in Greek and Roman art, and it was the artist, rather than the scientist, who later revived the study of human movement.

1.2.3 THE ITALIAN RENAISSANCE

The Italian Renaissance, which flourished from 1450 until the sack of Rome in 1527, was characterized by freedom of thought, as it saw the revival of ancient Greek philosophy, literature, and art. The political chaos of fifteenth century Italy provided the backdrop for the eruption of moral and intellectual freedoms. The authority of the ancients replaced the authority of the Church. Man became the measure of all things, as individual genius flourished in a time unparalleled since the golden age of anatomy in Alexandria. Encouraged by liberal patrons like the Medici in Florence, small numbers of scholars and artists including

Michelangelo, da Vinci, and Machiavelli, emerged as Renaissance men. With few exceptions, science played a limited role during the Renaissance, as an organized scientific community did not exist. The Renaissance, however, created the foundations for the Scientific Revolution in the seventeenth century. A revival of Platonic and Aristotelian philosophies broke down the rigid scholastic system of the Middle Ages, introducing the element of choice in intellectual pursuits. Most importantly, intellectual activity became a social adventure no longer subjugated by a predetermined orthodoxy and isolated by meditative study.

Leonardo da Vinci (1452-1519) was the illegitimate son of the notary Ser Pierro da Vinci and a peasant woman. Raised by his father's family in Vinci, he roamed the countryside, eagerly absorbing all that his senses could provide. Early education in Vinci provided only a rudimentary knowledge of reading, writing, and arithmetic. Never a man of letters, da Vinci was proud of his self-taught education, often criticizing his colleagues for relying too heavily on the words of others rather than ideas of their own. Realizing da Vinci's immense artistic talent, Ser Pierro sent him to Florence as an apprentice to Verrochio, an artist. With his contribution to Verrochio's painting *The Baptism of Christ*, da Vinci made his mark as an artist. Verrochio, however, vowed never to touch another brush as his young pupil had so far surpassed him. Best known today as a brilliant artist, da Vinci was primarily a military and civil engineer. Constant poverty throughout his life forced him to expand his talents, creating festival productions, and designing aqueducts and fortifications. The immense versatility of his mind, talents, and interests were reflected in the abundance of his inventions, which included distillation apparatus, water skis, a helicopter, a tank, a parachute, a steam cannon, and a hang glider. His personality was as well documented as his inventions. Strong, well-built, and handsome, his "...charm, his brilliance and generosity were not less than the beauty of his appearance" (Keele, 1983).

Da Vinci contributed substantially to the understanding of mechanics in his time. He described the parallelogram of forces, divided forces into simple and compound forces, studied friction, and questioned Aristotle's relationship between force, weight, and velocity. Da Vinci prepared Newton's third law in his analysis of the flight of birds "...an object offers as much resistance to the air as the air does to the object" (Keele, 1983). Da Vinci's mechanics, however, never reached the concepts of acceleration, inertia or mass as opposed to weight. By dissecting the movements of man, not merely a motionless dead cadaver, da Vinci's achievements in mechanics flourished with his application of mechanical principles to the anatomy of man. His unique ability to communicate dynamic movements in visual form shed light on the mechanics of human beings that is seldom as bright in modern anatomical illustration.

Da Vinci's mechanical analysis of human movement included joints, muscles, bones, ligaments, tendons, and cartilage. Examining the structures of movement, da Vinci discarded the theory of pneuma. Spiritual force replaced it, acting as a form of energy through the nerves or fleshy muscle fibres, "transmuting" their shape, "enlarging" or broadening them, and shortening their tendons.

Da Vinci's anatomical studies fused art and science, stressing the importance of perspective in producing pictures of reality. Ball and socket joints, such as the shoulder and hip joint, figured prominently in da Vinci's anatomical analysis. The "polo universale" allowed circular movement in many directions, with "polo" describing the fulcrum of balance for these rotating movements. Da Vinci illustrated the correct shape of the human pelvis for the

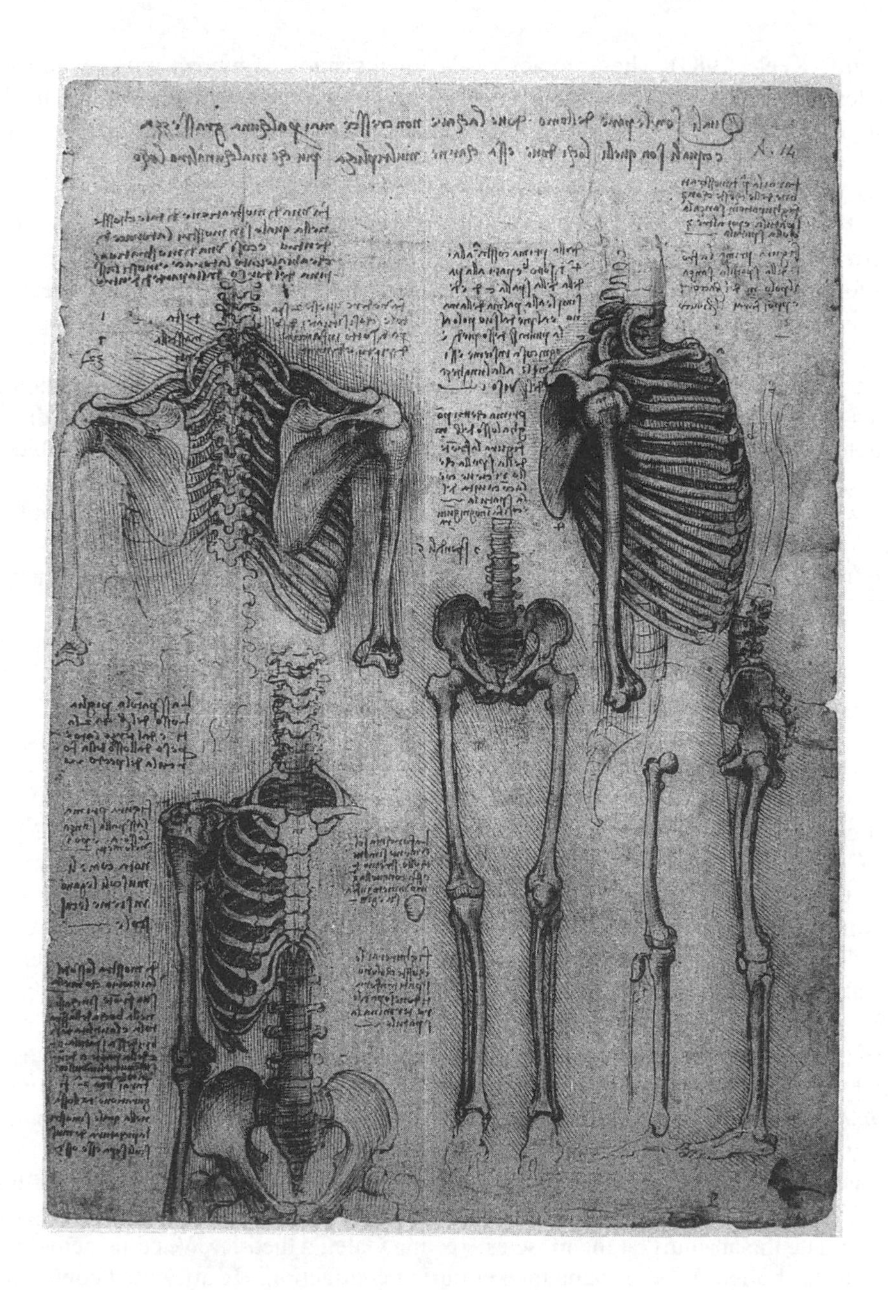

Figure 1.2.1 **Da Vinci's study of the human skeleton demonstrated the "polo dell' omo" and the shoulder joint (from Clayton, 1992, with permission).**

first time (Figure 1.2.1), describing the hip joint as the "polo dell' omo" (the pole of man).

Da Vinci depicted individual muscles as threads, by demonstrating the origin, insertion, relative position, and interaction of each muscle. Mechanical action, dependent upon the muscle's shape, was represented in his drawings, with forces acting along the line of muscle filaments (Keele, 1983). The accuracy of da Vinci's study of human musculature varied considerably. He was most successful illustrating the anatomy of the arm, elbow, and hand, but slightly less successful in illustrating the anatomy of the foot and lower leg, and least successful with the trunk and the thigh.

Da Vinci's ultimate goal was to integrate experience through geometrical analysis. As the first modern dissector and illustrator of the human body, he often realized contradictions and errors in Galen's comparative anatomy. With keen perception and an amazing aptitude for communicating movements pictorially, da Vinci presented a comprehensive mechanical analysis of the human body. Unlike Galen, verification and experimentation figured prominently in da Vinci's studies.

Currently, da Vinci's significance as a scientist is contested. The first publication of da Vinci's works occurred 200 to 300 years after his death. The eclectic nature of his drawings and notes, which were disorganized, often having several seemingly unrelated subjects on one page, discouraged some scholars from qualifying his contributions as entirely scientific. However, even if his notes had been organized, their influence would have been limited in his time by the lack of circulation. Furthermore, the absence of any formal scientific community during the Renaissance, prevented da Vinci's ideas from circulating among colleagues with similar interests.

Vesalius (1514-1564) was born during the final years of da Vinci's illustrious career. His career in anatomy contrasted sharply with that of his predecessor. Unlike da Vinci, Vesalius received formal training in medicine, taught his anatomical theories, and published his research and methodology. Born into a medical family, Vesalius trained as a physician at the University of Padua and graduated *magna cum laude* in 1537. By 1538 he had produced a dissection manual for his students. Originally a proponent of Galen, Vesalius began to notice contradictions in his dissections and those done by Galen. In 1539, Judge Marcantonio Contarini became interested in Vesalius' anatomical studies, encouraging him in further detailed investigation by making available the bodies of executed criminals. For the first time in history, enough cadavers were available to make, repeat, and compare detailed dissections. Vesalius became increasingly convinced that Galen's anatomy was actually an account of the anatomy of animals and that it was wrongly applied to the human body. In 1539, Vesalius distinguished himself by publicly challenging the authority of Galen.

Two revolutionary views of nature were published in the year 1543. Copernicus challenged the notion that the universe was heliocentric in his *De Revolutionibus Orbium Coelestium*. Vesalius' *De Humani Corporis Fabrica Libri Septem* revolutionized human anatomy by challenging a view that had dominated for 1300 years. Vesalius boldly declared that human anatomy could only be learned from the dissection and observation of the human body (Figure 1.2.2). Numerous human dissections provided Vesalius with the opportunity to reevaluate the anatomy of the muscles. Testing Galen's theories, Vesalius demonstrated that muscle shortened and became thicker during contraction. He attributed contraction to a property of the muscle, as it contracted when both ends were severed. Vesalius' experimentation stimulated scientific debate concerning the relationship between the nerves and muscles. Vesalius believed muscle to be "...composed of the substance of the ligament or tendon divided into a great number of fibres and of flesh containing and embracing these fibres" (Needham, 1971), which received branches of arteries, veins, and nerves. Vesalius' observations inspired the investigations of Fallopius (1523-1562), who was keenly interested in the fibrous tissue in connection with movement. Fallopius stated that "...motion requires a fibrous nature in the actual body that is moved; since whatever moves itself does

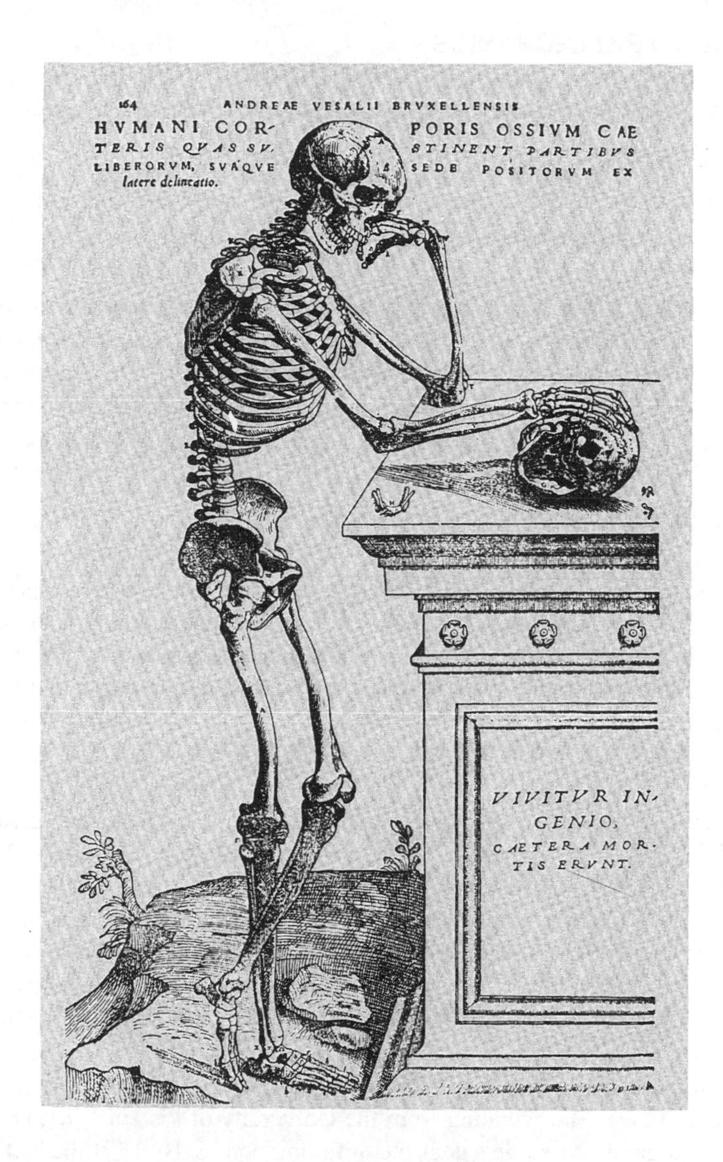

Figure 1.2.2 **Human skeleton from Vesalius' book *De Humani Corporis Fabrica Libri Septem* (from Singer, 1959, with permission of Oxford University Press).**

so by contraction or extension" (Needham, 1971).

Vesalius' demands on the development of human anatomy, his questioning of Galen's authority, and his detailed, descriptive anatomy laid the foundation for modern anatomy. Vesalius established the scientific principle of performing human dissections to understand human anatomy. Despite his criticism of Galen's anatomy, Vesalius adopted the theory of *spiritus animalis* as the cause of muscular contraction and carrier of the motor and spiritual function of the brain.

RELEVANCE TO BIOMECHANICS

The Italian Renaissance is relevant to biomechanics in three major aspects:

- Scientific work was revived,
- The foundations for modern anatomy and physiology were laid, and
- Movement and muscle action were studied as connected entities.

Scientific activities (the attempt to study the unknown), which remained dormant during the Middle Ages, experienced a revival. During the Italian Renaissance, da Vinci and Vesalius laid the foundations for modern anatomy by basing their teachings on observation and dissection of human cadavers. Parallel to the dissection of cadavers, human movements were discussed and factors contributing to human movement were studied.

1.2.4 THE SCIENTIFIC REVOLUTION

The environment in which the Scientific Revolution developed was similar to the environment of the Italian Renaissance. Men of science were supported by private and political institutions such as kings, counts, wealthy families, universities, and the Vatican in Rome. Intellectual freedom was highly respected and the interest in new ideas and findings was substantial. In addition, scientific societies began to emerge, encouraging the exchange of ideas and speculations, and intensive contact between scientists from different countries in Europe developed.

The Scientific Revolution of the seventeenth century, with which such great names as Galileo Galilei (1564-1642), Johannes Kepler (1571-1630), Rene Descartes (1596-1650), and Isaac Newton (1642-1727) are associated, saw a change in the understanding of nature and the way of doing scientific analysis. Led primarily by Galileo, the new scientific methodology revealed a picture of the universe that required updating interpretations of the Bible and abandoning Aristotelian mechanics. Scientists turned to new methods of research. Experimentation became the cornerstone of the new scientific method. Descartes, Galileo, and Newton advocated a methodology that questioned certain truths by experimentation.

Galileo Galilei (1564-1642) received his early education in Florence around the end of the Italian Renaissance. Matriculating from the University of Pisa in 1581, he began systematically investigating Aristotle's doctrine of falling bodies. By 1591, he had discovered the impossibility of maintaining that the rate of a fall was a function of the falling object's weight. One year later, Galileo was appointed the Professor of Mathematics at the University of Padua.

Galileo's Application of Mechanical theory went beyond the inorganic world as he proposed to publish a treatise on the mechanical analysis of animal movement. If published, this treatise, *De Animaliam Motibus* (The Movement of Animals), would have preceded Borelli's famous treatise *The Movement of Animals* (1680-81) by 40 years. Galileo's topics included the biomechanics of the human jump, an analysis of the gait of horses and insects, and a determination of the conditions that allowed a motionless human body to float. Galileo's *Discourses on Two New Sciences* (1638) contained some of his biomechanical observations. His analysis of the strength of materials, beam strength, discussion of the strength

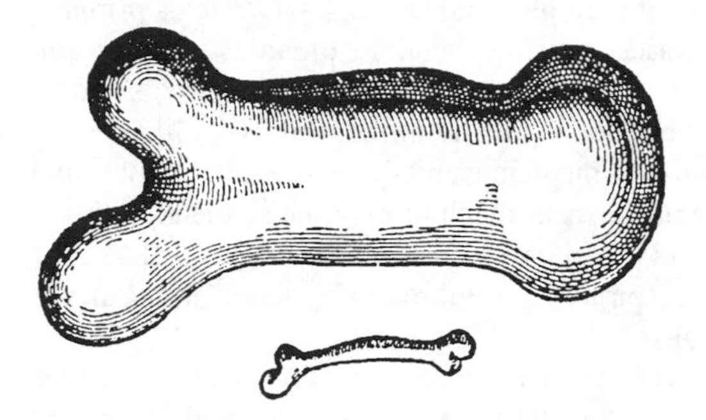

Figure 1.2.3 **Galileo's comparison of a normal femur with the femur needed to support an animal three times the size (from Singer, 1959, with permission of Oxford University Press).**

of hollow solids, and particularly his analysis of resistance of solids to fracture and the cause of their cohesion, had application in the dynamics of the structure of bone. Galileo also compared the effects of changing the structures of biological materials. One such comparison demonstrated the effect of increasing a normal femur to accommodate supporting an animal weighing three times the size (Figure 1.2.3). The following quote illustrates Galileo's theory: "You can't increase the size of structures indefinitely in art or nature. It would be impossible so to build the structures of men or animals as to hold together and function, for increase can be effected only by using a stronger material or by enlarging the bones and thus changing their shape to a monstrosity" (Singer, 1959).

Galileo applied this analysis to the predicament of enormous floating animals such as whales, in which he proposed that their weight was counterpoised by equivalent dispersal of water weight, altering the limit to their size. Ascenzi (1993) argues that Galileo's study on the mechanics of structure, function, and animal movements distinguishes him as the founder of biomechanics.

Galileo, was a short, stocky man with fiery red hair and a forceful character. A great detail in historical literature outlined his clash with the Church that led to the condemnation of the Copernican theory in 1616, and to his permanent house arrest in 1633. Galileo's personality played an immense role in the ultimate conflict with the Church. With the publication of *Letters on Sunspots* (1613), in which Galileo initially gave public commitment to the Copernican theory, he did not receive any opposition from the Church, and Cardinals Boroeo and Barberini (the future Pope Urban VIII) wrote letters to Galileo expressing their sincere admiration. In 1624, under the restraint of the 1616 decree, Galileo met with Pope Urban VIII to obtain permission to publish his analysis of the Copernican system. An admirer of Galileo, the Pope suggested a way to publish an analysis that avoided theological arguments, acknowledged the supremacy of God, and the inability of the human mind to understand all of God's mysteries. Galileo's arrogance sealed his fate as he placed the Pope's arguments in the mouth of a complete simpleton in his *Dialogue* (1631).

Galileo shaped the path of science for centuries to come. His theory of uniform motion, theory of projectiles, theory of the inclined plane, and definition of momentum provided foundations for Newton's three laws. Newton himself admitted: "If I have been able to see farther it was because I stood on the shoulders of giants" (Koestler, 1968). The giants Newton referred to were Galileo, Kepler, and Descartes. Galileo's primary contribution was the applicability of his theories to visible, tangible, organic, and inorganic objects.

Santorio Santorio (1561-1636) was a colleague of Galileo's and taught theoretical medicine at the University of Padua. Greatly influenced by Galileo, Santorio was the first scientist to apply mechanics to medicine, and the first to apply quantifiable methods to medicine. For 30 years, Santorio spent much of his time suspended from a steelyard weighing himself and his solid and liquid inputs and outputs (Figure 1.2.4). As he tried to deduce the amount of "insensible respiration" lost through his lungs and skin, his experiments laid the foundations of metabolism.

William Harvey (1578-1657), the eldest son of seven, was born into a family of yeoman farmers. He received his early education at Cambridge and completed his studies as a phy-

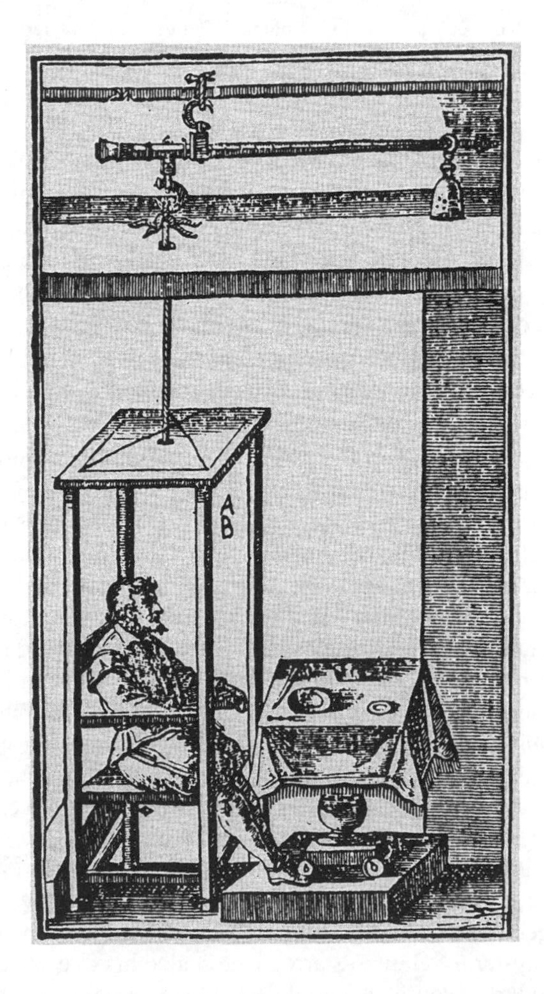

Figure 1.2.4 **Santorio in his balance measuring his solid and liquid input and output (from Millar et al., 1989, p. 343, Chambers Concise Dictionary of Scientists, with permission of Chambers Harrap Publishers Ltd., Scotland, UK).**

sician at the University of Padua. Moving back to London in 1602, Harvey was soon successful, becoming physician to James I and later Charles I. Harvey's main interest was medical research. Aristotle's theory of the primacy of the heart greatly influenced Harvey's discovery of circulation. By 1615, he had a clear conception of circulation, but did not publish his results until 1628 in the slim, poorly produced book, *De Motu Cordis*, which was considered a scientific classic.

Vivisection of dogs allowed Harvey to see the motion of the heart. According to Harvey's observations, the active motion was the heart's contraction – the systole – not the diastole as Galen had claimed. By measuring the capacity of the dissected heart, and estimating the number of heart beats every half an hour, Harvey demonstrated that the heart pumped more blood into the arteries in half an hour than the body contained. Therefore, the blood had to be returning to the heart by another route. Acting without the aid of a microscope, Harvey assumed that a system of capillaries connected the arterial and venous systems. He had discovered the circulation of blood. This discovery altered physiological thought and inspired generations to use his experimental methods. He applied mechanistic theory to the action of the heart, describing its function as a pump and recognizing the mechanical nature of the vascular system. Harvey became the first cardiac biomechanist and later died of a stroke at the age of 79.

Rene Descartes (1596-1650) was a philosopher and one of the dominant thinkers of the seventeenth century. Educated by the Jesuits, Descartes was exposed to the astronomical and mechanical observations of Galileo and rigorous mathematical training. Inheriting a modest wealth, Descartes spent his life travelling, serving as a soldier in Holland, Bohemia, and Hungary. In 1629 he settled in Holland, where he remained until persuaded to become the tutor to Queen Christina of Sweden in 1649. Descartes had the habit of sleeping late, claiming that his best thinking was done in the comfort of a warm bed. To his dismay, Queen Christina did not share his love for sleeping in, preferring to begin tutorials in philosophy at 5:00 a.m. in a freezing cold library. However, Descartes did not suffer too long from these untimely tutorials. Within five months of beginning to tutor the queen, Descartes died of lung disease. According to legend, Descartes invented the Cartesian coordinate system while lying in bed and observing the habits of a fly zooming around the room. As the fly flew into a corner of the room, Descartes realized the possibility of representing movement with a Cartesian coordinate system.

Descartes became one of the originators of the mechanical philosophy that stated: changes observed in the natural world should be explained only in terms of motion and rearrangements of the parts of matter. He represented the first wave of mathematical analysis of mechanics, which shaped the development of mechanics during the eighteenth century. Descartes' mechanical philosophy influenced physiology so that by 1670 all major physiologists had developed a mechanical approach to their investigations, e.g., William Harvey, Santorio Santorio, and Giovanni Borelli. Descartes' treatise *L'homme* (1664), written before 1637, applied mechanical principles to the human body. According to his mechanical philosophy, animals were organic machines running on autopilot, while humans, also a machine, were distinguished from animals by virtue of having a soul. The first modern book devoted to physiology, *L'homme* stressed the role of the nervous system in coordinating movement. However, Descartes had no extensive knowledge of physiology. Often grotesquely wrong in detail, his ideas about physiology quickly became obsolete.

Giovanni Borelli (1608-1679) was born in Naples. His family was under constant suspicion of conspiracy against Spain. In 1624, Borelli fled to Rome to escape persecution under such suspicions. Borelli was greatly influenced by Galileo, Harvey, Kepler, Santorio, and Descartes. He was a student and colleague of Galileo, holding degrees in medicine and mathematics. His own interests ranged from geometry, physics, mechanics, physiology, and astronomy to the study of volcanoes. Borelli served as the Professor of Mathematics at the universities of Messina, Pisa, and Florence. In 1624, Queen Christina, the eager pupil who may have hastened the death of Descartes with her enthusiasm for early mornings, renounced her crown to devote herself to the pleasures of culture. She became Borelli's patron, supporting his investigation of the mechanics of the human body with annual endowments. When he was robbed of all his possessions, Borelli was compelled to live in a building owned by the Society of the Scholae Piae of San Pantaleo. He remained there until his death in 1680. Due to his publication, *De Motu Animalium* (Figure 1.2.5), Borelli is often called the father of biomechanics.

Borelli's fundamental aim was to integrate physiology and physical science. *De Motu Animalium* (1680-81) demonstrated Borelli's geometrical method of describing complex human movements such as jumping, running, flying, and swimming (Figure 1.2.5 and Figure 1.2.6). His investigation of human movement included gait analysis and an analysis of the muscles. His work on the human gait and more complex movements was successful. However, he relied heavily on mechanical philosophy to explain the physiological nature of muscles. Borelli discussed several aspects of movement related to externally visible motion. He described muscle function, formulated mechanical lemmas to explain the movements produced by muscles, explained muscle function for the motion of the knee joint, and discussed the influence of the direction of muscle fibres for force production of the particular muscle in question. Borelli formulated his findings and hypotheses in the form of *propositions*. Examples from his propositions on jumping included:

Proposition CLXXVIII: **In jumping at an inclination to the horizon, the trajectory of the jump is parabolic.**

Proposition CLXXIX: **Why a jump during running is longer and higher.**

Borelli also discussed the mechanics of muscle contraction, the reason for fatigue, and offered some thoughts about pain. Some of his propositions for muscle contraction included:

Proposition II: **Muscle contraction does not consist of simple tension of the fibres similar to that exerted on a rope raising a weight.**

Proposition XIV: **Muscles do not contract by condensing the length of their fibres and bringing their extremities closer together. Muscle hardness and tightening results from swelling.**

Figure 1.2.5 **Cover of Borelli's book De Motu Animalium (from Borelli, 1680, translated 1989, with permission).**

Borelli's mechanical propositions were probably stronger than his propositions for muscle action. His mechanical propositions had a significant impact on the subsequent development of biomechanics.

Isaac Newton (1642-1727) was born on December 25th, 11 months after Galileo's death. A descendant of farmers, with no record of any notable ancestors, the young Newton quickly proved to be a poor farmer; often absent-minded and lazy. Newton read the works

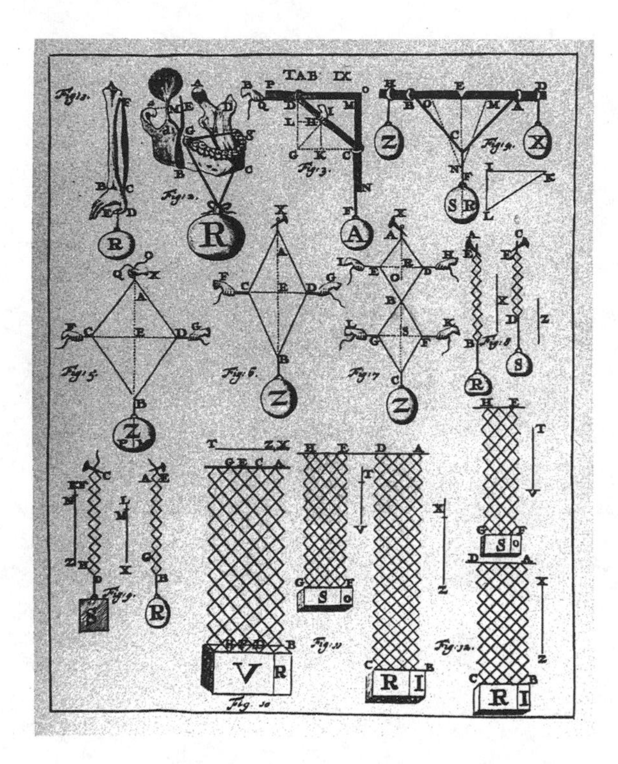

Figure 1.2.6 **Borelli's Table IX - The Movement of Animals - illustrating the models of muscular construction (from Borelli, 1680, translated 1989, with permission).**

of Kepler, Galileo, and Descartes to prepare for university. Although he was admitted to Trinity College, Cambridge, Newton returned to Woolsthorpe in 1666 to escape the plague. Conducting experiments in optics and roaming about the apple orchards of his home, Newton's fascination with mechanics peaked. As he explained: "I was in the prime of my age for invention and minded mathematics and philosophy more than at any other time" (Singer, 1959). Newton discovered the theory of gravitation during this time, but he misplaced it for 20 years when his interests turned to alchemy. He was a difficult man, remote and austere, often isolating himself from others in the pursuit of his science and secret metaphysical analysis. Newton gained recognition in his own time for his accomplishments. As the Master of the Royal Mint in 1696 and 1698, Newton reformed the currency and was knighted for his effort in 1705. He became President of the Royal Society in 1703, a position he held until his death. However, Newton was ridiculed and satirized at the end of his life in England as a scientist who concentrated on unimportant details.

Newton made several contributions that were important for science in general, and biomechanics specifically. However, his major contribution was the *synthesis* of many different pieces of the puzzle mechanics. Koestler (1968) described Newton as the conductor of an orchestra of individual players, each absorbed in tuning his or her instrument; a conductor who pulled the orchestra together and made the caterwauling of the single instruments into a beautiful sound of harmony.

The pieces of the puzzle that Newton established were:

- Kepler's laws of the motion of heavenly bodies.
- Galileo's law of falling bodies and projectiles.
- Descartes' law of inertia, which required straight motion if no external force acted.

However, the pieces did not fit together. For example:

- Kepler's forces responsible for the movement of planets did not apply to movement on earth.
- Galileo's laws of movement on earth did not provide an apparent possibility to explain the movement of the planets.
- Kepler's planets moved in elliptic pathways.
- Galileo's planets moved in circular pathways.
- Kepler's driving forces for the planets were spokes from the sun.
- Galileo did not need a driving force, since he proposed circular motion was self-perpetuating.
- Descartes' inertia of bodies required planets to move in a straight line. However, the planets certainly did not move in a straight line.
- There was disagreement on the forces responsible for the movement of planets.

Newton's contribution was to *synthesize* these puzzle pieces into the law of gravity and the laws of motion. It is not completely clear how Newton found the solution. He claimed later that he found the law of gravity by counterbalancing the moon's centrifugal force with the gravitational force. However, this explanation was not terribly convincing (Koestler, 1968). Some initial ideas (especially relating to the law of gravity) may have been present in 1666 when he was 24, and the concept may have developed in discussions with his colleagues at the Royal Society, e.g., Hooke, Halley, Wren, and continental Europe, e.g., Huygens in Holland. This developmental path can probably never be reconstructed. Koestler described an interesting aspect of this development (1968): "With true sleepwalker's assurance, Newton avoided the booby-traps strewn over the field: magnetism, circular inertia, Galileo's tides, Kepler's sweeping-brooms, Descartes' vortices - and at the same time knowingly walked into what looked like the deadliest trap of all: action-at-a-distance, ubiquitous, pervading the entire universe like the presence of the Holy Ghost".

Over 250 years after Newton's death, our view of the mechanical world is largely Newtonian. With the publication of these four basic laws in Newton's *Philosophiae Naturalis Principia Mathematica* in 1686, all motion in the universe could be described or predicted, as long as the movement was with relative speeds that were small compared to the speed of light:

- The law of inertia,
- The law of acceleration due to an acting force,
- The law of action and reaction, and
- The law of gravity.

Newton's second law of motion provided the fundamental tool for kinetic and kinematic analysis of movement. Newton was also credited with the first general statement of the parallelogram of force based on his observation that a moving body affected by two independent forces acting simultaneously moved along a diagonal equal to the vector sum of the forces (Rasch, 1958). When applied to muscular action, Newton's observation provided insight "...as two or more muscles may pull on a common point of insertion, each at different angles with a different force" (Rasch, 1958).

The improvement of the understanding of mechanics during the period of the Scientific Revolution was a quantum leap of scientific development. However, besides these exciting mechanical developments, other disciplines also made significant developments. Emerging biomechanical and physiological experiments benefited greatly from the invention of the microscope and the advent of experimental method. In 1663, the experiments of Jan Swammerdam (1637-1680) on frogs demonstrated the constancy of muscle volume during contraction. Nerve section experiments by William Croone (1633-1684) in 1664 illustrated that the brain must send a signal to the muscles to cause contraction. Between 1664 and 1667, Niels Stensen (1638-1686) provided precise descriptions of muscular structures. Francis Glisson (1597-1677) proposed the theory of irritability, the muscles' ability to react to stimuli. In 1691, Clopton Havers (1655-1702) used the microscope to conduct the first complete, systematic study of bone. He discovered that bone was composed of both inorganic and organic strings and plates arranged around cavities in tubular form.

RELEVANCE TO BIOMECHANICS

The relevance of the Scientific Revolution to biomechanics lies in two major aspects:

- Experiment and theory were introduced as complementary elements in scientific investigation, and
- Newtonian mechanics were established, providing a complete theory for mechanical analysis.

Experimentation became an accepted approach to understanding nature. Instruments improved the scope of experiments. As the telescope revolutionized the science of mechanics, the microscope revolutionized physiology. Separate fields of study began to emerge, and came fruition during the eighteenth century. The foundations for biomechanics were created during the seventeenth century, as scientists such as Galileo, Borelli, and Harvey used experimentation to understand the human body and its movements.

Newton's synthesis established a new way of studying motion. The emergence of the mechanical philosophy and Newton's laws provided the impetus for the study of human movement, and the tools to understand it. The human body became the focus of mechanical investigation. The emergence of the mechanical philosophy in the seventeenth century created a focus on movement and motion that persisted well into the eighteenth century.

1.2.5 THE ENLIGHTENMENT

The Scientific Revolution resulted in a quantum leap for science, specifically for the natural sciences. The contributions of Galileo, Kepler, Descartes, and especially Newton revolutionized science, and the former natural philosophy was replaced by a new foundation, the comprehensive mechanical paradigms. A new group of scientists developed: the mechanical philosophers. As is typical in such situations, some concepts were not yet understood. Much work was still needed, requiring the brilliant mathematical brains of the eighteenth century to clarify many unsolved parts of the puzzle; parts that would not change the basic foundation, but that would prove important for further development of the mechanical understanding of movement.

One area of disagreement between the leading thinkers of this time was the theory concerning the causes of motion. The mechanical philosophers argued whether matter was moved by external, internal or no force at all. The concept of force was not clear at this time. Descartes argued that there were no forces or power in matter. Newton argued that matter consisted of inertia particles, and that forces acted between every pair of particles. Leibnitz proposed that force was internal to matter. Typically, these positions were influenced by religious beliefs.

This period of increased understanding was called *le siècle des lumières* in France, *die Aufklärung* in Germany, and is known today in the Western world as the *Enlightenment*. The term was coined by mathematicians who were convinced that mathematics was the most important revolutionizing force of the Scientific Revolution. Mathematical analysis was proposed as the solution to the ills of society. In 1699, Fontanelle had argued that the new geometric spirit could improve political, moral, and literary works. D'Alembert believed that the Spanish Inquisition could be undermined by smuggling mathematical thinking into Spain (Hankins, 1985). It was in this atmosphere that natural philosophers were made into heroes. In France, Newton was the greatest hero of them all. He was, in the language of modern times, a star. By 1784, there were 40 books written on Newton in English, 17 in French, and three in German.

Eighteenth century science was not organized like it is today. Subdivision of science into the current disciplines began to develop during this time, as a reflection of the changing understanding of nature and the way its study progressed. Three mathematicians contributed significantly to the development of natural science during the Enlightenment as it related to biomechanics are Euler, d'Alembert, and Lagrange.

Newton's laws described the movement of mass points and could appropriately be applied to the movement of celestial bodies. However, they could not describe the motion of rigid bodies, the motion of fluids or the vibrations of a stretched spring. A group of mathematicians from Basel, Switzerland concentrated on the solution to these questions, the brothers Jakob and Johann Bernoulli, their nephew Daniel Bernoulli, and Johann's pupil Leonhard Euler (1707-1783). Euler was considered one of the ablest, most brilliant, and most productive mathematician and scientist of the eighteenth century, if not of all time. Euler developed mathematical theories, including theories to describe the motion of vibrating bodies, and the buckling of beams and columns. One of Euler's major contributions was the expansion of Newton's laws to the applications of rigid and fluid bodies on earth. Euler established the concept of conservation of energy on a solid mathematical background. This concept which was not recognized by Newton, but started to develop during the Enlightenment as *vis viva* (Latin for living force).

As a baby, Jean le Rond d'Alembert (1717-1783) was found on the steps of the St. Jean le Rond Church. As an adolescent, he began studying law and was to appear before the bar in 1741, but his interests strayed to medicine and finally to mathematics. In 1743, two years after his acceptance at the Academie des Sciences, d'Alembert published the *Traité de Dynamique*. Contained within was d'Alembert's principle stating "...that Newton's third law of motion holds not only for fixed bodies but also for those free to move" (Millar, 1989).

The implications of d'Alembert's work for the development of biomechanics included the application of his principles to kinetics.

Joseph Louis Lagrange (1736-1815) demonstrated a keen interest in mathematics at an early age. He studied mathematics at the Royal Artillery School, and in 1766 replaced Euler as the director of the Berlin Academy of Sciences. Moving to Paris in 1797, Lagrange became the Professor of Mathematics at the Ecole Polytechnique. A victim of periods of severe depression, Lagrange had virtually abandoned mathematics by his late 40s. He began work on his *Méchanique Analytique* (1788) at the age of 19, but did not complete it until the age of 52. Lagrange generally treated mechanical problems with the use of differential calculus. His treatise did not contain any diagrams or geometrical methods, unlike Newton's *Principia*. Lagrange's equations expressed Newton's second law in terms of kinetic and potential energy.

Physiologists of the eighteenth century adopted mechanical philosophy to explain the structures and functions of the human body. However, advances in chemistry began to provide a new approach to physiology. The study of the body's pumps, pulleys, and levers gave way to the investigation of growth, regeneration, nutrition, and the chemical functioning of the human body. Vitalistic theory began to challenge mechanistic theories. Marie Francois Bichat's (1771-1802) philosophy that form follows function began to influence physiology. Research in muscle physiology had been facilitated by the microscope in the seventeenth century. In the eighteenth century, discovery of electricity increased interest and understanding of the nature of muscles.

During the mid-eighteenth century, Albrecht von Haller (1708-1777) became a leading physiologist and a proponent of vitalism. Drawing upon Hermann Boerhaave's (1668-1738) theory that fibres were the basic structural element of the body, von Haller focused on the structure and function of muscles. He expanded Glisson's theory of irritability or contraction, suggesting that contractility was an innate property of muscle. Experiments demonstrated that the muscle retained its ability to contract even after death, and that contraction of the muscles could be provoked by mechanical, thermal, chemical or electrical stimuli. He believed that nerves played an important role in contraction. Electricity had been found to produce muscle contraction, and physiologists suggested that electricity in the body took a fluid form, flowing through the nerves carrying sense stimuli and motor commands (Hankins, 1985). However, von Haller urged caution in interpreting the electrical signal as the mysterious *spiritus animalius* that controlled movements. He believed the experimental technique of the eighteenth century was not capable of revealing the secrets of electro-chemical nerve impulses. This resistance to accepting chemical theory as an all encompassing explanation resulted from the failure of purely mechanistic interpretations to explain everything.

In the eighteenth century, muscle physiology began with the observations of Baglivi (1688-1706) who differentiated between the structures and functions of smooth and striated muscle in 1700. James Keill (1674-1719), a leading proponent of the mechanical approach to physiology, calculated the number of fibres present in certain muscles, and calculated the amounts of tension per fibre required to lift a given weight. In 1728, Daniel Bernoulli (1700-1728) developed a mechanical theory of muscular contraction, replacing Borelli's fermentation hypothesis. He also studied the mechanics of breathing and the mechanical work of the heart. Charles Dufay (1698-1739) believed that all living bodies had electrical properties. Nicholas Andre (1658-1742) coined the term *orthopedics* in 1741 and believed that muscular imbalances created skeletal deformities. In 1750, Jallabert was the first to reeducate paralyzed muscles with electricity, writing the first book on electrotherapy. The first demonstrations of reflex action in the spinal cord were done by Robert Whytt (1714-1766) in 1751. He also localized the sites of single reflexes. Between 1776 and 1793, John Hunter (1728-1793) produced a descriptive analysis of muscle functions stating that "...muscle was fitted for self motion and was the only part of the body so fitted" (Rasch, 1958). Hunter believed that muscle function should be studied with live subjects, but not cadavers.

RELEVANCE TO BIOMECHANICS

The relevance of the Enlightenment to biomechanics lies in four major aspects:

- The concept of force became clearly understood,
- The concepts of conservation of momentum and energy started to develop,
- A mathematical consolidation of the different mechanical laws took place, and
- Muscle contraction and action became an event influenced by mechanical, biochemical, and electrical forces.

The development of Newtonian mechanics and a Newtonian world view during the eighteenth century stimulated discussion and debate regarding what exactly was force and the effects of such force. The relationship between force and movement became important. From this debate, the laws of conservation of energy and momentum developed, forming the mechanical foundation of biomechanics.

Mathematical analysis during the eighteenth century advanced the study of mechanics. Lagrange and d'Alembert developed methods of analysis, based on Newton's mechanics, that facilitated the study of dynamic human movements. As the tools for studying human movements were becoming more sophisticated, physiology adapted a more holistic approach to understanding the human body. The structures and functions of biological material became clearer in terms of a combination of chemical and mechanical processes. A greater understanding of the mechanisms of movements improved analysis of their movements in future biomechanical investigations.

1.2.6 THE GAIT CENTURY

The development of science, as it related to human movement in the nineteenth century, was influenced substantially by three events in the second half of the eighteenth century. These events were:

- Jean Jacques Rousseau's novel *Emile* in 1762,
- The invention of the steam engine by James Watt in 1777, and
- The storming of the Bastille in 1789.

Jean Jacques Rousseau's (1712-1778) novel *Emile* (1762), revived the ancient idea of a complementary development of body and intellect by encouraging a return to nature and physical activity. Rousseau depicted movement and sport as an ideal form of human activity and fulfillment. The invention of the steam engine in 1777 by James Watt (1736-1819) heralded the start of the industrial revolution and the development and need for leisure time and recreation. The storming of the Bastille in 1789 signalled the beginning of the French Revolution, and an end of the monopoly on sport and leisure by the upper class. Developments in physiology paralleled this increased interest in physical activity, as Nicholas Andre proposed that exercise during childhood could prevent musculo-skeletal deformities in adulthood.

The development of sport and leisure during the late eighteenth century created a renewed scientific interest in human locomotion. The nineteenth century was characterized by the development of instruments and experimental methods to increase the understanding of how we move. The analysis of human gait occupied physiologists, engineers, mathematicians, and adventurers. The study of locomotion began as an observational science, and by the end of the nineteenth century, photography had revolutionized and quantified the study of human and animal movement.

In 1836, 150 years after Borelli published his mechanical analysis of the movement of animals, brothers Eduard (1795-1881) and Wilhelm (1804-1891) Weber published their treatise *Die Mechanik der menschlichen Gehwerkzeuge* (On the Mechanics of the Human Gait Tools). The treatise contained almost 150 hypotheses about human gait that were derived by the Weber brothers from observations or theoretical considerations, or both. As in Borelli's time, a lack of instrumentation prevented a quantified analysis of movement, forcing scientists to rely on their senses and intuition. Many of the Webers' hypotheses were incorrect, some were correct, and others were to be verified. However, the primary importance of these hypotheses lay – as proposed by Cavanagh (1990) – not in the accuracy and appropriateness of their statements, but rather in establishing an agenda for further research of the human gait.

Etienne Jules Marey (1838-1904) transformed the study of locomotion from an observational science to one based on quantification. Marey's numerous inventions were "...designed entirely for providing a quantified and unbiased description of movement" (Bouisset, 1992). With the assistance of the French government, Marey developed the most extensive facility ever devoted exclusively to biomechanics (Cavanagh, 1990). Built on what is presently the Roland Garros Tennis Courts in the Parc des Princes, the Station Physiologique included a 500 m circular track equipped with monitoring equipment. Here he analyzed the movements of adults and children during sport and work, and the movements of horses, birds, fish, insects, and even a jellyfish. Marey's study of locomotion was boundless and his analysis comprehensive.

Marey is often given more recognition as a pioneer of cinematography than of biomechanics. Yet his inventions correlated ground reaction forces with movement, using pneumatic devices such as shoes (Figure 1.2.7), hoof covers, and the dynamometric table, which was the first serious force plate. He quickly embraced the potential of the photographic plate "...to overcome the defectiveness of our senses and the insufficiency of our traditional language" (Tosi, 1992). Marey developed existing technology to record sequential motion at comparatively high speeds. His photographic rifle worked on a similar principle as the Kodak disk camera. In 1889, Marey also developed the *Chronophotographe a pellicule* or modern cinecamera. Marey preferred frame-by-frame analysis of movement (Figure 1.2.8), arguing that the screen portrayed images he could not see with his own eyes. Marey's data collection techniques quickly became in high demand in such fields as cardiology, microscopy, mechanics, music, civil engineering, and hydrodynamics.

Figure 1.2.7 **Marey's pneumatic analysis of human locomotion (from Centre National d'Art et de Culture Georges Pompidou, 1977, with permission of Ville de Beaune, Conservation des Musées).**

Marey greatly influenced the development of biomechanics, providing the ability to quantify movements and the rigorous scientific nature of his investigations. He was the first to combine and synchronize kinematic and force measurement, and inspired comprehensive analysis of locomotion. Marey's research also provided insight regarding the storage and re-use of elastic energy, variations in ground reaction forces and centre of gravity motion, and the dependency of physiological cost on movement characteristics. The variety and abundance of his data collection methods inspired others to adapt and create devices for the quantification of motion.

Edweard Muybridge (1830-1904) began his career in the study of locomotion based on the intuition of Mr. Leland Stanford. Stanford was an avid horseracing fan. While owning and training many of his own horses Stanford became interested in their anatomy and movements. Stanford believed that there was a period in which all four hooves were off the ground during a horse's trot. Marey had proven this with his pneumonic device at slower gaits, but was unable to do so at faster speeds. Stanford commissioned Muybridge to pho-

Figure 1.2.8 **Marey illustrated the movements he analyzed from film recordings with scientific drawings such as this (from Tosi, 1992, with permission).**

tograph his horse, Occident, trotting, to verify his belief in unsupported transit. Muybridge was successful and subsequently began a lifetime devotion to documenting the sequential motion of humans and animals. Initially, he continued his collaboration with Stanford, documenting the gaits of Stanford's horses on what is today the site of Stanford University. However, after some controversy over the rights of the publication of the research in *Horses in Motion*, Stanford and Muybridge ended their partnership.

Muybridge's contribution to biomechanics was the sheer quantity of pictures he produced to document movement. Between 1884 and 1885, Muybridge produced 781 plates for a total of 20,000 images. These images were eventually published in *Animal Locomotion*, *Animals in Locomotion,* and *The Human Figure in Motion*. However, recent analysis by Marta Braun indicates that "...the relationship in some 40% of the plates is not what Muybridge states it to be" (Braun, 1993). Muybridge's publications were plagued with several inaccuracies, but the richness of his pictures were testimony to the importance of photography as a new language of science. Marey's acknowledgment of the Stanford-Muybridge study accorded it the status of scientific research. Muybridge and Marey shared a fascination for locomotion and a passion for the new photographic language, but unlike Marey, Muybridge lacked a scientific methodology. Muybridge and Marey collaborated in Paris to document the flight of birds. Marey was interested in Muybridge's methods and adopted photography as a tool for the study of locomotion.

In 1891, Wilhelm Braune and Otto Fischer made precise mathematical analysis possible by conducting the first tri-dimensional analysis of the human gait. To complete a mathematical study, the centre of gravity and moments of inertia of the body and all its parts were required. The centres of gravity were determined experimentally with the use of frozen cadavers. Two cadavers were nailed to a wall with long steel spits, allowing dissection with a saw to intersect longitudinal, sagittal, and frontal planes of the centre of gravity. The points of intersection of the centre of gravity were then recorded on lifesize drawings, and compared photographically with those of over 100 soldiers. A subject with the same measurements was then dressed in a black suit with thin light tubes and passed by four cameras (Maquet, 1992). The light tubes left imprints on the sensitive plates with each discharge of a Rhumkorff coil. A network of coordinates was later photographed over the picture of the subject passing through the cameras. These experiments were conducted at night and took 10 to 12 hours of constant effort. The preparation of the subject alone took up to eight hours.

Braune and Fischer's *Der Gang des Menschen* (1895-1904) contained the mathematical analysis of "...3 transits of the human gait including 2 on free walking and 1 walking with army knapsack, 3 full cartridge pouches and 88 rifle in the shoulder arms position" (Cavanagh, 1990). The data analyzed was gathered in a single night of experimentation. Braune died soon after the experiments, leaving the arduous task of analysis to Fischer. Today, the calculations that took Fischer several years could be done by computer in a matter of hours or even minutes. Their methods are essentially the same as those used by today's investigators in the biomechanics of gait.

In the nineteenth century, biology became diversified into specialized fields of study. Darwin's theories of evolution coincided with the discovery and increased awareness of the development of the structures and their functions over time. As seen in the development of bone physiology, contributors from different fields of science, such as engineering, began to influence developments in biology.

The foundations of electromyography were laid by the pioneers Du Bois Reymond and Duchenne during the nineteenth century. Galvinism and the regeneration of animal electricity sparked the imagination of Mary Shelly to create Frankenstein in 1816. The popularity of electricity filtered into the nineteenth century as many charlatans sold it as a cure all to desperate aristocrats. It was in this environment that Du Bois Reymond (1818-1922) refined the methods for measuring currents in 1841, and traced electricity in contracting muscle to its independent fibres. His experimental methods dominated electromyography for a century and laid the foundations for future researchers. In 1866, Duchenne (1806-1875) published *Physiologie des Mouvements* which described the muscle action of every important superficial muscle. By developing electrodes for the surface of the skin, he was able to observe the actions of the superficial muscles on hundreds of normal and abnormal subjects. He was the first in this respect to use abnormal muscle function to analyze normal muscle function. Unfortunately, this work was not translated into English until 1949, but was believed to be "...one of the greatest books of all times" (Rasch, 1958).

Bone physiology experienced dramatic development in the nineteenth century. An emphasis on the importance of proper alignment of the skeleton for good health prompted increased awareness and interest in bone growth and function. Andrew Still (1828-1917) developed osteopathy as a separate school of medicine during this time. Mechanical forces were studied to increase the understanding of bone physiology. In 1855, Breithaupt (1791-1873) described stress fractures in military recruits of a Prussian unit, inspiring further analysis of mechanical forces and bone physiology. Volkmann described the effects of pressure on bone growth in 1862, indicating an inverse relationship between increasing pressure and bone growth. In 1867, van Meyer described the relationship between the architecture of bone and its function. Engineering principles were introduced to bone physiology by Culmann (1821-1881) in 1867, when he observed that the "...interior architecture of the femur coincided with the graphostatic determination of the lines of maximum internal stress in the Forborne Crane" (Rasch, 1958). Wolff synthesized many of the prevalent ideas about bone physiology in 1870 with the formulation of Wolff's law. Recognizing the interdependence between form and function, he proposed that physical laws strictly dictated bone growth.

RELEVANCE TO BIOMECHANICS

The relevance of the gait century to biomechanics lies in three major aspects:

- Measuring methods were developed to quantify kinematics and kinetics of movement, and applied extensively to human gait analysis,
- Measuring methods were developed to quantify electrical current during muscular activity, and
- Engineering principles were applied in biological and biomechanical analysis.

The transformation of biomechanics from an observational and intuitive science to one based on quantification and mathematical analysis created an exciting new perspective for the analysis of human movement and locomotion. Instrumentation developed by Marey, Muybridge, and Braune and Fischer allowed the quantification of movement for the first time. The developments in photography illustrated the subtle stages of motion that human eyes could not see. Additionally, muscle action could be quantified with EMG measurements, opening a new avenue of the understanding of muscle function. Biological materials were understood in greater detail, combining an interest in their mechanical functions and their growth.

1.2.7 THE TWENTIETH CENTURY

The twentieth century was characterized by several factors that, in turn, affected the development of biomechanics. These factors were the:

- Mechanical and technological developments resulting from two World Wars,
- Increased popularity, social and financial recognition of sport in society, and
- Explosion of financial support for medical and health care research.

In this environment, the discipline of *biomechanics* developed with respect to knowledge and understanding, and with respect to the number of researchers and research centres in biomechanics. The following sections discuss some milestones in biomechanics and some external indications for the development of the field.

In 1920, Jules Amar published an analysis of physical and physiological components of work in his book *The Human Motor* (1920), taking into consideration the worker's environment and individual movements. A product of the post World War I industrialization, Amar's study focused on the efficiency of human movements.

Nicholas Bernstein (1896-1966) descended from a family of physicians. After a year of service in a war hospital during 1914, Bernstein chose to pursue a medical career. Upon completion of his studies, he served on the Siberian front during the Russian Civil War. By 1921, Soviet science began to develop a new direction. The scientific organization of work psychophysiology dominated Soviet science from the early 1920s to 1940s, with Bernstein leading the way (Jansons, 1992). Bernstein developed a method for measuring movement based on mathematical analysis. His analysis of the human gait as "...a whole, unified complex of biological symptoms" (Jansons, 1992), was even more comprehensive than the analysis by Braune and Fischer (1895-1904). Bernstein's biodynamic studies also included:

- an analysis of the proper use of tools, such as the hammer and the saw, which saw the redesign of the driver's cab in Moscow's trams,
- an analysis of the movements of working women, and
- bridge dynamics.

Bernstein's analysis of the coordination and regulation of movement in both children and adults provided the basis for his theories of motor control and coordination. He also established that adults ran far more economically than children. Isolated from the west, Bernstein's work was not published in North America until 1967.

A.V. Hill (1886-1977) began his career as a mathematician at Cambridge in 1915, but switched to physiology after two years of studies in mathematics. In 1922, Hill received the Nobel Prize in Physiology or Medicine. His main research activities focused on the explanation of the mechanical and structural function of human muscle. Based on evidence from experiments using isolated frog sartorius muscles (Hill, 1970), Hill developed theories for mechanical and structural muscle action. Hill's classical paper (Gasser & Hill, 1924) discussed the dynamics of muscular contraction, signifying the beginning of an applicable understanding of muscle physiology to muscle functioning. Hill and his co-workers also studied human locomotion, contributing to the understanding of the efficiency of running.

Interest in human movement flourished during the twentieth century and was not confined to mechanical analysis. Although he was not typically considered a contributor to the field of biomechanics, Rudolph Laban (1879-1958) developed a method of representing a series of complex human movements still used in dance choreography today. Breaking down movements into symbols, Laban gained substantial knowledge regarding the use of movement as an expression of the individual. Escaping the Nazis and arriving in England in 1938, Laban found his knowledge of human movement could be used in industry. Laban increased the efficiency of individual factory workers through planned movement sessions with groups of workers. Teaching them to find the rhythm of their individual movements, Laban trained them to balance movements such that "...strong movements are compensated for light, and narrow ranging by wide..." (Hodgson & Preston-Dunlop, 1990). Productivity increased as the worker's movements became balanced and efficient. Laban's techniques were also used by the Air Ministry to create a more efficient way of parachute jumping.

In 1938 and 1939, Elftman estimated internal forces in muscles and joints. To perform these estimations, he developed a force plate to quantify ground reaction forces and the centre of pressure under the foot during gait. Elftman concluded that muscles act, regulating energy exchange, by using the strategies of transmission, absorption, release, and dissipation (Elftman, 1939).

A.F. Huxley (1924-) studied physics at Cambridge, worked on radar in World War II, and eventually applied his knowledge of physics to muscle physiology. Huxley also worked at the Massachusetts Institute of Technology, at Cambridge. Beginning in the 1950s, Huxley gained recognition for his work with the sliding filament model of muscle contraction, and the development of x-ray diffraction and electron microscopy. Huxley made a significant breakthrough in the study of muscle action when he proposed his Sliding Filament theory to explain muscle shortening in 1953. Huxley later expanded his work by proposing connecting mechanisms between both actin and myosin filaments, constituting his Cross-bridge theory.

The development and growth of biomechanics in the twentieth century may be illustrated by various political and professional activities, and by the numerical expansion of the biomechanical community. The first breakthrough of biomechanics into the curricula of universities was in sport related disciplines. In the early twentieth century, some universities started to teach biomechanics in faculties of physical education. At that time, there was discussion as to whether the field should be called kinesiology or biomechanics, a question which has been solved today, with kinesiology being the science dealing with various aspects of movement, e.g., mechanical, physiological, and neurological.

On August 21-23 of 1967, the First International Seminar on Biomechanics was held in Zürich, Switzerland. Initiated by E. Jokl, organized by J. Wartenweiler, and sponsored by the International Council of Sport and Physical Education of UNESCO. The congress attracted approximately 200 participants from Europe, Asia, and America for discussion on the following:

- Technique of motion studies,
- Telemetry,
- Principles of human motion studies: general aspects of coordination,
- Applied biomechanics in work,
- Applied biomechanics in sports, and
- Clinical aspects.

This first international seminar was repeated biannually. During the 1973 conference in Penn State, the International Society of Biomechanics (ISB) was founded. J. Wartenweiler became the society's first president. For the 1975 congress in Jyväskylä, Finland, the name was changed to International Congress of Biomechanics, a name still used today. The major trust of the International Society of Biomechanics and its congresses was and is locomotion, musculo-skeletal mechanics, ergonomics, sport biomechanics, and clinical biomechanics. The biannual congresses are ongoing, and in the early 1990s attracted between 500 and 1000 participants.

In the 1980s a movement started to emerge, allowing biomechanists from all sub-disciplines of biomechanics, e.g., locomotion, orthopedic, sport, muscle, material, tissue, dental, and cardiac, to meet periodically as one group to exchange ideas and findings. In 1989, under the chairmanship of Y.C. Fung, a group of biomechanists from all the sub-disciplines, in cooperation with most of the international, regional, and national biomechanic organizations, organized the First World Congress of Biomechanics in San Diego, USA. The congress attracted approximately 1200 participants and included all areas of biomechanics. A Second World Congress of Biomechanics was organized in Amsterdam, Netherlands in 1994, a third in Sapporo, Japan in 1998, a fourth in Calgary, Canada in 2002 and a fifth in Munich, Germany in 2006.

Overall, there are now thousands of biomechanists working on all continents. Biomechanics is a recognized discipline with many universities offering courses and graduate programs. Results from biomechanical research directly influence medicine, work, and sport equipment development and many other aspects of human life. Recently, biomechanical research became integrated into multi-disciplinary research projects, contributing the mechanical aspect of understanding the biological systems.

Research dealing with movement, exercise, and sport, of which biomechanics is a substantial part, now has a prestigious international prize, the IOC-Olympic Prize. This prize recognizes the importance of this research, and is awarded every two years in connection with the Olympic Games. The first IOC-Olympic Prize was awarded in 1996 in Atlanta, USA. The second IOC-Olympic Prize was awarded in 1998 in Nagano, Japan to a biomechanist, Dr. Woo, for his outstanding work on ligament characteristics and the effect of movement on ligament healing. This award to a biomechanist is an indication of the increasing importance of science studying human movement, and of the high quality of biomechanical research performed in many centres and laboratories around the world. The third award of $500,000 US was presented in Sydney in connection with the Sydney 2000 Olympic Games.

RELEVANCE TO BIOMECHANICS

The relevance of the twentieth century to biomechanics lies in three major aspects:

- Biomechanics developed as a discipline at universities with graduate programs and faculty positions,
- Biomechanical research results were increasingly used in practical, medical, and industry applications, and
- Biomechanics became a player in a multi-disciplinary attempt to understand human and animal movement, and the effects of movement on the musculo-skeletal system.

Today, most major universities have faculty members with the title Professor of Biomechanics. Hundreds of students are enrolled in graduate programs, studying for a masters or a Ph.D in biomechanics. Many biomechanical research centres have contracts with various industries, helping them to improve their products regarding performance or safety, or both. Many hospitals have gait laboratories to quantify medical treatment or to help in the decision making process for difficult cases. Many biomechanical research centres have moved into a multi-disciplinary cooperation with other disciplines to answer complex questions.

1.2.8 FINAL COMMENTS

The Selected Historical Highlights section, below, includes a multitude of facts. One may risk losing sight of the forest through the trees with such a lengthy exchange of information. This section attempts to provide a short overview and to synthesize the different facts into one picture. The graph in Figure 1.2.9 shows the intensity of the development of areas such as mechanics, mathematics, anatomy, muscle, and locomotion research. Figure 1.2.9 also illustrates the different phases of development.

An initial phase (650 B.C.-200 A.D.) was characterized by an awakening of Western scientific activity with initial developments in mechanics, mathematics, anatomy, muscle, and locomotion research. This initial phase was followed by a phase of little development

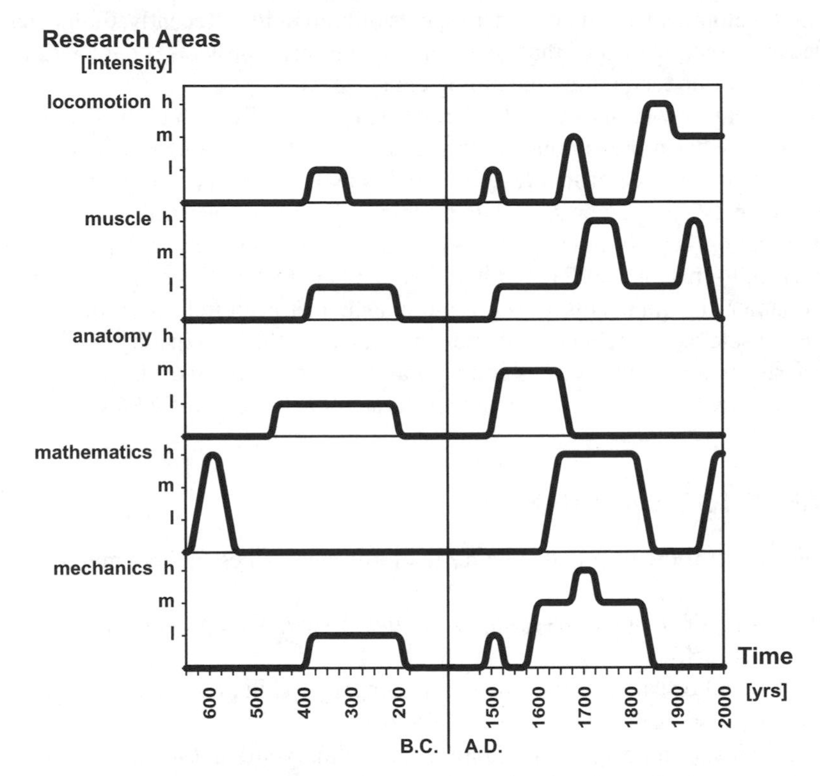

Figure 1.2.9 **Schematic illustration of time periods with significant developments in the areas of mechanics, mathematics, anatomy, muscle, and locomotion research as they relate to biomechanics.**

(200 A.D.-1450 A.D.). The Renaissance (1450 A.D.-1600 A.D.) changed the scientific picture, opening doors for further scientific development. Interest in the human and animal body awakened and the first locomotion studies were published.

The most decisive step for further biomechanical advancement occurred during the Scientific Revolution (1600 A.D.-1730 A.D.). This period was characterized by a new understanding of scientific research and a questioning of old concepts. It culminated in Newton's three laws of motion and a highly mechanistic world view. Some sources describe this period as the birthdate of biomechanics, with the publishing of the work *De Motu Animalium* by Borelli as the initiator of many future studies.

The mechanical understanding developed during the Scientific Revolution was ideal for celestial mechanics. However, when applied to terrestrial problems, many remained unsolved. The concept of force was not clearly understood, and the question of conservation of energy or momentum was debated intensively with no consensus. It was during the Enlightenment (about 1730 A.D.-1800 A.D.) that mathematicians such as d'Alembert, Lagrange, Leibnitz, and Euler worked intensively on these questions, providing the framework for mechanics which we use today. Conceptually, this was the period that changed mechanical thinking most dramatically.

The nineteenth century, the gait century, yielded the development of a variety of sophisticated techniques to analyze human and animal movement experimentally. The twentieth century provided an explosion of sophisticated experimental methods and, with the development of the computer, a wealth of numerical mathematical methods that could be applied to biomechanical research. Furthermore, it was characterized by an increased complex understanding of bone, cartilage, tendon, ligament, and especially muscle. Biomechanics developed into a discipline with university courses, departments, and graduate students. There are many thousands of biomechanical researchers today working in universities or in various industries. Results from biomechanical research contributed to the broadening of the understanding of the human body as well as to many practical applications in medicine, ergonomy, sport, and equipment.

1.3 MECHANICS

NIGG, B.M.

1.3.1 DEFINITIONS AND COMMENTS

Dynamics:

- Direct dynamics: Mechanical analysis of a system that determines forces from movement.
- Inverse dynamics: Mechanical analysis of a system that determines movement from forces.

Equation of motion: Set of mathematical equations for variables that describe the kinematics and kinetics of a system.

Particle: Matter that is assumed to occupy a single point in space.
Comment: in practical terms, this means that the volume of the body of interest is small compared to the space in which its behaviour is of interest.

Reference frame (RF): A system of coordinate axes relative to which positions can be defined.

- Inertial reference frame (IRF): A system of coordinate axes that have a constant orientation with respect to the fixed stars, and whose origin moves with constant velocity with respect to the fixed stars. An inertial reference frame, IRF, is a reference frame in which Newton's laws are valid.

Rigid body: Matter that is assumed to occupy a finite volume in space and that does not deform if subjected to forces.
Comment: a rigid body consists of a number of mass particles, with the distance between any two particles being constant. The rigid body concept may be valid where the deformations of the body are insignificant relative to the motion of the whole body.

Vector: A quantity which is defined by magnitude, direction, and sense of direction.
Comment: throughout this text, vectors are depicted with bold symbols.

1.3.2 SELECTED HISTORICAL HIGHLIGHTS

1686	Newton	An English philosopher, physicist, and mathematician who published the three laws of motion in his *Philosophiae Naturalis Principia Mathematica* (The Mathematical Principles of Natural Sciences).
1743	d'Alembert	A French mathematician who published an alternate form of Newton's second law of motion in his *Traité de Dynamique*.
1750s	Euler	A Swiss mathematician who expanded Newton's laws to the application of rigid and fluid bodies on earth.
1788	Lagrange	A French mathematician who published *Méchanique Analytique*, which included a formulation of Newton's second law in terms of kinetic and potential energy.
1834	Hamilton	An Irish mathematician and astronomer. His text *On a General Method in Dynamics* provided a different expression of Newton's second law by using coordinate position and energy.
1905	Einstein	Worked as a technical expert (class III) at the Swiss patent office in Bern where he developed the theory of relativity, which revolutionized the mechanical and philosophical thinking of humankind.

1.3.3 NEWTON'S LAWS OF MOTION

Newtonian mechanics is based on the three laws of motion, which were first formulated by Sir Isaac Newton in *Philosophiae Naturalis Principia Mathematica* in 1686. At least three other scientists, however, contributed significantly to the puzzle that Newton had to solve: Kepler, Galileo, and Descartes (Koestler, 1968). Kepler (who died about 30 years before Newton's synthesis) contributed the laws of motion of the heavenly bodies. Galileo (who died about 20 years before Newton's synthesis) contributed the laws of motion of bodies on earth. Descartes contributed the idea of inertia, which constrains bodies into straight motion if no external force is acting. Before Newton, these ideas did not appear to fit together. Some were even contradictory, e.g., planetary motion in ellipses or circles and the inertia concept. Newton's contribution was to *synthesize* these fragments into one concept, the three laws of motion, providing a framework for explaining them all. Newton's laws were formulated for single particles and were valid in an inertial frame of reference, a frame at rest or moving with uniform velocity relative to the distant, fixed stars. The three laws are:

First law: **A particle will remain in a state of rest or move in a straight line with constant velocity, if there are no forces acting upon the particle.**

If $\mathbf{F} = \mathbf{0}$ then $\mathbf{v} = \mathbf{constant}$

Second law: **A particle acted upon by an external force moves such that the force is equal to the time rate of change of the linear momentum.**

$$\mathbf{F} = k\frac{d(m\mathbf{v})}{dt}$$

where:

$\mathbf{F}$ = resultant force acting on the particle
m = mass of the particle
$\mathbf{v}$ = velocity of the particle
t = time
k = constant of proportionality where the constant of proportionality, k, is equal to unity for dynamically consistent unit systems, e.g., SI.

If the mass of a particle is assumed to be constant in time, Newton's second law reduces to:

$$\mathbf{F} = m\mathbf{a} = m\frac{d\mathbf{v}}{dt} = m\frac{d^2\mathbf{r}}{dt^2}$$

where:

$\mathbf{a}$ = acceleration of the particle

$\mathbf{r}$ = position of the particle

Note: velocity and linear momentum are measured relative to an inertial reference frame, IRF.

Third law: **When two particles exert force upon one another, the forces act along the line joining the particles and the two force vectors are equal in magnitude and opposite in direction.**

$$\mathbf{F}_{12} = -\mathbf{F}_{21}$$

where:

$\mathbf{F}_{ik}$ = force acting on particle i from particle k

Because of their importance for classical mechanics, Newton's laws are reproduced in their original Latin version (Cajori, 1960):

Lex I: **Corpus omne preservare in statu suo quiescendi vel movendi uniformiter in directum, nisi quatenus illud a viribus impressis cogitur statum suum mutare.**

Lex II: **Mutationem motis proportionalem esse vi motrice impressae, et fieri secundum lineam rectam qua vis illa imprimitur.**

Lex III: **Actioni contrariam semper et aequalem esse reactionem: sive corporum duorum actiones in se mutuo semper esse aequales et in partes contrarias dirigi.**

1.3.4 EQUATIONS OF MOTION FOR A RIGID BODY

This chapter attempts to summarize aspects of Newtonian mechanics as they relate to biomechanical applications. Derivation of formulas and detailed discussions of mechanics can be found in mechanics textbooks, e.g., Meirovitch, 1970, or in selected biomechanics textbooks, e.g., Fung, 1990. Consult these sources for further details. Specifically, the derivation of the equations from particle to rigid body mechanics is well documented in textbooks.

In this section, three-dimensional descriptions of translation and rotation are presented. Then, assumptions are made to reduce these general descriptions to less general ones or two-dimensional descriptions. In all cases, an outline of how the variables in the equations can be determined is provided.

GENERAL THREE-DIMENSIONAL CASE FOR TRANSLATION

The equation of motion for the centre of mass of a rigid body is:

$$\mathbf{F} = \frac{d\mathbf{p}_{CM}}{dt}$$

or for a rigid body with constant mass:

$$\mathbf{F} = m\mathbf{a}_{CM} = m\frac{d^2\mathbf{r}_{cm}}{dt^2} = m\ddot{\mathbf{r}}_{CM}$$

where:

$\mathbf{F}$ = resultant force acting on the rigid body
$\mathbf{r}_{CM}$ = position of the centre of mass of the rigid body
m = mass of the rigid body
$\mathbf{a}_{CM}$ = acceleration of the centre of mass of the rigid body
$\mathbf{p}_{CM}$ = linear momentum of the centre of mass of the rigid body

or in matrix form:

$$\begin{bmatrix} F_x \\ F_y \\ F_z \end{bmatrix} = m \begin{bmatrix} a_{CMx} \\ a_{CMy} \\ a_{CMz} \end{bmatrix} \tag{1.3.1}$$

Determination of variables in the equation

Mass: The mass of a rigid body as applied to biomechanical problems can be determined using methods described in Chapter 3. In general, experimental mass determination is appropriate for the required accuracies in biomechanical projects.

Acceleration: Accelerations can be measured with accelerometers or from film or video pictures using the second time derivative of position. Both approaches have methodological problems, as described in Chapter 3. However, both methods often provide accelerations with acceptable errors for movements with non-excessive accelerations, e.g., impacts.

SIMPLIFIED TWO-DIMENSIONAL CASE FOR TRANSLATION

Often it is sufficient and appropriate to discuss planar movement. In this special case, the accelerations in the third dimension are assumed to be zero.

Defining the plane of interest as the x-y-plane, the set of translational equations of motion are:

$$\mathbf{F}_x = m\mathbf{a}_{CMx}$$
$$\mathbf{F}_y = m\mathbf{a}_{CMy}$$

GENERAL THREE-DIMENSIONAL CASE FOR ROTATION

For the subsequent discussion, the following definitions and assumptions are used:

- X, Y, Z is an inertial reference frame with origin O.
- x, y, z is a body fixed reference frame with origin at the centre of mass, CM.

In the inertial reference frame, X, Y, Z, the rotational equation of motion is:

$$\mathbf{M}^{XYZ} = \dot{\mathbf{H}}^{XYZ} \tag{1.3.2}$$

where:

$\mathbf{H}^{XYZ}$ = the angular momentum of the body with respect to the inertial frame X, Y, Z, with:

$$\mathbf{H}^{XYZ} = \int_V (\mathbf{r} \times \dot{\mathbf{r}})\, dm$$

where:

$\mathbf{r}$ = vector from origin to a point in the body

$\mathbf{M}^{XYZ}$ = resultant moment (net moment) applied to the rigid body with respect to the origin of the inertial frame, X, Y, Z

Equation (1.3.2) may be split into two parts:

$$\mathbf{M}^{XYZ} = \mathbf{M}_O^{XYZ} + \mathbf{M}_{CM}^{xyz} = \dot{\mathbf{H}}_O^{XYZ} + \dot{\mathbf{H}}_{CM}^{xyz} = \dot{\mathbf{H}}^{XYZ} \tag{1.3.3}$$

with:

$$\mathbf{M}_{CM}^{xyz} = \dot{\mathbf{H}}_{CM}^{xyz} \tag{1.3.4}$$

$$\mathbf{M}_O^{XYZ} = \dot{\mathbf{H}}_O^{XYZ} \tag{1.3.5}$$

where:

$\mathbf{H}_O^{XYZ}$ = angular momentum due to the centre of mass revolving about the origin of the X, Y, Z inertial reference frame

$\mathbf{H}_{CM}^{xyz}$ = angular momentum due to the body rotating about the origin of the body fixed reference frame, x, y, z with:

$$\mathbf{H}_{CM}^{xyz} = \int_V (\mathbf{R} \times \dot{\mathbf{R}})dm$$

where:

$\mathbf{R}$ = vector from CM to a point in the body of interest

$\mathbf{M}_O^{XYZ}$ = moment about the origin of the X, Y, Z inertial reference frame

$\mathbf{M}_{CM}^{xyz}$ = moment about the origin of the body fixed reference frame, x, y, z

The expansion of $\mathbf{H}_{CM}^{xyz}$ in equation (1.3.4) yields:

$$\{\mathbf{M}_{CM}^{xyz}\} = [I]\{\alpha\} + \{\omega\} \times [I]\{\omega\}$$

where:

$\{\}$ = vector symbol

$[]$ = matrix symbol

$\{\mathbf{M}_{CM}^{xyz}\}$ = resultant moment vector about the CM expressed in components of x, y, z

$[I]$ = inertia tensor

$\{\alpha\}$ = angular acceleration vector with respect to the body fixed reference frame x, y, z

$\{\omega\}$ = angular velocity vector

or in an expanded version:

$$\begin{bmatrix} M_{CMx} \\ M_{CMy} \\ M_{CMz} \end{bmatrix} = \begin{bmatrix} I_{xx} & -I_{xy} & -I_{xz} \\ -I_{yx} & I_{yy} & -I_{yz} \\ -I_{zx} & -I_{zy} & I_{zz} \end{bmatrix} \begin{bmatrix} \alpha_x \\ \alpha_y \\ \alpha_z \end{bmatrix} + \begin{bmatrix} 0 & -\omega_z & \omega_y \\ \omega_z & 0 & -\omega_x \\ -\omega_y & \omega_x & 0 \end{bmatrix} \begin{bmatrix} I_{xx} & -I_{xy} & -I_{xz} \\ -I_{yx} & I_{yy} & -I_{yz} \\ -I_{zx} & -I_{zy} & I_{zz} \end{bmatrix} \begin{bmatrix} \omega_x \\ \omega_y \\ \omega_z \end{bmatrix} \quad (1.3.6)$$

where:

ω_i = angular velocity component relative to the X, Y, Z system expressed in components of x, y, z

α_i = angular acceleration component relative to the X, Y, Z system expressed in components of x, y, z

I_{ij} = $\begin{cases} i=j & \text{moment of inertia with respect to the principal axes} \\ i \neq j & \text{products of inertia} \end{cases}$

where for the moments of inertia:

$$I_{xx} = \int_V (y^2 + z^2)dm$$

$$I_{yy} = \int_V (x^2 + z^2)dm$$

$$I_{zz} = \int_V (x^2 + y^2)dm$$

and for the products of inertia:

$$I_{xy} = \int_V xy\,dm$$

$$I_{xz} = \int_V xz\,dm$$

$$I_{yz} = \int_V yz\,dm$$

Determination of variables in the equation

Moments of inertia: Experimental methods for determining the moments of inertia, I_{xx}, I_{yy}, and I_{zz} are discussed in Chapter 3. Several methods are available, and the moments of inertia can usually be determined with sufficient accuracy.

Products of inertia: The same experimental methods can be used to determine the products and moments of inertia. In most biomechanical applications, the body of interest is symmetrical enough to allow the principal axes, x, y, z, to be chosen such that the products of inertia are zero.
Note: principle axes can always be found and the products of inertia are exactly zero.

Angular velocity: Experimental methods for determining the angular velocity of a rigid body are described in Chapter 3. For typical biomechanical applications, these methodologies provide sufficiently accurate results.

Angular acceleration: Experimental methods for determining the angular acceleration of a rigid body are described in Chapter 3. For most biomechanical applications, these methodologies provide results that suffer from the typical problems of double differentiation.

SIMPLIFIED THREE-DIMENSIONAL CASE FOR ROTATION

If the body's fixed axes are aligned along the principal axes, the products of inertia are zero and the inertia tensor becomes diagonal. For this case, the rotational equations of motion are:

$$M_{CMx} = I_{xx}\alpha_x + (I_{zz} - I_{yy})\omega_y\omega_z$$

$$M_{CMy} = I_{yy}\alpha_y + (I_{xx} - I_{zz})\omega_x\omega_z$$

$$M_{CMz} = I_{zz}\alpha_z + (I_{yy} - I_{xx})\omega_x\omega_y$$

For the special case where the body fixed axis system is aligned along the principal axes, these equations are called the modified Euler equations. These equations are often used for three-dimensional biomechanical analysis.

SIMPLIFIED TWO-DIMENSIONAL CASE FOR ROTATION

If the motion is planar, e.g., in the x-y-plane, the rotational equations of motion take a further simplified form. The angular velocity vector has only one component (in this example, the z-component):

$$\omega_x = 0$$
$$\omega_y = 0$$
$$\alpha_x = 0$$
$$\alpha_y = 0$$

and the resulting rotational equation of motion is:

$$M_{CMz} = I_{zz}\alpha_z$$

This equation is used for most simplified rotational applications with typical forms such as:

$$I_{zz}\alpha_z = |\mathbf{r}_1 \times \mathbf{F}_1| + |\mathbf{r}_2 \times \mathbf{F}_2| + \ldots + |\mathbf{r}_n \times \mathbf{F}_n|$$

1.3.5 GENERAL COMMENTS

Mechanical analysis may concentrate on forces (kinetics) or on movement (kinematics), or both. Forces are the cause; movement is the result. Mechanical analysis can proceed from forces to movement or from movement to forces.

If the analysis starts with the cause (the force), the process is called direct or forward dynamics. A unique movement is the result of a defined set of forces. Consequently, the direct or forward dynamics approach has one solution and the approach is deterministic.

If the analysis starts with the result (the movement), the process is called inverse dynamics. A specific movement, however, can be the result of an infinite number of combinations of individual forces acting on a system. The inverse dynamics approach, therefore, has an infinite number of possible solutions and is not deterministic.

Newton's second law is essential for explaining and describing motion. Many researchers since Newton have centered their work on this law, and some of them have attempted to propose other (hopefully more elegant) mathematical procedures for applying Newton's second law of motion. The most popular ones are the approaches developed by Lagrange, d'Alembert, and Hamilton. This text concentrates primarily on Newtonian mechanics. The approaches of Lagrange, d'Alembert, and Hamilton are described in detail in advanced mechanics textbooks, and the reader should reference these sources for more specifics.

1.4 REFERENCES

Amar, J. (1920) *The Human Motor: Or the Scientific Foundations of Labour and Industry.* E.P. Dutton, New York.

Ascenzi, A. (1993) Biomechanics and Galileo Galilei. *J. Biomech.* **26 (2)**, pp. 95-100.

Asmussen, E. (1976) Movement of Man and Study of Man in Motion: A Scanning Review of the Development of Biomechanics. *Biomechanics V-A* (ed. Komi, P.V.). University Park Press, Baltimore. pp. 23-40.

Basmajian, J. and De Luca, C. (1985) *Muscles Alive: Their Functions Revealed by Electromyography* (5ed). Williams & Wilkins, London.

Bastholm, E. (1950) *The History of Muscle Physiology, from the Natural Philosophers to Albrecht von Haller.* Kopenhagen.

Bernstein, N. (1923) *Studies of the Biomechanics of the Stroke by Means of Photoregistration.* Res. Central Institute of Work, Moscow. **N1**, pp. 19-79. (In Russian).

Bernstein, N.A. (1967) *The Coordination and Regulation of Movements.* Pergamon Press, Oxford.

Borelli, G.A. (1680) *De Motu Animalium. Pars Prima* (trans. Bernabò, A., 1989). Springer Verlag, Berlin.

Borelli, G.A. (1681) *De Motu Animalium. Opus Posthumun. Pars Altera* (trans. Maquet, P., 1989). Springer Verlag, Berlin.

Borelli, G.A. (1734) *De Motu Animalium.* G. Bernoulli. *De Motu Musculorum et de Effervescentia et Fermentatione.* F. Mosca Publ., Naples, Italy.

Bouisset, S. (1992) *Etienne-Jules Marey. On When Motion Biomechanics Emerged as a Science.* ISB Series. pp. 71-87.

Braun, G.L. (1941) Kin. from Aristotle to 20th Century. *Research Quarterly.* **12 (2),** pp. 163-173.

Braun, M. (1993) *The Moment as it Flies.* University of Chicago Press, Chicago.

Braune, W. and Fischer, O. (1900) *Der Gang des Menschen* (trans. Hirzel, S.). Leipzig.

Braune, W. and Fischer, O. (1987) *The Human Gait* (trans. Maquet, P. and Furlong, R.). Springer Verlag, Berlin.

Cajori, F. (1960) *Sir Isaac Newton's MATHEMATICAL PRINCIPLES of Natural Philosophy and his System of the World.* University of California Press, Berkley, CA.

Cappozzo, A. and Marchetti, M. (1992) Borelli's Heritage. *Biolocomotion: A Century of Research Using Moving Pictures* (eds. Capozzo, A., Marchetti, M., and Tosi, V.). Rome, Italy. ISB Series - Volume I, Promograph. pp. 33-47.

Cavanagh, P.R. (1990) The Mechanics of Distance Running: A Historical Perspective. *Biomechanics of Distance Running* (ed. Cavanagh, P.R.). Human Kinetics Publishers, Champaign, IL. **1,** pp. 1-34.

Centre National d'Art et de Culture Georges Pompidou (1977) Museé National d'Art Moderne, Paris. *E.J. Marey. 1830/1904.* p. 22.

Clarys, P. and Lewille, L. (1992) *Clinical and Kin. Electromyography by Le Dr. Duchenne (de Boulogne).* ISB Series. pp. 90-112.

Clayton, M. (1992) *Leonardo da Vinci; Anatomy of Man.* Museum of Fine Arts, Houston.

Cooper, J.M. The Historical Development of Kinesiology with Emphasis on Concepts and People. *Kinesiology: A Natural Conference on Teaching* (eds. Dillman, C. and Sears, R.). University of Illinois Press, Urbana-Champaign. pp. 3-15.

d'Alembert, M. (1743) *Traité de Dynamique.* Chez David l'Aîné, Paris.

Dijksterhuis, E.J. (1961) *The Mechanization of the World Picture: Pythagoras to Newton.* Oxford University Press, Princeton.

Elftman, H. (1938) The Force Exerted by the Ground in Walking. *Arbeitsphysiologie.* **10,** pp. 485-491.

Elftman, H. (1939) Force and Energy Changes in the Leg During Walking. *Am. J. of Physiol.* **125 (2),** pp. 339-366.

Foster, M. (1901) *Lectures on the History of Physiology During the Sixteenth, Seventeenth, and Eighteenth Centuries.* Cambridge University Press, Cambridge.

Fung, Y.C. (1981) *Biomechanics.* Springer Verlag, New York.

Fung, Y.C. (1990) *Biomechanics: Motion, Flow, Stress, and Growth.* Springer Verlag, New York.

Furusawa, K., Hill, A.V., and Parkinson, J.L. (1927) The Energy Used in Sprint Running. *Proc. Roy. Soc.* **102 (B),** pp. 43-50.
Gasser, H.S. and Hill, A.V. (1924) The Dynamics of Muscular Contraction. *Proc. Roy. Soc.* **96 (B),** pp. 398-437.
Gillespie, C.C. (1973) *Dictionary of Scientific Biography.* Charles Scribner's Sons, New York.
Gray, Sir J. (1953) *How Animals Move.* Cambridge University Press, Cambridge.
Gray, Sir J. (1968) *Animal Locomotion.* Norton, New York.
Haas, R.B. (1976) *Muybridge: Man in Motion.* University of California Press, Berkeley.
Hamilton, W.R. (1834) On a General Method in Dynamics. *Phil. Trans. Roy. Soc.* pp. 247-308.
Hankins, T. (1985) *Science and Enlightenment.* Cambridge University Press, New York.
Hatze, H. (1971) Was ist Biomechanik. *Leibesübungen Leibeserziehung.* **25,** pp. 33-34.
Hay, J.G. (1973) *Biomechanics of Sports Techniques.* Prentice Hall, Inc., Englewood Cliffs, NJ.
Hill, A.V. (1970) *First and Last Experiments in Muscle Mechanics.* Cambridge University Press, Cambridge.
Hirt, S. (1955) What is Kinesiology?: A Historical Review. *Physical Therapy Reviews.* **35 (18),** pp. 419-426.
Hodgson, J. and Preston-Dunlop, V.M. (1990) *Rudolf Laban: An Introduction to his Work and Influence.* Northcote House, Plymouth, England.
Huxley, H.E. (1953) Electron Microscope Studies of the Organization of the Filaments in Striated Muscle. *Biochem. Biophys. Acta.* **12,** pp. 387-394.
Jansons, H. (1992) Bernstein: The Microscopy of Movement. *Biolocomotion: A Century of Research Using Moving Pictures* (eds. Capozzo, A., Marchetti, M., and Tosi, V.). Rome, Italy. ISB Series - Volume I, Promograph. pp. 137-174.
Kardel, T. (1990) Neils Stensen's Geometric Theory of Muscle Contraction: 1667: A Reappraisal. *J. Biomech.* **23 (10),** pp. 953-965.
Keele, K.D. (1983) *Leonardo da Vinci's Elements of the Science of Man.* Academic Press, New York.
Koestler, A. (1968) *The Sleepwalkers.* Hutchinson, London.
Lagrange, J.L. (1788) *Méchanique Analytique.* Veuve Desaint, Paris.
Lorini, G., Bossi, D., and Specchia, N. (1992) The Concept of Movement Prior to Giovanny Alphonso Borelli. *Biolocomotion: A Century of Research Using Moving Pictures* (eds. Capozzo, A., Marchetti, M., and Tosi, V.). Rome, Italy. ISB Series - Volume I, Promograph. pp. 23-32.
Maquet, P. (1992) *The Human Gait by Braune and Fischer.* ISB Series. pp. 115-125.
Marey, E.J. (1873) *La Machine Animale.* Librairie Germer Baillière, Paris.
Marey, E.J. (1878) *La Méthode Graphique.* G. Masson Editeur, Paris.
Marey, E.J. (1885) *La Méthode Graphique* (2ed). G. Masson Editeur, Paris.
Marey, E.J. (1972) *Movement.* Arno, New York. (Original 1895).
Marey, E.J. (1977) *La Photographie Mouvement*. Conservation des Musée de Beaune, France.
Meirovitch, L. (1970) *Methods of Analytical Dynamics.* McGraw-Hill, Toronto.
Millar, D., Millar, I., Millar, J., and Millar, M. (1989) *Chambers Concise Dictionary of Scientists.* W. & R. Chambers Ltd., Edinburgh.
Muybridge, E. (1887) *Animal Locomotion (1-11).* University of Pennsylvania, Philadelphia.
Muybridge, E. (1955) *The Human Figure in Motion.* New York, Dover. pp. 1-3. (Original 1887).
Needham, D. (1971) *Machina Carnes.* Cambridge University Press, Cambridge.
Nelson, R.C. (1976) *Contribution of Biomechanics to Improved Human Performance.* The Academy Pages. **10,** pp. 61-65.
Newton, J.S. (1686) *Philosophiae Naturalis Principia Mathematica.* Jussu Societatis Regiae ac Typis Josephi Streater, London.
Nussbaum, M.C. (1978) *Aristotle's De Moto Animalium.* Princeton University Press, Princeton, NJ.
Rasch, P.J. (1958) Notes Toward a History of Kinesiology (Parts 1, 2, and 3). *J. Amer. Osteo. Assoc.* pp. 572-574, 641-644, 713-715.
Sarton, G. (1953) *A History of Science: Ancient Science Through the Golden Age of Greece (1-2).* W.W. Norton & Company Inc., New York.
Singer, C.J. (1959) *A Short History of Scientific Ideas to 1900.* Clarendon Press, New York.
Symposia of the Society for Experimental Biology (1980) *Symposia XXXIV. The Mechanical Properties of Biologist Materials.* Cambridge University Press, Cambridge.

Tosi, V. (1992) Marey and Muybridge: How Modern Biolocomotion Analysis Started. *Biolocomotion: A Century of Research Using Moving Pictures* (eds. Capozzo, A., Marchetti, M., and Tosi, V.). Rome, Italy. ISB Series - Volume I, Promograph. pp. 51-69.

Wartenweiler, J. (1972) Zur Geschichte der Biomechanik. *Jugend und Sport.* **12,** pp. 323-324.

Weber, W. and Weber, E. (1836) *Die Mechanik der menschlichen Gehwerkzeuge.* Dietrichsche Buchhandlung, Göttingen.

Weisheipl, J. (1985) *Nature and Motion in the Middle Ages* (ed. Carroll, W.). The Catholic University of America Press, Washington, D.C.

Westfall, R.S. (1971) *The Construction of Modern Science.* Cambridge University Press, New York.

Wolf, J. (1986) *The Law of Bone Remodelling* (trans. Maquet, P.). Springer Verlag, Berlin. (Original 1892).

Zernicke, R.F. (1981) Emergence of Human Biomechanics. *Perspectives on the Academic Disciplines of Physical Education* (ed. Brooks, G.A.). Human Kinetics Publications, Champaign, IL. pp. 124-136.

2 BIOLOGICAL MATERIALS

The human body is constructed of bone, cartilage, ligament, tendon, muscle, and other connective tissues. The anatomical components of the human body can be divided into *active* and *passive* structures. Active structures produce force; passive structures do not. Muscles are active, while bones, cartilage, ligaments, tendons, and the remaining soft tissues are passive structures.

The musculo-skeletal system can influence the environment. Its actions can be interpreted mechanically, e.g., translation or rotation, biochemically, e.g., expression of joy or pain by facial movements, and in many other ways. This chapter provides insight into the *mechanical aspects* of construction and function of bone, cartilage, tendon, ligament, and muscle. However, the various biological materials also exist and function together in the joints, which allow relative movement between segments, transfer of forces from one segment to the next, and facilitate biochemical and physiological interaction between neighbouring biological materials. The synergy of the various components building a joint is described and discussed in a special section of this chapter. Finally, the mechanical characteristics are influenced by factors such as age, gender, exercise, immobility, and nutrition. Selected influences of these factors on the mechanical and biological integrity of biological materials are discussed in an overview section.

2.1 DEFINITIONS AND COMMENTS

2.1.1 ANATOMY

Bone

Bone modelling: Bone modelling is the process by which bone mass is increased.

Bone remodelling: Bone remodelling is the process by which bone mass is maintained or decreased.

Cancellous bone: Bone of reticular or spongy structure formed by thin spicules (trabeculae).
Comment: *cancellous* bone is also called *spongy* bone.

Cortex: An external layer of material.

Cortical bone: A solid external layer of material comprising the walls of diaphyses and external surfaces of bone.
Comment: *cortical* bone is also called *compact* bone.

Diaphysis: The elongated cylindric portion of a long bone between the ends (epiphyses).

Epiphysis: The end of a long bone.

Mineral content: The ratio of the unit weight of the mineral phase of bone to the unit weight of dry bone.

Osteoblast: A cell that arises from a fibroblast and that, as it matures, is associated with the formation and mineralization of bone.

Osteoclast: A large multinuclear cell associated with the absorption of bone.

Periosteum: Connective tissue covering the outer surface of bone.

Primary lamellar bone: Bone material that is arranged in circular form around the inner (endosteal) and outer (periosteal) circumference of a bony structure.
Comment: cancellous bone (at the end of long bones) belongs to the group of primary lamellar bone.

Primary osteon: A set of concentric lamellae together with associated bone cells (osteocytes) and a central vascular channel.

Trabecula: Little beam.

Trabecular bone: Bony spicules forming a meshwork of intercommunicating spaces that are filled with bone marrow. Comment: the terms trabecular and cancellous bone are used interchangeably.

Water content: The ratio of extracted water divided by the volume of the specimen (bone).

Articular Cartilage

Alymphatic: Not supplied with lymphoid tissue.

Aneural: Not supplied with nerves.

Anisotropic: Having physical properties that are not the same in all directions.

Articular cartilage: A thin layer of fibrous connective tissue on the articular surface of bones in synovial joints.

Avascular: Not supplied with blood vessels.

Cell: Basic unit of life within a tissue.

Chondrocyte: A mature cartilage cell embedded in a small cavity (lacuna) in the cartilage matrix.

Collagen: A protein substance of the matrix.

Homogeneous: Of a uniform quality throughout.

Isotropic: Having physical properties, e.g., wave conductivity, that are the same in all directions.

Lacuna: Small cavity.

Matrix: The intercellular substance of a tissue from which the structure develops.

- Interterritorial matrix: Matrix that lies outside the territorial matrix.
- Pericellular matrix: Matrix completely surrounding the chondrocytes (cells), which contains small amounts of collagen, proteoglycan, and other proteins.
- Territorial matrix: Matrix that surrounds the pericellular matrix. It contains fibrillar collagen, which adheres to the pericellular matrix.

Proteoglycans:	Group of glycoproteins found in connective tissue. Formed of subunits of disaccharides linked together and joined to a protein core.
Synovial fluid:	Transparent, alkaline, viscous fluid secreted by the synovial membrane contained in joint cavities.
Tropocollagen:	The molecular unit of collagen fibrils, about 1.4 nm wide and about 280 nm long. Tropocollagen is a helical structure consisting of three polypeptide chains. Each chain is composed of about a 1000 amino acids coiled around each other to form a spiral.
Zone:	Region or area with specific characteristics.
• Deep zone:	Layer of articular cartilage between the bone and the transitional zone.
• Superficial zone:	Thin part of articular cartilage situated near the surface.
• Transitional zone:	Layer of articular cartilage between the superficial and the deep zone.

Ligament

ACL:	Anterior cruciate ligament of the knee.
Elastin:	Elastic fibres present in various quantities in ligaments. Comment: elastin is found in small amounts (≈1.5%) in extremity ligaments. In elastic ligaments (e.g., lig. flavum), however, elastin fibres are about twice as common as collagen fibres.
Epiligament:	Loose envelope that encloses a ligament.
Fibroblasts:	Ligament cells, usually ovoid- or spindle-like and oriented longitudinally along the length of the ligament.
Fibronectin:	A protein composed of two 220-kD subunits joined by a single disulfide bridge and containing about 5% carbohydrate.
Insertion:	The area at which a ligament inserts into bone.
MCL:	Medial collateral ligament of the knee.
Matrix:	Body of a ligament; consists of water, collagen, proteoglycans, fibronectin, elastin, actin, and other glycoproteins.

Midsubstance:	Central part of a ligament, halfway between insertions.
Non-axial fibres:	Fibres not running parallel to the long axis of a ligament.
Proteoglycans:	Group of glycoproteins present in connective tissue. They are composed of subunits of disaccharides linked together and joined to a protein core.

Most ligaments are identified by the points where they attach to bone, e.g., talo-fibular, their shape, e.g., deltoid, their gross functions, e.g., capsular, their relationships to a joint, e.g., collateral, or their relationships to each other, e.g., cruciate. Some of these ligaments also have functional subdivisions, so there are probably at least twice as many functionally discrete ligaments as have so far been named. Only a few of these ligaments have been studied scientifically.

Tendon

Aponeurosis:	Tendinous expansion into which the muscle fibres insert, serving mainly to connect muscle and tendon.
Cell:	Basic unit of life within a tissue.
Chondrocyte:	A mature cartilage cell embedded in a small cavity (lacuna) within the cartilage matrix.
Collagen:	The protein substance of the white fibres (collagen fibres), composed of molecules of tropocollagen.
Matrix:	The intercellular substance of a tissue from which the structure develops.
Myotendinous junction:	Region where the muscle fibres join with the tendon.
Proteoglycans:	Group of glycoproteins present in connective tissue. Formed of subunits of disaccharides linked together and joined to a protein core.
Sheaths:	Tubular structure enclosing or surrounding some organ part.
• Endotenon sheath:	Connective tissue separating the fascicles of a tendon.
• Epitenon sheath:	Fibrous tissue sheath covering a tendon.
• Paratenon sheath:	Fibrous tissue sheath covering a tendon.
• Peritenon sheath:	Connective tissue investing larger tendons and extending as septa (partitions) between the fibres composing them.

Tendon: A fibrous, cord-like structure of connective tissue by which a muscle is attached to bone.

Muscle

Golgi tendon organ: A sensory organ consisting of nerve fibres located near the muscle-tendon junction. Golgi tendon organs lie in series with the muscle fibres and are sensitive to tension generated in the muscle.

Isokinetic contraction: A contraction of a muscle in which the rate of change in length (or the speed of contraction) is constant. The rate of change in length (or the speed) referred to may be associated with the joint angle, the muscle-tendon complex or the sarcomere, depending on the context.

Isometric contraction: A contraction of a muscle in which length remains constant. The length referred to may be that of the muscle-tendon complex, the fibre or the sarcomere, depending on the context.

Isotonic contraction: A contraction of a muscle in which the resistive force is kept constant throughout the contraction.

Motor neuron: A neural cell (neuron) that conveys impulses that initiate muscular contraction.

Motor unit: A set of muscle fibres that are innervated by the same motor neuron. It represents the smallest unit of control of muscular force.

Muscle: (*L. musculus*) A type of tissue composed of contractile cells or fibres, which effects movement of an organ or part of the body. The outstanding characteristic of muscular tissue is its ability to shorten or contract (Taber's Cyclopedic Medical Dictionary, 1981).

Muscle fibres: Threadlike cells of the muscle.

Muscle spindle: Sensory organ consisting of adapted muscle fibres enclosed in a capsule and lying parallel to the normal muscle fibres. Muscle spindles are sensitive to length changes and the rate of change in length of muscle fibres.

Myofibril: A tiny fibril containing myofilaments that runs parallel to the long axis of the fibre (cell).

Neuromuscular junction: Junction where the motor neuron meets the muscle and where impulses from the motor neuron are transmitted to the muscle.

Sarcomere: The basic contractile unit of striated muscle. It contains the region of a myofibril that lies between two adjacent dark lines called the Z-lines.

Thick filament: Filament predominantly made up of myosin proteins.

Thin filament: Filament predominantly made up of actin, tropomyosin, and troponin proteins.

Joints

Joint: Junction of two or more bones of the skeleton.
Comment: joints are also called articulations.

- Cartilaginous joint: Joints where the bones are connected by cartilage.
- Fibrous joint: Joints where the bones are connected by connective fibrous tissue.
- Synovial joint: Joints where the bones articulate relative to one another. They are connected by ligaments.

Synovial fluid: Transparent alkaline viscous fluid.
Comment: synovial fluid is also called synovium.

2.1.2 MECHANICS

Creep: Increase in deformation over time under a constant force or a force reached repetitively in a cyclic fashion.

Degrees of freedom (DOF): A minimum number of kinematic variables (coordinates) required to specify all positions and orientations of the body segments in the system.

Density: Mass per unit volume.

Elastic modulus (for a linearly elastic structure): The ratio of stress divided by strain.

$$E = \frac{\sigma}{\varepsilon}$$

where:

σ = stress

ε = strain

	E = elastic modulus
	with the units:
	$[E] = N/m^2 = Pa$
	$[\sigma] = N/m^2 = Pa$
	$[\varepsilon] = \%$
	Note: the unit percent is obtained by multiplying the measured relative length change by 100.
Force relaxation:	Decrease in force when a ligament is pulled to a particular deformation, either once or in cyclic succession.
Load:	Sum of all the forces and moments.
Properties:	
• Material properties:	Properties of a material which describe its general behaviour without including any information about its size and shape. Comment: material properties include ultimate stress and ultimate strain.
• Physical properties:	Properties of a material which relate to its physics. Comment: physical properties include density and specific weight.
• Structural properties:	Properties of a specific sample of a material which describe the behaviour of that sample, including the effects of its size and shape. Comment: structural properties include force to failure and deformation.
Rotation:	Movement of a body about an axis.
Strain:	Relative change of length.

$$\varepsilon = \frac{\Delta L}{L_o}$$

where:

ε = strain

ΔL = change in length $= L - L_o$

L_o = original length

Strain rate: Change of strain over time.

$$\dot{\varepsilon} = \frac{d\varepsilon}{dt}$$

with the unit:

$$[\dot{\varepsilon}] = \frac{1}{s}$$

Strength: Maximal force a material can sustain before failure.

- Compressive strength: Maximal force in compression a material can sustain before failure (= ultimate compressive strength).
- Tensile strength: Maximal force in tension a material can sustain before failure (= ultimate tensile strength).

Stress: Force per unit area.

$$\sigma = \frac{F}{A}$$

where:

σ = stress

F = force

A = area

- Compressive stress: Stress perpendicular to the surface that acts to compress an object.
- Shear stress: Stress parallel to the surface of an object.
- Tensile stress: Stress perpendicular to the surface that acts to elongate an object.
- Ultimate stress: The highest stress experienced by the tissue before complete failure.

Translation: Movement of a body along a rectilinear path.

2.2 SELECTED HISTORICAL HIGHLIGHTS

2.2.1 BONE

The following historical highlights have been selected from several sources (Steindler, 1964; Bouvier, 1989; Martin & Burr, 1989). Fossil records of bone, in the form of a dermal armour surrounding the heads of fish, have been found that date from the Paleozoic era, about 500 million years ago. Further fossil records suggest that about 50 to 100 million years later, bone had an established place in the evolutionary development of fish and land vertebrates. Bone structure and function has been discussed for several centuries. However, significant progress in the mechanical and morphological understanding of the construction, functioning, and growth of bone did not occur until the second half of the nineteenth century.

1678	van Leeuwenhoeck	Published microscopic observations in the Philosophical Transactions of the Royal Society of London. Reported observations of an extensive canal system in bone.
1691	Havers	Proposed that bone is composed of organic strings and inorganic plates, which are arranged around cavities in tubular form. He found pores between these plates.
1739	du Hamel	Demonstrated the lamellar nature of bone.
1742	Lieutaud	Suggested that bone is composed of laminae and compact fibres.
1754	Albinus	Recognized that the pores were used as vascular channels.
1776	Monro	Understood that bone resorption and formation occur throughout life.
1816	Howship	Found that interstitial bone could be removed by absorption.
1841	Burns	Reincarnated van Leeuwenhoeck's observation of osteons in bone.
1855	Breithaupt	Described stress fractures in military recruits of a Prussian military unit.
1856	Fick	Stated that bone is a passive structure and that the surrounding muscles determine the form of bone.
1856	Virchow	Stated that bone plays an active role in developing its form and structure.

1862	Volkmann	Suggested that pressure inhibits bone growth and the release of pressure promotes it.
1867	Culman	Stated the similarity between the trabecular arrangement in bone and the structural elements of a crane. In both cases, the principles of highest efficiency and economy are used.
1867	van Meyer	Suggested the relationship between architecture and function of bone.
1870	Wolff	Summarized various suggestions and statements about bone by stating that there is an interdependence between form and function of bone. Physical laws have strict control over bone growth.

The schematic construction of bone has often been depicted in ways that illustrate the similarity between theoretical considerations of three-dimensional trajectorial systems and the actual arrangement of the trabeculae in the human bone. One of the earliest examples of such an illustration, by Wolff (1870), is shown in Figure 2.2.1.

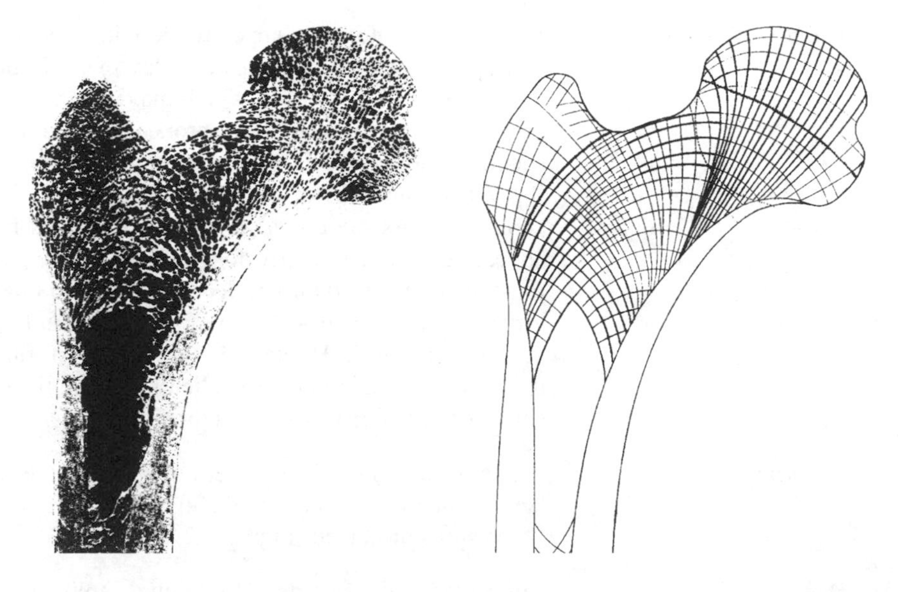

Figure 2.2.1 **Cross-section of the upper end of a femur of a 31-year-old male (left) and schematic representation of the same picture (right) (from Wolff, 1870, in the English translation by Maquet and Furlong, 1986, with permission).**

1883	Roux	Proposed that the orientation of the trabecular system corresponds to the direction of tension and compression stresses and is developed using the principle of maximum economy of use of material (as did Wolff). Architecture of bone follows good engineering principles.

1897 Stechow — Made first radiographic verification of stress fractures.

1920 Jores — Proposed that bone cells act as sensors for structural alterations.

1931 Greig — Suggested that microscopic local damage may stimulate bone remodelling.

1942 Maj — Showed, experimentally, that cortical tissue of bone becomes weaker with advancing age.

1984 Lanyon & Rubin — Demonstrated the influence of bone loading on bone remodelling.

1986 Frost — Proposed a controlling mechanism (mechanostat) for the structural and functional adaptations of bone.

2.2.2 ARTICULAR CARTILAGE

1742 Hunter — First to associate articular cartilage with mechanical concepts. Stated of cartilages that "by their elasticity, the violence of any shock, which may happen in running, jumping et cetera is broken and gradually spent... which must have been extremely pernicious, if the hard surfaces of bones had been immediately contiguous. As the course of the cartilaginous fibres appears calculated chiefly for this last advantage, to illustrate it, we need only reflect upon the soft undulatory motion of coaches, which mechanics want to procure by springs". Hunter also described a fibrous infrastructure radiating perpendicularly from the bone surface and deforming upon compression.

1898 Hultkrantz — Illustrated a surface fibre pattern by pricking the cartilage surface with a sharp tool and noting the direction of the splits produced (Figure 2.2.2).

1925 Benninghoff — Confirmed and expanded Hultkrantz's split line theory and investigated the deeper substance of the cartilage body. Introduced a detailed arcade theory, which confirmed Hunter's ideas and that described fibres radiating out from the bone and curving obliquely until parallel to the articular surface (Figure 2.2.3).

1920s and 1930s — Scientists attempted to measure elastic properties of articular cartilage (Bar, 1926; Göcke, 1927; Müller, 1939; Policard, 1936) to attempt to understand arthritic diseases.

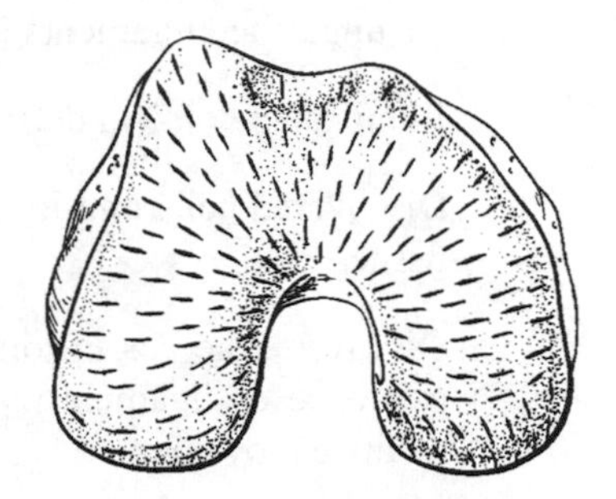

Figure 2.2.2 **Schematic diagram of Hultkrantz' split lines on the surface of articular cartilage. The split lines illustrate a surface fibre pattern (from Hultkrantz, 1898, with permission).**

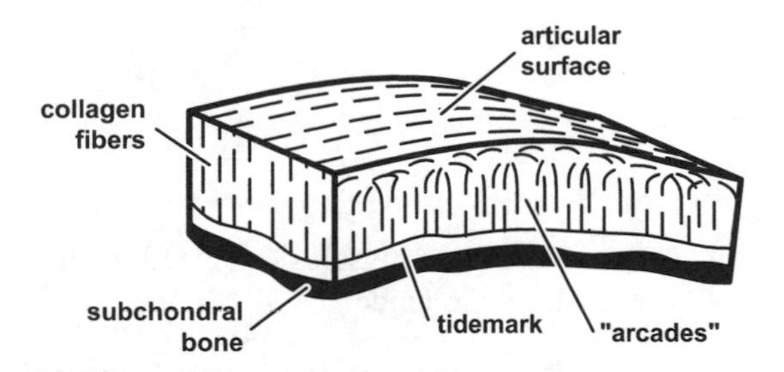

Figure 2.2.3 **Schematic diagram of Benninghoff's arcades (from Benninghoff, 1925, with permission).**

1960s		Ultrastructure of articular cartilage was described in great detail with the help of the newly developed polarized light microscope and scanning electron microscope (Goodfellow & Bullough, 1968; Hunter & Finlay, 1973).

2.2.3 LIGAMENT

3000 B.C.	Smith Papyrus	Described joint sprains.
400 B.C.	Hippocrates	Described treatments for ligament injuries.
300 B.C.	Herophilus	First person known to have performed extensive anatomical dissections of ligaments.
100 B.C.	Hegator	Provided the first anatomical definition of a ligament.

129-201 Galen	Provided further anatomical definitions of ligaments. Distinguished ligaments from tendons.
1514 Vesalius	Detailed anatomical definitions of joint tissues.
1830 Schleiden & Schwann	Discovered cells and long fibres in dense connective tissues.
1850s Rudinger & Hilton	Discovered nerves in joint tissues and postulated the existence of ligament-muscle feedback systems that mediate movement.
1911 Fick	Published the first biomechanical review of ligaments.
1980 Noyes	Provided the first definition of the complex structure-function relationships of ligaments.
1981 Woo	Reported on the viscoelastic properties of ligaments.

2.2.4 TENDON

129-201 Galen	Stated that muscles consisted of fibres that were connected to tendons.
1500s da Vinci	Stated that tendons were mechanical instruments playing a passive role in carrying out as much work as was put upon them.
1828 Bell	Observed that mechanical influences governed the growth and development of tendons.
1847 Wertheim	Noted that the stress-strain curve of tendon was non-Hookean.
1850 Kölliker	Stated that tendons were relatively avascular. In 1855, he noted that tendons contained waves (banding), which disappeared when the tissue was tensioned. He attributed these waves to the influence of elastic fibres upon collagen.
1949 Wickoff	Observed, from electron microscopic examination, that there was a helical (spiral) pattern within collagen fibrils.
1950 Rollhäuser	Stated that the extensibility of tendons increases with age.

2.2.5 MUSCLE

Milon of Kroton was probably the most famous athlete of the ancient Olympic games, winning six Olympic titles in wrestling in an athletic career that spanned more than 30 years. His incredible feats of strength are told in many stories, and one of these describes him exercising by carrying a young calf every day until it had grown up. In our present vocabulary of strength training, we may consider Milon of Kroton the first documented athlete who used variable resistance strength training, since the calf grew heavier as he became stronger.

384-322 B.C.Aristotle

The origins of discovery of muscles as the organ of force and movement production may be found in ancient Greece. Aristotle (De Motu Animalium) described the interrelation of breath, brain, and blood vessels in the production of movement. According to him, movement was associated with thrusting and pulling. Therefore, the organ producing movement must be capable of contracting and expanding, like the pneuma (spirits) (Farquharson, 1912; Needham, 1971).

129-201 Galen

The discovery that muscles are the true organs of voluntary movements must be accredited to Galen (De Tremore). Galen was born in Pergamon, the son of a mathematician. He received a broad philosophical education and started to study medicine at the age of 16, first in his home town, and then in Smyrna and Alexandria. His teachings and writings on medicine surpassed those of previous physicians and remained the foundation of medical practice for the following one and a half millennia. His detailed dissection of muscles, and his discovery that arteries contain blood (not air or spirits) were considered the first attempts of establishing a science of muscles (myology). It took the Renaissance to revive the sciences and the interest in how the human body functions (Daremberg, 1854-1857; Needham, 1971).

1543 Vesalius

(De Humano Corporis Fabrica) He was the first to break away systematically from the medical teachings of Galen. He discovered that the contractile power resides in the actual muscle substance, and he identified individual structural components of muscles. About 100 years after Vesalius, a series of scientists were concerned with the functioning of muscles.

1663	Swammerdam	Constancy of muscular volume during contractions was supported by elegant experiments on frog and human muscles (Needham, 1971).
1664	Croone	(De Ratione Motus Musculorum) Concluded from nerve section experiments that the brain must send a signal to the muscles to cause contraction.
1664	Stensen	(De Musculis et Glandulis Observationem Specimen, 1664; and Elementorum Myologiae Specimen sen Musculi Descripto Geometrica, 1667) Gave precise descriptions of muscular structures, showed that muscle fibres connect to tendons, and stated that contractions may occur without changes in muscular volume (Kardel, 1990).
1680	Borelli	Following these works concerned with the structure and function of muscles, Borelli (De Motu Animalium) incorporated the muscles systematically into the skeletal system to study the effects of external and muscular forces on the way the body works.
1682	van Leeuwenhoek	The major steps in identifying the structure of muscles and developing corresponding theories of muscular contraction were associated with the discovery of the light and electron microscopes. Van Leeuwenhoek performed (light) microscopic examinations and discovered the cross-striation of skeletal muscles. His descriptions of muscular structure dominated the following century (Needham, 1971; Pollack, 1990).
1939	Engelhardt	At about the same time that electron microscopes were first used for systematic studies of the structure of muscles, researchers in muscle biochemistry were able to associate actin proteins with thin myofilaments, associate myosin proteins with thick myofilaments, and, most importantly, discover the ATPase activity of myosin (Engelhardt & Lyubimova, 1939).
1954	Huxley, A.F. Huxley, H.E.	These biochemical discoveries, together with the structural findings using electron microscopy, led to the theory of sliding filaments during muscular contraction (Huxley, A.F. & Niedergerke, 1954; Huxley, H.E. & Hanson, 1954) and, finally, the Cross-bridge theory (Huxley, A.F., 1957; Huxley, A.F. & Simmons, 1971). The Cross-bridge theory has become the accepted paradigm for muscular force production, and, with few exceptions, has not been challenged or questioned seriously in the past three decades.

2.3 BONE

BOYD, S.K.
NIGG, B.M.

2.3.1 MORPHOLOGY AND HISTOLOGY

MORPHOLOGY AND FUNCTION

The musculoskeletal system is comprised of many bones and the connective tissues that hold them together. Bone differs from connective tissues in its rigidity and hardness, and is therefore often referred to as a hard tissue. It consists of an organic component (collagen fibres and non-collagenous proteins) and an inorganic mineral component. Bone is a dynamic, self-repairing tissue that can adapt its mass, shape, and properties in response to changes in load and physiological environments. Bone performs mechanical and physiological functions as described here:

Mechanical functions:
Bone provides the mechanical integrity for static posture, locomotion, and protection of internal organs. Specifically, bone has these mechanical functions:

- Providing *support* for the body against external forces, e.g., gravity,
- Acting as a *lever system* to transfer forces, e.g., muscular forces, and
- Supplying *protection* for vital internal organs, e.g., the brain.

Physiological functions:
Bone contributes to the functioning of the metabolic pathway associated with mineral homeostasis (Einhorn, 1996). Specifically, bone has these functions:

- Forming blood cells (*hematopoiesis*), and
- Storing calcium (*mineral homeostasis*).

Support

Both cortical and trabecular bone provide support for the soft tissues, skeletal construction, and shape of the body. The different positions of human and animal postures would not be possible without the bony structures.

Lever system

Bones provide points of attachment for skeletal muscle-tendon units. Bones and muscles function together to transfer forces in lever systems through muscular attachment and the articulation of bones at joints.

Protection

The flat bones of the skeleton, composed of a layer of trabecular bone wedged between two cortical plates, are largely responsible for protecting vital structures such as the heart, lung, and brain. The brain, for instance, is protected by the bones of the skull, the bladder and internal reproductive organs is protected by the pelvis, and the heart and lungs are protected by the rib cage.

Hematopoiesis

Hematopoiesis, the process of blood cell formation, occurs in red bone marrow (hematopoietic marrow). In adults, red marrow is found in bone regions composed largely of trabecular bone, e.g., the vertebrae, proximal femur, and iliac crest. In these regions, erythrocytes (red blood cells), leukocytes (white blood cells), and thrombocytes (platelets) are formed. In contrast to the red marrow, yellow marrow (or fatty marrow) is found primarily in regions of cortical bone such as the midshaft (diaphyseal) region of long bones such as the femur or tibia.

Mineral homeostasis

Bone is the body's largest reservoir of calcium, with 99% of total body calcium stored in the skeleton. Other critical minerals, such as phosphorus, sodium, potassium, zinc, and magnesium are also stored in bone. Since calcium is critical for many vital metabolic processes, maintaining serum calcium homeostasis always takes priority over the calcium requirements of bone. The hormones that regulate serum calcium balance include para thyroid hormone (PTH), calcitonin (CT), cholecalciferol (vitamin D), reproductive hormones, and growth hormones.

Table 2.3.1 Summary of the bones in the human body.

NAME	NUMBER OF BONES
VERTEBRAL COLUMN, SACRUM, AND COCCYX	26
CRANIUM	8
FACE	14
AUDITORY OSSICLES	6
HYOID BONE, STERNUM, AND RIBS	26
UPPER EXTREMITIES	64
LOWER EXTREMITIES	62
TOTAL	**206**

MORPHOLOGY AT ORGAN AND TISSUE LEVEL

The adult human skeleton consists of 206 distinct bones (Table 2.3.1). Although the bones vary considerably in size and shape, they are similar in structure and development. Bone may be classified into some basic shapes. The different bone types, selected examples, and their functions are given in shape (Table 2.3.2).

Table 2.3.2 Different bone types, classified according to shape.

SHAPE	EXAMPLES	FUNCTION
LONG	femur tibia radius	act as levers and to transmit longitudinal force
SHORT	carpal bone tarsal bone	provide strength and to transmit longitudinal force
FLAT	sternum ribs skull bones ilium scapula	provide protection and points of attachment for tendons and ligaments
IRREGULAR	ischium pubis bone vertebrae	various functions
SESAMOID	patella	improved lever situation

Despite the variety of external forms in the skeleton, the morphology of bone at the organ, tissue, and cellular levels is relatively consistent. It is assumed that the reader is familiar with basic anatomy, so no detailed discussion of skeletal anatomy is presented beyond the definitions of the structural components of a long bone.

The macroscopic structure of long bones can be described based on their centres of ossification. The *epiphyses* develop from seco ndary ossification centres and are found at the ends of long bones. The epiphyses are protected by a layer of hyaline cartilage, referred to as articular cartilage, and articulate with other bones. Between the two epiphyses is the shaft of the long bone, called the *diaphysis*, which develops from the primary ossification centre. The diaphysis is a hollow structure surrounding the *medullary cavity*. The medullary cavity, which is used as a fat storage site, is lined by a thin, largely cellular connective tissue membrane, the *endosteum*. There is no medullary cavity in a flat bone. The inner surface of the diaphysis is the *endosteal* surface, and the outer surface of the diaphysis is referred to as the *periosteal* surface. At the growth zone between the epiphyses and the diaphyses are the *metaphyses,* which are the flared ends of long bones. In a growing mammal, the epiphysis is separated from the metaphysis by a plate of hyaline cartilage named the growth plate, or epiphyseal plate, which is the region from which bone production and elongation of the cortex occurs. Epiphyseal and metaphseal bone supports articular cartilage in diarthrodial joints and is therefore wider than bone in the diaphyseal region (Figure 2.3.1).

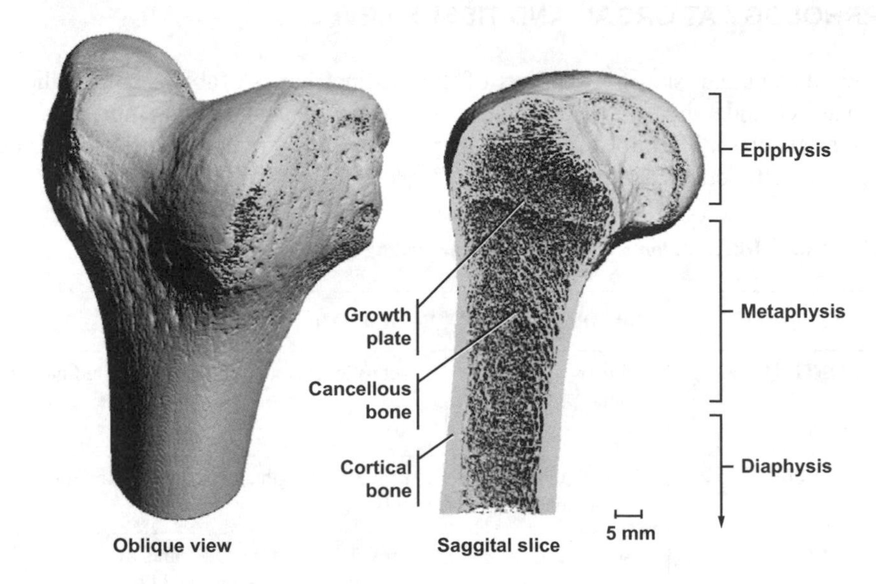

Figure 2.3.1 **The human distal femur (knee) measured by micro-computed tomography showing the three-dimensional architecture and a mid-sagittal slice.**

Surrounding and attached to the bone, except for areas covered by cartilage is a tough, vascular, fibrous tissue called the *periosteum*. The outer layer of the periosteum is well supplied with blood vessels and nerves, some of which enter the bone. The inner layer is anchored to the bone by collagenous bundles called *Sharpey's fibres*, which penetrate the bone. Some of the periosteum fibres are intertwined with fibres of tendons, which provide attachment for muscles. Projections of bone, called processes, act as sites of attachment for ligaments and tendons.

All bones in the adult skeleton have two basic structural components: cortical and cancellous bone. *Cortical* (or *compact*) bone is the solid, dense material comprising the walls of diaphyses and external surfaces of bones. This type of bone is solid, strong, and resistant to bending. The thickness of cortical bone varies between and within bones as a function of the mechanical requirements of the bone. *Cancellous* (or *spongy*) bone is named after the thin bony spicules that form it, called *trabeculae*. These trabeculae can be in the form of plates or rods, and have been observed to orient themselves primarily in the direction of the forces applied to the bone. Trabeculae are found in processes, the vertebral bodies, epiphyses of long bones, short bones, and sandwiched between the two layers of compact bone that make up the flat bones (called *diploe*). The rods and plates of trabecular bone are organized to withstand mechanical loads while minimizing the weight of the bone.

Red bone marrow, a hemopoietic tissue that produces red and white blood cells and platelets, is found in areas of trabecular bone. The yellow bone marrow found in the medullary canal of long bones consists of fat cells. Bones, therefore, provide an active site of hematopoiesis.

Bone consists of 65% mineral and 35% organic matrix, cells, and water. The *organic matrix* is approximately 95% collagen fibres. It is greatly strengthened by the *mineral com-*

ponent, which contains deposits of calcium and phosphate salts in the form of *hydroxyapatite*, $Ca_{10}(PO_4)_6(OH)_2$. The calcium and phosphate deposits give bone its strength, hardness, and rigidity, while the collagen fibres provide some flexibility.

HISTOLOGY

Three primary mature bone cell types are responsible for forming, resorbing, and maintaining bone. *Osteoclasts* are the cells responsible for resorbing bone, thereby releasing calcium into the serum. The bone forming cells are *osteoblasts*, which synthesize the collagen matrix (*osteoid*) and later deposit bone mineral (*hydroxyapatite*) within that matrix, producing mineralized bone. Once it has surrounded itself with mineralized bone tissue, the osteoblast is referred to as an *osteocyte*, the mature bone cell believed to be responsible for maintaining bone tissue. The processes of bone resorption followed by bone formation occur throughout life and comprise the process of bone *remodelling*. Remodelling can maintain existing bone mass or reduce bone mass.

Microscopic structures of bone types

In general, there are three broad categories of bone microstructure. These are determined by differences in organization or control, or both. Here, each is introduced, with emphasis placed on primary and secondary bone.

Woven bone:

Unlike the regularly oriented collagen fibres of lamellar bone, the collagen fibres in woven bone are randomly oriented. The result is a less dense bone, although there is generally no mineralization deficit. Woven bone further differs from other forms of bone in that it can be deposited *de novo*, i.e., without any previous hard tissue or cartilage model. The most characteristic form of woven bone is formed during embryonic life and at the growth plate during endochondral ossification. In adults, a form of woven bone is *callus* formation, which is important for fracture healing resulting from damage to, or tension on, the periosteum. Woven bone provides a relatively quick source of mechanical strength and the framework for the slower development of lamellar bone.

Primary bone:

There are several types of primary bone, and these differ in their development. Primary bone cannot be deposited *de novo*. It requires a pre-existing substrate, such as a cartilaginous model. The three main categories of primary bone differ both morphologically and by their mechanical and physiological properties.

(i) Primary lamellar bone

Primary lamellar bone or circumferential lamellar bone is arranged in circular rings around the endosteal and periosteal circumference of a bone. The trabeculae in the epiphyses of long bones are also primarily lamellar bone, and they are closely associated with marrow and vascular tissues. This proximity allows for rapid exchange of calcium between bone and serum and explains, in part, why regions of cancellous bone are the first to exhibit osteopenia (reduced bone mass). Primary lamellar bone is mechanically competent. Where there is a large surface area adjacent to marrow and blood, however, the requirements of hematopoiesis and mineral metabolism may override mechanical requirements.

(ii) Plexiform bone

Like primary lamellar bone, plexiform bone must be deposited on pre-existing surfaces. Like woven bone, it forms rapidly, but it has better mechanical properties. Structurally, plexiform bone has the morphology of highly oriented cancellous bone, with the trabecular plates thickening due to endosteal or periosteal surface apposition, or both. This type of bone is most commonly seen in rapidly growing large animals, such as the cow, whose rapid growth demands mechanical competence of the skeleton (Martin and Burr, 1989). However, it has also been observed in children during the growth spurt (Amprino, 1947).

(iii) Primary osteons

A set of lamellae arranged in concentric rings around a vascular channel (versus around the entire bone cortex) is called a primary osteon. Until recently, primary osteons were not distinguished from secondary osteons (see below). However, developmentally, morphologically, and possibly mechanically, they are quite different (Currey, 1984). The greatest distinction is that primary osteons do not have cement lines (reversal lines), because these lines are not developed through bone remodelling. As well, primary osteons appear to have smaller vascular channels and fewer lamellae than secondary osteons, so primary osteons may be stronger than secondary osteons (Martin & Burr, 1989). Primary osteons develop through the sequential filling in of vascular channels with layers of lamellar bone and are found in well-organized primary lamellar bone.

Secondary bone:

Secondary bone is the product of resorption of previously existing bone tissue and the deposition of new bone to replace it. In cortical bone, the result of bone resorption by osteoclasts, followed by apposition of bone by osteoblasts, is the *secondary osteon*. A secondary osteon's central vascular channel is larger than that found in primary osteons and called a *Haversian canal*. The Haversian canal is surrounded by concentric lamellae arranged in a circular fashion containing osteocytes. The two distinguishing morphological features of the secondary osteon are the presence of cement or reversal lines that separate the osteon from the extraosteonal bone matrix, and the organization of lamellae around the Haversian canal (Figure 2.3.2).

Each trunk in the cross-section of compact lamellar bone is referred to as a *Haversian system*. Haversian systems in healthy adult humans measure approximately 300 micrometers in diameter and are approximately 3 to 5 mm in length, with their long axes parallel to those of the long bone. Blood, lymph, and nerve fibres pass through the Haversian canal. Smaller *Volkmann's canals* pierce the bone tissue perpendicular to the periosteal and endosteal surfaces, linking the Haversian canals and creating a network for blood, nerve, and lymph supply to bone cells. Small spaces within the lamellae, called *lacunae*, contain mature bone cells (*osteocytes*), which receive nutrients via minute fluid-filled channels called *canaliculi* that radiate out from the Haversian canals.

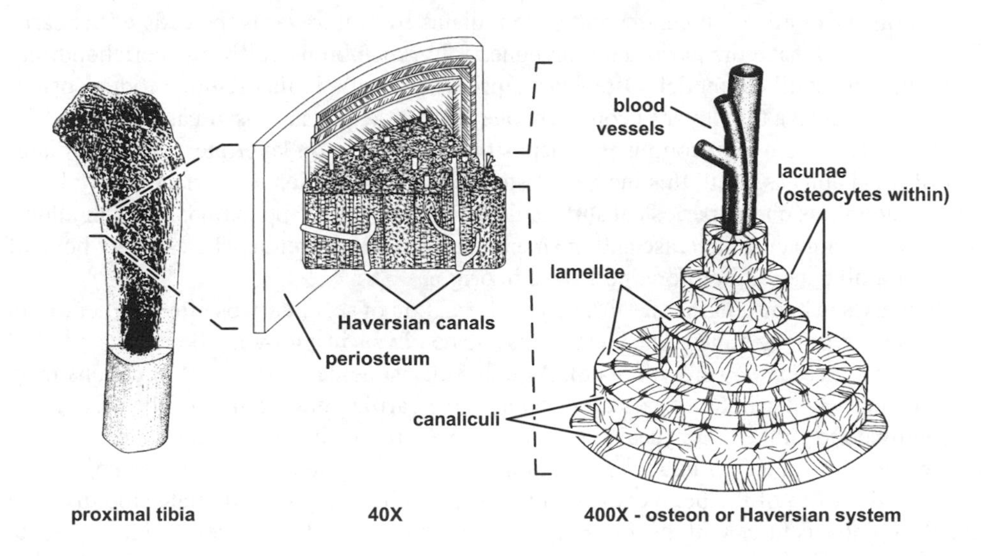

Figure 2.3.2 **Illustration of the structure of cortical bone (from White, 1991, with permission).**

Skeletal processes

There are four skeletal processes that occur at various stages of human life: growth (endochondral and intramembranous), modelling, remodelling, and repair. These processes are fundamentally distinct, operate under different controls, in different locations, and at different ages in the human, yet the same types of bone cells are involved in all. Each is discussed briefly. Consult books on anatomy, physiology, and the publications of H.M. Frost (1964, 1973a,b, 1986, 1987, 1989) for greater detail.

Bone growth:

(i) Intramembranous ossification
Intramembranous ossification refers to the growth of flat bones, such as the frontal and parietal bones of the skull. It occurs through the apposition of bone on tissue within an embryonic tissue membrane, and is the process by which woven bone is formed. Osteoblasts synthesize a bone matrix with collagen fibres that are not preferentially oriented and appear as irregular bundles. The osteocytes in the highly vascularized embryonic tissue are large and numerous, and calcification occurs in irregularly distributed patches. Woven bone trabeculae are formed with blood vessels incorporated between the structures. Later, remodelling of this structure results in the development of mature lamellar bone.

(ii) Endochondral ossification
The majority of bones in the skeleton grow through the process of endochondral ossification, in which bone is preceded by cartilage. Ribs, vertebrae, the cranial base, and bones of the extremities begin as cartilage models in utero, an environment in which the function of support is not necessary. Ossification occurs within the cartilage model at the centre of the diaphysis, as it is penetrated by blood vessels through the nutrient foramen. This region

is the primary ossification centre, and bone radiates from it towards the ends of the cartilaginous model. Like the periosteum on bone, a thin membrane called the perichondrium surrounds the cartilage model. Osteoblastic progenitor cells in this region produce osteoblasts that deposit a thin layer of compact bone around the primary ossification centre. The resulting periosteum then continues to deposit bone, layer upon layer increasing the diameter of the diaphysis. With this increase, osteoclasts on the endosteal surface resorb bone, while osteoblasts on the periosteal surface deposit bone. Thus, appositional growth allows diaphyseal diameter to increase and the medullary canal to develop. The compact bone of an adult limb bone is, therefore, periosteal in origin.

Increases in bone length arise through the formation of secondary ossification centres in the epiphyses of long bones. The metaphysis and epiphysis in a growing bone are separated by the *epiphyseal plate* (growth plate). As trabecular bone develops in all directions from the secondary and primary ossification centres, the cartilaginous centre of the plate is replaced by bone on the diaphyseal side. As the bone grows, the epiphyseal plate is pushed further from the bone's primary ossification centre, thereby lengthening the bone. Ossification and growth of the bone come to a halt when cells at the growth plate stop dividing and the epiphysis fuses with the metaphysis of the shaft. Most long bones develop two secondary centres in addition to the primary centres of ossification. A few long bones develop only one secondary centre, and the majority of carpal and tarsal bones develop completely from the primary ossification centre.

The growth of bone in a maturing skeleton is the only process that continually creates new trabeculae. There is no known counterpart in the adult skeleton. Noticeable changes in the size and shape of adult bones occur through the process of modelling.

Bone modelling and remodelling

Bone modelling is the process by which bone mass is increased.

Bone remodelling is the process by which bone mass is maintained or decreased.

Bone modelling is associated with the growth and development of trabecular and cortical structures where bone mass is added. Modelling can result in adaptation during growth where modelling drifts lead to thickened cortices and trabeculae in children, or adaptation to altered loading conditions. Modelling is predominant during skeletal development as the bone adapts to required loading conditions, and is less efficient in the adult skeleton.

Bone remodelling occurs at the surface of bone and involves a dynamic process of morphological adaptation in both cortical and trabecular bone. The maintenance of bone is achieved through bone remodelling by the ongoing replacement of old bone by new bone, thus repairing micro-cracks that occur during normal physiological activity. Resorption and formation are coupled, and in the normal adult skeleton formation normally only occurs where there has been previous resorption. The sequence of activation-resorption-formation (ARF) first described by H.M. Frost is the basic process for bone remodelling and occurs throughout a lifetime. Bone remodelling occurs at sites throughout the skeleton, and is achieved via a fundamental functional unit named a basic multicellular unit (BMU). The main stages of a BMU are activation following a period of rest, resorption of bone matrix by osteoclasts, a short intermediate (reversal) stage, and formation of new bone matrix by osteoblasts and subsequent mineralization. The complete remodelling cycle at each microscopic remodelling site takes 3 to 6 months.

The distinction between modelling and remodelling is critical to bone biology. Modelling differs from remodelling in many fundamental ways. The most profound differences are with respect to the temporal and spatial relationship of osteoblasts and osteoclasts. In modelling, resorption and formation occur on different bone surfaces. They have no spatial relationship, and no fixed temporal coupling exists between them. Osteoclastic and osteoblastic activity appears to be subject to independent regulation, and their activity, at respective surfaces, can occur for extended periods of time with no apparent interruption. In remodelling, osteoclastic and osteoblastic activities are coupled and occur in cycles of activation, resorption and formation (ARF, as described above). In remodelling, a negative balance between bone resorption and formation of the BMU can lead to remodelling-dependent bone loss. Finally, during the growth period, modelling processes involve almost the entire periosteal and endosteal surface at all times, while in remodelling only a fraction of the surface is active at any one time.

Bone repair

Bone repair occurs by different mechanisms at the microstructural and macrostructural levels. At the microstructural level, the remodelling process is responsible for constantly repairing microdamage that results from exposure to physiological loads. Repairing microcracks is a normal and necessary occurrence in the adult skeleton. If this repair mechanism is inhibited, for example, by the introduction of a pharmaceutical agent, microdamage can accumulate and fatigue fracture may result.

Fracture at the macrostructural level (organ level) is response to abnormal or excessive loads, or both, and requires the formation of woven bone (callus) to repair the damage. The process of repair is initiated with blood flow into the fracture region that normally coagulates to form a hematoma. The fracture ruptures the periosteum, stimulating the rapid formation of the callus or woven bone. It provides temporary strength and support for the fractured bone. Mineralization of the final callus takes approximately six weeks in the human adult. It is then gradually remodelled to produce lamellar bone. The final structure of the bone at the fracture site depends on the orientation of the broken bones and the loads applied during the healing process.

2.3.2 PHYSICAL PROPERTIES

The physical properties of bone depend on the tissue from which it is made, and how that tissue is organized to perform a mechanical function. The mechanical (structural) properties of bone are established in response to the loading to which it is normally subjected. The complex organization of cortical and trabecular architecture strikes a balance between maximizing strength while minimizing mass. Since bone is a dynamic tissue, its structure can alter in response to new loading patterns.

The relation between bone structure and function was first noted in the *Culmann and von Meyer drawings* where the principal stress trajectories in a crane-like curved bar exhibited similar patterns to the trabecular organization in the proximal human femur (von Meyer, 1867). This observed relation between the form and mechanical function of bone is first attributed to Roux (1885) and eventually established as the central concept of what is known today as *Wolff's law of functional adaptation* published by Wolff (1892, 1986).

WOLFF'S LAW OF FUNCTIONAL ADAPTATION

In his classic publication, Wolff wrote:

> **"The shape of bone is determined only by the static stressing..."**
>
> **"Only static usefulness and necessity or static superfluity determine the existence and location of every bony element and, consequently of the overall shape of the bone."**

Several comments seem appropriate in the context of these statements. First, a reference to Wolff's law should only be used as a philosophical statement that refers to the functional adaptation of bone. It is not a rigorous law in the sense of a mathematical law, and in fact, it is based on a *false premise* (Bertram & Swartz, 1991; Cowin, 1997) that the stress trajectories in a homogeneous elastic material coincide with the trabecular architecture in cancellous bone. Furthermore, Wolff's law does not account for differences between static and dynamic loading, which are known to have different influences on bone formation or remodelling processes, or both (Lanyon & Rubin, 1984), and it does not explain the so-called mechanostat concept, where a set point for remodelling occurs for either insufficient or excessive physiological loading (Frost, 1987). Finally, heredity may also play an important role in the development of bone architecture. Nevertheless, despite the weaknesses of Wolff's law as a rigorous mathematical law and in explaining all aspects of functional adaptation, it remains important terminology that has evolved to encompass the concept that form and function of bone are related. Considering the above points, the concept of functional adaptation could be restated in a more general way because the specific forms of loading are not the only factor that determines bone structure, although they are important ones. Adapted, Wolff's law may be worded simply as:

> **Physical laws are a major factor influencing bone modelling and remodelling.**

SELECTED PHYSICAL PROPERTIES FOR BONE

The physical properties of bone provide insight into its mechanical function, and how that function is impaired in diseases or augmented through therapeutic treatments. This information can be used, for example, for calculating stresses that occur during physiological loading, allowing us to improve our understanding of the functional adaptation of bone or to estimate fracture risk in patients suffering from diseases such as osteoporosis. Physical properties such as stiffness and strength can be measured directly through carefully applied biomechanical testing methods, including compression testing, tensile testing, or three/four-point bending. However, it is useful to relate the physical properties obtained through biomechanical testing to properties that are commonly measured in the clinic for diagnostic purposes, i.e., bone density. Before a discussion of the physical properties of bone, it is helpful to define some terminology.

The *structural* properties of bone are related to the complex organization of cortical and cancellous bone, and the *material* properties are related to the bone tissue itself. Structural properties are specific to a particular specimen, but material properties are independent of the specimen. For example, a common engineering material such as steel has particular material properties, but can be used to build structures with different physical properties. For example, a guitar string and the Eiffel tower may both be built from steel, but have vastly different structural properties. The terminology often used in bone biomechanics to describe the structural properties of bone is *apparent* properties, and the material properties are referred to as *tissue* properties. The distinction is particularly important for cancellous bone (Keaveny et al., 2001), as the tissue properties may be an order of magnitude higher than the apparent properties.

The characterization of the physical properties of bone typically requires using classical mechanical testing methods, i.e., axial and torsional testing. Testing a whole bone, the relation between the *load* applied and its *deflection* is used to generate a load-deflection curve. The linear region of this curve provides a measure of *stiffness* and the maximum value provides the *ultimate load*. These measurements provide useful physical properties. However, they depend on the geometry of the bone tested and the type of load applied, and are, therefore, specific to the particular bone tested. Alternatively, generalized results can be determined from similar tests using a machined specimen of bone (cube or cylinder) with known geometries. Under axial testing conditions (either compressive or tensile), loads applied to the test specimen can be divided over the area which they act to provide a measure of *stress*, σ, measured as force per unit area (N/mm^2 or Pa), and the displacements due to the loading are reported as *strain*, ε, which is a measure of the displacement relative to the original length (dimensionless units). Therefore, for a given specimen, a stress-strain curve can be determined, where the slope of the linear region is the *elastic modulus*, *E*, measured in the same units as stress (Pa), and the maximum point of the curve is the ultimate stress, σ_{ult}. Although the elastic modulus and ultimate stress are determined from machined specimens, these physical properties can be used in relatively basic mechanics of materials equations in combination with knowledge specific bone geometry to calculate either local stresses or failure loads. Examples of these calculations will be shown, and they illustrate that some basic mechanical testing can provide a powerful tool for estimating bone stress in complex loading configurations. The order of magnitude of the ultimate stress for apparent bone properties is provided for reference (Figure 2.3.3).

Material properties for bone are normally reported for either apparent-level or tissue-level properties. When determining the elastic modulus values using the methods described above, stress-strain curves are derived from testing a small machined specimen of solid bone tissue, i.e., a piece of cortical bone or a single trabeculae, and results provided are *tissue-level* material properties. They represent the properties of the tissue as a continuum.

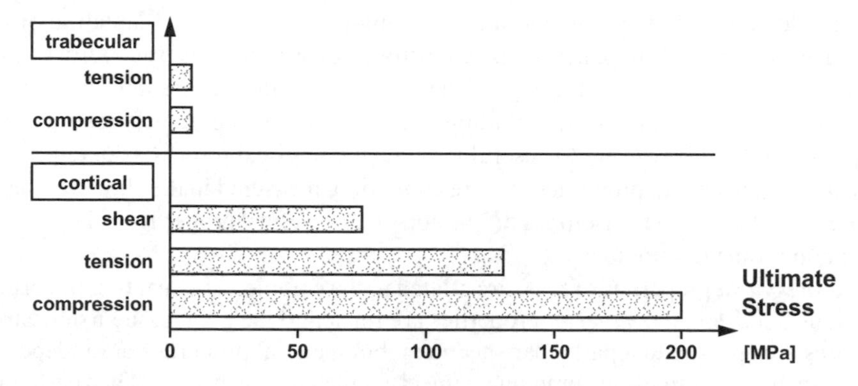

Figure 2.3.3 **Order of magnitude of ultimate stress for cortical and trabecular bone (based on data from Yamada, 1970, Steindler, 1977, Reilly & Burstein, 1975, and Martin & Burr, 1989, with permission of Williams & Wilkins, Baltimore, Maryland).**

However, a test based on a machined cube of cancellous bone containing several trabeculae provides *apparent-level* material properties. In this case, it is presumed that the cancellous bone behaves as a continuum, a reasonable assumption if the specimen is at least five trabeculae across (Harrigan et al., 1988). In this case, the apparent-level elastic modulus is influenced by tissue-level material properties, and organization of the trabecular tissue. Similarly, *bone density*, ρ, is an important physical property that should be distinguished as either apparent-level density, i.e., mass of bone tissue divided by the bulk volume, or tissue-level density, i.e., related to the degree of mineralization of bone tissue. Bone density at the apparent-level is a physical property that can be measured clinically for diagnosing diseases such as osteoporosis, using dual X-ray absorptiometry (DEXA).

Selected physical properties of bone (and, for comparison, selected other materials) are summarized in Table 2.3.3.

Table 2.3.3 **Selected physical properties of bone at the apparent level based on data from Yamada (1970), Burstein et al. (1976), Noyes et al. (1984), and Ascenzi and Bonucci (1972) (from Martin & Burr, 1989, with permission of Williams & Wilkins, Baltimore, Maryland).**

VARIABLE	COMMENT	MAGNITUDE	UNIT
DENSITY	cortical bone lumbar vertebra water	1700-2000 600-1000 1000	kg/m^3 kg/m^3 kg/m^3
MINERAL CONTENT WATER CONTENT	bone bone	60-70 150-200	% kg/m^3
ELASTIC MODULUS (TENSION)	femur (cortical)	5-28	GPa

VARIABLE	COMMENT	MAGNITUDE	UNIT
TENSILE STRENGTH	femur (cortical) tibia (cortical) fibula (cortical)	80-150 95-140 93	MPa MPa MPa
COMPR. STRENGTH	femur (cortical) tibia (cortical)	131-224 106-200	MPa MPa
	wood (oak) limestone granite steel	40-80 80-180 160-300 370	MPa MPa MPa MPa

The order of magnitude for typical values of the apparent-level elastic modulus, E, are:

- Trabecular (cancellous) bone 10^9 Pa = 1 GPa
- Cortical (compact) bone $2 \cdot 10^{10}$ Pa = 20 GPa
- Metals 10^{11} Pa = 100 GPa

EXAMPLE 1

Question

A hypothetical uniaxial compression test was performed on a machined cylinder of human femoral cortical bone, and the results of the force-displacement measurements are shown in Figure 2.3.4. Estimate the elastic modulus and the ultimate strength of this sample of bone.

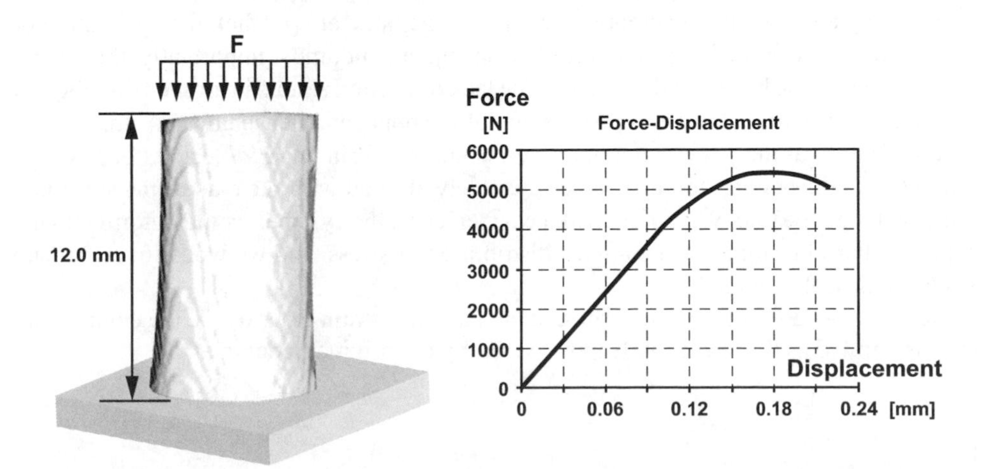

Figure 2.3.4 **A compression test of a machined core of femur cortical bone (12 mm length, 6 mm diameter) used to determine the material stiffness and strength. The force (N) applied and the change in length measured (mm) are represented in the force-length plot for the uniaxial test.**

Symbols

E	=	elastic modulus
A	=	cross-sectional area
L_o	=	unloaded length
ΔL	=	change in length from unloaded to loaded situation
σ_{ult}	=	ultimate stress

Assumptions

(1) The bone tissue is homogeneous and isotropic.
(2) The question can be treated as a one-dimensional problem.

Solution

In the first step, the force-displacement data must be converted into a stress-strain curve. We convert the force data into stress values by approximating the stress as the force applied divided by the area,

$$\sigma = \frac{F}{A}$$

where F is the applied force, and A is the cross-sectional area of the cylinder,

$$A = \frac{\pi d^2}{4}$$

where from the diagram $d = 6.0\ \text{mm}$, therefore,

$$A = 28.274\ \text{mm}^2.$$

The assumption that stress is simply the applied force divided by the area is reasonable because a *uniaxial* test has been performed. Some necessary conditions for uniaxial testing are that the platens applying compressive force to the bone are parallel, the application of force is aligned with the long axis of the test specimen, and, most importantly, there is no friction between the bone and the platens. The lack of friction results in an even distribution of the axial force throughout the cross-section of the bone, and hence uniaxial stress. Conversely, friction at the bone-platen interface would result in an *axial* test (as opposed to uniaxial) and the assumption that forces are solely aligned with the axis of the specimen would no longer be true. With friction at the platen ends, the cylinder would deform into the shape of a barrel resulting in a complex distribution of stress, and we would overestimate the elastic modulus of the bone.

Next, we need to convert the displacement data into strain data, where the engineering strain is defined as the change in length divided by the original length,

$$\varepsilon = \frac{\Delta L}{L_o}$$

where ΔL is the change in length in our force-displacement plot, and L_o = original length given as $L_o = 12.0\ \text{mm}$ from the diagram.

Conversion of our force-displacement data results in a new stress-strain plot (Figure 2.3.5), that represents the material behaviour of the human cortical bone specimen.

The linear region of the stress-strain plot represents that elasticity of the specimen, and the slope of this linear region provides our measure of the elastic modulus. Note that the linear region ends at approximately 0.75% strain (corresponding to 127 MPa), and we calculate the slope of that region as follows,

$$E = \frac{\Delta\sigma}{\Delta\varepsilon} = \frac{127}{0.0075} = 16933 \text{ MPA}, \text{ or approximately 17 GPa.}$$

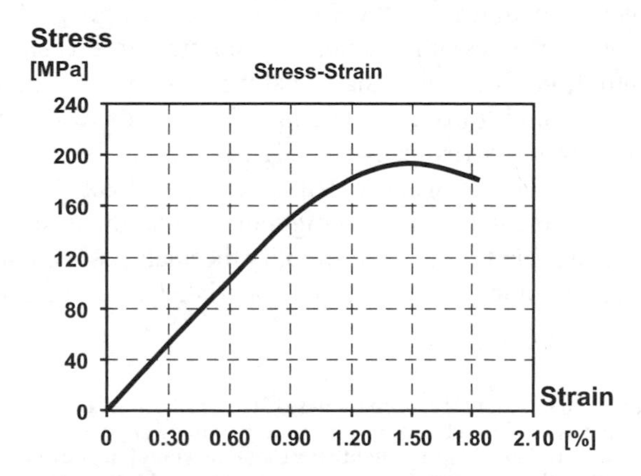

Figure 2.3.5 **The corresponding stress-strain curved based on force-displacement data. The ultimate stress is taken from the peak of the stress-strain curve and is approximately 193 MPa.**

The end of the elastic region is called the *elastic limit*. In the elastic region, the stress-strain relation is linear and obeys what is known as Hooke's law. If loading is maintained within this linear region, i.e., below the elastic limit, then the bone will return to its original length when released. Loading beyond the elastic limit results in a non-linear stress-strain relation, and the material undergoes plastic deformation where permanent changes to the tissue shape, or even failure, may occur. The normal physiological peak strain during daily activities is less than 0.1 to 0.2%, much less than the elastic limit of bone tissue.

The maximum point on the stress-strain curve represents the ultimate stress, σ_{ult}, which, taken from the plot, is approximately 193 MPa. Knowing the ultimate stress that the material can withstand, it is possible to approximate the failure load of a whole bone. For example, taking the cross-sectional area of a human femur at the midshaft (diaphysis) as 800mm^2, the ultimate load of the femur in axial compression would be approximately,

$$\sigma_{ult} = \frac{F_{ult}}{A} \rightarrow F_{ult} = \sigma_{ult}A = (193 \text{ MPA})(800 \text{ mm}^2) = 241 \text{ kN}.$$

The ultimate stresses for tension and compression are different. For cortical bone, the ultimate stress for compression, σ_{comp}, is about 30 to 50% higher than the ultimate stress for tension σ_{tens} (Reilly & Burstein, 1975; Steindler, 1977).

The ultimate stress for compression and tension of cancellous bone is at least one order of magnitude smaller than that of compact bone. The order of magnitude of averaged data from the literature is illustrated in Figure 2.3.3.

2.3.3 BONE MECHANICS AS RELATED TO FUNCTION

ELASTIC LOADING

Physiological loading of bone is normally within the elastic range. Therefore, using basic analytical tools for mechanics of materials, it is possible to estimate the stresses for a given loading configuration. The four fundamental types of loads that may occur are *axial loading*, *bending*, *torsion* and *transverse* loading. These loads result in combinations of normal stresses (either compressive or tensile) and shear stresses which vary throughout the bone structure. Using the principle of superposition, complex loads can be reduced to combinations of the four fundamental loading configurations, and the results are then superimposed to determine the stress in the bone. To simplify the analysis, it is common to make the assumption that bone is a homogenous isotropic material. This assumption is discussed in more detail later.

Table 2.3.4 Formulas to estimate stress on bone surfaces under loads of bending and torsion with associated moments of inertia for some common cross-sections. Maximum stresses occurring furthest from the neutral axis or torsional axis are given. These equations assume that the bone material is homogeneous and isotropic.

CROSS-SECTION	BENDING	TORSION	AREA MOMENT OF INERTIA
GENERAL	$\sigma = \frac{M_x \cdot y}{I_{xx}}$	$\tau = \frac{T \cdot c}{J}$	$I_{xx} = \int_x y^2 dx$ $J = \int_A \rho^2 dx$
SOLID CIRCLE (y, x, r)	$\sigma_{max} = \frac{4M_x}{\pi r^3}$ where $y = r$	$\tau_{max} = \frac{2T}{\pi r^3}$ where $c = r$	$I_{xx} = \frac{1}{4}\pi r^4$ $J = \frac{1}{2}\pi r^4$
HOLLOW CIRCLE (y, x, r_o, r_i)	$\sigma_{max} = \frac{4M_x r_o}{\pi(r_o^4 - r_i^4)}$ where $y = r_o$	$\tau_{max} = \frac{2Tr_o}{\pi(r_o^4 - r_i^4)}$ where $c = r_o$	$I_{xx} = \frac{1}{4}\pi(r_o^4 - r_i^4)$ $J = \frac{1}{2}\pi(r_o^4 - r_i^4)$

where:

σ = normal stress [N/mm^2 or Pa] either compressive or tensile
τ = shear stress [N/mm^2 or Pa]
M_x = bending moment about the x-axis [Nm]; similar for M_y
T = torsion [Nm]
I_{xx} = areal moment of inertia about the neutral x-axis [m^4]; similar for I_{yy}
J = areal polar moment of inertia about torsional axis [m^4]
r = radius [m] of inner r_i or outer r_o bone
y = distance [m] from the neutral axis of bending

The three most common types of loading in bone are axial, bending and torsion, while transverse loading plays a smaller role. For bending, both compressive and tensile normal stress is generated in different portions of the cross-section. The magnitude of the normal stress depends on the distance from the bending *neutral axis*, i.e., where the stress is zero, the border between compressive and tensile stress, and the *moment of inertia* of the cross-sectional area. Similarly, torsion generates shear stress and its magnitude depends on the distance from the torsional axis, i.e., zero at the axis, and the *polar moment of inertia* of the cross-sectional area. The moment of inertia is a measure of the distribution of solid mass and must be calculated for each cross-sectional area analyzed. For bending, the *maximal stress* occurs at the point furthest from the neutral axis, while for torsion, the *maximal stress* occurs at the point furthest from the torsional axis. A summary of the equations used to determine stresses for normal, bending and torsional loads is listed in Table 2.3.4. Consult a basic mechanics of materials text for more detail (Gere & Timoshenko, 1997; Beer et al., 2004).

EXAMPLE 2

Question

A mouse femur has the shape shown in Figure 2.3.6, as measured using high-resolution computed tomography at an isotropic resolution of 0.0322 mm, i.e., same resolution in all three dimensions. Given the loading configuration shown with F=15 N, determine the normal stress in the bone at the section *a-a*' of the mid diaphysis.

Assumptions

(1) The weight of the bone can be neglected.
(2) The bone tissue is homogeneous and isotropic.
(3) The problem can be treated two-dimensionally.

Solution

This is a case of eccentric loading, where the force F applied at the femoral head results in a combined loading condition of axial load plus bending at *a-a*'. It should be intuitive that the stresses due to bending result in compression on the medial side, and tension on the lateral side. The axial load will result in compression across the section a-a'. Since this is a

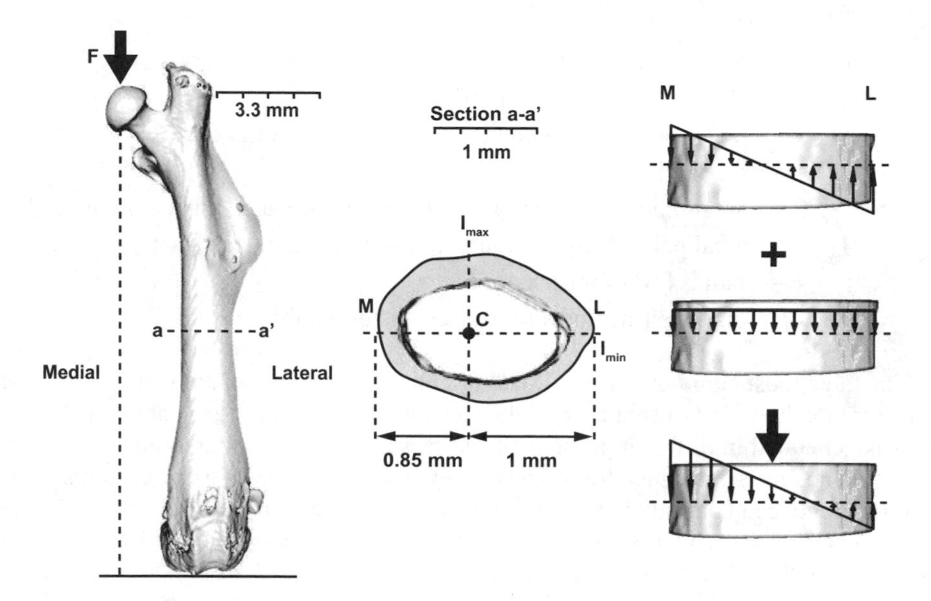

Figure 2.3.6 **Schematic illustration of eccentric loading of a mouse femur (left), a cross-section of the femur at a-a' (middle), and superposition of axial and bending loads due to force F to determine normal stress (right). M = medial, L = lateral.**

case of combined loading, we can superimpose the stresses due to bending with the stresses due to axial load.

The stresses due to bending can be calculated using the flexure formula described earlier (Table 2.3.4). The bending moment due to force F=15 N is,

$$M = F \cdot d = (15\ \text{N})(2.5\ \text{mm}) = 37.5\ \text{Nmm}$$

Knowledge of the centroid and the moment of inertia of the cross-sectional area is required before we can apply the flexure formula. However, it is difficult to calculate by hand for an irregular shape. For this example, we used a computer program that implements the moment of inertia calculation (shown in Table 2.3.4) to determine the centroid and the principal moments of inertia:

$$I_{max} = 0.2976\ \text{mm}^4 \text{ and } I_{min} = 0.1382\ \text{mm}^4.$$

Since the bending occurs about the vertical axis we will use I_{max} for our subsequent calculations. The maximum compressive stress due to bending will occur on the medial side of the femur (0.85 mm from the centroid), and the maximum tensile stress will occur on the lateral side (1.0 mm from the centroid). Therefore, from the flexure formula,

$$\sigma = \frac{My}{I}, \text{ we find,}$$

$$\text{Medial: } \sigma = \frac{(37.5\ \text{Nmm})(0.85\text{mm})}{0.2976\text{mm}^4} = 107.1\frac{\text{N}}{\text{mm}^2} = 107.1\ \text{MPa compressive}$$

$$\text{Lateral: } \sigma = \frac{(37.5\ \text{Nmm})(1.00\text{mm})}{0.2976\text{mm}^4} = 126.0\frac{\text{N}}{\text{mm}^2} = 126.0\ \text{MPa tensile.}$$

The stresses due to bending are illustrated in Figure 2.3.6, where the stress varies linearly from 107.1 MPa compressive stress medially to 126.0 MPa tensile stress laterally. The stress is zero at the neutral axis of bending.

The stresses due to axial loading are straightforward to calculate, but again the cross-sectional area requires some care due to the difficulty of making this calculation by hand. A computer program determined there were 922 bone voxels of 0.0322 mm by 0.0322 mm dimension, and, therefore, multiplying the number of voxels by their area resulted in a cross-sectional area of 0.956 mm^2. The axial stress is,

$$\sigma = \frac{F}{A} = \frac{15\text{N}}{0.956\text{mm}^2} = 15.7\frac{\text{N}}{\text{mm}^2} = 15.7\ \text{MPa compressive.}$$

The distribution of compressive stress on the cross-section *a-a'* is shown in Figure 2.3.6, and superposition of the stresses due to bending with the stresses due to compressive loading is shown (bottom). From these calculations, the maximum compressive stress is 122.8 MPa on the medial edge, and the maximum tensile stress of 110.3 MPa occurs on the lateral edge.

The stresses estimated due to bending are about one order of magnitude larger than the axial stresses. This illustrates that the geometry of the acting forces is extremely important. If the resulting force is not acting along the axis of a bone, i.e., eccentric loading versus concentric loading, the stress in the bone increases and can reach multiples of the stresses produced by the axial forces.

The results of this example have at least three practical implications:

(1) Joint forces (bone-to-bone contact forces in joints) that do not act along a bone axis are often compensated for by muscular forces that reduce the maximal stresses on the surface of the bone.

(2) Malalignment of the skeleton may necessitate increased muscular compensation to reduce the maximal stresses on the bone surfaces, e.g., at the diaphysis of a long bone. Muscular atrophy as a consequence of injury or aging may disturb the muscular balance and place excessive stresses on the bone.

(3) Movements in which external forces do not act along the bone axes in the human body, e.g., forces on the foot in different shoes, may produce high stresses on the bone (Nigg, 1985).

EXAMPLE 3

Question

A force of 4000 N acts at a distance 60 mm from the axis of the structure where we consider several idealized bone cross-sectional areas as illustrated (Figure 2.3.7). Determine the moment of inertia and the maximum bending stress for each section. The inner and outer radii for each section are given as follows:

A:	$r_i = 20\text{mm}, r_o = 25\text{mm}$	Base case
B:	$r_i = 15\text{mm}, r_o = 20\text{mm}$	Smaller diameter; same wall thickness
C:	$r_i = 25\text{mm}, r_o = 30\text{mm}$	Larger diameter; same wall thickness
D:	$r_i = 20\text{mm}, r_o = 28.3\text{mm}$	Periosteal thickening relative to base case
E:	$r_i = 15\text{mm}, r_o = 25\text{mm}$	Endosteal thickening relative to base case

Assumptions

The same assumptions as in the previous example are used.

Solution

The same procedure used in the previous example is applied here where we consider the superposition of stresses due to the bending and axial loads. The maximum stress is a function of the maximum compressive stress due to bending summed with the compressive stress due to the axial load,

$\sigma_{max} = \dfrac{M_x y}{I_{xx}} + \dfrac{P}{A}$, where y is equal to r_o, P is the internal normal force, and A is the cross-sectional area.

The moment of inertia is calculated based the formula provided in Table 2.3.4 for the hollow circle,

$$I_{xx} = \frac{1}{4}\pi(r_o^4 - r_i^4).$$

Taking section A as an example, $I_{xx} = 181132.5\ \text{mm}^4$ or $181.1 \cdot 10^3\ \text{mm}^4$. Knowing that the force F = 4000 N is acting with a moment arm of d = 60 mm, the resulting moment is $M_x = 240$ kNm, and the maximal stress for section A due to bending occurs at the edge, y = 25 mm. Superimposing the compressive stress due to the internal normal force $P = 4000\ \text{N}$ and $A = 706.9\ \text{mm}^2$, and the combined maximal normal stress, the maximum stress is $\sigma_{max} = 38.8\ \text{MPa}$. Similar calculations can be performed for all five cross-sectional areas, and taking section A as a base case, the results from the other calculations can be expressed as a relative percent difference. A summary of those calculations is presented in Table 2.3.5.

Some important observations should be made based on the cross-sectional area comparisons. Note that sections A, B and C represent three bones with identical wall thicknesses, but changing diameters. Comparing case A with cases B and C, it is evident that the per-

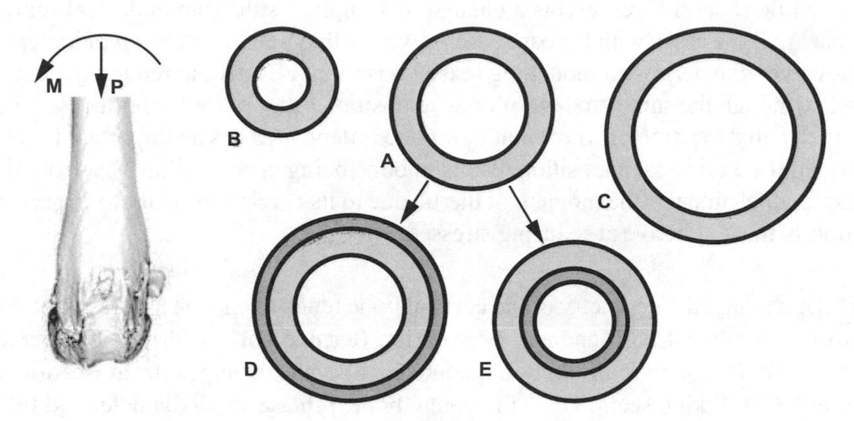

Figure 2.3.7 **A mouse femur with idealized cross-sectional areas subjected to a pure bending moment. Sections A, B and C have the same wall thicknesses, and sections D and E have periosteal and endosteal bone adaptation relative to section A.**

cent change in area is less dramatic than the change in moment of inertia. This occurs because the area is proportional to r^2, while moment of inertia is proportional to r^4. Thus, changes in bone diameter impact stresses due to bending disproportionally more than due to axial loading. Increasing the diameter of bone is an effective method to make bone more resistant to bending.

Table 2.3.5 **Area, moment of inertia and maximum stress calculated for five cross-sections of bone subjected to bending. Each cross-section is compared to the base case section A (shaded) and presented as a percent ratio.**

CROSS-SECTION DESCRIPTION	AREA		AREA MOMENT OF INERTIA		MAXIMUM STRESS	
	[mm^2]	[%]	[10^3 mm^4]	[%]	[MPa]	[%]
A: Standard cross-section Wall thickness, 5 mm. BASE CASE	706.9	100.0	181.1	100.0	38.8	100.0
B: Small cross-section Wall thickness, 5 mm	549.8	77.8	85.9	47.4	63.2	162.8
C: Large cross-section Wall thickness, 5 mm	863.9	122.2	329.4	181.8	26.5	68.3
D: Periosteal remodelling of section A. Wall thickness, 8.3 mm	1259.4	178.2	378.1	208.7	21.1	54.5
E: Endosteal remodelling of section A. Wall thickness, 10 mm	1256.6	177.8	267.0	147.4	25.7	66.1

Section D represents a change to wall thickness that might result from periosteal remodelling, while section E represents a change that might result from endosteal remodelling. Comparing these cases with the base case A, we see that both result in *equal* increases in area. However, the periosteal modelling (case D) is more effective at reducing the maximum stress. Although the same amount of bone apposition takes place for both cases, i.e., same area, the spatial location of where that new bone is deposited has an important impact on reducing stress. Periosteal apposition results in bone being deposited furthest from the bending axis, which impacts the moment of inertia due to its r^4 relation. Therefore, periosteal apposition is most effective at reducing stresses.

During aging, the geometry of the cortical bone tends to change as a result of periosteal apposition combined with endosteal resorption (termed cortical drift). This remodelling could result in changes of section geometry like the changes from section E to C (Figure 2.3.7). Taking section E as the young bone, it has a small diameter and thick walls while section C could represent the old bone with a larger diameter and thin walls. Notably, the total bone mass has decreased in the old bone (reduced area), yet the maximum stress is maintained through the change in cross-section resulting in increased bending stiffness that occurs due to the remodelling. This analysis is greatly simplified, of course, and ignores possible changes to the bone tissue that may occur. However, it demonstrates how the functional capacity of the bone to withstand loads can be sustained during age-related remodelling.).

The hollow structure of bone serves not only a physiological purpose (cavity for bone marrow), but it also is important for reducing metabolic costs while maintaining sufficient functional strength. The hollow structure has less mass and will, therefore, have a smaller inertial momentum. This is advantageous for bone segments involved in frequent, quick movements such as those in the extremities. In fact, most extremities in human and animal bodies have hollow bones, including femur, tibia, metatarsals, and radius. If the bone had a solid cross-section, it would result in a significant increase in mass (and therefore metabolic cost), yet there would be little advantage for increasing strength. Simple calculations comparing a hollow and solid cross-section, i.e., comparing case A of Figure 2.3.7 with and without a hollow core, can verify the marginal increase in strength. A high strength to mass ratio is advantageous from a metabolic and mechanical perspective, and this can be achieved through hollow bones. Different animal types have different strength-mass ratios, depending on the function of the bone (Oxnard, 1993).

BONE ANISOTROPY

Although bone is often considered an isotropic homogenous elastic tissue, this is a simplification of its mechanical properties. In fact, bone is an anisotropic inhomogeneous tissue that exhibits viscoelastic properties. For the purposes of calculations, the characterization of those mechanical properties may be simplified depending on the necessary level of complexity to solve the particular problem at hand. The simplifications, however, should be justified.

The *viscoelastic* characterization of cancellous and cortical bone refers to the components of its viscous and elastic properties. The elastic component suggests that bone returns to its original shape after applied loads have been removed. Since daily activities result in strains on the order of 2000 to 3000 $\mu\varepsilon$ it is often justified to consider bone as an *elastic* tissue, since these strains are well within the elastic limit. The viscous component is responsible for the dependency of stiffness on strain rate, the ability to dissipate energy within the elastic range, and for stress relaxation and creep behaviour of cancellous bone (van Rietbergen & Huiskes, 2001). The dependency of modulus of elasticity on strain rate is low for normal physiological strain rates (Carter & Hayes, 1977). Therefore, bone can be well described as an elastic material. When bone is loaded at strains and strain rates beyond normal physiological activities, i.e., error loads that may result in failure, clearly the elastic simplification of its material properties should be reconsidered.

The *inhomogeneity* of bone refers to the fact that its material properties vary spatially, i.e., properties vary depending on the region of skeleton considered. At the tissue level, cortical bone may vary depending on the degree of mineralization or changes in porosity. At the apparent level, cancellous bone may also depend on mineralization, but the organization of the cancellous structure can also play a significant role. For example, cancellous bone in the spine is organized to withstand primarily vertical loads, while its organization is more complex in the proximal femur. Sometimes the inhomogeneity of the tissue is not included simply due to a lack of available and accurate information. However, this problem is being rectified as more advanced three-dimensional imaging methodologies, which provide tissue density and organization, become available.

The *anisotropic* nature of bone refers to the fact that the material properties differ depending on the direction considered. At the apparent level, vertebral cancellous bone serves as a good example, as its stiffness in the vertical direction is greater than in the transverse directions. This difference is due to the organization of the underlying cancellous bone to withstand vertical loads. In cortical bone, anisotropy is largely due to variations in osteonal organization. The stiffness of human cortical bone in the longitudinal direction is approximately twice the stiffness in the transverse direction.

Hooke's law can describe the elastic relation between stress, strain and the material properties in three dimensions. There are six components of stress (three normal and three transverse/shear stresses) and, similarly, six components of strain. The relation between strain and stress for a linear elastic material is based on the elastic constants that form the most general form of Hooke's law, as follows:

$$
\begin{aligned}
\sigma_1 &= c_{11}\varepsilon_1 + c_{12}\varepsilon_2 + c_{13}\varepsilon_3 + c_{14}\varepsilon_4 + c_{15}\varepsilon_5 + c_{16}\varepsilon_6 \\
\sigma_2 &= c_{21}\varepsilon_1 + c_{22}\varepsilon_2 + c_{23}\varepsilon_3 + c_{24}\varepsilon_4 + c_{25}\varepsilon_5 + c_{26}\varepsilon_6 \\
\sigma_3 &= c_{31}\varepsilon_1 + c_{32}\varepsilon_2 + c_{33}\varepsilon_3 + c_{34}\varepsilon_4 + c_{35}\varepsilon_5 + c_{36}\varepsilon_6 \\
\sigma_4 &= c_{41}\varepsilon_1 + c_{42}\varepsilon_2 + c_{43}\varepsilon_3 + c_{44}\varepsilon_4 + c_{45}\varepsilon_5 + c_{46}\varepsilon_6 \\
\sigma_5 &= c_{51}\varepsilon_1 + c_{52}\varepsilon_2 + c_{53}\varepsilon_3 + c_{54}\varepsilon_4 + c_{55}\varepsilon_5 + c_{56}\varepsilon_6 \\
\sigma_6 &= c_{61}\varepsilon_1 + c_{62}\varepsilon_2 + c_{63}\varepsilon_3 + c_{64}\varepsilon_4 + c_{65}\varepsilon_5 + c_{66}\varepsilon_6
\end{aligned}
$$

where:

c_{ik} = elastic constants

with:

$c_{ik} = c_{ki}$

and

$\sigma_1, \sigma_2, \sigma_3$ and $\varepsilon_1, \varepsilon_2, \varepsilon_3$ are normal stresses and strains, respectively.

$\sigma_4, \sigma_5, \sigma_6$ and $\varepsilon_4, \varepsilon_5, \varepsilon_6$ are shear stresses and strains, respectively.

Not all of the elastic constants are independent. Anisotropy represents the most general relation between stress and strain for which there are 21 independent elastic constants. However, the determination of these elastic constants experimentally is impractical. Symmetries can be exploited to simplify the relation. *Orthotropy* exists when it is assumed that the mechanical behaviour differs along three orthotropic axes. The orthotropic nature of a material must often be approximated, and there are established methods available for making this approximation in bone (van Rietbergen et al., 1996). In this case, nine independent elastic coefficients must be determined that are directly related to three elastic moduli (E_1, E_2, E_3), three shear moduli (G_1, G_2, G_3), and three Poisson's ratios (v_{12}, v_{23}, v_{31}). Hooke's law can be expressed in matrix form $\{\varepsilon\} = [C]\{\sigma\}$, where $[C]$ is the six-by-six compliance tensor represented as:

$$C = \begin{bmatrix} c_{11} & c_{12} & c_{13} & 0 & 0 & 0 \\ c_{12} & c_{22} & c_{23} & 0 & 0 & 0 \\ c_{13} & c_{23} & c_{33} & 0 & 0 & 0 \\ 0 & 0 & 0 & c_{44} & 0 & 0 \\ 0 & 0 & 0 & 0 & c_{55} & 0 \\ 0 & 0 & 0 & 0 & 0 & c_{66} \end{bmatrix}.$$

Further symmetry can be exploited if material behaviour is identical in two axis directions, but different from the third axis. In this case we have *transverse isotropy* and five independent material coefficients must be determined. Finally, when the material behaviour is identical in all axis directions (most symmetric) we have *isotropy* for which there are only two independent elastic coefficients (E, v). There is a direct correspondence between the elastic coefficients and the material behaviour constants such as elastic modulus, shear modulus and Poisson's ratio (van Rietbergen & Huiskes, 2001).

Experimental testing of bone using standard mechanical testing approaches such as uniaxial compression or tension, torsion, and bending are often used to determine the mechanical behaviour of bone (Keaveny & Hayes, 1993). Normally, either orthotropy, transverse isotropy, or isotropy is assumed. An alternative approach is to use knowledge of the three-dimensional organization of the bone tissue, rather than performing mechanical tests, which are highly invasive, to derive the mechanical behaviour (Cowin, 1985; Kabel et al., 1999). This approach has been developed for cancellous bone where its mechanical properties at the apparent level are related to the organization of the underlying trabeculae (Odgaard et al., 1997). Once the trabecular organization is quantified using any number of methods, one of which is *mean intercept length* (MIL) method (Odgaard, 1997), that organization can be related to the mechanical properties. For example, in vertebral cancellous bone, MIL is used to quantify the preferred direction of the trabecular in the vertical direction, and this corresponds to the principal direction of the mechanical stiffness. The advantage of using the organization of the trabecular structure to estimate mechanical behaviour is that this measure can be performed based on medical imaging technologies such as magnetic resonance imaging or computed tomography.

MECHANICAL ADAPTATION

Mechanical adaptation is often used synonymously with the term bone remodeling. After primary bone has been established through modeling, there is a continuous process of remodeling that occurs. There is bone resorption by osteoclasts and formation by osteoblasts. Remodeling serves to maintain the integrity of bone structure by removing microcracks that accumulate during normal physiological activities, and it provides the ability to adapt to environmental changes in loading. If mechanical load is increased through exercise, an increase in bone mass can occur. Conversely, decreased load, as a result of space flight or bed rest, can result in decreased bone mass. Not only is the bone mass changed, but, particularly in the case of cancellous bone, the architecture may also adapt to the new mechanical environment. The mechanism that regulates remodeling is not well understood However, it is commonly postulated that cells such as osteocytes may act as so-called *mechanosensors* that regulate the remodeling process. The Mechanostat Theory was established by Frost (1986) to explain when modeling and remodeling processes will occur as a function of the strain environment. The theory suggests there is a physiological window of normal strains, but if strains reach a lower limit ($<200\,\mu\varepsilon$) a remodeling response will be evoked, and if strains reach an upper limit ($2500\,\mu\varepsilon$ compression or $1500\,\mu\varepsilon$ tension), a modeling response will occur (Martin & Burr, 1989). The Mechanostat Theory has been verified experimentally, and, although it is not perfect, it has proven to be a useful concept for understanding bone adaptation. The question remains, however, as to what type of loading is sensed by the bone to invoke a remodeling process.

The primary difficulty in determining the precise mechanical environment that leads to bone remodeling stems from the difficulty of measuring load in an in vivo environment. For example, exercise can result in increased mechanical loading, but its exact characterization at the tissue-level is difficult to determine, and other exercise-induced changes within the body may confound results. Currently, strain is considered the most important mechanical factor regarding bone remodelling, and there are many possible strain environments that play a role in bone adaptation. *Strain mode* can be either in compression or tension, both having osteogenic potential (Lanyon, 1974; Pauwels, 1980). For either strain mode, if eccentric loading is present, i.e., bending, then a *strain gradient* will result, meaning there is a spatial distribution of strain magnitudes (Lanyon, 1987). The regions of highest gradient may coincide with increased bone formation (Judex et al., 1997). The *strain direction* may control osteonal remodeling in cortical bone, so that alignment occurs with the principal strain directions (Lanyon & Bourn, 1979).

It is well documented, e.g., Liskova (1965), that bone responds differently to static and dynamic load. Dynamic loading causes an osteogenic response, and therefore *strain rate* and *strain frequency* have been extensively investigated regarding bone adaptation. Impact loading is an example of a high strain rate loading condition that induces bone formation. Vibration loading with a high frequency, and low magnitude strain load also induces bone formation (Rubin et al., 2001).

Finally, *strain energy* is a measure of the energy stored due to normal and shear strains that occur in the strain environment. It is easily calculated from methods such as finite element modeling, and can be expressed simply as a scalar that always has a positive value (a negative strain energy is not possible). For these reasons, strain energy per unit volume of tissue, or strain energy density, is a convenient method for relating the mechanical environment to adaptive processes.

Theories for the mechanical adaptation of bone have been implemented in computational models to provide predictive measures of bone density and to understand the processes of bone remodeling (Huiskes & van Rietbergen, 2005). Phenomenological models can simulate the outcome of the coordination of osteoclastic and osteoblastic activities based on an optimization of bone density distribution, usually using local mechanical loading, i.e., strain, strain energy density, as the optimization parameter. These phenomenological models can provide a reasonable prediction of bone density, for example, around implants, but they lack a basic underlying theory for the cell biology that controls the adaptive processes.

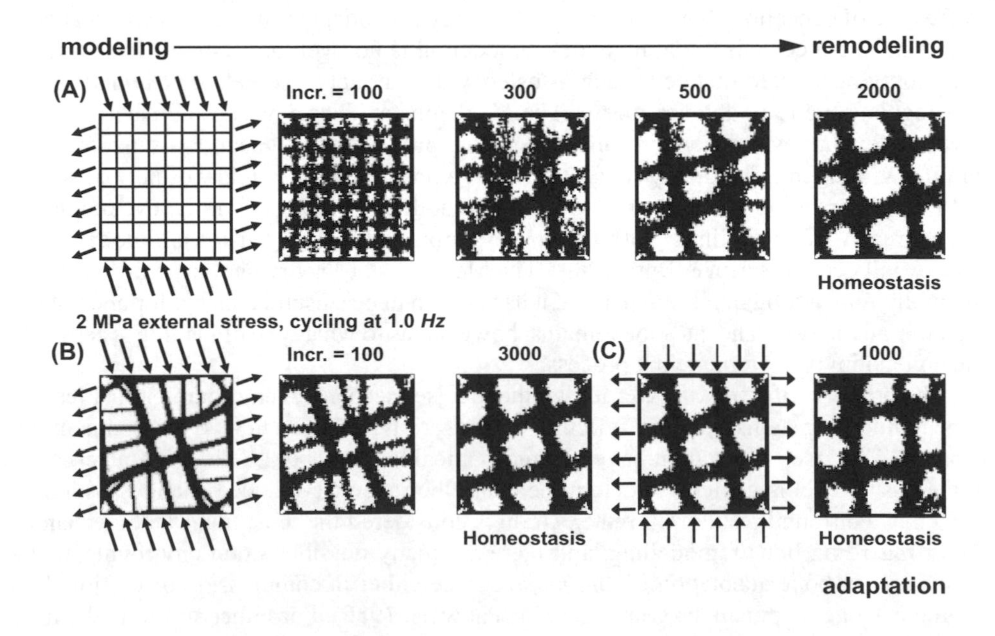

Figure 2.3.8 **Development of bone architecture in the simulation model starting from two different initial configurations (A) and (B). Homeostasis is obtained where only remodeling occurs, during which architecture no longer changes. When the load is reoriented (C), the trabecular orientation gradually adapts (adapted from Huiskes et al., 2000).**

A model that incorporates the underlying biological mechanisms for cancellous bone was proposed by Huiskes et al. (2000). The model presumes that osteocytes embedded within the bone tissue of cancellous bone act as mechanotransducers that send signals to the bone surface to attract BMUs, thus controlling the adaptive process. The mechanical signal is based on strain energy density, and the independent action of the osteoblasts and osteoclasts controlling formation caused modeling. Similarly, the independent action of the osteoblasts and osteoclasts controlling resorption caused remodeling. For a given dynamic load, a two-dimensional computer simulation demonstrated that a reasonable trabecular architecture would result following a modeling process as per the models' input parameters. Furthermore, once homeostasis was reached, the process of remodeling continued for bone maintenance. Subsequently, when a reoriented dynamic load was applied, the trabecular architecture was able to adapt to the new loading condition by realigning the trabeculae according to the load (Figure 2.3.8). The model is unique as it explains modeling (formation and adaptation) and remodeling (maintenance) based on a theoretical framework for the underlying biological processes. It has since been applied in three dimensions (Ruimerman et al., 2005), and shown to predict the remodeling cutting-cone behaviour inside the tissue matrix of cortical bone. Although it is still not clear why osteocytes may be particularly sensitive to strain-energy density, as opposed to other mechanical inputs, these types of issues will likely be addressed in the future. Nevertheless, these types of models will be important for learning about the physiological processes that control bone adaptation. As computational power increases to permit large three-dimensional models, they hold the potential to provide a means to predict the outcome of therapeutic interventions in diseases such as osteoporosis, and may provide insight into optimal timing of such interventions.

2.3.4 FAILURE OF BONE

FRACTURE RISK

Bone fractures are a major health concern, as they occur most frequently in the elderly where there is a devastating impact on morbidity and mortality. At least 20% of women are estimated to have suffered one or more fractures by age 65, and as many as 40% suffer such fractures after the age of 65 (Riggs et al., 1992; Melton et al., 1992). Increased *bone fragility* as a result of age- or disease-related changes leads to a higher risk of fracture. There are several possible causes for increased fragility, but the most prevalent is *osteoporosis*, which occurs most frequently in post-menopausal women and the aged. Osteoporosis is defined as a systemic skeletal disease characterized by low bone mass and deterioration of bone tissue. Osteoporosis results in a high turnover state leading to increased fragility and risk of fracture. Although osteoporosis is most often associated with postmenopausal women, the state of high turnover is also associated with immobilization osteoporosis, post-transplantation osteoporosis, secondary hyperparathyroidism, and Paget's disease. Other possible diseases leading to increased fragility are *osteopetrosis* resulting in hypermineralization of bone tissue, and *osteomalacia* associated with hypomineralization. There is a significant amount of research being performed on bone fragility, with the goal being to provide a clinical assessment method to estimate fracture risk, and, ultimately, to treat the disease leading to fragility.

Assessment of bone fragility requires identification of the risk factors associated with an increased possibility of fracture. In the clinic, bone density is normally measured since it is significantly correlated with mechanical parameters, i.e., stiffness and strength. In fact, the diagnosis of osteoporosis is defined by the World Health Organization (World Health Organization, 1994) on the basis of measured bone density compared to a normal population. Bone density below 2.5 standard deviations from the normal population is considered osteoporotic, and low bone density, i.e., between 1 and 2.5 standard deviations below average, is termed *osteopenic*.

Stochastic relations for the prediction of bone mechanical properties, based on apparent density, have been developed. By performing mechanical tests on a wide range of human and bovine cancellous and cortical bone, Carter and Hayes (1977) established the following relation between axial *elastic modulus* and apparent density:

$$E = 3790\dot{\varepsilon}^{0.06}\rho^{3},$$

where $\dot{\varepsilon}$ is the strain rate (s^{-1}), ρ is the apparent density ($(g)/cm^{3}$).

Other studies considering pure cancellous bone have applied the same general relation $E = a\rho^{p}$, but found that a power p of 2 provided a better fit (Rice et al, 1988). The power is not limited to integer values, and may be within the range of 2 to 3. Similar relations have been developed for estimating strength from bone density, where the *ultimate stress* was related to apparent density, as follows (Carter & Hayes, 1977):

$$\sigma_{ult} = 68\dot{\varepsilon}^{0.06}\rho^{2}$$

These stochastic relations provide a reasonably good prediction over a wide range of bone densities. On average, 70 to 80% of the variability of bone strength is determined by apparent density (Dalen et al., 1976). However, the predictive ability varies widely for individual specimens, as apparent density explains anywhere between 10 and 90% of the variation in the strength of trabecular bone (Ciarelli et al., 1991). This discrepancy illustrates an important problem of simply using bone density for fracture prediction. There must be other contributing factors, because not all people with decreased bone density develop fractures.

Several factors have been identified that influence fragility, and this has led to the emergence of the term *bone quality* as a general description of the physical changes associated with increased fragility. The precise definition remains elusive (NIH Consensus Development Panel, 2001). However, some general key components that comprise bone quality have been agreed upon within the research community (Bone quality, 2005):

- **Architecture**: Microarchitecture, including organization (disorganization) of trabecular microarchitecture and microporosity; macroarchitecture, including bone geometry and cross-sectional area.
- **Turnover**: Remodeling dynamics, increased rate of turnover or unbalanced turnover.
- **Damage Accumulation**: Accumulation of microcracks due to unrepaired fatigue damage, insufficient repair by remodeling.
- **Mineralization**: Tissue material properties, mineralization profile, extra-cellular matrix properties, and chemical consistency.

The assessment of three-dimensional microarchitecture can be achieved through high-resolution computed tomography (CT) (Rüegsegger et al., 1996) or magnetic resonance imaging (MRI) (Majumdar, 1998). Macroarchitecture can be assessed using clinical MR or CT. Bone turnover and micro-cracks can be assessed using techniques such as histomorphology, and mineralization measures are based on microradiography and fourier transform infrared spectroscopy. Many of these measures require bone biopsies, Therefore, it will be important to determine which features, or combinations, are most highly correlated to fracture risk. It may be possible that surrogate measures can be found for important factors, and that possibly the key factor has not yet been included in the list. Nevertheless, the bottom line is that there is a need for better assessment of fracture risk.

Imaging methods such as magnetic resonance imaging and computed tomography are promising technologies that can improve our understanding of bone fragility at both the basic research level (laboratory) and in the application to health care problems (clinic). In the laboratory, micro-computed tomography (micro-CT) has recently been developed that provides a non-destructive method to acquire high resolution three-dimensional images, i.e., less than 10 μm isotropic resolution of bone microarchitecture. This tool has had an enormous impact on the bone field, because a wide variety of bone microarchitectures can be examined in three dimensions with relative ease. Before micro-CT, a tedious serial-sectioning technique was required to evaluate architecture, and the method resulted in destruction of the specimens being examined.

Micro-CT has created novel opportunities to understand bone strength and stiffness in relation to its architecture. It was applied to provide a better understanding of the failure mechanisms of trabecular bone through multi-step compression tests within a micro-CT scanner (Nazarian & Müller, 2004), where it was demonstrated that for a given apparent strain there is a wide variation in local strains at the trabecular level. In addition, the three-dimensional data provided by micro-CT and high resolution MRI can be used to generate finite element (FE) models of complex bone architectures. Thus, without destroying the bone samples, the imaging technology combined with FE modeling can be used to provide a measure of the stiffness and strength of a bone specimen (Figure 2.3.9), and even back-calculate information about tissue properties (van Rietbergen et al., 1995).

Clinical applications of MRI and CT are becoming increasingly sophisticated. A typical clinical CT scanner, for example, does not provide sufficient resolution to assess bone microarchitecture. However, when calibrated, these scanners can provide estimates of apparent bone density that is then converted to stiffness based on stochastic relations. The resulting quantitative CT (QCT) data can be input into finite element models representing apparent-level bone properties to provide patient-specific measures of strength and assess biomechanical risk factors (Faulkner et al., 1991).

Clinical measurement of microarchitecture is more difficult, but is emerging as a potential tool. For example, it has recently been demonstrated that high resolution MR can distinguish the trabecular architecture, and subsequent strength from finite element models, of normal and osteopenic post-menopausal women (Newitt et al., 2002). In addition, a new generation of micro-CT scanners has been recently introduced that are capable of measuring bone microarchitecture in patients at resolutions less than 100 μm (Boutroy et al., 2005). The data from these new micro-CT scanners will be ideal for finite element models of bone microarchitecture, due to the high-contrast images that can be obtained from CT methodology. A limitation of both MR- and CT-based clinical methods for obtaining bone microar-

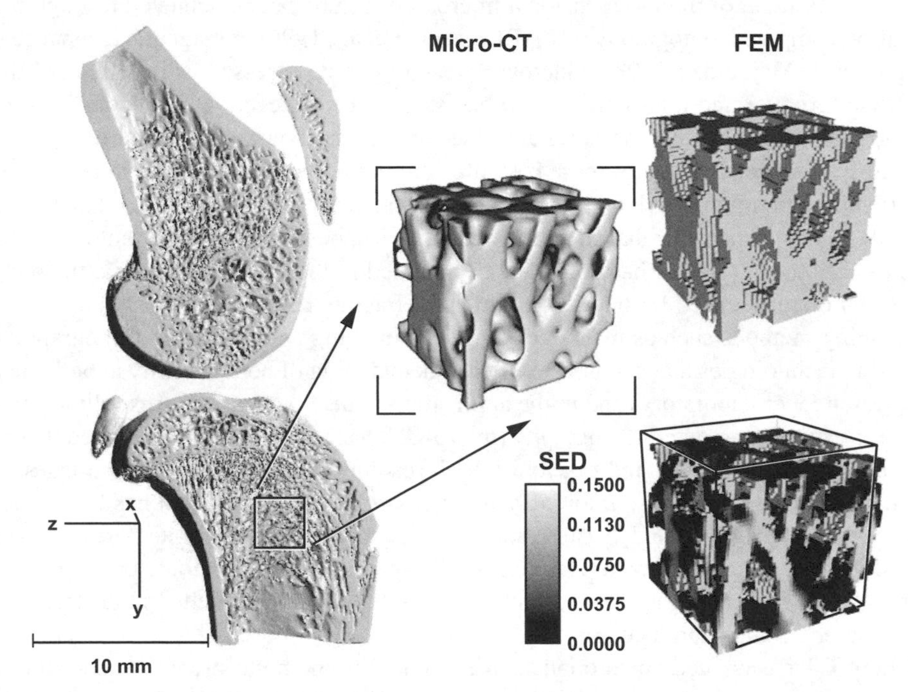

Figure 2.3.9 High resolution computed tomography (micro-CT) of a cat femur (left; adapted from Boyd et al., 2005) with a small region of interest extracted (middle). After conversion to a finite element model (top right), the deformation and the strain energy density can be determined for a given load (bottom right).

chitecture is that the measurements are restricted to peripheral limbs, i.e., distal radius or distal tibia, where signal-to-noise ratios can be maximized. Nevertheless, these tools offer exciting new potential for learning about bone quality in patients, and they are poised to become important options for improving bone-related health care.

2.4 ARTICULAR CARTILAGE

HERZOG, W.
FEDERICO, S.

2.4.1 INTRODUCTION

Articular cartilage is a thin (about 1 to 6 mm in human joints) layer of fibrous connective tissue covering the articular surfaces of bones in synovial joints (Figure 2.4.1). It consists of cells (2 to 15% in terms of volumetric fraction) and an intercellular matrix (85 to 98%) with 65 to 80% water content. Articular cartilage is a viscoelastic material that, in conjunction with synovial (joint) fluid, allows for virtually frictionless movement (coefficients of friction range from 0.002 to 0.05) of the joint surfaces. The primary functions of articular cartilage include:

- Transmitting force across joints,
- Distributing articular forces to minimize stress concentrations, and
- Providing a smooth surface for relative gliding of joint surfaces.

Osteoarthritis is a joint disease that is associated with a degradation and loss of articular cartilage from the joint surfaces and an associated increase in joint friction causing pain and disability, particularly in the elderly population. In most people, articular cartilage fulfills its functional role for decades, although the incidence of osteoarthritis in North America is about 50% among people of age 60 and greater.

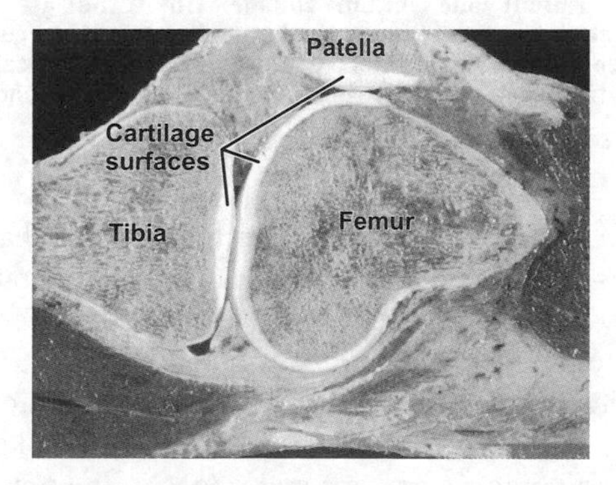

Figure 2.4.1 **Sagittal plane section through a human knee showing the femur, tibia, and patella, and associated articular cartilage.**

2.4.2 STRUCTURE

Articular cartilage is structurally heterogeneous, and material properties change as a function of depth. Although these changes are continuous, articular cartilage is typically divided into four structural zones.

The four structural zones of articular cartilage (Figure 2.4.2) are:

- Superficial zone,
- Middle (or transitional) zone,
- Deep (or radial) zone, and
- Calcified zone.

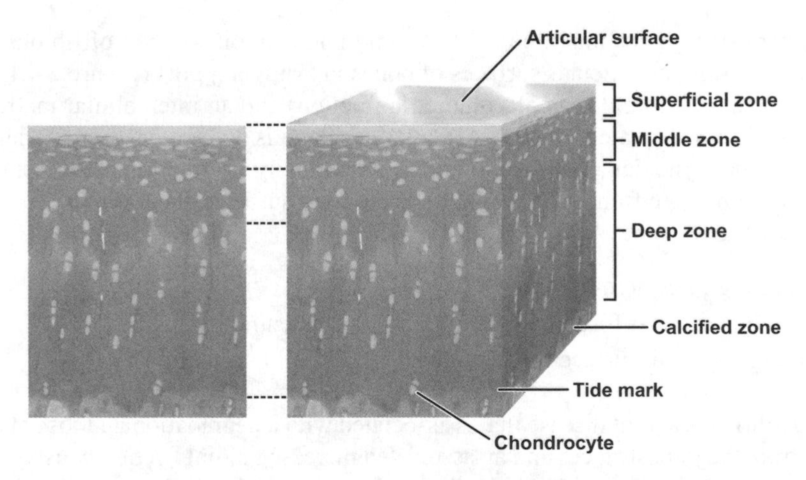

Figure 2.4.2 **The four zones of articular cartilage. The superficial zone provides the sliding surface of joints with collagen fibrils aligned parallel to the surface, and flat, relatively metabolically inactive cells. The middle (or transitional) zone contains collagen fibrils that are oriented randomly, and cells are nearly spherical. The deep (or radial) zone contains collagen fibrils that are oriented perpendicular to the subchondral bone (and articular surface), and the cells are typically aligned in radial columns. The calcified zone provides a mechanical transition that separates the relatively soft cartilage tissue from the stiff subchondral bone.**

The *superficial zone* is the thinnest, most superficial region that forms the gliding surface of joints. It contains a surface layer (lamina splendens) of about 2μm thickness, which is made up of randomly aligned collagen fibrils, and a deep layer consisting of collagen fibrils aligned parallel to the cartilage surface following the so-called "split line pattern" (Hultkrantz, 1898), which follows the direction of normal joint movement.

The collagen fibrils in the superficial zone show a wave-like pattern referred to as *c*rimp. This waving or crimping shows dips and ridges in the μm range (Dowson, 1990). The surface, although apparently smooth to touch and with a low friction coefficient, is not quite smooth at a microscopic level.

The deep layer of the superficial zone contains articular cartilage cells (chondrocytes) that are flat (Hunziker, 1992) and metabolically relatively inactive (Wong et al., 1996), as evidenced by a low content of mitochondria, Golgi organs, and endoplasmic reticula. The deep layer of the superficial zone also contains little proteoglycan, but has the highest water concentration (about 80%) of all zones, as water content decreases with depth to a value of about 65% in the deep zone (Maroudas, 1975; Torzilli, 1985).

The *middle (or transitional) zone* is typically thicker than the superficial zone. Collagen fibrils have a greater diameter in this zone than the superficial zone and are oriented in a nearly random fashion. Proteoglycan content is greater and aggregate complexes are larger than in the superficial zone. Chondrocytes are nearly spherical in this zone and contain great numbers of mitochondria, Golgi bodies, and a vast endoplasmic reticulum network, suggesting that cells in this zone are metabolically more active than those of the superficial zone.

The *deep (or radial) zone* contains the largest diameter collagen fibrils, which are oriented perpendicular to the subchondral bone and the cartilage surface. Water content is lowest and proteoglycan content is typically highest in this zone. The chondrocytes tend to be aligned in radial columns and contain intracytoplasmic filaments, glycogen granules, endoplasmic reticula, and Golgi bodies, suggesting great protein synthesis activity. Benninghof (1925) proposed that collagen fibres form arcades that extend from the deep to the superficial zone, and recent studies confirmed that collagen fibres, indeed, might be continuous through the various cartilage zones (Notzli & Clark, 1997).

The *calcified zone* provides a mechanical transition that separates the relatively soft cartilage tissue and the stiff subchondral bone. It is characterized by hydroxyapatite, an inorganic constituent of bone matrix. The calcified zone is separated from the deep (radial) zone by the *tidemark*, an undulating line that is a few μm thick. Collagen fibres from the deep zone cross the tidemark and anchor into the calcified zone, thereby strongly adhering cartilage to bone. The calcified zone contains metabolically active chondrocytes, serves for structural integration, and has been considered important for nutrition and cartilage repair arising from the underlying bone (Hunziker, 1992).

The structural differences across the various zones of articular cartilage have been thought to be the primary cause for the anisotropy of articular cartilage (e.g., Schinagl et al., 1997) and have motivated structural models of transverse isotropic cartilage models (Federico et al., 2005).

2.4.3 COMPOSITION

Articular cartilage consists mostly (85 to 98%) of matrix and a sparse population of cells (about 2 to 15% in terms of volumetric fraction). It is avascular, aneural, and alymphatic.

CELLS

Chondrocytes are metabolically active cells that are responsible for the synthesis and degradation of the matrix. They are isolated, lie in lacunae, and receive nourishment through diffusion of substrates that is thought to be facilitated by cycling loading and unloading, which is common for many articular joint surfaces. As described above, the volumetric fraction, shape, and metabolic activity of cells varies as a function of location, varies across cartilages in different joints, and even varies within the same joint at different locations, e.g., Stockwell (1979); Muir (1983); Clark et al. (2005). It is accepted that normal loading of articular cartilage produces deformations in chondrocytes and the corresponding cell nuclei (Guilak, 1995; Guilak et al., 1995). These deformations have been thought responsible for the biosynthetic activity of cells. In general, static, long-lasting loads have been associated with a tissue degrading response, while dynamic loading of physiological magnitudes has been related to positive adaptive responses.

Chondrocytes are softer than the surrounding extracellular matrix by a factor of about 1000x. Therefore, chondrocytes can be expected to be deformed much more during loading than the matrix, and might even be expected to collapse. However, chondrocytes are surrounded by a protective cover that consists of a pericellular matrix and a pericellular capsule (Poole et al., 1997). The chondrocyte and its pericellular matrix and capsule constitute the chondron, which is the primary functional and metabolic unit of cartilage. It has been suggested that the chondron acts hydrodynamically to protect the chondrocytes from excessive stresses and strains during cartilage compression. Chondrocytes in different types of cartilages have different functional and metabolic roles. In articular cartilage, chondrocytes specialize in producing type II collagen and proteoglycan.

MATRIX

The intercellular *matrix* of articular cartilage is largely responsible for the functional and mechanical properties associated with this tissue. It consists of structural macromolecules and tissue fluid. Fluid comprises between 65 to 80% of the wet weight of articular cartilage, and its volumetric fraction decreases from the superficial to the deep zone. Macromolecules, which are produced by the chondrocytes, comprise 20 to 40%. Of the macromolecules, collagens contribute about 50% of the tissue dry weight. Proteoglycans contribute approximately 30 to 35% and non-collagenous proteins/glycoproteins contribute 15 to 20% of the tissue dry weight (Buckwalter et al., 1991). The interactions of tissue fluid with the structural macromolecules give articular cartilage its specific mechanical and electro-static properties. As the distribution, orientation and density of these macromolecules changes with cartilage depth, so do the functional and material properties of the tissue.

Collagen

There are at least 18 different types of *collagen*. However, in articular cartilage, type II collagen is by far the most abundant (about 80 to 85% of all collagens). Other types of collagen (V, VI, IX, X and XI) are also found in articular cartilage and have been associated with specific functional roles (Thomas et al., 1994; Hasler et al., 1999). Collagen molecules are comprised of three α-chains that are interwoven in a helical configuration. Each chain has a high hydroxylysine content and covalently bound carbohydrates that make it adhere readily with proteoglycans.

Collagens have a high tensile stiffness (about 2 to 46 MPa for uncross-linked type I collagen from rat tail tendon and 380 to 770 MPa when cross-linked (Pins et al., 1997)). Because of their fibrillar structure, collagens are typically thought to have negligible compressive strength, as they are assumed to buckle when subjected to compressive loading (Li et al., 2000a; Soulhat et al., 1999). However, some scientists think of collagen fibrils as reinforced inclusions (similar to steel rods embedded in concrete) with appreciable compressive capabilities, and thereby contributing substantially to withstand compressive forces (Wu & Herzog, 2002). Collagens form a structural network that gives cartilage its tensile strength. Because of the characteristic orientation of the fibrillar network, collagens are associated with providing resistance to compressive loading, as fluid pressurization tends to load the collagen network in tension. Collagen fibrils are cross-linked for further strength (Pins et al., 1997), and are connected to proteoglycans via molecular chains arising from glycosaminoglycans and polysaccharides. Thus, collagens are intimately associated with other macromolecules and so make up a tough tissue that, despite its thinness, can withstand high repetitive loading for a lifetime.

Proteoglycans

Proteoglycans are large molecules composed of a central core protein with glycosaminoglycan side chains covalently attached. The protein core makes up about 10% of the molecular weight of proteoglycan, with the side chains making up about the remaining 90%. The glycosaminoglycan side chains contain sugars, which have a negative electrostatic charge, on a protein core. These negatively charged molecules repel each other and attempt to occupy as much volume as possible. Proteoglycans are kept from dissolving into the fluid and being swept away by their attachment to the stretched collagen network. The negative charges are thus forced to stay in close proximity When subjected to external compressive loads, proteoglycans are further compressed and the repulsive forces increase from their natural pre-tensed state.

Articular cartilage contains large aggregating proteoglycans (aggrecan and versican) and small interstitial proteoglycans (biglycan, decorin, fibromodulin, and lumican). The large proteoglycans contribute 50 to 85% of the total proteoglycan content. Aggrecan is the major proteoglycan. Aggrecan consists of a core protein and up to 150 chondroitin and keratin sulphate chains (Figure 2.4.3). The core protein's N-terminal G1 domain interacts with

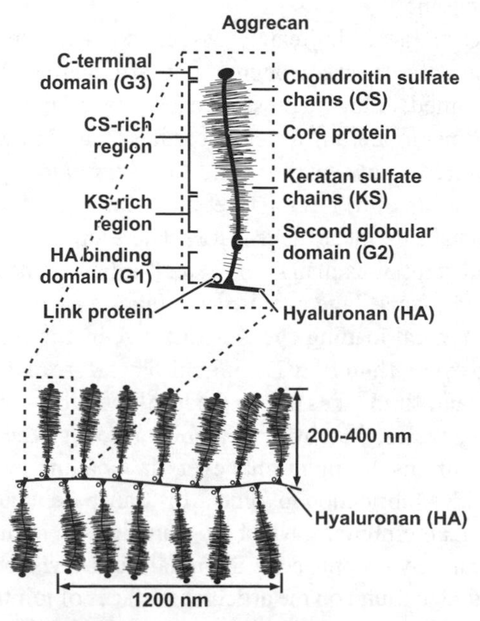

Figure 2.4.3 **Macromolecular aggregate formed by aggrecan molecules (inset-blown up) binding to a chain of hyaluronan through a link protein. The aggrecan molecule consists of a core protein with several domains: hyaluronan binding G1 domain, G2 domain, keratin sulphate rich region, chondroitin sulphate rich region, and C-terminal domain, G3.**

link proteins and hyaluronan, and these three components form stable macromolecular complexes (Figure 2.4.3).

Changes in proteoglycan structure and decreased density often accompany articular cartilage degeneration and aging (Lark et al., 1997; Buckwalter et al., 1985). These changes are typically accompanied by a corresponding increase in water content, decreased stiffness, and reduced resistance to withstand mechanical loading.

Non-collagenous proteins

Non-collagenous proteins play a role in the assembly and integrity of the extracellular matrix, and form links between the chondrocytes and the matrix. Non-collagenous proteins include adhesive glycoproteins such as fibronectin, thrombospondin, chondroadherin, and other matrix proteins such as the link protein, cartilage matrix oligomeric protein, cartilage matrix protein, and proline arginine-rich and leucine-rich repeat proteins. The detailed function of many of these non-collagenous proteins is, currently, not well understood.

Fluid

Articular cartilage tissue *fluid* consists of water and dissolved gas, small proteins and metabolites. Water is the most abundant component of cartilage and accounts for 65 to 80% of its wet weight. Fluid is moved around in articular cartilage, and is closely associated with the synovial fluid of the joint.

Fluid in articular cartilage is intimately associated with the proteoglycan network, which hinders its movement to a certain degree. This arrangement makes cartilage sponge-like in that water is restrained, but, with pressure caused by cartilage loading, will flow with the pressure gradient. When modelling articular cartilage, fluid flow is described by permeability, which is a measure of how easily a fluid flows through a porous material. Permeability is inversely proportional to the force required for a fluid to flow at a given speed through the tissue. Permeability in cartilage is low. Therefore, fluid flow in cartilage is typically small, and any substantial exchange of tissue fluid and synovial fluid, or any significant loss of fluid caused by cartilage compression takes a long time (in the order of minutes). Compared to the typical loading cycle of cartilage in a joint, e.g., about 0.5 seconds of loading during a step cycle, fluid flow is minimal. Therefore, cartilage maintains its stiffness well when loaded, and fluid takes up a big part of the forces in the tissue during physiological (quick) loading cycles, thereby presumably protecting the matrix (and, in turn, the cells) from stresses and strains during normal everyday loading cycles.

Synovial fluid provides lubrication to joints. It is transparent, alkaline and viscous, and is secreted by the synovial membranes, which are contained in joints. Synovial fluid is a dialysate of blood containing hyaluronic acid. Synovial fluid is viscous and contains a glycoprotein called lubricin that is found on the articular surfaces of joints, but not within the cartilage tissue. Cartilage surfaces are physically separated by a 0.5 to 1.0μm film of synovial fluid. Synovial fluid is essential for the mechanical sliding characteristics of joints, and permits the diffusion of gases, nutrients and waste products, thereby making it essential for tissue health.

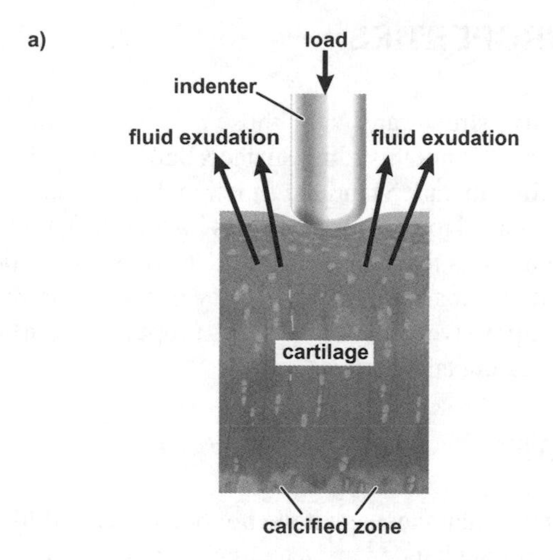

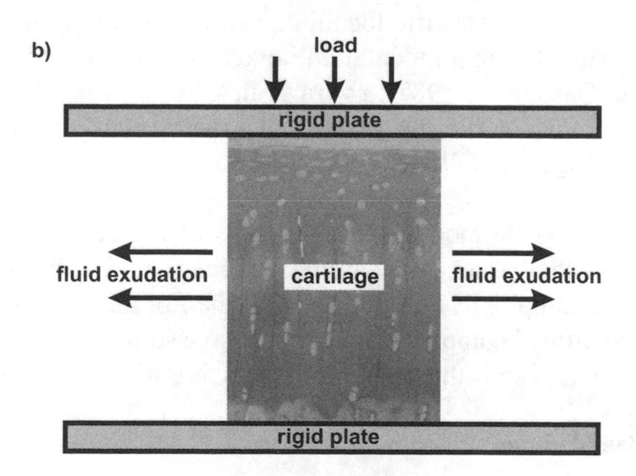

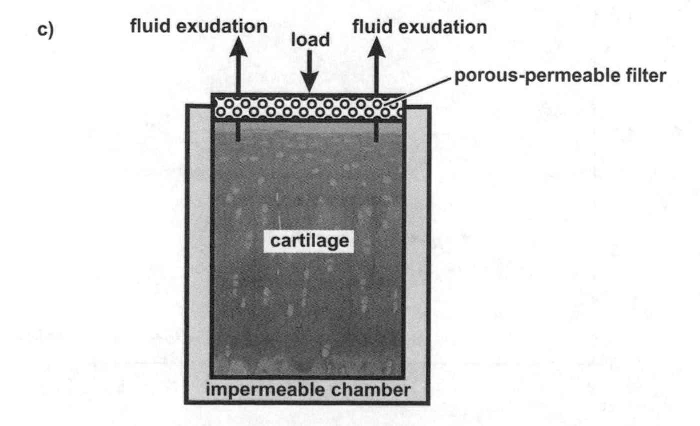

Figure 2.4.4 **Schematic illustration of the three most commonly used techniques for determining articular cartilage mechanical properties: (a) indentation testing, (b) unconfined compression, and (c) confined compression.**

2.4.4 MECHANICAL PROPERTIES

Articular cartilage is a passive structural tissue whose primary functions are mechanical. These functions are providing a smooth sliding surface, and transmitting and distributing forces across joints. Therefore, to understand the functional properties of this tissue, its mechanical (material properties) need to be known. This is insofar difficult as articular cartilage structure and composition varies from one joint to the next, its properties also change with location within a joint, and even across its depth. Nevertheless, many attempts have been made to elucidate the compressive, tensile and shear properties of articular cartilage using a variety of engineering testing approaches.

COMPRESSIVE PROPERTIES

Compressive properties of articular cartilage have not been obtained for the whole tissue attached to its native bone, although indentation testing (Figure 2.4.4a) has been used to evaluate compressive properties at specific locations using exposed joint surfaces, (e.g., Clark et al., 2003), or in patients using indentation devices that can be introduced arthroscopically into joints, e.g., Dashefsky (1987); Lyyra et al. (1995). Typically, material properties are determined for articular cartilage explants that are subjected to unconfined or confined compression testing (Figure 2.4.4b,c)

Compressive properties, elastic modulus and stiffness of a given sample of articular cartilage vary with depth in the tissue. For example, Schinagl et al. (1997) report a more than 20-fold increase in the compressive modulus across the full thickness of bovine articular cartilage tested in unconfined compression. Compressive strength is directly related to proteoglycan concentration, and so is the tissue's compressive stiffness (Figure 2.4.5).

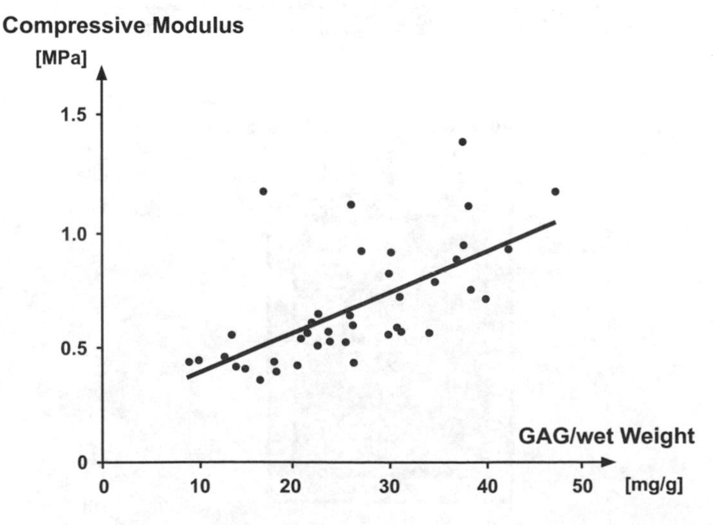

Figure 2.4.5 **Compressive modulus of elasticity as a function glycosaminoglycan (GAG) concentration. Since GAG concentration increases with cartilage depth, so does the tissue's compressive stiffness. Adapted from Sah et al. (1997), with permission.**

Since proteoglycan concentration is known to increase with depth, this result can be expected. However, even a five-fold increase in proteoglycan concentration is only associated with a doubling or tripling of the compressive modulus (Figure 2.4.5). Therefore, it appears that the great changes in stiffness observed experimentally in different cartilage layers must have an additional explanation. One such explanation might be the orientation of the collagen fibrillar network, assuming that the collagen network helps resist compressive loading.

TENSILE PROPERTIES

Articular cartilage functions as a compressive load absorber. However, when loaded, tensile forces are thought to act on the collagen network. Therefore, an understanding of the tensile properties might be important in understanding the functioning of cartilage in vivo. Tensile strength in articular cartilage is highest in the superficial zone and decreases continuously with increasing depth in the tissue (Figure 2.4.6), likely because of variations in the orientation of collagen fibrils, the cross-linking of collagen fibrils, and the ratio of collagen to proteoglycan. Tensile properties are also direction-dependent. When testing occurs along the split line, i.e., along the predominant orientation of the long axis of collagen fibrils, tensile strength and the tensile modulus are greater than when testing is performed perpendicular to the split line (Figure 2.4.6).

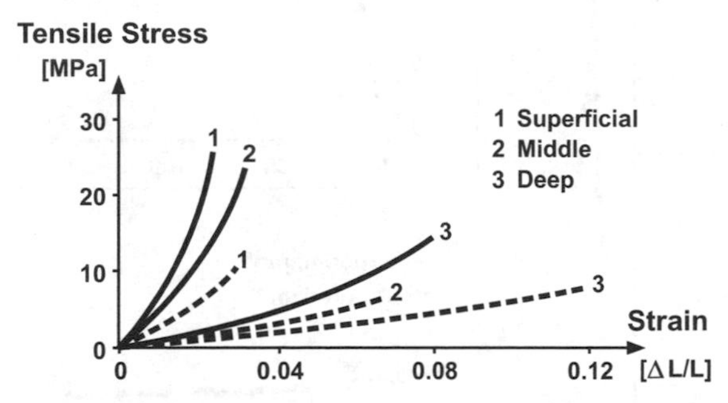

Figure 2.4.6 **Stress-strain curves for tensile testing of articular cartilage specimens from the superficial (1), middle (2), and deep zone (3), and along the long axis of the collagen fibrils (solid lines) and perpendicular to the long axis of the collage fibrils (dashed lines). Tensile strength decreases continuously from the surface to the deep zone and is greater along the collagen fibril direction than perpendicular to it. Adapted from Kempson (1972), with permission.**

SHEAR PROPERTIES

Shear properties in articular cartilage are not well understood, and shear testing is associated with numerous problems. For example, the smoothness of the surface layer makes it difficult to apply shear loads without slippage. Shear testing has become important for identifying flow-independent viscoelasticity of the tissue, as ideally, small deformation shear testing is accomplished with a constant volume and no fluid flow. Shear resistance is thought to come primarily from the collagen network. Consistent with that idea, shear modulus is decreased with decreasing collagen content, as may occur in cartilage diseases such as osteoarthritis. In addition, because the proteoglycan network is thought to pre-stress the collagen network, loss of pre-stress caused by loss of proteoglycans has also been found to decrease the dynamic and equilibrium shear properties (Zhu et al., 1993).

VISCOELASTICITY

If a material's response to a constant force or deformation varies in time, its mechanical behaviour is said to be viscoelastic, in contrast to an elastic material's response that does not vary in time. Articular cartilage is known to be viscoelastic (Figure 2.4.7), however, the origin for this property remains controversial. Most people agree that one aspect of the viscoelastic behaviour is associated with fluid flow, and some believe that there is also a flow-independent viscoelasticity that resides in the matrix proper. Since fluid flow is proportional to the pore pressure gradient in water, fluid flow can be described by the coefficient of hydraulic permeability. The inverse of this coefficient gives a measure of the material's resistance to fluid flow. Since proteoglycans attempt to restrain fluid flow, permeability is smallest in areas with high proteoglycan concentration, e.g., deep layers, and cartilage is most permeable in zones of low proteoglycan concentration, e.g., the surface layer.

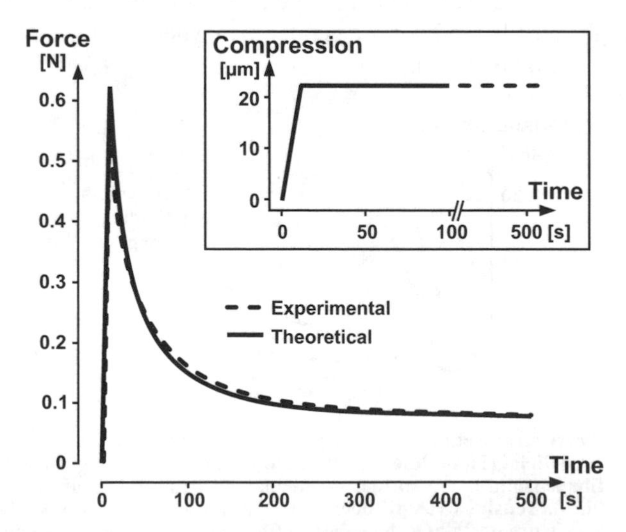

Figure 2.4.7 **Experimental and theoretical force-relaxation curve for articular cartilage exposed to unconfined ramp loading (as shown in the inset). Articular cartilage exhibits a typical viscoelastic response that is associated with fluid flow, and possibly also with an inherent viscoelasticity of the matrix component.**

2.4.5 BIOMECHANICS

The material properties of articular cartilage are of great importance because of articular cartilage's roles in transmitting and distributing force across joints, and providing a smooth surface for virtually frictionless gliding. However, articular cartilage is only one component of a joint, and to understand articular cartilage biomechanics, the functional requirements of articular cartilage within the "organ" joint need to be known. Malfunctioning of one component of a joint (for example, a breakdown of the articular cartilage, loss of a guiding ligament, or rupture of a meniscal inclusion) will affect all structures of a joint. Therefore, focusing on a single component might not do justice to the intricate functionality of that component in its native environment.

When a joint moves, the articular surfaces slide relative to one another, the size and location of the contact area changes, and the contact pressure distribution is affected by the changing surface geometry and the variable forces applied to the joint. In addition, for a given joint configuration, that is a given orientation of the articular surfaces, the contact area increases with increasing force transmission across the joint. For example, the cat patellofemoral joint contact area increases with increasing force potential of the quadriceps muscles, thereby providing a natural way for keeping average pressure between the contacting surfaces low when forces become big. Similarly, when forces transmitted across the patellofemoral joint are increased by a factor of five (from 100 to 500N), the contact area increases by a factor of about four (from about 7 to about 30 mm^2), while the mean contact pressure only increases by a factor of about 0.6 [from 7.7 to 12.9MPa (Figure 2.4.8)].

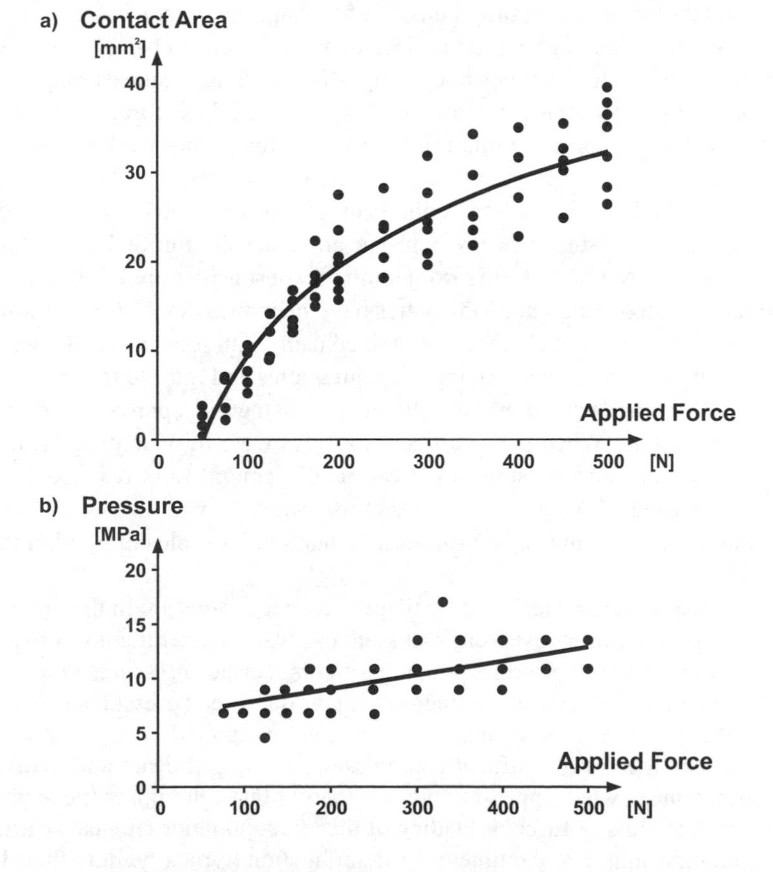

Figure 2.4.8 **Feline patellofemoral contact area (a) and peak pressure (b) transmitted across the joint. Contact area increases quickly with increasing force, especially in the range of physiological loading (0 to 200 N), while peak pressures are less sensitive to changes in joint loading.**

Therefore, contact pressures on the articular surfaces of joints are kept small with increasing forces by increasing:

- Contact areas for joint angles where large muscular forces are possible, and
- Contact area as a function of the applied forces through the viscoelastic properties of articular cartilage.

The loads transmitted across joint surfaces are not well known, because most of the loading in joints is caused by muscles. Muscle forces are hard to measure in vivo and cannot be estimated accurately and reliably. However, Rushfeldt et al. (1981a; 1981b) developed techniques to measure local hip joint contact pressures by mounting pressure transducers on an endoprosthesis that replaced the natural femoral head. They measured average articular surface pressures in the joint of about 2 to 3 MPa, with peak pressures reaching 7 MPa, for loads equivalent to the single stance phase of walking (2.6 times body weight). Peak hip joint pressures reached were in the range of 18 MPa and were obtained for walking downstairs and rising from a chair (Mann, 2005). Using telemetry-based force measurements from patients with total hip joint replacements, Bergmann et al. (1993) measured resultant hip forces of 2.8 to 4.8 times body weight for walking at speeds ranging from 1 to 5 km per hour. Very fast walking or slow running increased these forces to about 5.5 times body weight, while stumbling (unintentionally) gave temporary peak resultant hip joint forces of 7.2 and 8.7 times body weight in two patients. No corresponding joint pressure or force data are available for intact human joints during normal everyday movements.

Similarly, direct pressure measurements in an intact diarthrodial joint during unrestrained movements are not available from animal experimentation. However, quadriceps forces and the corresponding knee kinematics have been measured in freely walking cats. Joint pressure distributions could then be obtained using Fuji pressure-sensitive film in the anaesthetized animals while reproducing the joint angles and muscle forces observed during locomotion. Average peak pressures obtained by using this approach were 5.7 MPa in the patellofemoral joint. When the quadriceps muscles were fully activated (through electrical stimulation), median pressure in the cat patellofemoral joint reached 11 MPa, and peak pressure reached 47 MPa, indicating that joint pressures in normal everyday movements are much smaller than can be produced by maximal muscle contraction (Hasler and Herzog, 1998).

The forces transmitted by articular cartilage across leg joints are in the order of several times body weight for normal everyday tasks such as walking, getting up from a chair, and walking up or down stairs. The corresponding articular surface pressures range from about 2 to 10 MPa, but maximal muscle contraction can produce peak pressures in knee articular cartilage of almost 50 MPa. Such high pressures, when applied to articular cartilage explants through confined or unconfined compression, cause cell death and matrix damage, and also cause damage when applied to the whole joint through impact (peak pressures attained within 0.5 to 3 ms). Muscular loading of the intact joint does not cause articular cartilage damage, indicating that the time of load application (typically more than 100 ms for human muscles), joint integrity and articular surface movements during force production provide a safe environment for high cartilage loading that is lost when testing articular cartilage explants in vitro.

2.4.6 OSTEOARTHRITIS

Osteoarthritis is a joint degenerative disease that affects about 50% of all people above the age of 60 in North America. It is associated with a thinning and local loss of articular cartilage from the joint surfaces, osteophyte formations at the joint margins, swelling of the joint, and pain. The causes for osteoarthritis are not known, but risk factors include age, injury, and obesity. Osteoarthritis is often combined with a loss of muscle mass and movement control, and a decrease in strength.

It is well accepted that physiological loading of articular cartilage is essential for its health and integrity. Animal models of joint disruption through strategic cutting of ligaments or removal of menisci from knees have caused joint degeneration, presumably because of the altered loading conditions of the articular surfaces, although this is difficult to quantify in vivo. For example, when cutting the anterior cruciate ligament of the knee in a dog, cat, or rabbit unilaterally, there is an unloading of the experimental and an over-loading of the contralateral side that lasts for a few weeks (Figure 2.4.9). This intervention is associated with an instability of the knee, osteophyte formation at the joint margins, thickening of the medial aspect of the joint capsule and the medial collateral ligament, and increased thickness, water content, and softness of the articular cartilage (Herzog et al., 1993; Herzog et al., 1998). In the long-term, anterior cruciate ligament transection in the dog and cat have been shown to lead to bona fide osteoarthritis, as evidenced by full thickness loss of articular cartilage in some load bearing regions of the joint, leading to an increase in joint friction, decreased range of motion, pain, and a loss of muscle mass and strength in muscles crossing the knee (Brandt et al., 1991b; Brandt et al., 1991a; Longino et al., 2005; Clark et al., 2005).

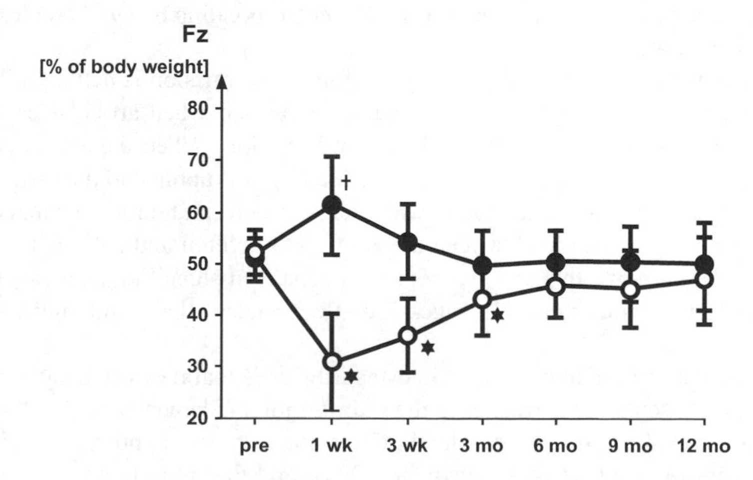

Figure 2.4.9 **Vertical ground reaction forces, Fz, normalized to body weight, as a function of time following unilateral anterior cruciate ligament (ACL) transection in cats (n = 7). Pre indicates testing before ACL transection, 1 wk, 3wk, ..., 12 mo indicate 1 week, 3 weeks, ...12 months post ACL transection. The vertical ground reaction forces are significantly (*) decreased in the ACL-transected (open symbols) compared to the contralateral (closed symbols) hind limb for 3 months post ACL transection. Thereafter, the differences are not statistically significant. One week following ACL transection, the vertical ground reaction force in the contralateral leg is greater (†) than in the pre-transection condition.**

Contact pressure measurements in intact cat patellofemoral joint that were loaded by muscular contraction showed that for a given amount of force transmitted through the joint, contact areas increased and peak contact pressures decreased by 50% in joints with early osteoarthritis compared to normal controls (Herzog et al., 1998). These results can be explained readily with the increased cartilage thickness and decreased stiffness observed in this phase of cartilage degeneration. Theoretical models with the altered geometries and functional properties have confirmed the experimental results. In late osteoarthritis, articular cartilage becomes thinner and harder than normal, and forces are transmitted across a decreased contact area and with peak pressures that are increased compared to the normal joint (Federico et al., 2004c)

It has been demonstrated that loading of articular cartilage affects the biological response of chondrocytes, which in turn affects the health and integrity of the cartilage matrix. The results are controversial, and there is no unified consensus on what loading is good or bad for cartilage health. However, it has been demonstrated that long, static loading is associated with an increase in cartilage matrix degrading enzymes, while intermittent physiological loading appears to strengthen the cartilage matrix through the formation of essential matrix proteins (e.g., Mow et al., 1994; Hasler et al., 1999.

One of the proposed pathways for transmitting load signals across the cell to the genome has been the cytoskeleton (specifically the integrin network). Integrins are transmembrane extra-cellular matrix receptors that have been shown to transmit forces from the matrix to the cytoskeleton of chondrocytes and vice versa. The mechanical signals transmitted by integrins may result in biochemical responses of the cell through force-dependent release of chemical second messengers, or by force-induced changes in cytoskeletal organization that activate gene transcription (Ingber, 1991; Ben-Ze'ev, 1991). However, a direct link between force transmission in integrins and a corresponding biosynthetic response has yet to be demonstrated.

Another potential cytoskeletal pathway for force transmission is actin. Guilak (1995) demonstrated that chondrocytes and their nuclei deformed when articular cartilage was loaded (15% compression) in an unconfined configuration. When the actin cytoskeleton was removed from the preparation and the same loading was applied to the cartilage, chondrocyte deformation remained the same, while nuclear deformation was changed, demonstrating an intimate connection between the actin cytoskeleton and cell nuclei. Although these results do not prove that actin works as a mechanical signalling pathway, they demonstrate that there are mechanical linkages that affect nuclear shape and might affect gene transcription.

A clinical observation in patients with osteoarthritis is that they often have diminished strength in the musculature surrounding the arthritic joint (Slemenda et al., 1997; Hurley, 1999; Slemenda et al., 1998), that maximal muscle contractions are not possible because of reflex inhibitions (e.g., Hurley & Newham, 1993), and that muscle contraction patterns change from normal for everyday movements (Herzog & Federico, 2005). Furthermore, it is generally accepted that muscles provide the primary loading of joints, with weight forces and external forces playing a relatively smaller role. Nevertheless, the role of muscles in the development and progression of osteoarthritis has largely been neglected, and some basic questions should be addressed in the future. These include the role of muscle activation patterns and muscle weakness as risk factors for osteoarthritis.

Only weak associations have been found between radiographic signs of joint disease and joint pain and disability in osteoarthritic patients (Claessens et al., 1990; McAlindon et al., 1993). Quadriceps weakness has been found to be one of the earliest and most common symptoms reported by patients with osteoarthritis (Fisher et al., 1991; Fisher et al., 1993; Fisher et al., 1997; Hurley & Newham, 1993). Furthermore, quadriceps weakness has been shown to be a better predictor of disability than radiographic changes or pain (Slemenda et al., 1997).

Therefore, there have been suggestions that muscle weakness might represent an independent risk factor for cartilage degeneration, joint disease, and osteoarthritis. Recent evidence from an animal model of muscle weakness supports this idea, showing that articular cartilage is degenerated following just four weeks of muscle weakness in knees that were otherwise not compromised (Figure 2.4.10). Based on the evidence that cartilage degeneration and osteoarthritis are associated with changed loading of joints, that mechanical loading influences the biosynthetic activity of chondrocytes, and that muscle coordination patterns and muscle strength are intimately associated with cartilage disease, conservative treatment modalities might represent promising approaches for preventing cartilage disease and osteoarthritis. These treatment modalities include appropriate movement biomechanics, exercise, and strengthening.

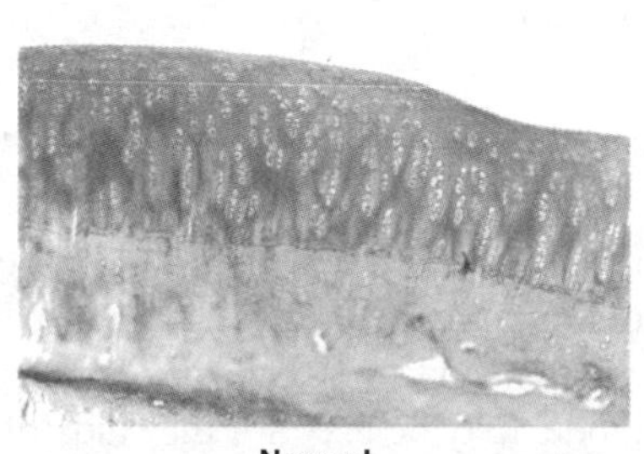

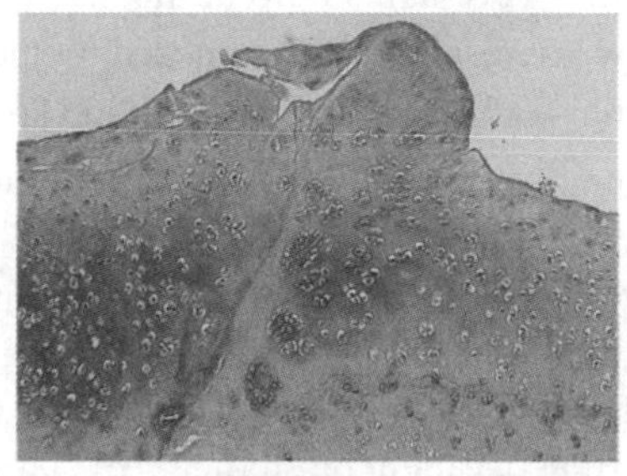

Figure 2.4.10 **Histological sections of articular cartilage from the femoral condyle of a normal and an experimental rabbit. The experimental rabbit received a single injection of botulinum toxin type-A into the knee extensor muscles causing weakness, and was sacrificed four weeks following the injection. The Mankin scores for the normal and experimental cartilage were 0 (normal) and 12 (signs of degeneration), indicating that muscle weakness may be a risk factor for joint degeneration.**

2.4.7 THEORETICAL AND NUMERICAL MODELS

Since it is virtually impossible at present to obtain in vivo joint mechanics (for example, instantaneous contact pressure distributions, or cartilage stresses and strains) much of our understanding of the local, internal joint biomechanics relies on joint contact and articular cartilage tissue models. These models have proven effective at estimating the contributions of fluid and matrix to the load sharing in the tissue, and describing stress-strain patterns in articular cartilage and chondrocytes with prescribed loading conditions. Selected cartilage modelling approaches are discussed below (Hasler et al., 1999; Federico et al., 2005; Herzog & Federico, 2005), ranging from single-phasic, homogeneous elastic models, to anisotropic, inhomogeneous, micro-structurally-based models of articular cartilage.

SINGLE PHASIC MODELS

The first theoretical models of articular cartilage were linearly elastic, homogeneous and isotropic. Hayes et al. (1972) related experimental data to the parameters of such models using results from indentation testing. Articular cartilage was represented as a linearly elastic material bonded to a rigid bone. A plane ended indenter was pressed with a force, P, into the tissue. Hayes et al. (1972) derived an equation for the shear modulus, μ, for this situation, assuming small strain theory and a frictionless indenter:

$$\mu = \frac{P(1-\nu)}{4aw\kappa(a/h, \nu)} \tag{2.4.1}$$

where:

ν = Poisson's ratio of the material
w = displacement imposed by the indenter
h = thickness of the cartilage layer
κ = scale factor depending on the aspect ratio a/h and the Poisson's ratio, ν

With this model, cartilage properties can be described by a single value for the elastic modulus, if a specific Poisson's ratio is assumed. Since the material is assumed to be elastic, its properties are independent of time. Therefore, creep and stress relaxation, inherent in articular cartilage, cannot be studied. An elastic model may be used for the study of instantaneous or equilibrium responses, when fluid flow is small or absent, and cartilage may behave almost like an elastic material. Both applications are confined to small loads to stay within the limits of the infinitesimal deformation assumption. Therefore, this approach is not appropriate for many physiological loading conditions.

In order to account for the experimentally observed creep and stress relaxation properties of cartilage, viscoelastic models have been proposed. The basis for viscoelastic models is the generalized Kelvin-Voigt-Maxwell spring-dashpot model (Figure 2.4.11). The linear springs produce instantaneous displacements proportional to the applied forces, while the dashpot produces velocities proportional to the applied forces. Using a viscoelastic model, Parsons and Black (1977) predicted the instantaneous and the long-term shear modulus for normal rabbit cartilage.

In summary, the use of single-phase models for the determination of the shear or Young's modulus requires that a specific Poisson's ratio for articular cartilage is assumed. It is not possible to separate the intrinsic cartilage properties associated with the solid matrix and fluid flow. Single-phase models are not suited for analysis of the internal mechanics of articular cartilage, or to estimate the loads acting on structural elements of cartilage. These models can merely provide an estimate of the external forces and the associated joint contact mechanics.

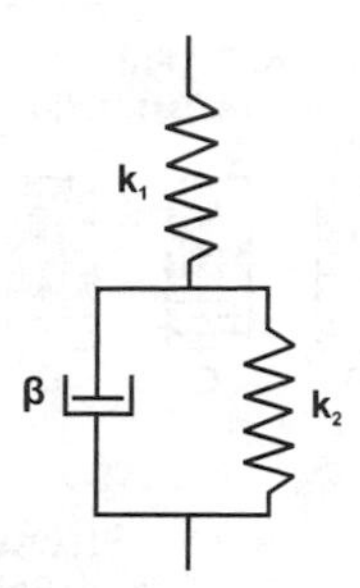

Figure 2.4.11 **Scheme of the generalized Kelvin-Voigt-Maxwell spring-dashpot system.**

BIPHASIC AND CONSOLIDATION MODELS

Interstitial fluid movement is thought to be the fundamental cause of the viscoelastic behaviour of articular cartilage. Thus, a structural model of articular cartilage should include at least two phases: a solid and a fluid phase. The behaviour of biphasic materials can be described in two ways

- A mixture theory, e.g., Fung (1993), that considers the material to be a continuum mixture of a deformable solid phase fully saturated by a fluid phase. At every point of the continuum, the two phases coexist, and the volumetric fraction fields obey the saturation condition, $\phi^s + \phi^f = 1$, at every point, where s indicates the solid phase, and f indicates the fluid phase.
- A consolidation approach (Oloyede & Broom, 1991), that considers the material to be a porous elastic solid saturated by a pore fluid that flows relative to the deforming solid.

Simon et al. (1996) proved that these two classes of models are equivalent if the fluid is assumed to be non-viscous.

BIPHASIC POROELASTIC MODELS

In biphasic models (Mow et al., 1980; Holmes & Mow, 1990), cartilage is represented as a mixture of a solid phase (collagen-PG matrix), which is assumed to be porous, homogeneous, isotropic, linearly elastic and with constant isotropic permeability, and a fluid phase (interstitial fluid), which is assumed to be non-viscous.

The solid and the fluid phases are assumed to be incompressible and immiscible. However, the mixture of an uncompressible solid and an uncompressible fluid can be globally compressible if fluid is allowed to escape through the boundaries of the system. Biphasic poroelastic models can explain the viscoelastic behaviour of cartilage in compression as the frictional drag associated with interstitial fluid flow. Movement of interstitial fluid determines the stress history for articular cartilage in confined compression (Figure 2.4.12). The stress rises with fluid exudation during compression because of the inherent resistance to fluid flow. During the relaxation phase, fluid redistribution decreases the stress in the compressed matrix. At equilibrium, the stress within the solid matrix represents the entire tissue stress.

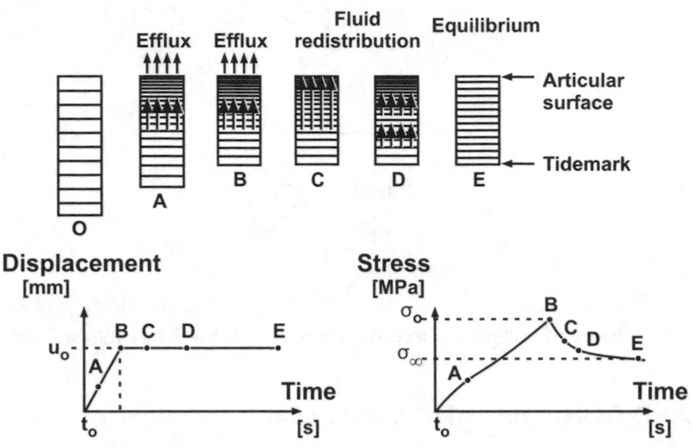

Figure 2.4.12 **Movement of the interstitial fluid in articular cartilage during a confined compression test, performed in displacement control. The displacement is applied with a linear ramp (A), and kept constant for the reminder of the test (B to E). The fluid starts to flow outwards as soon as the displacement is applied. The stress reaches its maximum value at the end of the displacement ramp (B). The fluid continues to flow, and this causes the stress to relax and tend asymptotically to the equilibrium value (E), which is reached when the fluid flow has ceased. From Mow and Rosenwasser; reproduced with permission from the American Academy of Orthopedic Surgeons.**

According to the biphasic constitutive law, the total Cauchy stress, σ, acting on the tissue is the sum of the stresses acting on the solid (superscript s) and fluid (superscript f) phases:

$$\sigma = \sigma^s + \sigma^f \tag{2.4.2}$$

The stress in the solid phase is expressed as the sum of the purely elastic stress, σ^{el}, plus a portion of the fluid pressure, p, proportional to the solid volumetric fraction, ϕ^s:

$$\sigma^s = -\phi^s pI + \sigma^{el} \tag{2.4.3}$$

where:

I = identity tensor

The reminder of the fluid pressure (proportional to the fluid volumetric fraction, $\phi^f = 1 - \phi^s$) represents the stress in the fluid phase:

$$\sigma^f = -\phi^f pI \tag{2.4.4}$$

In this way, the total stress (2.4.2) can be expressed as:

$$\sigma = -pI + \sigma^{el} \tag{2.4.5}$$

By defining the effective filtration velocity as $w = \phi^f(v^f - v^s)$, Darcy's law, describing the filtration of the fluid phase within the solid phase, reads:

$$w = -k \operatorname{grad} p \tag{2.4.6}$$

As the solid and the fluid phase are *intrinsically* incompressible, the continuity equation can be written in terms of volumetric fractions rather than mass densities (v^s is the velocity field in the solid phase, and v^f is the velocity field in the fluid phase):

$$\operatorname{div}(\phi^s v^s + \phi^f v^f) = 0 \tag{2.4.7}$$

Euler's equation for the whole system is written assuming negligible inertial effects, i.e., negligible acceleration, and negligible external volume forces:

$$\underbrace{\rho f}_{\approx 0} + \operatorname{div}\sigma = \underbrace{\rho a}_{\approx 0} \Rightarrow \operatorname{div}\sigma = 0 \tag{2.4.8}$$

However, Euler's equation for each of the two phases must take into account the frictional drag force (per unit volume) $\phi^f w/k$, that the fluid phase exerts on the solid phase, which is accounted for with the minus sign in the equilibrium equation for the solid phase:

$$\operatorname{div}\sigma^s = -\phi^f w/k \tag{2.4.9}$$

$$\operatorname{div}\sigma^f = \phi^f w/k$$

Summing the two equations in (Equation 2.4.9) retrieves Euler's equation for the whole system (Equation 2.4.8):

$$\operatorname{div}\sigma = \operatorname{div}(\sigma^s + \sigma^f) = -\phi^f w/k + \phi^f w/k = 0 \tag{2.4.10}$$

In the linear biphasic model (Mow et al., 1980), the permeability, k, is assumed to be a constant, so that Darcy's law (2.4.6) is linear, and the constitutive equation for the purely elastic stress, which features in Equations (2.4.3) and (2.4.5), is linear and isotropic:

$$\sigma_{ij}^{el} = \lambda \varepsilon_{kk} \delta_{ij} + 2\mu \varepsilon_{ij} \tag{2.4.11}$$

where:

ε = Green strain tensor for the solid phase

λ and μ = first and second Lamé's coefficients for the solid phase

Mow et al. (1980) applied the linear biphasic formulation to determine simultaneously the aggregate modulus, H_A, and Poisson's ratio, ν, of the solid phase, and the permeability, k, from a single indentation experiment. At equilibrium, interstitial fluid flow ceases and the entire load applied to the articular cartilage is carried by the solid phase, and the biphasic result reduces to the elastic solution found by Hayes et al. (1972). For indentation experiments, the fit between the theoretical solution and experimental data was poor for the early time response of creep. Mow et al. (1980) assumed that the poor fit was associated with the assumptions made to derive the biphasic indentation creep solution, which included negligible inertial effects, negligible intrinsic viscoelasticity of the solid phase, frictionless porous-permeable indenter tip, constant cartilage permeability, and isotropic and homogeneous tissue.

The non-linear model of Holmes and Mow (1990) includes a non-linear, hyperelastic constitutive law, and a strain-dependent permeability that can better represent the physiological behaviour of articular cartilage.

The purely elastic stress is obtained by derivation of the strain energy density, U, with respect to the components of the right Cauchy stretch tensor in the solid phase, C:

$$\sigma_{ij}^{el} = \frac{2}{J} F_{ik} \left[\frac{\partial U}{\partial C_{kl}}(C) \right] F_{jl} \tag{2.4.12}$$

where:

F = deformation gradient in the solid phase
J = determinant of F: J = det(F)

Holmes and Mow (1990) considered the material as isotropic. Therefore, the strain energy density function can be reduced to a function of the three principal invariants I_1, I_2, I_3 of the right Cauchy stretch tensor of the solid phase, C:

$$U(C) = W(I_1, I_2, I_3) \tag{2.4.13}$$

The strain energy given by Holmes and Mow (1990) is:

$$W(I_1, I_2, I_3) = \alpha_0 \exp[\alpha_1 (I_1 - 3) + \alpha_2 (I_2 - 3) - \beta \ln(I_3)] \tag{2.4.14}$$

where the coefficients α_0 (dimensions of energy per unit volume), α_1, α_2 (non-dimensional) are material parameters related to a material parameter β (non-dimensional) and to the homogenized values of the elastic Lamé's moduli for the solid phase, λ and μ, through the following equations:

$$\lambda = 4\alpha_0\alpha_2 \tag{2.4.15}$$

$$\mu = 2\alpha_0(\alpha_1 + \alpha_2) \tag{2.4.16}$$

$$\beta = \alpha_1 + 2\alpha_2 \tag{2.4.17}$$

α_0, α_1, and α_2 are fixed once λ and μ are known from small strain tests, and a value for β is chosen. The material is stable if α_0 is strictly positive, and α_1 and α_2 are non-negative, with at least one non-zero.

In the formulation of Holmes and Mow (1990), permeability depends on the volume fractions ϕ^f, ϕ^s and the third invariant of the right Cauchy tensor $I_3 = J^2$ for the solid phase, which is a measure of the apparent volume change (apparent because the local change of volume fractions makes the apparent density of the homogenized mixture change):

$$k = k_0\left[\frac{\phi_0^s\phi^f}{\phi^s\phi_0^f}\right]^{\kappa}\exp\left[\frac{M}{2}(I_3 - 1)\right] \tag{2.4.18}$$

where κ and M are non-dimensional material parameters to be determined experimentally.

Wu and Herzog (2000) expressed permeability as a function of the void ratio $e = \phi^f/\phi^s$, to implement the strain-dependent permeability into the Finite Element software ABAQUS. By setting $\phi^f/\phi^s = e$, $\phi_0^f/\phi_0^s = e_0$, $\phi_0^s/\phi^s = J$, and using the saturation condition, $\phi^f/\phi^s = 1$, one obtains:

$$k = k_0\left[\frac{e}{e_0}\right]^{\kappa}\exp\left[\frac{M}{2}\left[\left(\frac{1+e}{1+e_0}\right)^2 - 1\right]\right] \tag{2.4.19}$$

Qualitatively, the dependence of permeability on void ratio is intuitive. If the void ratio, $e = \phi^f/\phi^s$, decreases, the volume fraction of fluid decreases. In a fully saturated solid, the volume fraction of the fluid phase equals the volume fraction of the pores, thus, if the volume fraction of the pores decreases, permeability decreases as well.

The linear biphasic theory by Mow et al. (1980) was extended by Mak (1986) to the biphasic poroviscoelastic theory (BPVE), in which the solid phase was assumed to be intrinsically viscoelastic. This model allowed for the description of flow-dependent and flow-independent viscoelasticity in articular cartilage. The BPVE model increases the role of the solid phase and the deep layers in supporting applied loads within articular cartilage. In the BPVE model, predictions for confined compression during the early loading stage are improved compared to the linear elastic biphasic model, in situations in which loading is applied relatively "fast". Contributions from the intrinsic viscoelasticity of the solid matrix have been suggested to increase as permeability increases, as is the case for osteoarthritic cartilage with a damaged or fibrillated surface zone. In this case, intrinsic viscoelasticity of the solid matrix may alter the load-carrying behaviour of the tissue.

CONSOLIDATION MODELS

Oloyede and Broom (1991) wondered if cartilage could be described as a water-filled porous engineering material that undergoes mechanical consolidation like soils and clays. They performed simultaneous measurements of hydrostatic pore pressure and creep strain of the cartilage matrix during static compression in a one-dimensional consolidation configuration.

An idealized mechanism of consolidation (Figure 2.4.13) shows that the fluid initially carries the applied load (Figure 2.4.13b). After a critical hydrostatic pressure has been reached, fluid flows out of the porous matrix. Stress is then progressively transferred to the solid component of the tissue and consolidation starts. The tissue reaches the steady-state deformation when the hydrostatic pore pressure has decayed to zero. In the consolidated matrix, the final load is entirely carried by the compressed solid phase. The stress-strain relationship of the solid component, and therefore matrix stiffness, was shown to be non-linear for articular cartilage during consolidation. In addition, matrix stiffness increased progressively from low to medium to high strain-rate regimes. At very high strain rates, the fluid was locked within the matrix, thus greatly increasing tissue stiffness. The authors concluded that a model based on the consolidation theory conceptualizes cartilage on a phenomenological level without cumbersome assumptions in the analysis of the cartilage response to load.

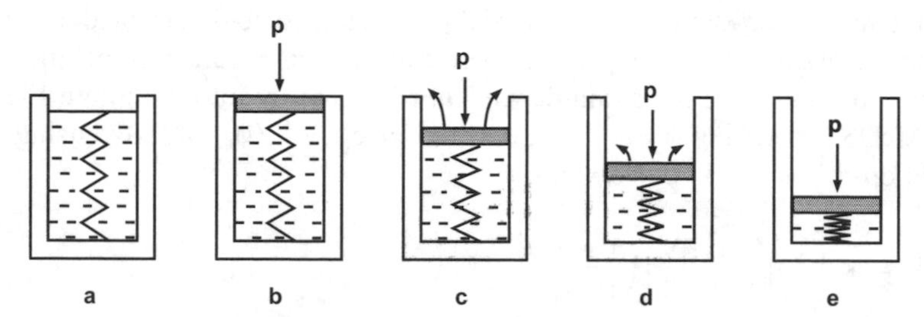

Figure 2.4.13 Scheme of the consolidation mechanism: the spring represents the stiffness of the solid matrix, and the fluid represents the interstitial fluid (a). In a load control test, when the load P is applied to the porous punch (b), the fluid starts to flow outwards. Equilibrium (e) is reached when fluid flow has stopped, and the elastic force exerted by the spring equals the applied load, P.

There is an equivalence of the biphasic and consolidation models for non-viscous fluids. As a result, consolidation models for non-viscous fluids can be applied to build finite element models to simulate problems with complex geometries and boundary conditions.

Van der Voet et al. (1992) modelled cartilage with poroelastic elements and compared their results with predictions obtained using the biphasic theory and results obtained experimentally. The authors concluded that, as predicted theoretically, results for cartilage compression from poroelastic models were comparable to those obtained using biphasic models. In the FE formulation of Suh and Bai (1998), the solid phase was considered viscoelastic, rather than elastic, as had been assumed in previous FE formulations. There are several other applications of the biphasic-poroelastic model to FE, which were used in a variety of problems of biomechanical interests. For example, Wu et al. (1999) and Wu and Herzog (2000) modelled the mechanics of chondrocytes embedded in the extracellular matrix, subjected to unconfined compression testing.

MULTIPHASIC MODELS

The main limitation of biphasic models is the impossibility to describe the structural elements that are responsible for the charged nature of articular cartilage, and to explain the origin of the physicochemical or electrochemical behaviour of the tissue, or both. Some constitutive theories that go beyond these limits are briefly discussed in this section.

Electromechanical Theory

Electromechanical models combine the laws for linear electrokinetic transduction in ionized media with the principles of the linear biphasic theory (Frank & Grodzinsky, 1987). When cartilage is compressed, mechanical-to-electrical transduction occurs, resulting in measurable electrical potentials. Deformation of the hydrated extracellular matrix causes a flow of interstitial fluid and entrained ions relative to the fixed charge groups of the extracellular matrix. Fluid flow tends to separate the mobile ions from the fixed charge groups on the proteoglycans, producing a local voltage gradient, or streaming potential, that is proportional to the local fluid velocity. Conversely, application of electrical current across the cartilage produces a current-generated stress.

Triphasic Theory

The triphasic theory is an extension of the biphasic theory, with the addition of a third phase, representing cations and anions (Lai et al., 1991). Triphasic models have a chemical expansion stress that can be computed from the physico-chemical activities in the matrix due to the mobile ions and the fixed, negatively charged proteoglycan aggregates. Gu et al. (1998) developed a generalized multiphasic model including 2 + n constituents (one charged solid phase, one non-charged fluid phase and n ion species) that enables analysis of the diffusion processes of different ion species. Adding ion species to the biphasic model has helped in reproducing selected experimental results. However, the model remains macroscopic (as it remains a continuum mechanics model), and is dissociated from the microscopic, molecular structure of articular cartilage.

A poroelastic transport-swelling (PETS) FE model was developed by Simon et al. (1996) to simulate the interaction between mechanical deformation, fluid flow, and transport of ions through the matrix. The PETS model was shown to be equivalent to the triphasic model (Lai et al., 1991), and it provides a basis for the analysis of transport and swelling in soft tissues.

MOLECULAR-BASED MODELS

Kovach (1996) proposed a structural model of articular cartilage that incorporates glycosaminoglycans which exert drag on the fluid phase. The collagen network and its bound water restrain fluid flow and the distribution of proteoglycans.

This model gives a theoretical explanation for why proteoglycans occupy a different volume in vivo compared to in vitro. The swelling pressure of the tissue is decomposed into two parts: a charge-independent component, caused primarily by the configurational entropy of the glycosaminoglycan chains, and a charge-dependent component, directly caused by the glycosaminoglycan charge. It has been suggested that the charge-dependent properties dominate the swelling pressure in articular cartilage. The time-dependent mechanical properties of cartilage may be explained in terms of friction between the structural components of the matrix and fluid.

ANISOTROPIC, INHOMOGENEOUS AND MICROSTRUCTURAL MODELS

All models described so far are characterized by the isotropy, i.e., invariance of properties for rotations, and homogeneity, e.g., invariance for translations, of the constitutive equations. However, it is well acknowledged that cartilage is anisotropic (e.g., Cohen et al., 1998; Bursac et al., 1999; Wang et al., 2003) and inhomogeneous, (e.g., Schinagl et al. 1997). Therefore, cartilage has sometimes been modelled as a transversely isotropic material, with the transverse plane assumed parallel to the articular surface, and the symmetry axis perpendicular to the articular surface. These models can account for the different elastic behaviour of cartilage in the axial and transverse directions, which has been observed experimentally.

Conversely, anisotropy and inhomogeneity of articular cartilage may be retrieved, to a certain degree, by modelling the structural components thought important in load bearing and force transmission across the tissue. Cells in articular cartilage can be approximated by revolution ellipsoids. Their shape changes from elongated in the deep zone to spherical in the middle zone to flattened in the superficial zone. Similarly, collagen fibres have a characteristic alignment within articular cartilage. Collagen fibres are perpendicular to the tidemark in the deep zone, randomly oriented in the middle zone, and parallel to the surface in the superficial zone (Hedlund et al., 1993). This arrangement of cells and fibres (Figure 2.4.14a) accounts for some of the anisotropy of the tissue. Assuming that the fibres in the superficial zone are parallel to the surface with no preferential direction on the superficial plane, the tissue can be represented as transversely isotropic (Figure 2.4.14b).

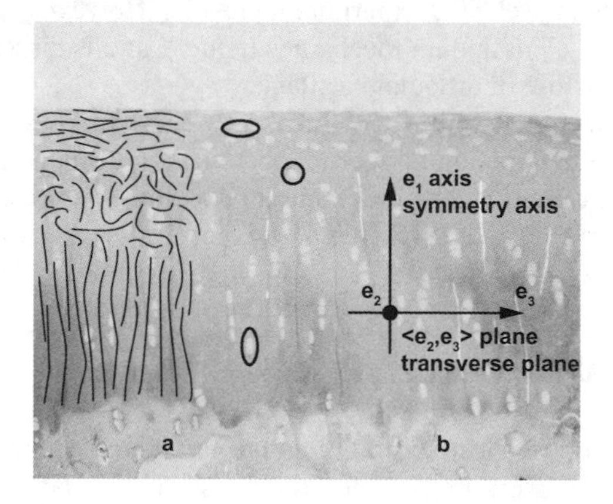

Figure 2.4.14 **Arrangement of the collagen fibres and cells in articular cartilage (a), and the consequent transverse isotropy (b).**

Farquhar et al. (1990) developed a fibril-reinforced micro-structural model of articular cartilage that accounts for the average effect of the collagen fibre network, but does not account for the spatial dependence of the arrangement of fibres. In the fibril-reinforced model of Li et al. (2000b), the depth-dependence of the mechanical properties of cartilage is taken into account, i.e., the model is inhomogeneous, and the collagen fibres are assumed to have tensile stiffness.

The models cited above do not account for the possible mechanical effects of cells on tissue properties. However, in small animals, cell volumetric concentration may exceed 20% (Clark et al., 2003), and, because their stiffness is much smaller than that of the surrounding matrix, they may affect the global elastic properties of the tissue (Federico et al., 2004b).

Based on a composite homogenization method (Walpole, 1981), Wu and Herzog (2002) built a microstructural three-layer model for the solid phase of the deep, middle, and superficial zone of articular cartilage, using zone-specific, structural arrangements of chondrocytes and collagen fibres. This model predicted the decrease of the axial elastic modulus from the deep zone to the surface, but could only give a separate description of the three cartilage layers, and not a global solution for the entire tissue, as it is a straightforward application of Walpole's (1981) method for aligned inclusions. Walpole (1981) calculated the elastic tensor, L, for a N+1-phasic composite, with an isotropic matrix (index 0) and N transversely isotropic, spheroidal inclusion phases aligned with the e_1 direction:

$$L = \left[\sum_{r=0}^{N} c_r L_r A_r\right]\left[\sum_{r=0}^{N} c_r A_r\right]^{-1} = \left[\sum_{r=0}^{N} c_r Z_r\right]\left[\sum_{r=0}^{N} c_r A_r\right]^{-1} \quad (2.4.20)$$

where, for the r-th phase:

c_r = volumetric concentration ($\Sigma_{r=0}^{N} c_r = 1$)
L_r = elasticity tensor
A_r = strain-concentration tensor
Z_r = $L_r A_r$

Generalizing Walpole's (1981) method, a homogenization method in which the inclusions can take any orientation can be derived (Federico et al., 2004a). The N-th phase is assumed to have a statistical orientation, governed by the normalized probability distribution density ψ, which is a function of the direction. The set of all directions in space is represented by the north unit hemisphere $\mathbf{S}^{2+}$, which is the set of all unit vectors w, lying in the positive hemi-space in the 1-direction. Vector w is conveniently represented as a function of the co-latitude angle, θ, and longitude angle, φ (Figure 2.4.15).

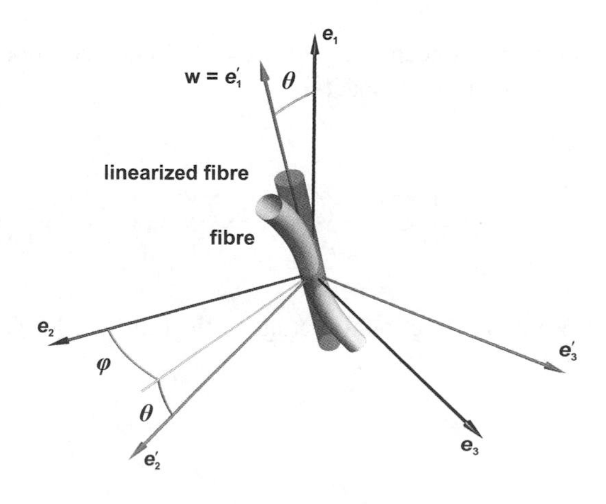

Figure 2.4.15 **Local reference frame for the representation of the direction, w, of the symmetry axis of an inclusion (a linearized fibre, in this example). The e'_1 axis of the local reference frame coincides with w, and its orientation with respect to the global reference frame (e_1, e_2, e_3) is given by the colatitude, θ, and the longitude, φ.**

With these assumptions on the N-th phase, Equation (2.4.20) can be generalized in the form:

$$L = \left[\sum_{r=0}^{N-1} c_r Z_r + \int_{S^{2+}} \psi c_N Z_N da\right]\left[\sum_{r=0}^{N-1} c_r A_r + \int_{S^{2+}} \psi c_N A_N da\right]^{-1} \qquad (2.4.21)$$

where tensors Z_N and A_N are explicit functions of the direction w of the generic inclusion, and the surface integrals are performed on the unit hemisphere $\mathbf{S}^{2+}$. By use of Walpole's (1981) formalism, tensors $Z_N(w)$ and $A_N(w)$ can be written as the linear combinations of Walpole's basis tensors B_α:

$$Z_N(w) = \bar{Z}_N^\alpha B_\alpha(w)$$
$$A_N(w) = \bar{A}_N^\alpha B_\alpha(w) \qquad (2.4.22)$$

where Walpole's components, $\bar{Z}_N^\alpha$ and $\bar{A}_N^\alpha$, *do not* depend on the direction, w (Federico et al., 2004a), as the dependence on w is accounted for by the basis tensors B_α. On the basis of these considerations, Equation (2.4.21), can be written as:

$$L = \left[\sum_{r=0}^{N-1} c_r Z_r + c_N \bar{Z}_N^\alpha \int_{S^{2+}} \psi B_\alpha da\right]\left[\sum_{r=0}^{N-1} c_r A_r + c_N \bar{A}_N^\alpha \int_{S^{2+}} \psi B_\alpha da\right]^{-1} \qquad (2.4.23)$$

after factorizing the volumetric concentration, c_N, and Walpole's components, $\bar{Z}_N^\alpha$ and $\bar{A}_N^\alpha$. Equation (2.4.23) can be written in compact form, as:

$$L = \left[\sum_{r=0}^{N-1} c_r Z_r + c_N \bar{Z}_N^\alpha H_\alpha\right]\left[\sum_{r=0}^{N-1} c_r A_r + c_N \bar{A}_N^\alpha H_\alpha\right]^{-1} \qquad (2.4.24)$$

where tensors H_α represent the directional average of the fourth order Walpole's (1981) basis tensors B_α:

$$H_\alpha = \int_{S^{2+}} \psi B_\alpha da \qquad (2.4.25)$$

This method can be extended to inhomogeneous materials and adapted to cartilage by including the dependence of the various parameters featuring in Equation (2.4.21) on tissue depth. This inclusion of the various parameters in Equation (2.4.21) gave rise to a transversely isotropic, transversely homogeneous (TITH) model of articular cartilage (Federico et al., 2005). Transverse homogeneity is defined in analogy with transverse isotropy. While isotropy denotes symmetry, i.e., invariance, with respect to rotations of the reference frame, homogeneity denotes symmetry with respect to translations. Therefore, if transverse isotropy is the invariance for rotations with respect to a given direction (and orthogonal to a given plane), transverse homogeneity is the invariance for translations parallel to a given plane. In the TITH model, the cell volumetric concentration and aspect ratio, and the fibre volumetric concentration and spatial arrangement vary as a function of depth (transverse homogeneity, i.e., invariance under translations of the reference frame parallel to the transverse plane). If cells are described as revolution ellipsoids (having the symmetry axis parallel to the direction of depth), and fibres are given a statistical distribution of orientation (symmetric with respect to the direction of the depth), the resulting model is transversely isotropic, with the symmetry axis parallel to the direction of depth.

Because of the depth-dependence of the above mentioned parameters, the elastic modulus in Equation (2.4.10) is an explicit function of the normalized depth ξ ($\xi = x_1/h$, where h is the thickness of the cartilage layer), and it is calculated as:

$$\begin{aligned} L(\xi) = {} & [c_0(\xi)L_0 + c_c(\xi)Z_c(\xi) + c_f(\xi)\bar{Z}_f^{\alpha}H_{\alpha}(\xi)] \\ & [c_0(\xi)I + c_c(\xi)A_c(\xi) + c_f(\xi)\bar{A}_f^{\alpha}H_{\alpha}(\xi)]^{-1} \end{aligned} \tag{2.4.26}$$

where:

0 = matrix
c = cells
f = fibres

A transversely isotropic, transversely homogeneous model can predict the decrease in axial modulus from the tidemark to the surface, and can describe non-uniformities in the displacement fields. The prediction of the elastic moduli as a function of depth depends on the choice of the probability distribution, ψ. The mathematical form of the latter has been based on the assumption that the collagen fibres in articular cartilage are orthogonal to the tidemark in the deep zone, randomly oriented in the middle zone, and parallel to the surface in the superficial zone. With this assumption only, and not by suitably adjusting the various parameters featuring in Equation (2.4.26), the TITH model compared well with the experimental results of Schinagl et al. (1997) on the variation of the axial aggregate modulus with depth (Federico et al., 2005). Both the experimentally and numerically evaluated plots of the axial aggregate modulus, normalized to its value at the tidemark (Figure 2.4.16) show a high value at the tidemark, a plateau corresponding to the middle zone, and a small value at the surface, where the fibres are parallel to the surface and cannot bear axial load.

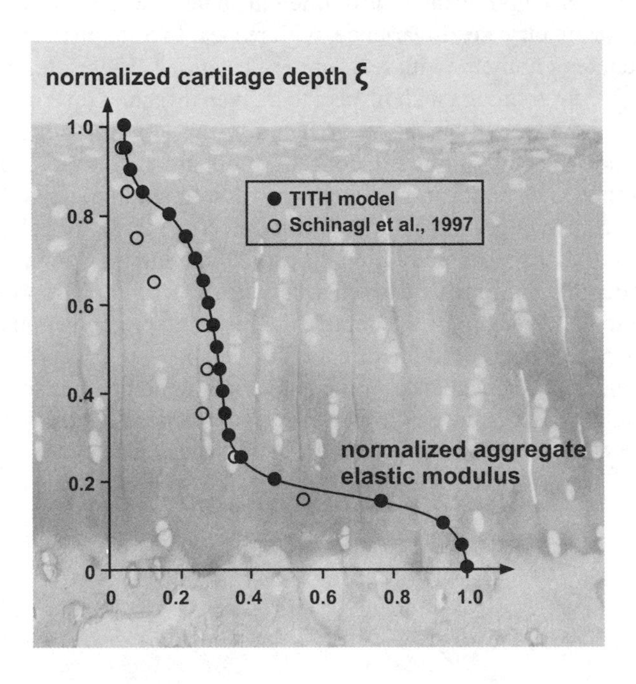

Figure 2.4.16 **Normalized aggregate elastic modulus (elastic modulus in uni-axial strain), as a function of the normalized cartilage depth, ξ, predicted by the TITH model, compared to the experiments performed by Schinagl et al. (1997).**

2.5 LIGAMENT

THORNTON, G.M.
FRANK, C.B.
SHRIVE, N.G.

2.5.1 MORPHOLOGY AND HISTOLOGY

MORPHOLOGY

The word ligament is derived from the Latin word *ligare*, which means "to bind". Ligaments consist of elastin and collagen fibres and attach one articulating bone to another across a joint. The major functions of ligaments are to:

- *Attach* articulating bones to one another across a joint,
- *Guide* joint movement,
- Maintain joint *congruency*, and
- Possibly act as a positional bend or strain *sensor* for the joint.

Collagen is the main protein present in ligaments. It is found chiefly in fibrillar form, and oriented between insertions to resist tensile forces. The hierarchical structure of the collagen in the ligament midsubstance includes fibres, fibrils, subfibrils, microfibrils, and tropocollagen (Figure 2.5.1). Tightly packed tropocollagen molecules, which are

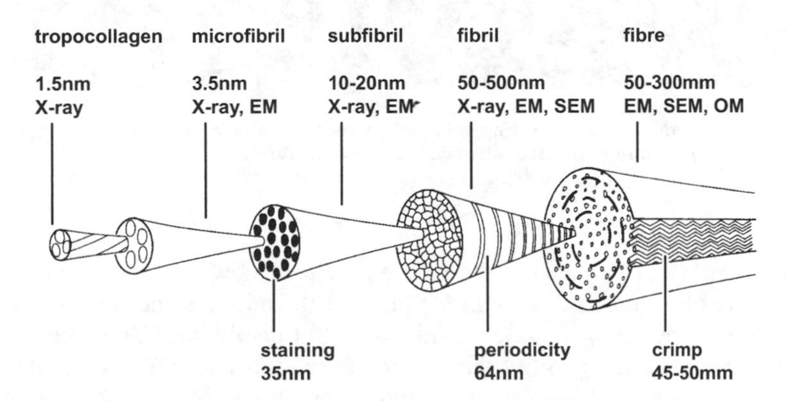

Figure 2.5.1 **Schematic illustration depicting the hierarchical structure of collagen in ligament midsubstance. EM = electron microscope; SEM = scanning EM; OM = optical microscope (from Kastelic et al., 1978, with permission of Gordon and Breach, Science Publishers Ltd.).**

approximately 1.5 nm in diameter, aggregate into groups of five, thereby becoming microfibrils of approximately 3.5 nm in diameter. The microfibrils group into subfibrils, which, in turn, aggregate to form fibrils. Fibrils are approximately 50 to 500 nm in diameter with a periodicity (spacing) of 64 nm. Fibres are an aggregation of fibrils and are 50 to 300 μ in diameter. They are the smallest unit of the collagen hierarchy that can be seen under a light microscope. Fibres have an undulating crimp with a distance between amplitudes of about 50 μ. The crimp period can vary quite dramatically in different locations within the ligament. In the rabbit medial collateral ligament (MCL), fibroblasts tend to align in rows between fibre bundles and are elongated along the long axis in the direction of normal tensile stress (see Figure 2.5.2). Fibres and fibre bundles may or may not aggregate into fascicles (Figure 2.5.3). In the MCL, fascicles are not as obvious as they are in the anterior cruciate ligament (ACL).

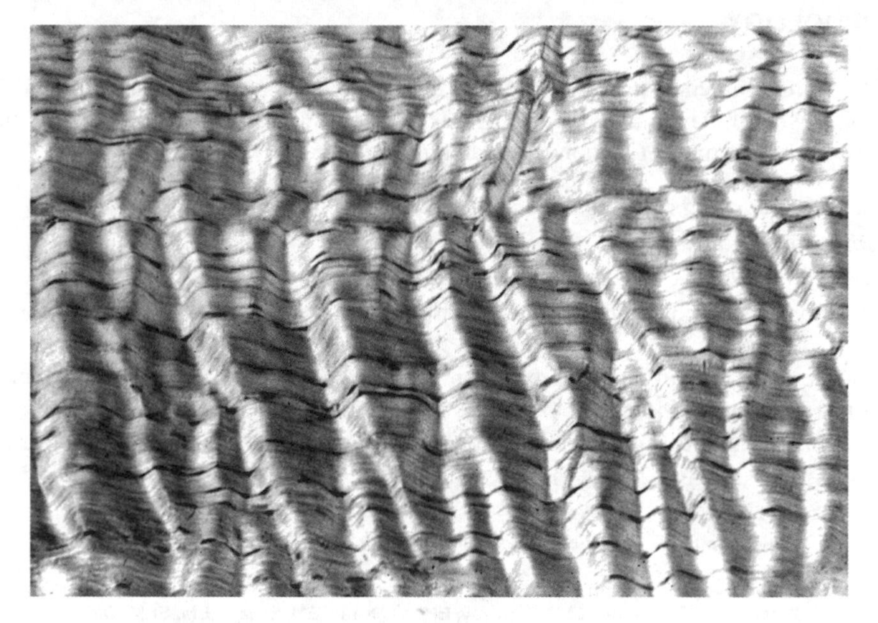

Figure 2.5.2 **Photograph illustrating crimped pattern of collagen in ligament. Fibroblasts may be seen interspersed between the collagen fibres.**

The surface of a ligament comprises a loose envelope known as the epiligament. Its collagen fibrils are of a smaller diameter than those of the midsubstance and are oriented in many directions. The epiligament contains a variety of cell types. These cells appear to have a greater proliferative ability than those of the midsubstance. Unlike the midsubstance, the epiligament encloses nerves and blood vessels that occasionally branch into the midsubstance. The function of the epiligament appears to be to:

- Protect the midsubstance of the ligament from abrasion,
- Support the neurovasculature,
- Control the water and metabolite flux, and
- Possibly act as a source for matrix, cells, and vasculature during maturation and healing.

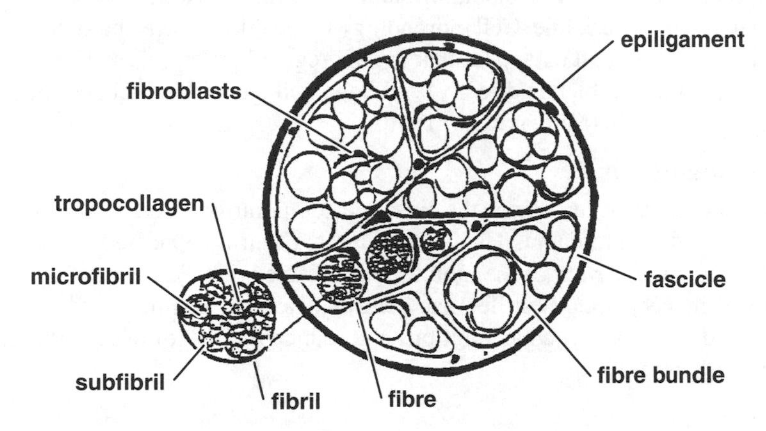

Figure 2.5.3 **Schematic diagram of a ligament in cross-section.**

Insertions

The structure of ligaments where they insert into bone is different from the midsubstance. Insertions anchor the ligament into the rigid, non-compliant bone. There are two general types of insertions: direct and indirect.

Direct insertions

Direct insertions, e.g., the femoral insertion of the MCL, occur where the ligament inserts directly into the bone. Direct insertions contain four different cellular zones, all of which occur within approximately 1 mm (Figure 2.5.4) of each other. The first zone is normal ligament midsubstance, with organized parallel collagen bundles, some elastin, and elongated fibroblasts. The second zone consists of non-mineralized fibrocartilage in which the cell numbers increase, become more ovoid, increase in size, and lie in rows. The collagen fibrils continue to extend into this region. The third zone may be characterized as min-

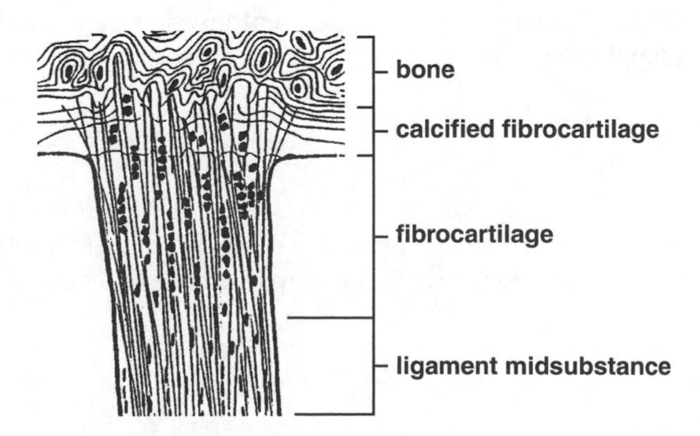

Figure 2.5.4 **Schematic diagram of a zonal ligament insertion into bone. The bone is at the top of the diagram, the ligament at the bottom (from Matyas, 1985, with permission).**

eralized cartilage and is clearly distinguishable from the previous zone by the *tidemark*, which is an undulating dark line. Cell morphology in the third zone is the same as in the second zone, but mineral crystals appear that are aggregated into masses. The fourth zone is where ligament collagen blends directly with bone collagen. The thickness of these zones is tissue and age specific (Matyas et al., 1990).

Indirect insertions

Indirect insertions occur where the ligament temporarily inserts into the periosteum during growth and development. The periosteum is, in turn, connected to bone. A typical example is the tibial insertion of the rabbit MCL (Matyas, 1990).

Insertions are mechanically stiffer near bone than near the ligament midsubstance. Such stiffening could lessen stress concentrations and reduce the risk of tearing due to shearing of the tissue at the interface.

Nerves

Recent years have seen much research into the role of the neural components of ligaments. As yet, the exact function of nerves in ligaments is not clear. It is speculated that they permit the sensing of joint position, the monitoring of ligament tension and integrity, and that they initiate protective reflexes. Some investigators have produced muscle activity by stimulating the nerves in ligaments (Sojka et al., 1989; Barrack & Skinner, 1990; Kraupse et al., 1992).

For the knee joint, two groups of nerves have been identified: one anterior and one posterior. These nerves respond to active and passive motion throughout the joint's range of motion. Their greatest responses occur at the extremes of knee motion, possibly warning of impending injury. Experimental stimulation of these receptors has been shown to cause reflex activation in the surrounding muscles (Adams, 1977; Kennedy et al., 1982).

Blood vessels

A fine network of blood vessels is in the epiligament of the MCL. A few vascular channels penetrate from the epiligament into the ligament substance and then course longitudinally between collagen fascicles. However, ligament insertions are poorly supplied with blood. The blood supply appears to nourish the fibroblasts, enabling them to remain metabolically active.

HISTOLOGY

The healthy ligament looks like a simple white band of homogeneous fibrous tissue, but it is actually highly complex and dynamic. It is composed of a few cells in a largely collagenous matrix.

Cells

Fibroblasts

Ligament cells are called fibroblasts or fibrocytes. Fibroblasts are not homogenous in ligament tissue and vary in size, shape, orientation, and number. Fibroblasts are usually ovoid or spindle-like and are generally oriented longitudinally along the length of the ligament body. Both metabolic and histological experiments suggest that there may be fibroblast subtypes within the ligament, but these remain to be defined (Frank & Hart, 1990). Fibroblasts are responsible for synthesizing and degrading the ligament matrix in response to various stimuli. Presumably, fibroblasts prevent or repair ongoing microscopic damage. Therefore, despite their relative scarcity, fibroblasts are crucial to maintaining the status quo of ligaments.

Matrix

The matrix comprises virtually the entire body of the ligament. It consists of water, collagen, proteoglycans, fibronectin, elastin, actin, and a few other glycoproteins.

Water

Water makes up approximately two-thirds of the wet weight of a ligament. Water can be associated with other ligament components in a variety of ways. It can be structurally bound to other matrix components. It can be bound to polar side chains, be so-called transitional water (loosely bound), or be freely associated with the interfibrillar gel. Most water in ligaments is freely bound or transitional. Although the exact function of water in ligaments is unknown, it appears to be crucial for at least three main reasons. First, its interaction with the ground substance, particularly the proteoglycans, influences the tissue's viscoelastic behaviour (Amiel et al., 1990; Bray et al., 1991). Second, it seems to provide lubrication and facilitate inter-fascicular sliding (Amiel et al., 1990; Bray et al., 1991). Third, it carries nutrients to the fibroblasts and removes waste substances.

Collagen

Collagen comprises approximately 70-80% of the dry weight of ligament (Amiel et al., 1983; Frank et al., 1983a; Frank et al., 1983b). For more information about these collagen types, the reader should consult Miller and Gay (1992).

Type I:	The majority of ligament collagen is fibrillar, type I collagen.
Type III:	The second most common type is type III, which is also fibrillar.
Type VI:	The third most common is type VI, which is beaded filament.
Types V, XI, XII, and XIV:	Most ligaments also have small quantities of types V, XI, XII, and XIV collagen (the latter nearer insertions).

Collagen fibres within ligaments vary in size (diameter) from 10 to 1500 nm, a range that appears to be age-, tissue-, and species-specific (Parry et al., 1978a; Parry et al., 1978b; Frank et al., 1989; Yahia & Drouin, 1989). Fibre size may affect the strength of the material. Ligaments with larger bimodal collagen distributions tend to be stronger and able to sustain higher stresses. Those with smaller unimodal diameters are more suited to resisting lower stresses. Collagen is enormously strong due to a combination of biochemical bonds known as molecular cross-links.

Proteoglycans

Ligament proteoglycans (mainly so-called small dermatan sulphate proteoglycans) comprise less than 1 percent of the dry weight of ligament, more than is found in tendon but considerably less than in cartilage (3 to 10 percent). The molecular structure of proteoglycans is described in Chapter 2.4. The role of proteoglycans in ligaments remains to be determined. However, ligaments (and tendons), because they experience primarily tensile forces, do not need the cushioning effect that proteoglycans give cartilage. Proteoglycans, with their pronounced hydrophilic properties, may instead be involved in regulating the amount and movement of water within the tissue. Therefore, proteoglycans would mainly influence the viscoelastic behaviour of the ligament.

Fibronectin

Fibronectin is a protein composed of two 220 kD subunits joined by a single disulfide bridge. This protein contains about 5 percent carbohydrate. In ligament, as in tendon (see section 2.6.1), fibronectin is found in small quantities in the matrix, usually in association with several other matrix components and blood vessels. Fibronectin also interacts with portions of the cell surface that are known to attach to intracellular elements, possibly forming part of an important matrix-cell feedback mechanism.

Elastin

Elastin is an elastic substance found in very small amounts in most skeletal ligaments (approximately 1.5 percent) in fibular form. However, in elastic ligaments, e.g., ligamentum flavum elastin fibres are about twice as common as collagen fibres. When unstressed, the insoluble globular proteinaceous elastin molecule takes on a complex, coiled arrangement, probably maintained in part by lysine derived cross-links. Elastin stretches into a more ordered configuration when it is stressed, reverting to its globular form when unstressed again. This behaviour probably accounts for part of the tensile resistance in ligament tissue and some of its elastic recoverability. The role of elastin is probably related to recovering ligament length after stress is removed. Elastin probably protects collagen, at least at low strains

Interactions of these components

Interactions between the various components of the extracellular matrix have, until recently, received little attention. A recent study (Bray et al., 1990) revealed, through a combination of cationic stains and enzymatic digestion, a network of electron-dense seams that connect cells and subdivide the matrix into compartments. The seams contain microfilaments of type VI collagen (Bray et al., 1990; Bray et al., 1993), microfibrils, and chondroitin sulphate-containing proteoglycan granules (Figure 2.5.5). The granules are suspended on the microfilaments and spiralled through the seams. The discovery of seams in the extracellular matrix of ligaments has led to the speculation that seams influence the viscoelastic behaviour of ligaments. The proteoglycans attached to the microfilament network may provide functional divisions within the matrix. As well, because of the hydrophilic nature of proteoglycans, water bound within the seams may act as a lubricant, facilitating sliding between adjacent collagen fascicles. Therefore, the seams would be an integral part of a ligament's viscous element.

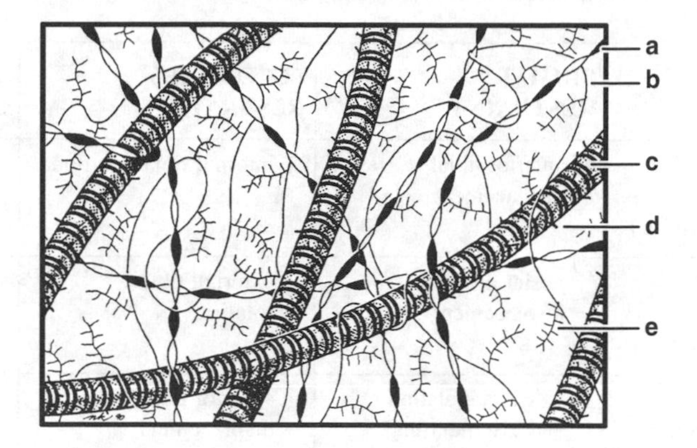

Figure 2.5.5 **Schematic diagram showing the proposed interrelationship between the elements of rabbit and human extracellular matrix. a = type VI collagen; b = non-beaded microfilaments; c = banded collagen fibrils; d = dermatan sulphate proteoglycan; e = chondroitin sulphate proteoglycan (from Bray et al., 1993, with permission).**

2.5.2 FUNCTION

The knee joint is used to illustrate ligament functions, because, like all weight-bearing joints, the knee is fundamentally concerned with stability over mobility. The four ligaments of the knee joint are illustrated in Figure 2.5.6. Their main passive restraining functions are summarized in Table 2.5.1.

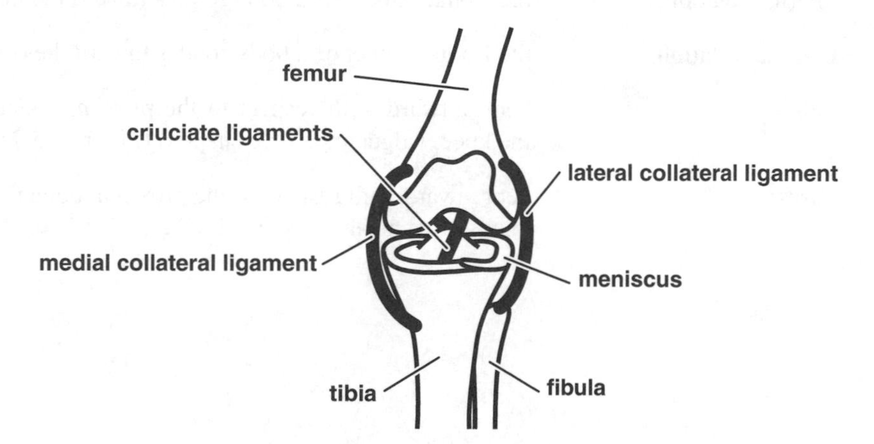

Figure 2.5.6 **Schematic diagram of the human knee joint from an anterior view. The cruciate (shaped like a cross) ligaments may be seen at the centre of the joint. The anterior cruciate ligament (ACL) attaches superiorly to the posterior lateral femoral condyle and inferiorly to the lateral anterior tibial spine. The posterior cruciate ligament (PCL) attaches superiorly to the medial femoral condyle and inferiorly to the posterior tibia. The collateral (parallel) ligaments lie lateral or to the side of the joint. The medial collateral ligament (MCL) is attached to the medial femoral condyle and the medial tibia, while the lateral collateral ligament (LCL) runs from the lateral epicondyle of the femur to the lateral fibular head.**

Table 2.5.1 Restraining functions of the four ligaments of the knee joint.

LIGAMENT	PRIMARY RESTRAINT	SECONDARY RESTRAINT	COMMENTS
ANTERIOR CRUCIATE LIGAMENT (ACL)	anterior tibial displacement	internal tibial rotation	no restraint to posterior tibial displacements
POSTERIOR CRUCIATE LIGAMENT (PCL)	posterior tibial displacement	external tibial rotation	no resistance to varus/valgus angulation
MEDIAL COLLATERAL LIGAMENT (MCL)	valgus angulation and external tibial rotation	anterior tibial displacement	
LATERAL COLLATERAL LIGAMENT (LCL)	varus angulation and internal tibial rotation	anterior and posterior tibial displacement	

The anatomical terms used in Table 2.5.1 are defined as follows:

Anterior: toward the front part.

Posterior: toward the back part.

Internal rotation: the frontal aspect of a body rotates toward the inside.

External rotation: the frontal aspect of a body rotates toward the outside.

Valgus: bent outward with respect to the proximal bone (for the knee, valgus means X-shaped) (Figure 2.5.7).

Varus: bent inward with respect to the proximal bone (for the knee, varus means O-shaped) (Figure 2.5.7).

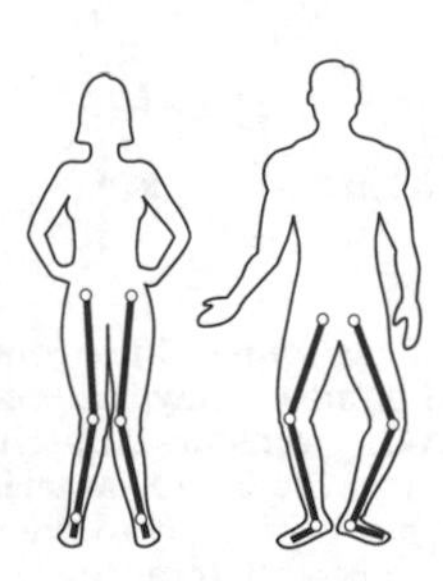

Figure 2.5.7 Schematic illustration of valgus knees (knock-knees) on the left and varus knees (bow-legged) on the right.

The knee joint has a complex ligament structure. In principle, the tibia has six possible degrees of freedom in relation to the femur (Figure 2.5.8): three translational (two sliding and one direct impingement and distraction) and three rotational. The ligaments, acting with the fibres of the joint capsule and the articular surfaces of the knee, restrict the range of motion, even when the muscles are completely relaxed.

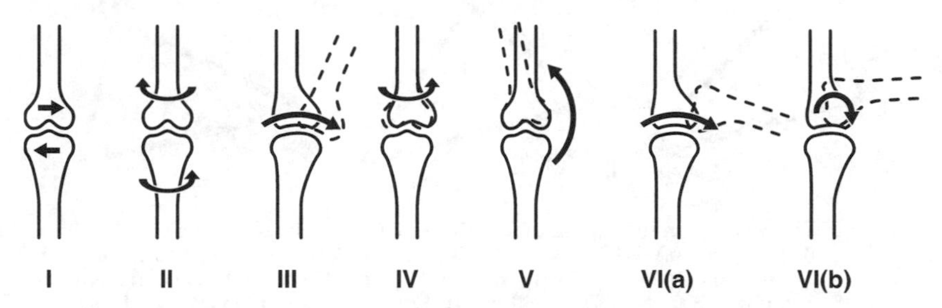

Figure 2.5.8 **Schematic diagrams illustrating the six possible degrees of freedom for the knee joint where (I) shows medial-lateral translation, (II) shows rotation, (III) shows anterior-posterior translation in the sagittal plane, (IV) shows tibial and femoral rotation, (V) shows varus-valgus rotation, and (VI) shows (a) the cruciate ligaments and (b) the collateral ligaments during flexion and extension in the sagittal plane.**

The joint may be flexed only through about 150°. The tibia can be rotated internally and externally relative to the femur through a range of about 35°. The range of varus and valgus movement is only about 5°. Only millimetres of translational movements parallel to the three axes are possible, and these are restricted by bone surface geometry and the ligaments. In normal physiological movements, however, it is rare for only one degree of freedom to be used. For example, knee flexion is associated primarily with rotation in the sagittal plane, but the tibia also translates along the frontal plane and rotates around the transverse plane while flexion occurs. Motions such as these are referred to as *coupled motions*. External loads seldom transmit only one component of force or moment across the joint. When examining the role of ligaments, however, we will simplify the possible combinations of the six degrees of motion.

CRUCIATE LIGAMENTS

Figure 2.5.9 shows a four bar linkage The tibial link AD and the femoral link CB are more or less parallel to the plateaux of the joints. The anterior cruciate AB and posterior cruciate CD are crossed (hence their name) and attach one condyle of one articulating bone to the opposite condyle of the apposing bone. Figure 2.5.9 shows how the shape of the four bar cruciate linkage changes during one particular degree of motion, flexion, and extension. Figure 2.5.9a shows the joint at full extension, (b) at 70° of flexion, and (c) at 140° of flexion. Between (a) and (c), the femoral link CB rotates 140° relative to the tibial link AD, and the cruciate links AB and CD rotate through 40° about their tibial attachments (A and D), and through 100° about their femoral attachments (B and C).

During this movement, the femoral condyles roll and slide on the tibia. The contact area moves backwards and outwards on the tibia in flexion, and forwards and inwards in extension. The femoral condyles diverge posteriorly and have reduced curvature. Thus, they sit in the menisci in an essentially anterior-posterior orientation in extension, but in flexion the diverging condyles force the menisci to assume a more medio-lateral curvature

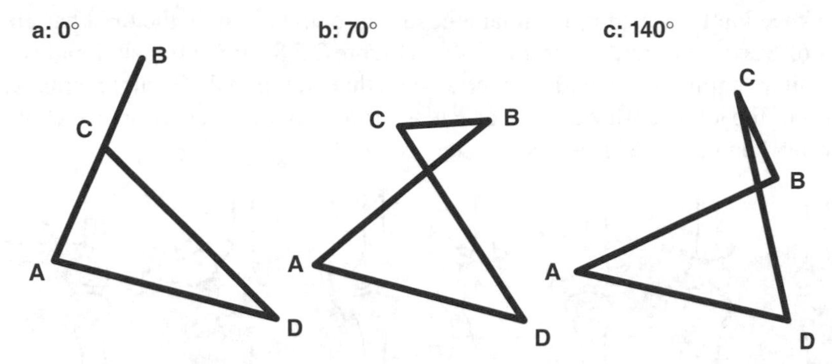

Figure 2.5.9 The cruciate four bar linkage ABCD at (a) full knee extension, (b) 70° of knee flexion, and (c) 140° of knee flexion. Between (a) and (c) the femoral link CB rotates through 140° relative to the tibial link AD, and the cruciate ligaments AB and CD rotate through 40° about their tibial attachments A and D, and through 100° about their femoral attachments B and C (from O'Connor et al., 1990, reprinted by permission of the Council of the Institution of Mechanical Engineers).

(Figure 2.5.10). It is through this mechanism that the condyles stay in contact with the menisci. The menisci are actually major load bearers in the knee, making the femur and tibia far more congruent than just a roller on a flat surface (Shrive et al., 1978). It takes about half

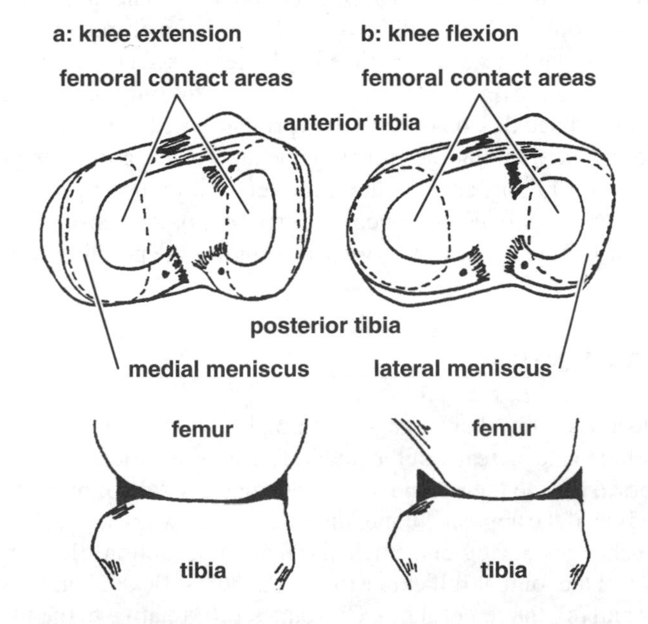

Figure 2.5.10 Schematic diagram of the knee joint showing distortion of the menisci during (a) extension and (b) flexion. When the menisci are viewed from above (top illustrations) the movement backwards and outwards from flexion to extension is apparent. The black dots show the points where each meniscus inserts firmly into the tibial eminence.

a body weight to make the femoral cartilage actually contact the tibial cartilage. At lower loads, typically all the load is carried by the menisci.

The cruciate ligaments play essential roles in controlling motion. The stresses that de-

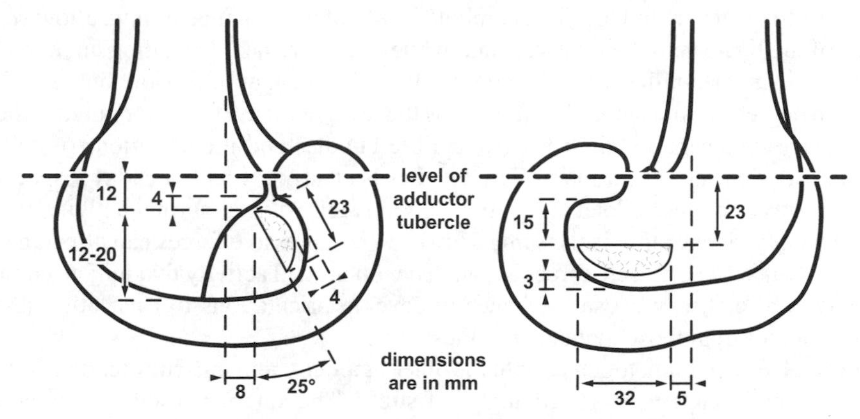

Figure 2.5.11 Average measurements and body relations of ACL (left) and the PCL (right) femoral attachment. The ACL attachment site is about 23 mm long and about half that wide. The PCL attachment site is about 32 mm in length and almost half that wide (from Girgis, 1975, with permission of J.P. Lippincott Co., Philadelphia, PA).

velop in the cruciate ligaments vary with joint angle. The twisted configuration of these ligaments and their wide insertion sites (Figure 2.5.11) appear well designed to deal with the varying load patterns.

Uniform stress is seldom achieved. The fact that the fibres are twisted suggests that the ACL is loaded in torsion as well as tension. The *screw home* mechanism as the knee comes

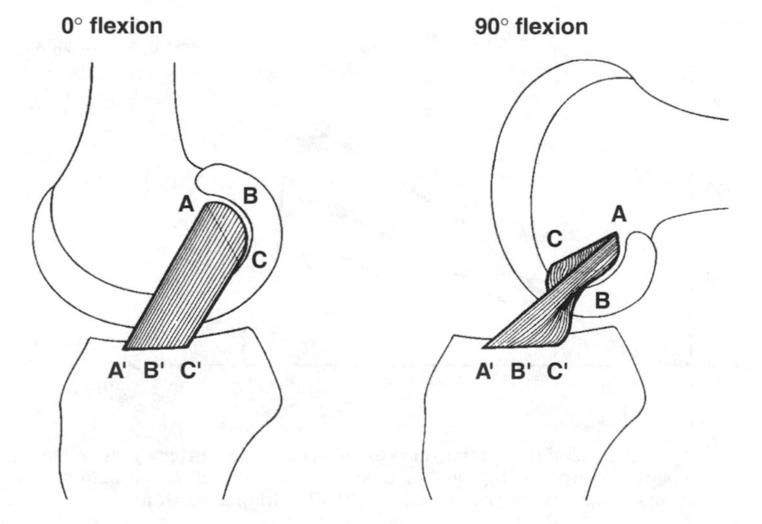

Figure 2.5.12 Schematic drawing of changes in shape and tension of ACL in extension (left) and flexion (right). The anteromedial band (A-A') lengthens and the posterolateral ligament aspect (C-C') shortens in flexion. Between (A-A') and (C-C') fascicles of an intermediate band (B-B') experience varying degrees of tension (from Arnoczky & Warren, 1988, with permission of J.P. Lippincott Co., Philadelphia, PA).

into extension, for example, involves a torsional rotation of the femur relative to the tibia. As the knee undergoes flexion and extension (and similarly internal, external, varus, and valgus rotations), the length and orientation of the fibres change (Figure 2.5.12). The broad attachments to the femur and the tibia, combined with inter-fascicular sliding, allow various portions of the ligament to be relatively taut while others are lax, depending upon the knee motion. For example, in the ACL, the anterior fascicles are taut in flexion, but are relaxed to a certain degree in extension. The situation is the reverse for the posterior bundles; hence, the advantage of separate sliding fascicular fibre bundles and the variations of collagen crimp. The tension and relaxation in fibre bundles is clearly evident in cadaver specimens missing gravity and muscle loads. The implication is that, during normal activity, the pattern of tension and relaxation is the same. However, compressive forces can affect inter-insertional distances. The muscle forces associated with normal activity that initiate compressive loads in a joint may cause the bones to develop orientations to each other that are slightly different from those that occur in most cadaver tests.

In normal joint motion, together with the other ligaments and musculo-tendon units, the cruciate ligaments cope with varied angles of stress. The anterior cruciate ligaments are stronger in directions similar to their predominant fibre orientation than in directions aligned with the tibial axis (Woo et al., 1990). Figure 2.5.13 shows the force-deformation curves obtained during tensile testing of ACLs from young donors. The specimens loaded along the axis of the ACL demonstrate a steeper linear stiffness and higher ultimate load than those deformed along the tibial axis. These results indicate that the cruciate ligaments are favourably oriented to take up load in flexion and extension, and varus and valgus movements that approximate the ligament's angle of alignment. Restraint of these movements is considered to be the prime function of this ligament complex in the knee.

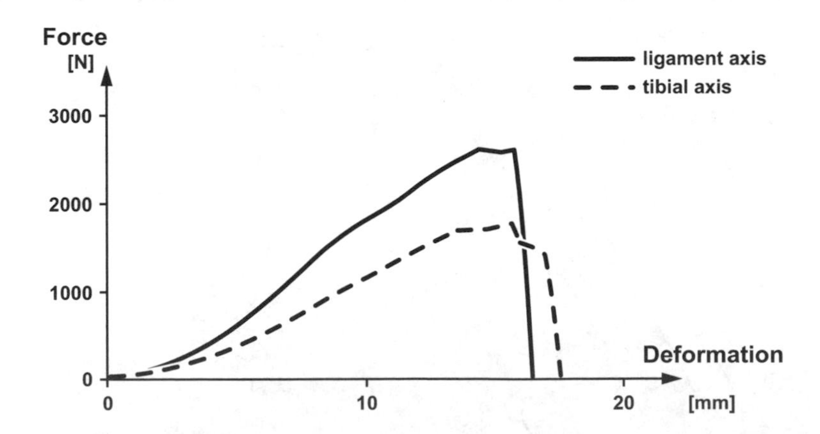

Figure 2.5.13 **Typical force-deformation curves for the femur-anterior cruciate ligament-tibia complex tested along the ACL axis (unbroken line) and along the tibial axis (broken line) (from Woo & Adams, 1990, with permission).**

COLLATERAL LIGAMENTS

The collateral ligaments are located outside the joint capsule (Figure 2.5.14), on the medial and lateral sides of the knee joint. Unlike the cruciate ligaments, the collateral

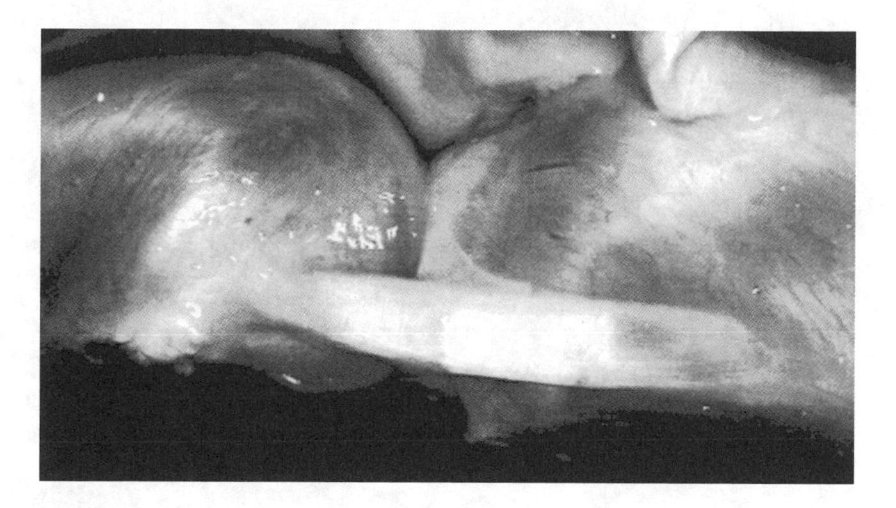

Figure 2.5.14 **Gross appearance of the rabbit MCL. The relatively parallel alignment of collagen fibres gives the MCL a cord-like appearance (from Bray et al., 1991, with permission).**

ligaments are not twisted. The main function of the collateral ligaments is to restrain varus and valgus angulation and rotation.

During varus angulation, the LCL takes up load, while the MCL relaxes. In valgus angulation, the roles of the collateral ligaments are reversed [Figure 2.5.8(V)]. During medial and lateral tibial rotation, the femoral condyles ride up the central eminence of the tibia, thus distracting the joint. The collateral ligaments resist the distraction, and consequently, the rotation. During rotation in the transverse plane, the collateral ligaments work together with the cruciate ligaments. During medial rotation, the MCL takes up a similar angulation to the ACL, and during lateral rotation, the LCL takes up a similar angulation to the PCL [Figure 2.5.8(IV)].

During flexion and extension, collateral ligaments play a secondary role to cruciate ligaments in restraining anterior and posterior displacement of femur and tibia, because alongside an increase in the joint angle during flexion, posterior bundles of the collaterals relax, while anterior bundles retain their length and take up some strain (Figure 2.5.15). This secondary role illustrates the advantage of sliding fascicles and the benefit of having different portions of ligaments able to take up stress at different joint angles.

2.5.3 PHYSICAL PROPERTIES AND MECHANICS

The tensile physical properties of all soft connective tissues can be classified into two general categories: structural and material. The structural properties of a ligament are derived from the behaviour of a bone-ligament-bone complex and thus involve the midsubstance of the ligament, the insertions, and the bone local to the insertions. The material properties describe the material irrespective of geometry. They are usually measured in the midsubstance of the ligament. In this section, both property types are discussed briefly as

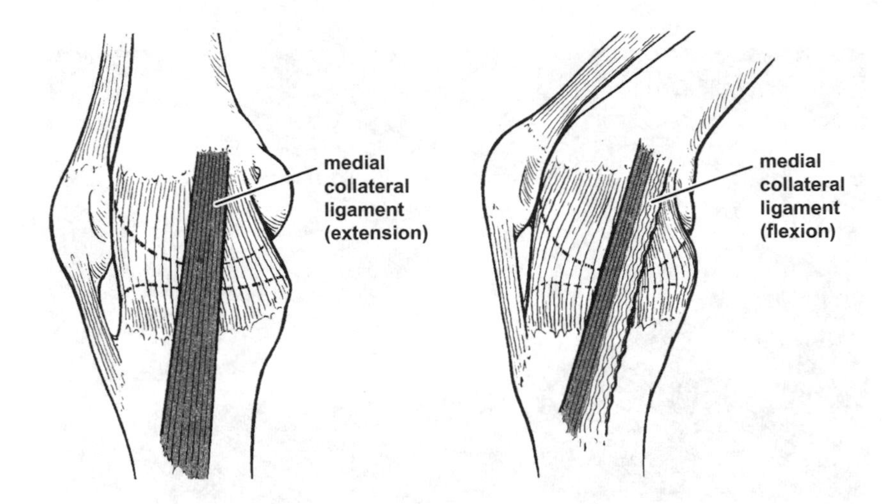

Figure 2.5.15 **The MCL at knee extension (left) and knee flexion (right). When the knee is fully extended, the fibres of the MCL are equally aligned and fully resist valgus angulation and tibial rotation. When the knee is in the flexed position, the anterior portion of the MCL remains taut while the posterior portion is relaxed. The anterior portion remains able to resist valgus and rotational stresses (from Indelicatio, 1988, with permission of Churchill Livingstone, New York).**

they pertain to the ligament model with which we have greatest experience, the rabbit MCL. Read the analogous section in Chapter 2.6 on tendon, as tendinous properties can be characterized similarly and are, at least in a gross sense, somewhat similar.

STRUCTURAL PROPERTIES

The non-linear force-deformation curve

Figure 2.5.16 shows a typical tensile force-deformation curve for ligaments. The stiffness (rate of change of force with deformation) of ligaments varies non-linearly with force. This non-linear behaviour allows ligaments to permit initial joint deformations with minimal resistance. The area under the curve in region I of Figure 2.5.16 is small compared, for cxample, to a straight-line relationship. Together with other ligaments, bone geometry and active muscles, ligaments work within their low-force range to guide bones through normal movement. At higher forces, ligaments become stiffer, thereby providing more resistance to increasing deformations. It is assumed that such stiffening protects the joint.

The non-linear force-deformation behaviour of ligaments occurs for at least two reasons: flattening out of collagen crimp and heterogeneous distribution of fibres. These reasons are described below.

Flattening out of collagen crimp

As described in more detail in Chapter 2.4, collagen (the main tensile-resisting substance in the ligament) is crimped. As with tendon, the crimp is thought to allow some extensibility of the ligament along its length under low forces. As tension is increased, the

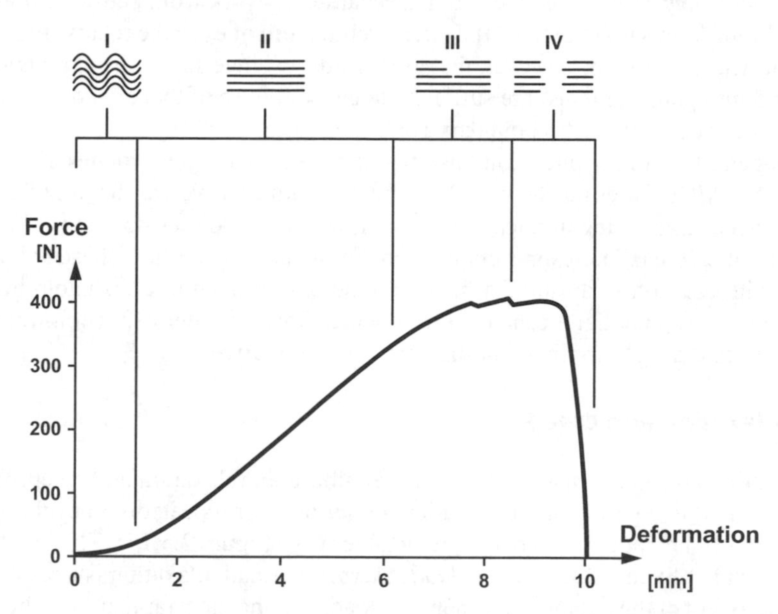

Figure 2.5.16 **A typical force-deformation curve for a typical rabbit ligament under monotonic loading. I = toe region; II = linear region; III = region of microfailure; IV = failure region. The top shows schematic representations of fibres going from crimped (I) through recruitment (II) to progressive failure (III and IV).**

crimp gradually flattens out. Once no more crimp can be removed, stiffness increases and an ever-increasing amount of force is required for further deformation. The toe region of the force-deformation curve (region I in Figure 2.5.16) is thought to correspond to the stretching out of crimp. With the crimp gone and the whole matrix under tension (the start of region II in Figure 2.5.16), a region with more constant linear stiffness (the linear region) begins.

Heterogeneous distribution of fibres

Neither fibres nor crimp are homogeneously distributed along the length of ligaments. As such, when ligaments are distracted, different fibres are recruited into load-bearing at different displacements. As noted above, when all the fibres have been recruited, the stiffness behaviour becomes more linear until some fibres (presumably those first recruited) fail. At this point, the net stiffness of the structure begins to drop, as shown in the fracture region of the force-deformation curve (region III in Figure 2.5.16). As some fibres fail, the load is redistributed onto the remaining fibres, increasing the load on them and the likelihood of their failure. It then takes little additional deformation to produce gross structural failure of the ligament through all the remaining fibres (region IV of Figure 2.5.16).

Microscopic examination shows some fibres crossing between parallel fibres in the ligament substance, some running perpendicular to the long axis and some at every angle between. While there are not many non-axial fibres (Liu et al., 1991) and they tend to be smaller than their longitudinally oriented partners, they should nonetheless contribute to non-linear stiffness behaviour as the ligament is loaded. Depending on how and where the non-axial fibres are connected, they could serve as tethers for longitudinal fibres. Alternatively, the crossing fibres could be connected in a separate network from end-to-end in some oblique fashion. At this point in time, the microarchitecture of even the relatively simple MCL is not known. It is known, however, that gross midsubstance strains from around 8 percent (in up to 5 mm gauge lengths) are sufficient to cause failure of that area of the rabbit MCL (at least under certain in vitro boundary conditions (Lam, 1988)).

Collagen fibres must spiral from insertion to insertion as a joint rotates. For example, at the knee, the MCL fibres must spiral along their longitudinal axis as the joint flexes and extends. This has interesting implications to interpreting two-dimensional histology slices in any plane. It also has interesting implications in understanding how these spiraling fibres interact with each other during that motion. One is reminded of a dishcloth being wrung out, suggesting a potential mechanism from water flow or water expulsion from the interfibrillar space during spiralling, but this remains speculative.

Load relaxation and creep

Like other connective tissues, ligaments exhibit load relaxation and creep. When a ligament is pulled to a particular deformation, either once or repeatedly in cyclic succession, the load in a ligament decreases in a predictable way (Figure 2.5.17). This predictable decrease in load at fixed deformation is *load relaxation*. Load relaxation is a result of the viscous component of the ligament's response to load, and the most rapid part of the load decay occurs immediately after loading. Decay continues non-linearly until a steady-state value of load is achieved: this is the elastic component of the response. The relative proportions of the viscous and elastic components in any one force depend on the rate of load application and the previous load history as well as in a variety of test-specific parameters, e.g., temperature and the solution in which the test is carried out. The faster a load is applied, the less time there is for the viscous component to dissipate. A ligament will appear stronger and slightly stiffer under rapid versus slow load application.

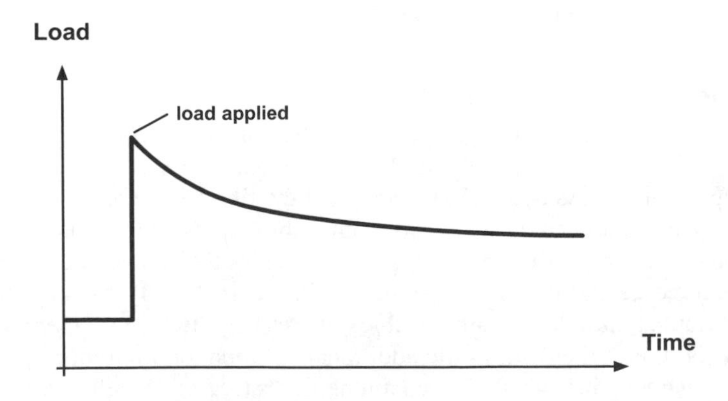

Figure 2.5.17 **Schematic load-relaxation curve for ligament.**

Creep is the analogous behaviour of a ligament under a fixed load when the load is either held or reached repetitively in a cyclic fashion. Creep is the increase in length over time under a constant load. With creep, as with load relaxation, manifestation of the viscous component through time-dependent load or deformation changes eventually ceases (Figure 2.5.18).

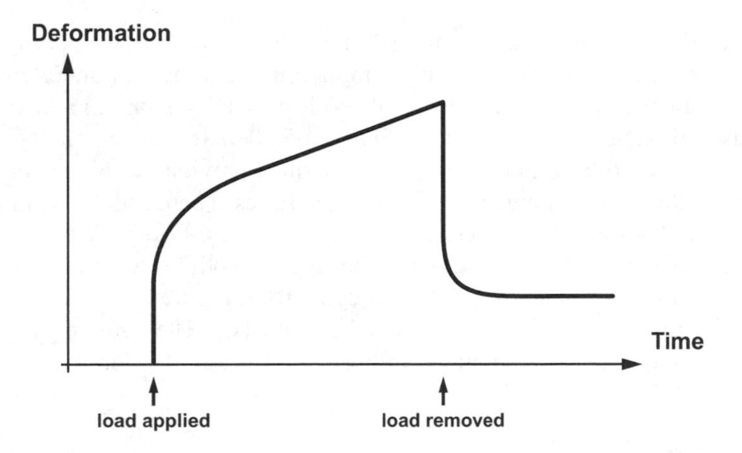

Figure 2.5.18 **Schematic creep curve for ligament.**

During cyclic load application, however, some of the viscous component can be recovered in each cycle (Figure 2.5.19). When the ligament is unloaded, the viscous component, while never recovering completely (at least during in vitro tests), can recover to over 90 percent of its original state after many hours in a relaxed condition. This recovery probably involves some combination of water influx, returning collagen crimp, elastin tensile force, and decreasing collagenous organization under unloaded conditions.

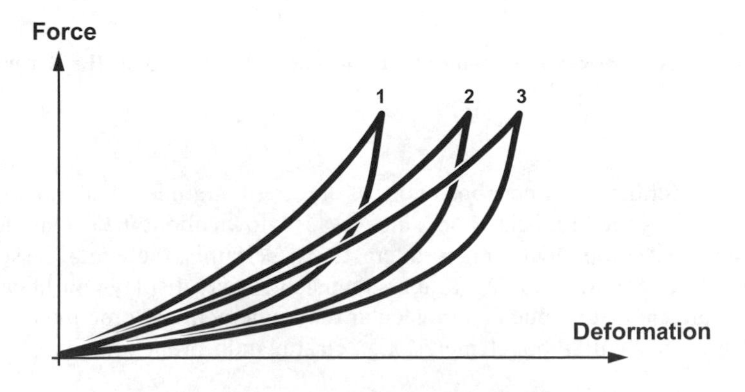

Figure 2.5.19 **Schematic force-deformation graph showing three successive cycles in displacement control to an upper load limit and a lower displacement limit, illustrating the viscoelastic creep effect of cycling upon a ligament.**

MATERIAL PROPERTIES

Non-linear behaviour

The material behaviour of ligaments, i.e., stress-strain behaviour, is also non-linear (Woo et al., 1982b; Lam, 1988), both under monotonic loading and under conditions in which viscoelastic behaviour is demonstrated. With increasing strain in a monotonic test, stress increases in the ligament, as shown in Figure 2.5.20. In the toe region (region I), there is little stress relative to the strain applied. As described previously, this region is thought to correspond to the straightening out of the collagen fibres. There follows a more linear region (region II in Figure 2.5.20), presumably as the collagen fibres take up load. A measurement of the tangent in the linear region is the tangent modulus which is often, but erroneously, called the elastic modulus. Finally, in region III of Figure 2.5.20, the curve flattens out, eventually dropping dramatically towards the strain axis. The flattening is presumably related to increasingly rapid microfailure, followed by catastrophic failure.

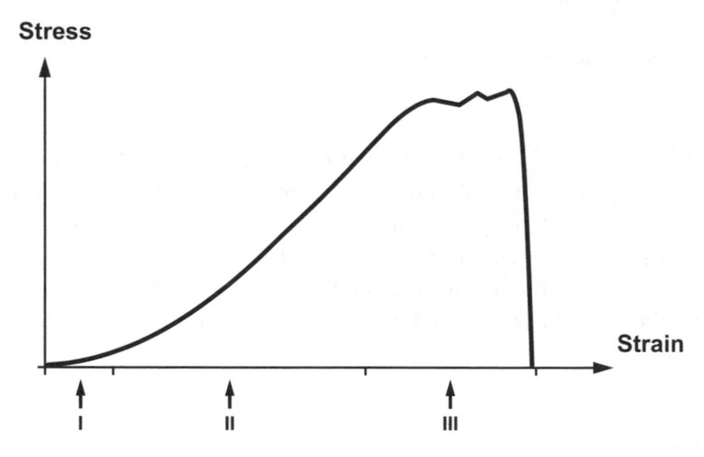

Figure 2.5.20 **Schematic stress-strain curve for ligament (I = toe region; II = linear region; III = failure region).**

The reasons for the non-linear behaviour at the stress-strain level are not as easy to explain as that of the structural behaviour, since less is known about the actual molecular nature of the ligament components or their interactions. Certainly, there must be some overlap with structural explanations. Collagen fibres themselves may display non-linear characteristics as they are elongated, due to a molecular rearrangement or some internal reordering of their relationships with other elements, e.g., elastin and fibronectin.

Water

Water makes up two-thirds of ligament composition which implies an important functional role. While these functions may be mainly biological, e.g., supply or removal of nutrients, water is known to contribute to the non-linear viscoelastic behaviour of the ligament in some significant way (Chimich et al., 1992; Thornton et al., 2001b). The amount of water in the interfibrillar space may influence collagen stiffness, or the distances at which fibres can interact in a physical or biochemical sense. Further, because of its attraction to the highly negatively charged proteoglycan molecules, water may induce pressure gradients in the tissue that are further manipulated by movement.

PROBLEMS IN MEASURING PHYSICAL PROPERTIES OF LIGAMENTS

Measuring the physical properties of ligaments is not as simple as might be expected. Selected problems with these procedures are discussed in the following paragraphs.

Source of the tissue sample

The source of the tissue sample may have a considerable effect on the mechanical properties of the sample. All the properties of ligaments noted above are to some extent boundary condition specific. The boundary conditions include:

- Species,
- Type of ligament,
- Gender,
- Age,
- Activity,
- Drugs, and
- Diet.

Considering the boundary condition of age, for example, immature, mature, and aging tissue samples have different properties. As tissue matures, it strengthens due to increases in the size of collagen fibres and the number of molecular cross-links. Aging tissues, by contrast, experience a gradual, limited reversal of the maturation process. The in vivo load history of a tissue is another factor that vastly influences its properties. In vivo, inactivity results in a weaker tissue, while mobility and exercise increase tissue strength.

Because many boundary conditions may influence the mechanical properties of ligaments, consensus about the precise biomechanical behaviour of ligaments is lacking. As a result, we prefer not to publish either structural or material properties in this text that could be misinterpreted as absolute.

Different aspect ratios and trouble securing the ends of ligaments

Testing isolated ligaments is complicated by the different aspect ratios (length and width) of different ligaments and by difficulties in effectively securing the cut ends. Putting the free ends in clamps induces end-effects and often results in stress concentration at the grips, which damages the tissue and may contribute to premature failure or defective data. A bone-ligament-bone preparation provides more secure clamping, but increases the difficulty in separating the properties of the ligament midsubstance itself from those of the insertion sites. Special devices such as buckle transducers and magnetic field (Hall Effect) displacement transducers have been used to estimate ligament forces during testing, but unfortunately they rely on direct contact with the tissue sample and may influence testing results. Measurements of ligament strains are also flawed. Optical analyzers that do not require contact with tissues have been used to quantify strains. However, the reported accuracy levels of these optical analyzers are known to be boundary condition specific (Lam et al., 1993), and the dye lines marked on the specimen may affect the strain being measured.

Measuring the cross-sectional area of a ligament

Measurements of stress have been compromised by the lack of an ideal method for measuring the cross-sectional area of a tissue sample. The irregular, complex shape and geometry of these tissues makes direct measurements difficult and errors large. Rigid calliper measurements, which are frequently used, require approximations, as the callipers are unable to take irregularities into consideration. Flexible callipers, which are better able to follow contours, are more accurate, but still require tissue contact. The only non-contact methods of measuring areas are optical, using either photography, or laser refraction to quantify distances. Laser refraction involves placing a specimen perpendicular to the path of a laser beam and rotating it by 180° or vice versa. The data of profile width and position are then recorded via a microprocessor. The centre of rotation and upper and lower boundaries are determined for each increment of rotation, and an iterative procedure is used to reconstruct the cross-sectional shape, from which the area may be calculated. Unfortunately, this technique misses depressions in an irregular surface. Like other techniques, this method is also unable to determine the circumference of the ligament with complete accuracy.

A *zero* strain position

The definition of a true *zero* strain position poses problems. The zero point on all the graphs used above is boundary-condition specific. The zero point depends on environment (including water content), load, and load history. The zero point is critical when properties are compared in absolute terms, i.e., at comparable forces and deformations, or comparable stresses and strains. A slight shift in the zero point of a test can drastically alter such comparisons (Figure 2.5.20). Changing water content by osmotic manipulation can affect this zero point (Thornton et al., 2001b). This has important implications for the need to control water content during in vitro tests, but also for in vivo length correction mechanisms in ligaments. Differences in environmental temperature can cause similar inaccuracies.

The necessity for well-documented biomechanical test procedures has been demonstrated. All the factors mentioned above must be taken into consideration before conclusions about ligament behaviour are drawn.

2.5.4 BIOLOGY AND FUNCTION

In vivo pressure recordings (used to estimate force) indicate that ligaments are exposed to repeated stress as part of daily activities (Holden et al., 1994). As such, ligaments are exposed to static and cyclic loading that causes creep. Examining the interplay between biology and function reveals potential mechanisms of creep in normal ligaments.

WATER

Water content plays a role in the viscoelastic behaviour of normal ligaments. In normal ligament, water flows out of the ligament during creep testing, resulting in a lower than normal water content at the end of the test (Thornton et al., 2000). The mechanisms driving this water exudation may involve a combination of fibre slack, Poisson's ratio, or osmotic pressure (Adeeb et al., 2004). If a period of *recovery* (unloading after loading) is permitted, water can flow back into the ligament to re-establish normal water content, but this process is stress- and time-dependent.

Decreasing the water content below normal values causes a decrease in relaxation and creep (Chimich et al., 1992; Thornton et al., 2001b). Correspondingly, an increase in water content above normal values causes an increase in relaxation and creep. Increasing the water content from normal creates a pre-stress in the ligament (Thornton et al., 2001b). Essentially, the functional length of the ligament decreases to accommodate the presence of greater than normal water content. The decrease in length is accompanied by an increase in cross-sectional area. These findings suggest that ligament viscoelastic behaviour could be affected by naturally occurring changes to water content in vivo, such as inflammation.

COLLAGEN

Despite similar responses to increased water content, relaxation and creep appear to be governed by different microstructural processes. When applying linear viscoelasticity within the linear framework of the quasi-linear viscoelastic theory (Fung, 1993), relaxation data were found to over-predict creep data for normal MCLs (Thornton et al., 1997). Fung (1993) speculated that creep is fundamentally more non-linear than relaxation, and the microstructural processes in a material undergoing creep could be quite different from relaxation.

The microstructural differences between creep and relaxation were investigated by recording the crimp pattern (marker of collagen fibre recruitment) at the beginning and end of relaxation and creep tests (Thornton et al., 2001a). Because relaxation is tested at a fixed deformation/strain, relaxation was found to be the behaviour of a distinct group of fibres recruited into load bearing at that fixed deformation with little or no change in the recruitment of fibres over the course of the test. Because creep is tested at a fixed load/stress and deformation is allowed to increase, creep was found to be the behaviour of fibres recruited initially and fibres recruited progressively over the course of the test.

The above observations were at stresses and strains in the toe region of the normal ligament stress-strain curve, thought to be within the normal, physiologic range. However, when ligaments are creep tested at stresses in the linear region of the stress-strain curve, the majority of the fibres are fully recruited (uncrimped) by the end of the test. Interestingly, the creep strains at the various toe region stresses were similar, despite doubling the stress. Yet, creep strain increased at a linear region stress. These findings indicate that fibre recruitment minimizes creep at toe region stresses, but this recruitment is limited at higher, linear region stresses.

A simple microstructural model demonstrated that the differences in creep and relaxation predictions could be accounted for by incorporating fibre recruitment (Thornton et al., 2001a). When the relaxation behaviour is taken to be that of fibres recruited into load-bearing, i.e. used to predict creep of recruited fibres, and coupled with progressive fibre recruitment, ligament creep behaviour is accurately predicted. The model also demonstrated that the load on the initially recruited fibres decreases as additional fibres are recruited into load-bearing.

This ability to redistribute stress is also demonstrated by the recovery of modulus following partial ligament failure (Thornton et al., 2002). In normal ligaments, modulus increases with successive loading cycles to toe region stresses. At a linear region stress, the ligament may have an increase in modulus for several cycles followed by an abrupt increase in strain with no corresponding increase in stress, i.e. a discontinuity, and a subsequent decrease in modulus. As the cyclic loading continues, the modulus increases, returning to the value achieved before the discontinuity. By evaluating the crimp pattern, the discontinuity was related to damage causing fibre rupture. Following fibre rupture, stress appears to be redistributed to the remaining intact fibres in order to avoid total ligament failure. From these results, different fibres in the ligament cross-section were at different stresses, with some stresses high enough to cause fibre rupture. Clearly, stress calculations based on total ligament cross-sectional area underestimate the stresses on the fibres that are recruited into load bearing.

Normal ligament mechanical behaviour is influenced by its most abundant component, water, and its major tensile load-bearing component, collagen. Examining the mechanical, morphological, and biochemical properties of ligaments healing from an injury further clarifies the importance of these and other components on ligament function.

2.5.5 FAILURE AND HEALING

Some of the most important factors that affect the normal integrity of ligaments include exercise, immobilization, aging and maturing, pregnancy, and injuries (see Chapter 2.8).

Accidental ligament ruptures in humans usually result from excessive loads. The rate of strain affects the type of ligament failure (Noyes et al., 1974). A study on primates revealed that there were only 29 percent ligamentous (or midsubstance) failures at a slow stretch rate, compared to 57 percent failures by tibial avulsion (insertion site at tibia). However, when the strain rate was increased a 100 fold, the percentage of ligament midsubstance tears increased to 66 percent and the percentage of tibial avulsions declined to 28 percent.

Ligaments that heal normally do so first by red blood cells and inflammatory white blood cells (disease preventing cells) entering the wound. Within days, a fragile fibrous scar composed mainly of blood-clot components and water appears in the wound. After about a week, metabolically active fibroblasts migrate into the wound to form an extracellular scar matrix. Eventually, these fibroblasts become the predominant cell type, as fluid and blood gradually dissipate.

When a ligament is injured and a gap is created between the torn ligament ends, the ligament heals with scar tissue that bridges that gap. This scar tissue is mechanically inferior to normal tissue, and several morphological and biochemical differences between normal and healing ligaments likely contribute to the inferior mechanical behaviour.

MECHANICAL

The failure properties of ligament scars are inferior to that of normal ligaments. Evaluating structural properties, MCL gap scars attain only 65 percent of the failure force of contralateral controls even after 40 weeks of healing (Chimich et al., 1991). Evaluating material properties (normalized for cross-sectional area), MCL gap scars achieve only 35 percent of the failure stress of controls at 40 weeks.

Like failure properties, viscoelastic properties improve with healing, but remain inferior to normal. Even after 14 weeks of healing, the creep strain of MCL scars is more than twice that of normal ligaments, when tested at the same stress (that corresponded to a physiologic stress in the toe region of the normal stress-strain curve) (Thornton et al., 2000). Also at 14 weeks, the modulus measured during cyclic loading of MCL scars is less than half of normal MCLs loaded to the same stress (Thornton et al., 2003).

MORPHOLOGICAL

The cross-sectional areas are larger for injured ligaments than for uninjured ligaments, even 40 weeks after the injury (Chimich et al., 1991). Because of this larger cross-sectional area, the failure properties of scar are worse when evaluated as a material property. The failure force was closer to that of normal, likely because of this larger cross-section of weak scar tissue.

Flaws (blood vessels, fat cells, hypercellular areas, and loose or disorganized matrix) account for a larger percentage of the area in MCL scar than in normal ligament (Shrive et al., 1995). The flaws decrease with healing, but remain greater than normal at 14 weeks. Stress concentrations arising from flaws likely affect both failure and viscoelastic properties.

Alignment of the scar collagen fibres along the longitudinal axis improves with healing and reaches normal values by 14 weeks (Frank et al., 1991). However, collagen fibril diameters are smaller than normal and do not change with healing by 40 weeks (Frank et al., 1992). Taking these collagen properties together, these small and initially poorly aligned fibres likely contribute to the mechanical inferiority.

BIOCHEMICAL

MCL scar water content is elevated early and decreases to normal values by 14 weeks (Thornton et al., 2000). Water content decreasing with healing may be related to the small improvement in viscoelastic behaviour over time, but cannot account for the persistent mechanical inferiority.

Proteoglycan (glycosaminoglycan) content of MCL scars decreases with healing, but remains elevated at 14 weeks (Frank et al., 1983a). This change in proteoglycan content may have altered the ratio of bound to free water and the resulting viscoelastic behaviour.

Collagen (hydroxyproline) concentration in MCL scars increases with healing and returns to normal values by 14 weeks (Frank et al., 1995). However, collagen (hydroxylysylpyridinoline) cross-link density increases with healing, but reaches about half of normal values at 14 and 40 weeks. Collagen fibres are small and initially poorly aligned, and never reestablish the same fibre interconnections as in normal ligament. These collagen properties as a whole likely contribute to both failure and viscoelastic inferiorities.

Remodelling occurs over many months or years. Scar flaws are removed, the number of collagen fibrils present increases, fibre alignment and cross-linking improves, and water and gylcosaminoglycan content decreases. However, there is no documented return to normal ligament.

2.6 TENDON/APONEUROSIS

HERZOG, W.

2.6.1 MORPHOLOGY AND HISTOLOGY

Tendon is a dense fibrous tissue that connects muscle to bone. Tendon is present in a wide variety of shapes and sizes, depending on the morphological, physiological, and mechanical characteristics of both the muscle and bone to which it is attached. Typically, tendon is glistening and pearly-white, and may take the shape of a thin cord or band, or a broad sheet. A variety of human tendons are well illustrated in Jozsa and Kannus (1997).

Usually, tendon consists of an external tendon, which is typically referred to as *tendon*, and an internal tendon, which is typically referred to as *aponeurosis*. The external tendon connects the muscle proper to bone. The aponeurosis provides the attachment area for the muscle fibres (Figure 2.6.1).

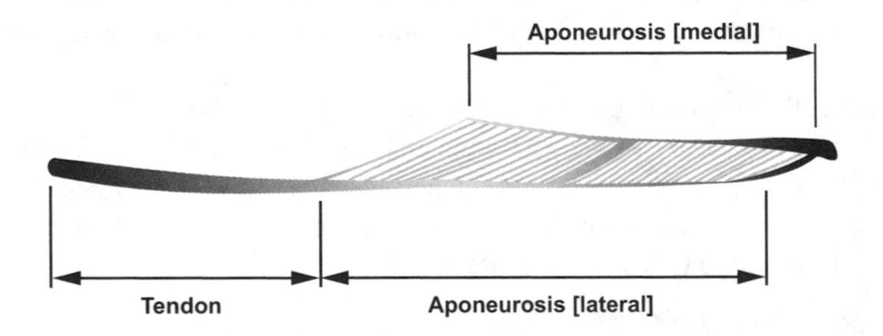

Figure 2.6.1 **Schematic illustration of a uni-pennate muscle with tendon and aponeuroses identified. The muscle is roughly based on the geometry of the cat medial gastrocnemius.**

A tendon may be constrained as it crosses a joint, either by bony prominences through which the tendon must pass or by specialized connective tissue sheaths, e.g., bicipital groove in the humerus and flexor retinaculum in the wrist. These constraints help to maintain the orientation of the tendon during joint motion. The retinacular sheaths are particularly important in the hands and feet, as tendons passing distally (towards the fingers or toes) over numerous small joints would be susceptible to injury if allowed to be displaced during finger or toe movements. A constrained finger flexor tendon is shown in Figure 2.6.2.

Despite the variable nature of tendon size and shape, every tendon has three distinct regions of organization: the muscle-tendon junction (myotendinous junction), the tendon *proper* (hereafter called tendon), and the bone-tendon junction (osteotendinous junction). Since the tendon region is often the most conspicuous region, it is the ideal place to begin a more detailed morphological and histological discussion.

Tendon is composed primarily of collagen fibres embedded in an aqueous gel-like ground substance. Since collagen comprises about 70 to 80% of the dry weight of tendon

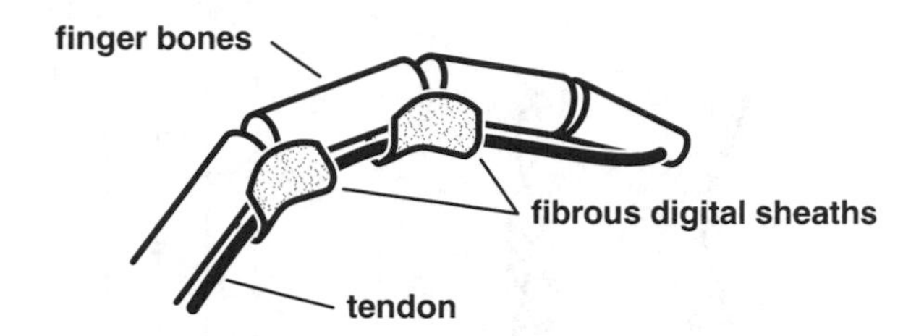

Figure 2.6.2 **Fibrous digital sheaths of the finger form a tunnel with the bone to keep the tendon close to the bone as the finger bends.**

(Elliot, 1965), a discussion of tendon morphology should first consider the biochemical organization of collagen.

Collagen is the main structural protein of animals. Although several different forms of collagen have been identified, e.g., type I collagen found in vertebrate tendon, collagen proteins are characterized, at least in part, by a predominance of the amino acids glycine, proline, hydroxyproline, and hydroxylysine. Hydroxyproline is unique to collagen.

The basic *unit* of collagen is tropocollagen, which consists of three non-coaxial helical polypeptide chains wound around each other to form a single closely-packed helix. The small glycine residues (a single proton), which occur approximately every third amino acid, allow for close packing between the three strands. The three strands are further stabilized by the hydrogen bonding from the relatively plentiful hydroxyproline residues. Each helical tropocollagen molecule is about 280 nm long. To make long strands of collagen fibres, tropocollagen molecules are covalently cross-linked to neighbouring tropocollagens at special amino acid sites where hydroxylysine is present. These amino acid sites give tropocollagen molecules the appearance of overlapping at approximately a quarter-length stagger. Five tropocollagen units in cross-section form the microfibril (Wainwright et al., 1982; O'Brien, 1992). The formation of a microfibril is schematically shown in Figure 2.6.3.

Microfibrils aggregate to form subfibrils. In cross-section, these subfibrils show 3.5 nm staining sites that correspond to the radii of the microfibrils. The subfibrils aggregate further to form fibrils. At this level of tendon organization, the electron microscope reveals a 64 nm longitudinal banding pattern. This banding pattern corresponds to the quarter-stagger overlap associated with the covalent cross-links between the tropocollagen units. Fibrils aggregate to form fibres and fibre bundles in which *crimping,* or longitudinal waviness of the collagen fibres, may first be apparent.

Fibre bundles aggregate to form fascicles. Fascicles are surrounded by endotenon, a connective tissue sheath composed of a well-ordered criss-cross pattern of collagen fibrils, proteoglycans, and elastin (Rowe, 1985). Endotenon contains the blood and lymphatic vessels, and nerves. Tendon cells, longitudinally-aligned among the fibre bundles, are visible at the fascicular level of tendon organization.

Fascicles aggregate into fascicle bundles and are surrounded by epitenon. Finally, several fascicular bundles are surrounded by paratenon, the outermost tendon sheath. Some tendons are enclosed by a further fluid-filled synovial sheath, which greatly reduces the friction associated with the movement of tendon against the surrounding tissue. A schematic diagram illustrating the structural hierarchy of tendon is shown in Figure 2.6.4.

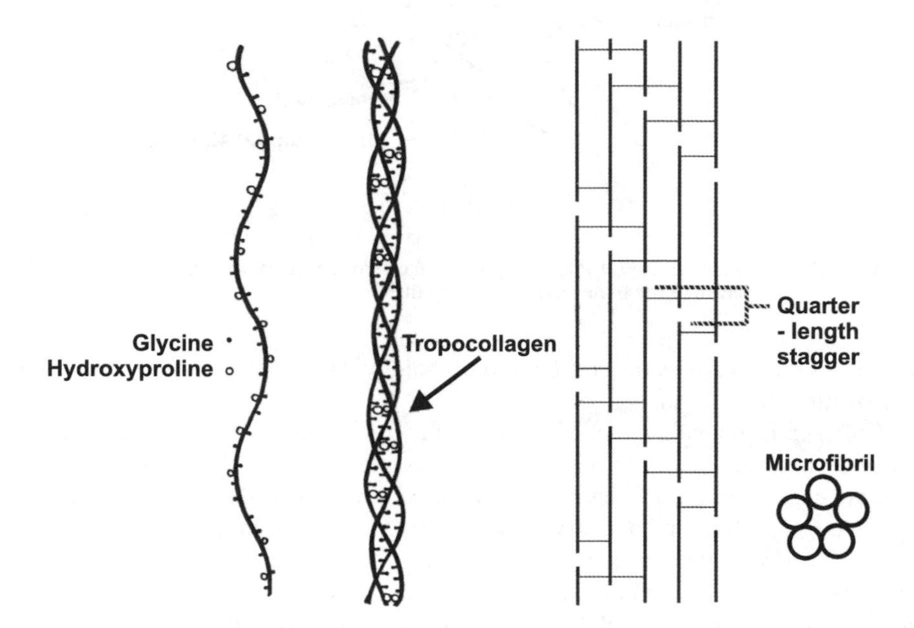

Figure 2.6.3 **Schematic diagram of tropocollagen and microfibril assembly.**

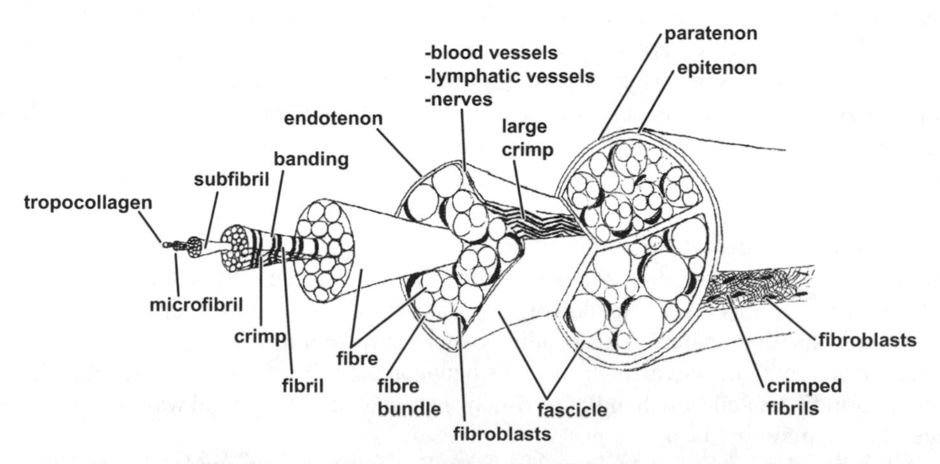

Figure 2.6.4 **The structural hierarchy of a tendon, from the tropocollagen molecule to the entire tendon. Connective tissue layers or sheaths envelop the collagen fascicles (endotenon), bundles of fascicles (epitenon), and the entire tendon (paratenon). Note that blood and lymphatic vessels and nerves are cut in the cross-section within the endotenon (from Kastelic et al., 1978, with permission).**

Golgi-tendon organs lie within the tendon and close to the myotendinous junction. Golgi-tendon organs are comprised of specialized nerve endings that lie in series with the contractile proteins of the muscle. Each golgi-tendon organ connects to the tendinous fascicles associated with approximately 10 muscle fibres, and sends a thick, myelinated afferent fibre to the spinal cord. The nerve endings extend many thin processes between the collagen fibres of their associated fascicles, partially encircling their associated fascicles. When the muscle contracts, tension is transferred to the tendon, causing the collagen fibres to compress the nerve endings and to initiate action potentials. As such, golgi-tendon organs are mechanoreceptors that monitor muscle tension. Afferent nerve fibres carry sensory information from the periphery to the spinal cord. The conduction velocities of the golgi afferents are enhanced by their relatively thick myelin and large size. Golgi-tendon afferents synapse with the internuncial neurons in the spinal cord and inhibit the alpha motor neurons of the corresponding muscle during isometric contraction. They may also excite the alpha motor neurons during cyclic movements, as has been shown for locomotion in the cat (Pearson, 1993).

The myotendinous junction is the region where the muscle joins with the tendon. Often, there is a myotendinous junction at both the origin and the insertion ends of the muscle belly. Transmission and scanning electron microscopy studies from a wide variety of vertebrate and invertebrate species show that the myotendinous junction is characterized by significant longitudinal surface folding (Figure 2.6.5). Several physical consequences are associated with this type of morphology, which greatly increases the area of contact between muscle and tendon (Tidball, 1991). The stresses transferred from the muscle to the tendon are greatly reduced in this region. The amplification in myotendinous contact area, calculated as the ratio between the myotendinous contact area and the physiological cross-sectional area of the muscle and its longitudinal end-folding area, has been estimated for several muscles, and appears to vary between 10- and 50-fold. Most of the values fall between 10- to 18-fold. Earlier studies by Trotter et al. (1985) suggested that the degree of infolding reflects the rate of tension development within the muscle, with slow twitch fibres having a larger junctional area than fast twitch fibres. However, more recent studies suggest that there is no consistent relationship between the amplification in myotendinous contact area and a physiological parameter, such as contractile speed or tonic versus phasic function (Trotter, 1993).

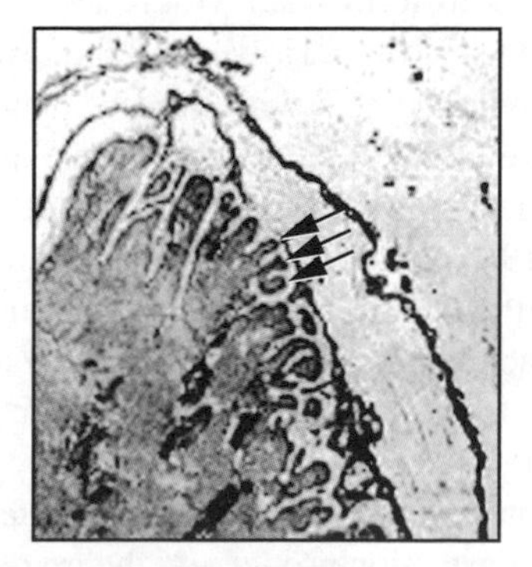

Figure 2.6.5 **The myotendinous junction (see arrows) as seen by transmission electron microscopy (from Jozsa & Kannus, 1997, with permission).**

Another consequence of the folded ultrastructure of the muscle-tendon junction is that the longitudinal folds facilitate the transfer of load by shear rather than by tension (Figure 2.6.6). The acute angle of contact between the muscle and tendon tissues ensures that the shear component of loading is greater than the tensile component. For adhesive joints between engineering materials, it is desirable to maximize the shear component of loading and to minimize the tensile component because engineering adhesives are stronger in shear than in tension. Biological materials that constitute the mechanical junctions at the ends of muscle fibres may also be stronger in shear than in tension. However, no direct evidence for this statement is available (Trotter, 1993).

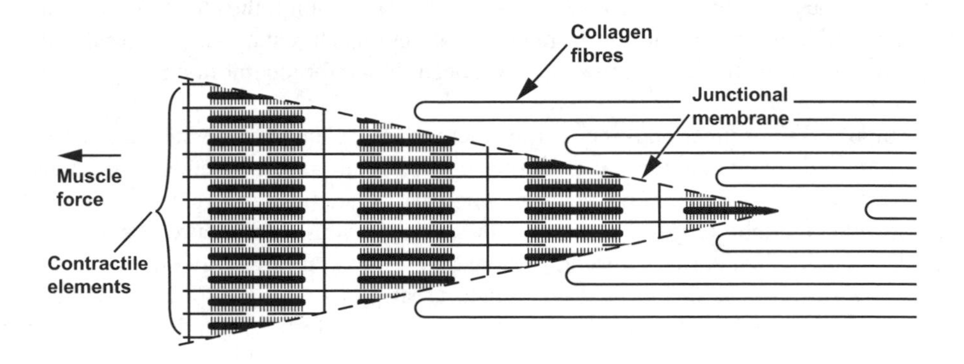

Figure 2.6.6 **Diagram of a myotendinous junction. Muscle force is applied parallel to the longitudinal axes of the myofilaments and the collagen fibres. The junctional membrane lies at an angle relative to the myofilaments. The acute angle with which the muscle and collagen fibrils meet creates a shear stress between the fibrils. If the fibrils met end-to-end, the junction would be loaded in tension.**

In vertebrate muscles, collagen fibres that form the areolar tissue of the endomysium (the muscle equivalent of the endotenon) become aligned near the muscle-tendon junction into a dense regular arrangement to form what Moore (1983) has termed a *microtendon*, which is one microtendon per muscle fibre. However, Ishikawa (1965) has pointed out that the collagen fibrils immediately associated with the fibre ends are not uniformly aligned with the fibre axis, but are more likely to have an isotropic arrangement, like those of the endomysium. The microtendon collagen fibrils, like those of the endomysium, have a uniform small diameter (Moore, 1983). At a short distance from the muscle-tendon junction, each microtendon blends with the tendon. The collagen fibrils of the tendon are heterogeneous in diameter, reaching much larger diameters than seen in the endomysium or microtendon. The microtendon thus forms a transitional structure for transmitting tension between muscle and tendon (Trotter, 1993).

The region where the tendon attaches to the bone, the osteotendinous junction, is morphologically distinct from both the tendon proper and the myotendinous junction. There are two types of attachment: one when the tendon becomes attached to epiphyseal bone and the other when the tendon approaches and attaches to bone at an acute angle.

When the tendon becomes attached to epiphyseal bone, the change takes place over four distinct, but well-blended, zones: fibrous tendon tissue; fibrocartilage tissue; calcified fibrocartilage tissue; and bone tissue. The transition from tendon to fibrocartilagenous tissue is characterized in part by a denser proteoglycan ground substance. Spindle-shaped tendon cells gradually change into rounded cartilage chondrocytes. The chondrocytes and cartilage ground substance lie among the collagen bundles that continue from the tendon proper. Blood vessels are generally absent from the zone of fibrocartilage (Benjamin et al., 1986). The amount of fibrocartilage present in an osteotendinous junction varies widely among tendons, and between regions of the same insertion. This newly formed fibrocartilage merges with articular cartilage, where present (Benjamin et al., 1986). The zone of calcified fibrocartilage is evident by the presence of one or more basophilic lines or tidemarks (visible by light microscopy), indicating the transition between the non-calcified and the calcified fibrocartilage. The calcified fibrocartilage zone is adjacent to the bone (Figure 2.6.7). Chondrocytes are less numerous on the bony side of the tidemark.

When tendon approaches bone at an acute angle, the morphology of attachment appears to be somewhat different than that described for attachment to epiphyseal bone. In this case, the superficial collagen fibres of the tendon blend in with the periosteum that covers the bony surface. The deep collagen fibres of the tendon penetrate the periosteum, and attach directly into the bone.

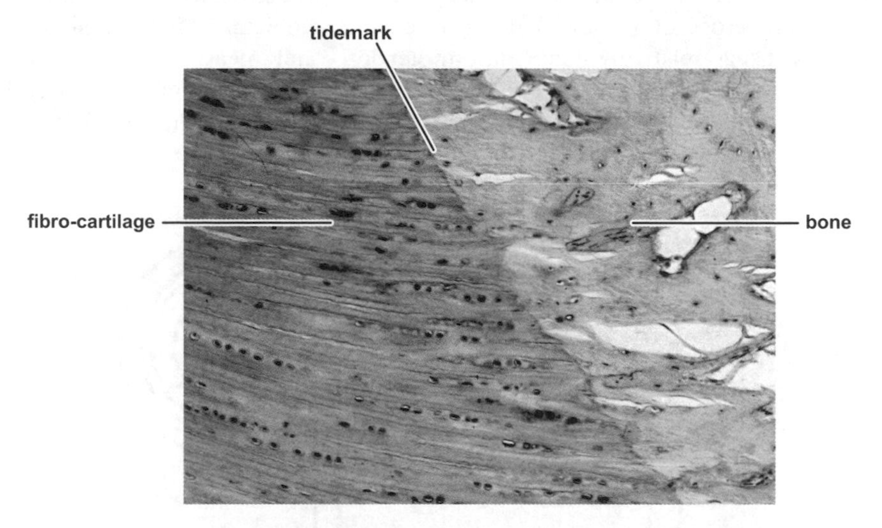

Figure 2.6.7 **Direct fibro-cartilagenous insertion into bone. The example shown is from a dog tendoachilles into the calcaneus (acknowledgement: Dr. J. R. Matyas).**

The functional significance of the epiphyseal-type of insertion is controversial. The gradual transition from fibrous tissue to bone may help to distribute forces over the attachment site, thereby, minimizing local stress concentration (Cooper & Misol, 1970; Noyes et al., 1974a; Woo et al., 1988). Benjamin et al. (1986) suggested that fibrocartilage is typical of epiphyseal attachments, because these tendons undergo a greater angular change during joint movement than do tendons inserting along the bony shafts.

2.6.2 PHYSICAL PROPERTIES

The primary role of tendon is to transmit the force of its associated muscle to bone. As such, tendon needs to be relatively stiff and strong in tension. Mechanical properties, such as tensile stiffness and ultimate strength, are typically determined using conventional engineering testing machines where one end of the test specimen is fixed to an actuator and the other end is fixed to a load cell.

Mechanical testing of tendons has proven to be a challenge. The main difficulty is that specimens tend to slip out of the standard testing machine grips. Alternatively, if tendon specimens are tightly squeezed between gripping surfaces, stress concentrations occur at the grip-specimen interface, and often cause premature failure of the specimen at the grip. Recently, cryo-grips have been introduced. In cryogrips, the specimen is frozen at the location where it is gripped. Frozen tissue is stronger than fresh tissue, therefore, this technique largely eliminates the failure problem at the specimen-grip interface. Similarly, the air-dried ends of tendon specimens (the remainders of which were kept moist) provided strong gripping material for measurements of stiffness, hysteresis, and ultimate strength (Wang & Ker, 1995; Wang et al., 1995). Specimens can also be tested by gripping the bone at one end, and the muscle or tendon at the other end. If the material properties of the tendon proper are sought, the structural effects involving the insertion must be separated from the material effects. Ker (1981) discussed potential problems associated with measuring the mechanical properties of tendon, and suggested that problems can be minimized, for example, by using relatively long uniform tendons and measuring length changes, specifically, near the mid-region of the tendon with a lightweight length transducer that is attached directly to the tendon. Possibilities for gripping tendon are shown in Figure 2.6.8.

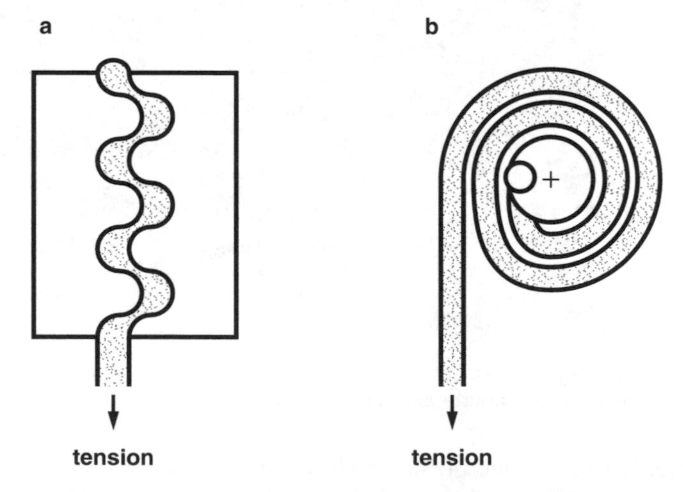

Figure 2.6.8 **Different grips for holding tendons. Serrated grips in (a) may have different shapes, but rely on friction between serration and specimen to help hold the specimen in place. Roller grips (b) are based on reducing the tension in the specimen as it is wrapped around the roller before the end is gripped.**

A typical stress-strain curve for tendon is shown in Figure 2.6.9. The curve illustrates the tensile response for a tendon specimen subjected to a single pull to failure. There are three distinct regions of the curve: toe, linear, and yield.

The toe region typically lies below 3% strain, a region in which specimen elongation is accompanied by very low stress. This low initial stiffness of tendon in the toe region is thought to be caused, in part, by the straightening of the collagen crimp. Rigby et al. (1959)

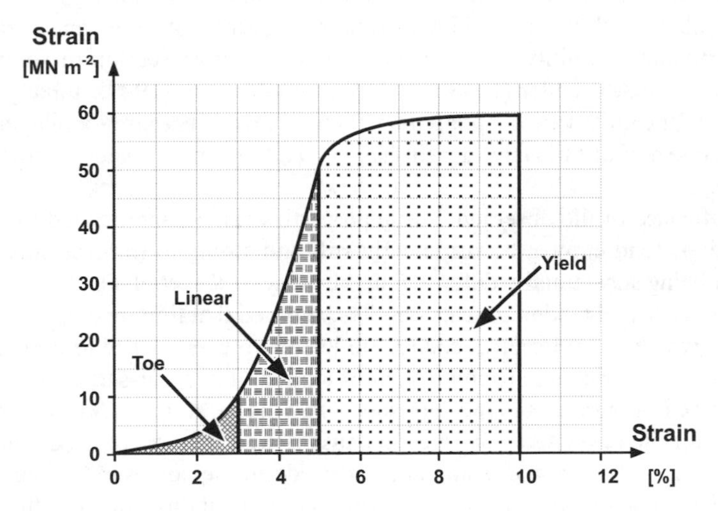

Figure 2.6.9 **Typical tendon stress-strain curve for a tensile test to rupture.**

studied rat tail tendon with polarized light and observed that the tendon lost its crimping pattern beyond strains of about 2 to 3%. However, Hooley and McCrum (1980) attributed the toe region to shearing action between the collagen fibrils and the ground substance of the tendon.

The linear region is evident beyond approximately 2 to 3% tensile strain. The slope of this linear portion of the curve has been used to define the Young's modulus of the tendon. This region of linear or *reversible* strain extends to about 4 to 5% (Wainwright et al., 1982). The Young's modulus for rat tail tendon is approximately 1.0 GPa (Rigby et al., 1959). The mean tangent moduli, derived from several aquatic and terrestrial mammalian species, ranges from 1.25 to 1.65 GPa with a mean across all species of 1.5 GPa (Bennett et al., 1986). Most of these measurements were obtained in dynamic tests at 1.0 to 2.2 Hz. Little variation in Young's modulus was found with frequencies in the range of 0.2 to 11 Hz.

Permanent deformation occurs beyond the region of linear or reversible strain. The ultimate or failure strain of tendon is about 8 to 10% (Rigby et al., 1959). There is a considerable yield region in which tendon deformation is accompanied by very little increase in stress. Tendon failure eventually results from collagen fibres pulling apart. The ultimate or failure stress of tendon is defined as tendon *strength*. The strength of any solid material depends on the presence of flaws. Since the number and severity of flaws likely varies between specimens, strength measurements for vertebrate tendons are also expected to be variable (Wainwright et al., 1982). Elliot (1965) reported tendon strengths ranging from 20 to 140 MPa. More recently, Bennett et al. (1986) measured the ultimate strengths of various mammalian digital and tail tendons using cryo-grips. Mean values ranged from 90 to 107 MPa, with a global mean across all species of about 100 MPa.

The high tensile stiffness and strength of tendon is attributed to its relatively high collagen content (70 to 80% dry weight; Elliot, 1965), and to its hierarchical organization into linear bundles. The majority of collagen fibres lie parallel to the long axis of the tendon, and the tropocollagen molecules in the heavily cross-linked regions of the fibres are crystallized, which together enhances the tensile properties of tendons. The multi-layered fibre-bundle organization of tendon allows for the maintenance of high tensile strength, with considerable flexibility in bending, in the same way that wire rope maintains high tensile strength and flexibility as compared to an equal cross-section of solid steel, e.g., Alexander (1988a); Alexander (1988b) and Wainwright et al. (1982). Finally, the helical interweaving of the units varies among the structural levels, e.g., tropocollagen has a left-hand helix, while microfibrils have a right-hand helix, to keep the material from unwinding when loaded.

The significance of the observed tensile properties can be appreciated by considering tendon function. Tendon must be sufficiently stiff and strong to transmit muscle force to bone without being substantially deformed in the process. Ker et al. (1988) studied the relative size of muscle and tendon dimensions. They argued that thin tendons require long-fibred muscles, which allow for great changes in length, to compensate for tendon deformation during muscle contraction. In contrast, thick tendons deform less than thin tendons, and may not need extra-long fibres in the corresponding muscle. Ker et al. (1988) calculated an optimal ratio for the cross-sectional area of muscles and their associated tendons, based upon the isometric properties of mammalian striated muscle, and minimizing muscle-tendon mass. They found that in many mammalian species, including humans, the muscle tendon units approximated their calculated optimal ratio (Ker et al., 1988; Cutts et al., 1991). Minimizing body mass is an important aspect of the functional design of living systems.

If a tendon specimen is cyclically stretched and allowed to recoil within the range of reversible strain, the resulting load-deformation measurements form a loop (Figure 2.6.10). This result means that a proportion of the energy expended during tensile deformation of the tendon is not recovered when the deforming force is removed. The lost energy, or hysteresis, is quantified by considering the area enclosed within the loop. It is usually expressed as a percentage of the deformation energy (the area below the elongation curve). The reciprocal quantity, the resilience, is often used to describe the fraction of energy returned following tendon recoil.

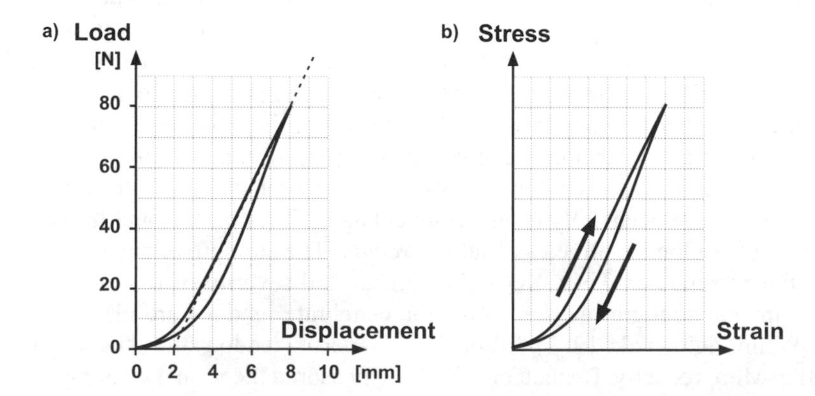

Figure 2.6.10 (a) Load-displacement curve for a wallaby tail tendon. The dashed line represents the tangent modulus. (b) Typical stress-strain curve for tendon showing loading, unloading, and hysteresis loop (see text for further explanation) (from Bennett et al., 1986, with permission).

Besides being relatively stiff and strong in tension, tendon is highly resilient. Ker (1981) reported that the hysteresis area of sheep plantaris tendon was about 6% of the area under the loading curve, and that this value was effectively independent of frequency in the range of 0.2 to 11 Hz. Bennett et al. (1986) showed that mean values of hysteresis of several vertebrate tendons ranged from 6 to 11%, and were virtually independent of the frequency of loading. More recently, Wang et al. (1995) demonstrated that hysteresis is virtually unchanged up to test frequencies of 70 Hz. Thus, about 89 to 94% of the energy associated with longitudinal deformation or stretch of a tendon is recovered when the load on the tendon is removed. For a biological material, tendon shows marked elastic behaviour, at least within the likely range of physiologically relevant frequencies of deformation.

Depending on the function of its associated muscle, a tendon may be subjected to prolonged static loads, such as those imposed by postural muscles or prolonged repetitive or cyclic loads, such as those measured during locomotion. A material that fails under prolonged constant stress is said to suffer from creep failure (or rupture). Fatigue failure (or rupture) is the term used to describe the failure of a material subjected to prolonged cyclic loading.

Wang and Ker (1995) investigated the creep properties of wallaby tail tendon by subjecting uniform sections of tendons to static stresses ranging from 10 to 80 MPa. The specimens were immersed in saline-saturated liquid paraffin to prevent dehydration during the prolonged experiments. They found that the tendons suffered from creep rupture at tensile stresses that were much lower than those that caused failure during a single dynamic pull (Figure 2.6.11). Between 20 and 80 MPa, the time to creep rupture, or *creep lifetime*, decreased exponentially with increasing stress. Tendon specimens subjected to 10 MPa, however, showed no signs of creep damage even after 15 days, at which time the experiments were abandoned. Based upon estimates of the physiological cross-sectional area of wallaby tail musculature and estimates of maximum skeletal muscle stresses, Wang and Ker (1995) concluded that the tail tendons would be subjected to tensile stresses of about 14 MPa, and that it was unlikely that tail tendons would suffer from creep damage in the living animal.

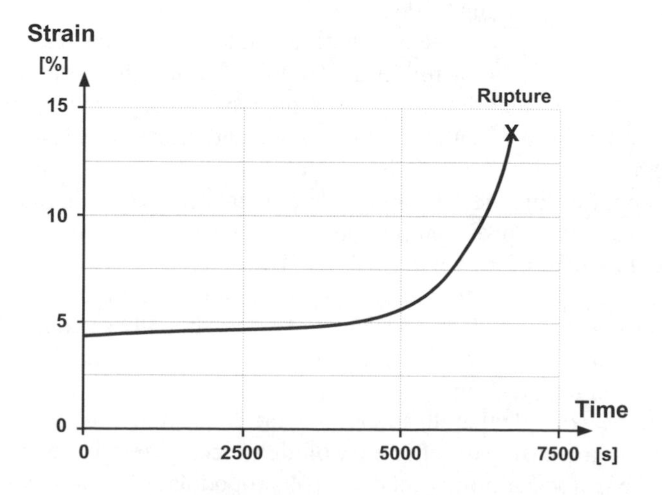

Figure 2.6.11 **Creep test. A load ramp was applied for the first 2 seconds. Thereafter, the stress was kept constant at 30 MPa until rupture. Wallaby tail tendon at 37°C. Specimen length was 150 mm (from Wang & Ker, 1995, with permission).**

In a companion study, Wang et al. (1995) subjected wallaby tail tendons to cyclic loading protocols to examine the potential for fatigue failure in these tendons. They found that the tail tendons suffered from fatigue rupture at stress levels that were much lower than those resulting in tendon rupture during a single dynamic pull. Additionally, dynamically-determined stiffness decreased gradually through each cycle of loading. They estimated that wallaby tail tendons would be subjected to about 14 MPa at a rate of about 3 Hz during moderately fast hopping (as the tail swings up and down in each stride), and calculated that the fatigue lifetime of the tail tendons would be about 56 hours. From these experiments, it seemed unlikely that tail tendons would fail due to fatigue damage. However, when Wang et al. (1995) tried to estimate the fatigue lifetime of the human Achilles tendon using their model and assumptions about typical stride frequencies and peak tensile stresses (about 1.4 Hz, as shown in Cavanagh & Kram, 1990), and 50 MPa (as shown in Ker et al., 1987), they concluded that fatigue failure should result after about one hour of continuous running. Clearly, Achilles tendons survive running times longer than one hour. Wang et al. (1995) suggested that despite the fact that mammalian tendons were virtually indistinguishable in terms of mechanical properties such as Young's modulus and resilience, highly stressed tendons, such as the Achilles tendon, may be more resistant to fatigue failure than tail tendons.

Because tendons are highly resilient structures, they are capable of storing and releasing significant amounts of elastic strain energy. This property of tendon is thought to be of considerable importance for the evolution of locomotor systems, particularly in high speed vertebrate locomotion such as that employed by ungulate mammals, i.e., those that walk on their toes, e.g., horses, deer, and camels. The proximal limb muscles of these animals are conspicuous for their large, multipennate, short-fibred structure, with long, cord-like tendons extending distally to the toes. When activated, these muscles are capable of generating large forces. However, because of their relatively short muscle fibre lengths, the active range of length changes in these locomotor muscles is limited. Therefore, part of the length changes in these muscle-tendon units is likely to occur within the tendons.

Alexander (1988a; 1988b) suggested that these types of locomotor muscles generate force to keep their corresponding tendons taut. Upon impact with the ground, these tendons are stretched and the energy associated with decelerating the mass of the animal during impact is stored as elastic strain energy in the tendons. During the propulsive phase of the stance phase, the tendons recoil, thereby, releasing part of the stored elastic strain energy. Alexander (1988b) proposed that the cyclic storage and release of elastic strain energy in tendon may be a mechanism for reducing the total amount of metabolic energy expended during locomotion. For this mechanism to work, metabolic energy is required to activate, and develop, tension in the proximal limb muscles. Short muscle fibres can exert as much force as an equal number of long muscle fibres. However, a muscle composed of short fibres is less massive, and, therefore, less costly to maintain and transport. Perhaps such a muscle is also less costly to activate. Alexander (1988b) hypothesized that high speed running in these animals has become more economical by virtue of having evolved muscles with short fibres and long tendons.

A more specific example that illustrates how the concept of elastic energy storage may be a mechanism for increasing the efficiency of high speed terrestrial locomotion is bipedal hopping of the kangaroos and their relatives (Macropodidae). Dawson and Taylor (1973) measured the oxygen consumption of kangaroos hopping on a treadmill. They showed that beyond a critical hopping speed, oxygen consumption leveled off, and in fact decreased slightly with increasing speed. Alexander and Vernon (1975) performed an analysis of the mechanics of hopping with a wallaby (Macropus rufogriseus), a small species of kangaroo, by taking cinefilm recordings of animals hopping across a force platform. They found that most of the energetic demand of hopping at high speeds was associated with the gravitational potential energy (of the centre of mass of the whole body) and the horizontal component of the external kinetic energy (of the centre of mass of the whole body). Each of these forms of mechanical energy decreased with each impact (braking) and increased with each subsequent take off (propulsive). Alexander and Vernon (1975) also found that the lengthening phase of the ankle extensors (gastrocnemius and plantaris) coincided with the impact phase, and the shortening phase coincided with the take off phase. By assuming that most of the changes in length of the ankle extensor muscle-tendon units occurred within the tendon, rather than the muscle, Alexander and Vernon (1975) suggested that elastic strain energy storage during impact, followed by the energy release from these tendons during take off, may help to save metabolic energy. Further, they suggested that Dawson and Taylor's discovery could be explained by the increased role of storage of elastic strain energy at higher speeds of locomotion.

The role of tendon elasticity in human locomotion seems less certain than that in animal locomotion. Cavagna et al. (1964) estimated the mechanical efficiency of running humans. They found that the efficiency of running was greater than that of contracting an isolated muscle in the absence of any prestretch. They argued that during impact and take off, muscle-tendon complexes were stretched and relaxed, thus storing and releasing elastic strain energy. However, they were unable to differentiate between the potential for muscle strain energy and tendon strain energy. Van Ingen Schenau (1984) calculated the potential for elastic strain energy storage during running in humans. His calculations were based, in part, on the tensile properties of tendons and measurements of the ground reaction forces during running. Van Ingen Schenau (1984) argued that, compared to the total energy required to run, the amount of energy that could be saved by means of tendon elastic energy storage was insignificant.

Voigt et al. (1995) considered stationary bipedal hopping in an attempt to estimate the potential role of tendon elastic energy storage in the efficiency of human locomotion. Oxygen consumption was measured during three different speeds of hopping. High speed film and force plate analyses were used to calculate net joint moments, and standardized anatomical data were employed to estimate the instantaneous muscle-tendon unit forces for the knee and ankle extensors. Corresponding tendon stresses and strain energies were calculated using a standardized tendon stress-strain function and morphological data. Absolute work rates of the knee and ankle extensors were compared to the absolute work rate associated with the whole body during the hopping movements. Voigt et al. (1995) calculated that the quadriceps femoris and triceps surae tendons performed 52 to 60% of the total absolute work associated with stationary bipedal hopping, and that these tendons made greater contributions to hopping efficiency at hopping speeds that exceeded those individually preferred by each participant. It must be emphasized that the 52 to 60% of total work performed by the tendons found by Voigt et al. (1995) is absolute work. In reality, of course, the negative work required to stretch a tendon always exceeds the positive work the tendon can return. Therefore, the total work performed by a tendon during a loading-unloading cycle is always negative. Thus, while the role of tendon elasticity appears well-established for at least some vertebrate species, the debate continues as to the relevance of this mechanism during human locomotion.

The question may arise as to which scenario is metabolically more efficient: a tendon that is stretched (while the muscle fibres remain at constant length) or muscle fibres that stretch (while the tendon remains at constant length). An isometrically contracting muscle fibre (no change in fibre length) consumes more metabolic energy than an eccentrically contracting muscle fibre (actively lengthening fibre), for the same output of force. However, an isometrically contracting muscle fibre consumes less metabolic energy than a concentrically contracting fibre (actively shortening fibre), for the same output of force. A systematic investigation of the relative merits of the two extremes of very compliant muscle and very stiff tendon, and very stiff muscle and very compliant tendon, and the intermediate combinations, therein, has yet to be conducted. Until such time, the role of elastic energy storage in the enhancement of locomotor efficiency remains an intriguing question.

2.6.3 PHYSIOLOGICAL PROPERTIES AND ADAPTIVE FUNCTION

The glistening, white appearance of tendon gives the impression that it is an avascular (without blood vessels) structure. Certainly, relative to intensely metabolic tissues, such as muscle and skin, tendon vascularity is minimal. However, despite its macroscopic appearance, tendon is a metabolically active structure.

Although the blood supply to tendon is variable in volume and organization, in general, tendon vascularity, like tendon morphology, involves three regions: tendon, myotendinous junction, and osteotendinous junction. If present, vessels that perfuse the tendon originate from the paratenon or synovial fluid. Small blood vessels in the paratenon run transversely toward the tendon, and branch several times before running parallel to the long axis of the tendon. Vessels enter the tendon along the endotenon. Capillaries loop from the arterioles to the venules, but do not penetrate the collagen bundles. At the myotendinous junction, vessels surrounding the muscle fibres continue around the junction into the endotenon. The vessels do not cross the actual junction. Similarly, the blood vessels do not cross the osteotendinous junction, but are seen coursing around it, continuous between the periosteum and the endotenon (O'Brien, 1992).

While the major role of the blood supply is to perfuse the tendon fibroblasts with nutrients, the cells may also obtain nourishment by diffusion directly from the synovial fluid (Lundborg & Rank, 1978; Lundborg et al., 1980; Manske & Lesker, 1982). Synovial diffusion appears to be particularly important in areas of the tendon that are enclosed within synovial sheaths. The effectiveness of nutrient uptake by diffusion was compared to that by perfusion in flexor tendon cells of chickens (Manske et al., 1978) and primates (Manske & Lesker, 1982). Tritiated proline, the traceable radio-isotope of proline (an amino acid necessary for the synthesis of collagen), was taken up by the cells more completely and quickly via diffusion than via perfusion, indicating that diffusion was an effective mechanism for the supply of nutrients to tendon cells.

Tendon cells (tenocytes are the inactive form, and fibroblasts are the active form) are responsible for maintaining the metabolic balance of tendon. Tendon cell populations differ, depending upon location. For example, fibroblasts isolated from the surface of tendon, and subsequently cultured, adhered less well to the culture substrate than fibroblasts isolated from the tendon interior. Additionally, surface fibroblasts synthesized less glycosaminoglycans (ground substance molecules), and a limited amount (10%) of a mixture of collagen. The interior fibroblasts, in contrast, synthesized great amounts of glycosaminoglycans

(30%) and type I collagen (the main tendon-fibre collagen) (Reiderer-Henderson et al., 1983).

The biosynthesis of collagen is, arguably, the primary role of the tendon cells. In normal, healthy tendon, the rate of metabolism of collagen is slow. The fibroblasts produce the ground substance, into which they synthesize and deposit the tropocollagen molecules. At this early stage, there are no cross-links between the tropocollagen molecules. Tropocollagen cross-linking takes place around 4 to 16 days following the initial deposition of the molecules. During the next 12 weeks or so, the collagenous tissue becomes organized. Collagen fibres are formed, at first assuming a random orientation. Gradually, the randomly-oriented fibres come to lie parallel to the applied tensile loads (O'Brien, 1992). This reorientation enhances further cross-linking between the collagen fibres thus increasing the tensile stiffness and strength of the newly formed tissue.

The ground substance is not simply a *filler*, but has many important functions within the tendon. It is composed primarily of water, proteoglycan and glycoprotein macromolecules, and inorganic salts. A proteoglycan is made up of a protein core, with many large glycosaminoglycan (amino-sugar) side-chains. The long sidechains of these molecules can interdigitate with one another and form a web-like network. Glycoproteins are also protein-carbohydrate macromolecules, but with a predominant protein portion to which many small sugar molecules are attached. The sugar groups allow the macromolecules to bind large amounts of water by hydrogen-bonding. The combined macromolecular potential of binding large numbers of water molecules and forming extensive networks is responsible for the hydrated, gel-like consistency of the ground substance.

The ground substance provides the necessary structural support for the collagen fibres, and also serves as a lubricating medium, enabling fibres to slide with respect to one another during tendon movement. Additionally, it is thought that the substantial water content of the ground substance allows for considerable dissipation of energy in the form of heat during tendon function. This would seem to be particularly important in a tissue with relatively low vascularity, and, therefore, relatively low capacity for heat exchange by blood perfusion. Finally, the ground substance functions as a medium for the diffusion of nutrients and gases, particularly for cells located well away from blood vessels or synovial fluid (O'Brien, 1992).

Some of the ground substance macromolecules have been implicated in more specific roles. For example, fibronectin, a large glycoprotein, plays an important role in cell-cell adhesion, fibroblast to collagen attachment, and cell migration. It may also play an indirect role in the control of ground substance production (Viidik, 1990). The proteoglycans, e.g., decorin and biglycan, seem to be associated with the regulation of fibril formation, as the concentration of these proteoglycan molecules appears to be inversely proportional to the ultimate size of the tropocollagen (O'Brien, 1992).

Tendons are capable of detecting and responding to mechanical and biochemical changes within their physiological environment. If the rates at which these changes occur are not too great, and the absolute changes are not too extreme, then the tendon can adapt to the new conditions. However, if the mechanical or biochemical changes are rapid, severe, or prolonged in nature, the tendon may not be capable of repair. A considerable number of studies have been aimed at trying to determine the mechanical and biochemical conditions associated with successful tendon remodelling (or adaptive repair) and tendon injury.

Flint (1982) found that if the Achilles tendon of a rabbit was completely severed, the glycosaminoglycan content, the fibroblast number, and the number of small collagen fibres, increased. The area of the tendon that was closest to the muscle belly showed the greatest changes, suggesting that tendon adaptation may be region-specific. If the severed tendon was sutured so that some tensile load could be transmitted, only minimal changes in glycosaminoglycan content and fibroblast and collagen fibre number, were observed. Thus, the removal of tension appeared to trigger tendon changes that were associated with collagen synthesis. In a study by Klein et al. (1977), the tensile forces normally associated with active muscle contraction were totally removed in the ankle extensor muscles of the rat by denervation (the surgical elimination of neural stimulation to the muscle). They observed dramatic increases in collagen turnover in the rat Achilles tendon, but did not observe changes in the relative amount of the various tendon components. It appeared that the tensile loads generated by the muscle had an important influence on collagen turnover.

Another way to change the mechanical environment of tendon is to increase the applied tensile load. Results from such studies are often difficult to compare because of the variable exercise regimes that were used to increase tendon load. For example, long-term running has been reported to increase the cross-sectional area and collagen content of swine extensor tendons. Similar changes were not observed in the flexor tendons of exercised animals (Woo et al., 1980; Woo et al., 1981). The differences in response of the extensor and flexor tendons to exercise were associated with different biochemical constituents and collagen concentrations, and different loading of flexor and extensor tendons during locomotion. Kiiskinen (1977) found that treadmill running appeared to cause an increase in the dry weight of the Achilles tendon in young mice. In senescent mice, however, treadmill running or voluntary exercise appeared to slow the age-related loss of glycosaminoglycans (Viidik, 1979; Vailas et al., 1985). In growing chicks, Curwin et al. (1988) observed that strenuous exercise slowed the maturation of collagen by increasing the turnover of ground substance and collagen. These results suggest that the response of a tendon to mechanical overload is tendon- and age-specific. Growth-related stimuli may conflict with exercise-related stimuli when growing animals are exercised. Furthermore, the stimuli responsible for tendon adaptation may change during maturation.

Over most of its length, tendon is loaded in tension. However, there are normal situations in which a tendon may be subjected to compressive loads. This occurs, for example, when a tendon passes around a bony prominence or turns through a digital sheath. Within this area, a specialized fibrocartilage *button* develops. The fibrocartilage button is like fibrocartilage that develops pathologically when a tissue that usually withstands tensile loads is experimentally subjected to compressive loads (Scapinelli & Little, 1970). Merrilees and Flint (1980) examined the ultrastructure of a rabbit flexor tendon. The rabbit flexor tendon is oriented such that tensile loads are sustained by the tissue along its superficial surface, while compressive loads are sustained by the tissue along its deep surface. The collagen in the tensile surface was oriented longitudinally, with typically elongated fibroblasts distributed among the collagen fibres. In contrast, in the deep zone of the tendon that was subjected to compressive loads, the collagen assumed a basket-weave orientation, and the associated cells resembled cartilage rather than tendon cells. The authors concluded that the observed differences in the morphology of the tendon between deep and superficial zones were caused by adaptive responses to the compressive and tensile loads.

Gillard et al. (1979) used the same rabbit tendon model as Merrilees and Flint (1980), however, they altered the loading pattern of the tendon by surgically relocating it, thereby, eliminating the compressive component of the load. Glycosaminoglycan content of the normally compressed region decreased by 60% within eight days of the removal of the compressive component. When the tendon was again subjected to the normal compressive component, the glycosaminoglycan content returned towards its previous level, albeit much more slowly than the rate at which the reduction occurred. These results support the notion that tendon is sensitive to the local loading environment, and that it is capable of responding to those changes in load by initiating changes in biochemical makeup and collagen orientation. In effect, tendon tissue becomes more cartilaginous in areas where it is subjected to compressive loads.

Tendon is capable of considerable self-healing, depending upon the type of injury. The most common tendon injuries are lacerations, ruptures, and tendinitis. The basic process of tendon healing may be considered to occur in three phases: inflammation with cellular infiltration, proliferation of new ground substance and collagen, and remodelling of the newly formed tissue. Lacerations and ruptures are considered to be acute traumatic injuries. Maintaining the blood supply to the healing tendon is critical to prevent necrosis of the collagen. However, maintaining blood supply may become problematic in areas of the tendon where vascularity is compromised because of friction, compression or torsion, e.g., in the Achilles tendon and the supraspinatus tendon. Another aspect of tendon healing that appears to affect the long-term result, particularly with injuries to the tendons in the hands, is the tendency for adhesions to form between the repair site and the tendon sheath. These adhesions may prevent the tendon from gliding easily within the sheath during joint motion. It has been suggested that maintaining contact between ruptured tendon ends (by prompt suturing), the early application of protected passive motion of the affected joint and hence, the injured tendon, and preventing adhesion formation between the injured tendon and its protective sheath, are the factors, which when combined, provide the best results for healing of tendon laceration and rupture (Gelberman et al., 1988).

Tendinitis (tendon inflammation), a common tendon injury most notably associated with pain, is often described as an overuse syndrome. Treatment protocols consist of pain management with the use of analgesics and anti-inflammatory agents to decrease the inflammation around the tendon. Physical agents, such as ice or ultrasound treatments, seem to be beneficial in the early stages of inflammation. It is suspected that the continuous cyclic loading of tendons causes microdamage, i.e., mechanical damage of the collagen fibres at the microscopic level. If the time frame between successive bouts of repetitive activity is sufficiently long, then the normal healing response can proceed and the fibroblasts can repair the microdamage. If the time frame is insufficient, the normal healing response will be initiated, but will not proceed beyond the inflammatory stage (Curwin & Stanish, 1984). While it is desirable to curtail activity that causes inflammation, it may be difficult to do so, particularly if the painful symptoms are work-related.

In an overload model of the rat plantaris tendon, Zamora and Marini (1988) found that there was a notable transformation of quiescent fibrocytes into active fibroblasts, without the presence of inflammatory cells. Collagen bundles in the tendon were disrupted, and empty longitudinal spaces were observed. Prompted by the observed ultrastructural changes, the authors suggested that the process of tendon remodelling may involve an initial transient period of mechanical weakness, and that the tendon breaks down structurally before remodelling. Unfortunately, the magnitudes of the overload, and the mechanisms by which the overloads contributed to the observed tendon changes, remain unknown (Archambault et al., 1995).

There seems to be a fine line between changes in the mechanical environment of a tendon (for example, magnitudes of the applied tensile loads and their respective time-histories) that initiate the normal healing response and result in successful completion of an adaptive and remodelling process, and changes in the mechanical environment of a tendon that initiate the normal healing response, but instead terminate in chronic tendon inflammation. Backman et al. (1990) presented the first potentially quantifiable experimental model of overuse tendon injury. The goal of their experiment was to standardize the conditions relating to the development of chronic Achilles paratenonitis (paratenon inflammation) with tendinosis (degeneration of the tendon). Anaesthetized rabbits were exercised using a kicking machine for two hours per day, three days per week, for five to six weeks. Four weeks into the exercise protocol, the rabbits had noticeable nodules and thickening about 0.5 to 1.0 cm above the calcaneus. At the end of the exercise program, the paratenon of the exercised legs was visibly thickened and there was an increase in the number of blood vessels, edema, and infiltration of inflammatory cells. Most of the degenerative changes were seen centrally in the tendons (Figure 2.6.12).

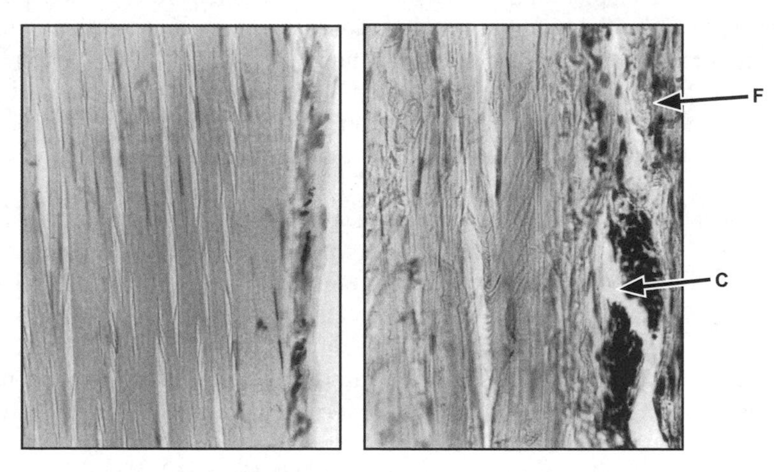

Figure 2.6.12 **Left: normal tendon and paratenon in control leg. Right: micrograph of exercised leg demonstrating fibrillation in tendon, dilated capillary (C), and increased number of fibroblasts (F) (from Backman et al., 1990, with permission).**

Archambault et al. (1997) developed a model for tendon overuse injury that built on the original model of Backman et al. (1990). Rabbits were exercised in kicking machines, and the tensile loads applied to Achilles tendons were recorded continuously (Figure 2.6.13). This model offers the possibility to correlate changes in the applied mechanical environment of the tendon with changes in the observed biochemical and physiological state of the tendon.

In the future, it may be possible to isolate the mechanical or temporal threshold, or both, of the tendon healing response, in which load-time histories below a threshold result in remodelling, and load-time histories above a threshold result in chronic inflammation.

2.6.4 TENDON-MUSCLE-APONEUROSIS INTERACTIONS

In section 2.6.2, we suggested that tendons play potentially important roles in modulating the length changes in fibres and fascicles, and in the storage and release of

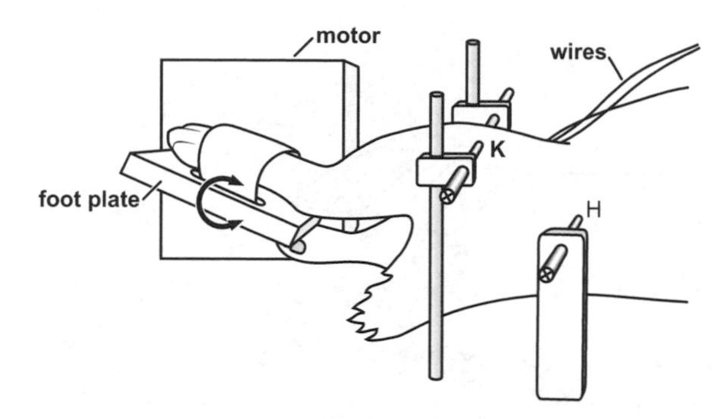

Figure 2.6.13 **Kicking machine (from Archambault et al., 1997, with permission).**

mechanical energy in cyclic movements. However, the examples only provided indirect evidence of the energy savings associated with the storage and release of energy, and did not directly tackle questions of how fibre length changes relate to the corresponding changes of the muscle tendon complex. Furthermore, the role of the aponeurosis as a potential site for energy storage was not considered.

When approaching the problem of energy savings through elastic elements in muscle, an important question that needs to be resolved is: are tendons and aponeuroses arranged in series, in parallel, or somewhere in between these two idealized states relative to the contractile muscle fibres? An argument frequently encountered in the biomechanics literature runs as follows: since the tendon and aponeurosis are structurally in series, they are also mechanically in series, and therefore sustain the same force, or at least, the forces they sustain are constantly in the same proportion. This reasoning has been used to measure the forces of a muscle at the distal end of the tendon, while recording elongations at some part of the aponeurosis, and then assuming that the two are related by a constitutive equation governing the (nearly) elastic behaviour of the aponeurosis, e.g., Magnusson et al. (2001) and Muramatsu et al. (2001). A good example of such reasoning, albeit made in a tacit fashion, has been put forward by Roberts et al. (1997). Similarly, van Ingen Schenau et al. (1997)) defined that the series elasticity of muscles is obtained simply by subtracting fibre length from the entire muscle tendon unit length, thereby explicitly including the tendon and aponeurosis as part of the series elasticity. Although there is little doubt that the tendon is arranged mechanically in series with the rest of the muscle, its arrangement relative to the aponeurosis is less obvious and cannot be easily discerned by casual analysis.

A frequently adopted model of series elasticity is shown in Figure 2.6.14, which represents a uni-pennate muscle with its associated contractile elements, tendon, and aponeuroses. Here, series elasticity (se) is depicted to reside in the fibres (f), the aponeuroses (a), and tendon (t). Although, there is little doubt that both tendon and aponeuroses might affect length changes of the fibres, for questions related to energy storage and release, as well as efficiency, the role of the aponeurosis in terms of force transmission needs clarification.

There are observations indicating that aponeuroses are not mechanically arranged in series with the muscle contractile elements or the tendon, or both. For example, Zuurbier et al. (1994) found that aponeurosis segment lengths decreased when a muscle was activated (and force increased), while Lieber et al. (2000) found that aponeurosis length depended on the

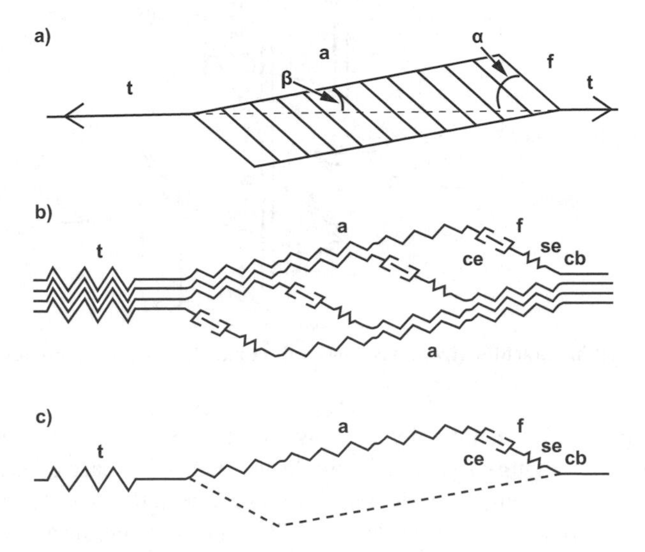

Figure 2.6.14 **Schematic illustration of a uni-pennate muscle with fibres (f), tendons (t), and aponeuroses (a). The angle of pennation between the line of action of the muscle and fibres (α) and the angle between the aponeurosis and the line of action (ß) are also shown. (b) Schematic illustration of the perceived series elastic elements (zig-zag lines) in the fibres, tendon and aponeuroses. (c) The corresponding series elasticity associated with a single fibre including the aponeurosis and tendon (from Ettema and Huijing, 1990, with permission).**

state of activation. Specifically, for a given amount of force acting on the distal tendon of a muscle, aponeurosis length was smaller when force was produced actively by the muscle compared to when force was passively applied to the muscle.

To clarify the role of aponeuroses in force transmission from fibres to the tendon, and to determine if aponeuroses are arranged mechanically in series with the tendon or muscle fibres, or both, we isolated the cat medial gastrocnemius muscle (MG) at its distal end with a remnant piece of the calcaneus bone still attached to the tendon. Otherwise, the muscle was left in its in situ configuration, with blood and nerve supply intact. The distal end of the muscle was attached to a muscle puller for computer controlled length changes of MG and for force measurement. The muscle was activated through tibial nerve stimulation (Herzog & Leonard, 1997) and tendon, aponeuroses and fibre length were measured using sonomicrometry (1997), with a crystal arrangement as shown in Figure 2.6.15.

When pacing the muscle through various work loop conditions (stretch-shortening cycles at various frequencies and magnitudes), it could be observed that some aponeurosis segments shortened when muscle force increased, and elongated when force decreased (Figure 2.6.16a - dashed vertical lines). This result indicates that aponeurosis length changes were not directly related to muscle force, and that force transmission from the muscle fibres to the aponeurosis was much more complex than could be explained with an idealized in series arrangement of fibres with the aponeurosis. When plotting the corresponding muscle force as a function of the aponeurosis segment length, the resulting loop ran in the counter-clockwise direction (Figure 2.6.16b). If this result was interpreted (incorrectly) as

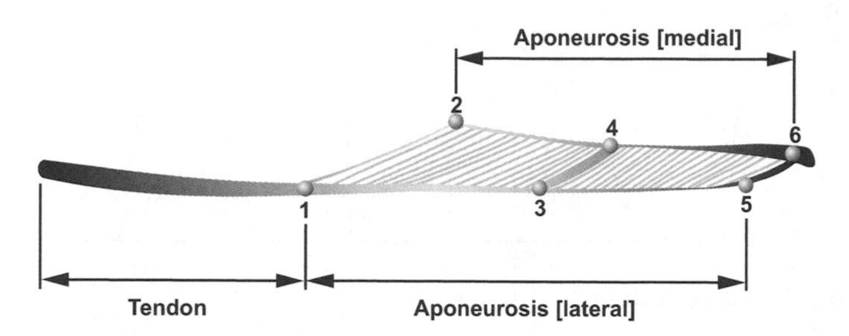

Figure 2.6.15 **Schematic illustration of the cat medial gastrocnemius muscle with the placement of six sonomicrometry crystals for measurement of fibre and aponeuroses segment lengths.**

a work loop of the aponeurosis, it would indicate that the aponeurosis produces net work during a full cycle, which is impossible for a (visco-) elastic tissue.

Roberts et al. (1997) studied the work contribution of the tendon and aponeuroses to the total work in the lateral gastrocnemius of running turkeys. They measured the total muscle length and fibre length and assumed implicitly that the difference between muscle and fibre length gave the length of an in series elastic element, as had been proposed by others (van Ingen Schenau et al., 1997). They found that the aponeurosis contributed as much as 60% of the total work in the shortening phase of the muscle, while the contractile contribution was about 40%. Their results cannot be properly evaluated, as data for full step cycles were not given. However, when applying identical methods to elucidate the role of the aponeurosis and tendon in the cat MG during free locomotion, we found that muscle force was primarily associated with the stance phase of gait, while the swing phase occurred essentially passively. However, muscle length changes during the stance phase of galloping were in the order of 10mm, while MG fibres remained at a nearly constant length (Figure 2.6.17a). If we now associate tendon/aponeurosis length changes with MG force by subtracting fibre lengths from muscle tendon unit length, as suggested by others (van Ingen Schenau et al., 1997; Roberts et al., 1997), while carefully accounting for any changes in the angle of pennation, the net mechanical work associated for a full step cycle (from zero force back to zero force) would be positive. In the example shown (Figure 2.6.17b), the net mechanical work associated for a full step cycle would account for more than 80% of the total work produced by MG, illustrating again that muscle force cannot be assumed to be transmitted along the MG aponeurosis in an in-series mechanical arrangement. This result was confirmed when the directly measured length of the distal segmet of the medial aponeurosis (crystals 2-4; Figure 2.6.15) was plotted as a function of the directly measured MG force. The resulting loops for all step cycles were counter-clockwise (Figure 2.6.18).

The conclusions drawn from these experiments are as simple as they are important. Aponeuroses of muscles must not be assumed to be arranged mechanically in series with either the muscle fibres or the tendon. Any such assumption needs careful justification in each case, and published literature on aponeuroses stiffness (or other mechanical properties) and the contributions of tendons/aponeuroses to the storage and release of elastic energy need careful evaluation, specifically if aponeurosis length changes (and, therefore, storage and release of energy in aponeuroses) are directly related to muscle forces.

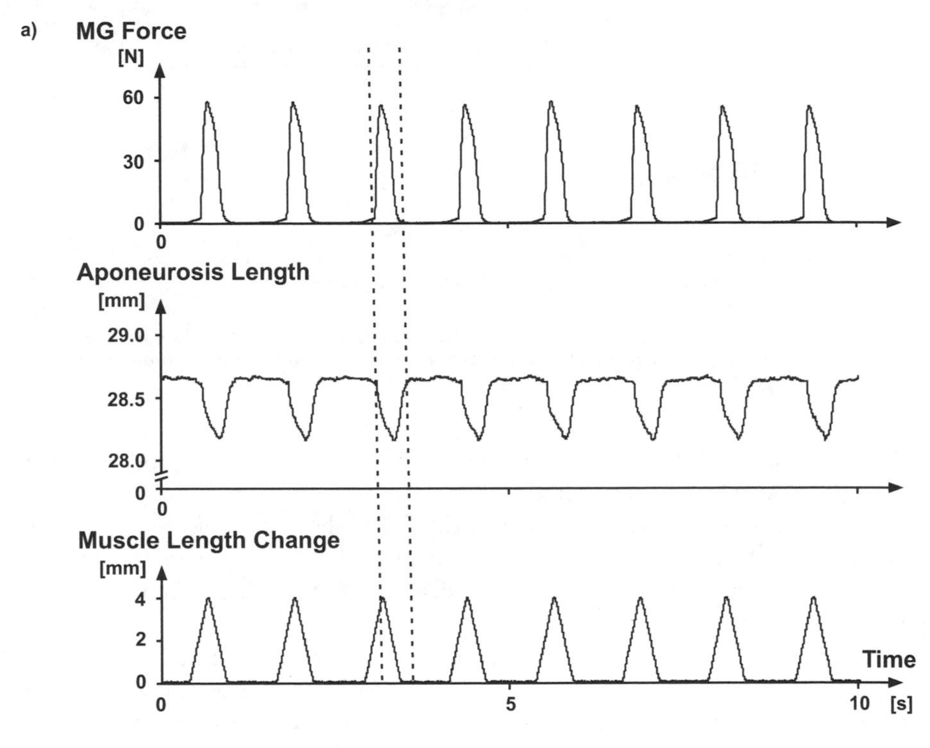

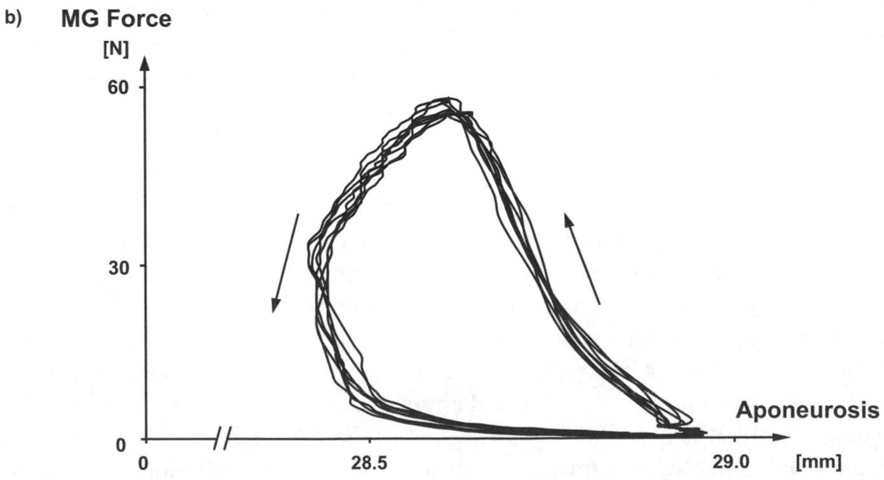

Figure 2.6.16 (a) Force-time, aponeurosis segment length-time, and muscle length change-time for work-loop experiments with cat medial gastrocnemius. Note that upon force production, the aponeurosis segment length decreases. (b) Medial gastrocnemius vs. aponeurosis segment length plots for the work loop experiments shown in (a). Note that the force-length plots run counter-clockwise, which would mean that the aponeurosis segment produces net mechanical work (which is impossible!), if it was assumed to be arranged mechanically in series with the tendon.

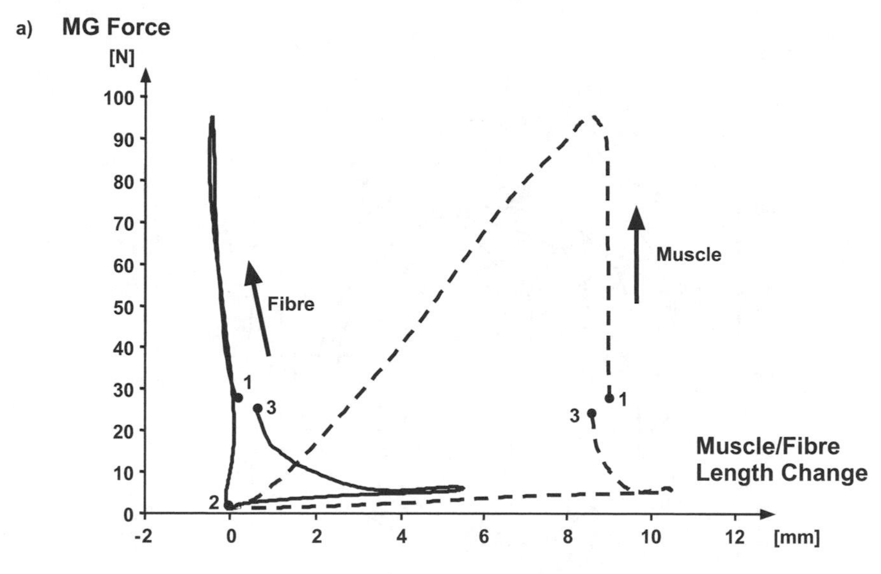

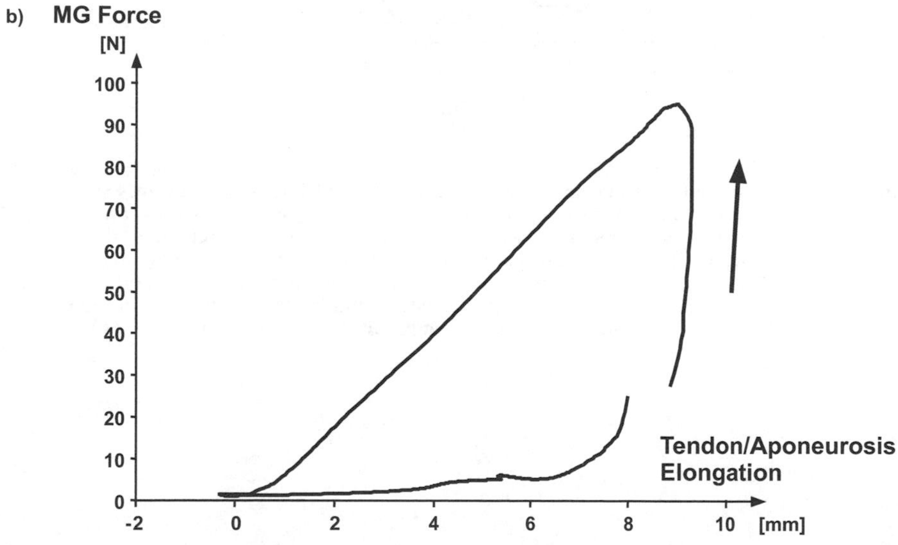

Figure 2.6.17 **(a) Medial gastrocnemius force as a function of muscle (dashed line) and fibre (solid line) length change for a single step cycle of a cat galloping at 4 m/s on a 10° uphill slope. Note that during the stance phase of the step cycle (from label 1 to 2), the muscle shortens by almost 10mm, while fibre length remains virtually constant. (b) If we now plot MG force as a function of tendon/aponeurosis elongation in the way proposed by van Ingen Schenau et al. (1997) and Roberts et al. (1997), we obtain a work-loop that runs counter-clockwise. Again, we have to realize that the tendon (where MG force is measured) is not mechanically arranged in series with the aponeurosis, otherwise, we would (erroneously) associate a great amount of net mechanical work with the aponeurosis segment.**

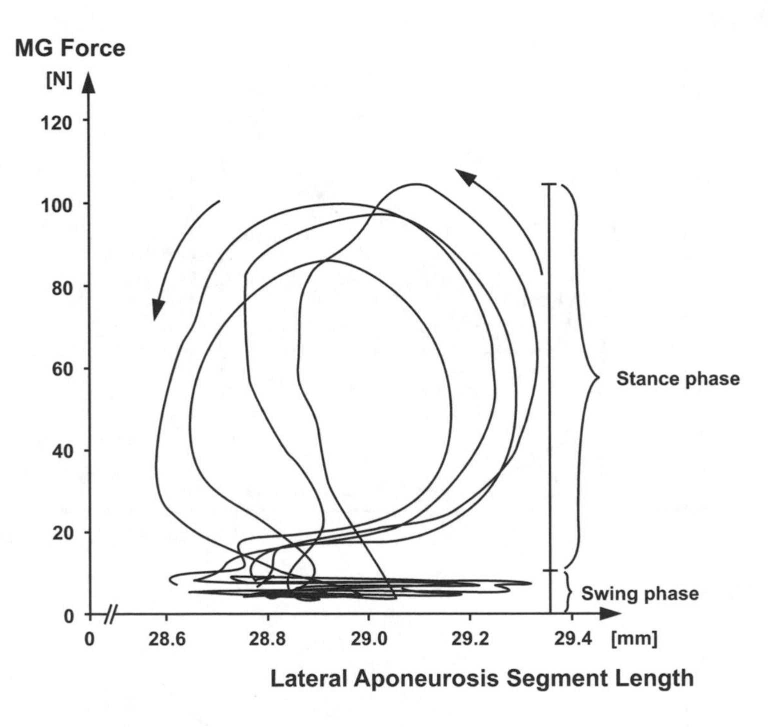

Figure 2.6.18 Medial gastrocnemius force as a function of lateral aponeurosis segment length changes for four step cycles during cat running at 4 m/s on a 10° uphill slope. Note again, the counter-clockwise loops that directly indicate that the aponeurosis segment cannot be mechanically arranged in series with the tendon. Note further that during the swing phase aponeurosis segment length changes are like those during the stance phase, despite much smaller forces, demonstrating that muscle force and aponeurosis segment length are not related in an in series fashion.

2.7 MUSCLE

HERZOG, W.

2.7.1 INTRODUCTION

Probably the most basic property of muscle is its ability to produce force. However, despite centuries of research on muscle and its contractile behaviour, some aspects of muscular force production have still not been resolved. For example, the precise mechanism of cross-bridge attachment and cross-bridge movement that are believed to cause relative movements of the myofilaments, and so produce force, are not clearly understood. Some scientists, e.g., Iwazumi (1978) and Pollack (1990), propose mechanisms of force production that do not agree with the most popular paradigm of muscular force production, the Cross-bridge theory (Huxley, 1957; Huxley & Simmons, 1971; Huxley, 1969; Rayment et al., 1993a). In this chapter, muscular force production and the mechanical properties of muscles are associated with the assumptions and predictions underlying the Cross-bridge theory. However, be aware, at all times, that not all of these assumptions and predictions have been tested and accepted unanimously in the scientific community.

Since this book is aimed at the student of biomechanics, muscles are viewed from a mechanical point of view. However, because of its contractile properties, muscle is less easily associated with strictly mechanical properties than are bones or ligaments, which have well-defined force-elongation or stress-strain relations. When dealing with the mechanics of muscle, muscle's physiological and biochemical properties must always be kept in mind.

Muscles exert force and produce movement and therefore may be considered the basic elements of movement mechanics in humans and animals. At the same time, force-time histories of muscles during movements are like small windows to the brain that may produce insight into the mechanisms of movement control. It is precisely the duality of mechanics and control function that splits biomechanists into two groups:

- Those who study the force output of muscles to determine the movement and loading effects on skeletal systems, in particular on joints, and
- Those who determine the force output of muscles for varying movement conditions to study the possible mechanisms that may be responsible for movement control.

This chapter focuses on muscle morphology and structure. In combination with the Cross-bridge theory, muscle morphology and structure determines, to a large extent, the mechanical properties of muscles.

2.7.2 MORPHOLOGY

Morphology is the science of structure and form without regarding function. This chapter demonstrates that muscle is a highly structured and organized material—every structure and each organization may be associated with specific functional properties of muscle.

Generally, muscles are grouped into striated and non-striated muscles. Striated muscles are further subdivided into skeletal and cardiac muscles. Cardiac and non-striated muscles are controlled by the autonomic nervous system, and in contrast to skeletal muscles, are not under direct voluntary control. In biomechanics, we are mostly (but by no means exclusively) concerned with skeletal muscles, and so, we will concentrate on skeletal muscles in this chapter. Although some of the force- generating mechanisms are similar for the different muscle types, there are also significant differences. Therefore, when studying cardiac or non-striated muscle, consult additional references that deal specifically with these types of muscles.

Skeletal muscle may be thought of in structural units of decreasing size. The entire muscle is typically surrounded by a fascia and a further connective tissue sheath known as the epimysium, which consists of irregularly distributed collagenous, reticular, and elastic fibres, connective tissue cells, and fat cells (Figure 2.7.1). The next smaller structure is the muscle bundle (or fascicle), which consists of several muscle fibres surrounded by a connective tissue sheath called perimysium. The muscle fibre, which is smaller than the muscle bundle, is an individual muscle cell surrounded by the endomysium, a thin sheath of connective tissue that consists principally of reticular fibres. The endomysium binds individual fibres together within a fascicle. Muscle fibres are cells within a delicate membrane, the sarcolemma.

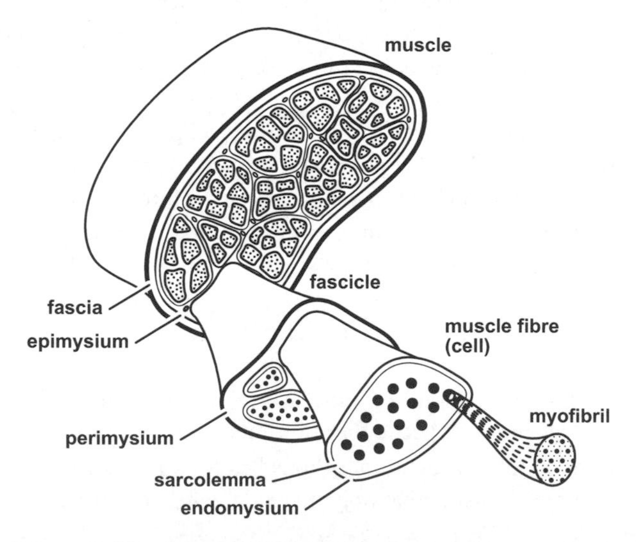

Figure 2.7.1 **Schematic illustration of different structures and sub-structures of a muscle.**

Muscle fibres consist of myofibrils lying parallel to one another. The systematic arrangement of the myofibrils gives skeletal muscle its typical striated pattern, which is visible under the light microscope (Figure 2.7.2). The repeat unit in this pattern is a sarcomere, which is the basic contractile unit of a muscle. Sarcomeres are bordered by the so-called Z-lines (named from the German Zwischenscheibe) and contain thin (actin) and thick (myosin) filaments. These filaments, which are mostly made up of the protein molecules that give them their names, lie parallel to one another (Figure 2.7.2).

Thick filaments are typically located in the centre of the sarcomere. However, evidence suggests that thick filaments may move from the centre towards the side of a sarcomere

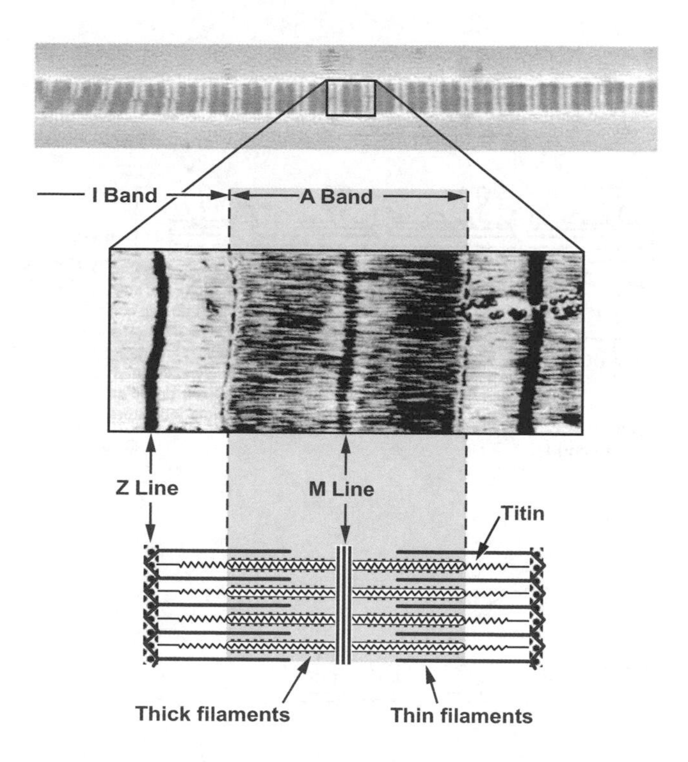

ʼigure 2.7.2 **Light microscope picture of sarcomeres within a myofibril, and the corresponding schematic illustration of a sarcomere with z-lines, contractile proteins (actin and myosin), and the structural protein titin.**

during prolonged contraction (Horowits, 1992). Thick filaments are responsible for the dark areas of the striation pattern, the so-called A bands (A = anisotropic) (Figure 2.7.2), and they are primarily composed of myosin molecules. A myosin molecule contains a long tail portion, which is composed of light meromyosin, and a globular head attached to the tail, which is composed of heavy meromyosin (Figure 2.7.3). The head portion extends outward from the thick filament. It contains a binding site for actin and an enzymatic site that catalyses the hydrolysis of ATP, which releases energy needed for muscular contraction. The myosin molecules in each half of the thick filament are arranged so their tail ends are directed towards the centre of the filament. Therefore, the head portions are oriented in opposite directions for the two halves of the filament, and when forming cross-bridges, i.e., when the myosin heads attach to the thin filament, the myosin heads are thought to pull the actin filaments towards the centre of the sarcomere.

The cross-bridges on the thick filament are offset by 14.3 nm in the axial direction and by 60° in the radial direction (Figure 2.7.4). Since cross-bridges are thought to come in pairs offset by 180°, two cross-bridges with identical orientation are 42.9 nm apart (3 x 14.3 nm).

The thin filaments are located on either side of the Z-lines within the sarcomeres (Figure 2.7.2). Thin filaments make up the light pattern of skeletal muscle striation, the so-called I-band (I = isotropic). The backbone of thin filaments is composed of two chains of serially-linked actin globules (Figure 2.7.5). The diameter of each actin globule is about 5-

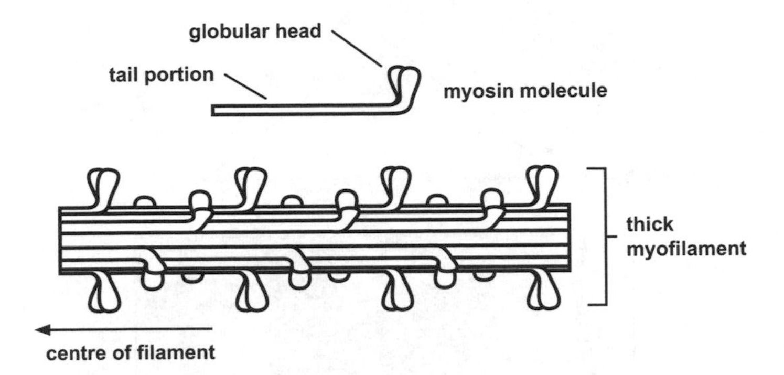

Figure 2.7.3 **Schematic illustration of the thick myofilament (from Seeley et al., 1989, with permission).**

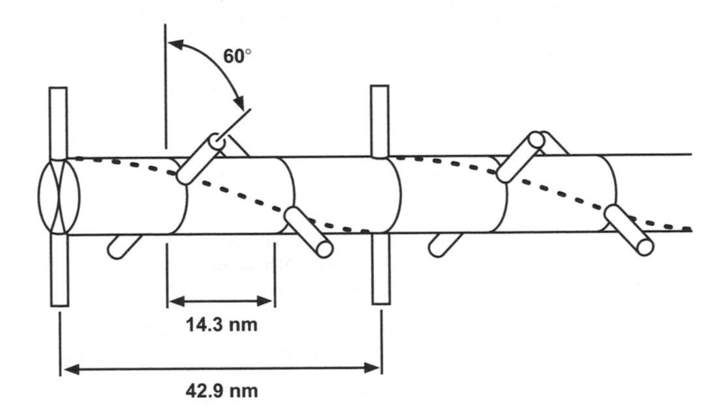

Figure 2.7.4 **Schematic illustration of the arrangement of the cross-bridges on the thick filament (adapted from Pollack, 1990, with permission).**

6 nm. The strands of serially-linked actin globules cross over one another every five to eight units in a somewhat random pattern. Thin filaments also contain tropomyosin and troponin. Tropomyosin is a long, fibrous protein that is believed to lie in the groove formed by the actin chains (Figure 2.7.5). Troponin is located at intervals of approximately 35-38.5 nm along the thin filament. Troponin is composed of three subunits: troponin C, which contains sites for calcium ion (Ca^{2+}) binding; troponin T, which contacts tropomyosin; and troponin I, which physically blocks the cross-bridge attachment site in the resting state, i.e., in the absence of Ca^{2+}. Actin and myosin are typically referred to as the contractile proteins, while tropomyosin and troponin are thought of as the regulatory proteins, because of their role in regulating cross-bridge attachment and force production.

Thick filament lengths appear to be remarkably constant among vertebrates (≈1.6 μm). However, thin filament lengths have been reported to vary from 0.925 μm in frogs to 1.27 μm in humans (Walker & Schrodt, 1973). These lengths correspond to 24 and 33

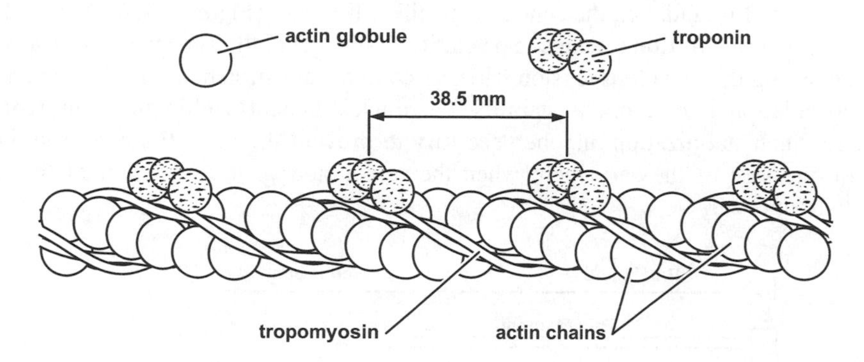

Figure 2.7.5 **Schematic illustration of the thin myofilament, composed of two chains of serially-linked actin globules (adapted from Seeley et al., 1989, with permission).**

periodicities of actin-binding proteins (38.5 nm intervals) on the thin filament. The influence of these differences in thin filament length on the force-length properties are discussed later in this chapter.

We have seen that skeletal muscle has an intriguing and systematic structural organization along the longitudinal axis of the muscle. This systematic organization is also observed in the cross-striation of the fibres and myofibrils, the regular arrangement of the cross-bridges on the thick filaments, and the periodicity of troponin on the thin filaments. The same regularity of structure may be observed in a cross-sectional plane (Figure 2.7.6). Each thick filament within a myofibril is surrounded by six thin filaments in the area of overlap. The cross-section of thick filaments is approximately 12 nm, and the cross-section of thin filaments is approximately 6 nm. The distance from thick to thick filament measures about 42 nm (Iwazumi, 1979).

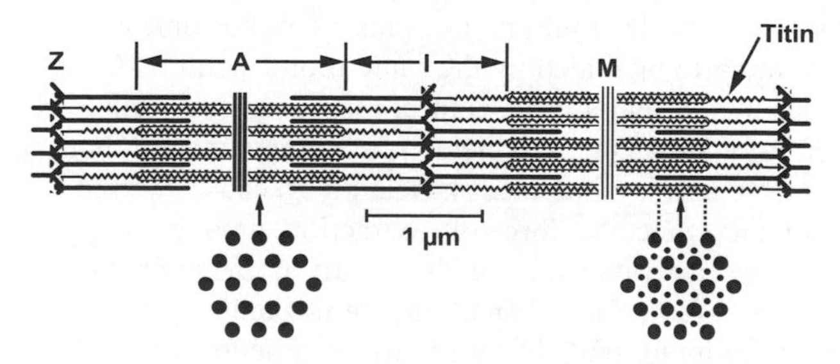

Figure 2.7.6 **Schematic illustration of thick and thin filament arrangement in a cross-sectional view in an area of thick filaments only and an area of thick-thin myofilament overlap.**

Aside from the contractile and regulatory proteins, skeletal muscle sarcomeres contain a variety of other proteins that are associated with structural and passive functional properties of the sarcomere, rather than active force production. The most important of these proteins, from a functional point of view, is *titin*.

Titin is a huge (mass ≈ 3 MDa) protein that is found in abundance in myofibrils of vertebrate (and some invertebrate) striated muscle. Within the sarcomere, titin spans from the Z-line to the M-band, i.e., the centre of the thick filament (Figure 2.7.7). Although the exact functional role of titin remains to be clarified, it is generally accepted that it acts as a molecular spring that develops tension when sarcomeres are stretched. Titin's location has led to speculation that it might stabilize the thick filament within the centre of the sarcomere. Such stabilization may be necessary to prevent the thick filament from being pulled to one side of the sarcomere when the forces acting on either half of the thick filament are unequal.

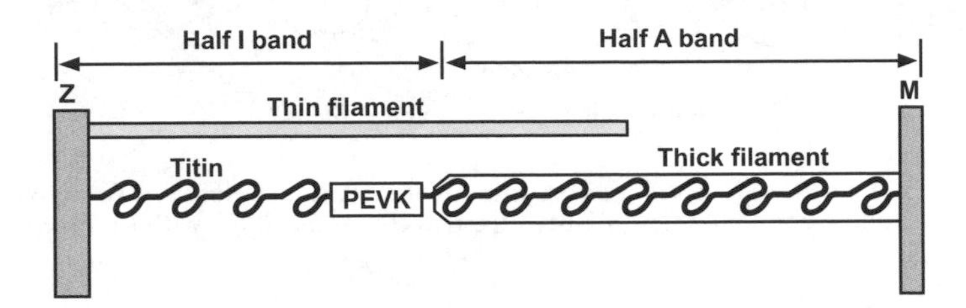

Figure 2.7.7 **Schematic illustration of a sarcomere, including titin, which runs from Z-line to M-band, and, through its elastic properties, keeps the thick myofilament centered in the sarcomere during contraction.**

Evidence of titin's role in centering thick filaments has been provided by Horowits and colleagues (Horowits & Podolsky, 1987; Horowits & Podolsky, 1988; Horowits et al., 1989), who showed, that upon prolonged activation in chemically skinned rabbit psoas fibres, the thick myofilaments could easily be moved away from the centre of the sarcomere at short (< 2.5 μm), but not at long (> 2.8 μm), sarcomere lengths when the titin *spring* presumably was tensioned and so helped centre the thick myofilament.

Motor units

Skeletal muscle is organized into motor units. A motor unit is defined as a group of muscle fibres that are all innervated by the same motor neuron. A small motor unit of a small muscle requiring extremely fine control may consist of only a few muscle fibres, while a motor unit of a large human skeletal muscle may contain thousands of muscle fibres. When a motor neuron is stimulated sufficiently to cause contraction, all fibres of the motor unit will contract. Since the force of contraction, among other parameters, depends on the number of fibres activated, a large motor unit (one that contains a large number of muscle fibres) can exert more force than a small motor unit.

A detailed study of motor units in cat soleus (McPhedran et al., 1965) and cat medial gastrocnemius muscles (Wuerker et al., 1965), showed large differences in motor unit size both from one muscle to the other, and within the same muscle. Differences in the contractile characteristics existed in parallel with these differences in motor unit size. Small motor units tended to be composed of slow twitch fibres, and large motor units were made up of fast twitch fibres.

The observation of differences in motor unit size, and that these differences are related to fibre type contents of a motor unit, is interesting from a morphological point of view. A series of two studies described by Henneman et al. (1965) and Henneman and Olson (1965) completely revealed the functional significance of this arrangement. These researchers showed that motor neurons with thin axons were the most excitable, i.e., were excited first in a graded muscular contraction, and belonged to small motor units, i.e., motor units with a small number of fibres that were typically of the slow type, and motor neurons with thick axons were the least excitable and belonged to large motor units. From these observations, they concluded that graded contractions of muscles are achieved by first recruiting small motor units and then progressively recruiting larger motor units as more force is required.

Most skeletal muscles are composed of slow and fast twitch motor units. The proportion of slow and fast motor units determines the force-velocity properties of a muscle, e.g., Hill (1970). In terms of energy requirements, fast contractions are more costly than slow contractions (Hill, 1938). The design and use of muscles may have evolved to account for the advantages and limitations of speed versus economy. In the cat triceps surae group, the functional tasks are divided, to a certain degree, among the muscles. The soleus is almost exclusively comprised of slow twitch fibres (Ariano et al., 1973) that are dominant in situations of low force requirements. In contrast, the gastrocnemius is composed of about 25% slow twitch and 75% fast twitch fibres, making it suitable to react to large and sudden force demands (Walmsley et al., 1978; Hodgson, 1983; Herzog et al., 1993).

Fibre arrangement within muscle

Depending on the arrangement of fibres within a muscle, muscles are referred to as parallel-fibred, fusiform or pennate. Among the pennate muscles, a further subdivision may be made to distinguish among unipennate, bipennate, and multipennate muscles (Figure 2.7.8). In a parallel-fibred and fusiform muscle, fibres run parallel to the line of action of the entire muscle, which is typically defined as the line connecting the distal and proximal tendons. In a unipennate muscle, fibres run at a distinct angle to the line of action of the entire muscle, and all fibres are approximately parallel to one another. In a bi or multi pennate muscle, fibres also run at a distinct angle to the line of action of the entire muscle, and there are two (or more) distinct directions of fibre alignment. Because the different

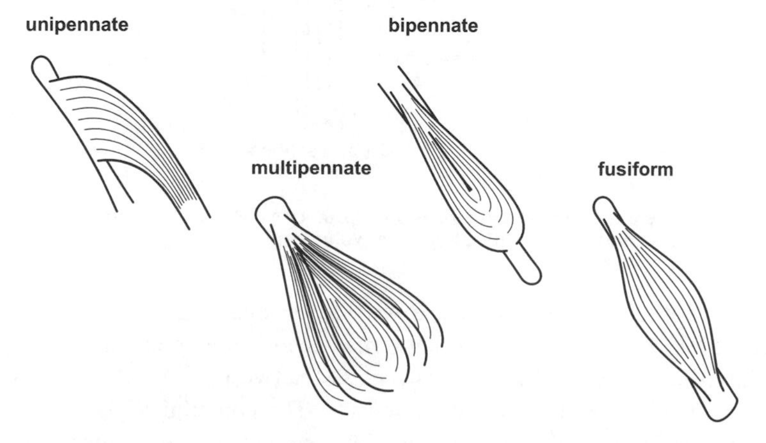

Figure 2.7.8 **Classification of muscles into fusiform, unipennate, bipennate, and multipennate, depending on the arrangement of fibres within a muscle.**

arrangements of fibres within a muscle influence some of the functional characteristics significantly, an index of architecture was proposed to quantify the structure of a muscle. This index is defined as the ratio of muscle fibre to muscle belly length at (an assumed) optimal length of all fibres (Woittiez et al., 1984).

The contractile properties of a muscle, as a function of the index of architecture, are described briefly in the form of a small example. Imagine that a muscle has a given volume. Assuming that all sarcomeres take up the same volume, there is a precisely defined number of sarcomeres that may be used to fill the available volume. However, these sarcomeres may be arranged to form long or short fibres. Long fibres, in contrast to short fibres, take up a large volume. Therefore, the number of long fibres that may be arranged in parallel in a given muscle volume is smaller than the corresponding number of small fibres. Assuming further that all sarcomeres can shorten and elongate by the same amount, the muscle with the long fibres can exert forces over a larger range of absolute muscle length than the short-fibred muscle (Figure 2.7.9). Assuming further that all sarcomeres produce the same

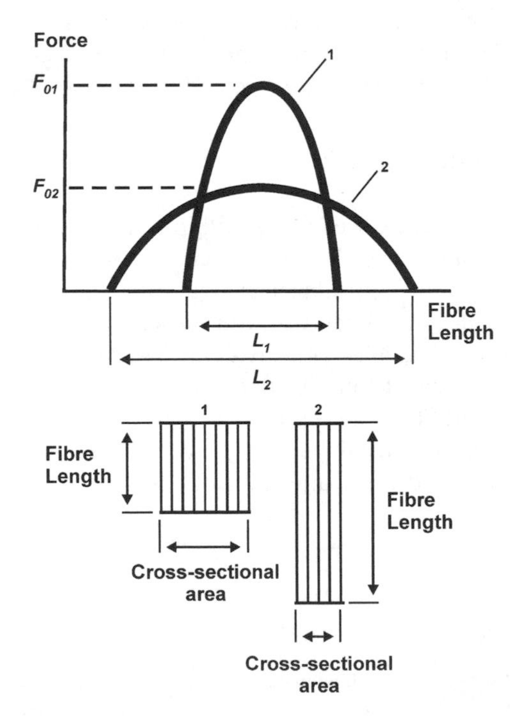

Figure 2.7.9 **Schematic force-length relationship of two muscles with different cross-sectional areas and fibre lengths, but equal volumes.**

amount of force, the short-fibred muscle will be stronger than the long-fibred muscle, because of the larger number of fibres that it may accommodate in parallel (Figure 2.7.9). Thus, the long-fibred muscle has a larger range of active force production, but a smaller peak force potential, than the short-fibred muscle. The potential of these two muscles to produce mechanical work is the same, because they contain the same number of sarcomeres (equal volume), and each sarcomere is assumed to have the same work potential.

2.7.3 MUSCULAR CONTRACTION

Skeletal muscles contract in response to electrochemical stimuli. Specialized nerve cells, called motor neurons, propagate action potentials to skeletal muscle fibres. When reaching the muscle, the axons of motor neurons divide into small branches, with each branch innervating one muscle fibre. Typically, the motor neuron reaches a muscle fibre near its centre, where it forms the neuromuscular junction or synapse (Figure 2.7.10).

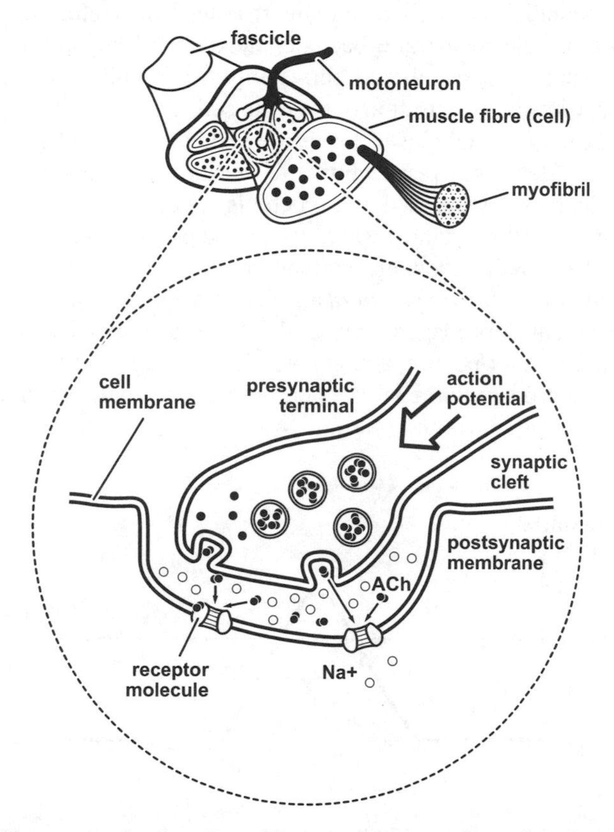

Figure 2.7.10 **Neuromuscular junction with motor neuron and muscle cell membrane.**

The neuromuscular junction is formed by an enlarged nerve terminal known as the presynaptic terminal, which is embedded in small invaginations of the cell membrane, called the motor endplate or postsynaptic terminal. The space between presynaptic and postsynaptic terminal is the synaptic cleft.

When an action potential of a motor neuron reaches the presynaptic terminal, a series of chemical reactions take place that culminate in the release of acetylcholine (ACh) from synaptic vesicles located in the presynaptic terminal. Acetylcholine diffuses across the synaptic cleft, binds to receptor molecules of the membrane on the postsynaptic terminal, and causes an increase in permeability of the membrane to sodium (Na^+) ions. If the depolarization of the membrane due to sodium ion diffusion exceeds a critical threshold, an action potential travels along the stimulated muscle fibre. To prevent continuous stimulation of muscle fibres, acetylcholinesterase rapidly breaks down acetylcholine into acetic acid and choline.

Excitation-contraction coupling

The process of excitation-contraction coupling involves the transmission of signals along nerve fibres across the neuromuscular junction (where the end of the nerve meets the muscle fibre) (Figure 2.7.10), and along muscle fibres. At rest, nerve and muscle fibres maintain a negative charge inside the cell compared to the outside, i.e., the membrane is polarized. Nerve and muscle fibres are excitable, which means that they can change the local membrane potential in a characteristic manner when stimuli exceed a certain threshold. When a muscle membrane becomes depolarized beyond a certain threshold, there is a sudden change in membrane permeability, particularly to positively charged sodium (Na^+) ions, whose concentration outside the cell is much higher than inside the cell. The resulting influx of Na^+ ions causes the charge inside the cell to become more positive. The membrane then decreases permeability to sodium, and increases permeability to potassium ions, which are maintained at a much higher concentration inside than outside the cell. The resulting outflow of the positively charged potassium ions causes a restoration of the polarized state of the excitable membrane. This transient change in membrane potential is referred to as an *action potential* and lasts for approximately 1 ms. In the muscle fibre, this action potential propagates along the fibre at a speed of about 5 to 10 m/s (Figure 2.7.11). In the motor neuron, action potentials propagate at speeds in proportion to the diameter of the neuron, with the largest neurons (in mammals) conducting at speeds of about 120 m/s.

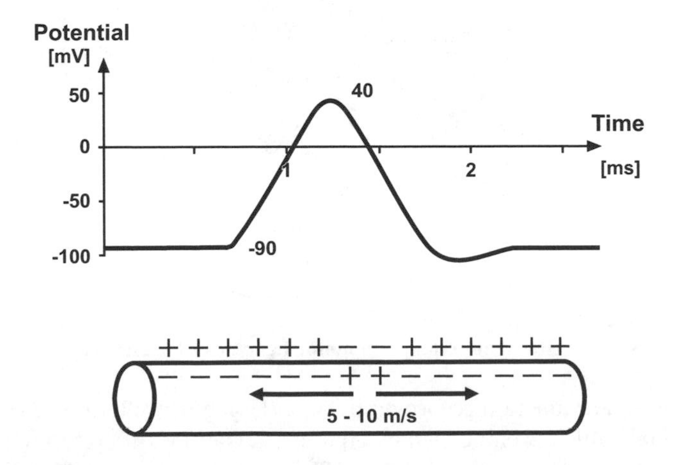

Figure 2.7.11 **Schematic illustration of a single muscle fibre action potential (top) and the corresponding propagation of the action potential along the muscle fibre (bottom).**

Once an action potential has been transmitted from the nerve axon to the muscle fibre at the neuromuscular junction, it is propagated along and around the fibre and reaches the interior of the fibre at invaginations of the cell membrane called T-tubules (Figure 2.7.12). Depolarization of the T-tubules causes the release of Ca^{2+} ions from the terminal cisternae of the sarcoplasmic reticulum (membranous, sac-like structure which stores calcium) into

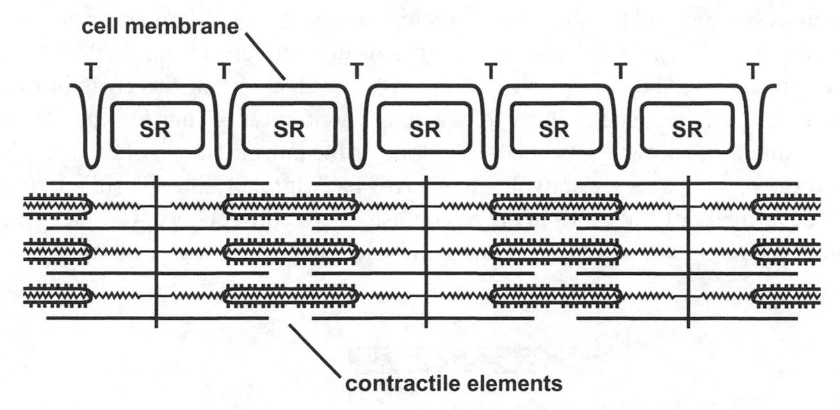

Figure 2.7.12 **Schematic illustration of T-tubules (T) in a section of a muscle fibre and its association with the sarcomplasmic reticulum (SR) and the contractile myofilaments.**

the sarcoplasm surrounding the myofibrils. Ca^{2+} ions bind to specialized sites on the troponin molecules of the thin myofilaments, and removing an inhibitory mechanism that otherwise prevents cross-bridge formations in the relaxed state (Figure 2.7.13). Cross-bridges then attach to the binding sites of the thin filaments and, through the breakdown of ATP into adenosinediphosphate (ADP) plus a phosphate ion (Pi), the necessary energy is provided to

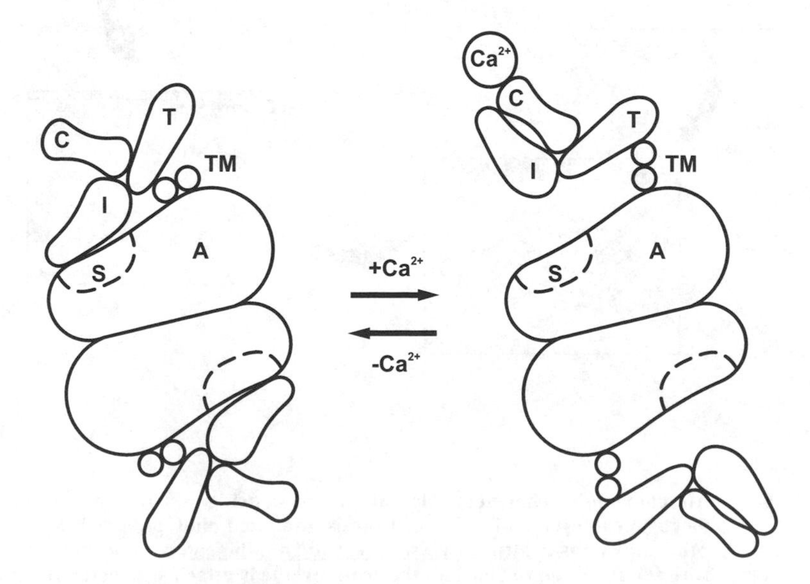

Figure 2.7.13 **Schematic illustration of the inhibitory/excitatory regulation of cross-bridge attachment on the actin filament (A). Without calcium (left), the tropomyosin (TM) and troponin complex (troponin T,C, and I) are in a configuration which blocks the cross-bridge attachment site (S). Adding calcium (Ca^{2+}) to the calcium binding site of the troponin (troponin C) changes the configuration of the tropomyosin-troponin complex so that the cross-bridge attachment site is exposed and cross-bridge attachment is possible (adapted from Gordon, 1992, with permission).**

cause the cross-bridge head to move and so attempt to pull the thin filaments past the thick filaments (Figure 2.7.14). At the end of the cross-bridge movement, an ATP molecule is thought to attach to the myosin portion of the cross-bridge so that the cross-bridge can release from its attachment site, go back to its original configuration, and be ready for a new cycle of attachment. This cycle repeats itself as long as the muscle fibre is stimulated. When stimulation stops, Ca^{2+} ions are actively transported back into the sarcoplasmic reticulum, resulting in a decrease of Ca^{2+} ions in the sarcoplasm. As a consequence, Ca^{2+} ions diffuse away from the binding sites on the troponin molecule and cross-bridge cycling stops.

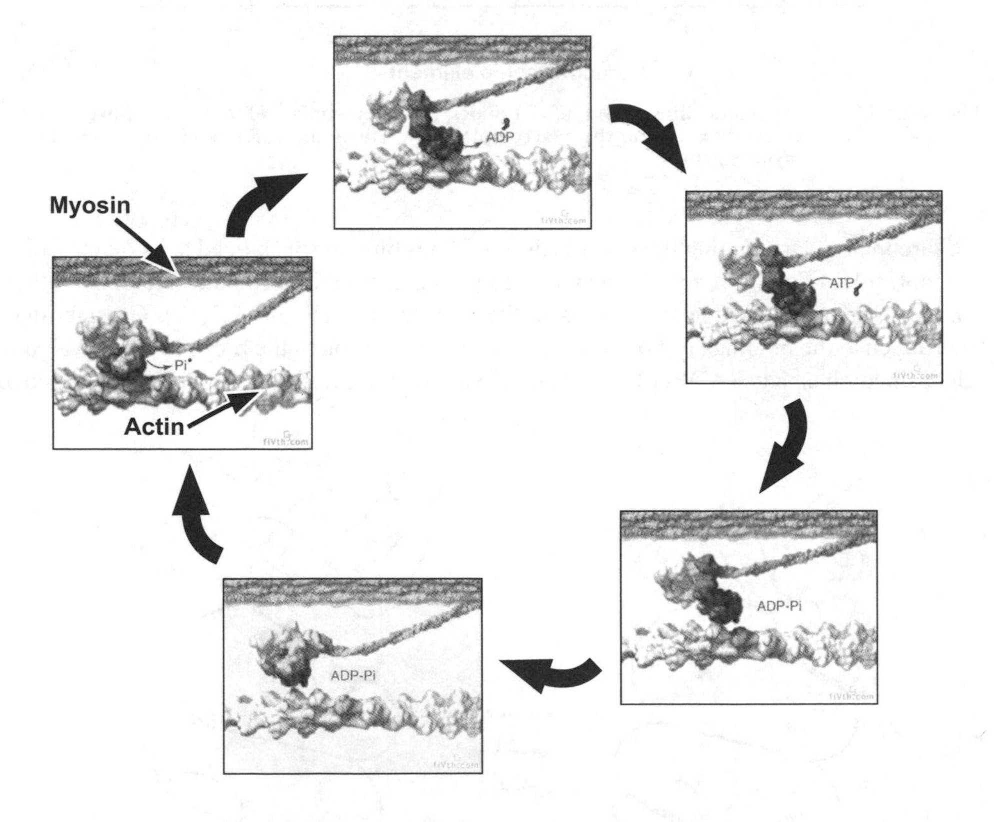

Figure 2.7.14 **Attachment/detachment cycle of a cross-bridge with the corresponding biochemical cycles of ATP hydrolysis (adapted and obtained from Dr. Ron Milligan's website http://www.scripps.edu/cb/milligan/, with permission). Starting with the figure on the far left, the cross-bridge is attached to actin and phosphate (Pi) is released, which initiates the power stroke. Following the arrows, the top figure shows the cross-bridge at the end of the power stroke and ADP is released. In the next figure (far right), ATP attaches to the nucleotide binding site of the cross-bridge, which allows for detachment of the cross-bridge from actin. With the cross-bridge detached (bottom right), ATP is hydrolyzed into ADP·Pi, and the cross-bridge head undergoes a conformational change (bottom left) and is ready for a new attachment to actin.**

Sensory organs

To understand the local control of muscle force, it is necessary to describe the muscle sensory organs that influence motor neurons. Local control is not only regulated by sending messages from the brain to the local level, but also by afferent fibres in muscles, tendons, skin, and joints, which may activate or inhibit motor neurons. The synaptic input to motor neurons, whether from afferent neurons or descending pathways, is typically channelled through interneurons rather than being direct. These interneurons can be thought of as switches that can turn signals on or off. Afferent fibres from muscle sensory organs connect with local interneurons, with the exception of the stretch reflex. Information carried by these fibres then influences the action of the muscle from which the afferent information is received (corresponding agonistic and antagonistic muscles).

The first sensory organ discussed is the muscle spindle. Muscles spindles consist of endings of afferent nerve fibres that are wrapped around modified muscle fibres, so-called spindle or intrafusal muscle fibres (Figure 2.7.15). Typically, several fibres are enclosed in a connective tissue capsule that makes up the muscle spindle. Muscle spindles are arranged in parallel with the muscle fibres and respond to stretch and the speed of stretch of the muscle.

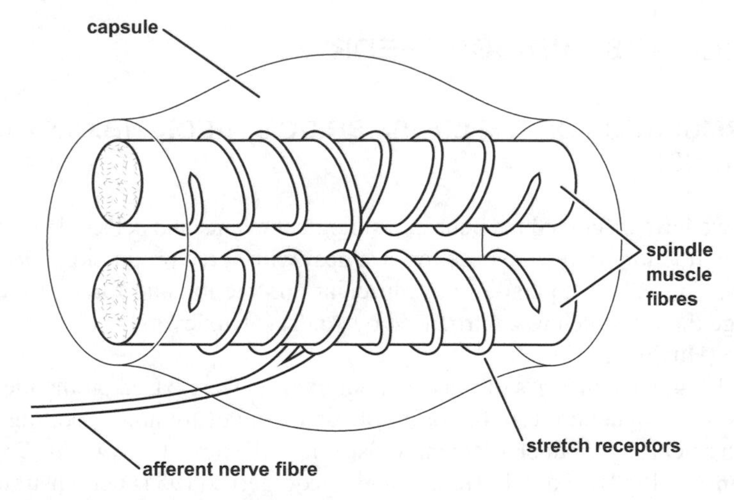

Figure 2.7.15 Schematic structure of a muscle spindle.

When a muscle is stretched, the afferent fibres of the muscle spindle send a signal to the central nervous system. This signal is divided into different branches that can take several pathways. One pathway directly stimulates the motor neurons going back to the muscle that was stretched. This particular pathway is known as the stretch reflex arc. A second pathway inhibits motor neurons of antagonistic muscles. A final pathway synapses with interneurons that convey information to centres of motor control in the brain.

Muscle spindles are able to contract in the end regions and can shorten with the muscle. This arrangement allows for transmission of information on muscle length and speed to higher centres of motor control.

The fibres of the muscle spindle have separate motor neurons from the actual muscle fibres, i.e., the extrafusal fibres. The motor neurons controlling muscle fibres are larger and are called alpha motor neurons. However, motor neurons that innervate the muscle spindles are known as gamma motor neurons. The interaction of the alpha and gamma systems during muscle shortening is important, because the middle of the intrafusal fibres must not become slack at any time. One way to prevent slackening is to link the activation of the alpha and gamma-motor neurons, e.g., Vallbo et al., (1979). However, experimental evidence suggests that there is a distinct difference between the control of gamma fusimotor neurons and alpha motor neurons, (Loeb, 1984; Murphy et al., 1984; Prochazka et al., 1985).

The second sensory organ associated tightly with muscular control is the golgi tendon organ. Golgi tendon organs are located near the muscle-tendon junctions. Golgi tendon organs consist of afferent nerve fibres wrapped around collagen bundles in the tendon, and they monitor tension. When the muscle contracts, forces are transmitted through the tendon, and afferent signals from the golgi tendon organs are transmitted to interneurons on the spinal level. Golgi tendon signals supply the motor control centres in the brain with continuous information about muscular tension. Branches of the afferent neuron inhibit motor neurons of the contracting muscle and activate motor neurons of antagonistic muscles during isometric contractions. This mechanism protects the muscle-tendon complex from injuries caused by excessive contractile forces.

2.7.4 THE CROSS-BRIDGE THEORY

1957 FORMULATION OF THE CROSS-BRIDGE THEORY (BIOLOGICAL CONSIDERATIONS)

So far, we have discussed the neurophysiological events and selected control issues of muscular contraction. In the next section, we deal with force production during muscular contraction. The accepted paradigm of muscular contraction and force production is the Cross-bridge theory, which was introduced by Andrew Huxley in his classic treatise half a century ago (Huxley, 1957).

Before 1954, most theories of muscular contraction were based on the idea that shortening and force production were the result of some kind of folding or coiling of the myofilaments (particularly the thick filaments) at specialized sites. However, in 1954 H.E. Huxley and Hansen (1954), and A.F. Huxley and Niedergerke (1954) demonstrated that contraction was not associated with an appreciable amount of myofilament shortening, and, therefore, postulated that muscle shortening is probably caused by a sliding of the thin past the thick myofilaments (the Sliding Filament theory). The mechanism that produces this myofilament sliding is referred to as the Cross-bridge theory.

In the classic description of the Cross-bridge theory, Huxley (1957) assumed that thick (myosin) filaments had side pieces that were connected via elastic springs to the thick filament. The side piece with its attachment site M (Figure 2.7.16) was thought to oscillate about its equilibrium position (O) because of thermal agitation. M was assumed to attach to specialized sites (A) on the thin (actin) filament. Attachment between thick and thin filaments at the M and A sites was thought to occur spontaneously and was restricted to occur asymmetrically only on one side of the equilibrium position, O, so that a combination of the M and A sites would produce a force (because of the tension in the elastic element constraining the side piece M) and movement that tended to shorten the sarcomere. Attachment of M to A was thought to be governed by rate function f, and detachment of M from A was thought to be governed by rate function g. In addition, f and g were modelled to be linear functions of x, where x is the distance from the equilibrium position of the side piece to the attachment site on the thin filament (Figure 2.7.16 and Figure 2.7.17). Since the combination of an M with an A site was taken to occur spontaneously, breaking the M-A connection had to be associated with an active, energy-requiring process. The energy for this process was assumed to come from splitting a high-energy phosphate compound.

For force production to occur smoothly, it was assumed that there were a number of M and A sites for combination of the thick and thin filaments, which were staggered relative to one another so that different sites would come into contact at different relative displacements of the two myofilaments. The M and A sites were further assumed to be so far apart that events at one site would not influence events at another site.

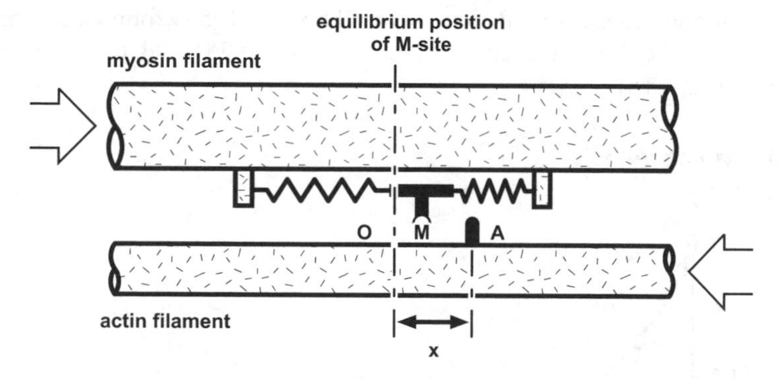

Figure 2.7.16 Schematic illustration of the 1957 model of the Cross-bridge theory (adapted from Huxley, 1957, with permission).

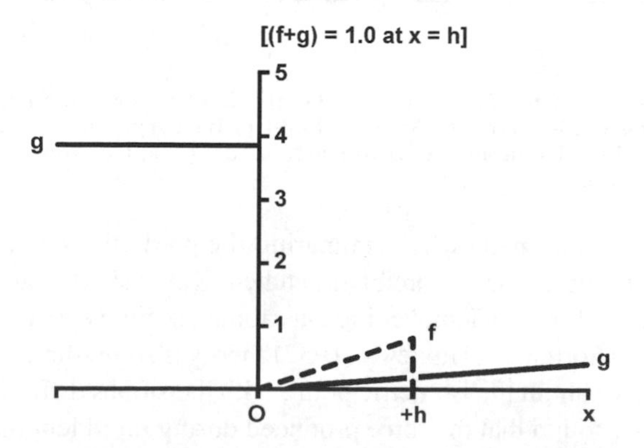

Figure 2.7.17 Rate functions for the formation, f, and the breaking, g, of cross-bridge links between the thick (myosin) and the thin (actin) myofilaments as a function of x, the distance from the attachment site on the thin filament to the equilibrium position of the cross-bridge (adapted from Huxley, 1957, with permission).

The Cross-bridge theory and its energetics are assumed to be associated with defined structures. The M sites are represented by the S1 subfragment of the myosin protein (the cross-bridge, Figure 2.7.3). The A sites are the attachment sites on the actin near troponin (Figure 2.7.5), and the high-energy phosphate supplying the energy for force production and cross-bridge detachment is associated with ATP. Typically, it is assumed that one ATP molecule is hydrolyzed for one cross-bridge cycle.

Since a thick myofilament in mammalian skeletal muscle is about 1600 nm in length and contains cross-bridges along its entire length (except for about the middle 160 nm), one-half of the thick filament contains about 50 (720 nm / 14.3 nm) pairs of side pieces offset by 180°. Each side piece is thought to contain two cross-bridge heads for possible attachment on the thin filament. Since neighbouring cross-bridge pairs are thought to be offset by 60° (Figure 2.7.4), there are about 16 (720 nm / 42.9 nm) side pieces available on each thick filament for interaction with a given thin filament.

To test the cross-bridge model of muscular contraction, Huxley (1957) compared the predictions of his theory with the experimental results obtained by Hill (1938) on frog striated muscle during tetanic stimulation at 0°C. Huxley (1957) found good agreement between the normalized force-velocity relation of Hill (1938) and his own theoretical predictions (Figure 2.7.18).

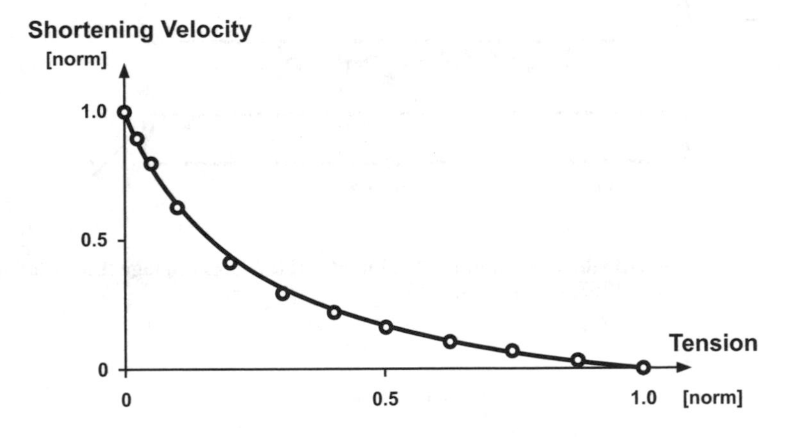

Figure 2.7.18 **Comparison of the force-velocity relationship obtained using Hill's characteristic equation (with a/Po = 0.25) (Hill, 1938) (solid curve), and obtained by Huxley (1957) (dots) based on the cross-bridge model (adapted from Huxley, 1957, with permission).**

Several observations were made when comparing the predictions of the theory to the properties of stimulated muscle that is forcibly stretched. Katz (1939) found that the slope of the force-velocity curve for slow lengthening was about six times greater than the corresponding slope for slow shortening. Huxley's (1957) theory also predicted this asymmetry in the force-velocity curve about the isometric point, with the slopes differing by a factor of 4.33. Katz (1939) further found that the force produced during rapid lengthening of a stimulated muscle was about 1.8 times the isometric force. Using the rate functions given by Huxley (1957), the force for increasing speeds of lengthening approaches asymptotically a value of 5.33 times the isometric force. This value is too large.

Similarly, Huxley's (1957) theory does not predict well the heat production of a muscle that is stretched. The theory predicts that the rate of heat liberation increases linearly with the speed of lengthening. This prediction vastly overestimates the actual heat production.

However, Huxley (1957) points out that the discrepancy between experiment and theory could be eliminated quite readily by assuming that, during lengthening, the cross-bridge connections were broken mechanically, rather than released via ATP splitting. This assumption has recently been implemented in various models to account for experimental observations made during concentric (Cooke et al., 1994) and eccentric contractions.

1957 FORMULATION OF THE CROSS-BRIDGE THEORY (MATHEMATICAL CONSIDERATIONS)

If we consider, as Huxley did in 1957, a great number of identical M-A pairs, i.e., cross-bridges that have at each instant in time the same value for x, the proportion of attached cross-bridges n(t) is a function of time alone. Since each cross-bridge has the same x-value, and thus the same force, n(t) directly reflects force, and the rate of change of n(t) can be calculated, based on the definitions of f(x) and g(x) as:

$$dn/dt = (1-n)f(x) - ng(x) \qquad (2.7.1)$$

Note that for a dynamic state of equilibrium, i.e., when dn/dt = 0, we obtain the following value for the proportion of attached cross-bridges:

$$n_{eq} = f(x)/f(x) + g(x) \qquad (2.7.2)$$

This result may be expected on intuitive grounds, as the proportion of attached cross-bridges at equilibrium (for M-A pairs with identical x-values at each instant in time) is governed by the probability of attachment and detachment. An interesting side issue is that, at equilibrium (or steady-state), the proportion of attached cross-bridges (and thus force) is independent of the contractile history. Therefore, the Cross-bridge theory predicts that the isometric steady-state force must be the same (for a given muscle length), independent of the contractile history. Later in this chapter, when discussing history-dependent effects, we show that this prediction is not correct. Experimental evidence was available before the formulation of the sliding filament and the classic Cross-bridge theory that showed convincingly that steady-state (equilibrium) forces depend directly on contractile history (Abbott & Aubert, 1952).

To solve equation 2.7.1 for n(t), we must specify the global relative motion x = x(t), and the initial condition $n_0 = n(0)$. It is sometimes convenient to provide the global relative sliding velocity v = v(t), instead of x(t), in which case x(t) can be obtained by integration as:

$$x(t) = x(0) + \int_0^t v(\tau)d\tau \qquad (2.7.3)$$

where sarcomere shortening is defined as negative.

If we assume now, more realistically, that M-A pairs are not identical (same x-value at each instant in time), but, for a great number of cross-bridges, are distributed randomly (and thus uniformly) over the range $[-0.5l_a, 0.5l_a]$, where l_a is the typical distance between actin sites. We now obtain a distribution function n(x,t) per unit length, such that the product n(x,t) dx represents, at time t, the proportion of attached cross-bridges whose distance to the nearest actin attachment site lies between x and x+dx. By the uniformity assumption, the proportion of detached cross-bridges in the same interval [x, x+dx] is given by:

$$\left[\frac{1}{l_a} - n(x,t)\right]dx \tag{2.7.4}$$

Again, we are interested in determining the rate of change of n(x,t) with time, as seen by an observer fixed at an actin site. For this situation, x varies as a function of time, and we obtain the material rate of change in time by:

$$\frac{Dn}{Dt} = \frac{\partial n}{\partial t} + \frac{\partial n}{\partial x}v \tag{2.7.5}$$

where v is again considered negative for shortening. Assuming that the actin and myosin filaments are rigid, the governing differential equation becomes:

$$\frac{\partial n}{\partial t} + \frac{\partial n}{\partial x}v = \left(\frac{1}{l_a} - n\right)f(x) - ng(x) \tag{2.7.6}$$

To solve equation (2.7.6), the initial conditions, n_0, of the unknown variable n must be specified, and the global motion, let's say in terms of the filament sliding velocity, v(t), must be known. For applications of equation (2.7.6) to practical problems in muscle mechanics, please see Zahalak and Ma (1990) and Epstein and Herzog (1998).

In the classical considerations of the Cross-bridge theory, e.g., Huxley (1957) and Huxley & Simmons (1971), it has been assumed that actin and myosin filaments are perfectly rigid. However, in the meantime it has been shown that not only are myofilaments not rigid, because of their great lengths compared to other structures, e.g., the cross-bridges, they contribute significantly (50 to 70%) to the total compliance of muscle fibres (Goldman & Huxley, 1994; Higuchi et al., 1995; Huxley et al., 1994; Kojima et al., 1994; Wakabayashi et al., 1994). If the filaments were not assumed rigid, Huxley's x-distance would not only depend on the relative position of actin and myosin filaments, but also on the extension of the filaments, and thus the cross-bridge attachment site on myosin and the host site on actin. To account for this complexity, an additional term containing the velocity gradient along the filaments would need to be added to equation (2.7.6). For a detailed solution of the Cross-bridge theory including compliant filaments, see Mijailovich et al. (1996); or for a simple non-cross bridge model, Forcinito et al. (1997).

If we now wish to calculate the force in a muscle or fibre based on the cross-bridge model, we only need to consider the contributions of all half-sarcomeres in a normal cross-section, since all half sarcomeres in series must produce the same force. If A is the area of such a cross-section, s the current average sarcomere length, and m the number of M sites per unit volume, then the number of M sites contained in all half sarcomeres of the cross-section considered is equal to mAs/2. The average force per site is calculated by the weighted average of the force in each cross-bridge:

$$f_{ave}(t) = \int_{-\infty}^{+\infty} kxn(x,t)dx \tag{2.7.7}$$

where the weighting function is the proportion of attached cross-bridges per unit length, n(x,t), and the total force is given by multiplying the average force per cross-bridge by the number of cross-bridges involved in the cross-section:

$$F(t) = \frac{mAs}{2}\int_{-\infty}^{+\infty} kxn(x,t)dx \tag{2.7.8}$$

By choosing his constants carefully, Huxley (1957) was able to nicely fit Hill's (1938) force-velocity data from frog skeletal muscles. Using a similar approach, energy and mechanical power can be calculated using the cross-bridge model (Huxley & Simmons, 1971; Zahalak & Ma, 1990; Zahalak & Motabarzadeh, 1997).

1969 FORMULATION OF THE CROSS-BRIDGE THEORY

So far, the cross-bridge head (M – Figure 2.7.16) was thought to be in a single configuration. However, Reedy et al. (1965) demonstrated that cross-bridges at rest were about perpendicular to the filament axis, while cross-bridges in rigor were tilted 45° to the perpendicular. Based on this information, H. Huxley (1969) proposed that force and filament transport would be generated by a tendency of the cross-bridge head to rotate about its attachment point (Figure 2.7.19). Therefore, subfragment 1 (S-1) was assumed to act as a crank, pulling the thin relative to the thick filament. Subfragment 2 (S-2), that connects subfragment 1 to the backbone of the thick filament, acts as a connecting rod, converting the ro-

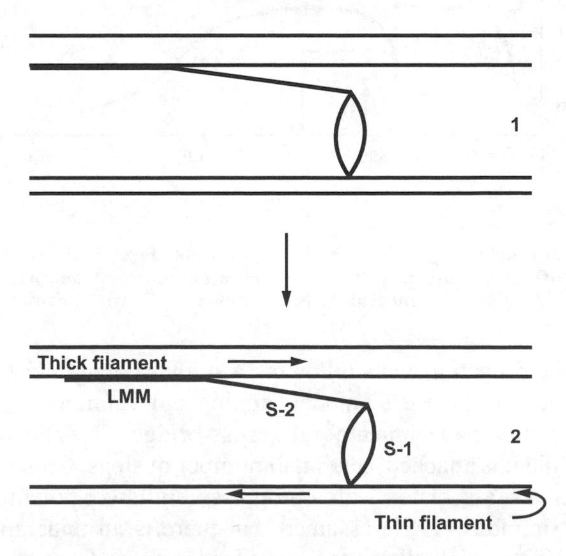

Figure 2.7.19 Force production and thick-thin filament sliding according to the 1969 Cross-bridge theory of H. Huxley (with permission). The cross-bridge head (S-1) was thought to rotate, as shown in the figure, and so produce the relative myofilament sliding.

tary movement of the cross-bridge head into a linear displacement of the filaments. The idea of the cross-bridge power stroke was born.

1971 FORMULATION OF THE CROSS-BRIDGE THEORY

A characteristic of muscular contraction that could not be predicted adequately with the 1957 theory is the force transients following a stepwise length change. When a muscle is shortened rapidly, the force drops virtually simultaneously with the length change and then recovers quickly (Figure 2.7.20). Two force parameters were defined by Huxley for describing these fast force transients: they are referred to as T_1 and T_2. T_1 is defined as the minimum force achieved during the rapid shortening; T_2 is the force at the end of the quick recovery phase (Figure 2.7.20). T_1 was found to vary almost linearly with the magnitude of the step, reaching zero force at a step amplitude of approximately 6 nm per half-sarcomere (Figure 2.7.21). T_2 remained nearly the same as the isometric force just before the length step, T_0, for step amplitudes of up to about 5 nm, but then started to become progressively smaller with increasing step amplitude, reaching zero force, i.e. the early, quick recovery was completely absent, at step amplitudes of about 13 nm per half-sarcomere and beyond.

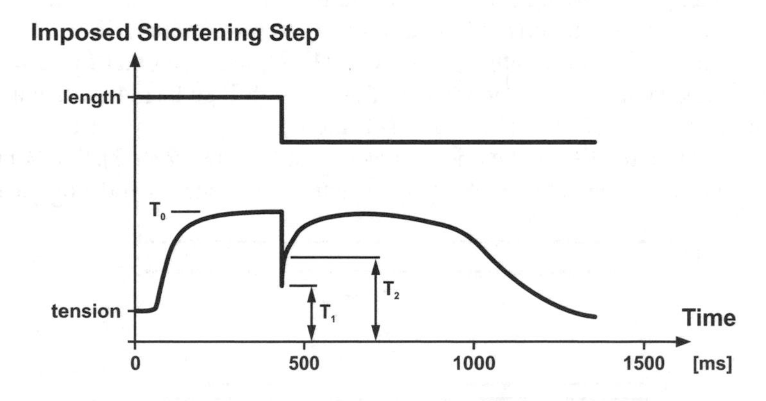

Figure 2.7.20 **Definitions of T_1 and T_2. T_1 is the minimal force value obtained during a rapid release of a muscle; T_2 is the force value achieved following the quick recovery period (adapted from Huxley & Simmons, 1971, with permission).**

To account for the force transients following a stepwise length change and to not lose the good predictive power of the 1957 model, Huxley and Simmons (1971) introduced the concept of different states of attachment for the cross-bridge. This allowed the cross-bridge to perform work (while it is attached) in a small number of steps. Going from one stable attachment to the next was associated with a progressively lower potential energy. Furthermore, Huxley and Simmons (1971) assumed that there is an undamped elastic element within each cross-bridge which allows the cross-bridge to go from one stable attachment state to the next without a corresponding relative displacement of the thick and thin filaments. A diagrammatic representation of the 1971 cross-bridge model is shown in Figure 2.7.22.

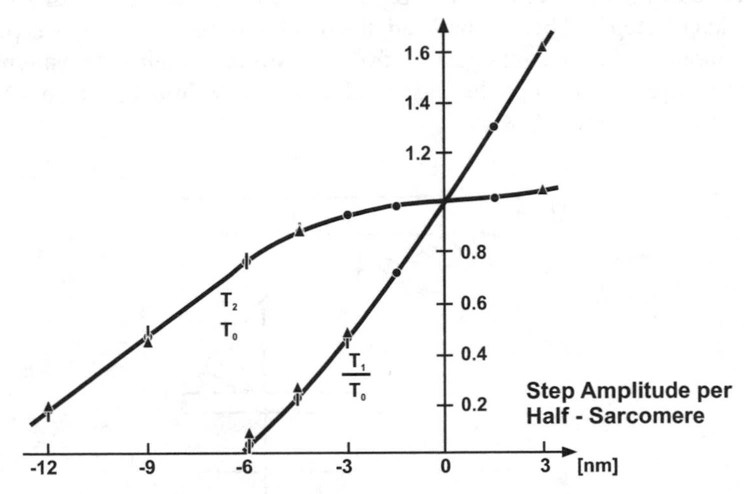

Figure 2.7.21 Values of T_1 (the extreme force reached after a sudden length step) and T_2 (the force reached during the early force recovery following the length step) as a function of the magnitude of the step release. Adapted from Ford et al. (1977).

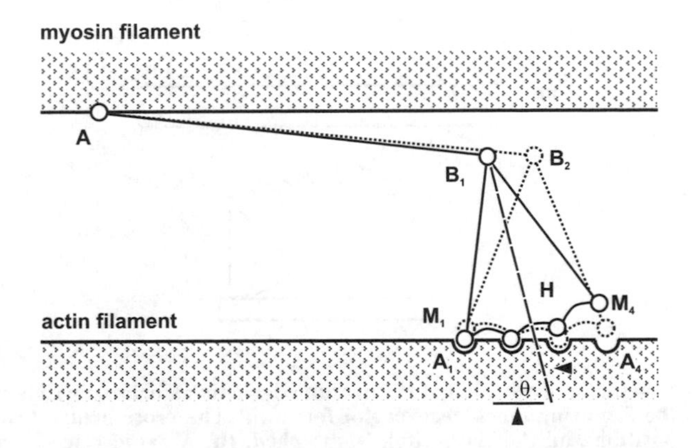

Figure 2.7.22 Schematic representation of the assumed interaction between the thick and thin filaments according to the 1971 Cross-bridge theory (Huxley & Simmons, 1971). The cross-bridge head is attached to the thick myofilament via an elastic spring. The cross-bridge head can rotate, and so produce different amounts of tension in the elastic link AB without relative movements of the myofilaments (adapted from Huxley & Simmons, 1971, with permission).

The force transients during a rapid length change are explained, as follows. If a muscle is released infinitely fast, there will be no rotation of the cross-bridge head (Figure 2.7.23a, b). Therefore, the drop in force observed during the length step (T_1) corresponds to the force-elongation property of the undamped elastic element within the cross-bridge. Since it had been argued that the relationship between the T_1-value and the distance of the length step was virtually linear (the experimentally observed non-linearity was associated with the beginning of the quick recovery during the large length steps), the cross-bridge elasticity was assumed to be linear as well (Huxley & Simmons, 1971, 2.3 x 10^{-4} N/m). Once the infinitely fast length step had been completed, the quick recovery of force was possible, because of a rotation of the cross-bridge head from a position of high to a position of low potential energy, thereby, stretching the elastic link in the cross-bridge, and so increasing the cross-bridge force (Figure 2.7.23c).

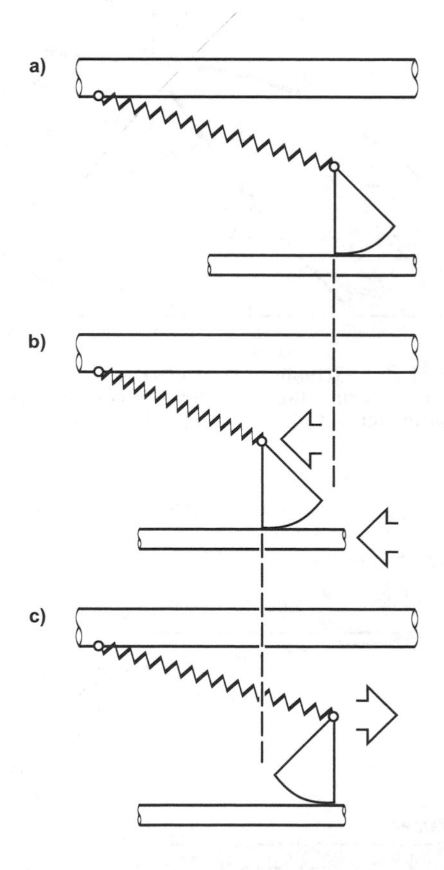

Figure 2.7.23 **Schematic illustration of the presumed events associated with a rapid release and the following quick recovery of force. (a) The cross-bridge head is in its initial position and the elastic link is stretched. (b) A rapid release has occurred. The cross-bridge head is in the same orientation as in (a) but the elastic link has shortened, because of the relative movement of the myofilaments. The cross-bridge force (carried by the elastic link) is smaller in (b) than in (a). (c) The cross-bridge rotates to a position of lower potential energy, thereby, stretching the elastic link and increasing the cross-bridge force without any myofilament movement.**

Huxley and Simmons (1971) discussed a cross-bridge model with three stable, attached states and derived equations for a system containing two stable states. Many further models with a variety of stable states have been proposed (Eisenberg & Greene, 1980; Eisenberg et al., 1980), but the basic ideas of these models can all be traced to the 1971 cross-bridge model (Huxley & Simmons, 1971).

For their theoretical model with two stable attached states Huxley and Simmons (1971) made the following assumptions:

(1) Detachment and attachment of cross-bridges for quick shortening experiments, and the subsequent force recovery are slow and can be ignored,

(2) Actin and myosin filaments are rigid,

(3) Relative filament sliding only occurs when the muscle fibre length is changed, i.e., there is no elasticity in series with the fibre and sarcomeres,

(4) The link AB (Figure 2.7.22) is linearly elastic, and it can exert negative and positive force,

(5) In the isometric state, cross-bridges spend equal time in the two stable positions, and

(6) The transition times between the two stable positions are negligibly small.

Using these definitions, the force in the link AB (Figure 2.7.22), when the cross-bridge is in state 1, F_1, and state 2, F_2, is given by:

$$F_1 = K(y + y_0 - h/2) \text{ and} \tag{2.7.9}$$

$$F_2 = K(y + y_0 + h/2) \text{ respectively} \tag{2.7.10}$$

Where K is the stiffness in link AB; y is the relative displacement of thin to thick filament, positive is defined as stretching; y_o is the extension of the elastic link AB when the cross-bridge is between the stable positions 1 and 2 (therefore if a shortening movement of magnitude y_o occurs from an isometric state, force drops to zero), and h is the increase in the length of link AB when the cross-bridge shifts from position 1 to position 2. If we now neglect the detached states of the cross-bridges, then the time averaged force, φ, becomes:

$$\begin{aligned} \varphi &= n_1 F_1 + n_2 F_2 \\ &= K(y + y_0 - h/2 + hn_2) \end{aligned} \tag{2.7.11}$$

Where n_1 corresponds to the fraction of attached cross-bridges in position 1, and n_2 corresponds to the fraction of attached cross-bridges in position 2, and
$n_1 + n_2 = 1$ (as we only consider transient phenomena of the attached cross-bridges).
In the isometric state, $y = 0$ and $n_2 = ½$. Therefore, the isometric force per attached cross bridge, φ_0, becomes:

$$\varphi_0 = K y_0 \tag{2.7.12}$$

Going from position 1 to position 2 is governed by the rate constant k_+, while going back from position 2 to position 1 is governed by the rate constant k_-. Rate constant k_+ is governed, in turn, by the energy barrier $B_2 = (E_2 + W)$, and rate constant k is governed by energy barrier $B_1 = E_1$ (Figure 2.7.24), where W is the work done in stretching the link AB when the cross-bridge goes from position 1 to 2. W is given by:

$$W = h\frac{F_1 + F_2}{2} \tag{2.7.13}$$

$$= K \cdot h(y + y_0)$$

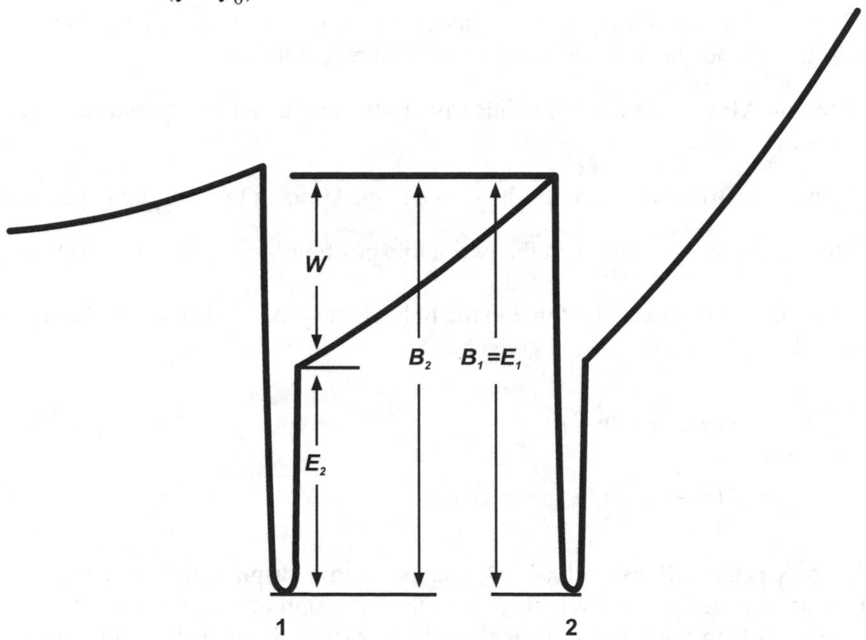

Figure 2.7.24 Potential energy diagram of a multi-state cross-bridge model with two stable states. See text for details.

Huxley and Simmons (1971) further assumed that the rate constants, k_+ and k_- are proportional to $\exp - B/k_BT$. Therefore, we have:

$$\begin{aligned} k_+ &= k_- \exp(B_1 - B_2)/k_BT \\ &= k_- \exp(E_1 - E_2 - W)/k_BT \\ &= k_- \exp(E_1 - E_2 - Kh(y + y_0))/k_BT \end{aligned} \tag{2.7.14}$$

Where k_B is the Boltzmann constant, and T is the absolute temperature. Note that k_- is constant, since B_1 is a fixed quantity that is independent of the force in the elastic link AB.

In the isometric state, Huxley and Simmons (1971) assumed that $n_1 = n_2$, thus, $k_+ = k_-$ and $y = 0$.Therefore, it follows from Equation (2.7.14) that:

$$E_1 - E_2 = Khy_0 \tag{2.7.15}$$

Thus Equation (2.7.14) becomes:

$$k_+ = k_- \exp - yKh/k_BT \tag{2.7.16}$$

If we now consider a quick stretch, or a quick release, experiment in which the fibre is held constant following the length change, the transfer of cross-bridge heads from one stable position to the other is given by:

$$dn_2/dt = k_+ n_1 - k_- n_2 \\ = k_+ - (k_+ + k_-)n_2 \tag{2.7.17}$$

If we define a rate constant, r, by:

$$r = k_+ + k_- \tag{2.7.18}$$

We can define the equilibrium situation as:

$$n_2 = k_+/(k_+ + k_-) \tag{2.7.19}$$

And by substitution from Equation (2.7.16) into Equation (2.7.18), we obtain:

$$r = k_-(1 + \exp(-yKh/k_B T)) \tag{2.7.20}$$

Huxley and Simmons (1971) found that Equation (2.7.20) approximated experimental results well if it was assumed that:

$$Kh = \alpha k_B T \tag{2.7.21}$$

Now, Equation (2.7.16) can be expressed as follows:

$$k_+ = k_- \exp{-\alpha y} \tag{2.7.22}$$

where α is a constant, and Equation (2.7.19) becomes:

$$n_2 = 1/2\left(1 + \tanh\frac{\alpha y}{2}\right) \tag{2.7.23}$$

The tension at the end of the quick recovery, φ_2, (corresponding to T_2 in the whole fibre) can now be calculated by using Equations (2.7.11), (2.7.21), and (2.7.23) as:

$$\varphi_2 = \frac{\alpha k_B T}{h}\left(y_0 + y + \frac{h}{2}\tanh\frac{\alpha y}{2}\right) \tag{2.7.24}$$

If we now assume that n_2 is always at its equilibrium, which would occur if the shortening speed was sufficiently slow, the work done during shortening can be calculated by integration of Equation (2.7.24):

$$\int \varphi_2 dy = k_B T\left(\frac{\alpha y}{h}\left(y_0 + \frac{y}{2}\right) - 1n\cosh\frac{\alpha y}{2}\right) \tag{2.7.25}$$

To match the experimentally observed results, α was taken as $5 \times 10^8 m^{-1}$. $(E_1 - E_2)$ was taken to be equal to $\alpha y_0 k_B T$, and y_0 was approximated as 8 nm, therefore $E_1 - E_2 = 4k_B T$. Finally, h was given a value of 8 nm for an optimal fit.

CURRENT THINKING

Cross-bridge rotation plays an important part in the 1969 and 1971 models of the molecular mechanisms of contraction. This rotation was thought to occur about the attachment point of the cross-bridge to actin. However, a series of structural studies (Rayment et al., 1993a; Rayment et al., 1993b) suggested that the connection between the cross-bridge head (subfragment S1) and actin does not allow for rotation, but that rotation takes place through a conformational change of the light-chain binding domain about a hinge within the myosin head.

Specifically, the following sequence of events was proposed for muscular contraction (Figure 2.7.25). Starting from the rigor conformation, Rayment et al. (1993a) suggested that the narrow cleft that splits the 50-kD segments of the myosin heavy chain sequence into two domains is closed (Figure 2.7.25A, horizontal gap, perpendicular to the actin filament axis). Addition of ATP and initial binding of ATP to myosin at the active site causes an opening of the narrow cleft between the upper and lower domains of the 50-kD segments. This event, in turn, disrupts the strong binding between actin and myosin, but still allows for a weak attachment (Figure 2.7.25B). The final ATP binding to myosin causes a closure of the nucleotide binding pocket and a corresponding change of the myosin molecule. Myosin now detaches from actin and ATP is hydrolyzed (Figure 2.7.25C). Rebinding of myosin to actin can now occur, presumably in multiple steps. The gap between the upper and lower domain closes in this process to produce strong binding, and phosphate, P, is released. This event starts the power stroke (Figure 2.7.25D). During the power stroke, the myosin S1 reverses its conformational change induced by ATP binding, the active site pocket is reopened, ADP is released, and the rigor conformation is established (Figure 2.7.25E). The cross-bridge cycle can now be restarted by ATP binding to myosin.

The detailed reactions of myosin S1 with actin and ATP are not known because not all of the rate constants have been determined. However, by combining the structural evidence (Rayment et al., 1993a; Rayment et al., 1993b), the biochemical data measured in solution (Bagshaw, 1993; Highsmith, 1976; Margossian & Lowey, 1978; Geeves, 1991; Millar & Geeves, 1983; Ma & Taylor, 1994; Greene & Eisenberg, 1980; Siemankowski et al., 1985), and the mechanical data (Huxley & Simmons, 1971; Finer et al., 1994), it is possible to infer the likely actin-myosin hydrolysis cycle (Table 2.7.1). Superimposing the mechanical cycle with a schematic structural model for the hydrolysis cycle (Figure 2.7.26) gives the current thinking of how muscular contraction might occur.

Starting from the left, top drawing in Figure 2.7.26, and following the arrows (the likely path of the actin-myosin hydrolysis cycle), we go through the following steps. The binding of ATP catalyzes the dissociation of myosin from actin (Szent-Györgyi, 1941; Eisenberg & Moos, 1968). When dissociated, ATP is hydrolyzed, and the cross-bridge head goes through a conformational change (the recovery stroke, Figure 2.7.26). Myosin attaches to actin, phosphate is released, and the power or working stroke is initiated. ADP is released, establishing the rigor conformation. ATP then attaches to the myosin head, myosin can now dissociate from actin, and the cycle may restart.

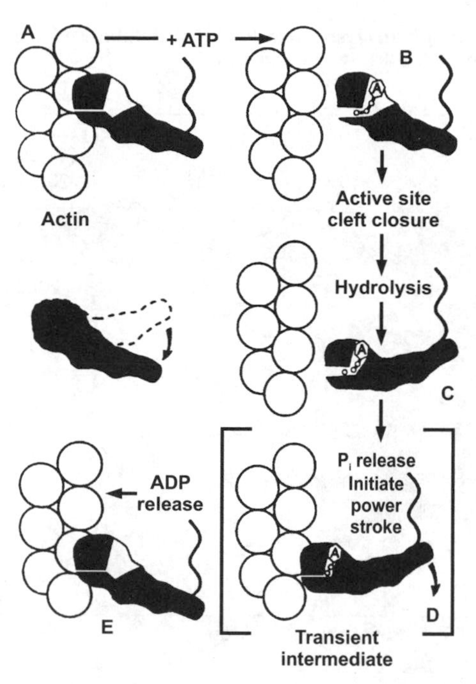

Figure 2.7.25 **Proposed molecular mechanism of contraction. A) Rigor conformation. The narrow cleft that splits the 50-kD segments of the myosin heavy-chain sequence into two domains is closed (horizontal gap, perpendicular to the actin filament axis). B) Addition of ATP, and initial binding of ATP to the active site, causes an opening of the narrow cleft and disrupts the strong binding between actin and myosin, but still allows for weak binding. The actin and myosin dissociate. C) The final binding of ATP to myosin causes a closure of the nucleotide binding pocket and a corresponding configurational change of the myosin molecule. ATP is now hydrolyzed. D) Myosin can now reattach to actin, presumably in multiple steps. The narrow cleft closes to produce strong binding. Phosphate, P, is released, and the power stroke starts. E) During the power stroke, the myosin molecule reverses its conformational change induced by ATP binding, and the active site pocket is reopened, establishing the rigor conformation. The cross-bridge cycle can now start all over again (reprinted with permission from Rayment et al., 1993a).**

Variations from the above proposed theory of contraction exist. However, Figure 2.7.25, Figure 2.7.26, and Table 2.7.1 represent the most commonly accepted thinking. An important aspect of the scheme discussed above is that the rate limiting step of the cross-bridge cycle occurs in the attachment of myosin to actin (Table 2.7.1), with the remaining steps occurring much faster .

The cross-bridge model, as discussed here, has dominated our thinking on muscular contraction for the past 50 years. It does not account for all observed phenomena. In fact, one might argue that the cross-bridge model neglects some basic phenomena, such as the long-lasting, history-dependent force production of muscle following stretch or shortening (Abbott & Aubert, 1952; Edman et al., 1978; Maréchal & Plaghki, 1979; Sugi & Tsuchiya, 1988; Granzier & Pollack, 1989; Edman et al., 1993). Therefore, it is quite conceivable that the cross-bridge model will evolve further, and, possibly, may be revised or replaced in the future. However, at present, it represents the paradigm of choice and strong experimental evidence and convincing theory will be required to replace the cross-bridge model.

Table 2.7.1 **Actin-myosin hydrolysis cycle (rabbit skeletal muscle). See text for details (adapted from Howard, 2001, with permission).**

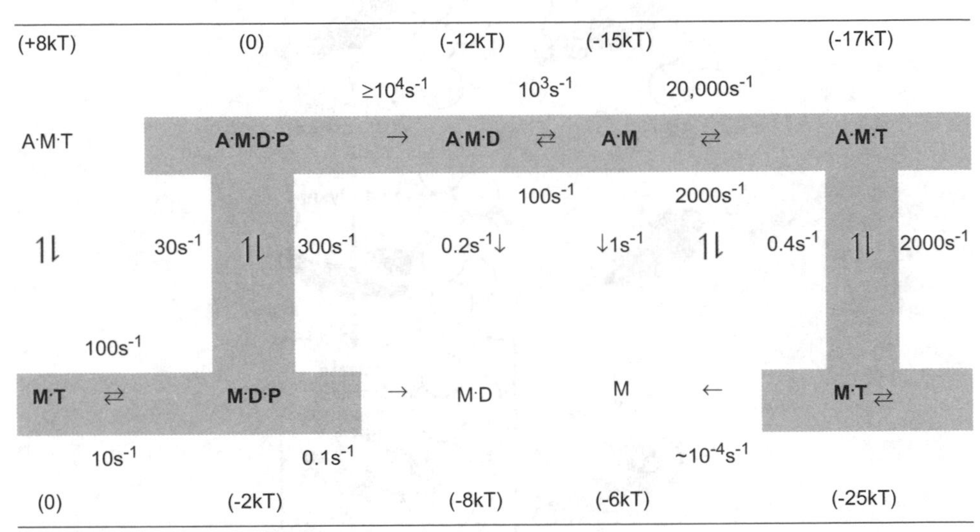

M = myosin, T = ATP, D = ADP, P = phosphate, A = actin

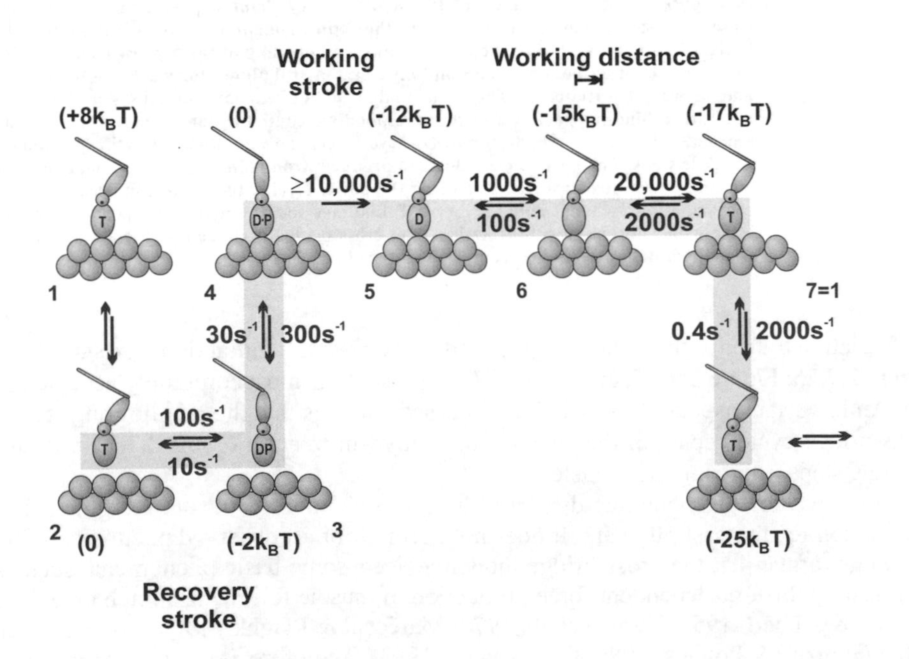

Figure 2.7.26 **Schematic illustration of the current thinking of the cross-bridge cycle, including a best estimate of the rate constants of the reactions. Adapted from Howard (2001), with permission.**

2.7.5 PHYSICAL PROPERTIES

Since muscles are active force-producing structures, one may argue that they do not have unique material or mechanical properties, or if basic material properties such as force-elongation relations are measured passively, they are not meaningful for the understanding of muscular function. In contrast, the force-elongation properties of passive musculo-skeletal structures such as ligaments, bones, and cartilage are essential to an understanding of their function in the intact biological system. Nevertheless, two properties of muscles are repeatedly used in biomechanical experiments involving muscles or the musculo-skeletal system. These two properties are the force-length and the force-velocity relation, discussed below. Another property of skeletal muscle is the history dependence of force production following active stretch or shortening. This property, although first described over half a century ago (Abbott & Aubert, 1952), has been virtually ignored in biomechanical research. In contrast to the force-length and force-velocity properties, which are well explained within the framework of the Cross-bridge theory, history-dependent force production is not part of the cross-bridge thinking, and is not trivial to account for without a conceptual adjustment of the cross-bridge model. History dependence of skeletal muscle force production is introduced and discussed following the considerations on the force-length and force-velocity properties.

Force-length and force-velocity relations of skeletal muscular tissues have been determined on the sarcomere, isolated fibre, isolated muscle, and the intact muscle level; and depending on the level of interest, these relations have to be interpreted differently. Furthermore, the terms force-length and force-velocity relations suggest an experimental procedure or a theoretical thinking governed by defined conditions. For example, it is implied here that force-length relations of a muscle are obtained under isometric conditions with the muscle maximally activated.

Maximal muscular activation is a term that may have to be treated liberally in this context. In an in vitro preparation of a single muscle fibre, activation can be adjusted until maximal activation, i.e., the highest possible force, is achieved. In experiments involving intact human skeletal muscles, maximal activation often is associated with maximal voluntary efforts, which, in the strictest sense, are not maximal in terms of absolute force production.

Force-length and force-velocity properties differ among muscles, and these differences likely reflect the functional demands of everyday activities. On occasion, we discuss muscular properties as they relate to the functional demands, and so go beyond just describing the material properties.

FORCE-LENGTH RELATION

Force-length relations describe the relation that exists between the maximal force a muscle (or fibre or sarcomere) can exert and its length. Force-length relations are obtained under isometric conditions and for maximal activation of the muscle. Isometric may refer to the length of the entire muscle, the length of a fibre or the length of a sarcomere, depending on the system that is studied.

Over a century ago, Blix (1894) noted that the force a muscle can exert depends on its length. In 1966, Gordon, Huxley and Julian published the results of a classic study in which they showed the dependence of force production in isolated fibres of frog skeletal muscle on sarcomere length. Their results were in close agreement with theoretical predictions of the Cross-bridge theory, and they helped establish the Cross-bridge theory as the primary paradigm of muscle contraction.

According to the Cross-bridge theory, cross-bridges extend from thick to thin filaments and cause sliding of the myofilaments past one another. Each cross-bridge is assumed to generate, on average, the same amount of force and work independently of the remaining cross-bridges. Since cross-bridges are believed to be arranged at equal distances along the thick filament, overlap between thick and thin filaments determines the number of possible cross-bridge formations and thus the total force that may be exerted.

For frog skeletal muscle, thick filament lengths are reported to be about 1.6 μm and thin filament lengths are reported to be about 0.95 μm (Page & Huxley, 1963; Walker & Schrodt, 1973). If the width of the Z-disc is about 0.1 μm and the H-zone (cross-bridge free zone in the middle of the thick filament) is 0.2 μm, a theoretical force-length relation for frog sarcomeres may be calculated (Figure 2.7.27). At long sarcomere lengths, thick and thin filaments cease to overlap and no cross-bridges can be formed. The corresponding force must be zero. For frog striated muscle, zero force is reached at a sarcomere length of about 3.6 μm (= thick filament length (1.6 μm), plus twice the thin filament length (1.9 μm), plus the width of the Z-disc (0.1 μm)). This sarcomere length is indicated in Figure 2.7.27 with label e.

Shortening of the sarcomeres increases the number of potential cross-bridge formations in a linear fashion with sarcomere length or correspondingly, thick and thin filament overlap, until a maximal number of cross-bridge formations are possible (Figure 2.7.27, label d). This optimal overlap corresponds to a sarcomere length of 2.20 μm in frog muscle, i.e., twice the thin filament length (1.9μm), plus the width of the Z-disc (0.1 μm), plus the width of the H-zone (0.20 μm). Further sarcomere shortening to 2.0 μm [Figure 2.7.27, label c; twice the thin filament length (1.9 μm), plus the width of the Z-disc (0.1 μm)] increases the area of overlap between thick and thin myofilaments, but does not increase the number of possible cross-bridge formations, since the middle part of the thick filament does not contain cross-bridges. Therefore, isometric force remains constant between 2.0 and 2.20 μm.

Sarcomere shortening below 2.0 μm has been associated with a decrease in force caused by interference of thin myofilaments as they start to overlap. Below 1.7 μm [(Figure 2.7.27, label b; thick filament length (1.6 μm), plus width of Z-disc (0.1 μm)], the rate of decrease in force becomes higher than between 1.7 and 2.0 μm. This steeper decline in force has been associated with the force required to deform the thick filament. At a sarcomere length of 1.27 μm, forces determined experimentally in frog striated muscle fibres became zero (Gordon et al., 1966).

The most important result in support of the Cross-bridge theory was the linear relation between force and length for sarcomere lengths between 2.2 and 3.6 μm (Figure 2.7.27). However, the strict linearity of this relation was questioned by several investigators who showed non-linear force-length behaviour on this so-called descending limb of the force-length relation, e.g. ter Keurs et al. (1978). The difference between studies showing linear and non-linear force-length behaviour was in the length control of the sarcomeres. Those studies showing a linear relation kept sarcomeres at a constant (controlled) length, whereas, those showing a non-linear relation kept fibre length at a constant length but allowed for

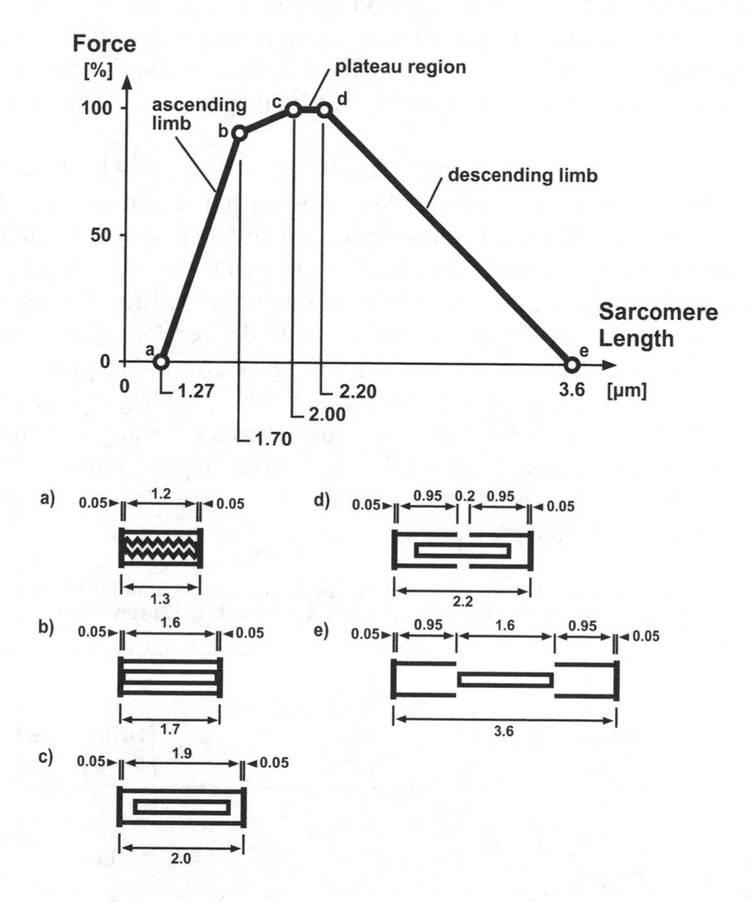

Figure 2.7.27 **Theoretical force-length relation of isolated fibres of frog skeletal muscle. Note the labelling of the schematic sarcomeres corresponds to the labels shown on the force-length graph. Adapted from Gordon et al., (1966), with permission.**

non-uniform changes in sarcomere length. This latter situation approaches actual physiologic conditions more appropriately and thus may be more relevant in studying intact skeletal muscles. The mechanisms that allow for the enhancement of fibre force when sarcomere lengths are changing in an isometric fibre preparation have not been elucidated in detail.

Traditionally, the decrease in external force of frog skeletal muscle below 2.0 μm has been associated with the double overlap of thin myofilaments and the corresponding interference caused by this arrangement. The decrease in force below 1.7 μm has been attributed to forces required to deform the thick myofilament. This view of force decrease on the so-called ascending limb of the force-length relation has been challenged by experiments that showed that Ca^{2+} release from the sarcoplasmic reticulum is length-dependent. Ruedel and Taylor (1971), in experiments on skeletal muscle, and Fabiato and Fabiato (1975), in experiments on cardiac muscle, showed that Ca^{2+} release from the sarcoplasmic reticulum was decreased at muscle fibre lengths shorter than optimal length. Adding caffeine to the activating solution enhanced Ca^{2+} release at short muscle fibre lengths and substantially increased maximal force. These results suggest that incomplete activation, i.e., decreased Ca^{2+} release from the sarcoplasmic reticulum, plays as much a role in decreased force production at fibre lengths below optimal length as factors typically associated with this decrease.

According to the Cross-bridge theory, force-length relations may be determined mathematically if the lengths of the thick and thin myofilaments are known. There is general agreement that thick myofilament lengths are nearly constant among many animal species (approximately 1.6 μm). However, thin myofilament lengths vary significantly among animals (Table 2.7.2) and sometimes even within the same animal. The influence of these differences in thin myofilament lengths on theoretically derived force-length properties of frog, cat, and human skeletal muscles is shown in Figure 2.7.28. The plateau regions and descending limbs of these curves were obtained strictly according to predictions of the Cross-bridge theory; the ascending limbs were determined assuming that interference of thin myofilaments at sarcomere lengths below the plateau region (Figure 2.7.27) had the same effect on the rate of force decrease, and that zero force was reached at sarcomere lengths of 1.27μm for all fibre preparations.

Table 2.7.2 Differences in thin myofilament length, L_{thin}, among animals (from three different sources: Page & Huxley, 1963; Walker & Schrodt, 1973; Herzog et al., 1992a).

ANIMAL	L_{thin} PAGE & HUXLEY [μm]	L_{thin} WALKER & SCHRODT [μm]	L_{thin} HERZOG ET AL. [μm]
CAT	-	-	1.12
RAT	-	1.04	1.09
RABBIT	1.07	-	1.09
FROG	0.975	0.925	-
MONKEY	-	1.16	-
HUMAN	-	1.27	

The plateau regions and the descending limbs (Figure 2.7.27) are identical for frog, cat, and human skeletal muscles, except for a shift along the sarcomere length axis. The width of the plateau corresponds to the width of the cross-bridge free zone in the middle of the thick myofilaments, i.e., the H-zone, here, assumed to be 0.17 μm. The length of the descending limb corresponds to the length of the thick myofilament minus the H-zone, i.e., 1.60 μm - 0.17 μm = 1.43 μm, and is identical for all three muscles shown. The shift of the plateau and descending limb regions along the sarcomere length axis between muscles from different animals is twice the difference in thin myofilament length. For example, thin myofilament lengths in human and frog skeletal muscle differ by 0.32 μm, i.e., 1.27 μm to 0.95 μm (Table 2.7.2), causing a corresponding shift of 0.64 μm.

Differences in thin myofilament lengths cause differences in the force-length properties. If we assume that during evolution muscular systems have optimized their properties to satisfy everyday functional requirements, one may speculate on the differences in functional requirements between, for example, human and frog skeletal muscle. The range of active force production appears to be larger for human compared to frog muscle. From this

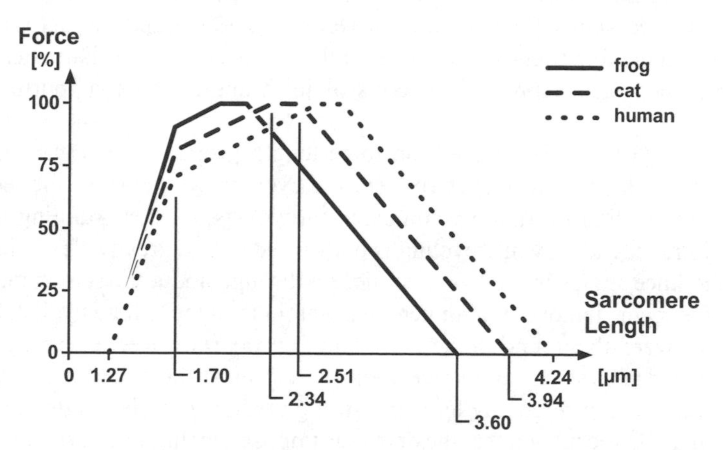

Figure 2.7.28 **Influence of differences in thin myofilament lengths on theoretically derived force-length properties of frog, cat, and human skeletal muscles.**

observation, it may be speculated that fibre lengths in human muscles are shorter relative to their normal, everyday operating range than frog muscles, and that this disadvantage is partially offset by a sarcomere design that can produce active force over a large range of lengths. A study in which muscle fibre lengths of many animals were related to the animal's size supported this speculation. Fibre lengths did not scale proportionally with animal size, but were relatively shorter for large animals compared to small animals (Pollock & Shadwick, 1994).

Another difference between human and frog sarcomere force-length relations is that the ascending limb of the frog curve is small compared to that for humans (Figure 2.7.28). Furthermore, there is a faster decrease in force per unit of sarcomere shortening for frog compared to human muscle. Therefore, it appears that human muscles may be better suited than frog muscles to operate on the ascending part of the force-length relation. Studies aimed at determining what part of the force-length relation is actually used by a muscle during normal, everyday movement tasks are rare and cannot be used as conclusive support for the speculation made above. However, the little data that are available tend to fit the speculation. For example, Mai and Lieber (1990) reported that the frog semitendinosus muscle works almost exclusively on the descending limb for a jumping movement (an everyday locomotor activity for a frog). However, Herzog and ter Keurs (1988) found force-length relations of intact human rectus femoris muscles that appear to be on the ascending and descending limbs of the force-length curves for anatomical ranges of knee and hip joint configurations. Intact human gastrocnemius muscles were found to operate virtually exclusively on the ascending limb of the force-length relation within the anatomical range of ankle and knee joint angles (Herzog et al., 1991b).

Force-length relations have typically been treated as a constant, invariant property of muscles. However, it has been speculated that force-length properties adapt to everyday functional demands, and that such adaptations occur within days or weeks. Support for the notion that force-length properties are associated with functional demands may be found in studies in which force-length relations have been determined for muscles comprising a functional unit. Such muscles satisfy similar functional demands, and thus, would be expected to have similar force-length properties. Studies on the triceps surae and plantaris group of the striped skunk (Goslow & Van DeGraaff, 1982) and the cat (Herzog et al., 1992b) show that the force-length properties of these muscles are similar when normalized to peak force and when expressed in terms of joint angles, thus supporting the above hypothesis.

Herzog et al. (1991a) determined the force-length properties of intact human rectus femoris muscles in high performance runners and cyclists. Runners use the rectus femoris in an elongated position in training, compared to cyclists, because running is performed with the hip joint extended, while cycling is performed with a greatly flexed hip joint. The corresponding knee angles in the two activities go through about the same range. Thus, the demands on the rectus femoris of a runner are different than those for a cyclist. Corresponding differences were also found in the force-length properties. Rectus femoris muscles of the runners tended to be strong at relatively long and weak at short muscle lengths, whereas, this muscle was weak at relatively long and strong at short muscle lengths in the cyclists. Therefore, it may be speculated that the demands imposed by high performance training are sufficient to alter the force-length properties of intact human skeletal muscles.

Since the force-length properties in the above study were obtained using maximal, voluntary effort contractions, it is not known if the differences in force-length properties between the two groups of athletes were caused by neurophysiological or mechanical mechanisms. A possible mechanical mechanism could be the addition or deletion of in series sarcomeres in muscle fibres.

FORCE-VELOCITY RELATION

Force-velocity relations are defined here as the relation that exists between the maximal force of a muscle (or fibre) and its instantaneous rate of change in length. Force-velocity properties are determined for maximal activation and are typically obtained at optimal length of the sarcomeres (Figure 2.7.29).

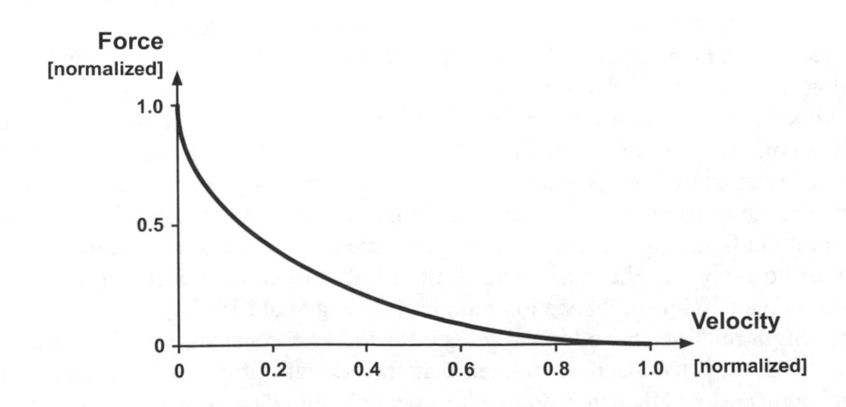

Figure 2.7.29 Normalized force-velocity relationship of skeletal muscle showing the well-known non-linear relationship between the speed of shortening of a muscle and its steady-state external force.

At the beginning of the twentieth century, it was recognized that the efficiency of human movement varied as a function of movement speed. At that time, it was found that for a given amount of work, the energy used increased with increasing speeds of muscular contraction, i.e., efficiency decreased. Fenn and Marsh (1935) were among the first to perform experiments on force-velocity properties of muscles. Their work was followed by the classic study of Hill (1938), who said that he stumbled over the force-velocity relation while working on problems of heat production in isolated frog muscles. Hill derived the force-velocity property of muscle by applying a force to a muscle and measuring the corresponding velocity of shortening, i.e., equation (2.7.26):

$$v = b(F_o - F)/(F + a) \tag{2.7.26}$$

where:

v = velocity of shortening
F_o = maximal force at zero velocity and optimal sarcomere length
F = instantaneous force
a = constant with unit of force
b = constant with unit of velocity

In biomechanics, one often thinks of the force-velocity relation as imposing the speed of movement and measuring the corresponding force. Many experiments on intact human skeletal muscles have been performed based on this idea using so-called isokinetic strength testing machines, e.g., Thorstensson et al. (1976) and Perrine & Edgerton (1978). In such cases, one may want to solve equation (2.7.26) for F, to illustrate that v is the independent and F the dependent variable:

$$F = (F_o b - av)/(b + v) \tag{2.7.27}$$

Setting v equal to zero in equation (2.7.27) corresponds to measuring the force under isometric conditions. For this situation F becomes equal to, F_o.

Setting the external force that is acting on the muscle (F) equal to zero, one can solve equation (2.7.27) for v, which, under these circumstances, corresponds to the maximal velocity of shortening (v_o):

$$v_o = b(F_o/a) \tag{2.7.28}$$

Equation (2.7.28) may be rearranged to yield:

$$a/F_o = b/v_o = \text{constant} \tag{2.7.29}$$

Typical values for a/F_o have been reported to be about 0.25 for skeletal muscles from a variety of animals including frog (Hill, 1938), rat (Close, 1964), and kittens (Close & Hoh, 1967).

Equations (2.7.26) or (2.7.27) can be obtained for in vitro fibre or muscle preparations by determining F_o, and then F and v, for a variety of different velocities of contraction. The constants a and b may then be determined in such a way that the corresponding equation gives a best fit to the experimental data.

In biomechanics, it is often of interest to describe force-velocity properties of intact human skeletal muscles. This description is by no means trivial, as it is virtually impossible to measure muscle forces in vivo, and velocities of shortening of the contractile elements can only be obtained for the most restricted experimental situations. However, in vivo force-velocity relations may be obtained by first estimating F_o and v_o, and then solving equation (2.7.29) for the constants a and b. Once a and b have been determined, equations (2.7.26) and (2.7.27) may be used with input of forces to calculate the corresponding velocities or with input of velocities to calculate the corresponding forces.

To estimate force-velocity properties of intact human skeletal muscles, it is necessary to know the physiological cross-sectional area (PCSA) and the average, optimal fibre length (l_o) of the muscle of interest. The physiological cross-sectional area (PCSA) is directly related to F_o and the average, optimal fibre length (l_0) of the muscle of interest is directly related to v_o.

Let us assume, for example, that we would like to estimate the force-velocity property of the human vastus lateralis muscle, and that average values from a variety of literature sources are 50 cm^2 for PCSA and 12 cm for l_o. Furthermore, research on mammalian skeletal muscles indicates that:

$$F_o \approx 25 N/cm^2 \cdot PCSA \tag{2.7.30}$$

and further, that:

$$v_o \approx 6 \cdot l_o/s \tag{2.7.31}$$

for muscles predominantly consisting of slow twitch fibres, and:

$$v_o \approx 16 \cdot l_o/s \tag{2.7.32}$$

for muscles predominantly comprised of fast twitch fibres, e.g., Spector et al. (1980). Therefore, $F_o = 1250$ N and $v_o = 72$ cm/s or 192 cm/s, depending on the fibre type content of the vastus lateralis. Since human skeletal muscles are typically of mixed fibre type composition, one may choose a statistical approach to derive force-velocity relations. Force-velocity properties of a muscle of mixed fibre composition may be calculated by separating the whole muscle into units of slow and fast fibres or better still, into a continuum of fibres ranging in properties from slow to fast, and weighing their contribution to the total force-velocity behaviour according to information available on the distribution of fibre types within the muscle (Hill, 1970).

From equations (2.7.30), (2.7.31), and (2.7.32), the constants a and b may be determined using:

$$a/F_o = b/v_o = 0.25 \tag{2.7.33}$$

for slow twitch fibres:

$a = 0.25 \cdot 1250\ \text{N}$

$b = 0.25 \cdot 72\ \text{cm/s}$

for fast twitch fibres:

$a = 0.25 \cdot 1250\ \text{N}$

$b = 0.25 \cdot 192\ \text{cm/s}$

Having determined the constants a and b as well as F_0 for the human vastus lateralis, its force as a function of the speed of shortening can be calculated using equation (2.7.27).

Since Hill's equation (2.7.26) was originally derived at muscle temperatures of 0°C, the question arises whether equation (2.7.33) also holds for physiological muscle temperatures, i.e., 37°C in humans. Values for a/F_0 were said to be largely temperature independent (Hill, 1938), although this is not strictly correct. However, it has been often assumed, as a first approximation, that the ratio of a/F_0 remains about constant over a range of muscle temperatures.

For many considerations, the power output of muscles is of prime importance. For example, in power sports, such as sprinting, or throwing and jumping, the peak power capability of muscles is considered a crucial determinant for success. Power (P) is an instantaneous scalar quantity, and muscle power is defined as the product of force and velocity ($F \cdot v$). Therefore, for a given force-velocity relation of a muscle, its instantaneous power ($P(v)$) may be determined throughout the range of shortening speeds (Figure 2.7.30). For many practical applications, it is of interest to calculate at what speed of shortening absolute maximal power, P_o, is reached.

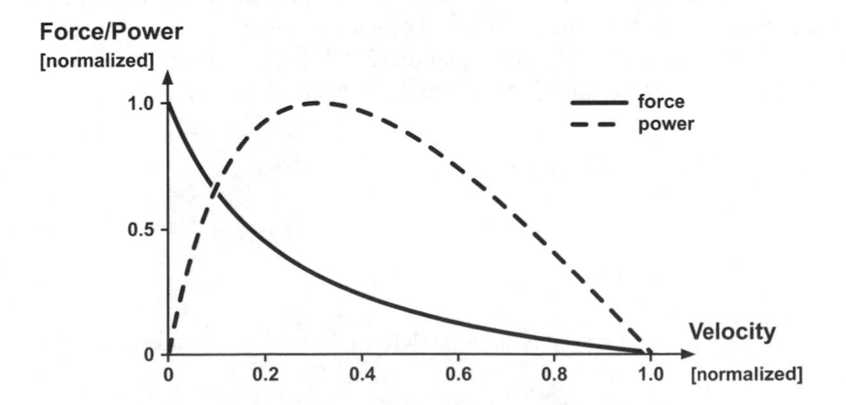

Figure 2.7.30 **Normalized force-velocity and power-velocity relationship of skeletal muscle. The power curve is obtained by multiplying force with the corresponding velocity. Typically, peak power is reached at a speed of muscle shortening of about 30-35% of the maximal, unloaded shortening speed.**

By definition:

$$P(v) = F(v) \cdot v \tag{2.7.34}$$

where:

$$\frac{dP(v)}{dv} = \frac{dF}{dv}v + F(v) \tag{2.7.35}$$

and using equation (2.7.27) gives:

$$\frac{dP(v)}{dv} = \frac{(F_o + a)b^2 - a(v+b)^2}{(v+b)^2} \tag{2.7.36}$$

Realizing that dP(v)/dv needs to be zero for P(v) to become maximal, i.e., P_o, one has:

$$0 = \frac{(F_o + a)b^2 - a(v+b)^2}{(v+b)^2} \tag{2.7.37}$$

Solving equation (2.7.37) for the velocity (v_m) at which P_o occurs yields:

$$v_m = b(\sqrt{(F_o/a)+1} - 1) \tag{2.7.38}$$

Solving equation (2.7.33) for a and b and substituting into equation (2.7.38) gives:

$$v_m = \frac{v_o}{4}(\sqrt{4+1} - 1) \tag{2.7.39}$$

or:

$$v_m \approx 0.31 v_o \tag{2.7.40}$$

which means that the speed of shortening at which maximal muscular power may be produced is about 31% of the maximal speed of shortening.

Substituting equation (2.7.39) into equation (2.7.27) and rearranging terms, it is possible to calculate the force produced at the maximal speed of shortening, v_m:

$$F(v_m) = \frac{F_o}{4}(\sqrt{4+1} - 1) \tag{2.7.41}$$

or:

$$F(v_m) \approx 0.31 F_o \tag{2.7.42}$$

The maximum power (P_o) may then be determined using equations (2.7.34), (2.7.39) and (2.7.41):

$$P_o = \frac{(\sqrt{4+1} - 1)^2}{16} F_o \cdot v_o \tag{2.7.43}$$

or:

$$P_o = 0.095 \cdot F_o \cdot v_o \tag{2.7.44}$$

Hill's (1938) force-velocity relation has been used for a long time. Recent evidence from experiments on single fibres of frog semitendinosus and tibialis anterior suggests that only the range of about 5 to 80% of the isometric force (F_o) may be approximated well with Hill's (1938) equation (Edman, 1979). Using Hill's equation, the maximal isometric force is typically overestimated and the maximal velocity of shortening is underestimated (Figure 2.7.31). Since v_o is typically not known, mathematical techniques have been used

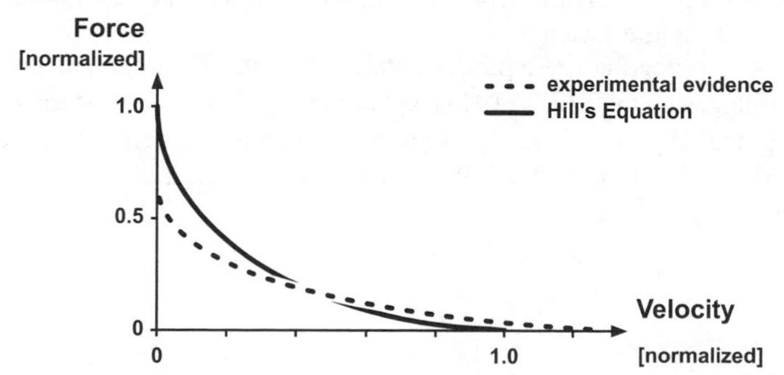

Figure 2.7.31 **Schematic illustration of the difference between theoretical estimation and experimental determination, e.g., Edman (1979) of the force-velocity relationships.**

to predict v_o theoretically. Based on the findings of Edman (1979), this mathematical approximation may underestimate the actual shortening velocity at zero load. The difference in theoretical and experimental results may be explained by arguing that v_o obtained experimentally is strictly a measurement of the fastest contracting fibres in a entire muscle, whereas, the theoretically determined value represents an estimate of the average maximal velocity of shortening of all fibres (Hill, 1970). However, this argument does not explain the discrepancy found by Edman (1979) when using single fibre preparations.

The difference between predicted and experimentally determined values for F_o in single fibre preparations has been explained by Edman (1979) as reflecting an inability of all possible cross-bridges to attach simultaneously during isometric contractions. This inability occurs because of the different repeat patterns of cross-bridge binding sites on the thin myofilaments and cross-bridges on the thick filaments. Edman further argued that this factor becomes less relevant at high velocities of shortening, because the proportion of attached cross-bridges is smaller at any given time. Therefore, more attachment sites are available for cross-bridge binding at high, compared to low, velocities of shortening.

The velocity of unloaded shortening (v_o) has been reported to remain virtually constant for a large range of sarcomere lengths, i.e., between about 1.65 and 2.7 μm for frog skeletal muscle (Edman, 1979). Since isometric tension varies considerably between 1.65 and 2.70 μm, this finding suggests that v_o is largely independent of the proportion of attached cross-bridges.

In frog skeletal muscle, maximal isometric force starts to decline below sarcomere lengths of about 2.0 μm (Figure 2.7.27). Traditionally, it has been assumed that this decrease in force is associated with some internal resistance, e.g., Gordon et al. (1966). However, if this were the case, one would expect v_o to decrease as well. Therefore, the loss of isometric force below 2.0 μm does not appear to be primarily a consequence of internal resistance, but possibly may be the result of an inability of the contractile elements to produce forces as high as F_o. This idea is supported by experiments on force-length properties in which Ca^{2+} release from the sarcoplasmic reticulum was enhanced artificially, and forces remained close to F_o for sarcomere lengths below 2.0 μm. This phenomenon has been described earlier in this chapter.

Maximal velocity of shortening decreases sharply below sarcomere lengths of 1.65 μm, and increases dramatically above about 2.70 μm. These two phenomena may safely be associated with increases in internal resistance because of thick filament deformation at the short sarcomere length, and with increases in muscular force caused by parallel elastic elements at the long sarcomere length.

Force-velocity properties are typically obtained at optimal sarcomere lengths. However, it has been suggested that Hill's (1938) equation may also be valid at sarcomere lengths other than optimal, by replacing F_o with the maximal isometric force at the target sarcomere length (Abbott & Wilkie, 1953). If F_o is replaced in this way, the velocity of unloaded shortening becomes:

$$v_o \quad = \quad bF_o l(x)/a \tag{2.7.45}$$

where:

$l(x)$ = a value between 0 and 1.0, representing the normalized force as a function of sarcomere length

Since the value for $l(x)$ in equation (2.7.45) decreases as soon as sarcomere lengths deviate from optimal, and since F_o, b, and a are all constants, v_o will be smaller for non-optimal compared to optimal sarcomere lengths. However, according to Edman's (1979) experiments on single frog fibres, v_o is not influenced within a range of sarcomere lengths of 1.65-2.70 μm. Thus, rather than replacing F_o by the maximal isometric force at the sarcomere length of interest ($= l(x) \cdot F_o$), the entire right hand side of equation (2.7.27) may be multiplied by $l(x)$:

$$F \quad = \quad [(F_o b - av)/(b + v)]l(x) \tag{2.7.46}$$

This equation appears to approximate experimental observations better than the one suggested above. In particular, maximal velocity of unloaded shortening remains as shown in equation (2.7.28), and the maximal isometric force at any sarcomere length F'_o would become $F'_o = F_o\ l(x)$.

When referring to velocity of shortening of a muscle or a fibre, it is typically implied that this is an average velocity. However, it has been argued that fibres in muscles may not shorten uniformly, and further, that sarcomeres within a fibre have a distinct maximal velocity of shortening (Edman & Reggiani, 1983). Therefore, the concept of uniform fibre or sarcomere shortening may not be adequate and may influence the force-velocity properties of a muscle. Further research in this area is needed.

So far, we have discussed the force-velocity properties of shortening muscle. When a muscle is stretched at a given speed, its force exceeds the maximal isometric force, F_o reaching an asymptotic value of about $2F_o$ at speeds of stretching much lower than the maximal velocity of shortening (Lombardi & Piazzesi, 1992). Also, using isotonic (Katz, 1939) or isokinetic stretches (Edman et al., 1978), there appears to be a discontinuity in the force-velocity relation across the isometric point: the rise in force associated with slow stretching is much greater than the fall in force associated with the corresponding velocities of shortening. In contrast to the force-velocity relation during shortening, the force-velocity relation during stretching is rarely described using a standard equation, such as the hyperbolic relation proposed by Hill (1938) for shortening. The primary reason for this discrepancy is that force-velocity properties during stretching have been investigated much less, and that these properties are not as consistent as those obtained during shortening. A mathematical description of the entire force-velocity relation (shortening and stretch) may be found in Epstein and Herzog (1998).

HISTORY DEPENDENCE

Skeletal muscle force production is history-dependent. This fact has been known for a long time, e.g. Abbott and Aubert (1952), but has not received much attention, probably for three primary reasons:

1) History dependence is not explicitly accounted for in the two basic models of muscle contraction used in biomechanics and motor control (the Hill model, and the Huxley or cross-bridge model), thus its implementation would be associated with modifications of accepted approaches,
2) History-dependent effects are often considered to be of insignificant magnitude, and
3) The mechanisms underlying history dependence are not known, and no generally accepted explanation has emerged.

History dependence of force production is defined by the effects of recent contractile conditions on muscle force. The word muscle is used here generically, and may mean an actual muscle, a single fibre, or a single myofibril. History dependence is hard to determine for dynamic movements with varying activation levels, although it likely occurs under such conditions. For the purpose of investigation and analysis, history dependence has been studied primarily for well-defined conditions, and, consequently, it is typically assessed by the decrease or increase of an isometric steady-state force at a given level of activation that is caused by shortening or lengthening, or a combination of shortening and lengthening. The increase or decrease of force is determined relative to the force produced by a purely isometric reference contraction at the corresponding length (Figure 2.7.32). Typically, force depression is produced by shortening of an activated muscle, and force enhancement by stretching of an activated muscle (Figure 2.7.32). Here, we only consider shortening-induced force depression, and stretch-induced force enhancement, although the combination of stretching and shortening produces some surprising results, e.g. Herzog & Leonard (2000) and Lee et al. (2001).

Force depression increases for increasing magnitudes of shortening (Figure 2.7.33a) (Abbott & Aubert, 1952; Maréchal & Plaghki, 1979; Meijer et al., 1998; Meijer, 2002; Josephson & Stokes, 1999; De Ruiter & de Haan, 2003), for increased forces during shortening (Figure 2.7.33b) (De Ruiter et al., 1998), and for decreased speeds of shortening (Maréchal & Plaghki, 1979; Ettema & Meijer, 2000; De Ruiter et al., 1998; Josephson & Stokes, 1999). Force depression has also been shown to be long-lasting, i.e. in excess of 20 s in mammalian skeletal muscle at physiologic temperatures, (Herzog et al., 1998), and force depression can be abolished instantaneously by deactivating the muscle for a period of

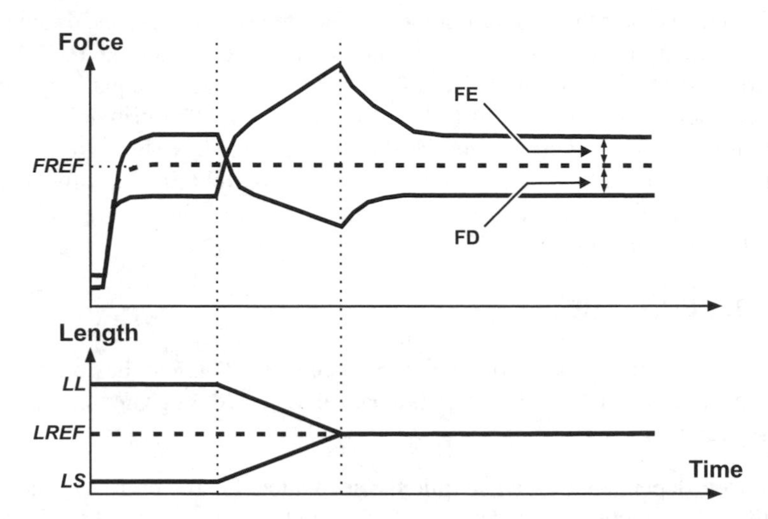

Figure 2.7.32 **Force enhancement (FE) following stretch, and force depression (FD) following shortening of an activated muscle. Force enhancement and force depression are measured at steady-state against a purely isometric reference contraction (FREF) at the final length of the stretch and shortening test contractions.**

time long enough for force to drop to zero (Figure 2.7.34) (Abbott & Aubert, 1952; Herzog & Leonard, 1997). These properties are well-accepted and have been observed in numerous muscles and single fibres. A slight controversy exists with respect to the relationship of force depression with the speed of shortening. A low speed of shortening (for a given activation level) produces a high force, and a high speed of shortening produces a small force (Hill, 1938). Therefore, it has not been easy to identify whether force depression is independently related to the speed of shortening, or if the speed of shortening is merely indirectly associated with the amount of force depression through the changing force, in accordance with the force-velocity property of muscle. We have shown that keeping the speed of shortening constant and changing the force during shortening produces a predictable relationship between the amount of force depression and the force during shortening (Herzog & Leonard, 1997). However, when keeping the force constant while changing the speed of shortening, force depression does not change appreciable (Leonard & Herzog, 2005), suggesting that the force during shortening, and not the speed, is important for decreasing the force following active shortening. Since force depression is related to the magnitude and force of shortening, it is obvious that it is also tightly related to the amount of mechanical work performed by the muscle during the shortening phase (Herzog et al., 2000; Josephson & Stokes, 1999) (Figure 2.7.35).

Four major mechanisms have been proposed to explain force depression of skeletal muscles following active shortening. The first of these is related to the idea that shortening produces a vast accumulation of free phosphate and hydrogen ions, thus essentially producing a fatigue-like effect that causes a reduction in force (Granzier & Pollack, 1989). One prediction arising from this proposed mechanism is that force depression must be long-lasting, as the fatigue-like processes would take minutes to be fully abolished. However, it has been shown in several investigations that force depression can be abolished virtually instantaneously by a deactivation of the muscle for a short period (Figure 2.7.34). Thus, this first mechanism is likely not causing force depression.

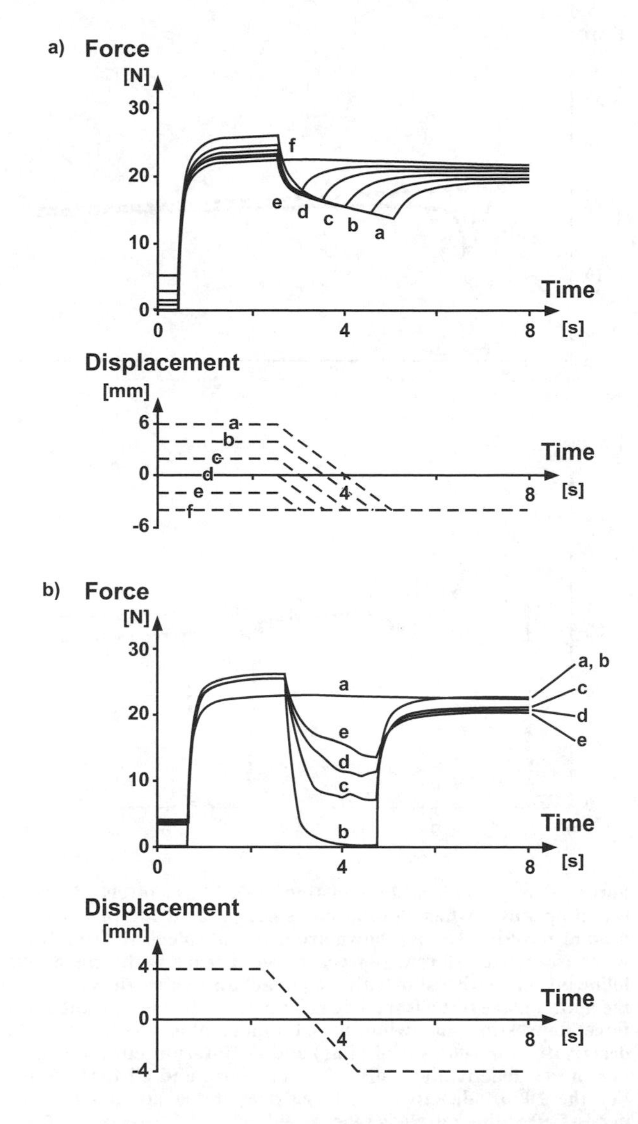

Figure 2.7.33 **Force depression increases with increasing magnitudes of muscle shortening (a) and with increasing force during shortening (b). Results shown are from cat soleus at 37°C. The different force levels in 2b were achieved by changing the stimulation voltage. Similar results were obtained when changing shortening forces by altering motor unit stimulation frequencies (not shown).**

The second mechanism is based on the idea that there is an inhibition of calcium affinity, and thus presumably cross-bridge attachments, during shortening compared to isometric conditions (Edman, 1996). Therefore, force should be depressed immediately following shortening, but then is predicted to recover quickly (within a second or two) to the normal, isometric, steady-state level. However, force does not completely recover following shortening (Abbott & Aubert, 1952; Herzog & Leonard, 2000), thereby eliminating this second mechanism as a serious candidate for explaining force depression.

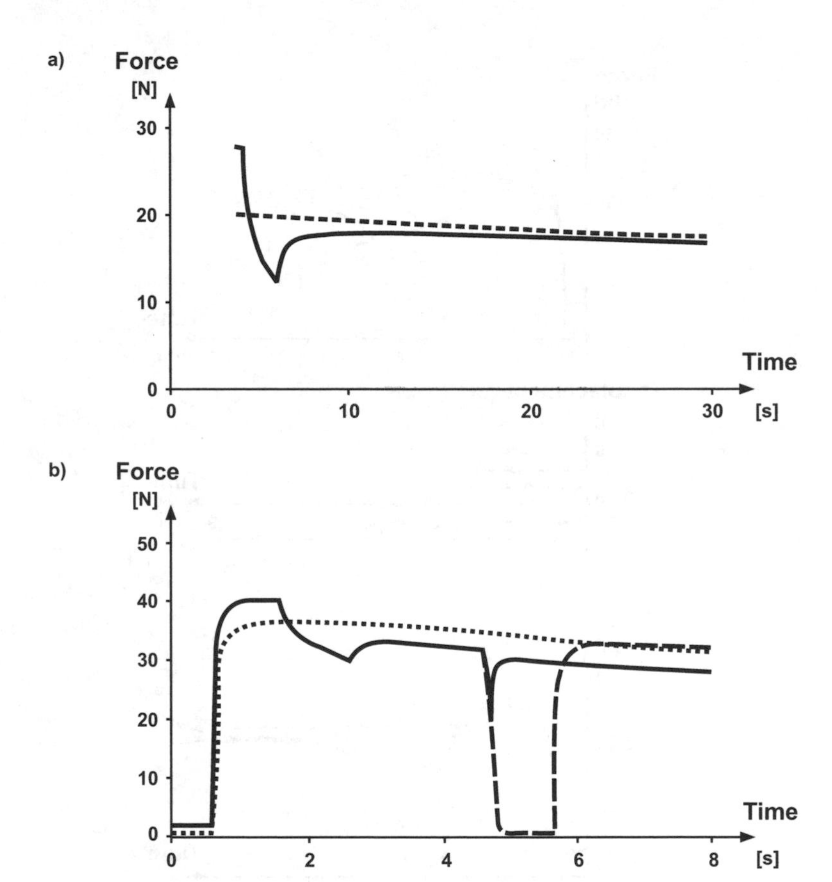

Figure 2.7.34 **Force depression is long-lasting (a), but force depression is abolished instantaneously when the muscle is deactivated for just long enough for force to drop to zero (b). Results shown are from cat soleus at 37°C. In (a), the dashed trace is the isometric reference force, the solid trace is the steady-state isometric force following the active shortening contraction (which finishes at ~ time = 7s). In (b), the dotted trace is the isometric reference contraction, while the solid line shows the force depression following muscle shortening. At time ~ 4.5s, the muscle is deactivated for 100ms (solid line) and 1s (intermittent line) and then is reactivated. For the 1s deactivation, force drops to zero, and all force depression is abolished. For the 100ms deactivation, force does not drop to zero, and force depression persists after the muscle is reactivated.**

The third mechanism is associated with the idea of the development of sarcomere length non-uniformities. These non-uniformities are assumed to occur because of an inherent instability of sarcomere lengths on the descending limb of the force-length relationship (Hill, 1953). Therefore, force depression has been said to only occur on the descending, but not the ascending, limb of the force-length relationship (Morgan et al., 2000; Zahalak, 1997). However, there is ample evidence demonstrating force depression on the ascending limb of the force-length relationship (Herzog & Leonard, 1997; De Ruiter et al., 1998). Furthermore, it has been demonstrated that sarcomere lengths in single myofibrils with all sarcomeres situated on the descending limb of the force-length relationship, can be perfectly stable (Rassier et al., 2003a). Together, these observations question the validity of the sarcomere length non-uniformity theory.

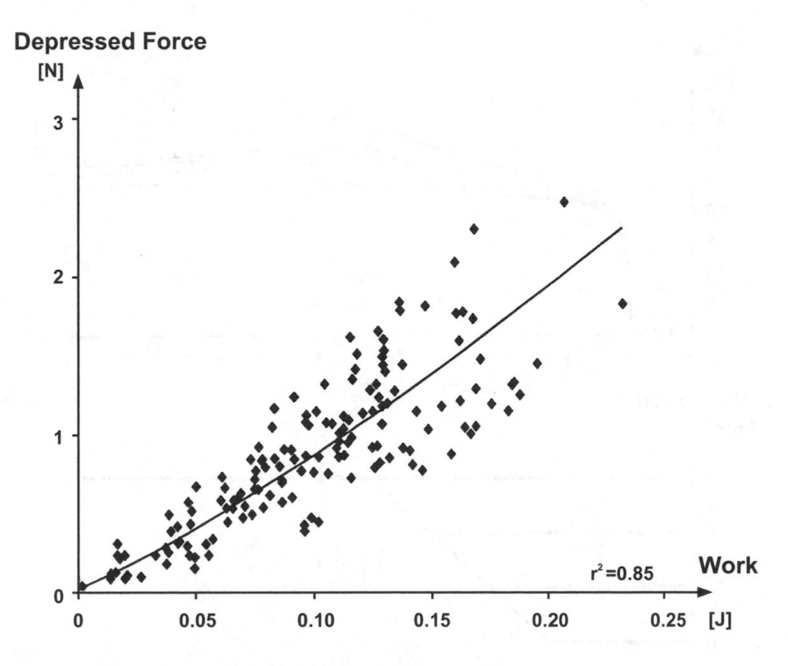

Figure 2.7.35 **Absolute force depression as a function of absolute work produced by cat soleus (37°C) during shortening. Individual data points are from six muscles and six shortening tests each (multiple trials). The coefficient of variation is extremely high (r^2 = 0.85) considering that no normalization was performed to account for differences in muscle strength.**

The final mechanism is associated with the idea that force depression is caused by a stress-dependent inhibition of cross-bridge attachments in the actin-myosin overlap zone that is formed during the shortening phase (Maréchal & Plaghki, 1979). Force production before shortening is assumed to cause an extension of the actin filaments (Huxley et al., 1994; Higuchi et al., 1995; Kojima et al., 1994; Wakabayashi et al., 1994). Upon shortening, these distorted actin filaments overlap with the myosin filaments, and cross-bridge attachment might be affected because of the distortion (Daniel et al., 1998). If so, force depression should increase with increasing overlap between distorted myofilaments, which is directly related to the amount of shortening, and furthermore, should be directly related to the stress acting on the myofilaments, which is related to the muscle force. Furthermore, force depression should last as long as stress on the myofilaments is maintained, and should be abolished instantaneously upon stress release (Abbott & Aubert, 1952). At present, all data on shortening-induced force depression conceptually agree with the idea of a stress-induced cross-bridge inhibition.

If force depression is caused by inhibition of cross-bridge attachments, and thus, a reduction in the number of cross-bridges in the force depressed compared to the normal isometric reference contraction, then stiffness of the muscle should decrease in direct proportion with the amount of force depression. Sugi and Tsuchiya (1988), for single fibres, and Lee and Herzog (2003), for human adductor pollicis, demonstrated that fibre and muscle stiffness decrease with the amount of force depression observed for tests performed for a variety of shortening magnitudes and speeds.

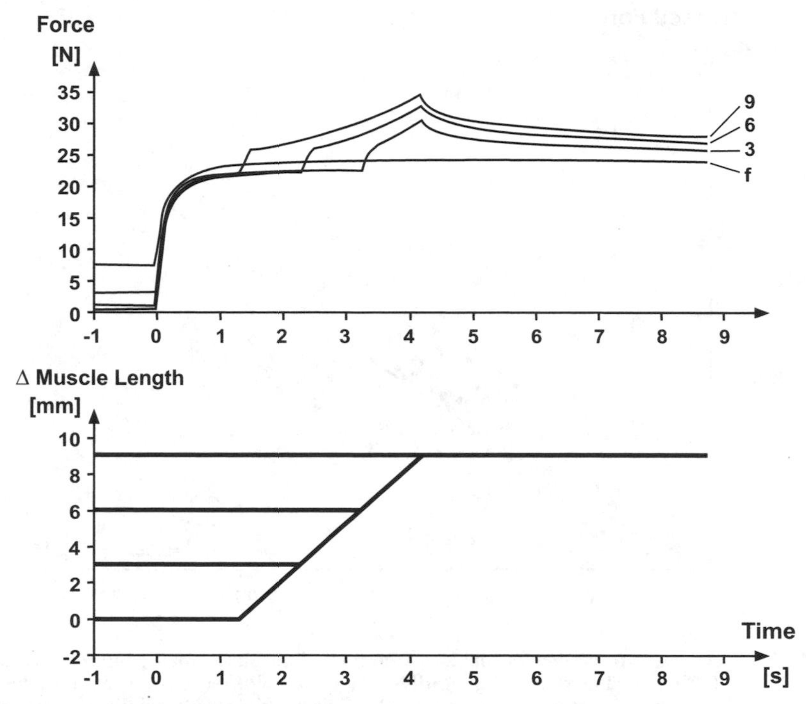

Figure 2.7.36 **Force enhancement increases with increasing magnitudes of stretch. Results shown are from cat soleus at 37°C and stretch magnitudes of 3, 6, and 9 mm. All stretches were performed on the descending limb of the force-length relationship.**

Force enhancement increases with increasing magnitudes of stretch (Figure 2.7.36), appears to be mostly unaffected by the speed of stretch, and has been associated with the descending limb of the force-length relationship exclusively (Edman et al., 1982; Morgan et al., 2000). Some of these properties need discussion, as they are not universally accepted. First, some scientists have found a relationship between the speed of muscle stretch and force enhancement (Abbott & Aubert, 1952; Rassier et al., 2003b; Sugi, 1972). However, the changes in force enhancement were small, suggesting that force enhancement may depend on the speed of stretching, but likely just to a small degree. Second, there has been some controversy if force enhancement could be obtained on the ascending limb of the force-length relationship. Proponents of the "sarcomere length non-uniformity theory" argued that force enhancement cannot occur on the ascending limb of the force-length relationship, e.g., Morgan et al. (2000) and Morgan (1990). However, these same investigators showed small but consistent force enhancement on the ascending limb in some of their own studies. For example, Figure 2.7.37 shows data by Morgan et al. (2000) (their Figure 3a). The force enhancement calculated for the eleven data points corresponding to the ascending limb of the force-length relationship ranged from 0 to 15.2%, with an average value across all 11 observations of 6.0%. These results are like those obtained in our own investigations [(5% force enhancement on the ascending limb of cat soleus (Herzog & Leonard, 2002)], and thus support the idea that force enhancement does exist on the ascending part of the force-length relationship. Experiments with single fibres confirmed the results obtained on whole muscles (Figure 2.7.38) (Peterson et al., 2004).

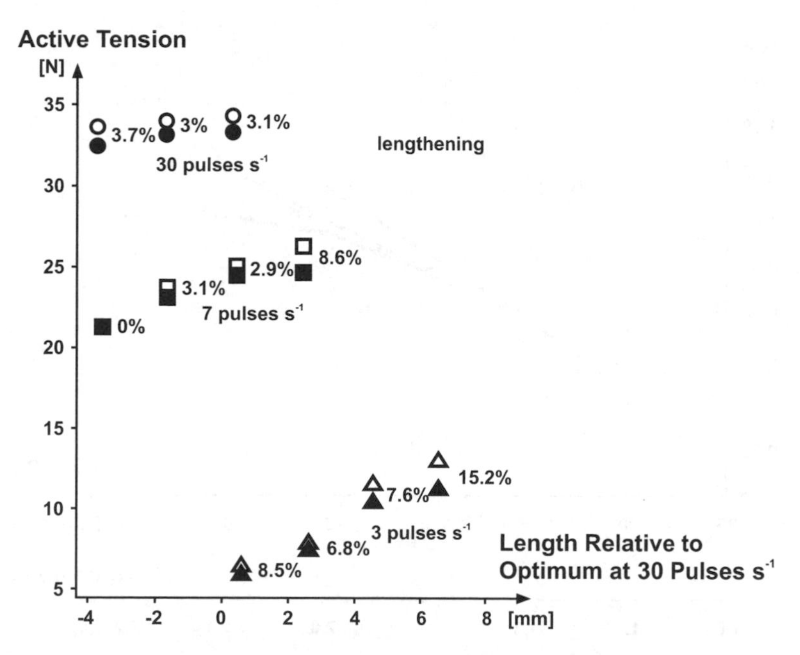

Figure 2.7.37 **Active tension during isometric reference contractions (filled symbols), and for isometric, steady-state conditions following stretch of activated cat soleus at physiological temperature (open symbols). All results are from the ascending limb of the force-length relationship and for stimulation frequencies of 30, 7, and 3Hz. The original data were presented by Morgan et al. (2000), and are reproduced here with permission from the Journal of Physiology. Note, that the force enhancement values were calculated by us from the original graph. Force enhancement was present in 10 of the 11 cases shown, ranging from 0 to 15.2%, with an average value across all observations of 6.0%. Note further that statistical analysis (non-parametric, repeated measures t-tests) reveals that force enhancement shown in this figure is statistically significant ($p < 0.01$).**

Only one mechanism has been proposed in the past to explain force enhancement following stretching of an activated muscle: the sarcomere length non-uniformity theory. This theory is based on the idea that there are instabilities in sarcomere lengths on the descending limb of the force-length relationship, as first proposed by Hill (1953). These instabilities are then thought to lead to substantial differences in individual sarcomere lengths. Some sarcomeres are thought to elongate until they are caught by passive forces, while others shorten onto the ascending limb of the force-length relationship (Allinger et al., 1996). This theory provides at least two testable predictions. First, there should be no force enhancement on the ascending limb of the force-length relationship. Second, force enhancement should not exceed the maximal steady-state isometric force at optimal muscle length. Both these predictions were found to be incorrect in single fibre and whole muscle experiments (Schachar et al., 2004; Peterson et al., 2004; Herzog et al., 2003; Rassier et al., 2003a; Morgan et al., 2000—their Figure 3a).

Therefore, there is no ready explanation for the force enhancement observed following stretching of an activated muscle. However, the following observations might provide hints to some contributing factors. When a muscle, or a single fibre, is stretched to a length at which there is a naturally occurring passive force, then passive forces are enhanced follow-

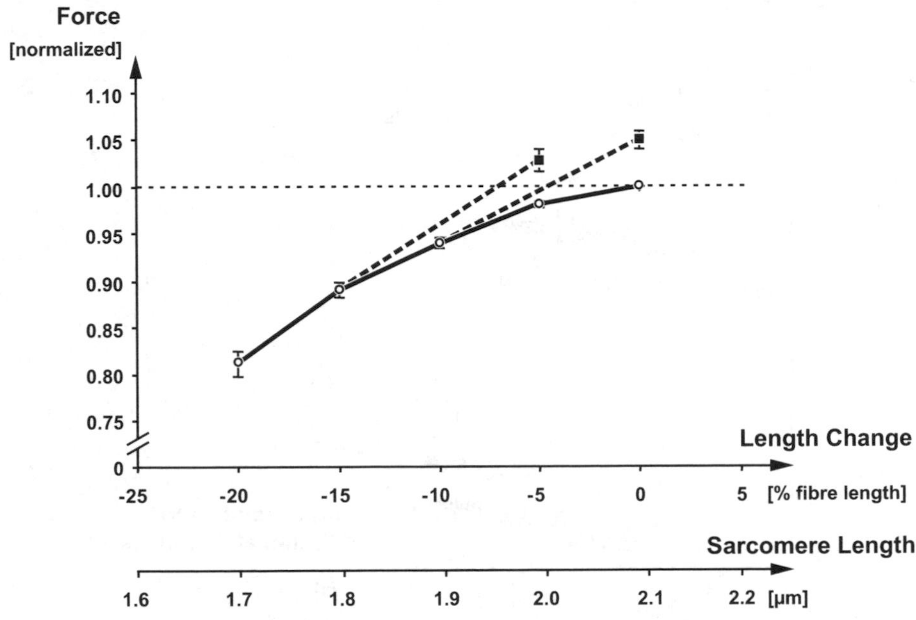

Figure 2.7.38 **Isometric forces (round symbols) and isometric forces following a 10% stretch (square symbols) as a function of fibre length and sarcomere length. Results were obtained from single frog fibres (lumbrical muscles) at about 8°C. Fibre lengths are normalized to optimal length (0% - about 3mm), and sarcomere lengths were obtained by assuming an optimal length of 2.1 μm and uniform sarcomere length changes proportional to the length changes of the fibres.**

ing active (but not passive) stretching (Figure 2.7.39). Therefore, it appears that at least part of the force enhancement might be explained by the engagement of a passive structural element. This idea has been proposed previously (Edman et al., 1982; Noble, 1992; De Ruiter et al., 2000; Herzog, 1998), but has only received experimental support recently (Herzog & Leonard, 2002; Lee & Herzog, 2002; Rassier et al., 2003b). We (Herzog, 1998) and others, e.g., Noble (1992), have suggested that the molecular spring titin might produce this passive force enhancement through a calcium induced increase in its stiffness (Labeit et al., 2003).

Muscle stiffness, in the force enhanced state, is greater than in the corresponding isometric reference state (Herzog & Leonard, 2000), and the rate of force decay upon deactivation is decreased (Rassier et al., 2003b). It is not clear how these two observations might be explained. However, the following scenario could account for the enhanced force, increased stiffness, and decreased rate of force decay in the force-enhanced compared to the isometric reference state. Imagine that the rate of cross-bridge detachment was decreased following stretch of attached cross-bridges. In the Cross-bridge theory, this would be equivalent to decreasing the rate constant of detachment [g(x), Equation 2.7.2] (Huxley, 1957). This would cause an increase in the proportion of attached cross-bridges, which in turn would be associated with an increase in force, an increase in stiffness, and a decrease in the rate of force decay. In other words, a stretch-induced decrease in Huxley's rate of cross-bridge detachment could explain many of the features observed in force enhancement following active muscle stretching.

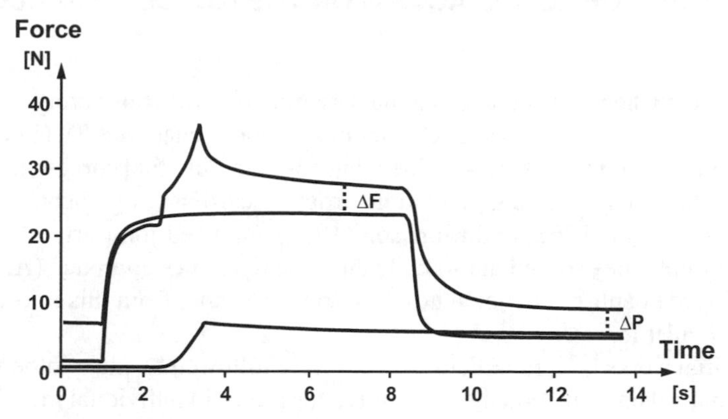

Figure 2.7.39 **Example of force enhancement (Δ F) and passive force enhancement (Δ P) in cat soleus following a 9 mm stretch on the descending limb of the force-length relationship, 37°C.**

2.7.6 APPLICATIONS

Muscle mechanics have countless practical applications. Experiments on muscles, single fibres or even single myofilaments are often aimed at studying the mechanisms underlying muscular force production.

The field of muscle mechanics could be extremely useful in sports biomechanics for predicting optimal performances of athletes or improving the outcome of an athletic activity. However, basic muscle mechanics are virtually non-existent in the field of sports biomechanics. For example, a large number of papers are related to the study of track and field, yet one would be hard-pressed to identify articles that attempt to maximize performance based on principles of muscle mechanics.

Bicycling is one of the few sports that has received systematic attention from a muscle mechanical point of view. Bicycling is a cyclic, virtually planar, and kinematically easy to describe activity, and offers itself better than most other sports to the study of muscle mechanics. Electromyographic work has given insight into muscular coordination for submaximal efforts of bicycling, e.g., Hull & Jorge (1985). Houtz & Fischer (1959), and Gregor et al. (1985). Furthermore, muscular action, and in particular inter-action and co-contraction of two-joint antagonist pairs of muscles, has been studied using bicycling as a model (Andrews, 1987). The performance of maximal effort sprint bicycling was optimized based on principles of muscle mechanics (Yoshihuku & Herzog, 1990).

However, most applied studies in muscle mechanics appear to be aimed at two major problems:

(1) The determination of loads acting on the musculo-skeletal system, in particular in joints of the human body, and

(2) The study of muscular force interactions during movement.

DETERMINATION OF LOADS ACTING ON THE MUSCULO-SKELETAL SYSTEM

The precise mathematics underlying the determination of loads acting on the musculo-skeletal system is discussed in a later chapter of this book (chapter 3.7). Here, it suffices to realize that one of the major forces influencing so-called internal forces (forces acting on the musculo-skeletal system, e.g., including articular cartilage, ligaments, and bones) are muscular forces. Paul (1965) and Morrison (1968) analyzed joint articular forces in the knee and hip joint. They solved the so-called inverse dynamics approach (Andrews, 1974), calculated forces of entire functional groups of muscles, and, from this information, calculated joint articular forces.

These initial works on joint articular forces were followed by many others, which often used mathematical optimization theory to solve for forces in individual muscles rather than in entire functional groups of muscles (Seireg & Arvikar, 1973; Penrod et al., 1974; Crowninshield, 1978; Crowninshield & Brand, 1981). Although individual muscle forces need not be known to calculate joint articular forces, they are required for determining local force and stress distributions in and around joints. Such information is typically relevant when approaching problems of artificial joint design, and tendon or ligament reconstructive surgery.

Since muscular forces cannot be measured directly in humans without significant ethical implications and serious problems of calibration, studies predicting muscle forces are typically theoretical or based on indirect measures of muscular activation, i.e., electromyography. We compared the predictions of a series of theoretical models aimed at estimating individual muscle forces to experimental muscle force measurements in an animal model, and concluded that none of the theoretical models agreed well with the experimental results (Herzog & Leonard, 1991). It will be a challenge to derive models that accurately predict individual muscle forces in an intact biological system.

STUDIES OF MOVEMENT CONTROL

Knowing the force-time history of a muscle (or even better, of all muscles in the system of interest) during movement, is like having a tiny window to the brain and its complex organization of voluntary movement. Thus, force-time histories of muscles have been predicted theoretically in the same way as described in the previous paragraphs, except with a different purpose in mind: the study of movement control, e.g., Pedotti et al., (1978), Dul et al. (1984), and Herzog (1987). Direct experimental force measurements have been performed to study interactions of muscles during movement. This type of work was pioneered by Walmsley et al. (1978), who studied the interaction between cat soleus and medial gastrocnemius for a variety of locomotor activities. This study was followed by a series of similar experiments, e.g., Hodgson (1983) and Whiting et al., (1984), probably because these two muscles are easily accessible to chronic force measurements, and because the force-time behaviour of the one-joint, predominately slow twitch, parallel fibred soleus muscle turned out to be quite different (and thus interesting) from the force-time histories of the two-joint, predominately fast twitch, pennate fibred medial gastrocnemius muscle. Abraham and Loeb (1985) measured forces from cat hindlimb muscles other than just soleus and medial gastrocnemius, but never for more than two muscles simultaneously.

To gain further insight into the force-sharing behaviour of muscles during locomotion, we decided to simultaneously study all the muscles in a functional group, plus one antagonist muscle. In our experimental protocol, forces from gastrocnemius, soleus, plantaris, and tibialis anterior muscles of the cat hindlimb were measured using standard tendon force transducers (Herzog & Leonard, 1991; Herzog et al., 1993). The first three of these muscles make up the Achilles tendon, and so are the major plantar flexors of the ankle joint in the cat; the tibialis anterior is one of the primary dorsiflexors of the ankle and may be considered a direct antagonist of the one-joint soleus.

Figure 2.7.40 shows representative plots of force-sharing between two muscles of the plantar flexor group for walking at nominal speeds of 0.4, 0.7, 1.2 m/s and trotting at 2.4 m/s. All force-sharing loops shown at each speed are averages over 10 consecutive step cycles. The arrows indicate the direction of force build-up and decay. Average peak gastrocnemius forces increase from about 15 N, and plantaris forces increase from about 18 N, at a walking speed of 0.4 m/s, to values of approximately 75 N and 45 N for trotting at a speed of 2.4 m/s. This change in force corresponds to a 5-fold increase for gastrocnemius and a 2.5-fold increase for peak plantaris forces from the slowest to the fastest experimental speeds shown. Corresponding average peak soleus forces remain between 17 to 20 N for all speeds. Therefore, it appears that increased force demands on the Achilles tendon, caused by increased speeds of locomotion, are accompanied by dramatic changes in force-sharing among the muscles involved.

To test if these changes in force-sharing among the cat ankle plantar flexor muscles are speed- or effort-dependent, muscle force measurements were also performed at a constant walking speed (0.7 m/s), but different external resistances. Changes in resistance were introduced by changing the slope of the walking surface from 10° downhill (D), to level (L), to 10° uphill (U). Conceptually, the results obtained using this protocol were like those obtained for level walking at different speeds (Figure 2.7.41). Therefore, changes in force-sharing among cat gastrocnemius, soleus, and plantaris muscles appear to be effort- rather than just speed-dependent..

From the previous two figures, it may appear that gastrocnemius and plantaris forces change significantly in response to changes in speed or resistance of locomotion, whereas, forces of the soleus remain relatively constant. This observation is not quite correct. When analyzing a series of consecutive steps at the same nominal speed, changes in soleus forces from step-to-step may vary considerably. For example, Figure 2.7.42 shows the mean force-time curves (M) for one animal walking at a nominal speed of 1.2 m/s on the treadmill. Means were obtained from 43 step cycles obtained during one experimental test, i.e., about 30 seconds. Also shown are the means of the three stride cycles yielding the highest (H) and lowest (L) peak forces in those same 43 steps. Gastrocnemius and plantaris peak forces range from approximately 30 to 45 N and from 27 to 33 N, much less than the range of peak forces measured for different walking speeds (Figure 2.7.40). However, peak soleus forces range from about 15 to 22 N, a range larger than corresponding changes of the average peak values for speeds of locomotion ranging from 0.4 m/s to 2.4 m/s. It appears, therefore, that soleus forces are sensitive to small variations in kinematics from one stride cycle to the next, whereas, corresponding average values are not sensitive to large changes in the speed of locomotion.

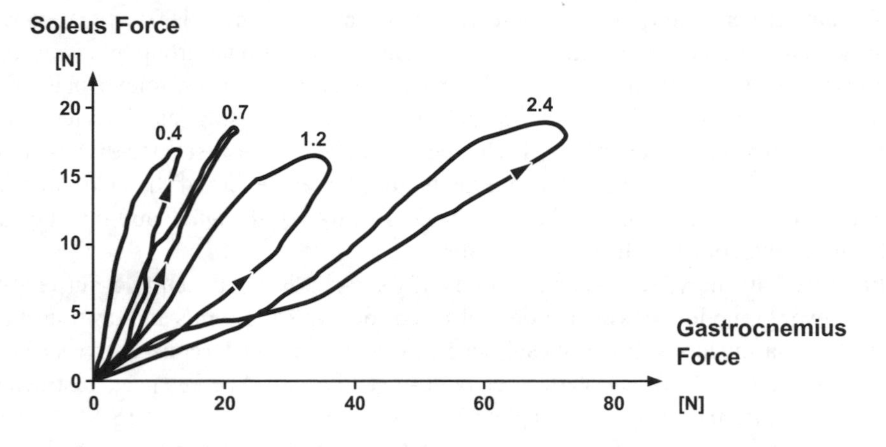

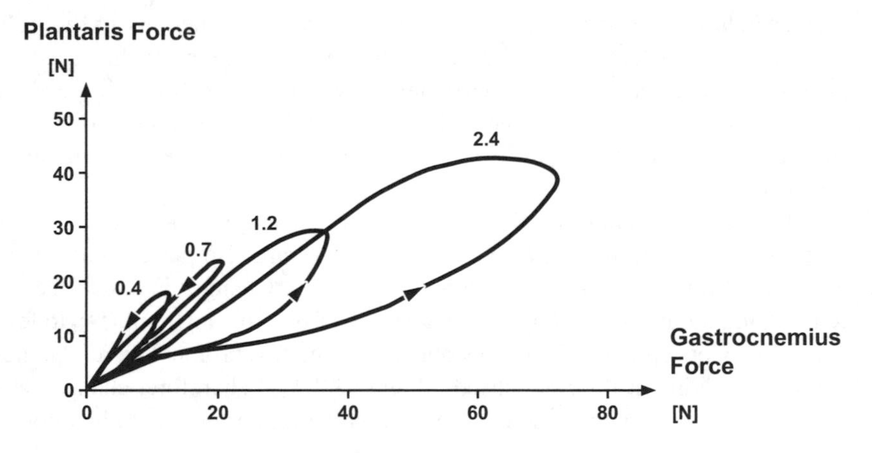

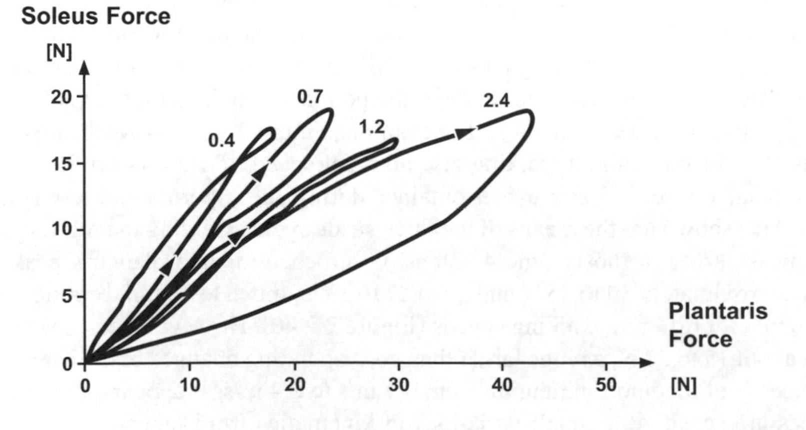

Figure 2.7.40 **Representative plots of force-sharing between muscles of the ankle plantar flexor group in the cat when walking at nominal speeds of 0.4, 0.7, 1.2 m/s and trotting at 2.4 m/s. All curves shown are averages obtained over 10 consecutive and complete step cycles.**

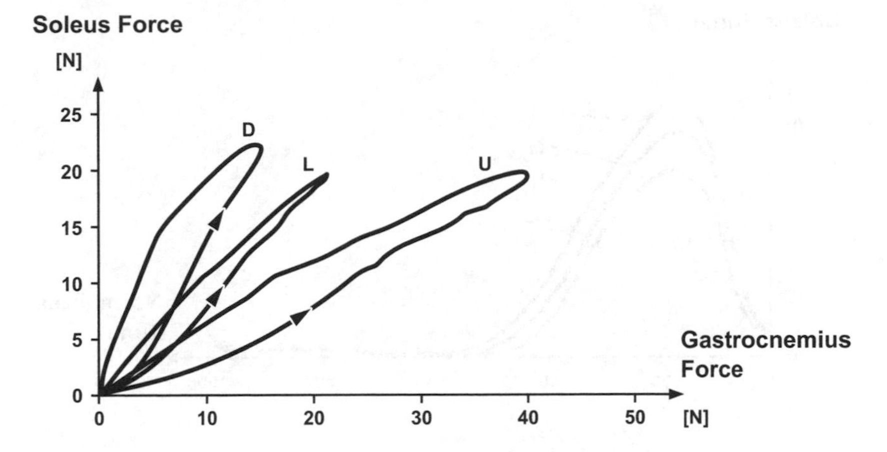

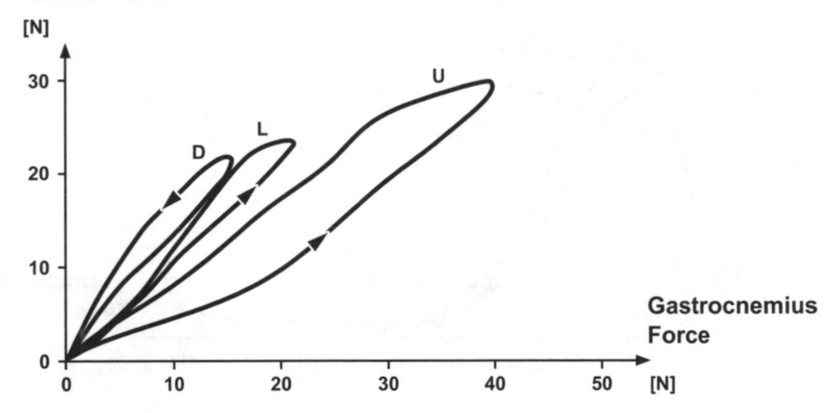

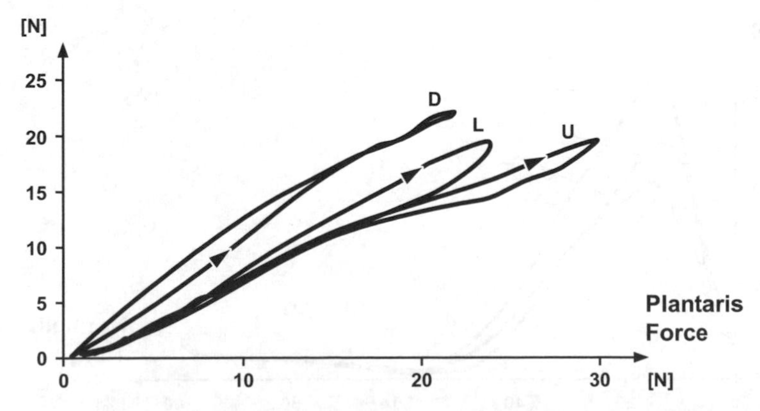

Figure 2.7.41 **Representative plots of force-sharing between muscles of the ankle-plantar flexor group in the cat when walking at a speed of 0.7 m/s at different external resistances. Changes in external resistance were introduced by changing the slope of the walking surface from 10° downhill (D), to level (L), to 10° uphill (U). All curves shown are averages obtained over 10 consecutive and complete step cycles.**

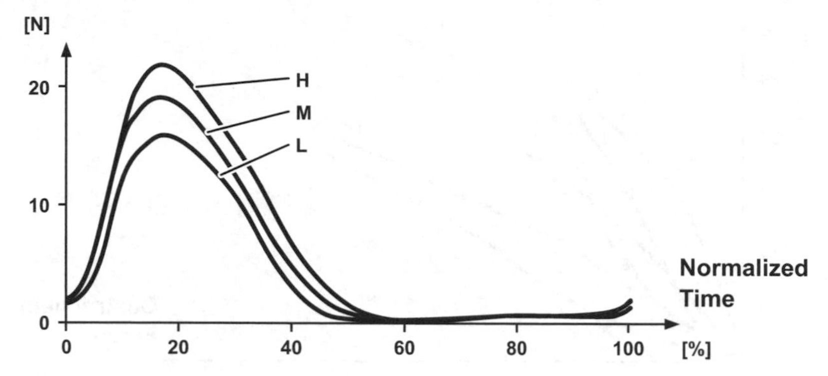

Plantaris Force

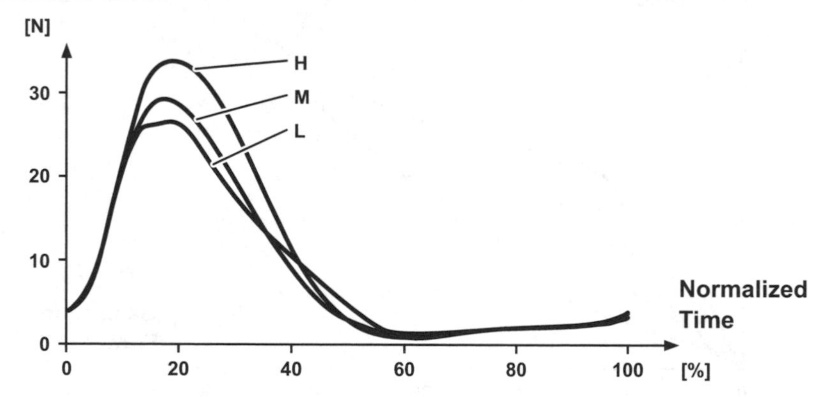

Gastrocnemius Force

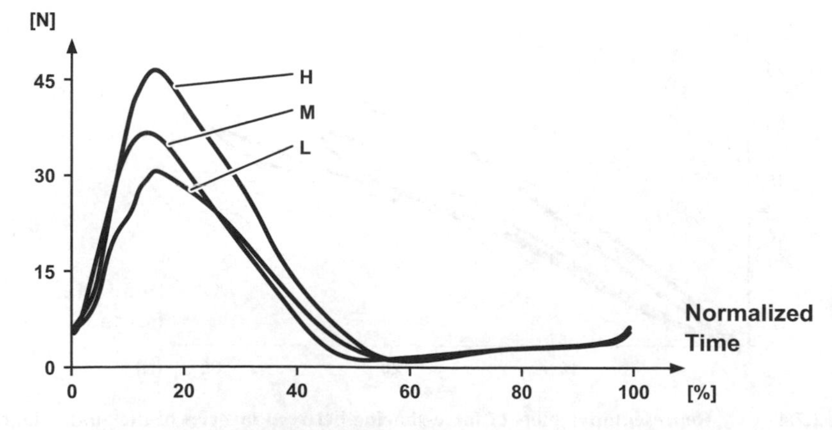

Figure 2.7.42 **Mean force-time curves (M) for one animal walking at a nominal speed of 1.2 m/s on a motor-driven treadmill. Means were obtained over 43 step cycles. Also shown are the mean force-time histories of the three steps yielding the highest (H) and lowest (L) peak forces in these 43 steps.**

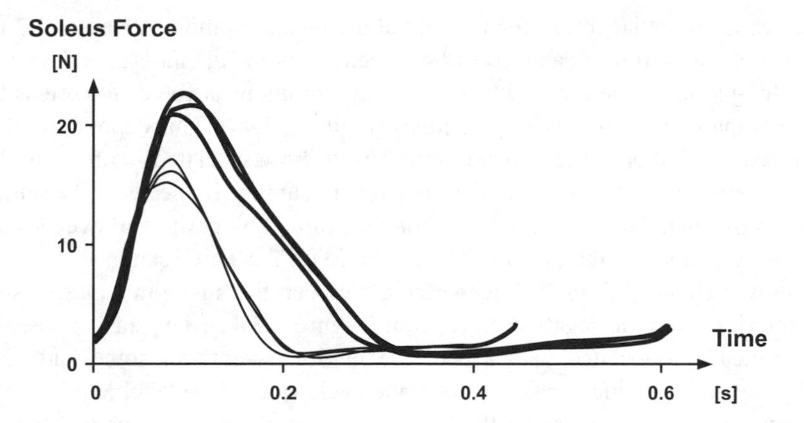

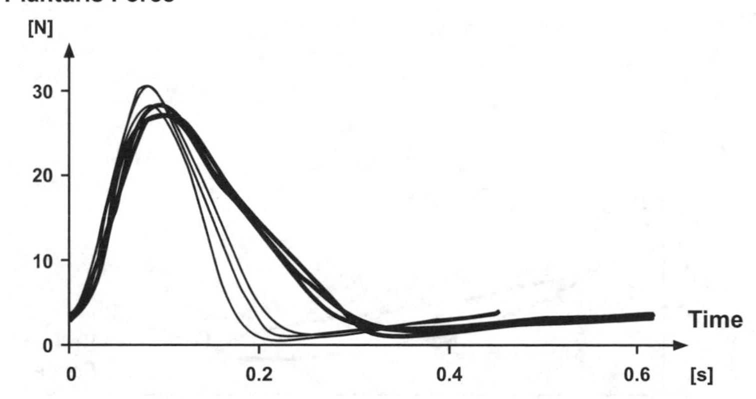

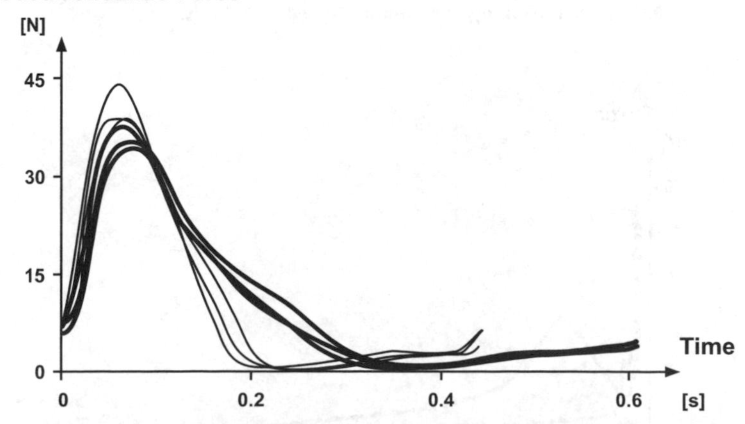

Figure 2.7.43 **Force-time curves for soleus, plantaris, and gastrocnemius muscles of the cat when walking at a nominal speed of 1.2 m/s. Variations in stride cycle time (horizontal axis) were associated with corresponding peak forces for the three muscles. Soleus forces tended to increase with increasing stride cycle times, whereas, forces in gastrocnemius and plantaris tended to decrease with increasing stride cycle times.**

Changes in muscular forces for walking at a constant nominal speed of 1.2 m/s were related to stride cycle time. Peak forces of gastrocnemius and plantaris muscles tended to be small for long stride cycle times, whereas, corresponding peak forces for soleus tended to be large, compared to short stride cycle times (Figure 2.7.43). This depression of the peak soleus forces for short stride cycle times must be associated with central control phenomena, since it is obvious from the force records at higher speeds of the same animal, i.e., 2.4 m/s trotting, that peak forces are not constrained by peripheral events such as the force-velocity relation or the time available to build up the active state.

For slow walking (0.7 m/s), force-sharing between the antagonist pair of soleus and tibialis anterior shows a negative correlation (Figure 2.7.44), i.e., an increase in soleus force is typically associated with a decrease in tibialis anterior force and vice versa, especially during the stance phase of the stride cycle (label S to label M_S, Figure 2.7.44). This negative correlation is exemplified by the low tibialis anterior force at the instant of peak soleus force (P_S), and the low soleus force at the instant of peak tibialis anterior force (P_{TA}, Figure 2.7.44).

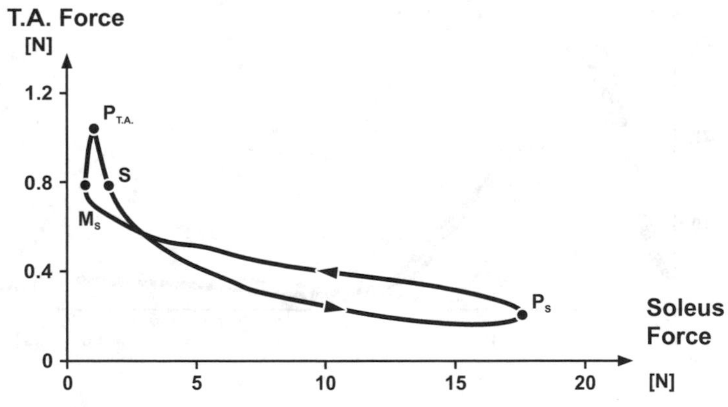

Figure 2.7.44 **Force-sharing between the antagonist pair of soleus and tibialis anterior muscles in the cat when walking at a nominal speed of 0.7 m/s.**

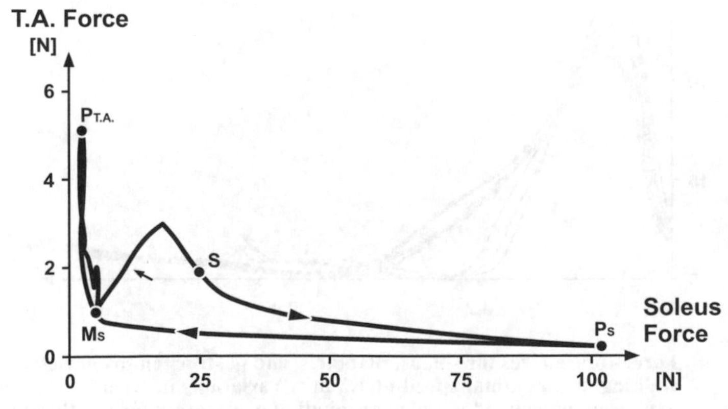

Figure 2.7.45 **Force-sharing between the antagonist pair of soleus and tibialis anterior muscles in the cat when trotting at a speed of 2.4 m/s.**

For trotting (2.4 m/s), there is a simultaneous increase in soleus and tibialis anterior forces just before paw contact with the ground (Figure 2.7.45, arrow), which is not seen as dramatically at the slow speeds of locomotion. It is hypothesized that this co-contraction of the antagonist pair causes an increase in ankle joint stiffness that may be necessary for accomplishing the increased force requirements for the trotting compared to the walking modes of locomotion. In the future, force-sharing between antagonist pairs of muscles should include even faster speeds of locomotion than shown here. It is speculated that co-contraction of soleus and tibialis anterior just before paw contact would be even more pronounced than the co-contraction shown for trotting (Figure 2.7.45).

2.8 ADAPTATION OF BIOLOGICAL MATERIALS TO EXERCISE, DISUSE, AND AGING

ZERNICKE, R.F.
JUDEX, S.
LORINCZ, C.

2.8.1 INTRODUCTION

Biological tissues are exquisitely tuned to their mechanical, biochemical, and electrical milieu. Load transmitting tissues, such as bone, cartilage, tendon, ligament, and skeletal muscle can modify their structure and composition in response to changes in their mechanical loading environment. When physical stimuli are increased through exercise, bone quality and quantity can be enhanced to accommodate the new loading levels. In contrast, when a person becomes bedridden, adopts a sedentary lifestyle, or experiences weightlessness, bone can be resorbed as the skeleton adapts to the diminished loading regime. Here, we provide an overview of the adaptive responses of bone, cartilage, ligament, tendon, and skeletal muscle to factors that generate adaptive tissue responses; namely, exercise, disuse, and aging.

2.8.2 BONE

The ability of bone to adapt functionally to stimuli is vital to meet the changing demands placed on the skeleton throughout life. If triggered, bone's response to stimuli can encompass remodeling or modeling events (Sommerfeldt & Rubin, 2001). Remodeling is resorption of bone tissue followed by subsequent bone formation. This sequence is often referred to as ARF (activation, resorption, and formation). Upon activation, large multi-nucleated osteoclasts perforate the bone matrix. Osteoblasts (bone forming cells) follow this cutting cone and deposit new bone behind the resorptive front. These groups of cells function as organized units and are called basic multicellular units (BMU) (Smit et al., 2002). Remodeling may reduce bone mass, but if activation frequency is low (as in healthy young adult bone) reductions in total bone mass are small. Remodeling plays an important role in maintaining stable serum calcium levels. If serum calcium levels are low, remodeling is initiated, and calcium stored in bone is released into the blood. Another vital function of remodeling is to maintain skeletal mechanical integrity, as it may replace bone material that has sustained activity-induced microcracks (Sommerfeldt & Rubin, 2001). If microcracks accumulate excessively in bone tissue, stress fractures, a common problem in long distance runners and gymnasts, may occur.

Remodeling is responsible for all events that occur within the bone cortex (intracortically). Modeling, on the other hand, acts exclusively on endocortical and periosteal surfaces, and, therefore, may change the shape of the bone (Rubin, 1984). During modeling, resorption and formation processes do not occur simultaneously on the same surface, i.e., bone resorption is not followed by formation. Modeling comprises activation-formation

(AF) and activation-resorption (AR) sequences. Until recently, modeling was only believed to be associated with changes during growth, and any changes after skeletal maturation were attributed to remodeling. That view has changed, and it is now generally accepted that limited modeling can be activated after skeletal maturation in response to stimuli such as increased mechanical usage (Burr et al., 1998).

Adequate mechanical usage is critical for maintaining skeletal integrity, as disuse causes rapid bone loss (Lang et al., 2004). Age-related changes in bone may lead to bone pathologies, e.g., osteoporosis, which have a catastrophic impact upon the quality of life in the elderly population. Exercise can be a means for moderately increasing bone mass or mitigating pathological bone loss, and a better understanding of bone's adaptive response to mechanical stimuli may help to accentuate these benefits (Judex et al., 1997a). It is important, however, to recognize the complexity of bone's physical and physiological environments and their potent interactions. Both local and systemic factors influence the processes by which bone adapts to physical stimuli. Some stimuli may negate exercise-derived positive effects, such as improper diet, e.g., high fat diet (Judex et al., 2000) or insufficient calcium, or hormonal imbalances, e.g., estrogen deficiency.

Exercise

Although heredity is the principal determinant of bone mineral density, up to half of the variance in bone mineral density is influenced by other factors (Krall & Dawson-Hughes, 1993), such as physical activity or exercise. Cross-sectional studies consistently reveal that bone morphology can change markedly in response to long-term exercise (Kontulainen et al., 2003; Ducher et al., 2004). In male tennis players, the cortical wall thickness of humeri in the playing extremity can be approximately 45% larger compared to humeri in the non-playing extremity (Hapaasalo et al., 1994). Similar bone hypertrophy has been reported in feet of classical ballet dancers (Kravitz et al., 1985).

Responses to exercise include alterations in bone shape, bone quality, and bone quantity. In animal models, molecular, histological, and biomechanical techniques can be used to detect exercise-related changes in bone. These methods, however, are generally highly invasive and not suitable for use in human studies. To assess bone status in humans, dual energy X-ray absorptiometry (DEXA) is currently the measurement standard. DEXA is non-invasive and quantifies bone mineral content (BMC [g]) and two-dimensional bone mineral density (BMD [$g \cdot cm^{-2}$]). Despite its advantages, DEXA does not differentiate changes in bone quality from changes in bone quantity or bone geometry. It underestimates actual BMD in people with small bones and overestimates BMD in people with large bones (Carter et al., 1992). Another potential limitation of DEXA is that the respective measure is averaged across the measuring field, and significant focal changes in bone morphology may remain undetected. High resolution quantitative computed tomography (QCT) can overcome these difficulties, as it can generate highly detailed three-dimensional images. This technique is now commonly used for peripheral sites of the skeleton (pQCT), but substantial exposure to ionizing radiation precludes its widespread use for routine BMD assessment of the axial skeleton (Syed & Khan, 2002).

Effect of exercise on bone

The skeletal response to exercise varies depending upon a person's age and physiological status. Because growing bone is more sensitive to exercise than adult bone, significant amounts of cortical and trabecular bone can be added in response to moderate exercise in growing bone. In premenarcheal girls, a 10 month high-impact aerobic exercise regimen significantly increased BMD at all measured regions, including the lumbar spine, leg, arm, pelvis, proximal femur, and femoral neck (Morris et al., 1997). Depending on the anatomical site, increases in BMD between the start and end of the exercise protocol were 2 to 7 times larger in the exercise group than in the control group (Figure 2.8.1). There is, however, an exercise threshold above which physical activity becomes detrimental to growing bone. In particular, exercise protocols that are high intensity and long in duration have been shown to suppress normal growth and maturation (Matsuda et al., 1986, Kiiskinen, 1977). Investigating the effect of exercise on growing bone is crucial because maximizing bone gains during growth may contribute significantly to preventing age-related bone pathologies.

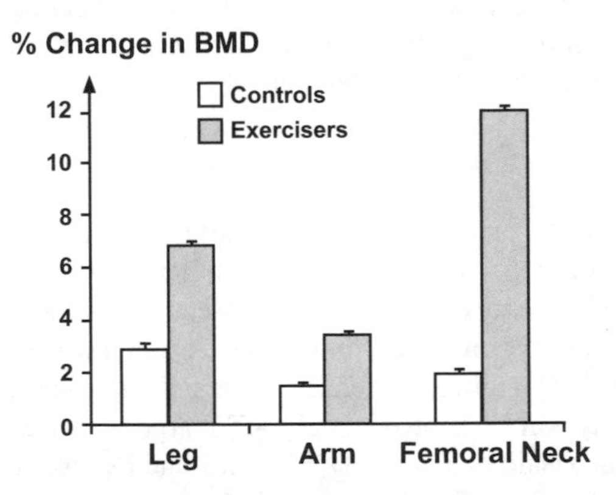

Figure 2.8.1 **Percent changes in mean BMD ± S.D. in premenarcheal girls that either followed a 10 month exercise program (n=38) or served as age-matched controls (n=33). Differences between groups were statistically highly significant. Data from Morris et al., 1997, with permission.**

In the young adult skeleton, exercise can moderately elevate BMD. Gains are typically less than 5% (Friedlander et al., 1995). Strenuous exercise can achieve slightly larger gains. After about age 40, there is significant bone loss at most skeletal sites. Studies involving older subjects suggest that the primary effect of exercise in the adult skeleton is conservation of bone mass and not acquisition of new bone. Nevertheless, some anatomical sites may gain moderate amounts of new bone in response to exercise (Bennel et al., 2002). The combination of diminished bone loss and slightly increased bone formation can have substantial effects on the skeleton, as demonstrated in a cross-sectional study of competitive runners (Lane et al., 1986; Bemben et al., 2004). Runners with a running history of 10 years had approximately 40% greater vertebral BMD than non-exercising, age-matched controls. Forwood and Burr (1993) summarized a host of studies related to exercise and bone, and their comprehensive review is recommended for more details.

The amount of bone mass that can be added to the skeleton is linked to the initial bone status. Therefore, individuals with extremely low initial bone mineral density have greater potential to add significant amounts of bone than individuals with average bone density (Forwood & Burr, 1993). Another critical question is if beneficial effects from exercise intervention can be maintained after ending the exercise program. Unfortunately, there is some evidence that gains are largely temporary and lost quickly if the exercise program is discontinued (Dalsky et al., 1988, Michel et al., 1991; Nordstrom et al., 2005).

How does bone sense exercise?

Contrary to cross-sectional studies that consistently reveal impressive differences between exercisers and sedentary controls, prospective longitudinal studies often fail to demonstrate significant increases in BMD or BMC. Some of these not-so-encouraging results from exercise studies are probably not associated with the inability of bone cell populations to respond to exercise but, at least in part, with our limited understanding of how exercise affects bone. Exercise is typically characterized by exercise mode, e.g., running, swimming, or weightlifting, duration, and intensity, e.g., percent of maximal heart rate. These terms, however, are non-specific to bone adaptation, and none of these parameters may be related directly to exercise-induced changes in bone. They define the mechanical environment of bone during exercise. Exercise can increase the forces acting on bone, and these force-related changes in the mechanical environment are primarily responsible for inducing morphological changes.

The relation between mechanical forces and skeletal morphology was identified long ago. Galileo (1638) was among the first to recognize this relation, and in 1892, Wolff promulgated his frequently cited law. In his treatise, however, Wolff only highlighted the form-function relation of bone, rather than a mathematical algorithm that explained mechanically induced bone adaptation.

The exercise-imposed mechanical environment may directly or indirectly affect bone cell populations, but the specific mechanical parameter or parameters that are the most potent for producing new bone or preventing bone loss remain to be identified. The mechanical measures used to quantify bone's mechanical environment include forces, stresses (forces normalized to unit area), strains (normalized deformations), strain frequency (number of strain cycles per unit time), strain rate (change in strain per unit time), and strain gradients (change in strain per unit length). Information about the osteogenically most potent mechanical parameter is important if designing an efficient or optimal exercise program to target and build bone. For instance, if strain magnitude is the most important parameter, then large loads should be imposed. If strain rate is more important, then loads should be applied rapidly. Traditionally, investigations have focused on strain magnitude and other parameters have received greater attention only recently. The current state of knowledge about how specific aspects of the mechanical environment affect bone is summarized in Table 2.8.1. These data are summarized from studies that analyzed the imposed mechanical environment. This knowledge, however, is not available from most exercise studies. Nevertheless, exercise studies consistently suggest that a few cycles of high-impact loading, e.g., weight training, are more effective for increasing BMD than repetitive aerobic low-impact loading, e.g., walking, and, consequently, support the major conclusions in Table 2.8.1.

Exercise may affect the mechanical environment of a bone and lead to changes in the physiological environment of bone. These changes can be systemic or local in nature and include alterations in blood flow or releases of cytokines. While the physiological stress response has limited potential to initiate bone (re)modeling on its own, it may influence the mechanically related adaptive process of bone.

The question of how bone senses exercise can be expanded to try to determine what bone attempts to achieve in response to mechanical usage. Intuitively, one might expect that bone seeks to minimize strains within its matrix, thereby reducing the risk of bone fracture (increasing safety factors). It has been demonstrated, repeatedly, that this may not be the full picture (Rubin, 1984). For instance, exercise increases, rather than diminishes,

Table 2.8.1 Relation between specific mechanical parameters and the adaptive responses in healthy mature bone. These relations were derived from studies involving animal models. Well-controlled clinical trials are required to confirm these results in humans.

MECHANICAL PARAMETER	EFFECT ON HEALTHY ADULT BONE
Normal Strain Magnitude	Strain magnitude is associated with the amount of new bone formation at the organ level. Applied at low frequencies, peak bone strains of 2000 to 4000 microstrain may initiate new bone formation while peak strains below 50 to 200 microstrain may induce bone resorption. This concept is expressed in the mechanostat theory (Frost, 1987).
Strain Frequency	When applied at higher frequencies (> 20Hz), even very small strains can elicit new bone formation or prevent bone loss (Rubin et al., 1997).
Shear Strain Magnitude	Several studies suggest that shear strains have less osteogenic potential than normal strains (Judex et al., 1997b), but shear strains may be important to minimize (intracortical) bone turnover (Rubin et al., 1996).
Strain Rate	Larger strain rates are more osteogenic (O'Connor et al. 1982; Turner et al., 1995). Static loading (strain rate=0) induces a state similar to disuse (Lanyon & Rubin, 1984a).
Loading Cycles	The maximal response appears to be elicited after a short number of loading cycles with further cycles producing little or no further gain (Rubin & Lanyon, 1984a). Extremely large numbers of loading cycles have been associated with stress fractures in bone.
Strain Distribution	Unusual strain distributions (different from distributions induced by walking or standing) are osteogenically more effective (Lanyon, 1996).
Strain Gradients	Strain gradients are related to fluid flow in bone and have been correlated with the specific sites of bone formation within the middiaphysis (Gross et al., 1997; Judex et al., 1997b).

curvature of long bones (Lanyon, 1980). An accentuated bone curvature is associated with higher strains in the middiaphysis. Also, the addition of bone mass within the middiaphysis does not always occur at sites where the largest strains are generated, but can happen at sites of least strains, but with greater strain gradients (Gross et al., 1997; Judex et al., 1997b). While larger bone curvatures and bone apposition close to the neutral axis do little to reduce activity-induced strains, they generate consistent loading conditions. It has been proposed that bone attempts to keep strains within a narrow and beneficial range (Rubin, 1984). Rubin's hypothesis is supported by experimental data demonstrating that peak bone strains are similar in a variety of species, e.g., buffalo, elephant, mouse, and turkey, including humans, ranging from 2000 to 3500 microstrain (Rubin & Lanyon, 1982; Burr et al., 1996). That phenomenon has been referred to as dynamic strain similarity (Rubin & Lanyon, 1984b). Alternatively, bone's response to mechanical stimuli may not follow a teleological algorithm, but instead may be governed by physiological processes without a specific goal from a structural engineering perspective.

Exercise effects are site-specific

Because the local mechanical loads induced during exercise produce changes in bone, it is not surprising that most exercise protocols do not produce general, systemic skeletal benefits. For instance, running in humans does not generate adaptive changes in the humerus, because running does not significantly alter the mechanical environment of this bone. Running, however, transmits large forces to the tibia that, in turn, may lead to adaptive changes. The principle of site specificity has been overlooked in some exercise studies and may explain, in part, why some investigators did not observe a significant skeletal response. The exercise-generated mechanical environment is highly variable both across different bones in the body and within a given bone. During many exercises, large compressive strains are generated on one side of the cortex, and large tensile strains are generated on the opposite side. It is likely that this non-uniform character of the mechanical environment contributes to the focal changes in bone morphometry commonly seen in response to exercise.

Disuse

Just as increasing aspects of the mechanical environment can lead to positive changes in bone geometry and composition, removing mechanical stimuli results in bone loss. For instance, immobilization, bed rest, or space flight may induce deleterious disuse-related changes in bone. In humans, the adverse effects of reduced loading have been highlighted by the substantial skeletal degeneration and calcium loss that can occur in space flight. Decreases in BMD of 3 to 8% have been observed during Apollo, Skylab, Spacelab, and Salyut-6/Soyuz missions (Zernicke et al., 1990). Increased urinary and fecal calcium excretion in astronauts causes a negative calcium balance and reflects bone loss during space flight (Figure 2.8.2). Consistent with the tenet that *the extent of adaptation is dependent upon the difference between the habitual and the newly imposed mechanical environment*, non-weight bearing bones seem to be less affected by disuse than weight bearing bones.

Because disuse changes can be substantial, and immobilization may be required in some rehabilitation instances, it is important to consider whether disuse related changes are reversed by remobilization. Rats immobilized for three weeks experience up to 12% decreased tibial and femoral ash weights compared to non-immobilized controls (Tukkanenen et al., 1991). After 9 weeks of remobilization, tibial bone mineral mass recovered by 62%, while the femur only regained 38% of the lost mineral mass. The Tukkanenen and coworker's study and others demonstrated that immobilization-related bone loss can be reversed to some extent, but that recovery does not occur as rapidly as the loss of bone. Recovery of cortical bone takes longer than recovery of cancellous bone (Lane et al., 1996). Additionally, the amount of recovery appears to be dependent upon the length of immobilization with greater and more efficient bone recovery after shorter immobilization periods (Jaworski & Uhthoff, 1986; Jarvin et al., 2003).

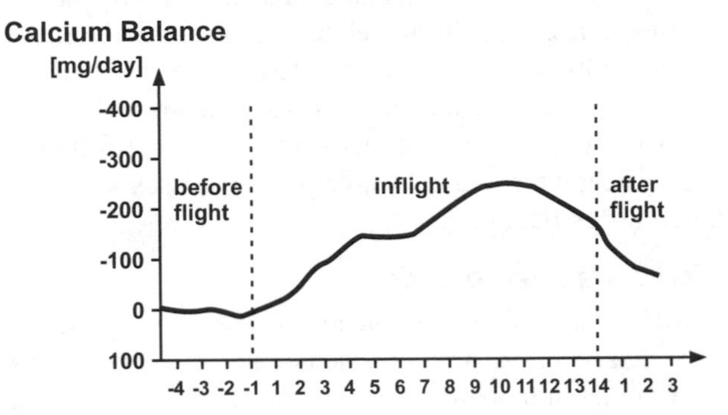

Figure 2.8.2 **Calcium balance of astronauts on Skylab missions undergoing immediate change upon entry into space. From Rambaut and Goode, 1985, (Figure 3), with permission.**

Age not only influences the magnitude of the disuse-related response, but also the location of the response. In quiescent, normal adult bone, disuse causes expansion of the endocortical cortex and an increase in intracortical porosity (Jaworski & Uhthoff, 1986; Gross & Rubin, 1995). In the normal immature skeleton, most periosteal surfaces are actively forming, while most of the endocortical surfaces are undergoing resorption. With disuse, growing bone manifests primarily impaired osteoblastic activity on the periosteal surface-with little changes in endocortical resorption (Jaworski & Uhthoff, 1986; Weinreb et al., 1991).

Aging

The exact time at which the human skeleton attains peak bone mass continues to be debated, but the consensus is that it occurs for most bones before age 30. In the following decade, bone mass remains relatively stable, and then begins to decline noticeably (Willet, 2005). The rate of bone loss is relatively steady in men. In women, there is acceleration in the rate of bone loss during menopause, after which the rate of bone loss returns to premenopausal levels. By age 90, women have lost 20% of their peak cortical bone mass and 40 to 50% of their peak trabecular bone mass. In contrast, men at that age have lost 5% of their cortical bone mass and 10 to 25% of their trabecular bone mass (Rigg et al., 1982).

Aging also changes the gross architecture of bone. Trabecular bone is particularly affected, as reflected in the dramatic decrease in connectivity between trabecular bone plates and struts. Additionally, bone geometry and composition may change with age. Osteons increase in number and change in size, and intracortical porosity increases (Martin et al., 1998). Many of these age-related changes are associated with alterations in the BMU-remodeling system, such as increases in BMU activation frequency and imbalances between bone resorption and formation.

Between 20 and 80 years of age, mechanical properties of bone are reduced by about 5 to 40% as long as no other influences improve or worsen the situation, with possible improvements resulting from physical activity. Possible negative effects may result from hormonal changes, e.g., menopause or a sedentary lifestyle.

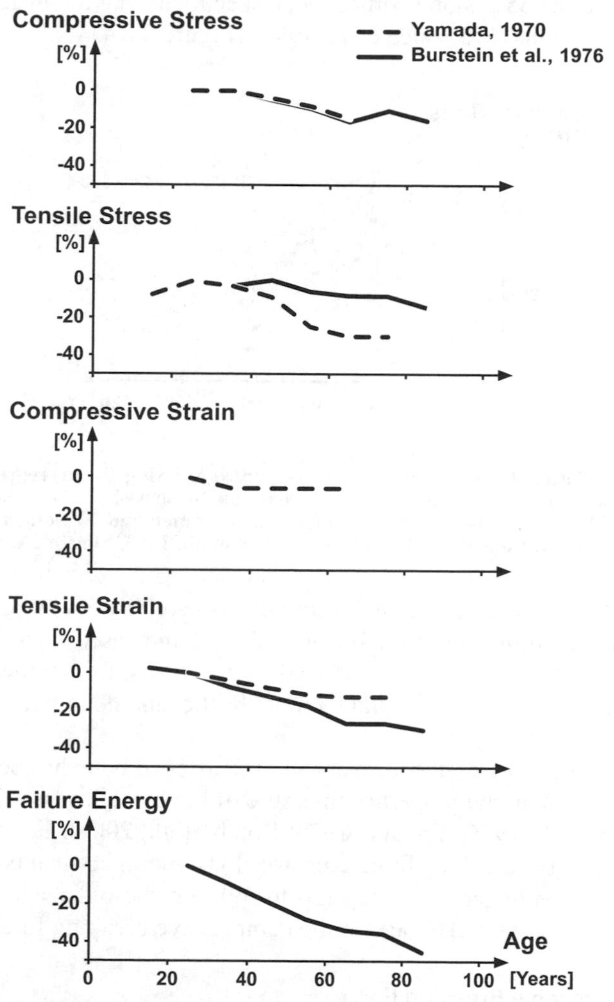

Figure 2.8.3 **Summarized results of the effect of age on mechanical properties of bone (based on data from Yamada, 1970, and Burstein et al., 1976, with permission of Williams & Wilkins, Baltimore, Maryland).**

Yamada (1970) and Burstein et al. (1976) measured mechanical properties of cortical bone specimens ranging from 20 to 90 years of age under various loading conditions. The percent differences in mechanical properties for femoral cortical bone between specimens from the 80 to 89 year old group and specimens from the 20 to 29 year old group show reductions of (Figure 2.8.3):

- Ultimate compressive stress ≈15%
- Ultimate tensile stress ≈30%
- Ultimate compressive strain ≈ 5%
- Ultimate tensile strain ≈30%
- Elastic modulus in compression ≈15%
- Elastic modulus in tension ≈10%

Observations from 235 femoral cortical bone specimens ranging in age from 20 to 105 years show similar results (McCalden et al., 1993) (Figure 2.8.4).

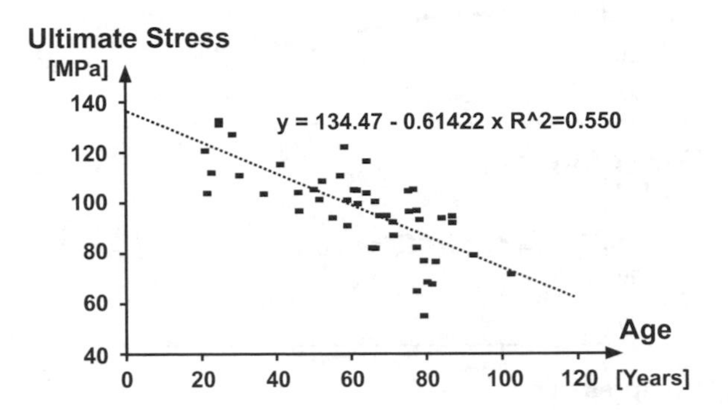

Figure 2.8.4 **Relation between ultimate stress and age. Using linear regression analysis, age accounted for 55% of the variation observed in mechanical stress data. Measurements were made on femora of 25 men and 22 women aged twenty to 105 years at time of death (from McCalden et al., 1993, Figure 2A, with permission).**

Ultimate stress, ultimate strain, and energy deteriorated by 5%, 9%, and 12% per decade, respectively. Additionally, the porosity of bone increased significantly with age, while mineral content of bone was not affected by age. Thus, from a mechanical perspective, cortical bone becomes weaker, slightly more brittle, and its impact energy absorption is impaired with aging.

The influence of gender on the mechanical properties of bone has been studied extensively. Reductions in material properties (measure of bone quality) are similar in men and women (Burstein et al., 1976; Yamada, 1970; Routh et al., 2005). For example, for both genders the compressive breaking force decreases for femoral cortical bone from about 15 to 20% between the younger age group (20 to 39) and the older age group (60 to 89) (Figure 2.8.5). The relative difference for the compressive breaking forces between males

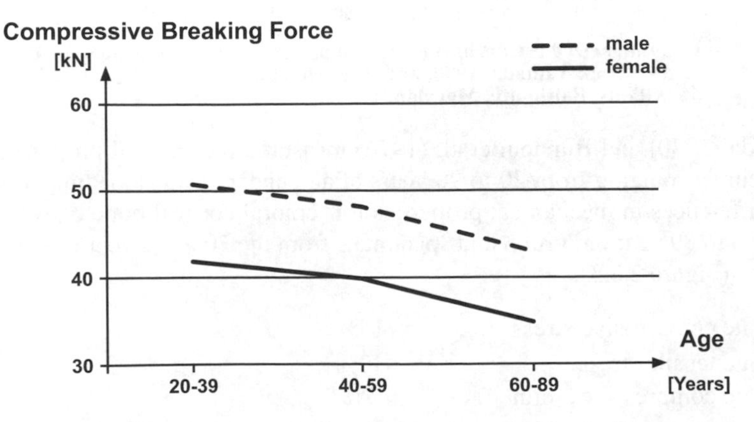

Figure 2.8.5 **Compressive breaking forces in the middle portion of wet femoral shafts in longitudinal direction (based on data from Yamada, 1970, with permission of Williams & Wilkins, Baltimore, Maryland).**

and females, however, remains constant for the two age groups (≈ 17%). Consequently, female bones fail at smaller forces compared to male bones.

The age-related reductions in material strength may be partially compensated by changes in the overall structure of long bone. This compensatory mechanism is more efficient in males than in females (Martin & Burr, 1989), as radial expansion of the cortex has been observed in aging men but not in women (Martin & Atkinson, 1977). Also, cortical bone width measured from iliac crest bone biopsies decreases with age for women but not for men (Christiansen et al., 1993). Taken together, these data explain, in part, why bone fractures happen more often in elderly women than in elderly men.

2.8.3 ARTICULAR CARTILAGE

Articular cartilage is especially well adapted to its function and is typically forgotten unless injured and osteoarthritis develops. As with bone and other load bearing connective tissues, normal use is important for articular cartilage health. If either excessive or too little loading, e.g., immobilization, occurs, cartilage will be adversely affected.

Exercise

Repetitive loading and unloading (exercise) can produce a swelling of cartilage (Walker, 1996). Chronic exercise can also produce increases in extracellular matrix and the size and number of chondrocytes (Engelmark, 1961; Munderman et al., 2005). There is a vigorous debate about the long-term effects of exercise on cartilage, with some investigators reporting positive effects and others reporting negative consequences of exercise. Saamanen and colleagues (1986) found that exercise enhanced articular cartilage matrix components, while others, e.g., Vasan (1983) and Little et al.(1997), reported that degenerative changes occurred in articular cartilage with excessive loading. Particularly with excessive loading, synthesis rate decreases and degradation increases, which may lead to osteoarthritis (Whiting & Zernicke, 1998).

Disuse

With a substantial decrease in repetitive loading or with immobilization, articular cartilage will atrophy and degenerate. When articular cartilage is immobilized for a long time, there is a marked reduction in the synthesis and amount of proteoglycan, a decrease in the size and number of aggregated proteoglycan, and an increase in the number of fibrillations in the surface of the cartilage. As a consequence of these changes in composition, cartilage mechanical properties are negatively affected. After immobilization, the free water in the matrix is more easily exuded as the cartilage is compressed. Mankin and colleagues (1994) noted, however, that after cartilage is remobilized (after a period of immobilization), these negative structural and functional changes may be reversed.

Aging

As articular cartilage ages, it becomes thinner, and it loses water and proteoglycan content (Mankin et al., 1994). Collagen concentration, however, tends to increase with age. In younger cartilage, the protein core and glycosaminoglycan chains are longer, but with advancing age, the protein cores become shorter. The changes in proteoglycan structure account for some of the decrements in mechanical behaviour and resiliency in articular cartilage.

Failure (acute/fatigue/chronic)

The incongruency of articular surfaces in synovial joints is crucial to the health of articular cartilage and the smooth lubrication of joint movements. Synovial fluid pooling in areas of incongruency permits nutrients to accumulate and provides enough fluid to create an elastohydrodynamic fluid film. Mechanical loading and unloading keep the tissue healthy by causing the influx of nutrients, the efflux of waste, and lubrication. As noted previously, disuse is linked to cartilage degeneration. Greenwald and O'Connor (1971), corroborating the work of Goodfellow and Bullough (1968), showed that areas of habitual non-contact matched areas of known degeneration in the hip joint. The dome of the acetabulum is a typical location for cartilage degeneration in older individuals. These persons are less likely to experience the high load conditions required to produce direct contact between synovial fluid and cartilage. Reduced mechanical stimulation results in a lack of nourishment, so the cartilage degenerates (Magnussen et al., 2005).

When cartilage is damaged, a remodeling response occurs. The remodeling of articular cartilage originates in the chondrocytes, which are responsible for the ongoing synthesis and degradation of the matrix. Investigators have concluded, however, that cartilage has a limited ability to remodel when damaged. The normal pattern of activity sees the clustering of chondrocytes around the site of damage. Their synthetic activity is high, but in general they fail to restore the matrix to normal, even if the defect is very small. Figure 2.8.6 shows

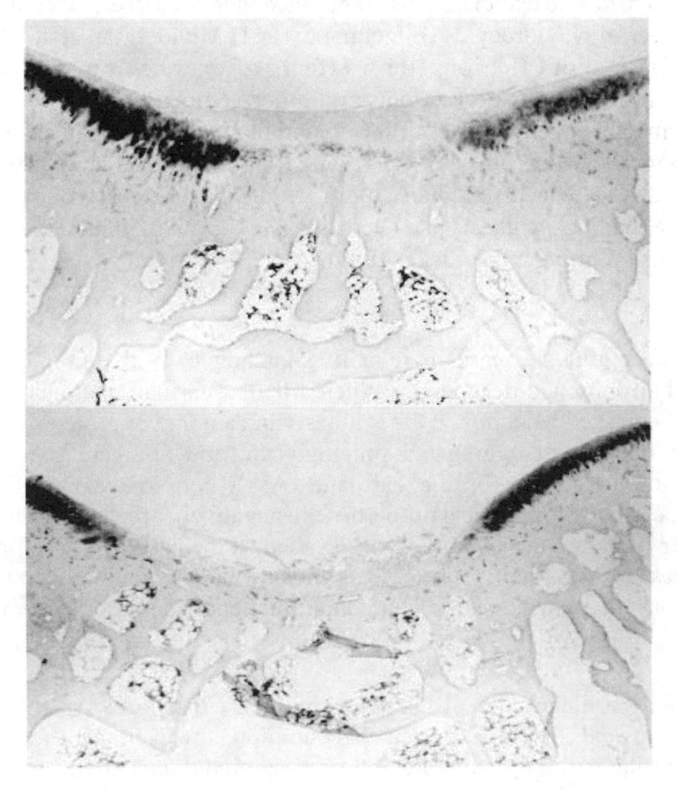

Figure 2.8.6 **Light micrographs of a full thickness articular cartilage lesion 2 mm in diameter six months after injury (top) and one year after injury (bottom) (Buckwalter et al., 1987, with permission).**

two light micrographs of healing articular cartilage, (a) at six months post injury and (b) at one year post injury. At six months, the tissue repair may clearly be seen in the centre. The area of subchondral bone has been restored below the obvious scar. The dark stain within the articular cartilage is Safranin O, which highlights proteoglycan. There is minimal staining within the scar. One year post injury, the staining remains much the same, and fibrillation (fraying and splitting) may be noted at the surface of the scar.

From a mechanical perspective, wear of cartilage and its subsequent failure may be caused by acute or chronic loading. Acute failure of cartilage results when local stress exceeds ultimate stress. This usually happens where there is a combination of high external forces and a small contact area between two adjacent bones. The external forces may be active or impact forces. Active forces that produce excessive local stress in a joint include forces from heavy lifting. By comparison, impact forces are produced by collisions, e.g., falls. It is speculated that acute cartilage failure is related to excessive impact loading.

Chronic failure of cartilage can result from interfacial problems or fatigue. The source of interfacial wear is a lack of lubrication at the bearing surface. It can be adhesive, where particles from the bearing surfaces adhere to each other and tear away from their surfaces, or abrasive, where a soft particle becomes scraped and damaged by a harder one. Interfacial wear should not happen in a normal joint with efficient lubricating mechanisms, but this type of wear may easily happen in a faulty or degenerative joint with impaired lubricating mechanisms. With increasing abrasive or adhesive wear, joint quality worsens.

Fatigue wear happens in articular cartilage tissue when the proteoglycan-collagen matrix is damaged by cyclic stressing. This stressing may produce *proteoglycan washout* from the repeated and massive exudation and imbition of the tissue fluid. Fatigue wear may result from the application of high active or impact forces over long epochs. The intensity, duration, frequency, and magnitude of stresses can cause structural changes in the properties of normal cartilage that affect its ability to resist mechanical wear. Such changes start at the molecular level, where macromolecules, cells, or collagen fibrillar network, or all three, are lost or damaged.

The destructive effects of arthritic diseases (diseases that involve inflammation of the joint) are self evident. Osteoarthrosis originates in cartilage and involves the fibrillation and softening (chondromalacia) of the articular tissue in the areas of greatest pressure and movement. Splitting happens when the fibres in the superficial zone cannot resist the lateral tension of the fluid drag. The cause of splitting is not precisely known, but it is thought to be related to age and wear. The damage caused by osteoarthrosis increases with wear.

2.8.4 LIGAMENT

Until a few decades ago, ligaments were considered to be relatively inert and unresponsive to loading. That attitude has changed dramatically, however, as these dense fibrous connective tissues have been shown to adapt to exercise and immobilization.

Exercise

The principal effects of exercise on ligaments are increased structural strength and stiffness. Frank (1996) noted that everyday activity is usually sufficient to maintain about 80 to 90% of a ligament's mechanical potential, and that exercise or training has the potential to enhance a ligament's normal potential by only about 10 to 20%. Nevertheless, studies of the effects of exercise on ligaments must be evaluated with care, because an exercise protocol may not change the specific stresses on a particular studied ligament. Stress, however, is a positive stimulus for connective tissues, including ligaments, and is needed for growth.

Studies with exercised dogs (Laros et al., 1971) failed to show morphological changes of the femoral and tibial insertions of the medial collateral ligaments of the knee, but the knee ligaments of exercised rats have more small-diameter collagen fibres than rats that have not been exercised (Binkley & Peat, 1986); Tipton et al., 1970). That change in collagen fibres may account for the increased stiffness of the exercised ligaments (Doschak & Zernicke, 2005).

Exercise can moderately influence the biochemical composition of ligaments. Activity-related increases can occur in collagen concentration, glycosaminoglycan, collagen turnover, and collagen non-reducible cross-links (Tipton et al., 1970). The physical properties of ligaments also adapt to exercise. Trained ligaments fail at somewhat greater forces (Tipton et al., 1967; Viidik, 1967; Tipton et al., 1970, 1974). Although the benefits of exercise may only be moderate for ligaments, it has been found consistently that injured ligaments that are exercised have significantly better physical, biochemical, and morphological properties than those that are not exercised (Eastlack et al., 2005).

Disuse

Joint immobilization or load deprivation produces rapid deterioration in ligament biochemical and mechanical properties, with concomitant losses in strength and stiffness (Frank, 1996). Morphologically, immobilization increases bone resorption and the number of osteoclasts at the insertion of ligaments to bone (Laros et al., 1971; Noyes, 1977; Noyes et al., 1974a; Woo et al., 1983). Furthermore, disuse increases the number of large diameter fibrils, decreases the number of small diameter fibrils, and decreases the density of collagen fibrils (Woo & Buckwalter, 1988).

Biochemically, immobilization decreases the glycosaminoglycan and the water content of ligaments. The rates of collagen synthesis and degradation increase, suggesting that there are more new collagen fibres. The total amount of collagen mass decreases, and the number of reducible cross-links increases with disuse (Akeson et al., 1977, 1980, 1987). The physical properties of ligaments are markedly altered with immobilization. Force to failure, stress to failure, energy absorbed to failure, and stiffness are reduced by immobilization. As well, immobilized ligaments tend to fail at the bony insertion rather than in mid-substance, a behaviour similar to immature ligaments (Doschak & Zernicke, 2005).

Changes induced by short term immobilization are reversible. The physical properties of primate anterior cruciate ligaments were studied after 8 weeks of immobilization and 5 and 12 months of remobilization (Noyes et al., 1974b). After 12 months, the force to failure compared to control values was 91%, the stiffness was 98%, and the energy absorbed to failure was 92%. Similar work with the rabbit medial collateral ligament (Woo et al., 1983) suggested that the longer the immobilization, the more harmful the effect on a ligament's physical properties.

Aging

During development and growth, the stiffness and modulus of elasticity increase until skeletal maturity. By middle age, the insertional points of ligaments begin to weaken, viscosity begins to decline, and the collagen fibres become more highly cross-linked and less compliant (Frank, 1996). With increasing age, the insertional junctions become weaker (related to increased numbers of osteoclasts), and the likelihood of avulsion fractures increases.

Morphologically, crimp periodicity of collagen fibres increases, suggesting that ligaments may take on different crimp patterns at different stages of life. Furthermore, the crimp angle decreases with age, a change that has been associated with the large toe region for ligaments of young specimens (Diamant et al., 1972). Cell density diminishes with age. The frequency of large collagen fibres is high in young ligaments, while there is a bimodal (small and large) distribution of collagen fibres in ligaments from older individuals (Parry et al., 1978; Frank et al., 1989; Bland & Ashurst, 2001).

Biochemically, aging increases the collagen concentration and reduces hexosamine and water in ligaments. The number of non-reducible cross-links in ligament increases throughout the aging process. The physical properties of ligaments also change with age, with structural properties changing to a greater extent than the material properties. During growth and maturation, the failure properties improve rapidly, leveling off in adulthood, at which time the failure properties begin to decline. Ultimate tensile stress of anterior cruciate ligaments was compared for young adults and for persons about 65 years old (Noyes & Grood, 1976). Their results showed a reduction of ultimate stress of more than 60% in the ligaments from older individuals. As well, the ligaments of younger individuals have a greater viscoelasticity than older individuals.

2.8.5 TENDON

Mature tendon is generally in metabolic balance, with resorption and deposition of fibres occurring as ongoing processes. This balance can be altered, however, by several factors, including exercise, disuse, and aging.

Exercise

Compared to ligaments, fewer quantitative data are available about exercise-related adaptations of tendon. The existing data suggest that exercise can increase the number and size of collagen fibrils and increase the cross-sectional area of tendons when compared with the tendons of sedentary controls (Michna, 1984). Exercise can lead to increased collagen synthesis in growing tendons (Curwin et al., 1988; Kjaer et al., 2005) and an increased number of fibroblasts in tendons (Zamora & Marini, 1988).

Changes in tendon that are attributed to increased loads have been equivocal, in part, because the exercise regimes that are typically used to increase tendon load are variable and likely result in different loads being applied to the tendon. Endurance running can increase the size and collagen content of swine extensor tendons, but prompts no change in flexor tendons (Woo et al., 1980, 1982). This may reflect the different recruitment patterns or contractile properties of flexor and extensor muscles. In young mice, treadmill running increased the dry weight of the Achilles tendon (Kiiskinen, 1977), while in senescent mice, treadmill running or voluntary exercise appeared to slow the age-related loss of glycosaminoglycans (Vailas et al., 1985; Viidik, 1979). In growing chicks, strenuous exercise slowed the maturation of collagen by increasing the turnover of matrix and collagen (Curwin et al., 1988). The response of tendon to overload appears to be specific to the tendon and the relative maturation of the animal. Growth- and exercise-related stimuli may conflict when growing animals are exercised, which suggests that the stimuli responsible for adaptation can change during maturation.

Disuse

Mechanically induced changes in tendon have been investigated in many studies. If the Achilles tendon of a rabbit is cut, glycosaminoglycan content and fibroblast number increase, and the number of small collagen fibres increases (Flint, 1982). Interestingly, the area of the tendon closest to the muscle belly shows the greatest change, which suggests that tendon adaptation is region-specific. If the tendon is sutured so that some tensile load is transmitted, only minimal changes occur (Wang, 2005).

Denervation or the elimination of neural impulses to the muscle eliminates the development of tension due to active muscle contraction and dramatically increases collagen turnover in the rat Achilles tendon (Klein et al., 1977), but produces no change in the relative amounts of the various tendon components. Tensile loads generated by the muscle, therefore, appear to exert an important influence over tendon collagen turnover.

The use of anabolic steroids to enhance athletic performance has become a concern among sports medicine specialists. Laboratory investigations focusing on the effects of anabolic steroids on tendon biomechanics and biochemistry suggest that long-term steroid use, particularly in combination with exercise, decreases tendon stiffness (Wood et al., 1988) and ultimate strength by stimulating collagen degeneration (Michna & Stang-Voss, 1983). In addition, anabolic steroids impair the healing of tendon injuries (Bach et al., 1987; Herrick & Herrick, 1987; Kramhoft & Solgaard, 1986). Ironically, substances used to enhance performance may ultimately predispose the individual to problems that will force the individual out of competition.

Aging

Tendon mechanical properties and composition are markedly influenced by age. Prior to skeletal maturity, tendons are more viscous and compliant (Frank, 1996). With increasing age, through middle age, the strength and stiffness of tendons are relatively steady. As muscular strength begins to diminish, there is a concomitant loss of strength in tendon. Maximal muscle strength in men and women is generally reached between the ages of 20 and 30 years, about the same time that the cross-sectional area of muscle is the greatest. The strength level tends to plateau through the age of 50, followed by a decline in strength that accelerates by 65 years of age.

2.8.6 SKELETAL MUSCLE

Skeletal muscle is a highly metabolic tissue capable of enormous adaptation to the physiological environment in which it is placed. Endurance training and resistance training both have profound but different effects on skeletal tissue and stand in marked contrast to muscle tissue that is aged, exposed to disuse (such as weightlessness), or sedentary lifestyle.

Exercise

Skeletal muscle adaptation is specific to the training demands placed upon the tissue. Endurance training can:

- Upregulate mitochondrial enzyme systems involved in energy production, such as in the Krebs cycle and the electron transport chain,
- Upregulate enzymes involved with fuel substrate availability, e.g., fatty acid activation, translocation, oxidation), and
- Increase transporter proteins involved in the regulation of glucose uptake into the muscle cells.

Shantz et al. (1983) found that in response to prolonged, low-intensity training (cross-country skiing, 6 d·wk^{-1} for 8 wk), the number of capillaries per muscle fibre in both quadriceps femoris and vastus lateralis were increased by 40%. There was also an increase in the number of mitochondria per slow twitch muscle fibre and myosin heavy chain isoform switching (from intermediate-fast twitch IIa, IIx isoforms to the more economical cross-bridge cycling kinetics of the IIb isoform) (Baldwin & Haddad, 2001). Endurance training, however, did not result in muscle cell hypertrophy, potentially because the level of force production needed for such training was relatively small when compared to maximal force-generation capacity of the muscle (Baldwin & Haddad, 2002).

To induce muscle cell hypertrophy, a threshold stimulus must be reached to induce adaptation in expression of contractile proteins such as actin and myosin, which will lead to subsequent muscle enlargement. That phenomenon is based on the training principle of overload, which states that muscles worked close to their force-generating capacity will increase in strength (McArdle, 1996). As such, resistance training programs are developed on the basis of exercise volume (number of repetitions and sets), frequency (number of times exercised per week), and intensity, e.g., a percentage of a one repetition maximum (Macaluso & De Vito, 2004). Importantly, loads must progressively increase to provide adequate stimulus throughout the entire training regime. Rapid improvement in the ability to perform the strength exercises can occur in the first and second weeks, due to a learning effect that involves improvements in motor skill coordination and level of motivation. During the third and fourth weeks, muscle strength increases without a corresponding increase in muscle size (Macaluso & De Vito, 2004). Those improvements are mainly due to neural adaptations, such as intermuscular coordination (efficient movements from agonist and antagonist muscle groups), intramuscular coordination (number of motor units recruited and synchronization of motor units), and an increased neural drive from the central nervous system. Beyond the sixth week, muscle hypertrophy occurs both within the whole muscle (5 to 8% increase in size) and within the muscle fibres themselves (25 to 35% increase) (MacDougal, 1992). This phenomenon occurs without hyperplasia (cell division) (McCall et al., 1999).

Muscle cell hypertrophy occurs through the activation of local support cells, called satellite cells, which are mitotically quiescent myoblasts located between the sarcolemma or the muscle fibre and its extracellular matrix (ECM). Upon physical stimulation, insulin-like growth factor-1 (IGF-1), found circulating in the ECM, is able to react with satellite cells, causing them to divide. The resultant daughter cells then fuse with the underlying muscle fibre, adding nuclei, cytoplasm, and proteins to the existing fibre. By increasing the number of nuclei within the muscle fibre, contractile protein synthesis can be upregulated and hypertrophy of the muscle cell results (Chakravarthy et al., 2000; Allen et al., 1989; Doumit et al., 1993). As a result of hypertrophy, muscle cross-sectional area increases, and capillary and mitochondrial density decrease as a result of a dilution effect (Baldwin et al., 1981). These processes are comparable in both sexes, but absolute strength in men exceeds that of women (Skelton et al., 1994).

Disuse

Chronic bedrest, spaceflight, or a sedentary lifestyle results in a decrease in protein synthesis and an increase in protein degradation within skeletal muscle. As a result, a rapid and marked degree of atrophy occurs with no decrease in muscle fibre number (Booth & Criswell, 1997). The amount of muscle atrophy depends on usage before immobility and the inherent function of the muscle. Antigravity muscles, such as the quadriceps, will experience greater atrophy than antagonist muscles (hamstrings) (Kasper et al., 2002). Lexell et al. (1988) reported that fast twitch fibres were lost preferentially with a sedentary lifestyle, as slow twitch fibres (found in most postural muscles) continued to be activated and maintained, even at relatively low forces. The potential to regain lost muscle mass depends on the length of immobility, age, nutrition, and rehabilitation program.

Aging

Human skeletal muscle strength peaks between the second and third decade, shows a slow decline until about 50 years of age, and then begins to deteriorate more rapidly at an approximate rate of 12 to 15% per decade, with greatly accelerated losses after age 65 (Larsson et al., 1979; Narici et al., 1991; Lindle et al., 1997). Quantitative loss of skeletal muscle tissue with age is called *sarcopenia* and comprises decreases in muscle cross-sectional area, fibre size, and fibre number (fast twitch fibres are more adversely affected) (Janssen et al., 2000). Loss of muscle tissue undoubtedly affects the generation of force with advancing age, especially with the accelerated loss of the stronger fast twitch muscles (Jones & Round, 1990). Both men and women experience a comparable decline in skeletal muscle mass with age when measured per unit of cross-sectional area. When quantified using absolute strength, however, women are weaker than men in various muscle groups throughout life. In many cases, the absolute strength of women may approach the minimal levels necessary to accomplish daily activities, therefore affecting independence and quality of life (Skelton et al., 1994).

Besides a quantitative loss in muscle mass, research suggests that declines in the *quality* of skeletal muscle can adversely affect strength production with age. Motor units within skeletal muscle decreased in both size and number with age (Doherty & Brown, 1993). Reduced motor unit firing rate and synchronization (Macaluso et al., 2003) and reduced number and diameter of motor axons results in slower conduction velocities in the aged (Kawamura et al., 1977). When compared to young adult skeletal muscle, fibre type grouping and a reduced number of fast twitch fibres existed in older humans (Lexell et al., 1988). The observed selective loss of fast twitch fibres and subsequent fibre type grouping may be linked to muscle fibre denervation and reinnervation. Initial denervation of fast twitch fibres is followed by reinnervation of these fibres by axonal sprouting from adjacent slow twitch motor units, resulting in large groups of slow twitch fibres. Aged skeletal muscle takes longer to reach maximal contraction and is slower to relax after excitation—properties likely related to the elevated slow twitch composition of the muscle (Narici et al., 1991).

Antagonist co-activation or co-contraction tends to be greater in 70-year-old men and women versus 20-to 40-year-olds (Macaluso et al., 2003). Antagonist co-activation during motion may help protect and stabilize the joint, but the net force exerted about a joint during a given action, e.g., knee extension, would be reduced in older people due to the greater simultaneous activation of the muscles exerting a torque in the direction opposite to that of movement (hamstrings). Thus, force production at the joint would be decreased.

Serum testosterone is a strong predictor of muscle mass in men, and a decline in testosterone production can lead to decreased contractile protein assembly and subsequent muscle atrophy with advancing age (Iannuzzi-Sucich et al., 2002). In large part, however, the role of steroidal hormones such as testosterone in men and estrogen in women—in relation to muscle strength and aging—are inconclusive at present. Nonetheless, research to date suggests that, although somewhat attenuated, skeletal muscle most likely retains a capacity to adapt favourably to adequate physical stimulus, even into advanced age.

2.9 JOINTS

NIGG, B.M.
HERZOG, W.

2.9.1 CLASSIFICATION OF JOINTS

Joints are classified into three subtypes: fibrous, cartilaginous, and synovial (Table 2.9.1).

Table 2.9.1 Description, function, movement, and examples for the three major joint types in the human and animal body, i.e., fibrous, cartilaginous, and synovial joints.

TYPE	DESCRIPTION	FUNCTION	MOVEMENT	EXAMPLES
FIBROUS	Bones connected by fibrous (connective) tissue	stability		
	• syndesmosis		small	tibia/fibula
	• suture		non	skull
CARTILAGINOUS	Bones connected by cartilage	bending		
	• synchondrosis		small	sterno costalis connection
	• symphysis		small	sym. pubica spinal vertebrae
SYNOVIAL	Bones connected by ligaments	movement	small translation large rotation	knee hip

In *fibrous joints,* the bones are connected by connective fibrous tissue. Fibrous joints have the appearance of hairline cracks and are designed for stability. The skull contains fibrous joints.

In *cartilaginous joints*, cartilage connects the bones. These joints allow for minimal movement, are relatively stiff, and are designed to provide stability and to transfer forces. Cartilaginous joints permit limited movement, but a combination of such joints may facilitate a large range of motion. The connection of spinal vertebrae is a typical example of a cartilaginous joint.

Synovial joints are the main focus of this section. The bones in synovial joints are in contact, are not connected by fibrous structures or cartilage, and are able to move with respect to each other. Synovial joints allow for virtually free movement within a defined, but limited, range, have little friction (see chapter 2.6), are designed to facilitate motion, and are able to transfer forces. A knee joint, for instance, does not heat up significantly during long movement exposure, e.g., during a marathon run, because of minimal joint friction. Knee joints transfer forces during normal walking of up to three times body weight, while hip joints transfer up to seven times body weight (Paul, 1976).

From a mechanical point of view, joints can be classified based on their shape, e.g., spherical, elliptical, hinge, saddle, and/or their translational and rotational movement possibilities, e.g., sliding or not sliding hinge joint.

2.9.2 FUNCTION

Functional, morphological, histological, and mechanical properties of biological materials have been discussed in chapters 2.3 to 2.7. However, joints should be considered as organs composed of several musculo-skeletal tissues. The integrity of a joint depends on the integrity of all its constituent parts. Failure of a single joint component, for instance, rupture of the anterior cruciate ligament in the knee, influences all tissues that make up the joint. Thus, understanding the mechanics and the function of joints requires (a) an understanding of the mechanics and function of muscles, tendons, ligaments, bones, and cartilage, and (b) a thorough understanding of the interaction of these tissues.

Selected aspects of joints are discussed in this section.

A joint is the junction between two or more bones of the human or animal skeleton.

The major functions of synovial joints include:

- To *permit limited movement*, and
- To *transfer forces* from one bone to another bone.

The skeletal system can move at joints. Cartilage covering the joint surfaces allows for relative movement of adjacent surfaces with minimal resistance. Furthermore, the joint surfaces are shaped to allow for restricted, e.g., tibio-femoral joint, or unrestricted, e.g., shoulder joint, movement. Each joint in the human or animal skeleton has structures that *limit* its range of motion, including ligaments, joint capsules, muscle-tendon units, or bony shapes, or all. The resistance to movement is typically small in the physiological range of motion, but increases dramatically towards the boundaries of this range of motion. Because of the rigidity of the contacting bones, synovial joints are well-suited to *transfer forces* between segments.

Joints are well-designed for their mechanical purposes. Human and animal joints, unlike their man-made equivalents, frequently function under demanding conditions for 70 or 80 years without malfunction.

2.9.3 DEGREES OF FREEDOM OF JOINTS

Relative movement between two adjacent segments of a joint can be described using six independent variables (degrees of freedom = DOF), three for translation, and three for rotation.

The degree of freedom (DOF) of a system of interest (a segment or a set of connected segments) is the number of independent variables (coordinates) needed to describe the motion of the system of interest.

The DOF of a system of rigid segments can be determined using the general relationship:

DOF = number of generalized coordinates – number of constraints

For three- and two-dimensional cases, the relationships read as:

3-D: DOF = 6N – C

2-D: DOF = 3N – C

where:

DOF = degree of freedom
N = number of segments
C = constraints

The degree of freedom provides information about the movement possibilities of one segment with respect to another segment of a joint, e.g., movement possibilities of the tibia with respect to the femur. Ideal *spherical joints* have three rotational and no translational DOF. The determination of the DOF for an ideal *elliptical joint* is complex. If the elliptical joint of interest has an elliptic cross-section for two principal axes, and a circular cross-section for one principal axis, the elliptical joint can rotate about an axis that is perpendicular to the circular cross-section. This corresponds to one rotational and no translational DOF. However, the two segments can also rotate about an axis that is perpendicular to an elliptical cross-section. In this case, the joint has two rotational and no translational DOF. *Sliding hinge joints* have one rotational and one translational DOF. *Saddle joints* have two rotational and two translational DOF.

Generally, joint axes have been found to change orientation during movement (Lundberg et al., 1989). This finding suggests that one degree of freedom is not sufficient for the description of actual joint motion. However, it has been proposed (Wilson et al., 1996) that unloaded joint motion at the human ankle and knee can be described using a one degree of freedom model, because there is a distinct movement path in these joints that offers minimal resistance. In the absence of other forces, the joint follows that minimal resistance movement path.

Human and animal joints are not ideal theoretical joints. Real joints allow for movements that are not discussed when assuming ideal theoretical joints. The surface geometry of opposing bones and the corresponding cartilage at joints is such that the two surfaces are not congruent. In the knee, for instance, the femoral condyles are curved, allowing for rolling, gliding, and twisting movements on the relatively flat tibial surfaces. The hip joint is a ball and socket joint, but its articular surfaces are not completely spherical and the two surfaces are incongruent (Bullough et al., 1968).

In biomechanics, joints are treated in two distinctly different ways. The first way corresponds to a purely mechanical thinking of a joint, e.g., section 2.9.3. In this case, the joint is defined as a point (coordinate) or a point and a directed line (axis) going through the point. This description is sufficient from a mechanical point of view to solve the inverse dynamics problem, to generate forward simulations or to attack the distribution problem. The second way corresponds to a biological thinking of a joint. In this case, joints are described as organs made up of contacting bones, their corresponding articular cartilage surfaces, and the structures providing stability or movement, or both, including ligaments, tendons, and muscles. These two approaches are virtually unrelated. Thus, discussion of a joint in biomechanics should clarify if the mechanical or the biological joint is of interest.

The first part of this chapter discussed joints in general terms. Every synovial joint in the human or animal body is specific and different from all the others. A full description of the anatomy, mechanics, and function of all joints goes well beyond the scope of this book. In the following sections, selected mechanical and functional aspects of joints are discussed using examples from the human ankle and the cat knee. The findings relate directly to the specific joints chosen and the comments might not be exactly appropriate for any other joint. However, general findings are often the same across joints. Thus, the specific examples may be considered in a broader sense than just as applicable for the joint discussed.

2.9.4 THE HUMAN ANKLE JOINT COMPLEX

Humans and animals have numerous joints with specific characteristics. In this section, the human ankle joint complex is used to discuss some general biomechanical aspects relevant to joints.

The ankle joint complex consists of the calcaneus, talus, tibia, fibula, and all ligamenteous and muscle-tendon structures crossing the joints between the four bones.

The subtalar joint is the joint between talus and calcaneus.

The ankle joint (or talo-crural joint) is the joint between tibia and talus.

ANATOMY OF THE ANKLE JOINT COMPLEX

The ankle joint complex consists of bones, ligaments, and muscle-tendon units. The bones of the ankle joint complex are the calcaneus, talus, tibia, and fibula. They are supported by several groups of ligaments (Figure 2.9.1).

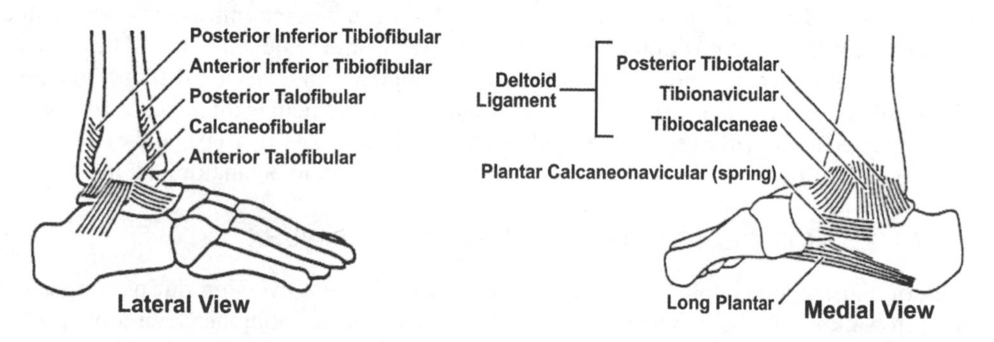

Figure 2.9.1 **Illustration of the medial (right) and lateral (left) ligaments of the ankle joint complex.**

The tibio-fibular ligaments (anterior and posterior) connect tibia and fibula on the lateral side. The talo-fibular (anterior and posterior) and the calcaneo-fibular ligaments connect talus and calcaneus with the fibula laterally. Medially, the ankle joint complex is supported by the long plantar ligament and by the plantar calcaneo-navicular (spring) ligament.

The muscle-tendon units of the ankle joint complex (Figure 2.9.2) include the triceps surae (gastrocnemius and soleus), the posterior compartment muscles (tibialis posterior, flexor digitorum longus, flexor hallucis longus), the lateral compartment muscles (peroneus brevis and longus), and the anterior compartment muscles (tibialis anterior, extensor hallucis longus, extensor digitorum longus, peroneus tertius). Each of these muscles has a specific function. The tibialis posterior tendon, for instance, originates at the lateral side of the foot, crosses the ankle joint complex on the medial side below the medial malleolus, and attaches to the navicular, the cuneiforms, and the cuboid bones. A contraction of the tibialis posterior may produce foot movement or compression, or both, of the bony structures of the rear- and mid-foot.

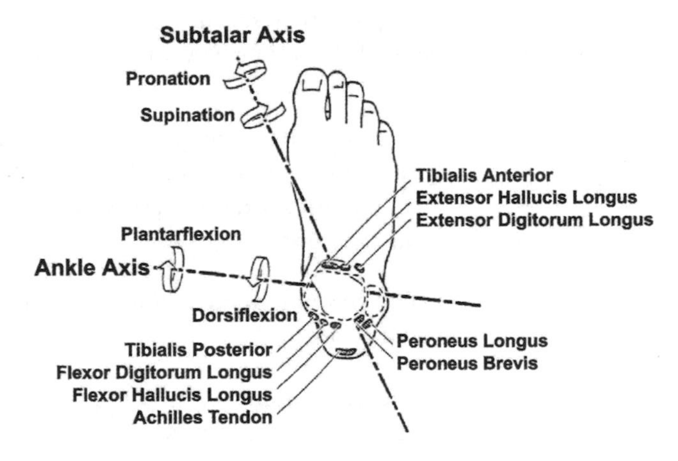

Figure 2.9.2 **Illustration of the extrinsic muscle-tendon units crossing the ankle joint complex and the functional axes of the subtalar and ankle joints.**

The ankle joint complex has two major functional axes, the subtalar joint axis, determined by talus and calcaneus, and the ankle joint axis, determined by talus and tibia (Figure 2.9.2). The position of the calcaneus and tibia can be determined with acceptable accuracy. However, the position of the talus is hidden from the outside view. Thus, orientation of, and movement about, the ankle and the subtalar joint axes are difficult to determine (van den Bogert et al., 1994). For this reason, foot movement is often quantified about an anterior-posterior (in-eversion), a medio-lateral (plantar-dorsiflexion), and an inferior-superior axis (ab-adduction), i.e., axes that do not correspond to an anatomical joint.

MOVEMENT COUPLING IN THE ANKLE JOINT COMPLEX

Foot and leg are coupled mechanically. For instance, foot eversion during the stance phase in walking results in internal leg rotation. Movement coupling between foot and leg or specifically, between calcaneus and tibia, has been described in numerous publications, e.g., Hicks, 1953; Wright et al., 1964; Lundberg, 1989. The subtalar and ankle joints were described as a combination of two hinge joints (Manter, 1941; Isman and Inmann, 1968; Mann, 1982), or as more complex joint structures (Sammarco et al., 1973; van Langelaan, 1983; Engsberg, 1987; Siegler et al., 1988; Nigg et al., 1993).

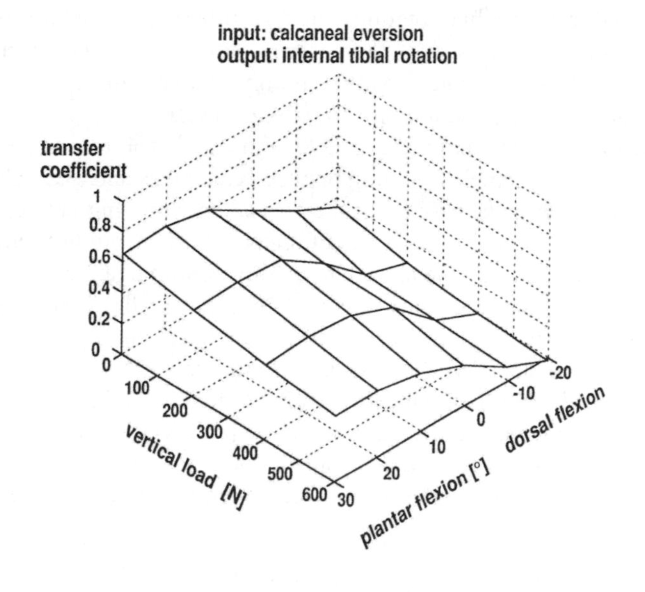

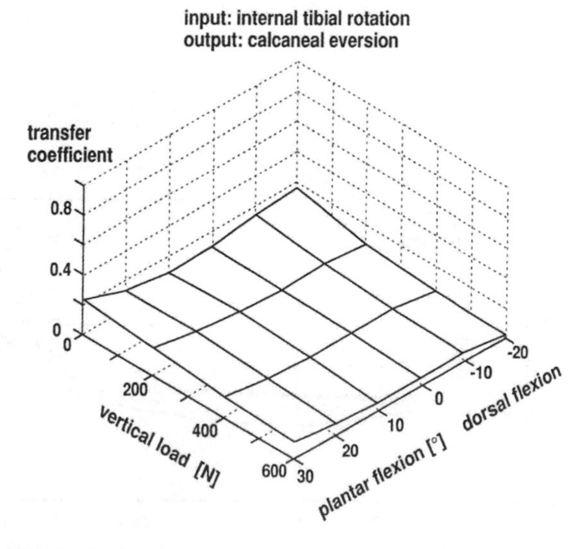

Figure 2.9.3 **Movement coupling between calcaneus and tibia for various axial tibial loads and foot positions. Movement input occurred through the calcaneus (top) or through the tibia (bottom). Summary based on data from Hintermann et al., 1994, with permission. The transfer-coefficient between calcaneus and tibia (foot and leg) was defined as T(FL) = $\Delta\rho : \Delta\beta$. The transfer-coefficient between tibia and calcaneus (leg and foot) was defined as T(LF) = $\Delta\beta : \Delta\rho$ with $\Delta\beta$ describing the change in calcaneal eversion (+) or inversion (-) and $\Delta\rho$ describing the change in internal (+) or external (-) tibial rotation.**

Movement coupling from calcaneus to tibia (bottom-up) or from tibia to calcaneus (top-down) has been studied in vitro with intact ligaments and without muscle-tendon forces (Hintermann et al., 1994). The results of these experiments (Figure 2.9.3) showed that the coupling between calcaneus and tibia was dependent on the input movement (calcaneal in-eversion or external-internal tibial rotation), the direction of the transfer (high for calcaneus to tibia and low for tibia to calcaneus), the position of the foot with respect to the leg (high for plantarflexion and low for dorsiflexion), and on the axial tibial loading (low coupling for high tibial loads and high coupling for low tibial loads).

Similar in vitro studies (Sommer et al., 1996) showed that cutting of selected ankle joint ligaments affected the movement coupling between calcaneus and tibia. Cutting the lateral ankle ligaments resulted in an *increase* of the movement coupling between calcaneus and tibia (indicated by the increased transfer coefficient), corresponding to an increased rotation of the ankle mortise around the talus. Cutting the deltoid ligament (on the medial side) resulted in a *decrease* of the calcaneus-tibia coupling, especially during foot eversion.

The range of motion between foot and leg was dependent on the ankle and subtalar joint integrity but also on the integrity of the neighbouring joints in the midfoot (Gellman et al., 1987; Hintermann et al., 1994). The most substantial reduction of the range of motion of the foot with respect to the leg occurred when the ankle joint was fused (Figure 2.9.4). Fusion of the talo-navicular joint (part of the midfoot) reduced the range of motion between foot and leg for all four movement directions (in-eversion and plantar-dorsiflexion) by about 10 to 20%.

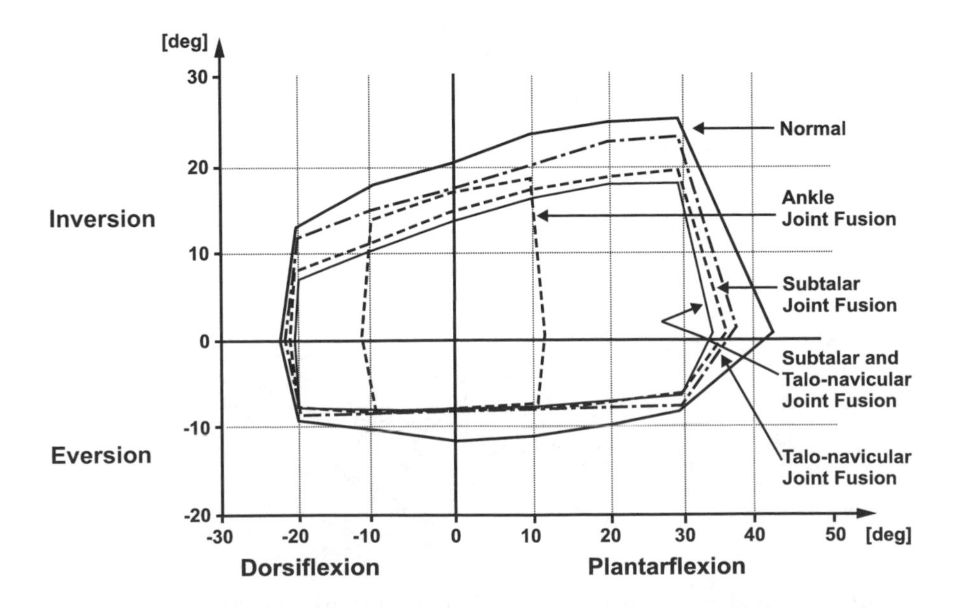

Figure 2.9.4 **Effect of selected joint arthrodeses on the range of motion for plantar-dorsiflexion and in-eversion between calcaneus and tibia. Average values for six specimens used in an in vitro setting (bold solid lines = normal; light dashed lines = ankle joint fusion; bold dashed lines = subtalar joint fusion; light solid lines = subtalar and talo-navicular joint fusion; dashed-dotted lines = talo-navicular joint fusion).**

In summary, movement between foot and leg or more specifically, between calcaneus and tibia is coupled. The coupling depends on:

- The input movement (calcaneal in-eversion or external-internal tibial rotation),
- The direction of the movement transfer (high for calcaneus to tibia and low for tibia to calcaneus),
- The position of the foot with respect to the leg (high for plantarflexion and low for dorsiflexion),
- The axial tibial loading (low coupling for high tibial loading and high coupling for low tibial loading), and
- The integrity of the ligaments (movement coupling increases when cutting ligaments on the lateral side and decreases when cutting ligaments on the medial side).

Ongoing research aimed at studying the effect of muscle-tendon forces on movement coupling in the ankle joint complex suggests that muscle-tendon forces increase the rotational stiffness of the ankle joint complex, thereby, reducing its range of motion. However, the movement coupling is influenced only minimally by muscle-tendon forces (Stähelin et al., 1997).

MOVEMENT AND MOVEMENT COUPLING IN VIVO

Most studies discussing the function of the ankle joint complex have used in vitro setups or in vivo studies with markers mounted to the skin. In one study, bone pins were used on five subjects during walking and running to study the in vivo ankle joint mechanics (Reinschmidt et al., 1997; Stacoff, 1998). Movement coupling between calcaneus and tibia during barefoot and shoed running in normal subjects was similar (illustrated for one subject in Figure 2.9.5). The maximal calcaneal eversion and the maximal internal tibial rotation were similar, suggesting that interventions with footwear, orthotics or inserts have a minimal effect on skeletal movement in the human ankle. Since the external and internal force conditions are substantially different between barefoot and shoed running, it is proposed that

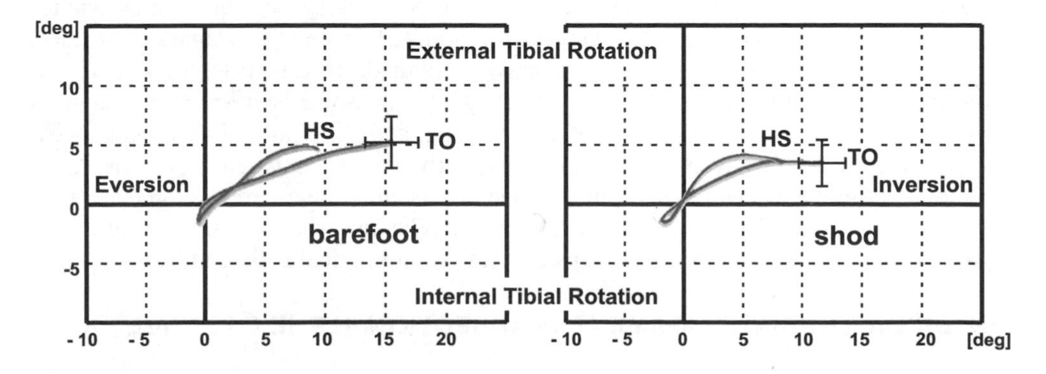

Figure 2.9.5 In vivo movement coupling between calcaneus and tibia for one subject running barefoot (left) and running with a running shoe (right) (from Stacoff, 1998, with permission).

changes in muscle activity are used to compensate for the interventions. Thus, changes in footwear or shoe inserts may not realign the skeleton, as is commonly thought, but may influence muscle activity in the lower extremities.

It is assumed that the mechanical range of motion and the mechanical movement coupling between segments of other joints follow similar patterns as those outlined here for the ankle joint complex. However, the actual relationships should be determined in detail for each joint and implemented in theoretical and experimental considerations.

2.9.5 JOINT ADAPTATION

Chapter 2.9.2 emphasized that the function of a joint can only be understood if (a) the function and properties of all components of a joint are known, and (b) if the relationships between the components comprising a joint are well understood. In other words, joints can only be understood completely if they are considered as *organs*.

In the previous section (2.9.4), the importance of the relationship between the components (bones, ligaments, muscles, cartilage) was emphasized on the example of ankle mechanics. The loss of a ligament or the fusion of two bones instantaneously influenced the mechanics of the entire joint.

This section, takes the idea of the joint as an *organ* one step further than in previous sections. Live materials can (and will) actively adapt when boundary conditions change. For example, in the previous section, joint ligaments were cut and the instantaneous effect on joint mechanics was measured. In this section, a joint ligament is cut and the mechanical and biological effects are observed over time. When studying the following paragraphs, focus on how complex and diverse the response of a joint can be to an intervention of a single joint component. The intervention described is a transection of the anterior cruciate ligament. The joint considered is the cat knee.

The mechanical and biological observations described in the following are unique to the intervention and joint chosen. However, the general observation that a disturbance of joint mechanics will cause adaptive responses in most or all tissues comprising the joint is likely correct for most interventions and joints. It should also be mentioned that the results from this example may not be directly translated to the human ACL-deficient knee. Although, it is well known that the design of knees across species has been quite consistent for about 300 million years (Dye, 1987), there are differences between the cat and human knee. Functionally, the cat uses its knee within a range of about 40° to 120° from full knee extension (0°) (Goslow et al., 1973; Herzog et al., 1992). Humans use their knees in a range of about 0° to 60° for activities such as walking, running, and climbing stairs. Also, the ACL is larger and the tibial plateau is tilted more posteriorly in cats compared to humans. However, the morphological components of the knee in cats and humans are the same and they have similar shapes. These components are: a bicondylar cam-shaped distal part of the femur, intra-articular ligaments, menisci, patella, and asymmetrical collateral ligaments.

ANTERIOR CRUCIATE LIGAMENT TRANSECTION IN THE CAT KNEE

Anterior cruciate ligament transections were performed either using an anterolateral capsulotomy (Herzog et al., 1993) or using an arthroscopic approach. Only adult, skeletally mature, outbred cats weighing a minimum of 30 N were used in any of these studies.

MECHANICAL CONSIDERATIONS

Movement and ground reaction forces

One week following ACL transection, ACL-transected animals exhibited a limping gait. This limping gait persisted for approximately three months. In these first three months following ACL transections, the knee was typically more flexed during walking, and final knee extension at the end of the stance phase was incomplete compared to normal control values (Hasler et al., 1998a). Static, vertical ground reaction forces and dynamic vertical and posterior ground reaction forces were also reduced in these initial months following intervention (Figure 2.9.6).

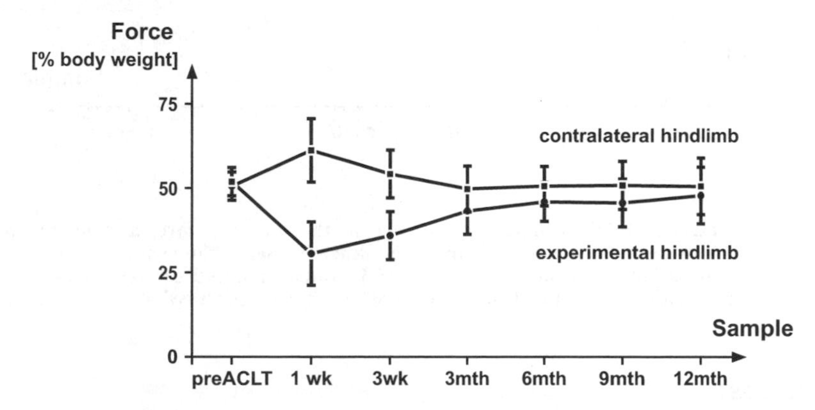

Figure 2.9.6 **Means ± standard deviations of the vertical ground reaction forces (normalized to % of body weight) for the ACL-transected (experimental) and the non-transected (contralateral) hindlimbs of seven cats followed over a one year period (from Suter et al., 1998, with permission).**

From three to 12 months following ACL transection, the visible limp disappeared in the experimental limb, and the static and dynamic ground reaction forces either returned to normal values or approached normal values closely (Suter et al., 1998). From 12 months to 3.5 years following ACL transection, quantitative gait analysis and qualitative examination of the animals' spontaneous movement behaviour appeared normal.

Stiffness

Anterior cruciate ligament loss in humans is associated with decreased knee stability. Specifically, for a given amount of anterior force on the tibia, the tibia translates anteriorly to a greater degree in the ACL-deficient, compared to the normal, human knee. Similarly, internal tibial rotation is increased following ACL rupture in humans.

In cats, as in humans, ACL transection caused a decreased stiffness in anterior translation and medial rotation of the tibia relative to the femur (Maitland et al., 1998). This result manifested itself in an increased anterior translation and medial rotation of the tibia relative to the femur immediately following ACL transection. However, over time, the knee became stiffer again. Anterior displacement of the tibia relative to the femur was substantially reduced three months compared to five weeks following ACL transection (Figure 2.9.7). Similarly, medial rotation of the tibia relative to the femur was decreased at two to four months following ACL transection compared to the values obtained immediately before and immediately after transection, indicating that secondary constraints more than compensated for the loss of the ACL (Maitland et al., 1998).

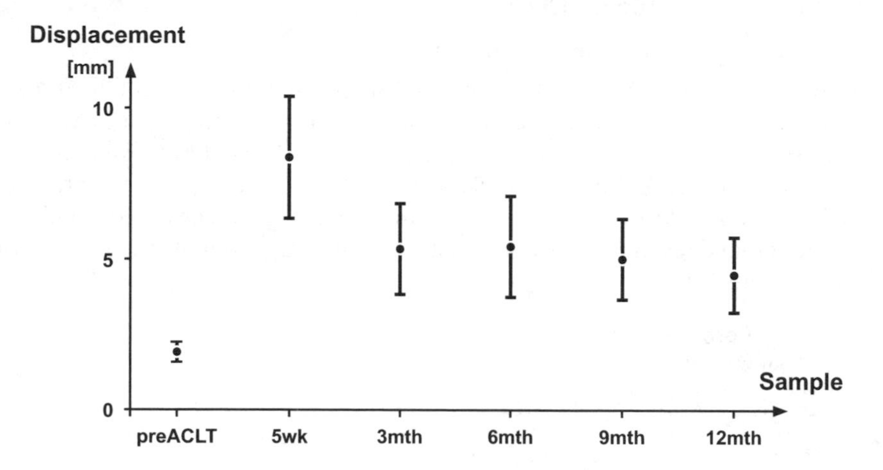

Figure 2.9.7 **Anterior tibial displacement relative to the femur before, and up to one year following, ACL transection in the cat knee. Observe the vast increase in anterior translation immediately following ACL transection and the reduction of anterior translation over time (from Suter et al., 1998, with permission).**

Internal forces

A complete and detailed understanding of joint mechanics requires that the internal forces are known. Internal forces are of importance for two primary reasons. First, the internal forces (particularly the muscular forces) tend to be much larger than the external forces. Second, the internal forces determine the amount of joint loading.

Following ACL transection, the forces in the cat quadriceps and gastrocnemius were reduced to about one-third of normal for the first two weeks (Figure 2.9.8). Also, the density, but not necessarily the magnitude, of knee extensor EMG signals was reduced following ACL transection compared to normal (Hasler et al., 1998a). Similarly, the patellofemoral joint contact forces were reduced in the ACL-transected compared to the normal knees. Quadriceps and gastrocnemius forces, knee extensor EMGs, and patellofemoral contact forces were all increased in the contralateral, non-operated hindlimbs compared to the corresponding values in normal animals (Hasler et al., 1998a; Hasler et al., 1998b). Follow-up results beyond two weeks of ACL transection are not available. We are not aware of any other direct muscle force measurements in animal models of joint or ligament injuries.

BIOLOGICAL CONSIDERATIONS

Anterior cruciate ligament transection in the cat caused fast, consistent, and in some tissues, dramatic adaptations. The biological adaptations described in the following were from observations made 12 to 16 weeks following ACL transection, except where specifically pointed out. However, it may be generally stated that the observations made at 12 to 16 weeks were present earlier (some as early as two to four weeks post transection) than at 12 to 16 weeks, and were also made for selected animals as late as 2.5 years following ACL transection. Therefore, the observations at 12 to 16 weeks post intervention represent one specific instant in time in a continuous adaptation process.

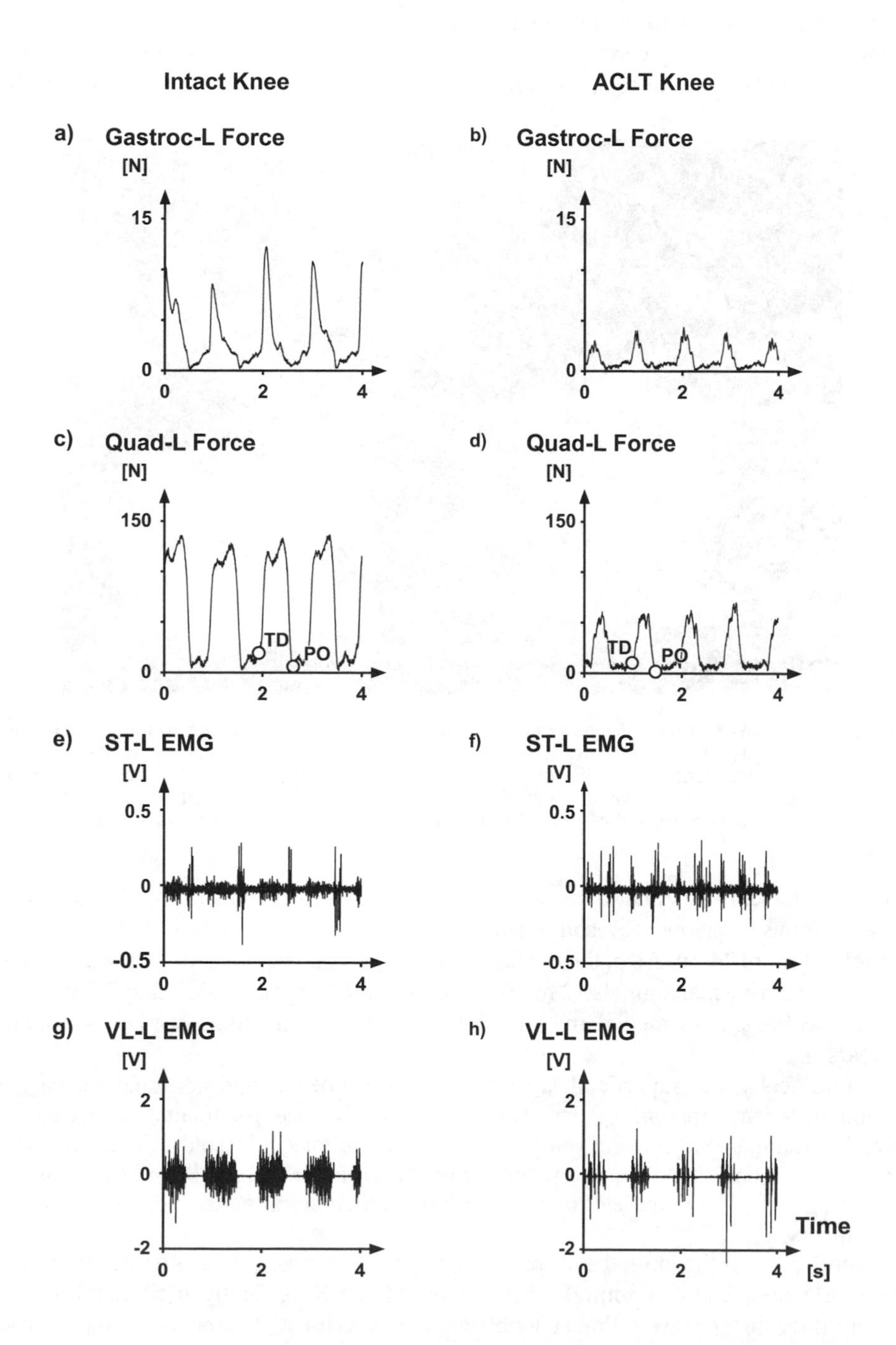

Figure 2.9.8 **Gastrocnemius (gastroc) and quadriceps (quad) forces, and semitendinosus (ST) and vastus lateralis (VL) EMGs, before (left panel) and seven days following ACL transection (right panel) (from Hasler et al., 1998a, with permission).**

Morphology

Twelve to 16 weeks following intervention, the ACL-transected joints revealed a hypertrophic and hemorrhagic synovium. The medial joint capsule and medial collateral ligament were thickened (Figure 2.9.9). The colour of the articular cartilage was tinged slightly brown in the experimental joints compared with the normal translucent blue of the contralateral joints. Osteophytes had formed around the joint margins and ligamentous insertions (Figure 2.9.9). Also, the medial menisci appeared ragged and they were torn in a few of the experimental animals 12 to 16 weeks post ACL transection. 1.5 to 2.5 years post ACL transections, most medial and lateral menisci were torn and showed buckle-handle-type tears.

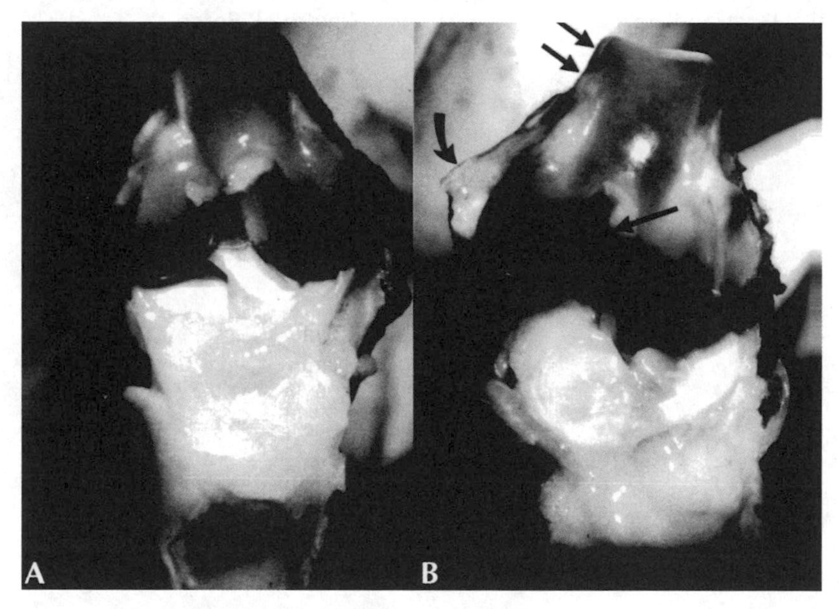

Figure 2.9.9 **ACL-intact (A) and ACL-transected (B) knees from one animal sacrificed 12 weeks post unilateral ACL transection. Observe the formation of osteophytes (straight arrows, B), and the increase in thickness of the medial capsule and medial collateral ligament (curved arrow, B) in the ACL-transected (B), compared to the non-transected control (A) (from Herzog et al., 1993, with permission).**

Histologically, the superficial layers of the experimental and contralateral cartilage were intact, but the apparent cell density was increased in the experimental compared to the contralateral joints. At 1.5 to 2.5 years post ACL transections, all tested animals (n = 4) had lesions of the articular cartilage either on the medial or lateral femoral condyle. In two cases, these lesions were complete to the underlying subchondral bone, indicating bona fide osteoarthritis.

In other parts of the joint, the articular cartilage was characteristically thicker in the experimental compared to the contralateral or normal knee. Specifically, in the patellofemoral joint, cartilage thickness was almost doubled in the experimental compared to the contralateral hindlimb at 16 weeks post ACL transection (Figure 2.9.10). Since the apparent material properties of the articular cartilage had not changed measurably within the first 16 weeks following ACL transection, the increased cartilage thickness was assumed to indi-

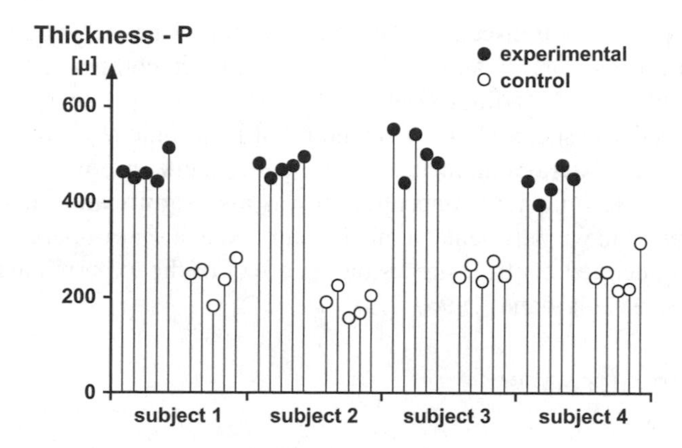

Figure 2.9.10 **Articular cartilage thickness (in microns) on the patella of four cats (subjects 1-4). Five measurements were made on each subject within a 4 mm^2 area in the centre of the patella. The experimental values were obtained from patellae of ACL-transected knees (16 weeks post transection). The control values were obtained from patellae of the corresponding non-transected contralateral knee.**

cate decreased cartilage stiffness and a corresponding change in force transmission across the patellofemoral joint. Measurement of the total contact area and peak pressure for a given force across the patellofemoral joint revealed that, in general, a given force was transmitted across a larger contact area and the peak pressure was smaller in the experimental compared to the contralateral knee (Figure 2.9.11).

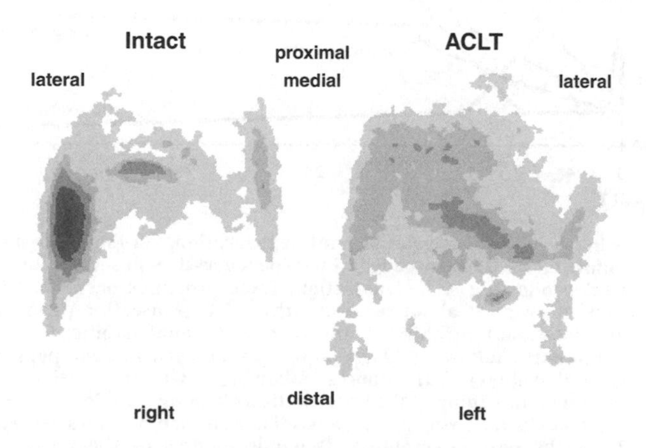

Figure 2.9.11 **Contact pressure distribution from an intact (non-transected, contralateral) patellofemoral joint and the corresponding ACL-transected (ACLT, ipsilateral) patellofemoral joint (16 weeks post transection). The darker the stain, the higher the pressure. Note, the higher peak pressure and the smaller total contact area in the intact, compared to the ACL-transected, joint for the same amount of total force transmitted across the joint.**

Biochemistry

At 12 weeks post ACL transection, the water content of the cartilage from the experimental hindlimb was elevated in almost all of the samples in comparison with that of cartilage from the contralateral hindlimb (Herzog et al., 1993). The tissue weight, both as fresh (wet) tissue and as dry tissue, and the total amount of hexuronic acid were elevated, especially in the medial and lateral femoral condyles and the patellar groove and patella (Herzog et al., 1993). The hexuronic acid concentration was also significantly higher in the contralateral hindlimb of the experimental animal compared to a sham-operated animal and its contralateral non-operated control, suggesting an effect of the experimental procedure on the contralateral knee (Maitland, 1996).

SUMMARY AND OUTLOOK

In summary, a simple intervention to a single structure of a joint (ACL transection in the cat knee) was shown to alter the internal and external mechanics of the joint and produce an adaptive response in all tissues comprising the knee: bones, ligaments, joint capsule, cartilage, and muscles. The changes in joint mechanics were shown to be time-dependent, and the effects on the joint tissues were progressive and continuous. It is likely that a similar intervention in any other joint would produce similar responses.

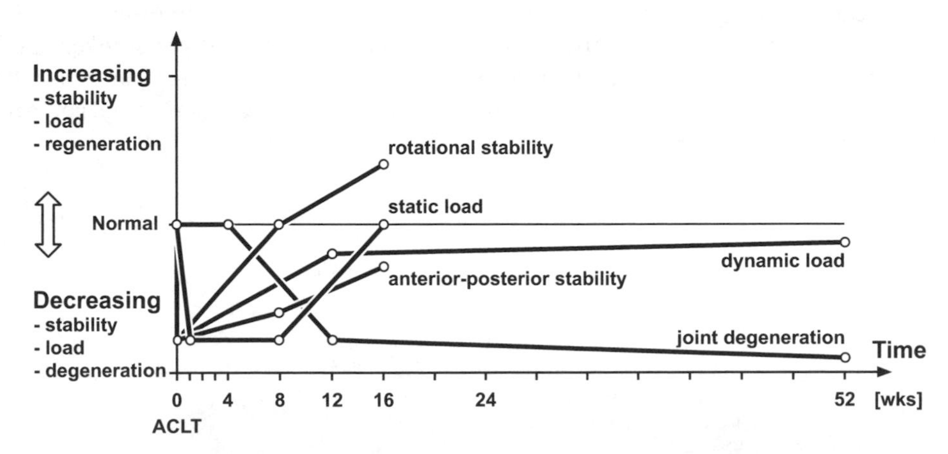

Figure 2.9.12 **Schematic time histories of joint degeneration, loading, and stability of the knee following ACL transection. First degenerative processes are seen about four weeks following ACL transection; these processes occur fast initially and then slow down after about three months. ACL transection produces immediate AP and rotational instability of the knee. Rotational stability is re-established about two months following ACL transection, and is hypercompensated (compared to normal values) four months following ACL transection. ACL transection produces unloading of the knee. Static unloading remains constant for about six to eight weeks following ACL transection and then becomes normal at around 16 to 18 weeks post intervention. Dynamic loading of the knee recovers relatively quickly in the first three months following ACL transection, and then remains virtually constant and below normal values for up to one year post intervention.**

Figure 2.9.12 shows a schematic summary of the mechanical and biological changes following anterior cruciate ligament transection for one year. In general, static and dynamic loading as well as translational and rotational stability are decreased immediately following

ACL transection. However, there are no obvious degenerative changes in the joint for the first two to four weeks following ACL transection. After a few weeks following ACL transection, the mechanical factors return to more normal values, loading of the ACL-transected joint increases, and translational and rotational stability increases. After about four weeks, biologic changes are observed in the joint that are generally considered degenerative. These degenerative processes occur rapidly initially and then progress more slowly up to one year (Figure 2.9.12) and beyond.

Following the ACL transection, the mechanics of the knee were altered dramatically. It is typically assumed that these alterations in joint mechanics are the primary stimulus for the biological adaptations observed. However, the relationship between joint mechanics and biological adaptations of joint structures is not well understood. Specifically, it is not known why ACL transection in the cat knee causes osteoarthritis, nor is it known why any of the tissues adapt the way they do. A future challenge will be to elucidate why and how musculo-skeletal tissues adapt to mechanical loads. Since tissue production is regulated on the cell level, it will be imperative (a) to determine the stress-strain states of cells in normal and experimentally perturbed joints, e.g., following the removal of a ligament, and (b) to quantify the changes in biosynthetic activity of the cells before and after perturbation. Such studies, combined with research on isolated cells aimed at elucidating the pathways of biosynthetic control, may ultimately lead to an understanding of adaptive and degenerative responses of joints, such as occur in osteoarthritis.

2.10 ADDITIONAL EXAMPLES

This section has been divided into basic and advanced examples. In general, the basic questions can be answered by studying the text or by performing some calculations based on information provided by the text. The advanced questions go beyond what can be found in the text.

BASIC QUESTIONS

Bone

1. Draw the force-deformation and stress-strain diagrams and determine the slopes for the listed cylindrical cross-sectional structures. Use the following assumptions:

    - The force-deformation and stress-strain relationships are linear, and
    - The force acts parallel to the symmetry axis.

Cylinder	Radius	Elastic Modulus	Height of Cylinder
A	r	E	h
B	2r	E	h
C	r	2E	2h
D	2r	2E	2h

2. A contact force is acting at point E on the drawing (Figure 2.10.1), representing the head of the femur of a human leg. Determine the magnitude and direction of additional muscle force acting at point A to minimize the magnitude of the local compression or tension stress in the point C or D. Discuss the solution with respect to human anatomy, and identify a muscle that corresponds to the proposed solution. Use the following assumptions:

    - The length-width ratio of the structure is large,
    - The femur is a homogeneous and isotropic structure, and
    - The shear influence is neglected (Bernoulli bending theory).

    Remember that muscle forces pull, but do not push.

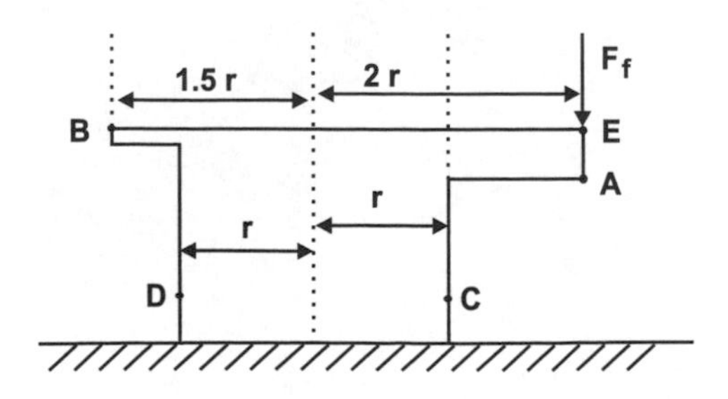

Figure 2.10.1 **Illustration of a contact force acting on the femur of a human leg.**

3. The two free body diagrams in Figure 2.10.2 are equivalent. Knowing that the contact force, $\mathbf{F}_c$, and the contact moment, $\mathbf{M}_c$, are the same for both diagrams, use the equations that govern each system to show that $\mathbf{F}_a = \mathbf{F}_b$, and calculate $\mathbf{M}_b$.

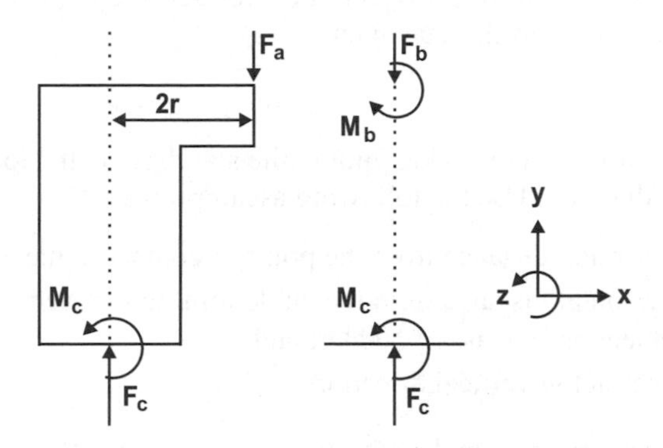

Figure 2.10.2 **Illustration of two equivalent free body diagrams.**

4. A cylindrical bone (illustrated in Figure 2.10.3) has the outer radius, R, and the inner radius, r. Determine analytically and graphically the maximal compression stress, σ_{comp}, at a cross-section C-D as a function of the relative wall thickness [(R-r) : R] for different wall thickness. Use the following assumptions:

- The structure is cylindrical,
- The outer radius is R = 3 cm,
- The force acts parallel to the symmetry axis at the edge of the cylindrical structure and has a magnitude of F = 1000 N,
- The structure is firmly anchored in the ground, and
- The length-width ratio of the structure is large.

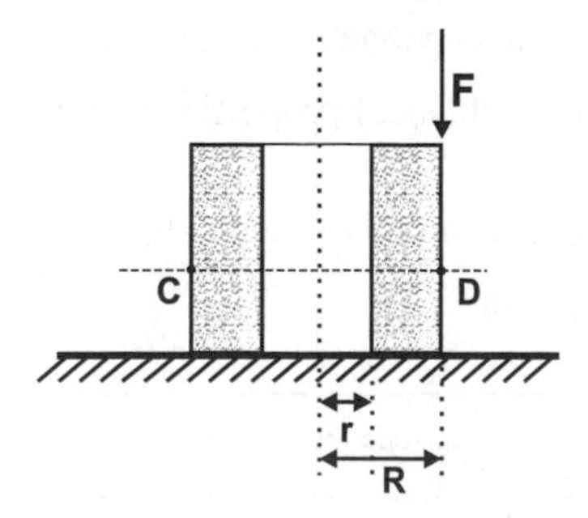

Figure 2.10.3 **Illustration of a cross-section through a cylindrical structure with an external force, F, acting on it.**

5. Determine the ultimate force in compression and tension for a cylindrical compact bone assuming a cross-sectional area of $A = 10\ \text{cm}^2$ and a modulus of elasticity of $E = 2 \cdot 10^{10}$ Pa.

6. It has been stated that bone is stronger in response to compressive forces than in response to tensile forces. It has been argued that this may be due to an evolutionary adaptation. Based on the actual loading pattern, is it reasonable to assume that all bones in the human body have evolved to be stronger in compression than in tension. Provide evidence to support this statement.

Articular Cartilage

1. (a) Estimate the forces in the ankle joint while standing on the toes of one foot using a free body diagram. Use the following assumptions:

   - The horizontal distance from the point of contact to the ankle joint is 25 cm,
   - The horizontal distance from the ankle joint to the Achilles tendon is 5 cm,
   - The subject has a mass of 100 kg, and
   - All forces act in vertical direction.

   (b) Estimate the average normal stress in the ankle joint if the idealized contact area is flat and has an area of 5 cm^2.

   (c) Compare and discuss the results of the findings in b with the ultimate compressive stress values published by Yamada (1973).

2. The idealized tensile stress–strain behaviours for the surface and the deepest layer of cartilage are illustrated in Figure 2.10.4.

   (a) Calculate the modulus of elasticity for the two layers.

   (b) Determine the strain in each of the two layers if a tensile force of 10,000 N acts on a piece of cartilage that has a flat area of 10 cm^2.

3. (a) Describe a stress-relaxation test.

   (b) Describe a creep test.

   (c) Which of these two tests simulations what might happen to the cartilage in the tibio-femoral joint when someone stands still for an extended period of time?

4. (a) When describing the mechanical properties of a material, what does anisotropic and inhomogeneous mean?

   (b) Describe the anisotropies and inhomogenieties of cartilage.

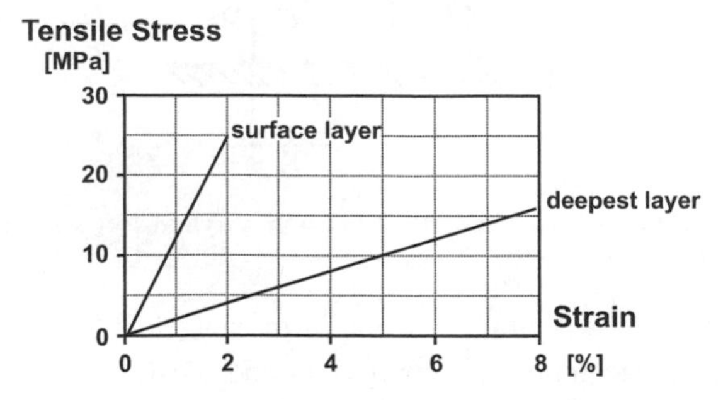

Figure 2.10.4 **Idealized tensile stress-strain diagram for the surface and the deepest layer of cartilage.**

Joints

1. Determine the degree of freedom (DOF) of a planar double pendulum fixed at a rigid structure (Figure 2.10.5).

2. Determine the DOF for a piston in a three-dimensional space (Figure 2.10.5, centre). Use the following assumptions:

   - The upper segment of the piston is fixed with a (non-sliding) hinge joint to a rigid structure, and
   - The joint between the two segments is a (non-sliding) hinge joint.

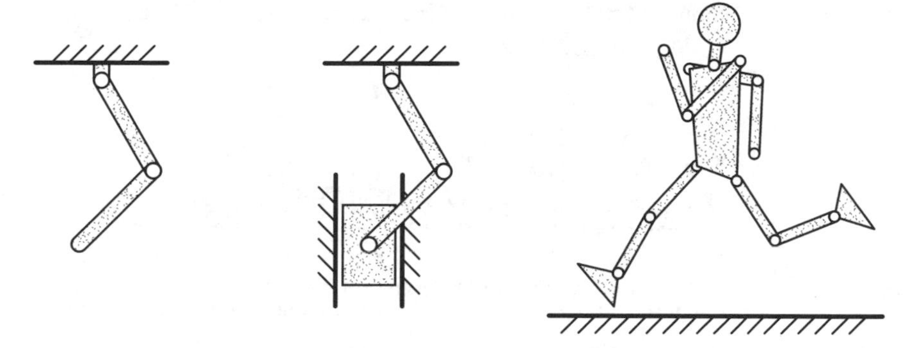

Figure 2.10.5 **Schematic illustration of three examples for the determination of the degree of freedom, a two-dimensional double pendulum fixed with a hinge joint on a rigid structure (left), a two-dimensional piston fixed with a hinge joint on a rigid structure (middle), and a schematic illustration of a human body with 12 segments in a three-dimensional space (right).**

3. Determine the DOF for a schematic representation of a human as illustrated in Figure 2.10.5, right. Use the following assumptions:

   - The 12 segments included are rigid,
   - The knee and elbow joints are (non-sliding) hinge joints, and
   - All the other joints are spherical joints.

ADVANCED QUESTIONS

Bone

1. Various suggestions have been made to explain the mechanism for bone formation or absorption, or both. List these possibilities, and provide supporting or contradicting experimental or theoretical evidence for the proposed mechanisms.

2. Determine the stress distribution for the firmly anchored structure illustrated in Figure 2.10.6. Specifically, determine the location of the neutral axis, the areas with compression and the areas with tension, and indicate where compression or tension stresses are low or high. Use the following assumptions:

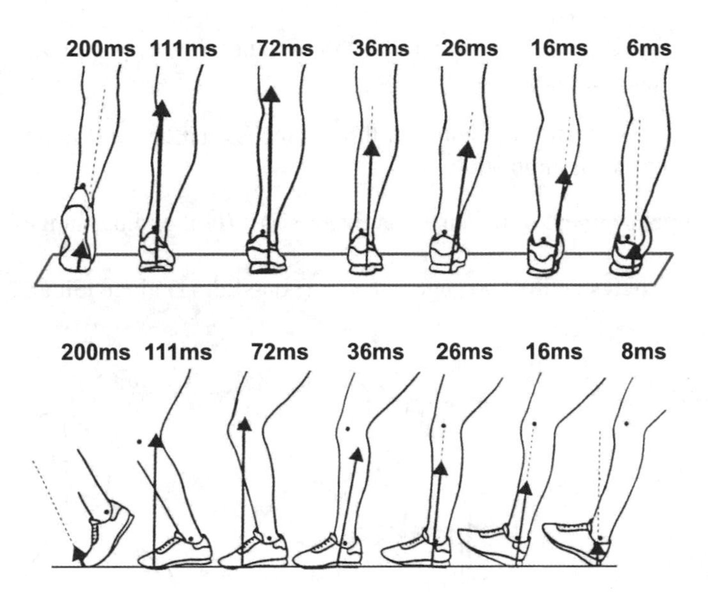

Figure 2.10.6 **Illustration of a human leg during ground contact in running, with the ground reaction forces acting on it for two views and seven different time points.**

- The structure is homogeneous,
- A general force acts with the components: **F** = (-314 N, -628 N, 0 N), and
- The information provided does not allow determining the stress in some boundary areas. Therefore, neglect those areas.

3. Estimate the stress distribution in a cross-section in the middle of the tibia during heel-toe running for the indicated time points, using an estimated centre of pressure of the resultant ground reaction force based on Figure 2.10.6. Use the following assumptions:

    - The ankle joint complex is stiff,
    - The bone material is isotropic and homogeneous,
    - The ground reaction forces act (contrary to the illustration) parallel to the tibia,
    - Muscle forces can be neglected,
    - The cross-section of the tibia is circular (with a radius of 2 cm) and completely filled with material, and
    - Other assumptions that allow you to solve this question with a reasonable amount of work.

4. Given a hollow cylindrical bone with a constant wall thickness, d, and a variable outside radius, r(x), where $r(0) = r_0$, and a moment, M, applied at x = 0 (Figure 2.10.7).

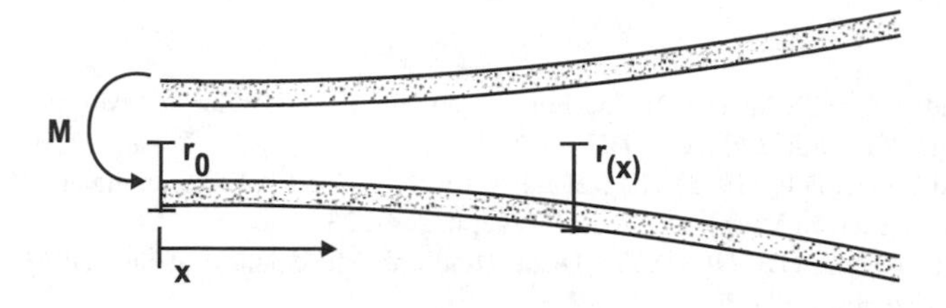

Figure 2.10.7 **Illustration of a hollow cylindrical bone and a moment acting on it.**

How must r vary as a function of x so that the maximal tensile strength at any point along the bone is constant?

5. Given:

Stress due to bending $= \sigma_{be} = M_{be} \cdot \frac{y}{I_{xx}}$

Moment of inertia $= \int_A y^2 dA$

Show that the bending stress for a hollow circle is $\sigma_{be} = \frac{4 \cdot M_{be} \cdot R}{\pi(R^4 - r^4)}$

6. It has been suggested that impact forces acting on the locomotor system can have a positive effect on bone mass and bone mineral density in adolescents and adults. It has also been suggested that bone mass and bone mineral density may be increased with a weight bearing exercise regime in postmenopausal women. Investigate and discuss the supporting evidence for these claims.

Articular Cartilage

1. The frequency of osteoarthritis is higher for the knee joint than for the ankle joint. Provide a list of possible reasons for this result and discuss the supporting evidence for these suggestions.

2. Compare the forces in the ankle and knee joint for the impact and active phases of ground contact during heel-toe running, using actual data from film or video and from force plates.

2.11 REFERENCES

Abbott, B.C. and Aubert, X.M. (1952) The Force Exerted by Active Striated Muscle During and After Change of Length. *J. Physiol.* **117,** pp. 77-86.

Abbott, B.C. and Wilkie, D.R. (1953) The Relation Between Velocity of Shortening and the Tension-length Curve of Skeletal Muscle. *J. Physiol.* **120,** pp. 214-223.

Abraham, L.D. and Loeb, G.E. (1985) The Distal Hindlimb Musculature of the Cat (Patterns of Normal Use). *Exp. Brain Res.* **58,** pp. 580-593.

Adams, J.A. (1977) Feedback Theory of How Joint Receptors Regulate the Timing and Positioning of a Limb. *Psychol. Rev.* **84,** pp. 504-523.

Adeeb, S., Ali, A., Shrive, N., Frank, C., and Smith, D. (2004) Modelling the behaviour of ligaments: a technical note. *Comput. Methods. Biomech. Biomed. Engin.* **7**(1), pp. 33-42.

Akeson, W.H., Amiel, D., Mechanic, G.L., Woo, S.L.-Y., Harwood, F.L., and Hamer, M.L. (1977) Collagen Cross-linking Alterations in Joint Contractures: Changes in the Reducible Cross-links in Periarticular Connective Tissue Collagen After Nine Weeks of Immobilization. *Connective Tissue Research.* **5,** pp. 15-19.

Akeson, W.H., Amiel, D., and Woo, S.L.-Y. (1980) Immobility Effects on Synovial Joints: The Pathomechanics of Joint Contracture. *Biorheology.* **17,** pp. 95-110.

Akeson, W.H., Amiel, D., Abel, M.F., Garfin, S.R., and Woo, S.L.-Y. (1987) Effects of Immobilization of Joint. *Clinical Orthopaedics and Related Research.* **219**, pp. 28-37.

Alexander, R.M. (1983) *Animal Mechanics* (2ed). Oxford, Blackwell, UK.

Alexander, R.M. (1984) Elastic Energy Stores in Running Vertebrates. *Amer. Zool.* **24**, pp. 85-94.

Alexander, R.M. (1988b) The Spring in Your Step: The Role of Elastic Mechanisms in Human Running. *Biomechanics XI-A* (eds. De Groot, G., Hollander, A.P., Huijing, P.A., and van Ingen Schenau, G.J.). Free University Press, Amsterdam. pp. 17-25.

Alexander, R.M. (1988a) *Elastic Mechanisms in Animal Movement.* Cambridge University Press, Cambridge, UK.

Alexander, R.M. and Vernon, A. (1975) The Mechanics of Hopping by Kangaroos (Macropodidae). *J. Zool.* **177**, pp. 265-303.

Alexander, R.M. and Vernon, A. (1975). The Dimensions of Knee and Ankle Muscles and the Forces They Exert. *J. Hum. Mvmt. Studies* **1**, pp. 115-123.

Alexander, R.M., Maloiy, G.M.O., Ker, R.F., Jayes, A.S., and Warui, C.N. (1982) The Role of Tendon Elasticity in the Locomotion of the Camel (Camelus Dromedarius). *J. Zool.* **198**, pp. 293-313.

Allen, R.E. and Boxhorn, L.K. (1989) Regulation of Skeletal Muscle Satellite Cell Proliferation and Differentiation by Transforming Growth Factor-beta, Insulin-like Growth factor-I, and Fibroblast Growth Factor. *Journal of Cellular Physiology.* **138**, pp. 311-5.

Allinger, T.L., Epstein, M., and Herzog, W. (1996) Stability of Muscle Fibers on the Descending Limb of the Force- length Relation. A Theoretical Consideration. *Journal of Biomechanics.* **29**, pp. 627-633.

American Academy of Orthopaedic Surgeons (1988) *Injury & Repair of the Musculo-skeletal Soft Tissues* (eds. Woo, S.L.-Y. and Buckwalter, J.A.). Amer. Academy of Ortho. Surg., Park Ridge, IL.

Amiel, D., Frank, C.B., Harwood, F.L., Fronek, J., and Akeson, W.H. (1983) Tendons and Ligaments: A Morphological and Biochemical Comparison. *J. Orthop. Res.* **1,** pp. 257-265.

Amiel, D., Billings, E., and Akeson, W.H. (1990) Ligament Structure, Chemistry, and Physiology. *Knee Ligaments: Structure, Function, Injury, and Repair* (eds. Daniel, D.D., Akeson, W.H., and O'Connor, J.J.). Raven Press, New York. pp. 77-91.

Amprino, R. (1947) La Structure du Tissu Osseux Envisagee Comme Expression de Differences dans la Vitesse de L'acroissement. *Arch. Bio.* **58,** pp. 315-330.

Andrews, J.G. (1974) Biomechanical Analysis of Human Motion. *Kinesiology.* Amer. Assoc. for Health, Phys. Ed., and Rec., Washington, D.C. **IV,** pp. 32-42.

Andrews, J.G. (1987) The Functional Roles of the Hamstrings and Quadriceps during Cycling. Lombard's Paradox Revisited. *J. Biomech.* **20,** pp. 565-575.

Archambault, J.M., Herzog, W., and Hart, D.A. (1997) Response of Rabbit Achilles Tendon to Chronic Repetitive Loading. *43rd Annual Meeting, Orthopaedic Research Society.* San Francisco, CA. pp. 28-35.

Archambault, J.M., Wiley, J.P., and Bray, R.C. (1995) Exercise Loading of Tendons and the Development of Overuse Injuries: A Review of Current Literature. *Sports Med.* **20 (2)**, pp. 77-89.

Ariano, M.A., Armstrong, R.B., and Edgerton, V.R. (1973) Hindlimb Muscle Fibre Populations of Five Mammals. *J. Histochem. Cytochem.* **21 (1),** pp. 51-55.

Armstrong, C.G. and Mow, V.C. (1980) Friction, Lubrication, and Wear of Synovial Joints. *Scientific Foundations of Orthopaedics and Traumatology* (eds. Owen, R., Goodfellow, J., and Bullough, P.). pp. 223-232.

Arnoczky, S.P. (1983) Anatomy of the Anterior Cruciate Ligament. *Clin. Orthop.* **172,** pp. 19-25.

Arnoczky, S.P. and Warren, R.F. (1988) Anatomy of the Cruciate Ligaments. *The Crucial Ligaments* (ed. Feagin Jr., J.A.). Churchill Livingstone, New York. pp. 179-195.

Ascenzi, A. and Bonucci, E. (1972) The Shearing Properties of Single Osteons. *Anat. Rec.* **172**, pp. 499-510.

Asmussen, E. and Bonde-Petersen, F. (1974) Apparent Efficiency and Storage of Elastic Energy in Human Muscles During Exercise. *Acta. Physiol. Scand.* **92 (4),** pp. 537-545.

Astrand, P. and Rodahl, K. (1977) Physiological Bases of Exercise. *Textbook of Work Physiology* (2ed). McGraw-Hill, New York.

Ateshian, G.A., Wang, H., and Lai, W.M. (1998) The Role of Interstitial Fluid Pressurization and Surface Porosities on the Boundary Friction of Articular Cartilage. *ASME J. Tribology.* **120,** pp. 241-248.

Bach Jr., B.R., Warren, R.F., and Wickiewicz, T.L. (1987) Triceps Rupture: A Case Report and Literature Review. *Am. J. Sports Med.* **15 (3),** pp. 285-289.

Backman, C., Boquist, L., Friden, J., Lorentzon, R., and Toolanen, G. (1990) Chronic Achilles Paratenonitis with Tendinosis: An Experimental Model in the Rabbit. *J. Orthop. Res.* **8,** pp. 541-547.

Bagshaw, C.R. (1993) *Muscle Contraction.* Chapman & Hall, New York.

Baldwin, K.M., Valdez, V., and Schrader, L.F. et al. (1981) Effect of Functional Overload on Substrate Oxidation Capacity of Skeletal Muscle. *Journal of Applied Physiology.* **50**, pp. 1272-1276.

Baldwin, K.M. and Haddad, F. (2001) Effects of Different Activity and Inactivity Paradigms on Myosin Heavy Chain Gene Expression in Striated Muscle. *Journal of Applied Physiology.* **90**, pp. 345-357.

Baldwin, K.M. and Haddad, F. (2002) Cellular and Molecular Responses to Altered Physical Activity Paradigms. *American Journal of Physical Medicine and Rehabilitation.* **81**, (Supplement) pp. 40-51.

Bar, E. (1926) Elastizitätsprüfungen der Gelenkknorpel. *Arch. F. Entwicklungsmech. D. Organ.* **108,** p. 739.

Barrack, R.L. and Skinner, H.B. (1990) The Sensory Function of Knee Ligaments. *Knee Ligaments: Structure, Function, Injury, and Repair* (eds. Daniel, D.D., Akeson, W.H., and O'Connor, J.J.). Raven Press, New York. pp. 95-114.

Beer, F.P., Johnston, E.R., and DeWolf, J.T. (2004) *Mechanics of Materials*, vol. 3rd edition SI. McGraw Hill, New York, NY.

Bell, C. (1828) *Animal Mechanics.* Baldwin & Craddock, Edinburgh. pp. 64-72.

Bemben, D.A., Buchanan, T.D., Bemben, M.G., and Knehans, A.W. (2004) Influence of Type of Mechanical Loading, Menstrual Status, and Training Season on Bone Density in Young Women Athletes. *Journal of Strength and Conditioning Research.* **18**, pp. 220-6.

Benjamin, M., Evans, E.J., and Copp, L. (1986) The Histology of Tendon Attachments to Bone in Man. *J. Anat.* **149,** pp. 89-100.

Bennel, K.L., Khan, K.M., Warmington, S., Forwood, M.R., Coleman, B.D., Bennet, M.B., and Wark, J.D. (2002) Age does not Influence the Bone Response to Treadmill Exercise in Female Rats. *Medicine and Science in Sports and Exercise*, 34(12), pp. 1958-65.

Bennett, M.B., Ker, R.F., Dimery, N.J., and Alexander, R.M. (1986) Mechanical Properties of Various Mammalian Tendons. *J. Zool.* **209 (A)**, pp. 537-548.

Benninghoff, A. (1925) Form und Bau der Gelenknorpel in ihren Beziehungen zur Funktion. Zeitsch. Zellforsch. *Mikrosk. Anat.* **2,** p. 814.

Ben-Ze'ev, A. (1991) Animal Cell Shape Changes and Gene Expression. *BioEssays* **13**, pp. 207-212.

Bergmann, G., Graichen, F., and Rohlmann, A. (1993) Hip Joint Loading during Walking and Running, Measured in Two Patients. *J. Biomech.* **26(8),** pp. 969-990.

Bertram, J.E. and Swartz, S.M. (1991) The 'Law of Bone Transformation': A Case of Crying Wolff? *Biol. Rev. Camb. Philos. Soc.* **66**, pp. 245-273.

Biot, M.A. (1941) General Theory of Three-dimensional Consolidation. *J. Appl. Physics*. **12,** pp. 155-165.

Binkley, J.M. and Peat, M. (1986) The Effects of Immobilization on the Ultrastructure and Mechanical Properties of the Medial Collateral Ligament of Rats. *Clin. Orthop. R. Res.* **203,** pp. 301-308.

Bland, Y.S. and Ashurst, D.E. (2001) The Hipjoint: The Fibrillar Collagens Associated with Development and Ageing in the Rabbit. Journal of Anatomy **198**, pp. 17-21.

Blix, M. (1894) Die Laenge und die Spannung des Muskels. *Skand. Arch. Physiol.* **5,** pp. 149-206.

Bone Quality: What is It and Can We Measure It? (2005) In: NIAMS-ASBMR Scientific Meeting. Bethesda, Maryland.

Booth, F.W. and Criswell, D.S. (1997) Molecular Events Underlying Skeletal Muscle Atrophy and the Development of Effective Countermeasures. *International Journal of Sports Medicine.* **18** (Supplement), pp. 265-269.

Borelli, A. (1680-1681) *De Motu Animalium*. Rome, Italy.

Boutroy, S., Bouxsein, M.L., Munoz, F., and Delmas, P.D. (2005) In Vivo Assessment of Trabecular Microarchitecture by High-Resolution Peripheral Computed Tomography. In: *Bone Quality: What is it and can we measure it?*, pp. 10-11. Bethesda, WA.

Bouvier, M. (1989) The Biology and Composite of Bone. *Bone Mechanics* (ed. Cowin, S.C.). CTC Press, Boca Raton, FL. pp. 1-13.

Boyd, S.K., Muller, R., Leonard, T., and Herzog, W. Long-Term Periarticular Bone Adaptation in a Feline Knee Injury Model for Post-Traumatic Experimental Osteoarthritis. *Osteoarthritis Cartilage.* **13**, pp. 235-242.

Brandt, K.D., Braunstein, E.M., Visco, D.M., O'Connor, B., Heck, D., and Albrecht, M. (1991a) Anterior (Cranial) Cruciate Ligament Transection in the Dog: A Bona Fide Model of Osteoarthritis, Not Merely of Cartilage Injury and Repair. *J. Rheumatol.* **18,** pp. 436-446.

Brandt, K.D., Myers, S.L., Burr, D., and Albrecht, M. (1991b) Osteoarthritic Changes in Canine Articular Cartilage, Subchondral Bone, and Synovium Fifty-Four Months after Transection of the Anterior Cruciate Ligament. *Arthritis Rheum.* **34,** pp. 1560-1570.

Bray, D.F., Frank, C.B., and Bray, R.C. (1990) Cytochemical Evidence for a Proteoglycan-associated Filamentous Network in Ligament Extracellular Matrix. *J. Orthop. Res.* **8 (1)**, pp. 1-11.

Bray, R.C., Frank, C.B., and Miniaci, A. (1991) Structure and Function of Diarthrodial Joints. *Operative Arthroscopy* (ed. McGinty, J.B.). Raven Press, New York. pp. 79-123.

Bray, D.F., Bray, R.C., and Frank, C.B. (1993) Ultrastructural Immunolocalization of Type VI Collagen and Chondroitin Sulphate in Ligament. *J. Orthop. Res.* **25 (10)**, pp. 1227-1231.

Buckwalter, J.A., Kuettner, K.E., and Thonar, E.J. (1985) Age-Related Changes in Articular Cartilage Proteoglycans: Electron Microscopic Studies. *J. Orthop. Res.* **3,** pp. 251-257.

Buckwalter, J., Rosenberg, L., Coutts, R., Hunziker, E., Hari Reddi, A., and Mow, V. (1987) Articular Cartilage Injury and Repair. *Injury and Repair of the Musculo-skeletal Soft Tissues* (eds. Woo, S.L.-Y. and Buckwalter, L.). American Academy of Orthopedic Surgeons, Park Ridge, IL. pp. 465-482.

Bullough, P.G. and Goodfellow, J. (1968) The Significance of the Fine Structure of Articular Cartilage. *J. Bone Jt. Surg.* **50 (B)**, pp. 852-857.

Bullough, P.G., Goodfellow, J., Greenwald, A.S., and O'Connor, J.J. (1968) Incongruent Surfaces in the Human Hip Joint. *Nature.* **217,** p. 1290.

Bursac, P.M., Obitz, T.W., Eisenberg, S.R., and Stamenovic, D. (1999) Confined and Unconfined Stress Relaxation of Cartilage: Appropriateness Of A Transversely Isotropic Analysis. *J. Biomech.* **32,** pp. 1125-1130.

Burr, D.B. and Martin R.B. (1989) Errors in Bone Remodelling: Toward a Unified Theory of Metabolic Bone Disease. *Amer. J. Anatomy.* **186**, pp. 186-216.

Burr, D.B., Martin R.B., and Sharkey, N.A. (1998) Mechanical Adaptability of the Skeleton. In: *Skeletal Tissue Mechanics* (ed. Robin Smith). Springer-Verlag, New York, NY, pp. 225-271.

Burr, D.B., Milgrom, C., Fyhrie, D., Forwood, M., Nyska, M., Finestone, A., Hoshaw, S., Saiag, E., and Simkin, A. (1996) In-vivo Measurement of Human Tibial Strains During Vigorous Activity. *Bone.* **18**, pp. 405-410.

Burstein, A.H., Reilly, D.T., and Martens, M. (1976) Aging of Bone Tissue: Mechanical Properties. *J. Bone Joint Surg. Am.* **58**, pp. 82-86.

Chakravarthy, M.V., Davis, B.S., and Booth, F.W. (2000) IGF-1 Restores Satellite Cell Proliferative Potential in Immobilized Old Skeletal Muscle. *Journal of Applied Physiology.* **89,** pp. 1365-1379.

Carter, D.R. and Hayes, W.C. (1977) The Compressive Behavior of Bone as a Two-Phase Porous Structure. *J. Bone Joint Surg. Am.* **59**, pp. 954-962.

Carter, D.R., Orr, T.E., Fyhrie, D.P., Whalen, R.T., and Schurman, D.J. (1987) Mechanical Stress and Skeletal Morphogenesis, Maintenance, and Degeneration. *Transactions of the Orthopaedic Research Society.* Adept Printing, Chicago, IL. **12,** p. 462.

Carter, D.R., Bouxsein, M.L., and Marcus, R. (1992) New Approaches for Interpreting Projected Bone Densitometry. *J. Bone and Mineral Res.* **7**, pp. 137-145.

Cavagna, G.A. (1970) The Series Elastic Component of Frog Gastrocnemius. *J. Physiol.* **206,** pp. 257-262.

Cavagna, G.A. (1977) Storage and Utilization of Elastic Energy in Skeletal Muscle. *Exercise and Sport Sci. Reviews* (ed. Hutton, R.S.). **5,** pp. 89-129.

Cavagna, G.A., Saibene, F.P., and Margaria, R. (1964) Mechanical Work in Running. *J. Appl. Physiol.* **19**, pp. 249-256.

Cavagna, G.A., Saibene, F.P., and Margaria, R. (1965) Effect of Negative Work on the Amount of Positive Work Done by an Isolated Muscle. *J. Appl. Physiol.* **20 (1),** pp. 157-158.

Cavanagh, P.R. and Kram, R. (1990) Stride Length in Distance Running. *Biomechanics of Distance Running* (ed. Cavanagh, P.R.). Human Kinetics, Champaign IL. pp. 35-63.

Chimich, D., Frank, C., Shrive, N., Dougall, H., and Bray, R. (1991) The effects of initial end contact on medial collateral ligament healing: a morphological and biomechanical study in a rabbit model. *J. Orthop. Res.* **9**(1), pp. 37-47.

Chimich, D.D., Shrive, N.G., Frank, C.B., Marchuk, L., and Bray, R.C. (1992) Water Content Alters Viscoelastic Behaviour of the Normal Adolescent Rabbit Medial Collateral Ligament. *J. Biomech.* **25 (18),** pp. 831-837.

Christiansen, P., Steiniche, T., Brockstedt, H., Mosekilde, L., Hessov, I., and Melsen, F. (1993) Primary Hyperparathyroidism: Iliac Crest Cortical Thickness, Structure, and Remodelling Evaluated by Histomorphometric Methods. *Bone.* **14**, pp. 755-762.

Ciarelli, M.J., Goldstein, S.A., Kuhn, J.L., Cody, D.D., and Brown, M.B. (1991) Evaluation of Orthogonal Mechanical Properties and Density of Human Trabecular Bone from the Major Metaphyseal Regions with Materials Testing and Computed Tomography. *J. Orthop. Res.* **9**, pp. 674-682.

Claessens, A.A., Schouten, J.S., van den Ouweland, F.A., and Valkenburg, H.A. (1990) Do Clinical Findings Associate with Radiographic Osteoarthritis of the Knee? *Annals of Rheumatic Diseases* **49,** pp. 771-774.

Clark, A.L., Barclay, L.D., Matyas, J.R., and Herzog, W. (2003) In-Situ Chondrocyte Deformation with Physiological Compression of the Feline Patellofemoral Joint. *J. Biomech.* **36,** pp.553-568.

Clark, A.L., Leonard, T.R., Barclay, L., Matyas, J.R., and Herzog, W. (2005) Opposing Cartilages in the Patellofemoral Joint Adapt Differently to Long-Term Cruciate Deficiency: Chondrocyte Deformation and Reorientation with Compression. *Osteoarthritis Cartilage.* **13**, pp. 1100-1114.

Close R. (1964) Dynamic Properties of Fast and Slow Skeletal Muscles of the Rat During Development. *J. Physiol.* **173,** pp. 4-95.

Close, R. and Hoh, J.F.Y. (1967) Force: Velocity Properties of Kitten Muscles. *J. Physiol.* **192,** pp. 815-822.

Cohen, B., Lai, W.M., and Mow, V.C. (1998) A Transversely Isotropic Biphasic Model for Unconfined Compression of Growth Plate and Chondroepiphysis. *J. Biomech. Eng.* **120,** pp. 491-496.

Cooke, R., White, H., and Pate, E. (1994) A Model of the Release of Myosin Heads from Actin in Repidly Contracting Muscle Fibers. *Biophys. J.* **66**, pp. 778-788.

Cooper, R.R. and Misol, S. (1970) Tendon and Ligament Insertion: A Light and Electron Microscope Study. *J. Bone Jt. Surg.* **52 (A),** pp. 1-20.

Cowin, S. (1985) The Relationship between the Elasticity Tensor and the Fabric Tensor. *Mechanics of Materials*. **4**, pp. 137-147.

Cowin, S. (1997) The False Premise of Wolff's Law. *Forma*, **12**, pp. 247-262.

Croone, W. (1664) *De Ratione Motus Musculorum*. London, England.

Crowninshield, R.D. (1978) Use of Optimization Techniques to Predict Muscle Forces. *J. Biomech. Eng.* **100,** pp. 88-92.

Crowninshield, R.D. and Brand, R.A. (1981a) A Physiologically Based Criterion of Muscle Force Prediction in Locomotion. *J. Biomech.* **14 (11),** pp. 793-801.

Crowninshield, R.D. and Brand, R.A. (1981b) The Prediction of Forces in Joint Structures: Distribution of Intersegmental Resultants. *Exercise and Sport Sciences Reviews* (ed. Miller, D.I.). The Franklin Institute Press, Philadelphia. **9**, pp. 159-181.

Currey, J.D. (1984) *The Mechanical Adaptations of Bones*. Princeton University Press, Princeton.

Curwin, S.L. and Stanish, W.D. (1984) *Tendinitis: Its Etiology and Treatment*. Collamore Press, Toronto.

Curwin, S.L., Vailas, A.C., and Wood, J. (1988) Immature Tendon Adaptation to Strenuous Exercise. *J. Appl. Physiol.* **65 (5)**, pp. 2297-2301.

Cutts, A., Alexander, R.M., and Ker, R.F. (1991) Ratio of Cross-sectional Areas of Muscles and Their Tendons in a Healthy Human Forearm. *J. Anat.* **176**, pp. 133-137.

Dalen, N., Hellstrom, L.G., and Jacobson, B. (1976) Bone Mineral Content and Mechanical Strength of the Femoral Neck. *Acta. Orthop. Scand.* **47**, pp. 503-508.

Dalsky, G.P., Stocke, K.S., Ehsani, A.I., Slatopolsky, E., Lee, W., and Birge, S.J. (1988) Weight-bearing Exercise Training and Lumbar Bone Mineral Content in Postmenopausal Women. *Annals of Internal Medicine.* **108**, pp. 824-828.

Daniel, T.L., Trimble, A.C., and Chase, P.B. (1998) Compliant Realignment of Binding Sites in Muscle: Transient Behaviour and Mechanical Tuning. *Biophysical Journal.* **74**, pp. 1611-1621.

Dashefsky, J.H. (1987) Arthroscopic Measurement of Chondromalacia of Patella Cartilage using a Microminiature Pressure Transducer. *Arthroscopy* **3,** pp. 80-85.

Daremberg, C.V. (1854-1857) *Oeuvres Anatomiques, Physiologiques et Medicales de Galen*. Paris.

Dawson, T.J. and Taylor, C.R. (1973) Energetic Cost of Locomotion in Kangaroos. *Nature*. London. **246,** pp. 313-314.

De Ruiter, C.J. and de Haan, A. (2003) Shortening-induced Depression of Voluntary Force in Unfatigued and Fatigued Human Adductor Pollicis Muscle. *Journal of Applied Physiology.* **94**, pp. 69-74.

De Ruiter, C.J., de Haan, A., Jones, D.A., and Sargeant, A.J. (1998) Shortening-induced Force Depression in Human Adductor Pollicis Muscle. *Journal of Physiology.* **507.2**, pp. 583-591.

De Ruiter, C.J., Didden, W.J.M., Jones, D.A., and de Haan, A. (2000) The Force-velocity Relationship of Human Adductor Pollicis Muscle during Stretch and the Effects of Fatigue. *Journal of Physiology.* **526.3**, pp. 671-681.

Diamant, J., Keller, A., Baer, E., Litt, M., and Arridge, R.G.C. (1972) Collagen: Ultrastructure and its Relation to Mechanical Properties as a Function of Aging. *Proc. R. Soc. London.* **180 (B),** pp. 293-315.

Doherty, T.J. and Brown, W.F. (1993) The Estimated Number and Relative Sizes of Thenar Motor Units as Selected by Multiple Point Stimulation in Young and Older Adults. *Muscle and Nerve.* **20** (Supplement), pp. 88-92.

Doschak, M.R. and Zernicke, R.F. (2005) Structure, Function, and Adaptation of Bone-tendon and Bone-ligament Complexes. *Journal of Musculoskeletal and Neuronal Interaction.* **5**, pp. 35-40.

Doumit, M.E., Cook, D.R., and Merkel, R.A. (1993) Fibroblast Growth Factor, Epidermal Growth Factor, Insulin-like Growth Factors, and Platelet-derived Growth Factor-BB Stimulate Proliferation of Clonally Derived Porcine Myogenic Satellite Cells. *Journal of Cellular Physiology.* **157**, pp. 326-332.

Dowson, D. (1990) Bio-tribology of Natural and Replacement Synovial Joints. *Biomechanics of Diarthrodial Joints* (eds. Mow, V. C., Ratcliffe, A., and Woo, S. L.-Y.). Springer Verlag, New York. **II,** pp. 305-345.

Dowson, D. (1992) Engineering at the Interface. *Proceedings of the Institution of Mechanical Engineers. Part C: Mechanical Engineering Science*. Mechanical Engineering Pubs. Limited, Suffolk. **206**

(3), pp. 149-165.

Ducher, G., Prouteau, S., Courteix, D., and Benhamou, C.L. (2004) Cortical and Trabecular Bone at the Forearm Show Different Adaptation Patterns in Response to Tennis Playing. *Journal of Clinical Densitometry.* **7**, pp. 399-405.

Dul, J., Johnson, G.E., Shiavi, R., and Townsend, M.A. (1984) Muscular Synergism - II: A Minimum Fatigue Criterion for Load Sharing Between Synergistic Muscles. *J. Biomech.* **17 (9),** pp. 675-684.

Dye, S.F. (1987) An Evolutionary Perspective of the Knee. *J. Bone Jt. Surg.* **69 (A)**, pp. 976-983.

Eastlack, R.K., Hargens, A.R., Groppo, E.R., Steinbach, G.C., White, K.K., and Pedowitz, R.A. (2005) Lower Body Positive-pressure Exercise after Knee Surgery. *Clinical Orthpeadic and Related Research.* **431**, pp. 213-219.

Edman, K.A.P. (1979) The Velocity of Unloaded Shortening and its Relation to Sarcomere Length and Isometric Force in Vertebrate Muscle Fibres. *J. Physiol.* **291**, pp. 143-159.

Edman, K.A.P. (1996) Fatigue vs. Shortening-induced Deactivation in Striated Muscle. *Acta Physiologica Scandinavica.* **156**, pp. 183-192.

Edman, K.A.P. and Reggiani (1983) Length-tension-velocity Relationships Studied in Short Consecutive Segments of Intact Muscle Fibres of the Frog. *Contractile Mechanisms of Muscle. Mechanics, Energetics, and Molecular Models* (eds. Pollack, G.H. and Sugi, H.). Plenum Press, New York. **II,** pp. 495-510.

Edman, K.A.P., Elzinga, G., and Noble, M.I.M. (1978) Enhancement of Mechanical Performance by Stretch During Tetanic Contractions of Vertebrate Skeletal Muscle Fibres. *J. Physiol.* **281,** pp. 139-155.

Edman, K.A.P., Elzinga, G., and Noble, M.I.M. (1982) Residual Rorce Enhancement After Stretch of Contracting Frog Single Muscle Fibers. *Journal of General Physiology.* **80**, pp. 769-784.

Edman, K.A.P., Caputo, C., and Lou, F. (1993) Depression of Tetanic Force Induced by Loaded Shortening of Frog Muscle Fibres. *J. Physiol.* **466,** pp. 535-552.

Einhorn, T.A. (1996) Biomechanics of Bone. *Principles of Bone Biology* (eds. Bilezikian, J.P., Raisz, L.G., and Rodan, R.A.). Academic Press, NY. pp. 25-37.

Eisenberg, E. and Greene, L.E. (1980) The Relation of Muscle Biochemistry to Muscle Physiology. *Annual Rev. Physiol.* **42,** pp. 293-309.

Eisenberg, E. and Moos, C. (1968) The Adenosine Triphosphatase Activity of Acto-heavy Mmeromyosin. A Kinetic Analysis of Actin Activation. *Biochemistry.* **7**, pp. 1486-1489.

Eisenberg, E., Hill. T.L., and Chen, Y.D. (1980) Cross-bridge Model of Muscle Contraction: Quantitative Analysis. *Biophys. J.* **29**, pp. 195-227.

Elliot, D.H. (1965) Structure and Function of Mammalian Tendon. *Biol. Rev.* **40,** pp. 392-421.

Engelhardt, V.A. and Lyubimova, M.N. (1939) Myosin and Adenosinetriphosphatase. *Nature.* **144**, p. 668.

Englemark, V.E. (1961) Functionally Induced Changes in Articular Cartilage. *Biomechanical Studies of the Musculo-skeletal System* (ed. Evans, F.G.). C.C. Thomas, Springfield, IL. pp. 3-19.

Engsberg, J.R. (1987) A Biomechanical Analysis of the Talocalcaneal Joint - In-vitro. *J. Biomech.* **20**, pp. 429-442.

Epstein, M. and Herzog, W. (1998) *Theoretical Models of Skeletal Muscle: Biological and Mathematical Considerations* John Wiley & Sons Ltd., New York.

Ettema, G.J. and Meijer, K. (2000) Muscle Contraction History: Modified Hill versus an Exponential Decay Model. *Biological Cybernetics.* **83(6)**, pp. 491-500.

Fabiato, A. and Fabiato, F. (1975) Dependence of the Contractile Activation of Skinned Cardiac Cells on the Sarcomere Length. *Nature.* **256,** pp. 54-56.

Farquhar, T., Dawson, P.R., and Torzilli, P.A. (1990) A Microstructural Model for the Anisotropic Drained Stiffness of Articular Cartilage. *J. Biomech. Eng.* **112,** pp. 414-425.

Farquharson, A.S.L. (1912) De Motu Animalium (Aristotle, translated). *The Works of Aristotle Volume 5* (eds. Smith, J.A. and Ross, W.D.). Clarendon Press, Oxford.

Faulkner, K.G., Cann C.E., and Hasegawa, B.H. (1991) Effect of Bone Distribution on Vertebral Strength: Assessment with Patient-Specific Nonlinear Finite Element Analysis. *Radiology.* **179**, pp. 669-674.

Federico, S., Grillo, A., and Herzog, W. (2004a) A Transversely Isotropic Composite with a Statistical Distribution of Spheroidal Inclusions: A Geometrical Approach to Overall Properties. *Journal of the Mechanics and Physics of Solids.* **52,** pp. 2309-2327.

Federico, S., Herzog, W., Wu, J.Z., and La Rosa, G. (2004b) A Method to Estimate the Elastic Properties of the Extracellular Matrix of Articular Cartilage. *J. Biomech.* **37,** pp. 401-404.

Federico, S., Herzog, W., Wu, J.Z., and La Rosa, G. (2004c) Effect of Fluid Boundary Conditions on Joint Contact Mechanics and Applications to the Modelling of Osteoarthritic Joints. *J. Biomech. Eng.* **126,** pp. 220-225.

Federico, S., Grillo, A., La Rosa, G., Giaquinta, G., and Herzog, W. (2005) A Transversely Isotropic, Transversely Homogeneous Microstructural-Statistical Model of Articular Cartilage. *J. Biomech.* **38,** pp. 2008-2018.

Fenn, W.O. and Marsh, B.S. (1935) Muscular Force at Different Speeds of Shortening. *J. Physiol.* **85,** pp. 277-296.

Finer, J.T., Simmons, R.M., and Spudich, J.A. (1994) Single Myosin Molecule Mechanics: Piconewton Forces and Nanometre Steps. *Nature.* **368**, pp. 113-119.

Fisher, N.M., Gresham, G.E., Abrams, M., Hicks, J., Horrigan, D., and Pendergast, D.R. (1993) Quantitative effects of physical therapy on muscular and functional performance in subjects with osteoarthritis of the knees. *Archives of Physical Medicine & Rehabilitation* **74,** 840-847.

Fisher, N.M., Pendergast, D.R., Gresham, G.E., and Calkins, E. (1991) Muscle Rehabilitation: Its Effect on Muscular and Functional Performance of Patients with Knee Osteoarthritis. *Arch. Phys. Med. Rehabil.* **72,** pp. 367-374.

Fisher, N.M., White, S.C., Yack, H.J., Smolinski, R.J., and Pendergast, D.R. (1997) Muscle Function and Gait in Patients with Knee Osteoarthritis Before and After Muscle Rehabilitation. *Disability and Rehabilitation* **19(2),** pp. 47-55.

Flint, M. (1982) Interrelationships of Mucopolysaccharides and Collagen in Connective Tissue Remodelling. *J. Embryol. Exp. Morph.* **27,** pp. 481-495.

Forcinito, M., Epstein, M., and Herzog, W. (1997) Theoretical Considerations on Myofibril Stiffness. *Biophys. J.* **72,** pp. 1278-1286.

Ford, L.E., Huxley, A.F., and Simmons, R.M. (1977) Tension Responses to Sudden Length Change in Stimulated Frog Muscle Fibers Near Slack Length. *Journal of Physiology.* **269**, pp. 441-515.

Forwood, M.R. and Burr, D.B. (1993) Physical Activity and Bone Mass: Exercise in Futility? *Bone and Mineral.* **21**, pp. 89-112.

Frank, C.B. (1996) Ligament Injuries: Pathophysiology and Healing. *Athletic Injuries and Rehabilitation* (eds. Zachazewski, J.E., Magee, D.J., and Quillen, W.S.). Saunders, Philadelphia. pp. 9-26.

Frank, C.B., Amiel, D., and Akeson, W.H. (1983a) Healing of the Medial Collateral Ligament of the Knee: A Morphological and Biochemical Assessment in Rabbits. *Acta. Orthop. Scand.* **54**, pp. 917-923.

Frank, C.B., Woo, S. L.-Y., Amiel, D., Harwood, F.L., Gomez, M.A., and Akeson, W.H. (1983b) Medial Collateral Ligament Healing: A Multi-disciplinary Assessment in Rabbits. *Am. J. Sports Med.* **11 (6)**, pp. 379-389.

Frank, C.B., Woo, S.L.-Y., Andriacchi, T., Brand, R., Oakes, B., Dahners, L., DeHaven, K., Lewis, J., and Sabiston, P. (1988) Normal Ligament: Structure, Function, and Composition. *Injury and Repair of the Musculo-skeletal Soft Tissues* (eds. Woo, S.L.-Y. and Buckwalter, J.A.). American Academy of Orthopaedic Surgeons, Rosemont, IL. pp. 45-101.

Frank, C.B., Bray, D.F., Rademaker, A., Chrusch, C., Sabiston, C.P., Bodie, D., and Rangayyan, R.M. (1989) Electron Microscopic Quantification of Collagen Fibril Diameters in the Rabbit Medial Collateral Ligament: A Baseline for Comparison. *Connect. Tissue Res.* **19,** pp. 11-25.

Frank, C.B. and Hart, D.A. (1990) The Biology of Tendons and Ligaments. *Biomechanics of Diathrodial Joints* (eds. Mow, V.C., Ratcliffe, A., and Woo, S.L.-Y.). Springer Verlag, New York. pp. 39-62.

Frank, C., MacFarlane, B., Edwards, P., Rangayyan, R., Liu, Z.Q., Walsh, S., and Bray, R. (1991) A quantitative analysis of matrix alignment in ligament scars: a comparison of movement versus immobilization in an immature rabbit model. *J. Orthop. Res.* **9**(2), pp. 219-227.

Frank, C., McDonald, D., Bray, D., Bray, R., Rangayyan, R., Chimich, D., and Shrive, N. (1992) Collagen fibril diameters in the healing adult rabbit medial collateral ligament. *Connect. Tissue Res.* **27**(4), pp. 251-263.

Frank, C., McDonald, D., Wilson, J., Eyre, D., and Shrive, N. (1995) Rabbit medial collateral ligament scar weakness is associated with decreased collagen pyridinoline crosslink density. *J. Orthop. Res.* **13**(2), pp. 157-165.

Frank, E.H. and Grodzinsky, A.J. (1987) Cartilage Electromechanics II. A Continuum Model of Cartilage Electrokinetics and Correlation to Experiments. *J. Biomech.* **20,** pp. 629-639.

Frankel, V.H. and Burstein, A.H. (1970) *Orthopaedic Biomechanics.* Lea & Febiger, Philadelphia, PA.

Friedlander, A.L., Genant, M.K., Sadowsky, S., Byl, N.N., and Gluer, C.C. (1995) A Two-year Program of Aerobics and Weight Training Enhances Bone Mineral Density of Young Women. *Journal of Bone and Mineral Research,* 10(4), pp. 574-85.

Frost, H.M. (1964) *The Laws of Bone Structure.* Charles C. Thomas, Springfield, IL.

Frost, H.M. (1973a) *Bone Modelling and Skeletal Modelling Errors.* Springfield, Charles C. Thomas.

Frost, H.M. (1973b) *Bone Remodelling and its Relationship to Metabolic Bone Disease.* Charles C. Thomas, Springfield, IL.

Frost, H.M. (1973) *Bone Modelling and Skeletal Modelling Errors.* Charles C. Thomas, Springfield, IL.

Frost, H.M. (1986) *Intermediary Organization of the Skeleton.* Vols. I and II. CTC Press, Boca Raton, FL.

Frost, H.M. (1987) The Mechanostat: A Proposed Pathogenic Mechanism of Osteoporoses and the Bone Mass Effects of Mechanical and Non-mechanical Agents. *Bone and Mineral.* **2**, pp. 73-85.

Frost, H.M. (1989) *Mechanical Usage, Bone Mass, Bone Fragility: A Brief Overview. Clinical Disorders in Bone and Mineral Metabolism* (eds. Kleerekoper, M. and Krane, S.). Mary Ann Liebow, New York. pp. 15-49.

Fung, Y.C. (1993) *Biomechanics: Mechanical Properties of Living Tissues*, 2nd ed, New York, Springer. pp. 41-48, 50-52, 277-287.

Galileo, G. (1638) *Discorsi e Dimonstrazioni Matematiche, Intorno a due Nuove Scienze Attentanti alla Meccanica ed a Muovementi Localli.* University of Wisconsin Press, Madison.

Geeves, M.A. (1991) The Dynamics of Actin and Myosin Association and the Crossbridge Model of Muscle Contraction. *Biochem. J.* **274**, pp. 1-14.

Gelberman, R.H., Goldberg, V.M., An, K.N., and Banes, A.J. (1988) Tendon. *Injury and Repair of the Musculo-skeletal Soft Tissues* (eds. Woo, S.L.-Y. and Buckwalter, J.A.). American Academy of Orthopaedic Surgeons, Park Ridge, IL. pp. 5-40.

Gellman, H., Lenihan, M., and Halikis, N. (1987) Selective Tarsal Arthrodesis: An In-vitro Analysis of the Effect on Foot Motion. *Foot and Ankle.* **8 (3)**, pp. 127-133.

Gere, J.M. and Timoshenko, S.P. (1997) *Mechanics of Materials*, 4th edition. PWS Publishing Company, Boston, MA.

Gillard, G.C., Reilly, H.C., Bell-Booth, P.G., and Flint, M.H. (1979) The Influence of Mechanical Forces on the Glycosaminoglycan Content of the Rabbit Flexor Digitorum Profundus Tendon. *Connect. Tissue Res.* **7,** pp. 37-46.

Girgis, F.G., Marshall, J.L., and Al Monajem, A.R.S. (1975) The Cruciate Ligaments of the Knee Joint. Anatomical, Functional, and Experimental Analysis. *Clin. Orthop.* **106,** pp. 216-231.

Göcke (1927) Elastizitätsstudien am jungen und alten Gelenkknorpel. *Verhandl. D. Deutsch. Orthop. Gesellsch.* pp. 130-147.

Goldman, Y.E. and Huxley, A.F. (1994) Actin Compliance: Are you Pulling my Chain? *Biophysical Journal.* **67**, pp. 2131-2136.

Goodfellow, J. and Bullough, P.G. (1968) Studies on Age Changes in the Human Hip Joint. *J. Bone Jt. Surg.* **50 (B)**, p. 222.

Gordon, A.M. (1992) *Regulation of Muscle Contraction: Dual Role of Calcium and Cross-bridges, in Muscular Contraction* (ed. Simmons, R.M.). Cambridge University Press, Cambridge, UK. pp. 163-179.

Gordon, A.M., Huxley, A.F., and Julian, F.J. (1966) The Variation in Isometric Tension with Sarcomere Length in Vertebrate Muscle Fibres. *J. Physiol.* **184,** pp. 170-192.

Goslow Jr., G.E. and van de Graaf, K.M. (1982) Hindlimb Joint Angle Changes and Action of the Primary Ankle Extensor Muscles During Posture and Locomotion in the Striped Skunk (Mephitis). *J. Zool.* **197,** pp. 405-419.

Goslow Jr., G.E., Reinking, R.M., and Stuart, D.G. (1973) The Cat Step Cycle: Hindlimb Joint Angles and Muscle Lengths During Unrestrained Locomotion. *J. Morphol.* **141,** pp. 1-42.

Granzier, H.L.M. and Pollack, G.H. (1989) The Transmission of Load Through the Human Hip Joint. *J. Biomech.* **4,** pp. 507-528.

Greene, L.E. and Eisenberg, E. (1980) Dissociation of the Actin Subfragment 1 Complex by Adenyl-5'-yl Imidodiphosphate, ADP, and PPi. *J. Biol. Chem.* **255**, pp. 543-548.

Greenwald, A.S. and O'Connor, J.J. (1971) Effect of Active Pre-shortening on Isometric and Isotonic Performance of Single Frog Muscle Fibres. *J. Physiol.* **415,** pp. 299-327.

Greenwald, A.S. and O'Connor, J.J. (1971) The Transmission of Load through the Human Hip Joint. *Journal of Biomechanics.* **4**, pp. 507-528.

Gregor, R.J., Cavanagh, P.R., and LaFortune, M. (1985) Knee Flexor Moments During Propulsion in Cycling: A Creative Solution to Lombard's Paradox. *J. Biomech.* **18 (5),** pp. 307-316.

Grimston, S.K. and Zernicke, R.F. (1993) Exercise-related Stress Responses in Bone. *J. Appl. Biomech.* **9 (1)**, pp. 2-14.

Grimston, S.K., Willows, N.D., and Hanley, D.A. (1993) Mechanical Loading Regime and its Relationship to Bone Mineral Density in Children. *Med. Sc. Sports Exercise.*

Gross, T.S. and Rubin, C.T. (1995) Uniformity of Resorptive Bone Loss Induced by Disuse. *J. Ortho. Res.* **13**, pp. 708-714.

Gross, T.S., Edwards, J.L., McLeod, K.J., and Rubin, C.T. (1997) Strain Gradients Correlate with Sites of Periosteal Bone Formation. *J. Bone and Mineral Res.* **12**, pp. 982-988.

Gu, W.M., Hou, J.S., and Mow, V.C. (1998) A Mixture Theory for Charged-Hydrated Soft Tissues Containing Multi-Electrolytes: Passive Transport and Swelling Behaviours. *J. Biomech. Eng.* **120,** pp. 169-180.

Guilak, F. (1995) Compression-induced Changes in the Shape and Volume of the Chondrocyte Nucleus. *J. Biomech.* **28,** pp. 1529-1541.

Guilak, F., Ratcliffe, A., and Mow, V.C. (1995) Chondrocyte Deformation and Local Tissue Strain in Articular Cartilage: A Confocal Microscopy Study. *J. Orthop. Res.* **13,** pp. 410-421.

Guralnik, D.B. (ed.) (1979) Webster's New World Dictionary. William Collins Publishers, Cleveland.

Haapasalo, H., Kannus, P., Sievanen, H., Heinonen, A., Oja, P., and Vuori, I. (1994) Long-term Unilateral Loading and Bone Mineral Density and Content in Female Squash Players. *Calcified Tissue International.* **54**, pp. 249-255.

Harrigan, T.P., Jasty, M., Mann, R.W., and Harris, W.H. (1988) Limitations of the Continuum Assumption in Cancellous Bone. *J. Biomech.* **21**, pp. 269-275.

Hascall, V.C. (1977) Interactions of Cartilage Proteoglycans with Hyaluronic Acid. *J. Supramol. Struct.* **7,** pp. 101-120.

Hasler, E.M. and Herzog, W. (1998) Quantification of In-vivo Patellofemoral Contact Forces before and after ACL Transection. *J. Biomech.* **31**, pp. 37-44.

Hasler, E.M., Herzog, W., Leonard, T.R., Stano, A., and Nguyen, H. (1998a) In-vivo Knee Joint Loading and Kinematics before and after ACL Transection in an Animal Model. *J. Biomech.* **31**, pp. 253-262.

Hasler, E.M., Herzog, W., Wu, J.Z., Muller, W., and Wyss, U. (1999) Articular Cartilage Biomechanics: Theoretical Models, Material Properties, and Biosynthetic Response. *Crit. Rev. Biomed. Eng.* **27(6),** pp. 415-488.

Hayes, W.C., Keer, L.M., Herrod, N.J., and Mockcros, L.F. (1972) A Mathematical Analysis for Indentation Tests of Articular Cartilage. *J. Biomech.* **5,** pp. 541-551.

Hedlund, H., Mengarelli-Widholm, S., Reinholt, F., and Svensson, O. (1993) Stereological studies on collagen in bovine articular cartilage. *Acta Pathologica, Microbiologica et Immunologica Scandinavica (APMIS)* **101,** pp. 133-140.

Henneman, E., Somjen, G., and Carpenter, D.O. (1965) Functional Significance of Cell Size in Spinal Motoneurons. *J. Neurophysiol.* **28,** pp. 560-580.

Henneman, E. and Olson, C.B. (1965) Relations Between Structure and Function in the Design of Skeletal Muscles. *J. Neurophysiol.* **28,** pp. 581-598.

Herrick, R. and Herrick, S. (1987) Ruptured Triceps in a Powerlifter Presenting as Cubital Tunnel Syndrome: A Case Report. *Am. J. Sports Med.* **15 (5),** pp. 514-516.

Herzog, W. (1987) Individual Muscle Force Estimations Using a Non-linear Optimal Design. *J. Neurosci. Methods.* **21,** pp. 167-179.

Herzog, W. (1998) History Dependence of Force Production in Skeletal Muscle: A Proposal for Mechanisms. *J. Electromyography and Kinesiology.* **8**, pp. 111-117.

Herzog, W. and Federico, S. (2006) Considerations on Joint and Articular Cartilage Mechanics. *Biomechanics and Modeling in Mechanobiology.* **5**, pp. 64-81.

Herzog, W. and ter Keurs, H.E.D.J. (1988) Force-length Relation of In-vivo Human Rectus Femoris Muscles. *Eur. J. Physiol.* **411,** pp. 642-647.

Herzog, W. and Leonard, T.R. (1991) Validation of Optimization Models that Estimate the Forces Exerted by Synergistic Muscles. *J. Biomech.* **24 (S1)**, pp. 31-39.
Herzog, W. and Leonard, T.R. (1997) Depression of Cat Soleus Forces Following Isokinetic Shortening. *J. Biomechanics.* **30(9)**, pp. 865-872.
Herzog, W. and Leonard, T.R. (2000) The History Dependence of Force Production in Mammalian Skeletal Muscle Following Stretch-shortening and Shortening-stretch Cycles. *J. Biomech.* **33**, pp. 531-542.
Herzog ,W. and Leonard, T.R. (2002) Force Enhancement Following Stretching of Skeletal Muscle: A New Mechanism. *J. Experimental Biology.* **205**, pp. 1275-1283.
Herzog, W., Adams, M.E., Matyas, J.R., and Brooks, J.G. (1993) Hindlimb Loading, Morphology, and Biochemistry in the ACL-deficient Cat Knee. *Osteoarthr. Cartil.* **1**, pp. 243-251.
Herzog, W., Read, L.J., and ter Keurs, H.E.D.J. (1991b) Experimental Determination of Force-length Relations of Intact Human Gastrocnemius Muscles. *Clinical Biomechanics.* **6**, pp. 230-238.
Herzog, W., Guimaraes, A.C., Anton, M.G., and Carter-Erdman, K.A. (1991a) Moment-length Relations of Rectus Femoris Muscles of Speed Skaters, Cyclists, and Runners. *Med. Sci. Sports Exerc.* **23 (11),** pp. 1289-1296.
Herzog, W., Guimaraes, A.C., Anton, M.G., and Carter-Erdman, K.A. (1991) Moment-length Relations of Rectus Femoris Muscles of Speed Skaters, Cyclists, and Runners. *Med. Sci. Sports Exerc.* **23 (11),** pp. 1289-1296.
Herzog, W., Leonard, T., Renaud, J.M., Wallace, J., Chaki, G., and Bornemisza, S. (1992) Force-length Properties and Functional Demands of Cat Gastrocnemius, Soleus, and Plantaris Muscles. *J. Biomech.* **25 (11)**, pp. 1329-1335.
Herzog, W., Leonard, T.R., and Wu, J.Z. (1998) Force Depression Following Skeletal Muscle Shortening is Long Lasting. *J. Biomech.* **31**, pp. 1163-1168.
Herzog, W., Leonard. T.R., and Wu, J.Z. (2000) The Relationship between Force Depression Following Shortening and Mechanical Work in Skeletal Muscle. *Journal of Biomechanics.* **33**, pp. 659-668.
Herzog, W., Kamal, S., and Clarke, H.D. (1992) Myofilament Lengths of Cat Skeletal Muscle: Theoretical Considerations and Functional Implications. *J. Biomech.* **8 (25)**, pp. 945-948.
Herzog, W., Leonard, T.R., and Guimaraes, A.C.S. (1993) Forces in Gastrocnemius, Soleus, and Plantaris Muscles of the Freely Moving Cat. *J. Biomech.* **26**, pp. 945-953.
Herzog, W., Schachar, R., and Leonard, T.R. (2003) Characterization of the Passive Component of Force Enhancement Following Active Stretching of Skeletal Muscle. *Journal of Experimental Biology.* **206**, pp. 3634-3643.
Herzog, W., Wu, J.Z., Leonard, T.R., Suter, E., Diet, S., Muller, C., and Mayzus, P. (1998) Mechanical and Functional Properties of Cat Knee Articular Cartilage 16 Weeks Post ACL Transection. *J. Biomech.* **31,** pp. 1137-1145.
Herzog, W., Zatsiorsky, V., Prilutsky, B.I., and Leonard, T.R. (1994) Variations in Force-time Histories of Cat Gastrocnemius, Soleus, and Plantaris Muscles for Consecutive Walking Steps. *J. Exper. Biol.* **191,** pp. 19-36.
Highsmith, S. (1976) Interactions of the Actin and Nucleotide Binding Sites on Myosin Subfragment 1. *Journal of Biological Chemistry.* **251**, pp. 6170-6172.
Higuchi, H., Yanagida, T., and Goldman, Y.E. (1995) Compliance of Thin Filaments in Skinned Fibers of Rabbit Skeletal Muscle. *Biophysical Journal.* **69**, pp. 1000-1010.
Hill, A.V. (1938) The Heat of Shortening and the Dynamic Constants of Muscle. *Proc. Royal Soc.* London. **126 (B),** pp. 136-195.
Hill, A.V. (1953) The Mechanics of Active Muscle. *Proceedings of the Royal Society London.* **141**, pp. 104-117.
Hill, A.V. (1970) *First and Last Experiments in Muscle Mechanics*. Cambridge University Press, Cambridge.
Hintermann, B., Nigg, B.M., Sommer, C., and Cole, G.K. (1994) Transfer of Movement Between Calcaneus and Tibia In-vitro. *Clin. Biomech.* **9,** pp. 349-355.
Hodgson, J.A. (1983) The Relationship Between Soleus and Gastrocnemius Muscle Activity in Conscious Cats: A Model for Motor Unit Recruitment. *J. Physiol.* **337,** pp. 553-562.
Hof, A.L., Goelen, B.A., and van den Berg, J. (1983) Calf Muscle Moment, Work and Efficiency in Level Walking; Roles of Series Elasticity. *J. Biomech.* **16 (7),** pp. 523-537.

Holden, J.P., Grood, E.S., Korvick, D.L., Cummings, J.F., Butler, D.L., and Bylski-Austrow, D.I. (1994) In vivo forces in the anterior cruciate ligament: direct measurements during walking and trotting in a quadruped. *J. Biomech.* **27**(5), pp. 517-526.

Hole, J.W. (1987) *Human Anatomy and Physiology* (4ed). Wm. C. Brown, Dubuque.

Holmes, M.H. and Mow, V.C. (1990) The Non-linear Characteristics of Soft Gels and Hydrated Connective Tissues in Ultrafiltration. *J. Biomech.* **23,** pp. 1145-1156.

Hooley, C.J. and McCrum, N.G. (1980) The Viscoelastic Deformation of Tendon. *J. Biomech.* **13 (6),** pp. 521-528.

Horowits, R. (1992) Passive Force Generation and Titin Isoforms in Mammalian Skeletal Muscle. *Biophys. J.* **61 (2),** pp. 392-398.

Horowits, R. and Podolsky, R.J. (1987) The Positional Stability of Thick Filaments in Activated Skeletal Muscle Depends on Sarcomere Length: Evidence for the Role of Titin Filaments. *J. Cell Biol.* **105,** pp. 2217-2223.

Horowits, R. and Podolsky, R.J. (1988) Thick Filament Movement and Isometric Tension in Activated Skeletal Muscle. *Biophys. J.* **54,** pp. 165-171.

Horowits, R., Maruyama, K., and Podolsky, R.J. (1989) Elastic Behaviour of Connectin Filaments During Thick Filament Movement in Activated Skeletal Muscle. *J. Cell Biol.* **109,** pp. 2169-2176.

Houtz, S.A. and Fischer, F.J. (1959) An Analysis of Muscle Action and Joint Excursion During Exercise on a Stationary Bicycle. *J. Bone Jt. Surg.* **41 (A),** pp. 123-131.

Howard, J. (2001) ATP Hydrolysis. In *Mechanics of Motor Proteins and Cytoskeleton*, ed. Howard J, pp. 229-244. Sinauer Associates, Inc., Sunderland, Massachusetts.

Huiskes, R., Ruimerman, R., van Lenthe, G.H., and Janssen, J.D. (2000) Effects of Mechanical Forces on Maintenance and Adaptation of Form in Trabecular Bone. *Nature.* **405**, pp. 704-706.

Huiskes, R., and van Rietbergen, B. (2005) *Biomechanics of Bone.* In: Basic Orthopaedic Biomechanics and Mechano-Biology, 3/e, pp. 123-179 (eds. V.C. Mow and R. Huiskes). Philadelhia, Lippincott Williams & Wilkins.

Hull, M.L. and Jorge, M. (1985) A Method for Biomechanical Analysis of Bicycle Pedalling. *J. Biomech.* **18 (19),** pp. 631-644.

Hultkrantz, W. (1898) Über die Spaltrichtungen der Gelenkknorpel. *Verh. Anat. Ges.* **12,** pp. 248-256.

Hunziker, E. (1992) Articular Cartilage Structure in Humans and Experimental Animals. In *Articular Cartilage and Osteoarthritis* (eds. Peyron, K.E., Schleyerback, J.G., and Hascall, V.C.) pp. 183-199. Raven Press, New York.

Hunter, J.A. and Finlay, B. (1973) Scanning Electron Microscopy of Connective Tissues. *Int. Rev. Connect. Tissue Res.* **6,** p. 218.

Hunter, W. (1742) Of the Structure and Diseases of Articulating Cartilages. *Phil. Trans.* **42,** pp. 513-521.

Hurley, M.V. (1999) The role of Muscle Weakness in the Pathogenesis Of Osteoarthritis. *Rheumatic Disease Clinics of North America* **25(2),** pp. 283-298.

Hurley, M.V. and Newham, D.J. (1993) The Influence of Arthrogenous Muscle Inhibition on Quadriceps Rehabilitation of Patients with Early Unilateral Osteoarthritic Knees. *Br. J. Rheumatol.* **32,** pp. 127-131.

Huxley, A.F. (1957) Muscle Structure and Theories of Contraction. *Prog. Biophys. Chem.* **7,** pp. 255-318.

Huxley, A.F. and Niedergerke, R. (1954) Structural Changes in Muscle During Contraction. Interference Microscopy of Living Muscle Fibres. *Nature.* **173,** p. 971.

Huxley, A.F. and Simmons, R.M. (1971) Proposed Mechanism of Force Generation in Striated Muscle. *Nature.* **233,** pp. 533-538.

Huxley, H.E. and Hanson, J. (1954) Changes in the Cross-striations of Muscle During Contraction and Stretch and Their Structural Interpretation. *Nature.* **173,** p. 973.

Huxley, H.E. (1969) The Mechanism of Muscular Contraction. *Science.* **164,** pp. 1356-1366.

Huxley, H.E., Stewart, A., Sosa. H, and Irving, T. (1994) X-ray Diffraction Measurements of the Extensibility of Actin and Myosin Filaments in Contracting Muscles. *Biophysical Journal.* **67,** pp. 2411-2421.

Iannuzzi-Sucich, M., Presetwood, K.M., and Kenny, A.M. (2002) Prevalence of sarcopenia and predictors of skeletal Muscle Mass in healthy, older men and women. *Journal of Gerontology.* **57A**, M772-M777.

Indelicatio, P.A. (1988) Injury to the Medial Capsuloligamentous Complex. *The Crucial Ligaments* (ed. Feagin Jr., J.A.). Churchill Livingstone, New York. pp. 197-216.

Ingber, D. (1991) Integrins as Mechanochemical Transducers. *Cell. Biol.* **3,** pp. 841-848.
Inmann, V.T. (1976) *The Joint of the Ankle.* Williams and Wilkins, Baltimore.
Ishikawa, H. (1965) The Fine Structure of Myo-tendon Junction in Some Mammalian Skeletal Muscles. *Arch. Histol.* Japan. **25,** pp. 275-296.
Isman, R.E. and Inmann, V.T. (1968) Anthropometric Studies of the Human Foot and Ankle. Biomechanics Laboratory, University of California. San Francisco and Berkley. *Technical Report 58.* The Laboratory, San Francisco.
Iwazumi, T. (1978) Molecular Mechanism of Muscle Contraction: Another View. *Cardiovascular System Dynamics* (eds. Baan, J., Noordergraaf, A., and Raines, J.). MIT Press, Cambridge. pp. 11-21.
Iwazumi, T. (1979) A New Field Theory of Muscle Contraction. *Cross-bridge Mechanism in Muscle Contraction* (eds. Sugi H. and Pollack G.H.). University of Tokyo Press, Tokyo. pp. 611-632.
Janssen, I, Heymsfield, S.B. Wang, Z., and Ross, R. (2000) Skeletal Muscle Mass and Distribution in 468 Men and Women Aged 18-88 Years. *Journal of Applied Physiology.* **89**, pp. 81-88.
Jarvin, T.L., Pajamaki, I., Sievanen, H., Voohelainen, T., Tuukkanen, J., Jarvinen, M., and Kannus, P. (2003) Femoral Neck Response to Exercise and Subsequent De-conditioning in Young and Adult Rats. *Journal of Bone and Mineral Research.* **18**, pp. 1292-1299.
Jee, W.S.S. (1983) *The Skeletal Tissues*. In: Histology, Cell and Tissue Biology, pp. 200-255. (ed. L Weiss). New York, Elsevier Biomedical.
Jones, D.A. and Round, J.M. (1990) Skeletal Muscle in Health and Disease. A Textbook of Muscle Physiology. Manchester University Press, Manchester, pp23-25, pp. 105-107.
Josephson, R.K. and Stokes, D.R. (1999) Work-dependent Deactivation of a Crustacean Muscle. *Journal of Experimental Biology.* **202(18)**, pp. 2551-2565.
Jozsa, L. and Kannus P. (1997) Human Tendons: Anatomy, Physiology, and Pathology. Human Kinetics. Champaign, IL.
Jozsa, L., Kannus, P., Balint, J.B., and Reffy, A. (1991) Three-dimensional Ultrastructure of Human Tendons. *Acta. Anat.* **142,** pp. 306-312.
Judex, S., Gross, T.S., and Zernicke, R.F. (1997b) Strain Gradients Correlate with Sites of Exercise-induced Bone Forming Surfaces in the Adult Skeleton. *J. Bone and Mineral Res.* **12**, pp. 1737-1745.
Judex, S., Gross, T.S., Bray, R.C., and Zernicke, R.F. (1997a) Adaptation of Bone to Physiological Stimuli. *Journal of Biomechanics*, 30(5), pp.421-429.
Judex, S., Wohl, G.R., Wolff, R.B., Leng, W., Gillis, A.M., Zernicke, R.F. Dietary fish oil supplementation adversely affects cortical bone morphology and biomechanics in growing rabbits. *Calcified Tissue International*, 66, pp 443-448, 2000.
Kabel, J., van Rietbergen, B., Odgaard, A., and Huiskes, R. (1999) Constitutive Relationships of Fabric, Density, and Elastic Properties in Cancellous Bone Architecture. *Bone*. **25,** pp. 481-486.
Kardel, T. (1990) Niels Stensen's Geometrical Theory of Muscle Contraction (1667): A Reappraisal. *J. Biomech.* **23**, pp. 953-965.
Kasper, C.E., Talbot, L.A., and Gaines, J.M. (2002) Skeletal Muscle Damage and Recovery. *Clinical Issues.* **13**, pp. 237-247.
Kastelic, J., Galeski, A., and Baer, E. (1978) The Multicomposite Structure of Tendon. *Connect. Tissue Res.* **6,** pp. 11-23.
Katz, B. (1939) The Relation Between Force and Speed in Muscular Contraction. *J. Physiol.* **96**, pp. 45-64.
Kawamura, Y., Okazaki, H., O'Brian, P.C., and Dyck, P.J. (1977) Lumbar Motoneurons of Man. I. Numbers and Diameter Histograms of Alpha and Gamma Axons and Ventral Roots. *Journal of Neuropathology and Experimental Neurology.* **36,** pp. 853-860.
Keaveny, T.M., and Hayes, W.C. 1993. A 20-year Perspective on the Mechanical Properties of Trabecular Bone. *J. Biomech. Eng.* **115**, pp. 534-542.
Keaveny, T.M., Morgan, E.F., Niebur, G.L., and Yeh, O.C. (2001) Biomechanics of Trabecular Bone. *Annu. Rev. Biomed. Eng.* **3**, pp. 307-333.
Kempson, G.E. (1972) The Tensile Properties of Articular Cartilage and Their Relevance to the Development of Osteoarthrosis. *Orthopaedic Surgery and Traumatology. Proceedings of the 12th International Society of Orthopaedic Surgery and Traumatology.* Tel Aviv. Excerpta Medica, Amsterdam. pp. 44-58.
Kempson, G.E. (1979) Mechanical Properties of Articular Cartilage. *Adult Articular Cartilage* (ed. Freeman, M.A.R.). Elsevier Science Publishers BV. pp. 333-414.

Kempson, G.E., Muir, H., Freeman, M.A.R., and Swanson, S.A.V. (1970) Correlations Between Stiffness and the Chemical Constituents of Cartilage on the Human Femoral Head. *Biochimica et Biophysica Acta*. Amsterdam. **215**, pp. 70-77.

Kempson, G.E., Muir, I.H.M., Pollard, C., and Tuke, M. (1973) The Tensile Properties of the Cartilage of Human Femoral Condyles Related to the Content of Collagen and Glycosaminoglycans. *Biochim. Biophys. Acta*. Elsevier Science Publishers BV. **297,** pp. 456-472.

Kennedy, J.C., Alexander, I.J., and Hayes, K.C. (1982) Nerve Supply of the Human Knee and its Functional Importance. *Am. J. Sports Med.* **10 (6),** pp. 329-335.

Ker, R.F. (1981) Dynamic Tensile Properties of Sheep Plantaris Tendon (Ovis Aries). *J. of Exp. Biol.* **93**, pp. 283-302.

Ker, R.F., Dimery, N.J., and Alexander, R.M. (1986) The Role of Tendon Elasticity in a Hopping Wallaby (Macropus Rufogriseus). *J. Zool.* **208 (A)**, pp. 417-428.

Ker, R.F., Bennett, M.B., Bibby, S.R., Kester, R.C., and Alexander, R.M. (1987) The Spring in the Arch of the Human Foot. *Nature.* **325,** pp. 147-149.

Ker, R.F., Alexander, R.M., and Bennett, M.B. (1988) Why are Mammalian Tendons so Thick? *J. Zool.* **216**, pp. 309-324.

Kiiskinen, A. (1977) Physical Training and Connective Tissues in Young Mice - Physical Properties of Achilles Tendons and Long Bones. *Growth.* **41,** pp. 123-137.

Kjaer, M., Langberg, H., Miller, B.F., Boushel, R., Cramei, R., and Koskinen, S. et al. (2005) Metabolic Activity and Collagen Turnover in Human Tendon inResponse to Physical Activity. *Journal of Musculoskeletal and Neuronal Interaction.* **5**, pp. 41-52.

Klein, L., Dawson, M.H., and Heiple, K.G. (1977) Turnover of Collagen in the Adult Rat after Denervation. *J. Bone Jt. Surg.* **59 (A),** p. 1065.

Kojima, H., Ishijima, A., and Yanagida, T. (1994) Direct Measurement of Stiffness of Single Actin Filaments With and Without Tropomyosin by in vitro Nanomanipulation. *Proc. Natl. Acad. Sci. USA.* **91**, pp. 12962-12966.

Kölliker, R.A. (1850) *Mikroskopische Anatomie oder Gewebelehre des Menschen.* Leipzig.

Kölliker, R.A. (1855) *Handbuch der Gewebelehre des Menschen.* Leipzig.

Komi, P.V. and Bosco, C. (1978) Utilization of Stored Elastic Energy in Leg Extensor Muscles by Men and Women. *Med. Sci. Sports Exerc.* **10,** pp. 261-265.

Kontulainen S., Sievanen, H., Kannus, P., Pasanen, M., and Vuori, I. (2003) Effect of Long-term Impact Lading on Mass, Size, and Estimated Strength of Humerous and Radius of Female Racquet-sports Players: A Peripheral Quantitative Computed Tomography Study between Young and Old Starts and Controls. *Journal of Bone and Mineral Research* **18**, 352-9.

Kovach, I.S. (1996) A Molecular Theory of Cartilage Viscoelasticity. *Biophysical Chemistry* **59,** pp. 61-73.

Krall, E. and Dawson-Hughes, B. (1993) Heritable and Lifestyle Determinants of Bone Mineral Density. *J. Bone and Mineral Res.* **8**, pp. 1-9.

Kramhoft, M. and Solgaard, S. (1986) Spontaneous Rupture of the Extensor Pollicis Longus Tendon after Anabolic Steroids. *J. Hand Surg.* **11 (B)**, pp. 87.

Kraupse, R., Schmidt, M.B., and Schaible, H.G. (1992) Sensory Innervation of the Anterior Cruciate Ligament. *J. Bone Jt. Surg.* **74 (A),** pp. 390-397.

Kravitz, S.R., Fink, K.L., Huber, S., Bohanske, W., and Cicilioni, S. (1985) Osseous Changes in the Second Ray of Classical Ballet Dancers. *J. Amer. Podiatric Med. Assoc.* **75**, pp. 103-147.

Labeit, D., Watanabe, K., Witt, C., Fujita, H., Wu, Y., Lahmers, S., Funck, T., Labeit, S., and Granzier, H.L. (2003) Calcium-dependent Molecular Spring Elements in the Giant Protein Titin. *Proceedings of the National Academy of Sciences of the United States of America.* **100**, pp. 13716-13721.

Lai, W.M., Hou, J.S., and Mow, V.C. (1991) A Thriphasic Theory for the Swelling and Deformation Behaviours of Articular Cartilage. *J. Biomech. Eng.* **113,** pp. 245-258.

Lam, T.C. (1988) *The Mechanical Properties of the Maturing Medial Collateral Ligament.* Ph.D. Thesis, University of Calgary. Nat'l Library of Canada, Ottawa.

Lam, T.C., Frank, C.B., and Shrive, N.G. (1993) Changes in the Cyclic and Static Relaxations of the Rabbit Medial Collateral Ligament Complex During Maturation. *J. Biomech.* **26 (1)**, pp. 1-8.

Lane, N.E., Bloch, D.E., Jones, H.H., Marshall, W.H., Wood, P.D., and Fries, J.F. (1986) Long-distance Running, Bone Density and Osteoarthritis. *J. Amer. Med. Assoc.* **255**, pp. 1147-1151.

Lane, N.E., Kaneps, A.J., Stover, S.M., Modin, G., and Kimmel, D.B. (1996) Bone Mineral Density and Turnover Following Forelimb Immobilization and Recovery in Young Adult Dogs. *Calcified Tissue Inter.* **59**, pp. 401-406.

Lang, T., LeBlanc, A., Evans, H., Lu, Y., Genant, H., and Yu, A. (2004) Cortical and Trabecular Bone Mineral Loss from the Spine and Hip in Long-term Spaceflight. *Journal of Bone and Mineral Research*, 19, pp.1006-1012.

Lanyon, L.E. (1974) Experimental Support for the Trajectorial Theory of Bone Structure. *J. Bone Jt. Surg.* **56 (B)**, pp. 160-166.

Lanyon, L.E. (1980) The Influence of Function on the Development of Bone Curvature. *J. Zool.* **192**, pp. 457-466.

Lanyon, L.E. (1987) *Strain and Remodelling*. Hard Tissue Workshop, Sun Valley, ID.

Lanyon, L.E. (1996) Using Functional Loading to Influence Bone Mass and Architecture: Objectives, Mechanisms, and Relationship with Estrogen of the Mechanically Adaptive Process in Bone. *Bone*. **18**, pp. 37S-43S.

Lanyon, L.E. and Bourn, S. (1979) The Influence of Mechanical Function on the Development and Remodeling of the Tibia. An Experimental Study in Sheep. *J. Bone Jt. Surg.* **61 (2)**, pp. 263-273.

Lanyon, L.E. and Rubin, C.T. (1984) Static vs Dynamic Loads as an Influence on Bone Remodelling. *J. Biomech.* **17,** pp. 897-905.

Lark, M.W., Bayne, E.K., Flanagan, J., Harper, C.F., Hoerrner, L.A., Hutchinson, N.I., Singer, I.I., Donatelli, S.A., Weidner, J.R., Williams, H.R., Mumford, R.A., and Lohmander, L.S. (1997) Aggrecan Degradation in Human Cartilage. Evidence for both Matrix Metalloproteinase and Aggrecanase Activity in Normal, Osteoarthritic, and Rheumatoid Joints. *J. Clin. Invest.* **100,** pp. 93-106.

Laros, G.S., Tipton, C.M., and Cooper, R.R. (1971) Influence of Physical Activity on Ligament Insertion in the Knees of Dogs. *J. Bone Jt. Surg.* **53 (A),** pp. 275-286.

Larsson, L., Grimby, G., and Karlsson, J. (1979) Muscle Strength and Speed of Movement in Relation to Age and Muscle Morphology. *Journal of Applied Physiology.* **46**, pp. 451-456.

Lee, H.D. and Herzog, W. (2002) Force Enhancement Following Muscle Stretch of Electrically and Voluntarily Activated Human Adductor Pollicis. *Journal of Physiology.* **545**, pp. 321-330.

Lee, H.D. and Herzog, W. (2003) Force Depression Following Muscle Shortening of Voluntarily Activated and Electricallystimulated Human Adductor Pollicis. *Journal of Physiology.* **551**, pp. 993-1003.

Lee, H.D., Herzog, W., and Leonard, T.R. (2001) Effects of Cyclic Changes in Muscle Length on Force Production in situ Cat Soleus. *Journal of Biomechanics.* **34**, pp. 979-987.

Lekhnitskii, S.G. (1963) *Theory of Elasticity of an Anisotropic Elastic Body.* Holden-Day, Inc., San Francisco, CA.

Leonard, T.R. and Herzog, W. (2005. Does the Speed of Shortening affect Steady State Force Depression in Cat Soleus Muscle? *Journal of Biomechanics.* **38**, pp. 2190-2197.

Lexell, J. Taylor, A.W., and Sjostrom, M. (1988) What is the Cause of the Aging Atrophy? Total Number, Size, and Proportion of Different Fiber Types Studies in Whole Vastus Lateralis Muscle from 15-to-83 Year Old Men. *Journal of Neurological Science.* **84,** pp. 275-283.

Li, L.P., Buschmann, M.D., and Shirazi-Adl, A. (2000a) A Fibril Reinforced Nonhomogeneous Poroelastic Model for Articular Cartilage: Inhomogeneous Response in Unconfined Compression. *J. Biomech.* **33,** pp. 1533-1541.

Li, L.P., Buschmann, M.D., and Shirazi-Adl, A. (2000b) A Fibril Reinforced Nonhomogeneous Poroelastic Model for Articular Cartilage: Inhomogeneous Response in Unconfined Compression. *J. Biomech.* **33,** pp. 1533-1541.

Lieber, R.L., Leonard, M.E., and Brown-Maupin, C.G. (2000) Effects of Muscle Contraction on the Load-strain Properties of Frog Aponeurosis and Tendon. *Cells Tissues Organs.* **166**, po. 48-54.

Lindle, R.S., Metter, E.J., Lynch, N.A., Fleg, J.L., Fozard, J.L., Tobin, J., Roy, T.A., and Hurley, B.F. (1997) Age and Gender Comparisons of Muscle Strength in 654 Women and Men Aged 20-93 Yr. *Journal of Applied Physiology.* **83**, pp. 1581-1587.

Lipshitz, H., Etheredge III, R., and Glimcher, M.J. (1976) Changes in the Hexosamine Content and Swelling Ratio of Articular Cartilage as Functions of Depth from the Surface. *J. Bone Jt. Surg.* **58 (A-8)**, pp. 1149-1153.

Liskova, M. (1965) The Thickness Changes of the Long Bone after Experimental Stressing During Growth. *Plzensky Lekarsky Sbornik.* **25,** pp. 95-104.

Little, C.B., Ghosh, P., and Rose, R. (1997) The Effect of Strenuous versus Moderate Exercise on the Metabolism of Proteoglycans in Articular Cartilage from Different Weight-bearing Regions of the Equine Third Carpal Bone. *Osteoarthritis and Cartilage.* **5**, pp. 161-172.

Liu, Z.Q., Rangayyan, R.M., and Frank, C.B. (1991) Statistical Analysis of Collagen Alignment in Ligaments by Scale-space Analysis. *I.E.E.E. Trans. Biomed. Eng.* **38 (6)**, pp. 580-588.

Loeb, G.E. (1984) The Control and Responses of Mammalian Muscle Spindles During Normally Executed Motor Tasks. *Exerc. Sports Sci. Rev.* (ed. Terjung, R.L.). D.C. Health & Co., Lexington. **12,** pp. 157-204.

Lombardi, V. and Piazzesi, G. (1992) Force Response in Steady Lengthening of Active Single Muscle Fibres. *Muscular Contraction* (ed. Simmons, R.M.). Cambridge University Press, Cambridge and New York. pp. 237-255.

Longino, D., Butterfield, T., and Herzog, W. (2005) Frequency and Length Dependent Effects of Botulinum Toxin-Induced Muscle Weakness. *J. Biomech.* **38,** pp. 609-613.

Lundberg, A., Svensson, O.K., Bylund, C., Goldie, I., and Selvik, G. (1989) Kinematics of the Ankle/foot Complex - Part 2: Pronation and Supination. *Foot and Ankle.* **9,** pp. 248-253.

Lundborg, G. and Rank, F. (1978) Experimental Intrinsic Healing of Flexor Tendons Based upon Synovial Fluid Nutrition. *J. Hand Surg.* **3,** pp. 21-31.

Lundborg, G., Holm, S., and Myrhage, R. (1980) The Role of the Synovial Fluid and Tendon Sheath for Flexor Tendon Nutrition: An Experimental Tracer Study on Diffusional Pathways in Dogs. *Scand. J. Plast. Reconstr. Surg.* **14,** pp. 99-107.

Lyyra, T., Jurvelin, J., Pitkänen, P., Väätäinen, U., and Kiviranta, I. (1995) Indentation Instrument for the Measurement of Cartilage Stiffness under Arthroscopic Control. *Med. Eng. Phys.* **17,** pp. 395-399.

Ma, Y.Z. and Taylor, E.W. (1994) Kinetic Mechanism of Myofibril ATPase. *Biophys. J.* **66,** pp. 1542-1553.

Macaluso, A., Young, A., Gibb, K.S., Rowe, D.A., and De Vito, G. (2003) Cycling as a Novel Approach to Resistance Training Increases Muscle Strength, Power, and Selected Functional Abilities in Healthy Older Women. *Journal of Applied Physiology.* DOI 10.1152/japplphysiol.00416.2003.

Macaluso, A. and De Vito G. (2004) Muscle Strength, Power, and Resistance Training in Older People. *European Journal of Applied Physiology.* **91**, pp. 450-72.

MacConaill, M.A. (1951) The Movements of Bones and Joints - 4. The Mechanical Structure of Articulating Cartilage. *J. Bone Jt. Surg.* **33 (B-2)**, pp. 251-257.

MacDougal, J.D. (1992) Hypertrophy or Hyperplasia? In: *Strength and Power in Sports: The Encyclopedia of Sports Medicine* (ed. Komi P.). Oxford, Blackwell, (pp. 230-8).

Magnusson, S.P., Aagaard, P., Dyhre-Poulsen, P., and Kjaer, M. (2001) Load-displacement Properties of the Human Triceps Surae Aponeurosis in vivo. *J. Physiology.* **531**, pp. 277-288.

Magnussen, R.A., Guilak, F., and Vail, T.P. (2005) Cartilage Degeneration in Post-collapse Cases of Osteonecrosis of the Human Femoral Head: Altered Mechanical Properties in Tension, Compression, and Shear. *Journal of Orthopedic Research.* **23**, pp. 576-583.

Mai, M.T. and Lieber, R.L. (1990) A Model of Semitendinosus Muscle Sarcomere Length, Knee, and Hip Joint Interaction in the Frog Hindlimb. *J. Biomech.* **23 (3),** pp. 271-279.

Maitland, M.E. (1996) *Longitudinal Measurement of Tibial Motion Relative to the Femur During Passive Displacements and Femoral Nerve Stimulation in the ACL-deficient Cat Model of Osteoarthritis.* Ph.D. Dissertation, Faculty of Medical Science. University of Calgary.

Maitland M.E., Leonard, T.R., Frank, C.B., Shrive, N.G., and Herzog, W. (1998) Longitudinal Measurement of Tibial Motion Relative to the Femur During Passive Displacements in the Cat before and after Anterior Cruciate Ligament Transection. *J. Orthop. Res.* **16**, pp. 448-454.

Majumdar, S. (1998) A Review Of Magnetic Resonance (Mr) Imaging of Trabecular Bone Micro-Architecture: Contribution to The Prediction of Biomechanical Properties and Fracture Prevalence. *Technol Health Care.* **6**, pp. 321-332.

Mak, A.F. (1986) Apparent Viscoelastic Behaviour of Articular Cartilage – Contributions from Intrinsic Matrix Viscoelasticity and Interstitial Fluid Flow. *J. Biomech. Eng.* **102,** pp. 73-84.

Mankin, H.J., Mow, V.C., Buckwalter, J.A., Iannotti, J.P., and Ratcliffe, A. (1994) Form and Function of Articular Cartilage. *Orthopaedic Basic Science* (ed. Simon, S.R.). *Amer. Acad. Orthop. Surg.* Park Ridge, IL. pp. 1-44.

Mann, R.A. (1982) Foot Problems in Adults. Part I: Biomechanics of the Foot. *Amer. Acad. Orthop. Surg.* (ed. Frankel, V.H.). pp. 167-180.

Mann, R.W. (2005) Comment on "an articular cartilage contact model based on real surface geometry", Han Sang-Kuy, Salvatore Federico, Marcelo Epstein and Walter Herzog, Journal of Biomechanics, 38 (2005) 179-184. *.J Biomech.* **38,** pp. 1741-1742.

Manske, P.R., Bridwell, K. and Lesker, P.A. (1978) Nutrient Pathways to Rlexor Tendons of Chickens using Tritiated Proline. *J. Hand Surg. [Am.].* **3**, pp. 352-357.

Manske, P.R. and Lesker, P.A. (1982) Nutrient Pathways of Flexor Tendons in Primates. *J. Hand Surg.* **7**, pp. 436-457.

Manter, J.T. (1941) Movements of the Subtalar and Transverse Tarsal Joints. *Anat. Rec.* **80**, pp. 397-400.

Maréchal, G. and Plaghki, L. (1979) The Deficit of the Isometric Tetanic Tension Redeveloped After a Release of Frog Muscle at a Constant Velocity. *J. Gen. Physiol.* **73,** pp. 453-467.

Margossian, S.S. and Lowey, S. (1978) Interaction of Myosin Subfragments with F-actin. *Biochemistry.* **17**, pp. 5431-5439.

Maroudas, A. (1975) Biophysical Chemistry of Cartilaginous Tissues with Special Reference to Solute and Fluid Transport. *Biorheology.* **12,** pp. 233-248.

Maroudas, A. (1979) Physiochemical Properties of Articular Cartilage. *Adult Articular Cartilage* (ed. Freeman, M.A.R.). Pitman, London. pp. 215 - 290.

Martin, R.B. and Atkinson, P.J. (1977) Age and Sex Related Changes in the Structure and Strength of the Human Femoral Shaft. *J. Biomech.* **10 (4),** pp. 223-231.

Martin, R.B. and Burr, D.B. (1989) *Structure, Function, and Adaptation of Compact Bone.* Raven Press, New York.

Martin, R.B., Burr, D.B., and Sharkey, N.A. (1998) Skeletal Biology. In: Skeletal Tissue Mechanics (pp.29-77) (ed. Smith, R.). New York, NY: Springer-Verlag New York Inc.

Matsuda, J.J., Zernicke, R.F., Vailas, A.C., Pedrini, V.A., Pedrini-Mille, A., and Maynard, J.A. (1986) Structural and Mechanical Adaptation of Immature Bone to Strenuous Exercise. *J. Appl. Physiol.* **60**, pp. 2028-2034.

Matyas, J.R. (1985) *The Structure and Function of Tendon and Ligament Insertions into Bone.* M.Sc. Thesis, Cornell University, New York.

Matyas, J.R. (1990) *The Structure and Function of the Insertions of the Rabbit Medial Collateral Ligament.* Ph.D. Thesis, University of Calgary. Nat'l Library of Canada, Ottawa.

Matyas, J.R., Bodie, D., Andersen, M., and Frank, C.B. (1990) The Developmental Morphology of a "Periosteal" Ligament Insertion: Growth and Maturation of the Tibial Insertion of the Rabbit Medial Collateral Ligament. *J. Orthop. Res.* **8 (3)**, pp. 412-424.

McAlindon, T.E., Cooper, C., Kirwan, J.R., and Dieppe, P.A. (1993) Determinants of Disability in Osteoarthritis of the Dnee. *Annals of Rheumatic Diseases.* **52,** pp. 258-262.

McArdle, W.D., Katch, F.I., and Katch, V.L. (1996) *Exercise Physiology.* Williams and Wilkins, Baltimore, MD., pp. 350, 393.

McCalden, R.W., McGeough, J.A., Barker, M.B., and Court-Brown, C.M. (1993) Age-related Changes in the Tensile Properties of Cortical Bone. *J. Bone Jt. Surg.* **75 (A-8)**, pp. 1193-1199.

McCall, G.E., Byrnes, W.C., and Fleck, S.J. (1999) Acute and Chronic Hormonal Responses to Resistance Training Designed to Promote Muscle Hypertrophy. *Canadian Journal of Applied Physiology.* **24**, 96-107.

McCutchen, C.W. (1962) The Frictional Properties of Animal Joints. *Wear.* **5,** p. 1.

McPhedran, A.M., Wuerker, R.B., and Henneman E. (1965) Properties of Motor Units in a Homogeneous Red Muscle (Soleus) of the Cat. *J. Neurophysiol.* **28,** pp. 71-84.

Meijer, K. (2002) History Dependence of Force Production in Submaximal Stimulated Rat Medial Gastrocnemius Muscle. *J. Electromyogr. Kinesiol.* **12**, pp. 463-470.

Meijer, K., Grootenboer, H.J., Koopman, B.F.J.M., van der Linden, B.J.J.J., and Huijing, P. (1998) A Hill Type Model of Rat Medial Gastrocnemius Muscle that Accounts for Shortening History effects. *Journal of Biomechanics.* **31**, pp. 555-563.

Melton, L.J., 3rd, Chrischilles, E.A., Cooper, C., Lane, A.W., and Riggs, B.L. (1992) Perspective. How Many Women have Osteoporosis? *J. Bone Miner Res.* **7**, pp. 1005-1010.

Merrilees, M.J. and Flint, M.H. (1980) Ultrastructural Study of Tension and Pressure Zones in a Rabbit Flexor Tendon. *Am. J. Anat.* **157,** pp. 87-106.

Michel, B.A., Lane, N.E., Bloch, D.A., Jones, H.H., and Fries, J.F. (1991) Effect of Changes in Weightbearing Exercise on Lumbar Bone Mass After Age Fifty. *Annals of Medicine.* **23**, pp. 397-401.

Michna, H. (1983) Organization of Collagen Fibrils in Tendon: Changes Induced by an Anabolic Steroid II: A Morphometric and Stereologic Analysis. *Int. J. Sports Med.* **4**, p. 59.

Michna, H. (1984) Morphometric Analysis of Loading-induced Changes in Collagen-fibril Populations in Young Tendons. *Cell and Tissue Res.* **236**, pp. 465-470.

Michna, H. and Stang-Voss, C. (1983) The Predisposition to Tendon Rupture After Doping with Anabolic Steroids. *Int. J. Sports Med.* **4,** p. 59.

Mijailovich, S.M., Fredberg, J.J., and Butler, J.P. (1996) On the Theory of Muscle Contraction: Filament Extensibility and the Development of Isometric Force and Stiffness. *Biophysical Journal.* **71**, pp. 1475-1484.

Milgrom, C., Giladi, M., Simkin, A., Rand, N., Kedem, R., Kashtan, H., Stein, M., and Gomor, M. (1989) The Area Moment of Inertia of the Tibia: A Risk Factor for Stress Fractures. *J. Biomech.* **22 (11-12),** pp. 1243-1248.

Millar, N.C. and Geeves, M.A. (1983) The Limiting Rate of the ATP-mediated Dissociation of Actin from Rabbbit Skeletal Muscle Myosin subfragment 1. *FEBS Lett.* **160**, pp. 141-148.

Miller, K. (1998) Modelling Soft Tissue Using Biphasic Theory-A Word of Caution. *Computer Methods in Biomech. and Biomedical Eng.* **1**, pp. 261-263.

Miller, E.J. and Gay, S. (1992) Collagen Structure and Function. *Wound Healing: Biochemical and Clinical Aspects* (eds. Cohen, I.K., Diegelmann, R.F., and Lindblad, W.J.). W.B. Saunders, Philadelphia. pp. 130-151.

Morgan, D.L. (1990) New Insights into the Behavior of Muscle during Active Lengthening. *Biophysical Journal.* **57**, pp. 209-221.

Morgan, D.L., Whitehead, N.P., Wise, A.K., Gregory, J.E. and Proske, U. (2000) Tension Changes in the Cat Soleus Muscle following Slow Stretch or Shortening of the Contracting Muscle. *Journal of Physiology.* **522.3**, pp. 503-513.

Moore, M.J. (1983) The Dual Connective Tissue System of Rat Muscle. *Muscle and Nerve.* **6,** pp. 416-422.

Morris, F.L., Naughton, G.A., Gibbs, J.L., Carlson, J.S., and Wark, J.D. (1997) Prospective Ten-month Exercise Intervention in Premenarchal Girls: Positive Effects on Bone and Lean Mass. *J. Bone Mineral Res.* **9**, pp. 1453-1462.

Morrison, J.B. (1968) Bioengineering Analysis of Force Actions Transmitted by the Knee Joint. *Biomed. Eng.* **3,** pp. 164-170.

Mow, V.C., Bachrach, N.M., Setton, L.A., and Guilak, F. (1994) Stress, Strain, Pressure and Flow Fields in Articular Cartilage and Chondrocytes. In *Cell Mechanics and Cellular Engineering* (eds. Mow, V.C., Guilak, F., Tran-Son-Tray, R., and Hochmuth, R.M.) pp. 345-379. Springer Verlag, New York.

Mow, V.C., Kuei, S.C., Lai, W.M., and Armstrong, C.G. (1980) Biphasic Creep and Stress Relaxation of Articular Cartilage in Compression: Theory and Experiments. *J. Biomech. Eng.* **102,** pp. 73-84.

Mow, V.C., Holmes, M.H., and Lai, W.M. (1984) Fluid Transport and Mechanical Properties of Articular Cartilage: A Review. *J. Biomech.* **17 (5),** pp. 377-394.

Mow, V.C., Ateshian, G.A., and Spilker, R.L. (1993) Biomechanics of Diarthrodial Joints: A Review of Twenty Years of Progress. *J. Biomech. Eng.* **115,** pp. 460-467.

Mow, V.C. and Rosenwasser, M. (1987) Articular Cartilage: Biomechanics. *Injury and Repair of the Musculo-skeletal Soft Tissues* (eds. Woo, S.L.-Y. and Buckwalter, L.). American Academy of Orthopaedic Surgeons, Park Ridge, IL. pp. 427-463.

Mow, V.C. and Proctor, C.S. (1989) Biomechanics of Articular Cartilage. *Basic Biomechanics of the Musculo-skeletal System* (eds. Nordin, M. and Frankel, V.H.). Lea & Febiger, Baltimore, Maryland. pp. 31-58.

Munderman, A., Dyrby, C.O., Andriacchi, T.P., and King, K.B. (2005) Serum Concentration of Cartilage Oligomeric Protein (COMP) is Sensitive to Physiological Cyclic Loading in Healthy Adults. *Osteoarthritis and Cartilage.* **13**, pp. 34-38.

Muramatsu, T., Muraoka, T., Takeshita, D., Kawakami, Y., Hirano, Y., and Fukunaga, T., (2001) Mechanical Properties of Tendon and Aponeurosis of Human Gastrocnemius Muscle in vivo. *J. Applied Physiology.* **90**, pp. 1671-1678.

Muir, I.H.M. (1978) The Chemistry of the Ground Substance of Joint Cartilage. *The Joints and Synovial Fluid* (ed. Sokoloff, L.). Academic Press, New York. pp. 27-94.

Muir, I.H.M. (1983) Proteoglycans as Organisers of the Intercellular Matrix. *Biochem. Soc. Trans.* **11 (6),** pp. 613-622.

Müller, W. (1939) *Biologie der Gelenke*. Leipzig.

Murphy, P.R., Stein, R.B., and Taylor, J. (1984) Phasic and Tonic Modulation of Impulse Rate in γ Motoneurons During Locomotion in Premammilliary Cats. *J. Neurophysiol.* **52,** pp. 228-243.

Myers, R. and Mow, V.C. (1983) Biomechanics of Cartilage and its Response to Biomechanical Stimuli. *Cartilage. Volume I. Structure, Function and Biochemistry.* (ed. Hall, B.K.). Academic Press, New York, NY. pp. 313-341.

Narici, M.V., Landoni, L., and Minetti, A.E. (1991) Effect of Aging on Human Adductor Pollicis Muscle Function. *Journal of Applied Physiology.* **71**, pp. 1277-1281.

Nazarian, A. and Müller, R. (2004) Time-lapsed Microstructural Imaging of Bone Failure Behavior. *J. Biomech.* **37**, pp. 55-65.

Needham, D.M. (1971) *Machina Carnis,* Cambridge University Press, Cambridge.

Newitt, D.C., Majumdar, S., van Rietbergen, B., von Ingersleben, G., Harris, S.T., Genant, H.K., Chesnut, C., Garneo, P., and MacDonald, B. (2002) In Vivo Assessment of Architecture and Micro-Finite Element Analysis Derived Indices of Mechanical Properties of Trabecular Bone in the Radius. *Osteoporos Int.* **13**, pp. 6-17.

Nicholas, J.A. (1986) *The Lower Extremity and Spine in Sports Medicine.* C.V. Mosby Company, St. Louis.

Nigg, B.M. (1985) Biomechanics, Load Analysis, and Sports Injuries in the Lower Extremities. *Sports Med.* **2,** pp. 367-379.

Nigg, B.M., Cole, G.K., and Nachbauer, W. (1993) Effects of Arch Height of the Foot on Angular Motion of the Lower Extremities in Running. *J. Biomech.* **26 (8),** pp. 909-916.

NIH Consensus Development Panel. (2001) Osteoporosis prevention, diagnosis, and therapy. *JAMA.* **285**, pp. 785-795.

Noble, M.I.M. (1992) Enhancement of Mechanical Performance of Striated Muscle by Stretch During Contraction. *Experimental Physiology.* **77**, pp. 539-552.

Nordstrom, A., Karlsson, C., Nyquist, F., Olsson, T., Nordstrom, P., and Karlsson, M. (2005) Bone Loss and Fracture Risk after Reduced Physical Activity. *Journal of Bone and Mineral Research.* **20**, pp. 202-207.

Noyes, F.R. (1977) Functional Properties of Knee Ligaments and Alterations Induced by Immobilization. *Clin. Orthop. Rel. Res.* **123,** pp. 210-242.

Noyes, F.R., Butler, D.L., Grood, E.S., Zernicke, R.F., and Hefzy, M.S. (1984) Biomechanical Analysis of Human Ligament Grafts Used in Knee-Ligament Repairs and Reconstructions. *J. Bone Joint Surg. Am.* **66**, pp. 344-352.

Noyes, F.R., DeLucas, J.L., and Torvik, P.J. (1974a) Biomechanics of Anterior Cruciate Ligament Failure: An Analysis of Strain-rate Sensitivity and Mechanisms of Failure in Primates. *J. Bone Jt. Surg.* **56 (A-2)**, pp. 236-253.

Noyes, F.R., Torvik, P.J., Hyde, W.B., and DeLucas, J.L. (1974b) Biomechanics of Ligament Failure II: An Analysis of Immobilization, Exercise, and Reconditioning Effects in Primates. *J. Bone Jt. Surg.* **56 (A),** pp. 1406-1418.

Noyes, F.R. and Grood, E.S. (1976) The Strength of the Anterior Cruciate Ligament in Humans and Rhesus Monkeys. *J. Bone Jt. Surg.* **58 (A-6),** pp. 1074-1082.

O'Brien, M. (1992) Functional Anatomy and Physiology of Tendons. *Clin. Sports Med.* **11(3),** pp. 505-520.

O'Connor, J.A., Lanyon, L.E., and MacFie, H. (1982) The Influence of Strain Rate on Adaptive Bone Remodelling. *J. Biomech.* **15**, pp. 767-781.

O'Connor, J.J., Shercliff, T., Fitzpatrick, D., Bradley, J., Daniel, D.M., Biden, E., and Goodfellow, J. (1990) Geometry of the Knee. *Knee Ligaments: Structure, Function, Injury, and Repair* (eds. Daniel, D.M., Akeson, W.H., and O'Connor, J.J.). pp. 163-199.

Odgaard, A.. (1997) Three-dimensional Methods for Quantification of Cancellous Bone Architecture. *Bone.* **20**, pp. 315-328.

Odgaard, A., Kabel, J., Van Rietbergen, B., Dalstra, M., and Huiskes, R. (1997) Fabric and Elastic Principal Directions of Cancellous Bone are Closely Related. *J. Biomech* **30**, 487-495.

Oloyede, A. and Broom, N. (1991) Is Classical Consolidation Theory Applicable to Articular Cartilage Deformation? *Clin. Biomech.* **6,** pp. 206-212.

Oxnard, C.E. (1993) Bone and Bones, Architecture and Stress, Fossils and Osteoporosis. *J. Biomech.* **26 (1),** pp. 63-79.

Page, S.G. and Huxley, H.E. (1963) Filament Lengths in Striated Muscle. *J. Cell Biol.* **19,** pp. 369-390.

Parry, D.A.D., Barnes, G.R.G., and Craig, A.S. (1978a) A Comparison of the Size Distribution of Collagen Fibrils in Connective Tissues as a Function of Age and a Possible Relation Between Fibril Size Distribution and Mechanical Properties. *Proc. R. Soc. Lond.* **203 (B),** pp. 305-321.

Parsons, J.R. and Black, J. (1977) The Viscoelastic Shear Behaviour of Normal Rabbit Articular Cartilage. *J. Biomech.* **10,** pp. 21-29.

Parry, D.A.D., Craig, A.S., and Barnes, G.R.G. (1978b) Tendon and Ligament from the Horse: An Ultrastructural Study of Collagen Fibrils and Elastic Fibres as a Function of Age. *Proc. R. Soc. Lond.* **203 (B),** pp. 293-303.

Paul, J.P. (1965) Bioengineering Studies of the Forces Transmitted by Joints - II. *Biomechanics and Related Bioengineering Topics* (ed. Kenedi, R.M.). Pergamon Press, Oxford.

Paul, J.P. (1976) Approaches to Design. Force Actions Transmitted by the Joints in the Human Body. *Proc. R. Soc. Lond.* **192 (B),** pp. 163-172.

Pauwels, F. (1980) *Biomechanics of the Locomotor Apparatus: Contributions on the Functional Anatomy of the Locomotor Apparatus* (trans. Maquet, P. and Furlong, R.). Springer Verlag, Berlin.

Pearson, K.G. (1993) Common Principles of Motor Control in Vertebrates and invertebrates. *Annu. Rev. Neurosci.* **16,** pp. 265-297.

Pedotti, A., Krishnan, V.V., and Stark, L. (1978) Optimization of Muscle Force Sequencing in Human Locomotion. *Math. Biosci.* **38,** pp. 57-76.

Penrod, D.D., Davy, D.T., and Singh, D.P. (1974) An Optimization Approach to Tendon Force Analysis. *J. Biomech.* **7 (3),** pp. 123-130.

Perrine, J.J. and Edgerton, V.R. (1978) Muscle Force-velocity and Power-velocity Relationships Under Isokinetic Loading. *Med. Sci. Sports Exerc.* **10 (3),** pp. 159-166.

Peterson, D., Rassier, D., and Herzog, W. (2004) Force Enhancement in Single Skeletal Muscle Fibres on the Ascending Limb of the Force-length Relationship. *Journal of Experimental Biology.* **207**, pp. 2787-2791.

Pins, G.D., Huang, E.K., Christiansen, D.L., and Silver, F.H. (1997) Effects of Static Axial Strain on the Tensile Properties and Failure Mechanisms of Self-Assembled Collagen Fibers. *J. Applied Polymer Science.* **63,** pp. 1429-1440.

Podolsky, R.J. (1960) Kinetics of Muscular Contraction: the Approach to the Steady State. *Nature.* **188,** pp. 666-668.

Policard, A. (1936) *Physiologie Générale des Articulations a l'État Normal et Pathologique*. Masson et Cie, Paris.

Pollack, G.H. (1990) *Muscles and Molecules: Uncovering the Principles of Biological Motion.* Ebner & Sons, Seattle, WA.

Pollock, C.M. (1991) *Body Mass and Elastic Strain in Tendons*. M.Sc. Thesis, University of Calgary. Nat'l Library of Canada, Ottawa.

Pollock, C.M. (1991) *The Relationship Between Body Mass and the Capacity for Storage of Elastic Strain Energy in Mammalian Limb Tendons.* Ph.D. Thesis, University of Calgary. Nat'l Library of Canada, Ottawa.

Pollock, C.M. and Shadwick, R.E. (1994) Relationship between Body Mass and Biomechanical Properties of Limb Tendons in Adult Mammals. *American Journal of Physiology.* **266**, pp. R1016-R1021.

Poole, C.A., Gilbert, R.T., Herbage, D., and Hartmann, D.J. (1997) Immunolocalization of Type Ix Collagen in Normal and Spontaneously Osteoarthritic Canine Tibial Cartilage and Isolated Chondrons. *Osteoarthritis Cartilage.* **5,** pp. 191-204.

Prochazka, A., Hulliger, M., Zangger, P., and Appenteng, K. (1985) Fusimotor Set: New Evidence for α Independent Control of γ Motoneurons During Movement in the Awake Cat. *Brain Res.* **339,** pp. 136-140.

Rambaut, P.C. and Goode, A.W. (1985) Skeletal Changes During Space Flight. *Lancet.* **2**, pp. 1050-1052.

Rassier, D.E., Herzog, W. and Pollack, G.H. (2003a) Dynamics of Individual Sarcomeres during and after Stretch in Activated Single Myofibrils. *Proceedings of the Royal Society London B* **270**, pp. 1735-1740.

Rassier, D.E., Herzog, W., Wakeling, J., and Syme, D. (2003b) Stretch-induced, Steady-state Force Enhancement in Single Skeletal Muscle Fibers Exceeds the Isometric Force at Optimal Fibre Length. *Journal of Biomechanics.* **36**, pp. 1309-1316.

Rayment, I., Holden, H.M., Whittaker, M., Yohn, C.B., Lorenz, M., Holmes, K.C. and Milligan, R.A. (1993a) Structure of the Actin-myosin Complex and its Implications for Muscle Contraction. *Science.* **261**, pp. 58-65.

Rayment, I., Rypniewski, W.R., Schmidt-Bäse, K., Smith, R., Tomchick, D.R., Benning, M.M., Winkelmann, D.A., Wesenberg, G. and Holden, H.M. (1993b) Three-dimensional Structure of Myosin Subfragment-1: A Molecular Motor. *Science.* **261**, pp. 50-58.

Recker, R.R. (1983) *Bone Histomorphometry: Techniques and Interpretation.* CTC Press, Boca Raton, FL.

Reedy, M.K., Holmes, K.C., and Tregear, R.T. (1965) Induced Changes in Orientation of the Cross-bridges of Glycerinated Insect Flight Muscle. *Nature.* **207**, pp. 1276-1280.

Reiderer-Henderson, M.A., Gauger, A., and Olson, L. (1983) Attachment and Extracellular Matrix Differences Between Tendon and Synovial Fibroblastic Cells. *In-vitro.* **19 (2),** pp. 127-133.

Reilly, D.T. and Burstein, A. (1975) The Elastic and Ultimate Properties of Compact Bone Tissue. *J. Biomech.* **8 (6),** pp. 393-406.

Reinschmidt, C., van den Bogert, A.J., Murphy, N., Lundberg, A., and Nigg, B.M. (1997) Tibiocalcaneal Motion During Running, Measured with External and Bone Markers. *Clin. Biomech.* **12 (1),** pp. 8-16.

Rice, J.C., Cowin, S.C., and Bowman, J.A. (1988) On the Dependence of the Elasticity and Strength of Cancellous Bone on Apparent Density. *J. Biomech.* **21**, pp.155-168.

Rigby, B.J., Hirai, N., Spikes, J.D., and Eyring, H. (1959) The Mechanical Properties of Rat Tail Tendon. *J. Gen. Physiol.* **43**, pp. 265-283.

Riggs, B.L. and Melton, L.J., 3rd. (1992) The Prevention and Treatment of Osteoporosis. *N. Engl. J. Med.* **327(9)**, pp. 620-627.

Rigg, B.L., Wahner, H.W., Seeman, E., Offord, K.P., Dunn, W.L., Mazess, R.B., Johnson, K.A., and Melton, L.J. (1982) Changes in Bone Mineral Density of the Proximal Femur and Spine with Aging: Differences Between the Postmenopausal and Senile Osteoporosis Syndromes. *J. Clin. Investigation.* **70**, pp. 716-723.

Roberts, T.J., Marsh, R.L., Weyand, P.G., and Taylor, C.R. (1997) Muscular Force in Running Turkeys: The Economy of Minimizing Work. *Science.* **275**, pp. 1113-1115.

Rollhäuser, H. (1950) Konstruktions-und Altersunterschiede in Festigkeit kollagener Fibrillen. *Gegenbauers Morph. Jahrb.* **90,** pp. 157-179.

Routh, R.W., Rumancik, S., Pathak, R.D., Bursheil, A.L., and Nauman, E.A. (2005) The Relationship between Bone Mineral Density and Biomechanics in Patients with Osteoporosis and Scoliosis. *Osteoporosis International* (Epub ahead of print).

Roux, W. (1885) Beitrage zur Morphologie der Funktionellen Anpassung. *Arch. Anat. Physiol. Anat. Abt.*

Rowe, R.W.D. (1985) The Structure of Rat Tail Tendon. *Connect. Tissue Res.* **14,** pp. 9-20.

Rubin, C.T. (1984) Skeletal Strain and the Functional Significance of Bone Architecture. *Calcified Tissue International.* **36**, pp. S11-S18.

Rubin, C.T. and Lanyon, L.E. (1982) Limb Mechanics as a Function of Speed and Gait: A Study of Functional Strains in the Radius and Tibia of Horse and Dog. *J. Exp. Biol.* **101**, pp. 187-211.

Rubin, C.T. and Lanyon, L.E. (1984a) Regulation of Bone Formation by Applied Dynamic Loads. *J. Bone Jt. Surg.* **66 (A)**, pp. 397-402.

Rubin, C.T. and Lanyon, L.E. (1984b) Dynamic Strain Similarity in Vertebrates: An Alternative to Allometric Bone Scaling. *J. Theo. Biology.* **107**, pp. 321-327.

Rubin, C.T. (1984c) Skeletal Strain and the Functional Significance of Bone Architecture. *Calcified Tissue Int.* **36**, pp. S11-S18.

Rubin, C.T., Gross, T.S., Qin, Y.X., Fritton, S., Guilak, F., and McLeod, K.J. (1996) Differentiation of the Bone-tissue Remodelling Response to Axial and Torsional Loading in the Turkey Ulna. *J. Bone Jt. Surg.* **78 (A)**, pp. 1523-1533.

Rubin, C., Turner, A.S., Bain, S., Mallinckrodt, C., and McLeod, K. (2001) Anabolism. Low Mechanical Signals Strengthen Long Bones. *Nature.* **412**, pp. 603-604.

Rubin, C.T., Turner, A.S., Mallinchrodt, C., Fritton, J., and McLeod, K. (1997) Site-specific Increase in Bone Density Stimulated Non-invasively by Extremely Low Magnitude Thirty Hertz Mechanical Stimulation. *Trans. Ortho. Res. Soc.* **22**, p. 110.

Ruedel, R. and Taylor, S.R. (1971) Striated Muscle Fibres: Facilitation of Contraction at Short Lengths by Caffeine. *Science.* **172,** pp. 387-388.

Rüegsegger P, Koller B, and Müller R. (1996) A Microtomographic System for the Nondestructive Evaluation of Bone Architecture. *Calcif. Tissue Int.* **58**, pp. 24-29.

Ruimerman, R., Hilbers, P., van Rietbergen, B., and Huiskes, R. (2005) A Theoretical Framework for Strain-Related Trabecular Bone Maintenance and Adaptation. *J. Biomech.* **38**, pp. 931-941.

Rundgren, Å. (1974) Physical Properties of Connective Tissue as Influenced by Single and Repeated Pregnancies in the Rat. *Acta. Physiol. Scand. Supplement.* **417,** pp. 1-138.

Rushfeldt, P.D., Mann, R.W., and Harris, W.H. (1981a) Improved Techniques for Measuring in vitro the Geometry and Pressure Distribution in the Human Acetabulum--i. uUltrasonic Measurement of Acetabular Surfaces, Sphericity and Cartilage Thickness. *J. Biomech.* **14,** pp. 253-260.

Rushfeldt, P.D., Mann, R.W., and Harris, W.H. (1981b) Improved Techniques for Measuring in vitro the Geometry and Pressure Distribution in the Human Acetabulum. II Instrumented Endoprosthesis Measurement of Articular Surface Pressure Distribution. *J. Biomech.* **14,** pp. 315-323.

Saamanen, A.M., Tammi, M., Kiviranta, I., Helminen, H., and Jurvelin, J. (1986) Moderate Running Increases but Strenuous Running Prevents Elevation of Proteoglycan Content in Canine Articular Cartilage. *Scandinavian Journal of Rheumatology.* **60** (Supplement), p. 45.

Sah, R.L., Yang, A.S., Chen, A.C., Hant, J.J., Halili, R.B., Yoshioka, M., Amiel, D., and Coutts, R.D. (1997) Physical Properties of Rabbit Articular Cartilage after Transection of the ACL. *J. Orthop. Res.* **15,** pp. 197-203.

Sammarco, G.J., Burstein, A.H., and Frankel, V.H. (1973) Biomechanics of the Ankle: A Kinematic Study. *Orthop. Clin. of North America.* **4**, pp. 75-95.

Scapinelli, R. and Little, K. (1970). Observations on the Mechanically Induced Differentiation of Cartilage from Fibrous Connective Tissue. *J. Pathol.* **101,** pp. 85-91.

Schachar, R., Herzog, W., and Leonard, T.R. (2004) The Effects of Muscle Stretching and Shortening on Isometric Forces on the Descending Limb of the Force-length Relationship. *Journal of Biomechanics.* **37(6),** pp. 917-926.

Schmidt, R.F. (1978) Motor Systems. *Fundamentals of Neurophysiology* (ed. Schmidt, R.F.). Springer Verlag, New York. pp. 158-161.

Seeley, R.R., Stephens, T.D., and Tate, P. (1989) Anatomy and Physiology. Times Mirror/Mosby College Publishing, Toronto. Mosby-Year Book, Inc., St. Louis.

Seireg, A. and Arvikar, R.J. (1973) A Mathematical Model for Evaluation of Force in Lower Extremities of the Musculo-skeletal System. *J. Biomech.* **6 (3),** pp. 313-326.

Schinagl, R.M., Gurskis, D., Chen, A.C., and Sah, R.L. (1997) Depth-Dependent Confined Compression Modulus of Full-Thickness Bovine Articular Cartilage. *J. Orthop. Res.* **15,** pp. 499-506.

Shantz, P., Henriksson, J, and Jansson E. (1983) Adaptation of Human Skeletal Muscle to Endurance Training of Long Duration. *Clinical Physiology.* **3**, pp. 141-151.

Shrive, N.G., O'Connor, J.J., and Goodfellow, J.W. (1978) Load-bearing in the Knee Joint. *Clin. Orthop.* **131**, pp. 279-287.

Shrive, N., Chimich, D., Marchuk, L., Wilson, J., Brant, R., and Frank, C. (1995) Soft-tissue "flaws" are associated with the material properties of the healing rabbit medial collateral ligament. *J. Orthop. Res.* **13**(6), pp. 923-929.

Siegler, S., Chen, J., and Schenck, C.D. (1988) The Three-dimensional Kinematics and Flexibility Characteristics of the Human Ankle and Subtalar Joints. Part I: Kinematics. *J. Biomech. Eng.* **110**, pp. 364-373.

Siemankowski, R.F., Wiseman, M.O, and White, H.D. (1985) ADP Dissociation from Actomyosin Subfragment 1 is Sufficiently Slow to Limit the Unloaded Shortening Velocity in Vertebrate Muscle. *Proc. Natl. Acad. Sci. USA.* **82**, pp. 658-662.

Simon, B.R., Liable, J.P., Pflaster, D., Yuan, Y., and Crag, M.H. (1996) A Poroelastic Finite Element Formulation Including Transport and Swelling in Soft Tissue Structures. *J. Biomech. Eng.* **118,** pp. 1-9.

Skelton, D.A., Greig, C.A., Davies, J.M., and Young, A. (1994) Power and Related Functional Ability of Healthy People Aged 65-89 Years. *Age Ageing.* **23,** pp. 371-377.

Slemenda, C., Brandt, K.D., Heilman, D.K., Mazzuca, S., Braunstein, E.M., Katz, B.P., and Wolinsky, F.D. (1997) Quadriceps Weakness and Psteoarthritis of the Knee. *Ann. Intern. Med.* **127,** pp. 97-104.

Slemenda, C., Heilman, D.K., Brandt, K.D., Katz, B.P., Mazzuca, S., Braunstein, E.M., and Byrd, D. (1998) Reduced Quadriceps Strength Relative to Body Weight. A Risk Factor for Knee Osteoarthritis in Women? *Arthritis Rheum.* **41,** pp. 1951-1959.

Smit, T.H., Burger, E.H., and Huyghe, J.M. (2002) A Case for Strain-induced Fluid Flow as a Regulator of BMU-coupling and Osteonal Alignment. *Journal of Bone and Mineral Research,* 17(11), pp.2021-2028.

Sojka, P., Johansson, H., Sjolander, P., Lorentzon, R., and Djupsjobacka, M. (1989) Fusimotor Neurones can be Reflexly Influenced by Activity in Receptors from the Posterior Cruciate Ligament. *Brain Res.* **483 (1),** pp. 177-183.

Sommer, C., Hintermann, B., Nigg, B.M., and van den Bogert, A.J. (1996) Influence of Ankle Ligaments on Tibia Rotation: An In-vitro Study. *Foot and Ankle.* **17 (2),** pp. 79-84.

Sommerfeldt, D.W., and Rubin, C.T. (2001) Biology of Bone and How it Orchestrates the Form and Function of the Skeleton. *European Spine Journal*, 10, pp. S86-S95.

Soulhat, J., Buschmann, M.D., and Shirazi-Adl, A. (1999) A Fibril-Network-Reinforced Biphasic Model of Cartilage in Unconfined Compression. *J. Biomech. Eng.* **121,** pp. 340-347.

Spector, S.A., Gardiner, P.F., Zernicke, R.F., Roy, R.R., and Edgerton, V.R. (1980) Muscle Architecture and Force-velocity Characteristics of Cat Soleus and Medial Gastrocnemius: Implications for Motor Control. *J. Neurophysiol.* **44,** pp. 951-960.

Spence, A.P. (1987) *Human Anatomy and Physiology* (3ed). Benjamin Cummings, Menlo Park, CA.

Stacoff, A. (1998) *The Effect of Shoe and Insert Interventions on Skeletal Movement of the Lower Extremities*. Unpublished Thesis, University of Calgary.

Stähelin, T., Nigg, B.M., Stefanyshyn, D.J., van den Bogert, A.J., and Kim, S.-J. (1997) A Method to Determine Bone Movement in the Ankle Joint Complex In-Vitro. *J. Biomech.* **30 (5),** pp. 513-516.

Steindler, A. (1977) *Kinesiology of the Human Body Under Normal and Pathological Conditions*. Charles C. Thomas, Springfield, IL.

Stockwell, R.S. (1979) *Biology of Cartilage Cells*. Cambridge University Press, Cambridge.

Storey, E. and Feik, S.A. (1982) Remodelling of Bone and Bones: Effects of Altered Mechanical Stress on Analages. *British Journal of Experimental Pathology.* **6,** pp. 184-193.

Sugi, H. (1972) Tension Changes during and after Stretch in Frog Muscle Fibers. *Journal of Physiology.* **225**, pp. 237-253.

Sugi, H. and Tsuchiya, T. (1988) Stiffness Changes During Enchancement and Deficit of Isometric Force by Slow Length Changes in Frog Skeletal Muscle Fibres. *J. Physiol.* **407,** pp. 215-229.

Suh, J.-K. and Bai, S. (1998) Finite Element Formulation of Biphasic Poroviscoelastic Model for Articular Cartilage. *J. Biomech. Eng.* **120,** pp. 195-201.

Suter, E., Herzog, W., Leonard, T.R., and Nguyen, H. (1998) One Year Changes in Hindlimb Kinematics, Ground Reaction Forces, and Knee Stability in an Experimental Model of Osteoarthritis. *J. Biomech.* **31**, pp. 511-517.

Syed, Z., and Khan, A. (2002) Bone Censitometry: Applications and Limtations. Journal of Obstetrics and Gynaecology Canada, 24(6), pp.476-84.

Swammerdam, J. (1758) *The Book of Nature. The History of Insects* (trans. Flloyd, T.). Seyffert, London.

Szent-Györgyi, A.G. (1941) Contraction of Myosin Threads. *Stud. Inst. Med. Chem. Univ. Szeged.* **1**, pp. 17-26.

Taber, C.W. (ed.) (1981) Taber's Cyclopedic Medical Dictionary. Davis, Philadelphia.

ter Keurs, H.E.D.J., Iwazumi, T., and Pollack, G.H. (1978) The Sarcomere Length-tension Relation in Skeletal Muscle. *J. Gen. Physiol.* **72,** pp. 565-592.

Thomas, J.T., Ayad, S., and Grant, M.E. (1994) Cartilage Collagens: Strategies for the Study of Their Organisation and Expression in the Extracellular Matrix. *Ann. Rheum. Dis.* **53,** pp. 488-496.

Thornton, G.M., Oliynyk, A., Frank, C.B., and Shrive, N.G. (1997) Ligament creep cannot be predicted from stress relaxation at low stress: a biomechanical study of the rabbit medial collateral ligament. *J. Orthop. Res.* **15**(5), pp. 652-656.

Thornton, G.M., Leask, G.P., Shrive, N.G., and Frank, C.B. (2000) Early medial collateral ligament scars have inferior creep behaviour. *J. Orthop. Res.* **18**(2), pp. 238-246.

Thornton, G.M., Frank, C.B., and Shrive, N.G. (2001a) Ligament creep behavior can be predicted from stress relation by incorporating fiber recruitment. *J. Rheol.* **45**(2), pp. 493-507.

Thornton, G.M., Shrive, N.G., and Frank, C.B. (2001b) Altering ligament water content affects ligament prestress and creep behaviour. *J. Orthop. Res.* **19**(5), pp. 845-851.

Thornton, G.M., Shrive, N.G., and Frank, C.B. (2002) Ligament creep recruits fibres at low stresses and can lead to modulus-reducing fibre damage at higher creep stresses: a study in rabbit medial collateral ligament model. *J. Orthop. Res.* **20**(5), pp. 967-974.

Thornton, G.M., Shrive, N.G., and Frank, C.B. (2003) Healing ligaments have decreased cyclic modulus compared to normal ligaments and immobilization further compromises healing ligament response to cyclic loading. *J. Orthop. Res.* **21**(4), pp. 716-722.

Thorstensson, A., Grimby, G., and Karlsson, J. (1976) Force-velocity Relations and Fibre Composition in Human Knee Extensor Muscles. *J. Appl. Physiol.* **40 (1)**, pp. 12-16.

Tidball, J.G. (1983) The Geometry of Actin Filament-membrane Associations can Modify Adhesive Strength of the Myotendinous Junction. *Cell Motility.* **3,** pp. 439-447.

Tidball, J.G. (1984) Myotendinous Junction: Morphological Changes and Mechanical Failure Associated with Muscle Cell Atrophy. *Exp. and Mole. Pathol.* **40,** pp. 1-12.

Tidball, J.G. (1991) Myotendinous Junction Injury in Relation to Junction Structure and Molecular Composition. *Exerc. Sport Sci. Rev.* **19,** pp. 419-445.

Tipton, C.M., Schild, R.J., and Tomanek, R.J. (1967) Influence of Physical Activity on the Strength of Knee Ligaments in Rats. *Am. J. Physiol.* **212,** pp. 783-787.

Tipton, C.M., James, S.L., Mergner, W., and Tcheng, T.K. (1970) Influence of Exercise on Strength of the Medial Knee Ligaments of Dogs. *Am. J. Physiol.* **218,** pp. 894-902.

Tipton, C.M., Matthes, R.D., and Sandage, D.S. (1974) In-situ Measurement of Junction Strength and Ligament Elongation in Rats. *J. Appl. Physiol.* **37 (5)**, pp. 758-761.

Torzilli, P.A. (1985) Influence of Cartilage Conformation on its Equilibrium Water Partition. *J. Orthop. Res.* **3,** p. 473.

Torzilli, P.A., Rose, D.E., and Dethmers, D.A. (1982) Equilibrium Water Partition in Articular Cartilage. *Biorheology.* **19,** pp. 519-537.

Trotter, J.A. (1993) Functional Morphology of Force Transmission in Skeletal Muscle. *Acta. Anat.* **146,** pp. 205-222.

Trotter, J.A., Hsi, K., and Samora, A. (1985) A Morphometric Analysis of the Muscle-tendon Junction. *Anat. Rec.* **213,** pp. 26-32.

Turner, C.H., Owan, I., and Takano, Y. (1995) Mechanotransduction in Bone: Role of Strain Rate. *Amer. J. Physiol.* **269,** pp. E438-442.

Vailas, A.C., Pedrini, V.A., Pedrini-Mille, A., and Holloszy, J.O. (1985) Patellar Tendon Matrix Changes Associated with Aging and Voluntary Exercise. *J. Appl. Physiol.* **58,** pp. 1572-1576.

Vallbo, A.B., Hagbarth, K.E., Torebjork, H.E., and Wallin, B.G. (1979) Somatosensory, Proprioceptive, and Sympathetic Activity in Human Peripheral Nerves. *Physiol. Rev.* **59,** pp. 919-957.

van B. Cochran, G. (1982) Biomechanics of Orthopaedic Materials. *A Primer of Orthopaedic Biomechanics* (ed. van B. Cochran, G.), Churchill Livingstone, New York. pp. 71-142.

van den Bogert, A.J., Smith, G.D., and Nigg, B.M. (1994) In-vivo Determination of the Anatomical Axes of the Ankle Joint Complex: An Optimization Approach. *J. Biomech.* **27 (12)**, pp. 1477-1488.

van der Voet, A.F. (1992) *Finite Element Modelling of Load Transfer Through Articular Cartilage*. Ph.D. Thesis, University of Calgary. Nat'l Library of Canada, Ottawa.

van der Voet, A.F., Shrive, N.G., and Schachar, N.S. (1992) Numerical Modelling of Articular Cartilage in Synovial Joints: Poroelasticity and Boundary Conditions. *Proceedings of the International Conference on Computer Methods in Biomechanics and Biomedical Engineering* (eds. Middleton, J. and Pande, G.N.). Books and Journals International, Swansea. pp. 200-209.

van Ingen Schenau, G.J. (1984) An Alternative View of the Concept of Utilisation of Elastic Energy in Human Movement. *Human Movement Science.* **3**, pp. 301-336.

van Ingen Schenau, G.J., Bobbert, M.F. and de Haan, A. (1997) Mechanics and Energetics of the Stretch-shortening Cycle: A Stimulating Discussion. *J. Applied Biomech.* **13**, pp. 484-496.

van Langelaan, E.J. (1983) A Kinematic Analysis of the Tarsal Joints. *Acta. Orthop. Scand.* **54** (Supplement) **204,** pp. 22-56.

van Leeuwenhoeck, A. (1678) Microscopical Observations on the Structure of Teeth and Other Bones. *Phil. Trans. Royal Soc.* London. **12,** pp. 1002-1003.

van Rietbergen, B. and Huiskes, R. (2001) *Elastic Constants of Cancellous Bone*. In: Bone Mechanics Handbook, 2/e, Chapter 15, pp. 1-24. Ed by SC Cowin. Boca Raton, CRC Press.

van Rietbergen, B., Odgaard, A., Kabel, J., and Huiskes, R. (1996) Direct Mechanics Assessment of Elastic Symmetries and Properties of Trabecular Bone Architecture. *J. Biomech.* **29**, pp.1653-1657.

van Rietbergen, B., Weinans, H., Huiskes, R., and Odgaard, A. (1995) A New Method to Determine Trabecular Bone Elastic Properties and Loading using Micromechanical Finite-Element Models. *J. Biomech.* **28**, pp. 69-81.

Vesalius, A. (1543) *De Humani Corporis Fabrica.* Hoffman La-Roche, Basel.

Vasan, N. (1983) Effects of Physical Stress on the Synthesis and Degradation of Cartilage Matrix. *Conn. Tissue Res.* **12**, pp. 49-58.

Viidik, A. (1967) The Effect of Training on the Tensile Strength of Isolated Rabbit Tendons. *Scand. J. Plast. Reconstr. Surg.* **1,** pp. 141-147.

Viidik, A. (1979) Connective Tissues: Possible Implications of the Temporal Changes for the Aging Processes. *Mech. Aging Dev.* **9,** pp. 267-285.

Viidik, A. (1990) Structure and Function of Normal and Healing Tendons and Ligaments. *Biomechanics of Diathrodial Joints* (eds. Mow, V. C., Ratcliffe, A., and Woo, S.L.-Y.). Springer Verlag, New York. pp. 3-38.

Voigt, M., Bojsen-Moller, F., Simonsen, E.B., and Dyhre-Poulsen, P. (1995) The Influence of Tendon Young's Modulus, Dimensions and Instantaneous Moment Arms on the Efficiency of Human Movement. *J. Biomech.* **28 (3),** pp. 281-291.

von Meyer, G.H. (1867) Die Architektur der Spongiosa. *Arch. Anat. Physio.l Wiss Med. Reichert DuBois-Reymonds Arch.* **34**, pp. 615-628.

Wainwright, S.A., Biggs, W.D., Currey, J.D., and Gosline, J.M. (1982) *Mechanical Design in Organisms.* Princeton University Press, Princeton, NJ. pp. 88-93.

Wakabayashi, K., Sugimoto, Y., Tanaka, H., Ueno, Y., Takezawa, Y., and Amemiya, Y. (1994) X-ray Diffraction Evidence for the Extensibility of Actin and Myosin Filaments during Muscle Contraction. *Biophysical Journal.* **67**, pp. 2422-2435.

Walker, J.M. (1996) Cartilage of Human Joint and Related Structures. *Athletic Injuries and Rehabilitation* (eds. Zachazewski, J.E., Magee, D.J., and Quillen, W.S.) Saunders, Philadelphia. pp. 120-151.

Walker, S.M. and Schrodt, G.R. (1973) I-segment Lengths and Thin Filament Periods in Skeletal Muscle Fibres of the Rhesus Monkey and the Human. *Anat. Rec.* **178,** pp. 63-82.

Walpole, L.J. (1981) Elastic Behaviour of Composite Materials: Theoretical Foundations. *Advances in Applied Mechanics.* **21,** pp. 169-242.

Walmsley, B., Hodgson, J.A., and Burke, R.E. (1978) Forces Produced by Medial Gastrocnemius and Soleus Muscles During Locomotion in Freely Moving Cats. *J. Neurophysiol.* **41 (5),** pp. 1203-1216.

Walsh, S. (1989) *Immobilization Affects Growing Ligaments*. M.Sc. Thesis, University of Calgary. Nat'l Library of Canada, Ottawa.

Walsh, S., Frank, C.B., and Hart, D.A. (1992) Immobilization Alters Cell Metabolism in an Immature Ligament. *Clin. Orthop.* **277,** pp. 277-288.

Wang, C.C., Chahine, N.O., Hung, C.T., and Ateshian, G.A. (2003) Optical Determination of Anisotropic Material Properties of Bovine Articular Cartilage in Compression. *J. Biomech.* **36,** pp.339-353.

Wang, J.H. (2005) Mechnobiology of Tendon. *Journal of Biomechanics* (Epub ahead of print).

Wang, X.T. and Ker, R.F. (1995) Creep Rupture of Wallaby Tail Tendons. *J. Exp. Biol.* **198,** pp. 831-845.

Wang, X.T., Ker, R.F., and Alexander, R.M. (1995) Fatigue Rupture of Wallaby Tail Tendons. *J. Exp. Biol.* **198,** pp. 847-852.

Weineck, J. (1986) *Functional Anatomy in Sports*. Year Book Medical, Chicago, IL.

Weinreb, M., Rodan, G.A., and Thompson, D.A. (1991) Depression of Osteoblastic Activity in Immobilized Limbs of Suckling Rats. *J. Bone and Mineral Res*. **6,** pp. 725-731.

Wertheim, M.G. (1847) Mémoire sur l'élasticité et la Cohésion des Principaux Tissues du Corps Humain. *Ann. Chim. (Phys)*. **21,** pp. 385-414.

Whalen, R.T., Carter, D.R., and Steele, C.R. (1987) The Relationship Between Physical Activity and Bone Density. *Trans. Orthopaedic Research Society*. Adept Printing, Chicago, IL. **12,** p. 463.

White, T.D. (1991) *Human Osteology*. New York, Academic Press.

Whiting, W.C., Gregor, R.J., Roy, R.R., and Edgerton, V.R. (1984) A Technique for Estimating Mechanical Work of Individual Muscles in the Cat during Treadmill Locomotion. *Journal of Biomechanics*. **17**, pp. 685-694.

Willet, A.M. (2005) Vitamin D Status and its Relationship with Parathyroid Hormone and Bone Mineral Status in Older Adolescents. *Proceedings from the Nutrition Society*. **64**, pp. 193-203.

Woittiez, R.D., Huijing, P.A., and Rozendal, R.H. (1984) Twitch Characteristics in Relation to Muscle Architecture and Actual Muscle Length. *European Journal of Physiology*. **401**, pp. 374-379.

Wolff, J. (1892) *Das Gesetz der Transformation der Knochen*. Berlin, A Hirschwald.

Wolff, J. (1986) *The Law of Bone Remodelling*. Berlin, Springer-Verlag.

Wong, M., Wuethrich, P., Eggli, P., and Hunziker, E. (1996) Zone-Specific Cell Biosynthetic Activity in Mature Bovine Articular Cartilage: A New Method using Confocal Microscopic Stereology and Quantitative Autoradiography. *J. Orthop. Res*. **14(3),** pp. 424-432.

World Health Organization (WHO). (1994) *Assessment of Fracture Risk and Its Application to Screening for Postmenopausal Osteoporosis*. World Health Organ. Tech. Rep. Ser. Geneva, World Health Organization.

Whiting, W.C., Gregor, R.J., and Edgerton, V.R. (1984) A Technique for Estimating Mechanical Work of Individual Muscles in the Cat During Treadmill Locomotion. *J. Biomech*. **17 (9),** pp. 685-694.

Whiting, W.C. and Zernicke, R.F. (1998) *Biomechanics of Musculo-skeletal Injury*. Human Kinetics, Champaign, IL.

Wickoff, R.G.W. (1949) *Electron Microscopy: Techniques and Applications*. Interscience Publications, New York.

Williams, P.F., Powell, G.L., and Laberge, M. (1993) Sliding Friction Analysis of Phosphatidylcholine as a Boundary Lubricant for Articular Cartilage. *Proc. Inst. Mech. Eng., J. Eng. Med.* **207 (H1),** pp. 59-66.

Wilson, D.R., Feikes, J.D., Zavatsky, A.B., Bayona, F., and O'Connor, J.J. (1996) The One Degree-of-freedom Nature of the Human Knee Joint - Basis for a Kinematic Model. *Proceedings of the Ninth Biennial Conference*. Canadian Society for Biomechanics, Vancouver.

Woittiez, R.D., Huijing, P.A., Boom, H.B.K., and Rozendal, R.H. (1984) A Three-dimensional Muscle Model: A Quantified Relation Between Form and Function of Skeletal Muscles. *J. Morphol.* **182,** pp. 95-113.

Woledge, R.C., Curtin, N.A., and Homsher, E. (1985) *Energetic Aspects of Muscle Contraction*. Academic Press, London.

Wolff, J. (1986) *The Law of Bone Remodelling* (trans. Maquet, P. and Furlong, R.). Springer Verlag, Berlin. (Original 1870).

Woo, S.L.-Y., Ritter, M.A., Amiel, D, .Sanders, T.M., Gomez, M.A., Kuei, S.C., Garfin, S.R., and Akeson, W.H. (1980) The Biomechanical and Biochemical Properties of Swine Tendons: Long Term Effects of Exercise on the Digital Extensors. *Connect. Tissue Res*. **7,** pp. 177-183.

Woo, S.L.-Y, Gomez, M.A., Amiel, D., Ritter, M.A., Gelberman, R.H., and Akeson, W.H. (1981) The Effects of Exercise on the Biomechanical and Biochemical Properties of Swine Digital Flexor Tendons. *J. Biomech. Eng.* **103 (2),** pp. 51-56.

Woo, S.L.-Y., Gomez, M.A., Woo, Y.K., and Akeson, W.H. (1982) Mechanical Properties of Tendons and Ligaments: II. The Relationships of Immobilization and Exercise on Tissue Remodelling. *Biorheology.* **19,** pp. 397-408.

Woo, S.L.-Y., Gomez, M.A., Woo, Y.K., and Akeson, W.H. (1982a) Mechanical Properties of Tendons and Ligaments II: The Relationships of Immobilization and Exercise on Tissue Remodelling. *Biorheology.* **19,** pp. 397-408.

Woo, S.L.-Y., Gomez, M.A., Woo, Y.K., and Akeson, W.H. (1982b) Mechanical Properties of Tendons and Ligaments. I. Quasi-static and Non-linear Viscoelastic Properties. *Biorheology.* **19,** pp. 385-396.

Woo, S.L.-Y., Gomez, M.A., Seguchi, Y., Endo, C.M., and Akeson, W.H. (1983) Measurement of Mechanical Properties of Ligament Substance from a Bone-ligament-bone Preparation. *J. Orthop. Res.* **1 (1),** pp. 22-29.

Woo, S.L.-Y. and Buckwalter, J.A. (1988) *Injury and Repair of the Musculo-skeletal Soft Tissues.* American Association of Orthopedic Surgeons, Park Ridge, IL.

Woo, S.L.-Y., Maynard, J., Butler, D.L., Lyon, R., Torzilli, P.A., Akeson, W.H., Cooper, R. R., and Oakes, B. (1988) Ligament, Tendon, and Joint Capsule Insertions into Bone. *Injury and Repair of the Musculo-skeletal Soft Tissues* (eds. Woo, S.L.-Y. and Buckwalter, J.A.). American Academy of Orthopaedic Surgeons, Park Ridge, IL. pp. 133-166.

Woo, S.L.-Y. and Adams, D.J. (1990) The Tensile Properties of Human Anterior Cruciate Ligament (ACL) and ACL Graft Tissues. *Knee Ligaments: Structure, Function, Injury, and Repair* (eds. Daniel, D.M., Akeson, W.H., and O'Connor, J.J.). Raven Press, New York. pp. 279-289.

Wood, T.O., Cooke, P.H., and Goodship, A.E. (1988) The Effect of Exercise and Anabolic Steroids on the Mechanical Properties and Crimp Morphology of the Rat Tendon. *Am. J. Sports Med.* **16,** pp. 153-158.

Wright, D.G., Desai, S.M., and Henderson, W.H. (1964) Action of the Subtalar and Ankle Joint Complex During the Stance Phase of Walking. *J. Bone Jt. Surg.* **46 (A),** pp. 361-375.

Wuerker, R.B., McPhedran, A.M., and Henneman, E. (1965) Properties of Motor Units in a Heterogeneous Pale Muscle (M. Gastrocnemius) of the Cat. *J. Neurophysiol.* **28,** pp. 85-99.

Yahia, L.H. and Drouin, G. (1989) Microscopical Investigation of Canine Anterior Cruciate Ligament and Patellar Tendon: Collagen Fascicle Morphology and Architecture. *J. Orthop. Res.* **7 (2)**, pp. 243-251.

Yamada, H. (1970) *Strength of Biological Materials* (eds. Evans, F. and Gaynor, F.). Williams & Wilkins, Baltimore, Maryland. p. 49.

Yoshihuku, Y. and Herzog, W. (1990) Optimal Design Parameters of the Bicycle-rider System for Maximal Muscle Power Output. *J. Biomech.* **23 (10),** pp. 1069-1079.

Zahalak, G.I. (1997) Can Muscle Fibers be Stable on the Descending Limbs of their Sarcomere Length-tension Relations? *Journal of Biomechanics.* **30**, pp. 1179-1182.

Zahalak, G.I. and Ma, S.P. (1990) Muscle Activation and Contraction: Constitutive Relations based Directly on Cross-bridge Kinetics. *Journal of Biomechanical Engineering.* **112**, pp. 52-62.

Zahalak, G.I. and Motabarzadeh, I. (1997) A re-examination of Calcium Activation in the Huxley Cross-bridge Model. *Journal of Biomechanical Engineering.* **119**, pp. 20-28.

Zajac, F.E. (1989) Muscle and Tendon: Properties, Models, Scaling, and Application to Biomechanics and Motor Control. *Crit. Rev. Biomed. Eng.* **17 (4)**, pp. 359-411.

Zamora, A.J. and Marini, J.F. (1988) Tendon and Myo-tendinous Junction in an Overloaded Skeletal Muscle of the Rat. *Anatomy and Embryology.* **179**, pp. 89-96.

Zernicke, R.F., Vailas, A.C., and Salem, G.J. (1990) Biomechanical Response of Bone to Weightlessness. *Exercise and Sport Sci. Rev.* **18**, pp. 167-192.

Zuurbier, C.J., Everard, A.J., van der Wees, P., and Huijing, P.A. (1994) Length-force characteristics of the aponeurosis in the passive and active muscle condition and in the isolated condition. *J.Biomech.* **27(4)**, pp. 445-453.

3 MEASURING TECHNIQUES

Many biomechanical research activities concentrate on experimental work that needs measuring techniques. Several measuring methods were established and further developed at the end of the nineteenth and during the twentieth century. The methodologies developed and described by Marey at the end of the nineteenth century are still a milestone in biomechanical measuring techniques. Their principles still apply for many of today's methodologies.

Today, many measuring techniques are highly developed and allow the quantification of kinetic, kinematic, and other aspects of human or animal movement. This chapter describes some of the most important methods for movement analysis used in biomechanical research.

3.1 DEFINITIONS AND COMMENTS

Acceleration: The time derivative of the velocity (vector) or the second time derivative of the position (vector).

$$\mathbf{a} = \frac{d\mathbf{v}}{dt} = \frac{d^2\mathbf{x}}{dt^2}$$

Accelerometer: A device used to measure acceleration.

Adaptive filtering system: A system whose structure (or weights of the filters) is automatically adjustable in accordance with some algorithm to produce an estimate of the desired response.

Attitude matrix: A matrix that describes the orientation of a rigid body in three dimensions.

Bi-, multi-polar electrodes: Two or more electrodes measuring EMG signals relative to a common ground electrode. The difference between the signals received by each of the electrodes, relative to the common ground electrode, is amplified and gives the EMG records.

Calibration frame: A rigid frame with markers of known positions distributed throughout its volume.

Comment: the position of markers on the calibration frame has typically been measured with high accuracy. Ideally, markers are evenly distributed throughout the volume of calibration.

Calibration rod: A rigid body with markers at each end for which the distance is known.

Comment: the distance of the two markers on the rigid rod has typically been measured with high accuracy. The positions of the rod are determined while moving the rod throughout the calibration space.

Cardan angles: A sequence of rotation angles about the i, j, and k axes, respectively.

Centre of force: Point in a plane of interest through which the line of action of the resultant force vector passes.
Comment: the plane of interest is, for instance, the force plate. Centre of force is also called *centre of pressure.*

Centre of gravity: The centre of gravity of an object is the centroid of all the gravitational forces, G, acting on its mass elements, dm, with:

$$\mathbf{r}_{CG} = \frac{\int \mathbf{r}\, dG}{\int dG}$$

Centre of mass: The centre of mass of an object is the centroid of all mass elements, dm, with the position vector:

$$\mathbf{r}_{CM} = \frac{\int \mathbf{r}\, dm}{\int dm}$$

Centroid: The centroid of an assemblage of n similar quantities, $\Delta_1, \Delta_2, ..., \Delta_n$, situated at points $P_1, P_2, ..., P_n$, and having the position vectors $\mathbf{r}_1, \mathbf{r}_2, ..., \mathbf{r}_n$, has a position vector $\mathbf{r}$.

$$\mathbf{r} = \frac{\sum_{i=1}^{n} \mathbf{r}_i \Delta_i}{\sum_{i=1}^{n} \Delta_i}$$

Compound motor unit action potential: Also referred to as CMUAP. The algebraic sum of MUAP, obtained when several motor units are stimulated during a contraction.

Critical damping: The amount of damping required to return a displaced elastic system to its initial position in minimal time without oscillation or overshoot.

Crosstalk: A signal at the output of a transducer caused by a variable not allocated to this particular output.
Comment: the common crosstalk in force sensors occurs in all channels. It is, for instance, possible for the vertical force channel to show a signal when only horizontal forces are acting on a sensor.

DLT: Direct linear transformation (DLT) is a mathematical procedure used to determine three-dimensional coordinates of markers from two sets of two-dimensional images of these markers.

Damping: The dissipation of energy with distance or time.

Damping factor: The ratio of actual damping to critical damping.

Digitizing: Determination of the location of markers on the body of interest in a frame of film, video or any other image.

Drift: Change in output of the transducer over time that is not a function of the measurand.

Electromyogram(EMG): Recording of the electrical potential, at some distance from the muscle fibres, which was generated by a superposition of motor unit action potentials.

Electromechanical delay (EMD): EMD is typically defined as the time interval between the onset of the EMG signal and the onset of the corresponding muscular force.

Euler angles: A sequence of rotation angles about the i, j, and i axes, respectively.

f/stop setting: Quantification of lens opening with:

$$f/stop = \frac{f_L}{\text{diameter of lens opening}}$$

Field of focus: Also referred to as *depth of field*. The range of distance within which objects in a picture appear in focus (sharp).

Focal length: Focal length is defined by the lens equation:

$$\frac{1}{f_L} = \frac{1}{a} + \frac{1}{b}$$

where:

f_L = focal length
a = distance from object to lens
b = distance from lens to film (image)

Force: Force cannot be defined. However, effects produced by forces can be described.

- Active force: Active forces in human locomotion are forces generated by movement that is entirely controlled by muscular activity.

- Dynamic force: If something is able to accelerate a mass, this something is called a force in a dynamic sense.

- Impact force: Impact forces in human locomotion are forces that result from a collision of two objects and that reach their maximum earlier than 50 milliseconds after the first contact of the two objects.
  Comment: the most often used application of impact forces is the landing of the heel of the foot on the ground.

- Normal force: A force perpendicular to the surface of interest.

- Shear force: A force parallel to the surface of interest.

- Static force: If something is able to keep a spring deformed, this something is called a force in a static sense.

Frequency: Number of oscillations per time unit.

- Natural frequency: A frequency of vibration that corresponds to the elasticity and mass of a system under the influence of its internal forces and damping.
  Comment: the lowest natural frequency of a system is known as the fundamental frequency. Higher natural frequencies are harmonics of the fundamental value. Some systems exhibit several coupled or uncoupled modes of vibration.

- Resonance frequency: The frequency of a forced vibration input to a system that corresponds to a natural frequency of the system.

Frequency space or domain: A signal of duration T can be viewed as a vector. If sine and cosine waves used in the Fourier transform represent the base of the vector space containing the signal (signal space), then the space is called frequency space. In contrast, if the base vectors are vectors of individual time points the same space is called time space. The length of the vector represents the square root of the energy of the signal.

Frustrum: The solid figure formed when the top of a cone or pyramid is cut off by a plane parallel to the base.

Gauge factor: Quantity with which an initial signal output is multiplied to receive the measurand of interest.

Hysteresis: The biggest difference in the output signal for any value of the measured variable within the specified range when this value is reached with increasing or diminishing measurand.

Indwelling electrodes: Electrodes placed inside the muscle to measure EMG signals.

Innervation zone: The volume containing the neuromuscular junctions of all nerves activating the muscle.

Integration: Mathematical integration of previously rectified EMG signals over specified periods of time, or until the integrated EMG value reaches a preset limit.

Isometric contraction: A contraction of the muscle without changing its length.

Joint Coordinate System: One approach to quantify relative segmental motion with the first axis embedded in the proximal segment, the second axis embedded in the distal segment, and the third axis, which is orthogonal to the first and the second axis, floating.

Kinematics: The study of the geometry of motion.

LMS algorithm: Least mean square (LMS) gradient search adaptive algorithm. The LMS algorithm is widely used for implementing the adaptive filtering process because of its simplicity and robustness.

Linearity: Amplitude of output signal in direct proportion to amplitude of input signal.
Comment: linearity is not necessary to obtain accurate measurements.

Measurand: Quantity to be measured.

Membrane capacity: Amount of charge, Q, stored by the membrane per volt of membrane potential. This amount of charge has to be moved across the membrane to build up or collapse the membrane potential.

Membrane potential: Potential measured across the membrane. In the present case, this is across the membrane of the muscle fibre.

Microstrain: Unit that is typically used for strain measurements.

$$1 \text{ microstrain} = 10^{-6} = 1\mu\varepsilon$$

Comment: $10^4\mu\varepsilon = 1\%$

Moment: Turning effect of a force, **F**, applied at point, P, about a base point, O, with:

$$\mathbf{M}_O = \mathbf{r}_{PO} \times \mathbf{F}$$

where:

$\mathbf{M}_O$ = moment (or moment of force) about point O

$\mathbf{r}_{PO}$ = position vector from O to P

$\mathbf{F}$ = force

Moment of inertia: The moment of inertia, I, of a mass, m, is the sum of the axial moments of all its elements, dm.

$$I = \int r^2 dm$$

Monopolar electrodes: A single electrode measuring EMG signals relative to a ground electrode.

Motor unit: A motor unit (MU) consists of a group of muscle fibres that are stimulated by one nerve with multiple neuromuscular junctions.

Motor unit action potential: Also referred to as MUAP. Electrical potential measured over time or space at some distance from the motor unit, with reference to the electrical potential at another place outside the muscle fibres. Usually, the reference electrode is at ground potential. In a bipolar recording setup, the reference electrode is an equivalent electrode placed close to the first one.

Muscle fibre action potential: Change of electrical potential measured inside a muscle fibre, with reference to a ground electrode placed at some distance from the outside of the muscle fibre.

Neuromuscular junctions: Contact point between the nerve and the muscle. These are the points where the muscle gets activated.

Noise: Any unwanted disturbance of the original signal in a frequency band of interest.
Comment: a researcher distinguishes between *background noise* and *random noise*.

Normal force: Force perpendicular to the surface.

Overload: The highest value of a measured variable that the transducer can sustain without its measuring properties being changed beyond the specified tolerances.

Phase angle: If two signals are of identical shape, but time shifted with respect to one another, they are called out of phase by Δt. The phase angle measures this phase shift if the signal is sinusoidal or has a wavelet base. The notation angle stems from the fact that in Fourier transform a sinusoidal wave is represented by an amplitude, a circular frequency and a phase, $\psi(a \cdot \exp(i\omega t + \psi))$. In this notation, the phase is represented by an angle in the complex plain. The phase angle is always between $-\pi$ and $+\pi$. For two signals with equal frequency, but different phases, ψ_1 and ψ_2, the difference in phase angle is $\psi_2 - \psi_1$ and the time shift is $\Delta t = (\psi_2 - \psi_1)/\omega$.

Potential: A scalar quantity representing the possibility to gain energy or perform work.

- Electrical potential: Indicates the energy that has to be used to move an electric charge or charged chemical from a point with zero electrical potential to a point with some electrical potential. Electrical potentials are measured in volts. Energy is measured in electron volts.

- Chemical potential: Indicates the energy used to bring a molecule or atom (chemicals) to an area of a different concentration. The chemical potential is the driving force for balancing the concentration of chemicals.
- Electrochemical potential: A chemical at some point can simultaneously have an electrical and a chemical potential. Thus, the total potential is of electrical and chemical nature.
- Power: In a mechanical sense, the energy used or gained per unit time. In signal analysis, the energy, e.g. of an EMG, is the integral over the signal squared, and the power is this energy of the signal per unit time. If the signal is considered a voltage source (no drop in voltage if a current is drawn) then power represents a measure for the possibility to draw energy from that source. The energy of a signal of length T is equal to the integral over the power spectral density obtained by a Fourier transform.

Power density spectrum: Mathematical conversion of EMG signals from the time to the frequency domain for analysis of the frequency content of the signal.

Pressure: Force perpendicular to the surface of a sensor per unit area of this sensor where:

p = pressure
F = force perpendicular to the surface of a sensor
A = area of this sensor

with the unit:

$$[p] = \text{Pa} = \text{N} \cdot \text{m}^{-2}$$

Pressure is a scalar quantity.

Range: The upper and lower limits of a variable for which a transducer is intended.

Raw EMG: The unprocessed (typically amplified) EMG signal displayed in the time domain.

Rectification: Elimination of all negative values from the raw EMG signal (half-wave rectification) or multiplying all negative values from the raw EMG signal by -1 (full-wave rectification).

Resting potential: Electrical potential across the muscle fibre membrane at rest (about -90 mV inside the muscle fibre).

Rigidity: Measured deflection as a result of a defined acting force.

Rotation matrix: A 3x3 matrix used to convert three-dimensional coordinates from one to another reference frame.
Comment: this matrix can also be called *transformation* matrix.

Sampling frequency: Number of measurements of the same variable made during a one second time interval, or the ratio of the number of measurements of a variable to the time interval in which the measurements were taken.

Scalogram: A representation of the intensity extracted by the wavelet transform.

Sensitivity: The ratio between the change in the transducer output signal and the corresponding change in the measured value.
Comment: sensitivity can change over the possible range of measurements.

Shock rating: Highest acceleration for which the transducer is not damaged.

Shutter: Mechanical device that allows light to pass through the lens onto the film for a specified period of time.

Shutter factor: Fraction of the total circle of a circular shutter that is available to let the light pass through the lens to the film.

Smoothing: Mathematical procedures aimed at reducing the high-frequency content of signals.

Strain:

- Engineering strain:

$$\varepsilon = \frac{\Delta L}{L_o} = \frac{L - L_o}{L_o}$$

where:

ε = engineering strain
ΔL = change in length
L_o = original length
L = current length

- True strain:

$$\varepsilon_{true} = \left(\int_{L_o}^{L} \frac{dl}{l} \right) = \ln\left(\frac{L}{L_o}\right)$$

Comment: strain is non-dimensional. Units of strain are typically % or microstrain ($\mu\varepsilon$).

Surface electrodes: Electrodes placed outside (on the surface) of the muscle to measure EMG signals.

Three-dimensional reconstruction: A method by which a minimum of two two-dimensional images are used to determine each marker's three-dimensional position (coordinates).

Threshold: The smallest change in the measured variable that causes a measurable alteration of the transducer output signal.

Time/frequency analysis: A signal that fluctuates over time can be represented as a linear superposition of sine and cosine waves of various frequencies. The weights used for this linear superposition are obtained by a Fourier transform. The properties of the signal can thus be obtained by an analysis of the temporal aspects and of the frequency related aspects.

Time space or domain: See frequency space. A space is a means of representing a signal of length T, whereby, a vector points to the position in that space which represents the signal. The components of the vector represent the amplitudes at the individual time points. The vectors with only component being 1 form the base vectors of time space. The length of the vector of the signal represents the square root of the energy of the signal.

Torque: Result of a force couple. Ideally, its use should be confined to situations where it is producing rotation or torsion. Otherwise, the clear and useful distinction between torque producing twisting and moment producing bending is lost.

Wavelet: Short wave of finite width. Its integral over time is 0. Its power spectrum is band limited.

Wavelet spectrum: A representation of the intensity extracted by the wavelet transform plotted against the scale or center frequencies of the wavelets.

Wavelet transform: A set of scaled and shifted wavelets form the base vectors of a space containing the signal (signal space) (see time space and frequency space). The wavelet transform represents the projection onto these base vectors.

The force plates described in Chapter 4 measure a resultant ground reaction force. This resultant ground reaction force is a summation of all the forces acting from the ground on the test subject's feet.

Assume a test subject standing (bipedal stance) on a force plate with the inside of the feet 20 cm apart. For this example, the force plate indicates a resultant force vector with the magnitude of body weight, acting in a vertical direction, with the line of action of the force vector passing through a point on the surface of the force plate between the two feet (Figure 3.1.1). This point on the force plate is called point of application, centre of pressure,

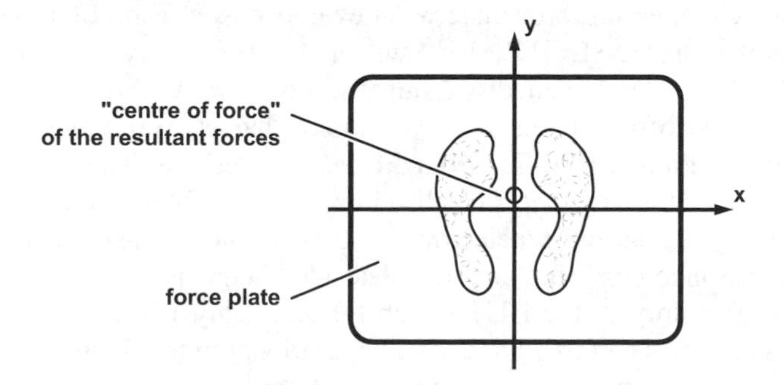

Figure 3.1.1 **Illustration of the possible location of the centre of force in bipedal standing.**

or centre of force. The resultant force is obviously not a real force that can be felt by a specific part of the test subject's body. It is the vector sum of all the local forces acting from the force plate onto the feet. Similarly, the point where the resultant force vector intersects with the force plate is not a point where an actual force is acting but the point through which a theoretical resultant force would pass. Figure 3.1.1 shows that this point may not even be a contact point between test subject and force plate. Consequently, the use of the term *centre of force* would be more appropriate than the use of the term *point of application*.

The use of a resultant force with a centre of force is often appropriate in the analysis of human motion or behaviour, or both. However, in some applications, information about the local forces acting between the force measuring device and parts of the contacting body may be of importance. Therefore, information about local pressure distribution is needed. This chapter discusses selected aspects relevant to pressure distribution measurements.

3.2 SELECTED HISTORICAL HIGHLIGHTS

Techniques for quantifying forces were developed in previous centuries. However, the first attempts to quantify forces related to biomechanical questions were only made at the end of the nineteenth century, and force measuring devices for biomechanical purposes have only been commercially available since the late 1960s.

Today, commercial force plates are used as a basic tool for biomechanical tests and measurements. However, it has only been a few years since the first commercial force plate was built. The first commercial force plate was developed as a result of an initiative of J. Wartenweiler, a professor at the ETH Zürich, Switzerland. When serving as an officer in the Swiss cavalry, Wartenweiler began discussing the comparative advantages of various horseshoes with another officer in his unit. As a biomechanist, Wartenweiler focussed on the external and internal forces resulting from different horseshoes. The other officer (H.C. Sonderegger) owned a company that specialized in precision force and acceleration measurements for industry that used piezoelectric elements for their measurements. As a result of their discussions, Sonderegger built a force plate and charge amplifiers that were sold to the biomechanics laboratory of the ETH Zürich for gait analysis measurements with humans and animals (1969). The force plate is still operational in that laboratory.

Force plates and force sensors for biomechanical applications are currently commercially available. Piezoelectric and strain gauge sensors are the two most commonly used force measuring sensors in biomechanical and clinical-biomechanical applications.

Attempts to quantify pressure distribution began at the end of the nineteenth century. However, significant progress in the development of commercially available measuring devices was not made until between 1980 and 1990.

Accelerometry, the measurement of acceleration with electronic equipment, is relatively new in biomechanics. Acceleration measurement methods were developed parallel to force measuring techniques. Today, accelerometers can be bought commercially. However, the use of accelerometers is not widespread in biomechanical applications, possibly due to the problems associated with these measurements, as discussed in the following sections.

1664	Croone	(De Ratione Motus Musculorum) Concluded from nerve section experiments that the brain must send a signal to the muscles to cause contraction.
1686	Newton	Published the three laws in "Philosophiae Naturalis Principia Mathematica" (The Mathematical Principles of Natural Science) that were the cornerstone of all subsequent force measurements.
1743	Borelli	Published the first part of "De Motu Animalium" (The Movement of Animals), which included suggestions on how to determine forces on, and in, a biological system.
1791	Galvani	Showed that electrical stimulation of muscular tissues produces contraction and force.

1849	DuBois-Reymond	The first to discover and describe that contraction and force production of skeletal muscles were associated with electrical signals originating from the muscle.
1856	Lord Kelvin	Demonstrated that the resistance of copper and iron wire change when subject to strain.
1860	Harless	Used the *water immersion method* to determine the mass, the location of the centre of mass, and the moment of inertia of body segments.
1867	Duchenne	The first to perform systematic investigations of muscular function using an electrical stimulation approach.
1872	Carlet	Developed a pneumatic measuring device to quantify the forces between foot and ground.
1873	Marey	Developed a portable pneumatic device to quantify forces between foot and ground of a human subject.
1876	Janssen	Developed a sequential photographic instrument for taking pictures of the movement of the planet Venus in front of the sun.
1878	Muybridge	Published photos of movement in the journal *Nature*.
1880	Curie	The Curie brothers (Pierre and Jacques) discovered the piezoelectric effect, which was later used to quantify forces. The piezoelectric effect was put to practical use in the 1920s and 1930s.
1882	Beely	Used subjects standing on a sack of plaster of Paris. Impressions were assumed to be related to maximal local forces.
1882	Marey	Photographic documentation of movement with the photographic rifle (la fusil photographique), an instrument based on the ideas of Janssen. Published the first chronophotographic pictures of a marching human and a horse jumping an obstacle.
1887	Muybridge	Published an 11-volume photographic work, Animal Locomotion, considered a treasure of early photography of human and animal movement.
1889	Braune & Fischer	Published data about the centre of gravity of the human body as a contribution to the design of military equipment.
1927	Abramson	Used steel shot underneath a soft lead sheet. Penetration of steel shot into lead sheet indicated amount of maximal local force.

1927	Basler	Used force plate beams suspended on metal wires. Frequency of wires was assumed to be proportional to the force acting on the beam.
1930	Morton	Used a rubber mat with longitudinal ridges on top of an inked ribbon. The colouring of inked ribbon was proportional to the maximal local forces.
1931	Carlson	Constructed the first unbonded resistance strain gauge.
1934	Elftman	Used a rubber mat with pyramidal projections on its undersurface mounted on a rigid glass plate, with reflecting fluid in the spaces between the mat and the glass plate. Filming the deformation of the rubber pyramids provided information on the dynamic pressure distribution.
1935	Bloach	Constructed the first bonded carbon film strain gauge.
1938	Elftman	Published measurements of the ground reaction forces and the centre of force for human locomotion. The measurements were obtained using a force plate.
1938	Weinbach	Used volume contour maps to determine the location of the centre of mass and moment of inertia.
1945	Eberhart & Inman	Revolutionized gait analysis with the development of three-dimensional cine photography, the first useful 6-quantity force platform, and the 6-quantity force transducer for the leg prosthesis design.
1947	Bernstein	Used weight measurements in different positions to determine the mass of body segments if the location of the CM was known or to determine the location of the CM if the relative mass of the body segments was known.
1952	Cunningham & Brown	Developed two devices for measuring the forces acting on the human body during walking.
1954	Barnett	Developed a *plastic pedobarograph*, an instrument with 640 vertically mounted beams on a rubber support. The movement of the beams was filmed to determine the dynamic pressure distribution.
1955	Dempster	Identified mass and inertia values for all major human body segments using various methods.
1957	Hertzberg	Used a *photogrammetric method* to determine the mass, the location of the centre of mass, and the moment of inertia of body segments.
1963	Baumann	Used capacitive force transducers for selected local forces between foot and shoe.

1964	Contini	Reported the use of accelerometers in gait analysis.
1964	Hanavan	Presented a mathematical model of the human body for determining inertial properties using symmetric mathematical volumes as a base.
1967	Bouisset & Pertuzon	Determined moments of inertia for limb segments experimentally by using the so-called quick release method.
1967	Gage	Used accelerometers to determine the vertical and horizontal accelerations of the trunk as well as the angular acceleration of the shank analyzing human gait.
1969	Clauser	Contributed cadaver data for weight, volume, and centre of mass of human limb segments.
1969	Gurfinkel & Safornov	Used a *mechanical vibration method* to determine the moments of inertia of body segments.
1969	KISTLER	The KISTLER Company constructed the first commercially available piezoelectric force plate for gait analysis for the biomechanics laboratory of the ETH Zürich.
1972	Hutton	Developed a strain gauge force plate divided into beams.
1972	Prokop	Used accelerometers mounted in the shoe sole of spike shoes on various track surfaces.
1973	Morris	Used five accelerometers to quantify the three-dimensional movement of the shank assuming that the transverse rotations of the shank are small and can be neglected.
1973	Nigg	Used accelerometers mounted at the head, hip, and shank during alpine skiing.
1974	SELSPOT	A SELective light SPOT recognition system which became commercially available using LED's as active markers (Lindholm & Oeberg, 1974).
1974	Unold	Used accelerometers mounted at the head, hip, and shank during walking and running with different footwear on various surfaces.
1976	AMTI	The AMTI Company constructed the first commercially available strain gauge force plate for gait analysis for the biomechanics laboratory of the Boston Children's Hospital.
1976	Arcan	Used a sandwich of reflective and polarizing material with interference techniques to assess the plantar pressure distribution in standing individuals.

1976	Nicol	Developed and used capacitive force transducers and multiplexing techniques for the assessment of pressure distribution with flexible mats.
1976	Scranton	Developed a transducer to quantify local plantar forces using shear sensitive liquid crystals.
1976	Theyson	Used an array of pins on high precision springs and video analysis for clinical gait analysis.
1977	Saha	Studied the effect of soft tissue in vibration tests using skin-mounted accelerometers.
1978	Chao	Proposed an experimental protocol for the quantification of joint kinematics using systems of accelerometers.
1979	Light	Measured skeletal accelerations at the tibia using bone-mounted accelerometers.
1979	Ziegert	Studied the effect of soft tissue on skin-mounted accelerometer measurements.
1980	Denoth	Used acceleration measurements in an effective mass model to determine bone-to-bone impact forces in the ankle and knee joint. Used accelerometers to determine in vivo force deformation diagrams of human heels.
1980	Hatze	Presented a mathematical model for the computational determination of inertial properties of anthropometric segments.
1980	Light	Compared results from bone and skin-mounted accelerometers to find a loss of high frequency content and a phase shift for skin-mounted accelerometers.
1982	Hennig	Used piezoceramic transducers in a rubber matrix and applied them for insoles.
1982	VICON	A television-based system for movement analysis (VICON), which became commercially available. Since 1982, many other systems have been developed that currently on the market.
1983	Aritomi	Developed force detecting sheets (Fuji foils) for maximal local pressure measurements.
1983	Voloshin	Studied the shock (impact force) absorbing capacity of the leg using skin-mounted accelerometers in conjunction with force plate measurements.
1983	Zatziorsky	Used a *gamma-scanner method* to determine the mass, the location of the centre of mass, and the moment of inertia of body segments.

1985	Basmajian & De Luca	Summarized the existing knowledge and research on muscle function, as revealed by electromyographic studies.
1985	Diebschlag	Used inductive sensors embedded in a resilient material for pressure distribution measurement in an insole.
1987	Maness	Developed ultra-thin conductive type transducers to quantify pressure distribution in various applications.
1987	Valiant	Estimated a magnification factor for skin-mounted accelerometers using a linear spring damper model.
1989	Martin	Used MRI to determine the mass, the location of the centre of mass, and the moment of inertia of body segments.
1990	Nicol	Developed strain gauge sensors on a silicon membrane for pressure distribution measurements on soft surfaces.
1991	Lafortune	Described the contribution of angular motion and gravity to tibial acceleration during walking and running.

Electrical resistance strain gauges

1870s	Unbonded wire gauges were invented.
1936-38	Bonded wire gauges developed separately and simultaneously at Caltech, MIT, and the GEC labs. The first commercial gauge was based on Professor A.C. Ruge's gauge from MIT, with circuitry based on the Caltech technology (Stein, 1992).
1952	Foil strain gauges were patented in the UK by Peter Jackson, using the concept of printed circuit technology (Stein, 1992).
1960s	Semi-conductor strain gauges evolved from microelectronics industry.
1970s	Liquid mercury strain gauges introduced in biomechanics.

Mechanical strain gauges

1900s	Huggenberger	Developed a mechanical extensometer.
1953	Demec	Introduced a demountable extensometer for measurements on concrete (Morice and Base, 1953).
1960s		Development of strain gauge based cantilever extensometers.
1992	Shrive	Described soft tissue extensometers.

Optical methods

1815	Brewster	Discovered double refraction (Hendry, 1948).
1906	Coker	Used celluloid for photoelasticity.
1930	Mesnager	Described photoelastic coatings.
1935	Solakian	Three-dimensional analysis by stress freezing was described.
1953	Zandman	First practical results from a photoelastic coating were reported (Redner, 1968).
1970s		Introduction of holography to biomechanics measurements.

Other techniques

1925	Sauerwald & Wieland	Brittle lacquers.
1983	Woo	Video dimension analysis was developed to determine soft tissue strain.
1984	Noyes	Used high speed film analysis to determine strain in ligaments and tendons.
1985	Arms	Developed the Hall Effect displacement transducer to measure ligament strain.
1989		Introduction of SPATE to biomechanics (Friis et al., 1989; Kohles et al., 1989).

3.3 FORCE

NIGG, B. M.

3.3.1 MEASURING POSSIBILITIES

Force cannot be defined. However, effects produced by forces can be described:

If *something* is able to keep a spring deformed, this *something* is called a force in a static sense.

If *something* is able to accelerate a mass, this *something* is called a force in a dynamic sense.

Various mechanical and electrical devices can quantify the effects of a force. They include:

- Air balloon,
- Capacitor,
- Conductor,
- Inductive sensor,
- Photonic sensor,
- Piezoelectric sensor,
- Pyramid system,
- Spring, and
- Strain gauge.

Capacitor, conductor, piezoelectric, and strain gauge sensors are often used in biomechanical force measurements, and are discussed in detail in this section. Other types of sensors are less frequently used in biomechanical force measurements, and are discussed in less detail.

Transducers used for force measurements are typically designed to measure the stress in the material under load, or the deflection or deformation of the material as a function of the applied load (Kain, 2004). The Institute of Measurement and Control (1998) published a guide to assist users of force measurement systems with the principles, methods, applications, calibration and operation of various systems.

CAPACITOR SENSORS

An electric capacitor consists of two electrically conducting plates that lie parallel to each other, separated by a distance that is small compared to the linear dimensions of the plates. The space between the plates is, for the purpose of biomechanical force measurements, filled with a *dielectric* non-conducting elastic material. If a normal force is applied to a capacitor, the distance between the plates changes. This process can be described mathematically for a parallel capacitor with a normal force, F, acting as:

$$C = \frac{Q}{U}$$

and:

$$C = \varepsilon_o \cdot \varepsilon_1 \cdot \frac{A}{d}$$

where:

C = capacitance of the capacitor
Q = magnitude of the total charge on each plate
U = potential difference between the two plates
A = area of one of the two identical capacitor plates
ε_o = permittivity of the free space
ε_1 = permittivity of the dielectric
d = distance between the plates

consequently:

$$Q \cdot d = \varepsilon_o \cdot \varepsilon_1 \cdot A \cdot U$$

Assuming that the relationship between an acting force, F, and the resulting distance, d, can be experimentally determined, for instance:

$$F = c_o \cdot d_o + c_1 \cdot d$$

the relation between an acting force and the resulting change in charge, Q, can be quantified:

$$F = c_o \cdot d_o + \frac{c_1}{Q} \cdot \varepsilon_o \cdot \varepsilon_1 \cdot A \cdot U$$

where:

d_o = constant
c_o, c_1 = constants

A change in force produces a change in the thickness of the dielectric material that is inversely proportional to a current that can be measured.

CONDUCTOR SENSORS

Sensors based on the conductivity principle consist of two layers of conductive material (top and bottom) and a conductive material in between, with construction like that of a capacitor. However, the material between the two plates is electrically conductive. Resistive sensors rely on conductive elastomers as material between the plates. As a force is applied to the two plates, the elastomer between the plates is deformed, thereby reducing the electrical resistance between the two plates. By examining the current output, the applied force can be determined using Ohm's law:

$$V = I \cdot R$$

where:

V = voltage
I = current
R = resistance

PIEZOELECTRIC SENSORS

The piezoelectric effect was discovered by the Curie brothers in 1880. A piezoelectric material is a non-conducting crystal that generates an electrical charge when subjected to mechanical strain. Several materials have piezoelectric properties, including quartz, which is the most stable of the piezoelectric materials. The output of quartz is low compared to the output of ceramic materials, but adequate for biomechanical measurements. The electric charge that appears on the surface of quartz, when strained, is a result of the relative movement of electric charges where the positive and negative charges line up on opposite sides of the material. This movement is illustrated in a simplified form in Figure 3.3.1.

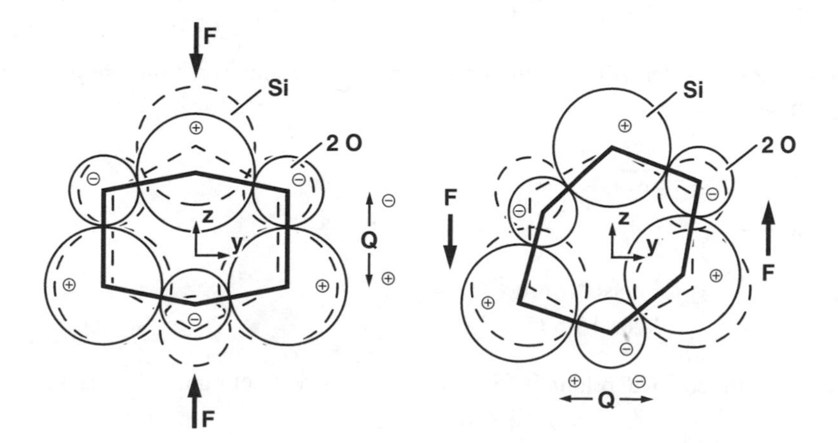

Figure 3.3.1 **Schematic illustration of the piezoelectric effect (from Martini, 1983, with permission of the Instrument Society of America). Compressive forces produce a change in the electric charges on the surfaces where the force has been applied (left). Skew forces produce a change in charges on the surfaces perpendicular to the applied skew force direction (right).**

For biomechanical force sensors, quartz crystals are commonly cut into disks perpendicular to their crystallographic x or y axis. If such disks are subjected to force, they yield an electrical charge of 2.26 pC/N (pico Coulombs per Newton) on the x axis and 4.52 pC/N on the y axis. These charges are temporarily stored in the element's inherent capacitance, CT. As with all capacitors, however, the charge dissipates with time due to leakage. Electronic circuits have been developed to minimize this shortcoming. Modern charge amplifiers overcome this problem, and allow dynamic and quasi-static measurements. Additional information about piezoelectric force measuring sensors can be found in Gohlke (1959), Cady (1964), Tichy and Gautschi (1980), and Martini (1983).

One major advantage of piezoelectric force plates is their wide range of applications. The same force plate can be used to measure the forces between the ground and the hoof of a horse trotting or galloping, or to measure human microvibrations (Nigg, 1977) such as forces on the ground from a heartbeat. Piezoelectric force plates have typically a high stiffness, which provides a high frequency response to minor changes in geometric properties, making them suitable for dynamic measurements.

STRAIN GAUGE SENSORS

Strain measurements are discussed, in detail, in chapter 3.7. In this section, only selected aspects of measuring strain are discussed. All structures deform to some extent when subjected to external forces. The deformation results in a change in length that is called strain when normalized.

$$\varepsilon_a = \frac{(L_2 - L_o)}{L_o} = \frac{\Delta L}{L_o}$$

where:

ε_a = average axial strain
L_o = original length
L_2 = final (strained) length

Lateral strain, as described by the Poisson ratio, n, results if a structure is subjected to simple uniaxial stress:

$$n = \frac{-\varepsilon_L}{\varepsilon_a}$$

where:

n = Poisson's ratio
ε_L = lateral strain

The three-dimensional relations for simultaneous stress acting in all three axis direction are:

$$\varepsilon_x = \left[\frac{1}{E}\right][\sigma_x - \nu(\sigma_y + \sigma_z)]$$

$$\varepsilon_y = \left[\frac{1}{E}\right][\sigma_y - \nu(\sigma_z + \sigma_x)]$$

$$\varepsilon_z = \left[\frac{1}{E}\right][\sigma_z - \nu(\sigma_x + \sigma_y)]$$

The stress strain relationship for uniaxial loading condition is:

$$\sigma_a = E \cdot \varepsilon_a$$

where:

σ_a = average uniaxial stress
ε_a = average strain in direction of the stress
E = Young's modulus

Most strain measurement methods used in biomechanics are based on electrical methods. Electrical type strain gauges use resistive, piezoresistive, piezoelectric, capacitive, inductive, or photoelectric principles. The following discussion concentrates on resistive and piezoresistive methods.

Electrical resistance transducers use the principle that the application of stress to a structure changes its geometry. For example, if a wire is stretched, the cross-sectional area of the wire is reduced, which changes the electric conductivity. The change in resistance of the stretched structure can be calibrated.

Piezoresistive transducers function, in principle, the same way as electrical resistance transducers. The difference lies in the material used. Piezoresistive transducers use semiconductor materials, such as silicon, that have a higher sensitivity than normal electrical resistance transducers (gauge factor > 100). However, piezoresistive transducers are non-linear with strain, their sensitivity is temperature dependent (low temperatures must be avoided to prevent operational failure), and their strain range is significantly lower than the strain rate for conventional resistive strain gauge sensors.

Transducers are mounted on structures that deform if subjected to stress. Different structures are usually used for different axes' directions, since crosstalk would be too big if the transducers were applied to a single structure. Every transducer has crosstalk, but the user is probably unaware of the crosstalk in a one-channel transducer. Provided that the deformation of the transducer does not sensibly affect the values of load and force that it is measuring, crosstalk should be taken into account. The restriction on deformability implies that the crosstalk will be second order and generally linear in behaviour, allowing calibration to produce a 6x6 matrix of sensitivities and cross-sensitivities.

Transducers are arranged to maximize electric output. Strategies to achieve this maximization include the application of one or more sensors on the top and bottom of a deformable structure. The sensors are electrically connected by a bridge arrangement, as explained and illustrated in chapter 3.7, Figure 3.7.2.

OPTIC FIBRE SENSORS

Photonics technology has developed rapidly in recent years, with an increase in the availability and use of fibre optic sensors. A typical optic fibre consists of a silica glass (or similar material) central core that is surrounded by a dielectric silica or plastic cladding-coated in a buffer that protects the fibre (Figure 3.3.2). The sensors also consist of a light source, a photodetector, demodulation, processing and display optics, and necessary electronics.

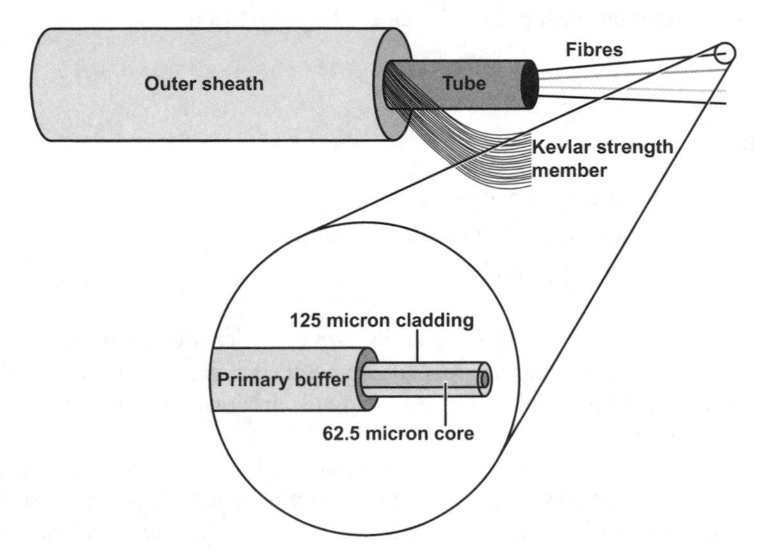

Figure 3.3.2 Schematic diagram of a fibre optic sensor (from Network Cabling Help – Fibre Optics, 2005, reproduced by permission of Ian Patrick - www.datacottage.com).

The principle behind the use of optic fibres is that the geometrical properties of the fibre, and therefore the optical properties of the light beam, experience a change that is dependent on the nature and the extent of the perturbation (Selvarajan & Asundi, 1995).

Fibre optic sensors that are based on the modulation of the wave properties can be intensity (amplitude), phase, frequency, or polarization sensors. Changes in phase and frequency are measured using interferometric techniques such as Mach-Zehnder, Michelson, Fabry-Perot or Sagnac forms. They use coherent detection and have a complex design, but have better sensitivity and resolution than intensity modulated sensors, which are incoherent in nature and have a simple construction (Tohyama et al., 1998; Selvarajan & Asundi, 1995). The advantages of fibre optic technology include (Tohyama et al., 1998):

- High resolution and accuracy,
- Small geometry of the sensing element,
- Immunity to electromagnetic interference, and
- Suitability for biomedical applications.

Another sensor using a Fibre Bragg Grating (FBG) has been proposed for strain measurement. This consists of a spatially periodic (or quasi-periodic) modulation of the refractive index created along the desired length of the core of an optical fibre (Botsis et al., 2005). The fibre gratings act as a wavelength selective mirror for incoming light. The reflected light is of narrow wavelength range and shows up as a dip in the spectrum. The remaining light is transmitted through the grating without any change to the wavelengths (Figurc 3.3.3) (Maaskant et al., 1997; Botsis et al., 2005). The centre wavelength, λ_o o, of the reflected spectral band is defined by the Bragg condition:

$$\lambda_o = 2n_{eff} \cdot \Lambda$$

where:

λ_o = centre wavelength
n_{eff} = the effective core refraction index
Λ = pitch length

The response of the fibre optic Bragg gratings are due to either the induced change in the pitch length (Λ) of the grating, or the perturbation of the effective core refractive index (n_{eff}) (Maaskant et al., 1997).

This system has advantages in that the gratings are intrinsic to the optic fibres, thereby making them small and reliant on the spectral content of the optical signal. They are also inexpensive, provide high durability, have long-term measurement stability, and allow mass construction and multiplexing of many sensors along a single optical fibre (Maaskant et al., 1997).

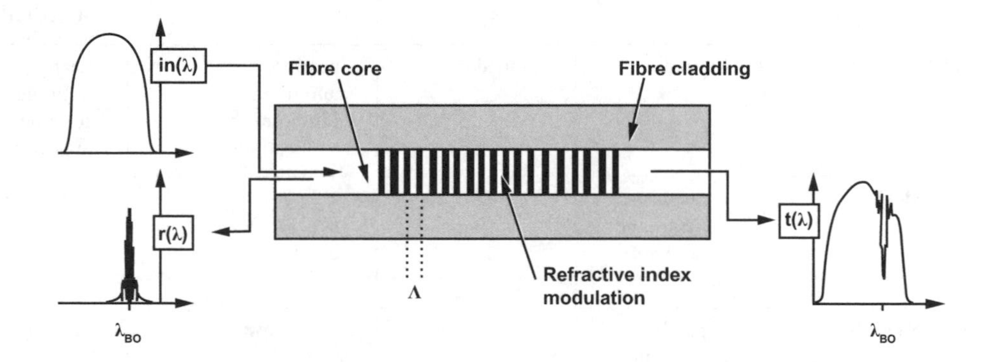

Figure 3.3.3 **Schematic illustration of the functioning of a fibre optic sensor. The broadband light in (λ) is launched into the fibre core. The term $r(\lambda)$ is the reflected signal, and $t(\lambda)$ is the transmitted signal around the Bragg wavelength, λ_{BO}. Λ is the grating period (from Botsis et al., 2005).**

SELECTED CHARACTERISTICS OF FORCE SENSORS

Table 3.3.1 summarizes and compares selected characteristics of force measuring sensors used in biomechanical force measuring instruments. Specific characteristics of force sensors that are sold commercially can be found in company brochures and information sheets. Data in the table is summarized from the information provided by major companies and our practical research experience. Table 3.3.1 concentrates on aspects associated with the principles of each methodology that do not depend on short technical developments.

Sensors can be divided into two groups: capacitance and conductive sensors, and piezoelectric and strain gauge sensors.

- **Capacitance and conductive sensors.** Force measuring devices using capacitance and conductive sensors are advantageous for measuring forces on soft or uneven surfaces and for pressure distribution measurements. They are usually less accurate, and display a lower frequency response than piezoelectric or strain gauge sensors. Force measuring devices using capacitance or conductive sensors are typically used in pressure measuring devices, e.g., in shoes or on sports equipment.

- **Piezoelectric or strain gauge sensors.** Force measuring devices using piezoelectric or strain gauge sensors are advantageous for measuring forces on rigid structures, e.g., force plates and for measurements where the sensor can be rather stiff. They are usually of relatively high accuracy. Force measuring devices using piezoelectric or strain gauge sensors are typically used for force plates.

Table 3.3.1 Summary of selected characteristics of force sensors typically used in biomechanical experiments. Specific data are extracted from sales brochures of major suppliers or are based on our own experience, or both. FSO = full scale output. If corrected with software. If not corrected, errors may be substantially bigger.

CHARACTERISTIC	SENSOR TYPE			
	CAPACITOR	CONDUCTOR	PIEZOELECTRIC	STRAIN GAUGE
RANGE	limited	limited	nearly unlimited for biomech. applications	nearly unlimited for biomech. applications
LINEARITY	depends on dielectricum mostly not linear		> 99.5% of FSO highly linear	> 96% of FSO good linearity
CROSSTALK	depends on construction can be high		rather low	rather low
HYSTERESIS	small < 3%	small < 3%	very small < 0.5%	small < 4%
THRESHOLD	a few N	a few N	< 5 mN	50 to 100 mN
TEMPERATURE SENSITIVITY	small	small	small	small
ACCURACY	errors up to 20% depending on application	errors up to 20% depending on application	errors up to 5%	errors up to 5%
DRIFT	none	none	[a] 0.01 N/s	none
COST	high	low	high	low

OTHER METHODS

If a balloon filled with fluid or air is subjected to a force, a pressure in the balloon is necessary to balance the force. The pressure in the balloon can be calibrated for the applied external force and the system can be used for force assessment. Carlet (1872) initially used this principle in a circular walkway with an air balloon under a foot. The balloon was connected to a central recording instrument by long rubber hoses. Marey (1873) modified Carlet's apparatus by making it portable, allowing the apparatus to quantify the forces under a horse's hoof. Currently, this principle of force measurement is used in internal force/pressure measurements in animals, e.g., between heart and surrounding tissue. In biomechanical and clinical research and monitoring, hydro-fluid balloons are used to quantify the forces between soft tissue and the environment, e.g., between a foot and shoe. These sensors show high linearity, small hysteresis, little drift, and high frequency response.

Forces, especially local forces, may be quantified by using the pyramid system. A mat, which is flat at the top and has small pyramid-type elements at the bottom, is placed on a glass plate (Figure 3.3.4). If a force is applied to the top of the mat, the pyramids will be deformed to a degree that depends on the magnitude of the force applied. The deformation results in an increased contact area between pyramid and glass plate (Figure 3.3.4), which can be quantified optically by video. This force measuring principle was used to assess pressure distribution under the foot during walking (Miura et al., 1974).

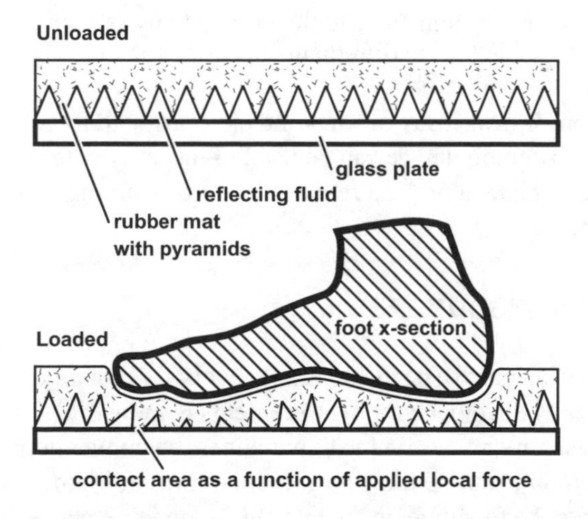

Figure 3.3.4 **Illustration of local force assessment with the pyramid system as used by Miura et al. (1974), with permission.**

Inductive type force transducers measure the relative movement of one or several coils with respect to a core. This type of force transducer has seen limited use in biomechanics.

Force measuring devices may have some mechanical elastic parts like *springs*. Spring force measuring devices are not widely used in biomechanical research.

A force applied to the elastic part of a spring force measuring device results in a deflection that is usually linear. The deflection can be used as a measure of the applied force using, in the linear case:

$$F = -k\,x$$

where:

F = applied force
k = deflection (spring) constant
x = deflection

3.3.2 APPLICATIONS

CAPACITANCE TRANSDUCERS

The most frequent application of capacitance transducers in the biomechanics of locomotion is in assessing pressure distribution (see chapter 3.4) using insoles or plates, or both. Capacitance transducers are also used to assess forces between two soft surfaces (Nicol & Hennig, 1976)

CONDUCTIVE TRANSDUCERS

Transducers based on conductor technology are currently used in two main applications:

- Measuring pressure distribution close to the human body (see chapter 3.4). The sensor is in the form of an insole for measurements between foot and shoe, or in a special shape for specific applications.
- Combining this sensor in an array of rows and columns with force plates. The fine grid of this pressure distribution device can be used to instantly determine the position of the foot on the force plate, a procedure that is time-consuming and often not accurate with optical methods.

PIEZOELECTRIC TRANSDUCERS

Piezoelectric transducers have been used extensively for dynamic force measurements. The sensors were initially developed for measurements in industrial applications, including force or pressure measurements in combustion engines, on conventional train shoe brakes, on turbine blades, and on machine tools during operation. In biomechanics, piezoelectric force sensors were used the first time, in the late 1960s, on force plates designed to quantify ground reaction forces during human locomotion. Currently, force plates based on the piezoelectric principle are available in different sizes and made from different materials. Piezoelectric force transducers are also used in many biomechanical applications, including transducers for force-deformation assessment or drop tests, or both.

STRAIN GAUGE TRANSDUCERS

Strain gauge transducers have played an important role in developing biomechanical knowledge in the last few decades (Canderale, 1983). Assessing the ground reaction forces during human locomotion has often been performed using force plates with strain gauge transducers. Additionally, in vitro stress and strain measurements on bone, specifically the femur and the stem, have been performed (Weightman, 1976; Breyer et al., 1979; Calderale et al., 1979; Huiskes & Slooff, 1979; Jacob & Huggler, 1979; Huiskes, 1980). Similar stress and strain measurement experiments have been performed for the tibia (van Campen et al., 1979). Asang (1974) took mechanical properties of the human tibia, as related to alpine skiing, and assessed these properties using strain gauge transducers.

In vivo forces have also been quantified on various animal and human bones and tendons. Forces in the human hip joint have been quantified by implanting a femoral prosthesis with strain gauge transducers, (English & Kilvington, 1979; Kotzar et al., 1991; Bergmann et al., 1993; Graichen et al., 1999; Bergmann et al., 2001; Stansfield et al., 2003). Semiconductor transducers have been used to quantify the forces in the thoracic vertebrae of sheep (Lanyon, 1972). Graichen et al. (1996) described a telemetry system with semiconductor strain gauges for measuring loads on spinal fixation devices. Similar procedures have been used to examine these loads in vivo (Rohlmann et al, 1997, 2002) and in vitro (Rohlmann et al, 1997).

Because strain gauge transducers can easily be placed on human or animal bone, assessing forces for bones is simpler than tendons or ligaments. Tendon or ligament forces are more difficult to measure because these tissues are usually not accessible. With the help of buckle transducers, measuring multiple muscle forces has been quantified on animal tendons (Walmsley et al., 1978). Buckle transducers have also been used to quantify forces in the human Achilles tendon during walking, running, and other sporting activities (Komi et al., 1987). Finally, strain gauge transducers have been used in evaluating sporting activities, such as taking force measurements from the oar in rowing (Nolte, 1979; Schneider, 1980).

FORCE MEASUREMENTS IN TENDONS AND LIGAMENTS

Several sensors have been developed to quantify forces in tendons and ligaments in vivo or in vitro. They include:

- Foil strain gauge transducer,
- Liquid metal strain gauge transducer,
- Buckle transducer,
- Hall Effect strain transducer, and
- Implantable force transducer.

Ravary et al. (2004) and Fleming and Beynnon (2004) provide extensive reviews of the devices available to measure tendon and ligament forces in vivo. The majority of research has taken place in animals, with only more recent developments suitable for human studies. In humans, in vivo force in the tendons of the hand (Schuind et al., 1992 cited in Fleming & Beynnon, 2004), the Achilles tendon (Finni et al., 1998, 2000), the patellar tendon (Finni et al., 2000, 2003) and the ACL (Henning et al., 1985, Roberts et al., 1994, both cited in Fleming & Beynnon, 2004) have been measured.

Force measuring transducers are commonly classified as either direct, meaning that force is measured directly, or indirect, meaning that force is measured indirectly, e.g., by strain. Buckle transducers and implantable force transducers are generally considered direct force transducers. Strain gauge transducers, video dimension analysis, and Hall Effect transducers are generally considered indirect force transducers. This categorization, however, seems artificial, since every possible force-measuring device must quantify force indirectly. A buckle transducer, for instance, uses deformation, and therefore strain, to determine force. For further information, see chapter 2.7.

Foil strain gauge transducers typically consist of (White & Raphael, 1972):

- A foil gauge, and
- A stainless steel shim that is sutured to the tendon or ligament.

Theoretically, these transducers are highly accurate, but problems sometimes arise because the sutures are unable to act as rigid displacement transducers between the tissue and the shim (Shrive et al., 1992).

Liquid metal gauge transducers consist of (Brown et al., 1986):

- A gauge body of silastic or silicone tubing,
- A liquid metal (purified mercury, indium, or gallium),
- Two insulated wires plugging the ends of the gauge tubing, and
- Two end seals of silicone heat shrink tubing that attach and seal the gauge tube to the wires.

Axial strain in the tendon or ligament causes the gauge to deform by increasing the column length of the mercury and decreasing the cross-sectional area, thus creating a net increase in electrical resistance. The liquid metal strain transducer is small enough to fit into individual fibre bundles. Because it is not very stiff, the liquid metal strain transducer interferes minimally with the mechanics of the bone. Early models were limited, as implantation, measurement and calibration needed to be carried out within one day. However, later developments increased this maximum period to five days. Jansen et al. (1998) used these devices to measure in vivo tendinous interrosseous muscle forces in ponies.

Buckle transducers consist of (Salmons, 1969):

- A rectangular, oval, or E-shaped stainless steel frame element for supporting a portion of the tendon or ligament,
- A stainless steel beam mounted transversely between the tendon and the frame element, and
- Two conventional foil strain gauges mounted on the buckle for measuring deformation (Figure 3.3.5).

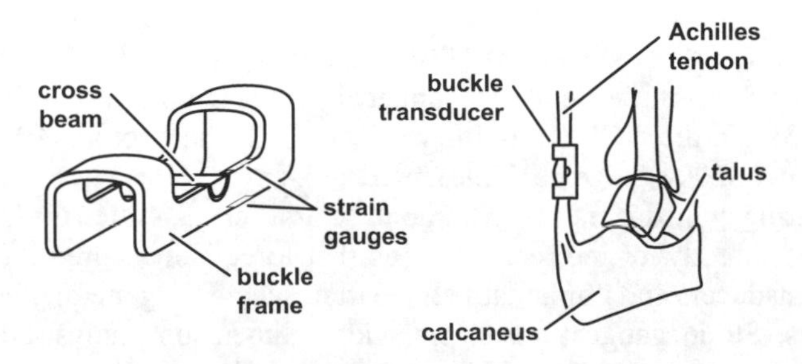

Figure 3.3.5 **Schematic illustration of a buckle transducer (left) and a possible arrangement on an Achilles tendon (right) (from Salmons, 1969, with permission).**

Under axial loading, the tendon or ligament lengthens, producing a transverse force and a resulting deformation of the beam element, both of which can be quantified. The buckle transducer is a relatively stable measuring device for chronic in vivo experiments.

Typical insertion times are three to seven days (Komi et al., 1987; Komi, 1990; Herzog et al., 1993), but Komi et al. (1996) found that a recovery period of two to three weeks may be necessary for subjects to return to normal walking after implantation of a buckle transducer in the Achilles tendon (Fleming & Beynnon, 2004). Buckle transducers require a finite tendon or ligament length and distance between bony structures and the sensor. They also require quite a large incision, and, as they shorten the tendon length, they may affect the length-tension relationship of the muscle. These requirements make buckle transducers applicable only to a few structures in human and animal bodies.

Hall Effect strain transducers consist of (Arms et al., 1983):

- A magnetized cylindrical rod, and
- A tube mounted with a Hall generator.

The device is fixed to the tendon or ligament by barbs attached to the magnet and the tube. As the tendon or the ligament is strained, the rod displaces with respect to the tube, and the Hall generator produces a proportional voltage. The Hall Effect strain transducer may be firmly anchored to the tissue and does not load the tissue during strain monitoring. It is small, and minimal surgical intervention is required to affix it to tissue. Hall Effect strain transducers measure local forces. Typical insertion times are about seven days (Platt et al., 1992). Marlkof et al. (1998a) used a variation of a Hall Effect strain transducer with a metal rod sliding within a small cylindrical electrical coil (Ravary et al., 2004).

Two *implantable force transducers* have been described in the literature. One type is based on the deflection of a metallic element (Xu et al., 1992), and the other is based on a modification of a miniature, circular pressure transducer (Holden et al., 1991).

The implantable force transducer, based on the deflection of a metallic element, consists of (Ravary et al., 2004):

- A curved steel spring with foil strain gauges mounted on the upper and lower side (Figure 3.3.6), or
- A C-shaped tube with a longitudinal slit and a strain gauge mounted opposite the slit.

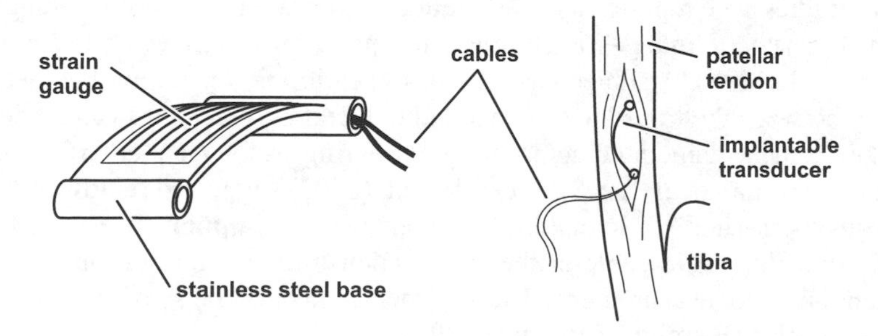

Figure 3.3.6 **Schematic illustration of an implantable force transducer (left) and a possible arrangement in a patellar tendon (right) (from Xu et al., 1992, with permission).**

This transducer is inserted directly into a longitudinal slit in the tendon or ligament mid-substance and is typically held in place with sutures through the adjacent tissue fibres and at the slit edges. The tissue fibres are deflected around the sensor and exert a compressive load when the ligament or tendon is in tension (Fleming & Beynnon, 2004). Implantable force transducers theoretically allow an absolute force and an absolute zero line to be determined if they can be calibrated in situ. However, there are some limitations for their application (Herzog et al., 1996).

Arthroscopically implanted force transducers have been used to measure forces in the animal and human tissue. Specifically, these transducers have been used to measure forces in goat ACL (Holden et al., 1994, 1995), and human ACL (Roberts et al., 1992 cited in Fleming & Beynnon, 2004) and subscapularis tendon (Reilly et al., 2003 cited in Fleming & Beynnon, 2004).

Komi et al. (1996) proposed the use of implantable force transducers based on fibre optic technology as an alternative to buckle transducers for measuring in vivo tendon forces. These transducers are inexpensive, minimally invasive, have a reduced risk of injury and infection to the subject (Arndt et al., 1998), and cause minimal disturbance to the gross tendon geometry (Fleming & Beynnon, 2004). These devices have been used to measure forces in the Achilles tendon (Komi et al., 1996; Finni et al., 1998, 2000; Arndt et al., 1998; Erdemir et al., 2002; 2003), and the patellar tendon (Finni et al., 2000, 2003).

Implantable force transducers based on fibre optic technology consist of optic fibres of 265 or 500 μm to 0.5 mm diameter (Ravary et al., 2004). The procedure involves passing a hollow needle through the tendon under local anaesthetic. The optic fibre is inserted into the needle, which is then removed, leaving the fibre in situ. When used in the Achilles tendon, the fibre is aligned mediolaterally through the tendon, transverse to the loading axis. Both ends are then connected to a transmitter-receiver unit, which comprises a light emitting diode and a photodiode receiver (Finni et al., 2000). The geometric properties of the optic fibre are changed by bending induced by tensile and compressive loading of the tendon. Tensile loading on the tendon causes the tendon to lengthen and increases the compressive loading on the optic fibre by adjacent tendon fibres (Fleming & Beynnon 2004). This increases bending of the fibre, which decreases the intensity of the transmitted light. The modified light signal travels in the core of the fibre to the unit, where it is converted into an analog signal and is given as a voltage output of the transducer (Ravary et al., 2004; Erdemir et al., 2003).

Several studies have reported a linear relationship between increased loading of the tendon and the intensity of light passing through the optic fibre (Komi et al., 1996; Arndt et al., 1998; Finni et al., 1998). A subject-specific in situ calibration is required to determine the relationship between the transducer output and the tendon force. This can be done by calculating the moments generated across the joint during isometric plantarflexion using an ankle ergometer (Finni et al., 1998). Erdemir et al. (2003) suggested removing the skin and subcutaneous tissues and retracting the skin at entry and exit points to reduce the chances of bending the fibre optic cable at the skin-tendon interface and ensure more accurate Achilles tendon force measurements. Measurements may also be sensitive to joint position and cable migration (Fleming & Beynnon, 2004).

FORCE PLATES

Force plates have been used in biomechanics for many years for quantifying external forces during human and animal locomotion. Publications over the last 50 to 100 years illustrate that force plates have been one of the most important measuring devices in biomechanics. Marey (1895) built the first force plate, which consisted of spirals of India rubber tubes installed on a wooden frame. Elftman (1938) used a plate moving vertically against four springs upon which it was supported, with the vertical displacement recorded optically. This quantified the vertical component of the resulting ground reaction force and the path of the centre of force. Cavagna (1964) used force plates to quantify external vertical impact force peaks, a quantity that was frequently used in the late 1970s, and thereafter, in sport shoe and sport surface research. With the developing commercially available force plates, the number of researchers working in the field of locomotion analysis increased significantly, and resulting ground reaction forces were assessed for many different movements. A comprehensive discussion of ground reaction forces in distance running was published in 1980 (Cavanagh & Lafortune).

Force plates currently on the market use a construction in which a rectangular plate is supported at four points (Figure 3.3.7). Such a measuring system is mechanically overdetermined, which may lead to accuracy problems in some cases (Bobbert & Schamhardt, 1990). Force transducers for each axis direction are mounted in each corner.

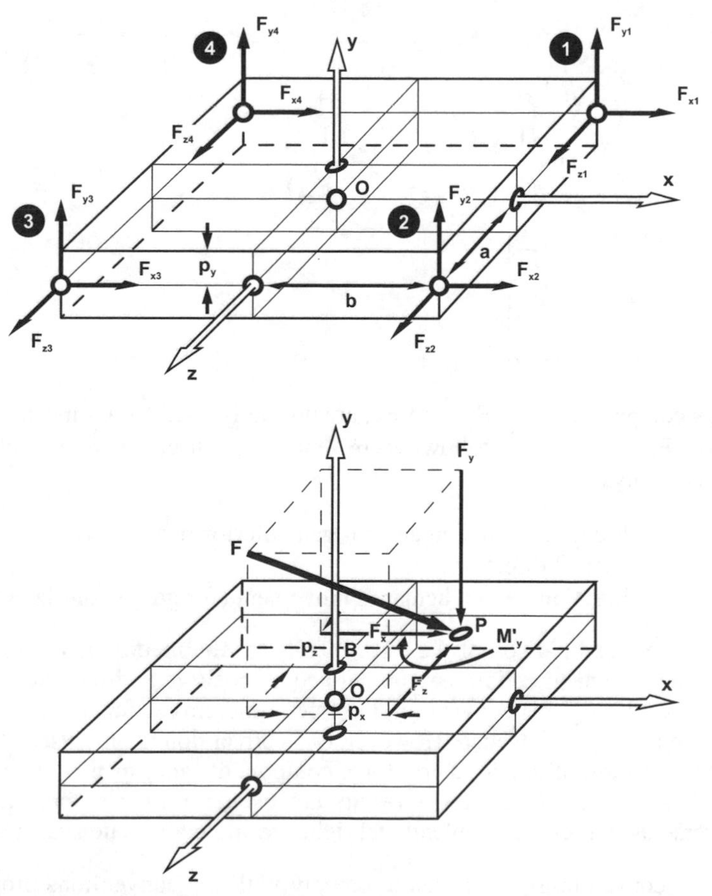

Figure 3.3.7 **Illustration of the forces and moments applied to the force plate (top) and the forces registered by the force transducers in the four corners of the force plate. The force transducers are assumed to be located in the corners of the indicated frame.**

The signals from these transducers are:

F_{x1}	corner 1
F_{x2}	corner 2
F_{x3}	corner 3
F_{x4}	corner 4
F_{y1}	corner 1
F_{y2}	corner 2
F_{y3}	corner 3
F_{y4}	corner 4
F_{z1}	corner 1
F_{z2}	corner 2
F_{z3}	corner 3
F_{z4}	corner 4

where:

$$F_x = F_{x1} + F_{x2} + F_{x3} + F_{x4}$$

$$F_y = -(F_{y1} + F_{y2} + F_{y3} + F_{y4})$$

$$F_z = -(F_{z1} + F_{z2} + F_{z3} + F_{z4})$$

$$\mathbf{F} = \begin{bmatrix} F_x \\ F_y \\ F_z \end{bmatrix}$$

The force components, F_x, F_y, and F_z, are normally used for biomechanical analysis of locomotion. For this text, the following conventions are used when force plates are used for locomotion analysis:

(a) x = direction of movement (for gait anterior-posterior)
y = vertical direction
z = direction perpendicular to movement (for gait medio-lateral)

(b) To make a comparison of the force results for the medio-lateral direction possible, the medial direction may be defined as positive and the lateral direction as negative. Using this convention, the force time curves for the left and right foot can be averaged if needed. However, this convention is only used for the graphical illustration of medio-lateral force components and for the numerical analysis of these curves. If the force results are used for further force analysis, e.g., inverse dynamics, the conventional right-turning coordinate system is used.

(c) These conventions are chosen to comply with the conventions proposed by the International Society of Biomechanics (ISB). However, the results are similar if another convention is used, as long it is a right turning coordinate system. Therefore, the convention should not be overemphasized.

Force plates provide a resulting force that is an integral quantity. The vertical component describes the change in momentum of the centre of mass of the test subject in the vertical direction, and the a-p and the m-l components correspond in the two horizontal directions. This can be used for speed control of test movements. If a person must run or walk with constant speed, the integral of the a-p force component, over time, must be zero.

$$\int_{t_1}^{t_2} F_{ap}(t)dt \quad = \quad mv_2 - mv_1 \quad = \quad \int_{t_1}^{t_2} F_x(t)dt$$

where:

t_1 = first contact with the force plate
t_2 = last contact with the force plate
m = mass of the test subject
v_1 = horizontal speed of the centre of mass of the test subject before landing
v_2 = horizontal speed of centre of mass of the test subject after take-off

The measurements of the force sensors in the four corners of a force plate may be used to determine the centre of pressure, and the free moment of rotation about a vertical axis through the centre of pressure. In determining these variables, the following conventions are used:

O = origin of the force plate, arbitrarily assumed in the centre of the force plate
P = point where the force, F, applies on the top surface of the force plate
B = origin of the surface of the force plate
a = distance of the force transducers from the x axis
b = distance of the force transducers from the z axis
p_y = distance between O and B
M'_y = external moment component applied to the force plate with respect to a vertical axis through P

For simplicity, the horizontal force components can be combined this way:

$$F_{x14} = F_{x1} + F_{x4}$$

$$F_{x23} = F_{x2} + F_{x3}$$

$$F_{z12} = F_{z1} + F_{z2}$$

$$F_{z34} = F_{z3} + F_{z4}$$

In determining the mathematical function for the x and z coordinates of the centre of force, p_x and p_z, two sets of moment equations can be formulated. The first set describes the moments that the forces measured with the transducers produce with respect to the laboratory coordinate system with origin in O (Figure 3.3.7):

$$M_x = a(F_{y1} - F_{y2} - F_{y3} + F_{y4})$$

$$M_y = b(-F_{z12} + F_{z34}) + a(-F_{x14} + F_{x23})$$

$$M_z = b(F_{y1} + F_{y2} - F_{y3} - F_{y4})$$

The second set of equations includes the moments of the force components with respect to the laboratory coordinate system with origin in O (Figure 3.3.7):

$$M_x = -p_z F_y - p_y F_z$$

$$M_y = -p_z F_x + p_x F_z + M_y'$$

$$M_z = -p_x F_y - p_y F_x$$

The free moments M_x' and M_z' are zero as long as the shoe or foot is not sticking to the ground, e.g., for locomotion. This free moment applies at all levels in the leg, while its value changes in offset positions due to the shear forces. The equations can be solved for the coordinates of the centre of force, p_x and p_z, and the free moment of rotation, M_y':

$$p_x = \left[\frac{-1}{F_y}\right][b(F_{y1} + F_{y2} - F_{y3} - F_{y4}) + p_y(F_{x14} + F_{x23})]$$

$$p_z = \left[\frac{-1}{F_y}\right][a(F_{y1} - F_{y2} - F_{y3} + F_{y4}) - p_y(F_{z12} + F_{z34})]$$

and:

$$M_y' = -b(-F_{z12} + F_{z34}) + a(-F_{x14} + F_{x23})$$

$$+p_z(F_{x14} + F_{x23}) + p_x(F_{z12} + F_{z34})$$

The output of the force transducers on the four corners can be used to calculate the centre of force and the free moment of rotation applied to the force plate. However, determining the coordinates of the centre of force has technical problems. The coordinates are calculated by dividing forces by forces. The calculation is sensitive to noise when these forces are small (division of a small quantity by a small quantity). As a result, the coordinates of the centre of force, $p_x(t)$ and $p_z(t)$, are typically inaccurate for small forces at the beginning and the end of the stance phase.

An example on how shoe construction can influence the path of the centre of force is illustrated in Figure 3.3.8. Figure 3.3.8 shows results of the centre of force for the initial phase of ground contact during heel-toe running ($v = 4.5$ m/s) for one test subject in three different running shoes. The shoes differed only in one aspect of construction, the shape of the heel on the lateral side of the rear foot. The heel construction of one shoe was flared, one was neutral, and the third was rounded. The shoes were identical other than their differences in heel construction. For illustration purposes, the outline of the shoe sole and an arbitrarily defined line, which may represent a projection of the subtalar joint axis, are drawn. The example illustrates how even small alterations in shoe construction can be used to influence the centre of force path.

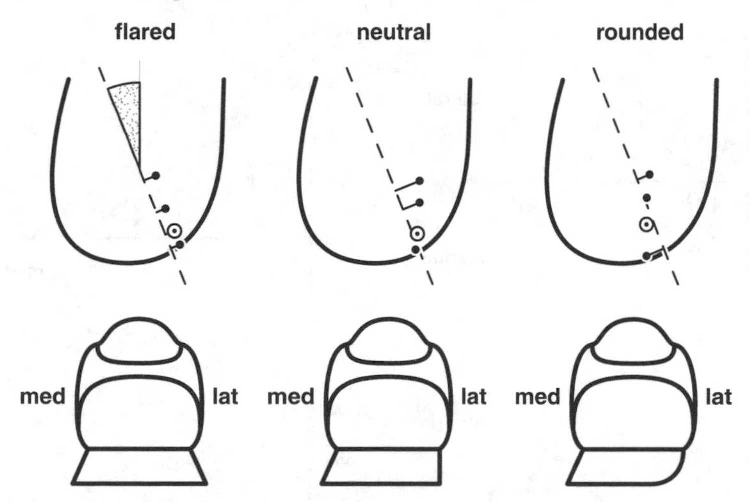

Figure 3.3.8 **Illustration of changes in the centre of force path for one subject, running heel-toe (v = 4.5 m/s) in three different running shoes; a shoe with a flared (left), a neutral (centre), and a rounded lateral heel (right). The outline of the shoe during stance and an arbitrary selected projection of a possible subtalar joint axis are drawn to allow an easy comparison between the three situations.**

Locomotion studies often quantify the force between the subject or animal and the ground (ground reaction forces). For humans, the most frequently analyzed locomotion is walking and running. In the following pages, ground reaction forces for running and walking are depicted, and the typical characteristics are discussed. Ground reaction forces for heel-toe running (Figure 3.3.9) are used to define impact and active forces.

The vertical force component for heel-toe running usually has two peaks. The first vertical force peak occurs about 5 to 30 milliseconds after first ground contact and is referred to in this text as F_{yi}. Similar impact force peaks may occur in the two horizontal force time curves.

In general, impact forces can be defined as follows:

Impact forces in human locomotion are forces that result from a collision of two objects reaching their maximum earlier than 50 milliseconds after the first contact of the two objects.

In running, impact forces occur when the foot lands on the ground. The time, occurrence and magnitude of the impact force peaks depend on various factors, including running speed, material properties of heel and shoe sole, geometrical construction of the shoe sole (Nigg et al., 1987) and running style (Cavanagh & Lafortune, 1980). The vertical impact force peaks are smaller for toe landing than for heel landing (see the discussion of effective mass in chapter 4). The vertical impact force peaks are earlier for barefoot running (5 to 10 milliseconds after first contact) than for running with shoes, and earlier for running with

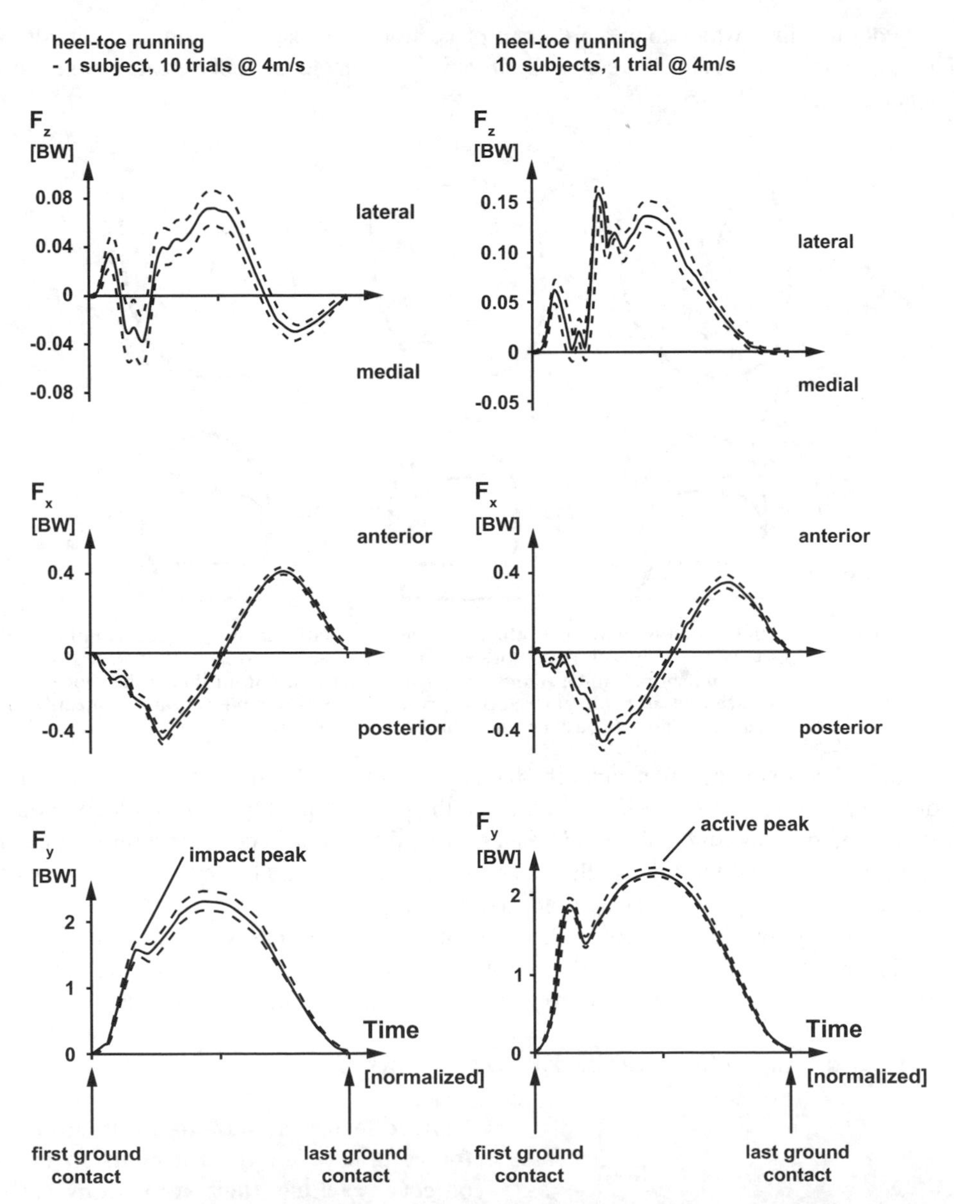

Figure 3.3.9 **Mean and standard deviation of the components of ground reaction force in vertical (bottom), a-p direction (middle) and m-l direction (top) for heel-toe running for one subject and 10 trials (left), and 10 subjects with one trial each (right) in units of body weight (BW).**

harder shoe soles than running with softer shoe soles. Some individuals show more than one impact peak, one resulting when the heel hits the ground and one when the forefoot does. Impact forces in walking are not always evident. Some subjects show impact peaks, but some do not.

The second peak in the vertical force time curve is the *active vertical force peak* and is referred to in this text as F_{ya}. Similar peaks may occur in the horizontal force time curves.

Active forces can be defined as follows:

Active forces in human locomotion are forces generated by movement that is entirely controlled by muscular activity.

Vertical active force peaks from running at a speed of 4 m/s are about two to three times body weight. In running, F_{ya} occurs in the middle of the stance phase, which is about 100 to 300 milliseconds after first ground contact (Figure 3.3.9 and Figure 3.3.10).

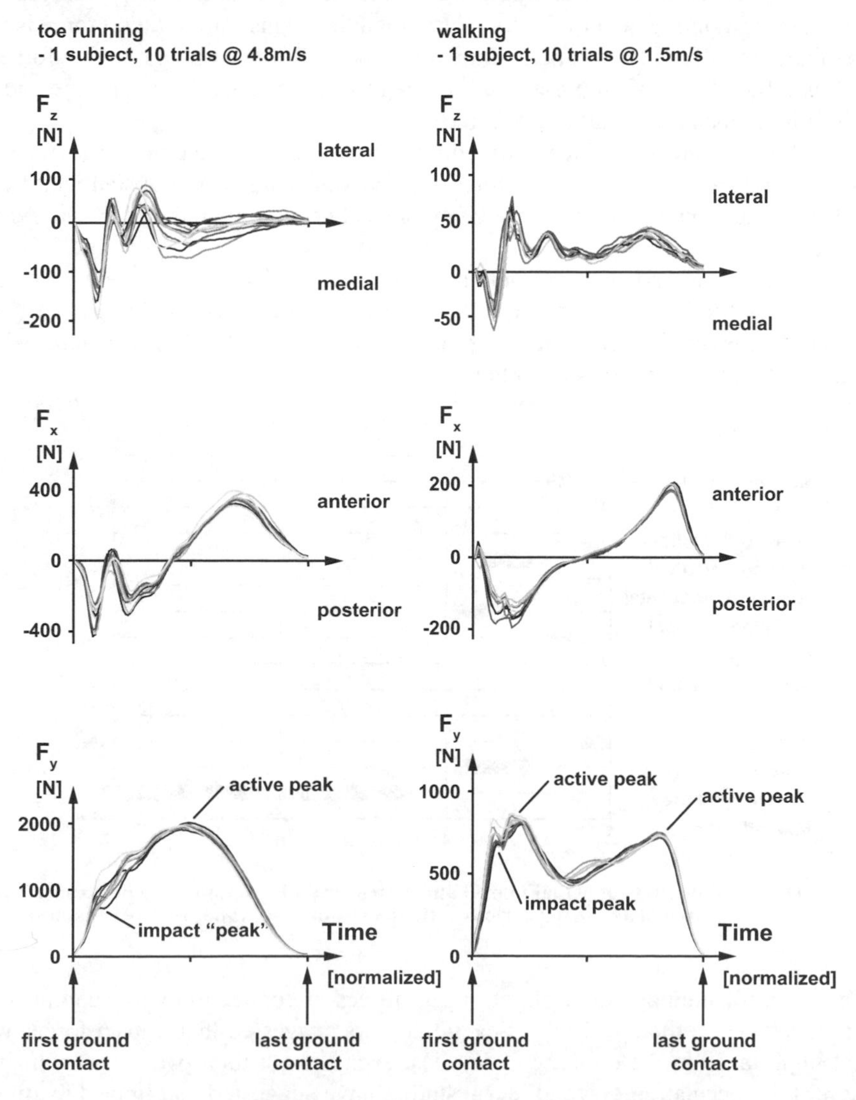

Figure 3.3.10 **Results for individual trials for ground reaction force components in vertical (bottom), a-p direction (middle) and m-l direction (top) for toe-running (left) and walking (right) in Newtons for one subject and 10 trials.**

In walking (Figure 3.3.10), two active force peaks are present, the first being associated with deceleration and the second with the acceleration phase. The first peak is associated with deceleration, and the second is associated with the acceleration phase. The force component in the a-p direction has two active parts, which are similar in walking and running. In the first half of ground contact, the foot pushes in the anterior direction. Consequently, the reaction force from the force plate is directed in the posterior direction (backwards). In the second half of ground contact, the foot pushes in the posterior direction. Consequently, the reaction force from the force plate is in the anterior direction. The force component in the m-l (medio-lateral) direction is less consistent intra- and inter-individually. The medio-lateral component often shows an initial reaction force in lateral direction that results from a medial (inward) movement of the foot during landing. This initial lateral force is usually shorter than 20% of the total contact time. It is usually followed by a reaction force in the medial direction that is often present during the rest of the ground contact time and that is usually smaller than the initial lateral force.

The intra- and inter-individual variability is much bigger for the medio-lateral than for the vertical and the anterior-posterior force-time curves. In addition, substantial differences may exist in the ground reaction force components between the left and the right foot fall for one subject.

Values of impact and active ground reaction force peaks during different movements have been reported by various authors. A summary of these results (Figure 3.3.11) illustrates that the maximal external impact forces can exceed 10 BW (body weight) while the maximal external active forces do not exceed 5 BW.

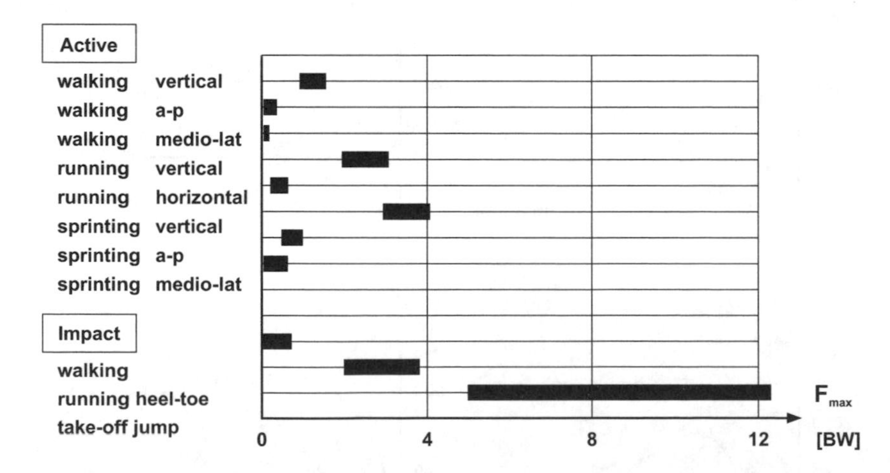

Figure 3.3.11 **Magnitude of actual impact and active ground reaction force peaks measured with a force plate during various activities (from Nigg, 1988, with permission).**

The quantification and analysis of impact forces in connection with running shoe research developed in the late 1970s. Several authors suggested that impact forces were of special importance in the etiology of injuries, even though there was no conclusive evidence for this speculation. Several recent studies have suggested that impact loading positively affects the development and maintenance of bone if it is not excessive and if the body has enough time to repair and recover. Recent research speculates that additional work used for muscle tuning to minimize soft tissue vibrations may be an important contribution to impact forces (Nigg, 2001). Selected external impact forces, as reported in the literature, are summarized in Table 3.3.2.

Table 3.3.2 Selected experimental results for external vertical impact force peaks in walking and running (from Nigg, 1985, with permission). Values shown with an * have been calculated assuming a body weight BW = 700 N.

MOVEMENT	FOOTWEAR	v [m/s]	F(max) [N]	F(max)/BW	AUTHOR	YEAR
WALKING	barefoot	1.3	385	0.55	Cavanagh	1981
	army boots	1.3	259	0.37	"	"
	street shoes	1.3	189	0.27	"	"
RUNNING (HEEL)	run. shoe	4.5	1540	2.2	Cavanagh	1980
	run. shoe hard	4.0	2000	2.9	Nigg	1980
	run. shoe soft	4.0	1100	1.6	"	"
	run. shoe	3.4	1365	2.0	Frederick	1981
	run. shoe	3.8	1590	2.3	"	"
	run. shoe	4.5	1963	2.9	"	"
	run. shoe	3.0	1345	2.0	Nigg	1987
	run. shoe	4.0	1521	2.2	"	"
	run. shoe	5.0	1799	2.6	"	"
	run. shoe	6.0	2070	3.0	"	"
RUNNING (TOE)	run. shoe	4.0	300	0.4	Denoth	1980
TAKE-OFF FOR JUMP	spikes	2.0	1000	1.4	Nigg	1981
	spikes	4.0	2300	3.3	"	"
	spikes	6.0	3700	5.4	"	"
	spikes	8.0	5700	8.3	"	"
	run. shoe	2.0	1400	2.0	"	"
	run. shoe	4.0	2000	2.9	"	"
	run. shoe	7.0	2900	4.2	"	"

The results indicate that several factors influence the magnitude of the external impact force peaks including footwear, speed of movement, and type of movement. Externally measured forces are also used to estimate internal forces (see chapter 4, Modelling).

3.4 PRESSURE DISTRIBUTION

NIGG, B. M.

3.4.1 MEASURING POSSIBILITIES

The possibilities for quantifying forces were discussed in chapter 4 and are not repeated here. However, methods used in pressure distribution measurements, and their characteristics as they relate to pressure distribution measurements, are discussed in this chapter. This section's main emphases are on measuring devices used in pressure distribution plates and devices that allow the quantification of pressure distribution between two surfaces, e.g., foot and shoe, teeth, and leg and shoe shaft. Single force sensors used for the assessment of local pressures are discussed in less detail.

The measuring elements used in pressure distribution measurements (in alphabetical order) are:

- Capacitor,
- Conductor,
- Critical light reflection,
- Force sheet (Fuji foils),
- Inductive sensor,
- Piezoceramic element,
- Reflecting/polarizing sheet,
- Rod/spring elements, and
- Strain gauge element.

Their main characteristics of these measuring elements are discussed in the following paragraphs. The discussion is grouped into two parts.

- A detailed discussion, including specific characteristics of measuring techniques that have been used frequently in biomechanical research (capacitor, conductor, force sheet, and piezoceramic element).
- A discussion of remaining methods, which work in principle, but are less frequently used in biomechanical research.

CAPACITOR SENSORS

Pressure distribution plates or insoles, or both, on the basis of capacitor elements consist of n stripes of conducting material arranged in rows on the top, m stripes made of conducting material arranged in columns on the bottom, and a resilient dielectric material between two stripe layers. The stripes of row i and column k overlap and form a condenser, C_{ik}. This construction consists of m times n condenser elements (Nicol & Hennig, 1976).

The dielectric between the condenser plates is an elastic material, and its elastic material properties are of critical importance for sensor quality. Ideally, dielectric materials that allow compression to about 50% of the unloaded thickness under maximal force are required. Dielectric materials with viscoelastic properties produce hysteresis in the force signals and should be avoided.

In most applications, a multiplexing technique is used to quantify the forces acting on each element C_{ik}. Alternating current is sequentially connected to each row by a de-multiplexer switch, and columns are connected to a resistor by a multiplexer switch (Nicol & Hennig, 1976). Thus, each measuring capacitor, C_{ik}, is an element of a voltage divider. The advantage of such a multiplexer circuit is that only (m + n) channels and the same number of cables are required for m times n sensors. For example, a pressure distribution insole with 10 rows and 30 columns would have 40 cables and 300 sensors, which may provide adequate information on pressure distribution underneath the foot, and provides a technical solution with acceptable interference to the test subject. However, if each sensor required a cable connection, 300 cables would be needed for the same arrangement, which could encumber the test subject substantially.

CONDUCTOR SENSORS

Force transducers based on the principle of conductivity are described in chapter 4. Pressure distribution measuring devices, based on conductivity, use the same construction and multiplexing concepts as pressure distribution measuring devices based on capacitance. Rows and columns of conductive material are used in the top and bottom layer of the system. However, the material between the rows and columns functions as a resistance (not a dielectric, as before), and consists of force-sensitive ink (Maness et al., 1987) or carbon-impregnated conductive polyurethane foam (Shereff et al., 1985). As force is applied to a sensor, the electrical resistance of the material between the rows and columns is changed. The change in resistance corresponds to the applied force. Polymer film sensors and discrete force sensing resistors are also based on these principles.

PRESSURE SHEET (FUJI FOIL)

Pressure sheets consist of two sheets: A and C (Figure 3.4.1). The upper side of sheet A is in contact with the force applying object, e.g., the human foot. Microcapsules, which contain a colour producing agent, are coated onto the lower side of sheet A. Sheet A lies on

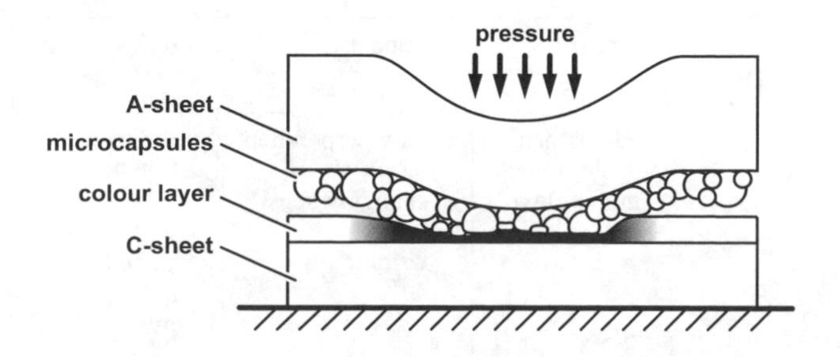

Figure 3.4.1 **Illustration for the force sheet method for maximal pressure distribution measurement as presented by Aritomi et al. (1983), with permission.**

top of sheet C. When force is applied locally, capsules are ruptured and colour-producing agents colour sheet C. The colour intensity corresponds to the maximal force applied locally, and the intensity of the coloured sheet C provides a picture of maximal local forces. With the optical density:

$$D = \log(1/\text{transmission}) \quad (3.4.1)$$

the colour intensity can be quantified. The correlation between optical density, D, and local force, F_{loc}, is linear in the middle part, but non-linear in the end part of the applicable range of the foil (Hehne, 1983). In the case of a dynamic measurement, the maximal forces at different locations may not occur at the same time (Aritomi et al., 1983).

PIEZOCERAMIC SENSORS

Hennig et al. (1981) used 499 piezoceramic transducers in their first attempt to develop a measuring device to quantify pressure distribution between foot and shoe.

The elements had a surface area of 4.78 mm² and were embedded in a 3 to 4 mm-thick layer of highly resilient silicon rubber. The embedment was electrically isolating and impermeable to moisture. Each sensor was connected through a cable to a data collection unit. In the first attempt, 499 cables connected the measuring insole with the data collection unit.

Table 3.4.1 Selected characteristics for pressure distribution sensors (based on data from Nicol & Hennig, 1976; Hennig et al., 1981; Aritomi et al., 1983; Cavanagh et al., 1983; Hehne, 1983; Maness et al., 1987; with permission), and on personal experiences from experimental projects. For general characteristics of force sensors, see chapter 4.

CHARACTERISTIC	SENSOR TYPE			
	CAPACITOR	CONDUCTOR	PRESSURE SHEET	PIEZO-CERAMIC
FLEXIBILITY (INSOLES)	limited due to dimensions	high but may wrinkle	high but may wrinkle	good
TEMPERATURE SENSITIVITY	not sensitive	not sensitive	sensitive	must be taken into account
DYNAMIC MEASUREMENTS	standard	standard	not possible yet	standard
MAJOR LIMITATIONS	only perpendicular forces limited flexibility	only perpendicular forces changing sensitivity wrinkle	force direction not known only static wrinkle only max. pressure	only perpendicular forces temperature sensitivity
COSTS	high	low	low	high

The use of piezoceramic transducers has specific problems. Since each transducer is connected directly with the outside, special shoe adaptations to provide room for cables, or holes to take the cables out of the shoe at different locations, are required. Consequently, the number of cables limits the number of measuring cells in piezoceramic measuring devices.

Piezoceramic transducers were used for the first comprehensive pressure distribution measurements in shoes (Hennig et al., 1981; Cavanagh et al., 1983), providing initial insight into foot-shoe interaction. However, this technology was not extensively developed, and has been used rarely in biomechanics research. These devices have the disadvantage of being relatively fragile, with a poor fatigue resistance (Urry, 1999).

A comparison of selected characteristics of the most important pressure distribution sensors is summarized in Table 3.4.1.

OTHER METHODS

In addition to the above-mentioned methods for quantifying pressure distribution, other methods were developed that are applied less often in current biomechanical research and testing.

The *critical light reflecting technique* was used by Chodera and Lord (1979) to assess pressure distribution between the foot and the ground. The hardware consisted of a plastic sheet, a glass plate, a video camera, and a microcomputer. The plastic sheet was flat on the upper side, was in contact with the foot, and had an array of small knobs on the bottom side. The bottom side was in contact with the glass plate. The single knobs had a point contact with the glass when no force was applied to the plastic. However, the knobs flattened against the glass surface when a local force was applied. The glass plate was illuminated from the side. The light rays passing through the glass were totally reflected if the neighbouring medium on top of the glass was air. The light was partially reflected if the neighbouring medium was plastic. If the forces applied to the plastic increased, the surface area that was in contact with plastic increased, and the intensity of the reflected light could be used as an indicator of the locally applied force. The system used by Chodera and Lord (1979) has the limitation that only pressure distribution components perpendicular to the measuring surface can be quantified.

Inductive sensors were used by Diebschlag et al. (1985). In their construction, a spring suspended metal pin was inserted into a coil with pin and coil embedded into a flexible rubber material. If force was applied, the pin penetrated the coil and inductive current was measured. A second order multiplexer technique was used to scan the sensors.

Arcan and Bull (1976) used a sandwich of a *reflective and polarizing material* placed onto a rigid transparent plate. Depending on local forces, optical interference patterns were generated and used for the qualitative analysis of dynamic pressure distribution in locomotion (Lord, 1981).

Further developments in *optic fiber technology* led to the design of a high resolution sensor plate to measure static and dynamic plantar pressure distribution (Hughes et al., 2000). A coherent light source was applied through an optically clear material such as Perspex. The beams of light combined and interfered at a certain point, producing an interferogram. Applying pressure deformed the surface, resulting in a change in the interferogram, the magnitude of which could be measured and evaluated. The device had a resolution of 70 pixels/cm^2, which is much higher than other pressure measuring systems. However, as only a small area of 6cm x 9cm could be examined in this prototype, future developments need to address this issue.

Totsu et al. (2005) developed an ultra-miniature fibre optic pressure sensor system of 125 μm in diameter. Two mirrors separated by a cavity formed a Fabry-Perot cavity at the end of the optic fibre. Pressure-induced change to the diaphragm of the cavity varied the cavity length. Changes in the spectrum of the reflected light were detected by a commercial spectrometer, which allowed the cavity length and applied pressure to be determined. This system has been used to measure blood pressure in the heart and arteries of a goat, thus making it suitable for small vessel pressure measurements.

Theyson et al. (1979) suspended about 1000 pins on high precision springs (*rod and spring elements*), recorded the shadows of the springs on videotape, and performed a computer analysis. Cavanagh and Ae (1978) further developed this methodology to provide qualitative (visual) and quantitative information on pressure distribution during locomotion. The rod and spring system is used currently in routine gait analysis in Germany, and is built in small numbers commercially.

Nicol (personal communication) used *strain gauge sensors* on a silicon membrane to quantify dynamic pressure distribution on soft surfaces, which provided improved flexibility when compared to earlier methods. These sensors seem more accurate than comparable single foil capacitor sensors used for soft surfaces.

New developments include *piezosheets* or *piezofoils*, which are extremely thin measuring sheets that allow static and dynamic quantification. One major application seems to be pressure distribution measurements in joints.

Other devices to measure pressure include *force-balanced sensors and resonant pressure sensors*. In a force-balanced sensor, applied pressure causes displacement of the membrane. The force required to maintain a non-deflected position, and therefore the output, vary according to this pressure. Some devices use capacitors that are used to produce electrostatic forces. These electrostatic forces are also inversely proportional to the distance between the capacitors, so they increase as pressure is applied. This cancels the pressure-induced deflection of the membrane. Resonant pressure sensors are based on changes in the resonant frequency of the membrane with applied pressure. These devices are not commonly used in biomechanics research at the current time, partly due to their complicated manufacture (Novotny & Kilpi, 2005).

THREE-DIMENSIONAL PRESSURE DISTRIBUTION

Pressure distribution is typically quantified for forces normal to the surface of interest, e.g., the plantar surface of the foot. However, it has been proposed that local shear forces are also an important component of mechanical stress and may be functionally more important to assess load and excessive load on the human body. A review of recent developments in this area (Urry, 1999) outlined some of the approaches used to develop sensors that allow simultaneous quantification of normal and shear stresses.

Initial studies used magneto-resistive devices, which consisted of two thin stainless steel discs to measure two-dimensional stresses (Laing et al., 1992; Lord et al., 1992; Perry et al., 1995). A magnet was attached to the top disc and a magneto-resistive sensor attached to the bottom disc. Adjacent discs contained a ridge or a corresponding groove to restrain the relative lateral movement between them. Movement of the magnet away from the central position was detected as a proportional change in the resistance and the electrical signal of the magneto-resistive element. The discs were bonded with silicone rubber, which provided a restoring force when the upper disc was displaced. The discs could be aligned to measure either longitudinal or transverse shear force (Urry et al., 2002). In a three-dimensional device, a third disc was added and the normal force was measured by a strain gauge diaphragm attached to the shear section (Urry et al., 2002). The shear section alone was used to measure plantar shear forces (Lord et al., 1992; Hosein & Lord, 2000; Lord & Hosein, 2000), while the triaxial device was used for stump socket shear pressure measurement (Williams et al., 1992 cited in Urry, 1999; Zhang et al., 1998).

Pressure transducers using piezoelectric film were used to measure shear stresses in two and three orthogonal directions (Akhlaghi & Pepper, 1996; Razian & Pepper, 2003). The device consisted of a 10x10 mm element of copolymer film sandwiched between three layers of 0.7 mm-thick double-sided circuit boards. The circuit boards provided mechanical stiffness and electrodes to which the cables could be attached. The layers were bonded together and the transducers embedded in an insole. The piezoelectric coefficient was defined as the ratio of electric charge developed along a specific axis to the mechanical stress applied along a specific axis of the piezoelectric element. The following relationship was used to calibrate the device:

$$\frac{d_i j = QA_f}{FA_e}$$

where:

$d_i j$	=	piezoelectric coefficient
F	=	applied force
Q	=	charge output
A_f	=	area of applied force distribution
A_e	=	electroded area (Razian and Pepper, 2003)

Several devices have used strain gauge technology (Davis et al., 1998) or combined it with other technologies, such as capacitance, to measure the vertical and shear stresses simultaneously (Christ et al., 1998 cited in Cavanagh, 2000; Heywood et al., 2004). Other methods used include using light emitting diodes (Lebar et al., 1993 cited in Urry et al., 2002) and fibre optic sensors (Lebar et al., 1996 cited in Urry, 1999; Koulaxouzidis et al., 2000).

Sensor size is the main limitation of currently available methods, although recent developments have allowed the manufacture of smaller devices, which are more suitable for in-shoe measurements. These initial studies provide the methods to measure shear pressures. Future studies using this technology to compare clinical groups will provide additional clinical information.

3.4.2 APPLICATIONS

Pressure distribution measuring devices are currently used in several major biomechanical and clinical fields of application.

Plates measuring pressure distribution are currently on the market with a sensor size of less than 0.5 cm^2. Plates of different construction were used in gait analysis (Elftman, 1934;

Grundy et al., 1975; Cavanagh & Ae, 1978; Clarke, 1980; Hutton & Dhanendran, 1981), and in assessing the diabetic foot (Stokes et al., 1975; Rhodes et al., 1988). General gait analysis concentrated on the analysis of temporal plantar pressure distribution patterns for different foot regions, the movement of the centre of force, and the plantar pressure distribution patterns for specific foot types (Clarke, 1980). Additionally, pressure distribution plates can be used to describe foot types dynamically.

Research by sport shoe manufacturers was partially responsible for the development of *shoe insole* devices that assess pressure distribution between the plantar aspect of the foot and the shoe insole (Hennig et al., 1981; Cavanagh et al., 1983). The devices were used to study the effect of specific shoe constructions on pressure distribution at the shoe-sole interface (Soames, 1985), the pressure distribution at the ski boot-shaft interface (Hauser & Schaff, 1987), or the estimation of internal forces in the human foot (Morlock and Nigg, 1991). Additionally, insoles were used to study the effect of foot type or orthotics, or both, on pressure distribution (Figure 3.4.2). Inshoe devices based on capacitive and force sensing resistor principles are currently commercially available.

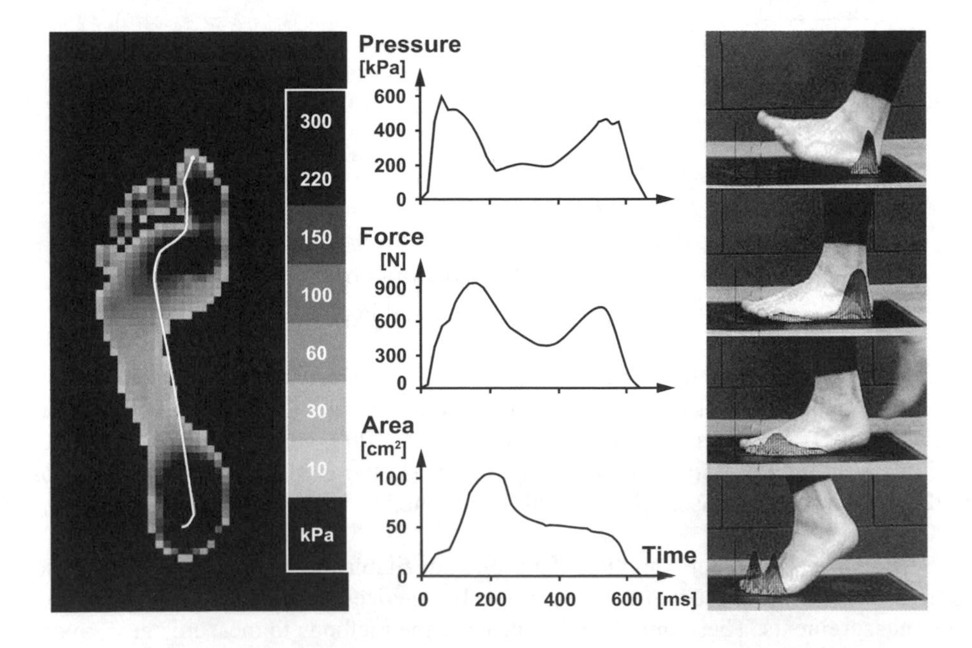

Figure 3.4.2 **Illustration of pressure distribution measurements between foot and ground measured with a pressure distribution platform during walking for one subject. The left picture shows the pressure in gray shades over the whole foot and the path of the center of pressure (COP). The numbers show the scale for the gray shades, with each number indicating that the gray shade corresponds to pressure values between the indicated number and the next higher number in kPa. The middle graphs indicate the pressure (top) the total force (middle) and the area with pressure (bottom) as a function of time (NOVEL GmbH, with permission).**

Pressure distribution measuring devices have also been used for *local pressure distribution* measurements in specific applications. Hehne (1983) used Fuji foils to experimentally determine maximal pressure distribution between the patella and femur in cadaver knees. Studies have examined pressures at the stump socket interface in subjects with femoral and transtibial prostheses to aid socket design (Lee et al., 1997; Sanders et al., 1998, in press). Many studies use strain gauges, although other technologies, such as force sensing resistors and printed circuit sheet sensors, have been used. The devices include individual sensors to measure pressures at discrete sites and sensor mats to measure pressure distribution at an interface (Mak et al., 2001). Stump socket pressure sensor systems are commercially available, but are used more for research than in clinical settings (Shem et al., 1998; Polliack et al., 2000). Capacitive or conducting devices have also been used to monitor pressure changes in the points of tooth contact during jaw closure (Podoloff and Benjamin, 1989; Liu & Herring, 2000).

An indirect application of pressure distribution measuring devices is combining thin conductive type transducer sheets with force plates, where the sheets are mounted on top of the force plate. In this application, the pressure distribution sensor is primarily used to determine the location of the foot on the force plate. This combination may prove powerful in clinical applications in which the combination of force and foot is of interest.

Several devices based on piezoresistive, force sensing resistor, capacitive, hydraulic, and pneumatic principles have been developed to measure pressure distribution and related comfort levels during sitting, lying, and positioning. Other applications include measuring the effects of material selection and ergonomic position during driving, assessing the forces exerted between the saddle and horse, and determining the pressure distribution of the rider.

In addition to the extensively discussed pressure distribution plates and insoles, the literature provides many studies that have quantified local forces with single sensors. These sensors were often held in place with adhesive tape (Schwartz & Heath, 1937; Bauman and Brand, 1963; Shereff et al., 1985; Soames, 1985), or embedded within an insole (Hennig & Milani, 1995; Sterzing & Hennig, 1999; Sanderson, 2000). Eight discrete piezoceramic transducers, placed under anatomical landmarks of the foot, were used to measure inshoe plantar pressure distribution during running to examine various types of footwear (Hennig & Milani, 1995) and fatigue effects (Sterzing & Hennig, 1999). A similar arrangement of twelve transducers was used to measure inshoe pressure during cycling (Sanderson, 2000).

The main advantage of these techniques is the limited electronic and computer power required. The main criticism of this technology is the potential variation of the positioning of the transducers during movement. This concern may be important for clinical and sport applications.

Methods that use plates or insoles for pressure distribution measurements can also be used to determine the path of the centre of force during one ground contact. Calculating the centre of force using this method is less sensitive to noise than the method described for force plates in chapter 4. An example for the path of the centre of force during a walking stride is illustrated in Figure 3.4.3. It is measured with a pressure distribution insole and shows the influence of an orthotic in a running shoe on the path of the centre of force.

The methods for quantifying pressure distribution between foot and shoe are relatively new, and reliable instrumentation has only been on the market for a few years. Current research comparing these systems has generally found that capacitive systems are more accurate than resistance insoles and mats (Hsiao et al., 2002; Quesada et al., 1996; Quesada & Rash, 2000). However, such devices can potentially be applied to various tasks, since they provide easily understandable on-line information. The findings may have implications for footwear selection, discomfort, lower extremity pain, and walking ability (Hennig, 2002). Groups that may use such devices include gait analysis laboratories in hospitals, sport medicine centres, podiatry centres, sport shoe stores, and sport training centres, e.g., golf. However, specific research is needed to provide the necessary background for these applications.

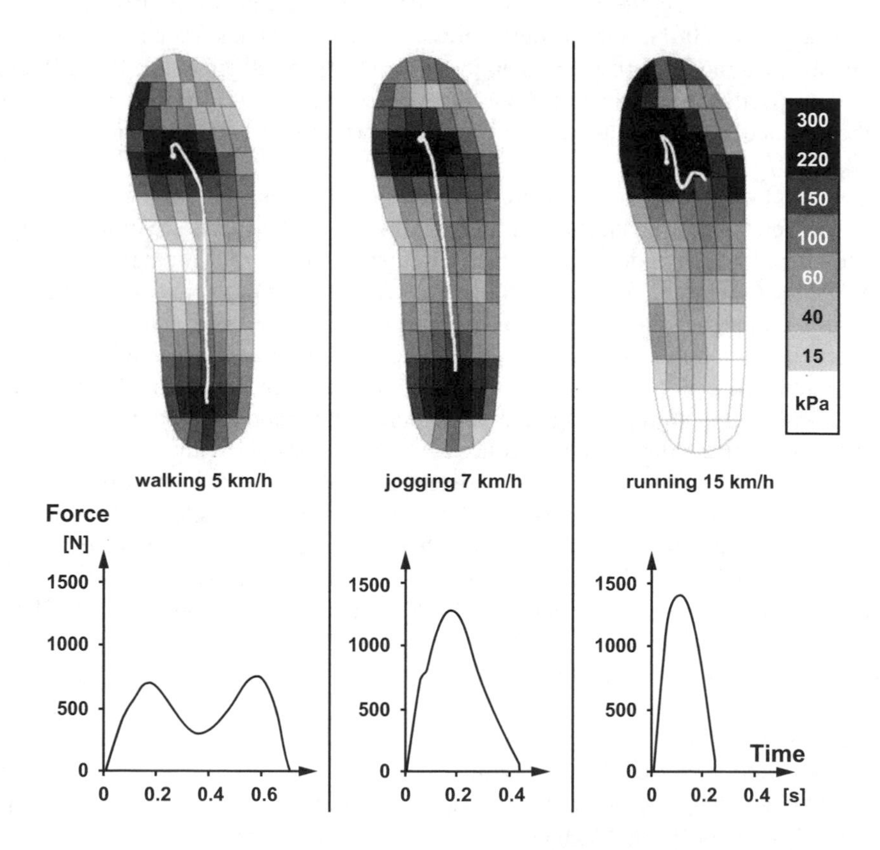

Figure 3.4.3 **Illustration of pressure distribution and path of center of pressure for walking (left), jogging (center) and forefoot running (right). The numbers on the right side of the graph show the scale for the gray shades with each number indicating that the gray shade corresponds to pressure values between the indicated number and the next higher number in kPa (NOVEL GmbH, with permission).**

The techniques for quantifying pressure distribution discussed in this section have primarily been used for applications related to the human foot, which is where biomechanical research applications have concentrated in the last few decades. However, the methods are applicable to most other biomechanical studies in which pressure distribution is the appropriate technique. Therefore, the use of these devices for other biomechanical applications has increased in recent years.

3.5 ACCELERATION

NIGG, B.M.
BOYER, K.A.

3.5.1 MEASURING POSSIBILITIES

Acceleration can be determined by using the measuring techniques described in this chapter, or by using positional information (from film or video measurements) and calculating the second time derivative. Sensors described as *accelerometers* belong to the first category. They include strain gauge, piezoresistive, piezoelectric, and inductive transducers.

STRAIN GAUGE AND PIEZORESISTIVE ACCELEROMETERS

A strain gauge accelerometer consists of several strain sensitive wires (often four) attached to a cantilevered mass element that is mounted on a fixed base. The wires are connected to an electric Wheatstone bridge circuit. If the base is accelerated, the mass element causes a deformation due to its inertia. Deformation of the mass element causes a change of the strain in the wires, therefore changing their resistance, and consequently changing the balance of the bridge circuit. The result is an electric output proportional to the acceleration of the base.

A piezoresistive accelerometer is based on the same principle as a strain gauge accelerometer, except that piezoresistive, instead of wire strain gauge elements, are used. The transducers are typically solid state, single silicon crystals that change their electric resistance in proportion to the applied mechanical stress. The design of a piezoresistive accelerometer is illustrated in Figure 3.5.1. For an acceleration in the illustrated direction, which corresponds to an axis of measurement in this example, and for the corresponding displace-

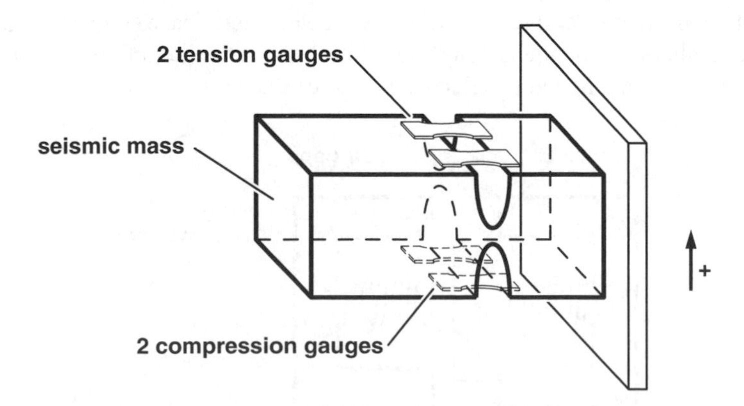

Figure 3.5.1 **Illustration of a piezoresistive accelerometer (from Instruction Manual for Endevco Piezoresistive Accelerometers, 1978, with permission).**

ment of the mass element, the two top gauge elements operate in tension and the two bottom gauge elements in compression (Figure 3.5.2).

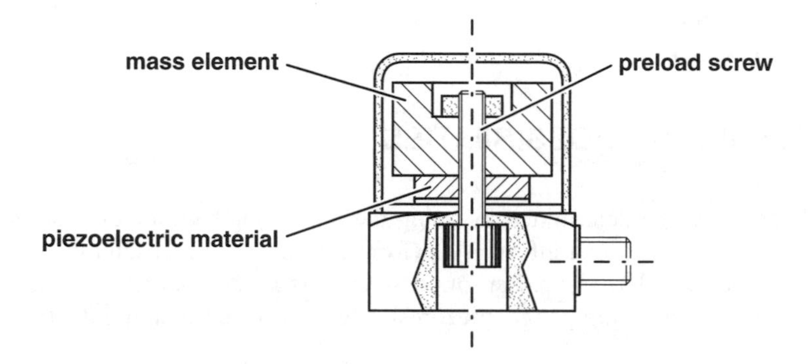

Figure 3.5.2 **Schematic illustration of the construction of a piezoelectric accelerometer (from Bouche, 1974, with permission).**

PIEZOELECTRIC ACCELEROMETERS

Piezoelectric accelerometers work on the same principles as outlined in chapter 3.3. They often use ceramic materials with piezoelectric properties, producing an electric output in response to a stress. In a typical setup, the piezoelectric material is positioned and preloaded in compression between the base of the accelerometer and a mass element (see chapter 3.2). Vibration of the base and the inertia of the mass element create dynamic stress and deform the piezoelectric material, resulting in an electric output.

INDUCTIVE ACCELEROMETERS

An inductive accelerometer consists of a mass element positioned and magnetically coupled between a pair of coils attached to the accelerometer base (Figure 3.5.3). Accelerating the mass alters the magnetic coupling. This changes the inductive current of the coils and is measured as a change in the electric output of the coils.

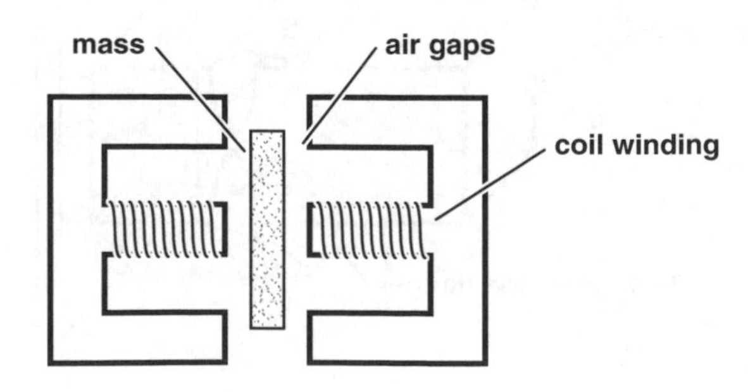

Figure 3.5.3 **Schematic illustration of the construction of an inductive accelerometer.**

The characteristics of accelerometers correspond to the characteristics of the force sensors constructed on the same principles. However, due to the specific use of accelerometers in biomechanics, accelerometer-specific characteristics can be discussed. Factors of specific interest for accelerometers include mass, natural frequency, frequency response, shock rating, and range (Table 3.5.1).

Table 3.5.1 Summary of selected characteristics of accelerometers typically used in biomechanical experiments. Specific data was extracted from sales brochures of major suppliers or from our own experience, or both.

CHARACTERISTIC	SENSOR TYPE			
	STRAIN GAUGE	PIEZO-RESISTIVE	PIEZO-ELECTRIC	INDUCTIVE
MASS	low 1-2 g	low 1-2 g	low 1-2 g	higher several grams
NATURAL FREQUENCY	2000 Hz to 5000 Hz	2000 Hz to 5000 Hz	20,000 to 30,000 Hz	200 to 400 Hz
FREQUENCY RESPONSE	0 to 1000 Hz	0 to 1000 Hz	0 to 5000 Hz	low no shock measurements
SHOCK RATING	one magnitude higher than upper limit of range	one magnitude higher than upper limit of range	several magnitudes higher than upper limit of range	several magnitudes higher than upper limit of range
RANGE	limited to about one magnitude	limited to about one magnitude	0.01 to 10,000 g	about 30 g
ADVANTAGES	measure static and dynamic accelerations	measure static and dynamic accelerations	excellent range	accuracy in low frequency accelerations
DISADVANTAGES	limited range	limited range	limited use for static measure-ments	limited range

Most accelerometers available for the measurements in biomechanics are extremely light and weigh only a few grams. The natural frequency is typically high and the frequency response is sufficiently high for the frequencies relevant for biomechanical measurements (an exception are the inductive accelerometers). The shock rating for strain gauges and piezoresistive accelerometers is typically about one magnitude higher than the upper limit of the range, and may be a safety problem for accelerometers with low measuring ranges (dropping of such accelerometers usually results in the destruction of the sensor). Piezoelectric accelerometers have the advantage that their range is excellent, while the range for strain gauges and piezoresistive accelerometers is typically about one magnitude. Accelerometers used for biomechanical measurements apply typical strain gauge techniques with conventional or piezoresistive strain wires or piezoelectric accelerometers. Strain gauges using the piezoelectric principle are advantageous for highly dynamic acceleration measurements. However, they are not an ideal solution for static or quasi-static measurements. Inductive accelerometers are rarely used in biomechanical applications.

OTHER METHODS

The methods described above quantify the linear acceleration of an idealized mass point. However, in some specific applications, angular acceleration may be of interest. A magneto-hydrodynamic acceleration sensor has been proposed for assessing angular acceleration. The principle of operation was described by Laughlin. This description is a summary of his publication (Laughlin, 1989) and the general description that appeared in the journal *Sensor* (September 1989, pp. 32-37). A permanent magnet near a constrained annulus of conductive fluid produces a constant magnetic field with flux lines perpendicular to the fluid annulus (Figure 3.5.4). As the sensor moves, the fixed permanent magnet moves

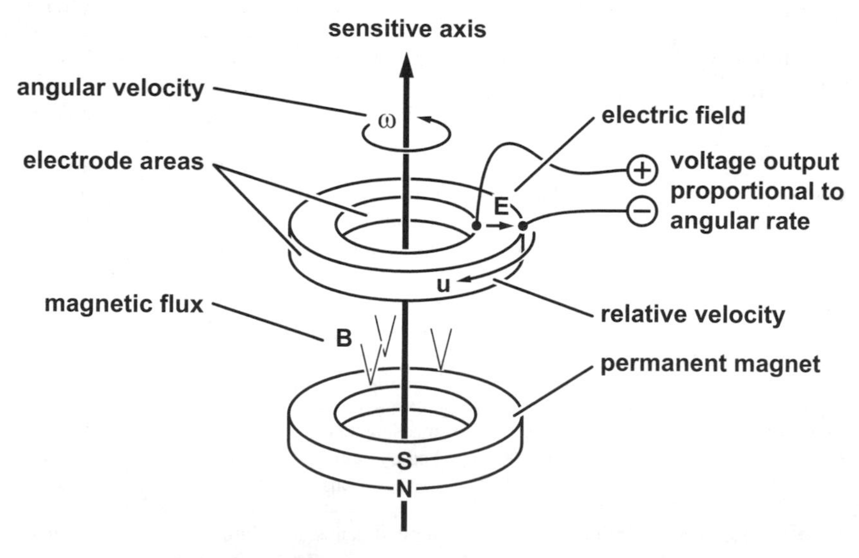

Figure 3.5.4 **Schematic illustration of a magneto-hydrodynamic angular acceleration sensor (from Laughlin, 1989, with permission).**

with the case. The fluid tends to remain inertially stable about its rotational sensitive axis, but tends to move with the sensor in translation and cross-axial rotation. A rotation of the sensor about its sensitive axis results in a velocity difference, **u**, between the fluid and the applied magnetic flux density, **B**. This induces an electric field, **E**, that is radial in direction and proportional in amplitude to, **u**, and, **B**:

$$\mathbf{E} = \mathbf{u} \cdot \mathbf{B} \tag{3.5.1}$$

The circumferential velocity, **u**, is determined by:

$$\mathbf{u} = r \cdot \omega \tag{3.5.2}$$

where:

r = root mean square radius of the channel

ω = input angular velocity of the sensor

The voltage, **V**, generated radially across the annulus of fluid is the integration of the induced electric field, **E**, over the width of the fluid channel:

$$V = \int_{r_i}^{r_o} \mathbf{E}(r)dr \tag{3.5.3}$$

where:

V = voltage across channel

r_i = inner channel wall radius

r_o = outer channel wall radius

The voltage is electronically amplified and is used to quantify angular acceleration. A special arrangement of three such instruments can be used to quantify three-dimensional angular acceleration and has been used in car crash experiments.

MEASURING SEGMENTAL ACCELERATION WITH ACCELEROMETERS

In experiments with human test subjects where accelerations of selected body segments ought to be measured, accelerometers may be mounted in various ways and at different locations to the segment of interest. However, this segment consists of rigid and soft tissue. For such measurements, several questions are of interest, including:

- Which acceleration should be determined: the acceleration of a specific rigid part of the segment, the acceleration of a specific soft tissue part or an average acceleration of rigid and soft tissue?
- How well does the measured acceleration correspond to the actual acceleration of interest?

Relevant acceleration

The acceleration at different locations of a bone is usually different, as demonstrated with axial in vitro acceleration measurements on a tibia and a femur mounted on a shaker (Chu et al., 1986). Additionally, rotational aspects may become important. Consider, for instance, a steel beam landing in an inclined position on a steel plate. The accelerations of the contacting end, the centre of mass, or the other end of that beam are different. Consequently, it is important to decide which acceleration is relevant to answer the question of interest for a specific research project.

This question may become important when acceleration measurements are used to determine impact ground reaction forces. In principle, the (impact) ground reaction force can be determined by adding all the mass times acceleration terms for each segment of the body and subtracting the body weight. However, the rigid and the non-rigid masses of a segment do not have the same acceleration. One may solve the problem by introducing an *effective acceleration*, $\mathbf{a}_{\mathbf{eff}}$, which is the acceleration with which the segmental mass would have to be multiplied to describe the actual inertial contribution of the specific segment. The equation of motion would then read:

$$\sum m_i \cdot (\mathbf{a}_{\mathbf{eff}})_i \quad = \quad \mathbf{F}_{\mathbf{ground}}(t) - \sum m_i \cdot \mathbf{g} \tag{3.5.4}$$

However, this would not solve the question of how this effective acceleration should or could be measured.

Error in the measured acceleration

The magnitude of the acceleration measured with an accelerometer depends on:

- Bone acceleration,
- Mounting interaction,
- Angular motion,
- Gravity, and
- Cross-sensitivity.

The influence of these factors is discussed in the following paragraphs.

Bone acceleration

Accelerometers may be mounted by screwing them onto the bone of interest (Light et al., 1980; Lafortune & Cavanagh, 1987), or by strapping them to the segment of interest at a location with minimal soft tissue between the accelerometer and the bone of interest. In any mounting case, the acceleration measured does not represent the bone acceleration, but instead represents the acceleration of a specific mass element at the surface of the bone or even of a point outside the bone (if the accelerometer is screwed onto the bone). Acceleration of a specific bone location can then be determined mathematically from several acceleration measurements or from additional measurements (with film, for instance).

Mounting interaction

Accelerometers screwed onto the bone measure the acceleration outside the bone at the location of the accelerometer. Since the connection of the screw to the bone is usually rigid, resonance problems are minimal. The only correction that must be included is the reduction of the signal to the location of interest, e.g., the proximal end of the tibia.

However, accelerations measured with accelerometers strapped to the segment of interest with soft tissue between the mounted accelerometer and the underlying bone may not reflect the acceleration of the underlying bone. The following simple experiment illustrates this statement. Two accelerometers were mounted on a wooden rod. The first accelerometer was screwed to the rod, providing the true acceleration of the rod, a_{true}. The second accelerometer was strapped, with rubber bands, to the rod, with a water bag between the accelerometer and the rod (Figure 3.5.5), providing the acceleration measured with the skin-mounted accelerometer, a_{skin}. The strapping of this second accelerometer was arbitrarily

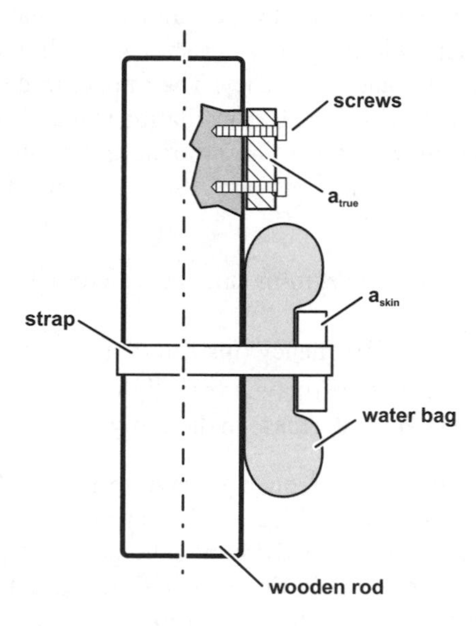

Figure 3.5.5 **Schematic illustration of an experiment using two accelerometers mounted to wooden rod. The first accelerometer is screwed to the rod, and the second strapped to the rod with a water bag between the accelerometer and the rod.**

defined as light, medium, and strong, based on the subjective feeling of the experimenter. The rod was dropped from several different heights onto three different surfaces: linoleum with the thickness of 0.3 cm, artificial sport surface (tartan) with the thickness of 1.2 cm, and foam rubber with the thickness of 6 cm, all mounted on concrete. During ground contact, the maximal accelerations, a_{true} and a_{skin}, were determined for each drop and the two measured values were represented as one data point in a a_{skin} - a_{true} diagram (Figure 3.5.6). Additionally, the line $a_{skin} = a_{true}$ is indicated with a dotted line. For the ideal case where the two accelerations provided identical results, all experimental data points would be on this dotted line.

The results of this experiment can be summarized as follows:

(1) The two actually measured acceleration amplitudes, a_{true} and a_{skin}, were only identical in a few cases. The biggest difference between a_{true} and a_{skin} was about 700%.

(2) A specific result for a_{skin} can be bigger, equal or smaller than the corresponding result for a_{true}. The acceleration amplitudes for a_{skin} were smaller than or equal to a_{true} for the light strapping, and bigger as well as smaller for the strong strapping.

(3) The measured skin accelerations, a_{skin}, were influenced by the surface on which the rod was dropped. For the hardest surface (linoleum) all amplitudes of a_{skin} were smaller than the amplitudes of a_{true}. For the softest surface (foam rubber) about 50% of all amplitudes of a_{skin} were larger than the amplitudes of a_{true}.

The results may, simplistically, be explained as follows: the accelerometer strapped onto the rod with a water bag between the rod and the accelerometer is, mechanically speaking, a vibrating system with the characteristics of such a system such as stiffness, damping, and natural and resonance frequency. The impact force at the moment when the rod contacts the surface may be considered an excitation signal. Due to this impact force the strapped-on accelerometer may perform translational and rotational movements with respect to the rod. Several assumptions are used for the following considerations (Figure 3.5.6):

- The relative movement between skin-mounted accelerometer and the rod is translational,
- The excitation signal has one frequency (instead of a frequency spectrum), and
- An increase in the tightness of the fixation from light to medium to strong corresponds to an increase in stiffness and a decrease in damping.

The idealized relative amplitude-frequency diagram for the three fixations used, with these assumptions in effect, is shown in Figure 3.5.7. The forces during one specific contact of the rod with the surface for one selected dropping height correspond to excitation signals with different frequencies. The signal for linoleum has the highest frequency, and foam rubber has the lowest frequency. For illustration purposes, results with similar input signals a_{true} from Figure 3.5.6 are illustrated in Figure 3.5.7. The selected experimental result for linoleum for the light fixation is indicated with a number 1, the medium fixation is indicated with a number 2, and the strong fixation with a number 3. The results for the tartan and foam rubber surfaces are numbered in a similar way. The schematic illustration in Figure 3.5.7 contains the following messages that are relevant for acceleration measurements:

- The tightness with which an accelerometer is strapped to a human or animal segment influences stiffness and damping, which, in turn, affect the amplitude of the measured acceleration.
- Accelerations measured with strapped-on accelerometers, a_{skin}, for high excitation frequencies are often lower than the accelerations measured with screwed on accelerometers, a_{true}.
- Accelerations measured with strapped-on accelerometers depend on the excitation frequency of the impact force responsible for the acceleration.

An increase in the excitation frequency relates to a decrease in the acceleration measured with a strapped-on accelerometer, a_{skin}, if the excitation frequencies are above the natural frequency of the mounted accelerometer system.:

As mentioned earlier, this is a simplistic way to describe the actual situation. In reality, the situation is more complex. Two important differences are: 1) the movement of the strapped-on accelerometer is translational and rotational, and 2) the excitation signal has a frequency spectrum. However, the basic message does not change for the more complex, actual situation.

The discussed problems may suggest that acceleration measurements with strapped-on accelerometers should be treated with caution, and one may wonder what strategies could be applied to allow a restricted use of acceleration results in biomechanical experiments. Some suggestions are discussed in the following.

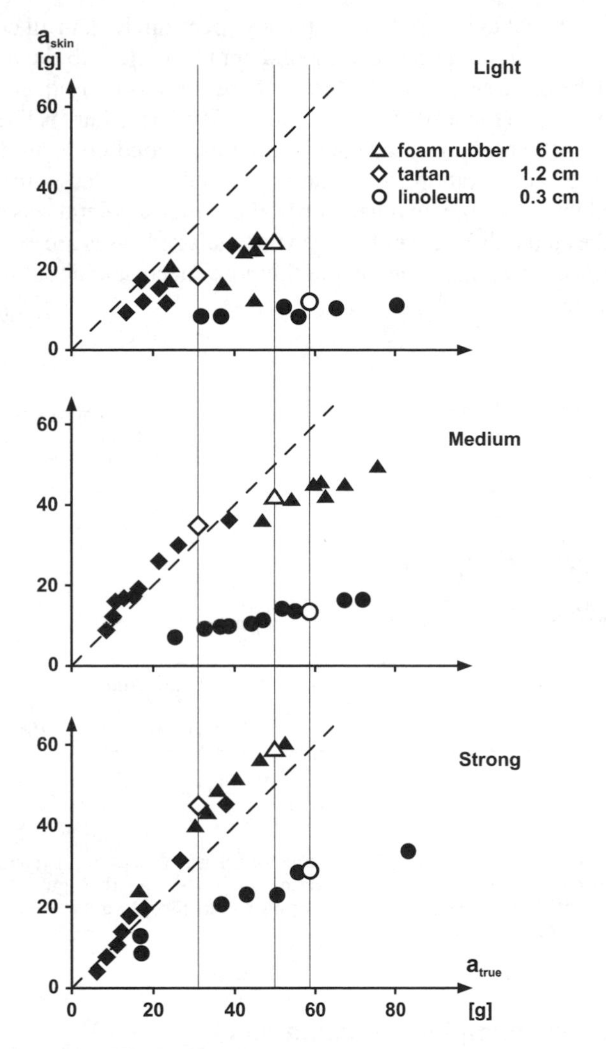

Figure 3.5.6 **Experimental acceleration results for a_{true} and a_{skin} for three different strapping procedures, light, medium, and strong, for drops onto three different surfaces.**

Understanding the direction of the error

The knowledge of the tendency in which the error of the measured acceleration develops may help to support acceleration results in applications where accelerations are compared for different situations.

EXAMPLE 1

Let us assume that acceleration measurements at the crest of the tibia have been made for running with shoes with a soft, a medium, and a hard midsole. Let us assume that the mean results were 10 g for the soft, 13 g for the medium, and 16 g for the hard midsole and that the differences were significant. Let us further assume that the natural frequency of the mounting was 5 Hz and that the excitation frequency spectrum had, in all cases, a maximum higher than 10 Hz. These assumptions correspond approximately to the medium strapping in Figure 3.5.7 and the three surfaces. The three measurements consequently correspond in principle to three points: 8 (for soft), 5 (for medium), and 2 (for hard). Therefore, the measured acceleration for the medium shoe sole is underestimated compared to the soft shoe sole, and the measured acceleration for the hard shoe sole is underestimated compared to the soft and the medium shoe sole. In other words, the actual accelerations of the crest of the tibia are such that the actual differences between the analyzed cases are even bigger than the ones shown in the measurements. Note that nothing has been said about the biological relevance of this statement.

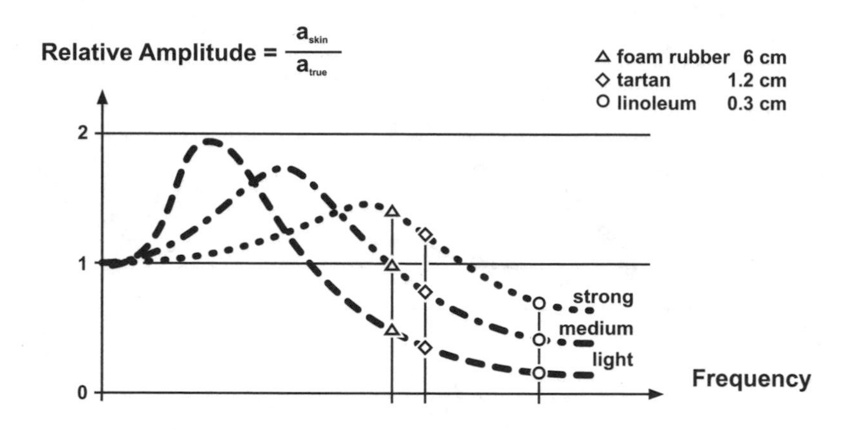

Figure 3.5.7 **Schematic illustration of a relative amplitude (a_{skin}: a_{true})-frequency diagram for different fixation modes of a setup, as illustrated in Figure 3.5.5. The idealized excitation frequencies for landing on the different surfaces have been arbitrarily selected.**

Appropriate stiffness/damping combinations

Use stiffness and damping combinations with a relatively flat relative amplitude-frequency diagram for the dominant frequency content of the excitation signal (Neukomm & Denoth, 1978).

EXAMPLE 2

Assume that the dominant frequency content of the acceleration signal of interest is between 20 and 40 Hz. A mounting technique with a damping coefficient of $c = 200\ s^{-1}$, and a natural frequency of 50 Hz provides a relative amplitude-frequency diagram that is relatively flat in the frequency region between 20 and 40 Hz (Figure 3.5.8). Consequently, the error is constant (about + 10 to 15%) and comparisons may be less critical.

Similar experimental and theoretical considerations were presented by Valiant et al. (1987). Using a simplistic linear model, they estimated a magnification factor for the mea-

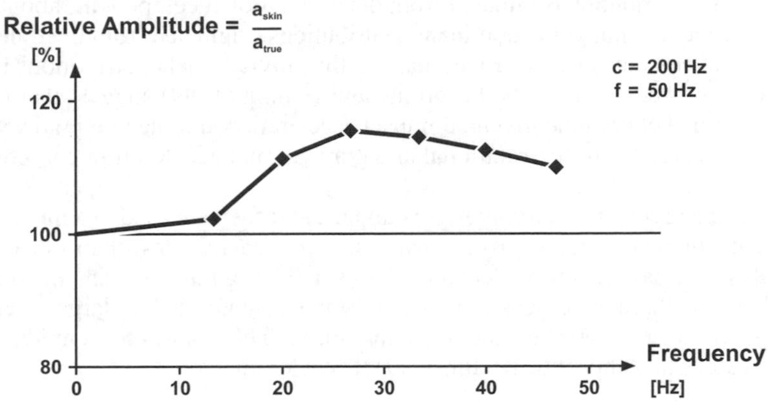

Figure 3.5.8 **Schematic relative amplitude-frequency diagram for a damping coefficient of c=200 s^{-1}, and a natural frequency of 50 Hz (from Neukomm & Denoth, 1978, with permission).**

sured acceleration amplitudes and concluded that in their specific case (a damping ratio of 0.4 for an accelerometer mass of 4.4 g) the acceleration amplitudes had been overestimated by 20 to 30%, and that the overestimation was higher for the barefoot measures than for the measures with running shoes. This result is in agreement with the results of the general considerations of this book.

Translational, rotational, and gravitational components of acceleration

Acceleration measurements on a segment of the human body provide a signal which is composed of a translational, a rotational, and a gravitational component (Winter, 1979; Lafortune & Hennig, 1989). A point at the tibia during landing, e.g., the tibial tuberosity, experiences a rotational acceleration due to the rotation near the ankle joint and a translational acceleration due to the deceleration of the foot and lower leg during initial contact with the ground.

$$\mathbf{a}_{tot} = \mathbf{a}_{tr} + \mathbf{a}_{rot} + \mathbf{a}_{gra} \tag{3.5.5}$$

where:

$\mathbf{a}_{tot}$ = total acceleration measured with an accelerometer mounted to the rigid structure

$\mathbf{a}_{tr}$ = contribution to the total acceleration due to the translational acceleration of a point with no rotation, e.g., contact point

$\mathbf{a}_{rot}$ = contribution to the total acceleration due to the rotation of the rigid structure

$\mathbf{a}_{gra}$ = contribution to the total acceleration due to gravity

Accelerations measured during human or animal locomotion have different combinations of the three acceleration components, depending on the actual movement. Actual contributions of translational, rotational, and gravitational acceleration components have been presented for human walking and running (Lafortune & Hennig, 1989). Accelerations at the human tibia were measured with triaxial strain gauge accelerometers secured to a traction pin which was inserted in the tibia 3 cm below the proximal articular surface. During impact in walking, the dominant contribution originated from the translational component. A second, smaller, contribution originated from the gravitational component (about 15 to 20%). During impact in running, the dominant contributions originated from the translational and the rotational components. The rotational and the gravitational contributions in this part of the movement were about 45%. Lafortune and Hennig (1989) suggest that the relatively low correlation between the maximal impact acceleration and the maximal vertical impact force (0.76) is related to the rotational and gravitational acceleration components during impact in running.

Cross-sensitivity of accelerometers is important if they are used in a three- dimensional movement situation, although cross-sensitivities are generally less than 1%. Commonly, in gait analysis we have major movement patterns in the sagittal plane and the minors ones in other planes, so that a cross-sensitivity from the large signal can be a large percentage of the output of an accelerometer sensing in another plane. This applies to individual accelerometers as well as to compact three-dimensional accelerometers.

3.5.2 APPLICATIONS

ORDER OF MAGNITUDE OF MEASURED ACCELERATIONS

The peak accelerations measured during selected activities are summarized in Table 3.5.2. The results illustrate several factors that influence the acceleration or deceleration of a body segment.

Table 3.5.2 Order of magnitude of peak acceleration values measured in selected activities (based on data from Nigg et al., 1974; Unold, 1974; Voloshin & Wosk, 1982; Lafortune & Hennig, 1991, with permission). Acceleration values are not corrected for mounting artifacts, or for gravitational or rotational components, or both.

MOVEMENT	SPECIFIC COMMENTS	ORDER OF PEAK MAGNITUDE ACCELERATIONS OF		
		HEAD [g]	HIP [g]	TIBIA [g]
SKIING:				
POWDER SNOW	10 m/s	1	2	4-6
PACKED RUN	10 m/s	2	3	30-60
PACKED RUN	15 m/s	-	-	60-120
PACKED RUN	25 m/s	-	-	100-200
WALKING		1	1	2-5
RUNNING:				
HEEL-TOE	on asphalt	1-3	2-4	5-17
HEEL-TOE	on grass	1-3	2-4	5-10
TOE	on asphalt	1-3	2-4	5-12
GYMNASTICS:				
LANDING FROM 1.5M	on 7 cm mat	3-7	8-14	25-35
LANDING FROM 1.5M	on 40 cm mat	2	5	8
ROUND OFF SOMERSAULT	on 7 cm mat	3	14	24
STRADDLE DISMOUNT	on 40 cm mat	3	8	10
TAKE OFF MINITRAMP		3	7	9

The results indicate that accelerations seem to be smaller:

- For increasing body masses,
- With increasing number of joints between the input force and the measurement location, and
- For soft than for hard surfaces.

Acceleration values of 200 g during alpine skiing may be impressive and indicate high loading. However, the magnitude of the acceleration is not the only important factor in the analysis of a load to which the human body is exposed. Knocking with the knuckle of the index finger at a door may correspond to accelerations of the finger joint of 100 to 200 g. However, this action is not considered dangerous. Consequently, one may dispute the relevance of simple acceleration measurements in the context of load analysis.

SOFT TISSUE VIBRATIONS AND MUSCLE TUNING DURING LOCOMOTION

When quantifying skeletal movement, the vibrations of soft tissue packages are considered an artifact. However, vibrations of soft tissue may contain information of interest for biomechanical considerations. For example, soft tissue vibrations have been studied in the context of impact forces and muscle tuning (Nigg, 1997; Nigg & Wakeling, 2001; Wakeling et al., 2002; Wakeling et al., 2003; Boyer & Nigg, 2004; Boyer & Nigg, 2006).

To understand soft tissue vibrations and related muscle tuning, the human body must be considered as a mechanical system with a rigid skeleton and attached wobbling masses (Garg & Ross, 1976; von Gierke, 1971; Yue & Mester, 2002; Liu & Nigg, 2000; Nigg & Liu, 1999). Following an excitation pulse, a single degree of freedom system vibrates at its natural frequency. The natural frequency is determined by the mass, m, and spring stiffness, k, with:

$$\Omega_n = \sqrt{\frac{k}{m}} \tag{3.5.6}$$

or

$$f_n = \frac{\Omega_n}{2\pi} \tag{3.5.7}$$

The time interval to complete one cycle is the period:

$$T = \frac{2\pi}{\Omega_n} \tag{3.5.8}$$

For a simple damped system, the damped natural frequency is:

$$\Omega_d = \Omega_n(\sqrt{1-\zeta^2}) \tag{3.5.9}$$

$$\zeta = \frac{c}{c_c} \tag{3.5.10}$$

$$c_c = 2\sqrt{km} \tag{3.5.11}$$

where:

Ω_n = natural circular frequency of the system
Ω_d = damped natural circular frequency of the system
f_n = natural frequency of the system
k = spring stiffness
m = mass of the system
ζ = fraction of critical damping
c = damping coefficient
c_c = critical damping coefficient

The magnitude of the vibration response is a function of the shape and frequency of the input signal and the natural frequency of the system. If the frequency of the input signal is close to the natural frequency of the system, then the magnitude of the response is greatest (Figure 3.5.9). For dynamic activities such as walking, running, or jumping, the input signal can be modeled in a first approximation as a sine or a versed sine pulse. In this case, the frequency is of the input signal is the inverse of the pulse length.

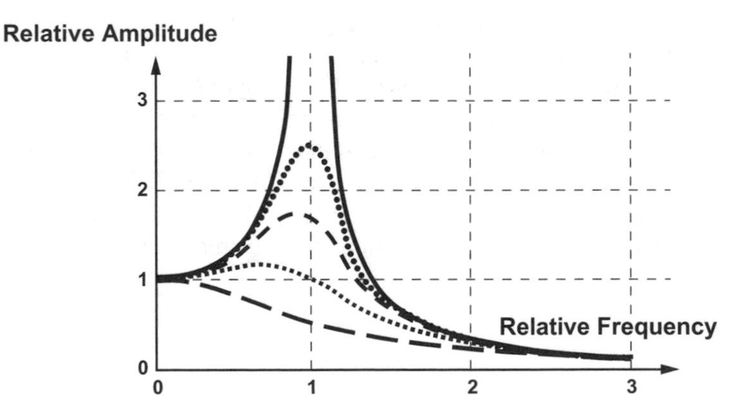

Figure 3.5.9 **Response spectrum for a simple vibration system with a sinusoidal input. A relative frequency of 1 corresponds to the condition when the frequency of the input signal equals the natural frequency of the vibrating mass. The response spectrum for a shock-like input signal shows, in principle, the same behaviour.**

The concept of muscle tuning as a reaction to impact forces suggests that the human locomotor system attempts to avoid a mechanical resonance situation using a muscle adaptation. Resonance can be avoided by changing the mechanical characteristics of a soft tissue package or by changing the input signal, or both.

Changing the mechanical characteristics

Changes in natural frequencies of soft tissue packages, as a result of changes in muscle activity for the triceps surae and quadriceps, have been assessed experimentally (Wakeling & Nigg, 2001). Changes ranged between 10 and 50 Hz. Corresponding changes in damping coefficients were up to 400%. Similar changes were found for the tibialis anterior and the hamstrings.

Changing the input signal

The original input signal of an impact force into the human leg is the ground reaction force. However, the relevant input signal for vibrations of the soft tissue package of the triceps surae is the acceleration of the tibia. The input signal for the soft tissue packages of the quadriceps and the hamstrings is the acceleration of the femur. For heel-toe running, the input signal for the triceps surae can be quantified with an accelerometer fixed to the heel-cup of the shoe, or by using a motion analysis system and reflective skin markers placed just proximal to the lateral malleolus.

Soft tissue accelerations are measured relative to an inertia reference frame, and are therefore a convolution of the input signal, e.g. bone acceleration, and the relative soft tissue acceleration. To quantify the mechanical response of the soft tissue package to different input signals, a biodynamic response function is calculated, which is a transfer function of the input signal through the body. Typical biodynamic response functions include (Griffin, 2001):

- Apparent mass (force per unit acceleration),
- Mechanical impedance (force per unit velocity),
- Transmissibility (transfer of acceleration through structure), and
- Apparent stiffness (force per unit displacement).

The *transmissibility* of the input acceleration to the soft tissue package describes its mechanical response for a specific activity. In regions of resonance frequencies, the transmissibility reaches a local maximum and the phase angle of the transmissibility function shows a steep slope. The relative magnitude of a peak in the transmissibility curve indicates the magnitude of damping within the system for a particular experimental condition.

The acceleration transmissibility is determined from the ratio of the frequency spectrums of the input and output acceleration signals. For running or jumping, a Hanning window function is used to take the data around heel-strike. The windowed data is then padded with zeros on both ends to increase the data length to an nth power of 2. Using a Discrete Fourier transformation, the data is transformed into the frequency domain. A combination of the auto and cross power spectrums are used to minimize the effects of signal noise on the transfer function result (Harris, 1995).

The auto power spectrum is calculated with the formula:

$$S_{II} = I(f) \cdot \overline{I(f)} \tag{3.5.12}$$

where:

$I(f)$ = input signal amplitude as a function of frequency
$\overline{I(f)}$ = complex conjugate of input signal frequency spectrum

The cross power spectrum is determined as:

$$S_{AI} = A(f) \cdot \overline{I(f)} \tag{3.5.13}$$

where:

$A(f)$ = acceleration signal amplitude as a function of frequency f
$\overline{I(f)}$ = complex conjugate of input frequency spectrum

The transmissibility is calculated using average auto and cross power spectrums from all landings for a specific experimental condition. Using the average of ten or more landings reduces noise effects and smoothes measurements, producing a more reliable biodynamic response of the system. The transfer function, transmissibility, is calculated using:

$$H(f) = \frac{\overline{S_{AI}(f)}}{\overline{S_{II}(f)}} \tag{3.5.14}$$

where:

$\overline{S_{AI}(f)}$ = mean cross power sectrum of the input signal and output signals

$\overline{S_{II}(f)}$ = mean auto spectrum of the input signal

A coherence function should be calculated to check if the input signal used is related to the resulting vibrations measured at the soft tissue package. The coherence function is defined as the ratio of two biodynamic response estimates. At a given frequency, the coherence indicates if the vibration recorded at a given frequency is due to the input signal. The coherence is zero when the vibration is not due to the applied input force/acceleration, and one if the input and output signal are related. The coherence function is determined as:

$$\gamma_{AI}^{2} = \frac{\left|\overline{S_{AI}\langle f\rangle}\right|^{2}}{\overline{S_{II}\langle f\rangle S_{AA}\langle f\rangle}} \tag{3.5.15}$$

where:

$S_{AA}\langle f\rangle$ = mean auto spectrum of the output signal

$S_{AI}\langle f\rangle$ = mean cross power spectrum of the input and output signals

$S_{II}\langle f\rangle$ = mean auto spectrum of the input signal

Muscle tuning

The concept of muscle tuning suggests that a muscle response occurs in reaction to an input signal with a frequency near the resonance frequency of the related soft tissue package. If a muscle response to such an input signal does not occur, then an increase in the magnitude of vibration is expected (Nigg, 1997; Nigg & Wakeling, 2001). The effect of this muscle response is a change in the mechanical properties of the soft tissue packages (natural frequency or the damping properties, or both) to minimize resonance responses.

Muscle tuning reactions can be quantified by comparing changes of (a) EMG pre-activation, (b) mean EMG frequency at landing, and (c) acceleration transmissibility for two different input signals. If a muscle tuning reaction has occurred, a greater EMG intensity or mean frequency, or both, is expected, along with a decrease in the magnitude of the transmissibility at the resonance frequency.

The transmissibility and EMG pre-activation intensity results for one subject exposed to repetitive impacts with two different shoe conditions (elastic and visco-elastic) on a pendulum apparatus shows a muscle tuning reaction for the visco-elastic shoe (Figure 3.5.10). The muscle response shows a greater EMG pre-activation intensity occurring in response to the visco-elastic shoe condition. The main peak in the transmissibility for both shoes occurred at the same frequency (about 12 Hz). However, the magnitude of this peak differs, indicating no change in resonance frequency, but an increase in damping for the visco-elastic shoe condition. Thus, an increase in the EMG intensity occurred to increase the damping of the soft tissue package, which is a muscle tuning reaction. Similar reactions were quantified for most subjects tested.

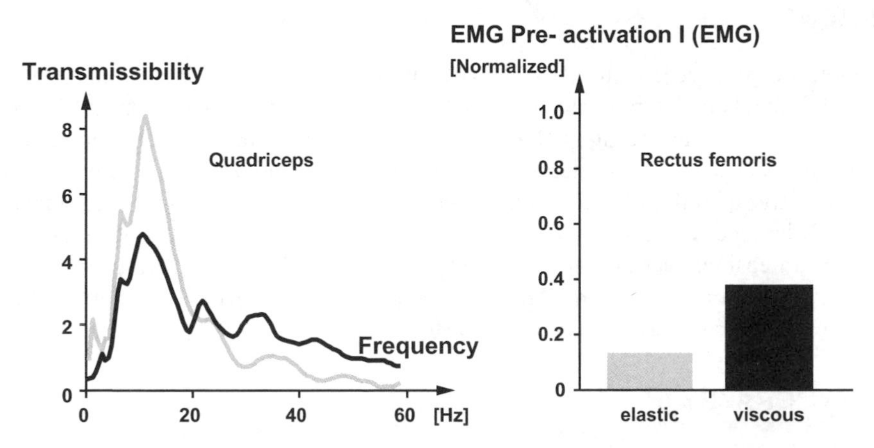

Figure 3.5.10 **Transmissibility (left) for the quadriceps acceleration for one subject for an elastic (grey) and a visco-elastic (black) shoe condition. Corresponding EMG pre-activation intensity (right) for the rectus femoris for the elastic (grey) and visco-elastic (black) shoe conditions.**

Final comments

To quantify a muscle tuning response for a given shoe or surface, a measure of both the input and soft tissue accelerations, and the muscle activity are needed. A biodynamic response function can then be calculated to determine the effect of a measured change in the EMG signal intensity or mean frequency, or both, on the mechanical properties of the soft tissue package and vibration magnitude.

Assigning muscle activity to a specific task is difficult due to the complexity of motor control aspects in landings. Based on the current knowledge, muscle activity during the 100 ms surrounding impact is used to (a) prepare the leg geometry and (b) leg stiffness for landing, and (c) to possibly control vibrations of soft tissue packages (Boyer & Nigg, 2004; Boyer & Nigg, 2005; Boyer & Nigg, 2006; Wakeling et al., 2003; Wakeling & Liphardt, 2005; Mündermann et al., 2005). This application example illustrates that acceleration measurements were instrumental in providing insight into possible mechanisms of soft tissue vibrations and related muscle tuning.

SHOCK ABSORPTION IN THE HUMAN BODY

Comparison of acceleration measurements during locomotion at selected locations of the human body have been used to discuss possible etiology of pain or injuries, or both. Acceleration amplitudes at the medial femoral condyle and at the forehead during walking have been compared for patients with low back pain and with no pain (Voloshin & Wosk, 1982). The quotient of the femoral and the head acceleration has been determined as a measure for shock attenuation, and it has been suggested that the shock attenuation was better for healthy subjects than for subjects with low back pain. Similar studies have been performed on landing from a jump (Gross & Nelson, 1988) and of artificial shock absorbers in shoes (Voloshin & Wosk, 1981).

IMPACT ACCELERATIONS IN THE ASSESSMENT OF PROTECTIVE EQUIPMENT

Accelerometers have been used extensively in the assessment of appropriate head protection with helmets for various human activities. Aircrew helmets (Norman et al., 1979), American football helmets (Bishop et al., 1984), and bicycling helmets (Bishop & Briard, 1984) have all been studied. Accelerations of standardized head dummies were used to assess the protective qualities of such helmets. Similar procedures are applied in car crash experiments. Tolerances of head impacts with respect to brain injuries have been developed using three-dimensional acceleration input into a mathematical model (King et al., 1995). Additionally, acceleration measurements have been used to compare the vibrational characteristics of tennis racquets (Elliott et al., 1980).

VIBROMYOGRAPHY

Vibromyography (VMG) represents a simple, non-invasive technique to assess a mechanical signal in the low frequency range that is produced by the contracting muscle. Grimaldi was the first to report muscle sound (Grimaldi, 1665). Wollastone recognized that muscle sound increases with increasing muscle activity (Wollaston, 1810). Recently, vibrations due to muscle contraction have been studied by many investigators (Barry, 1987; Oster & Jaffe, 1980; Stokes & Dalton, 1991; Zhang et al., 1992; Orizio, 1993). Vibromyography measurements use miniature accelerometers because of their small size, small mass (less than 1.5 g), good frequency response, and appropriate sensitivity. VMG frequencies lie in the range of the infrasonic (< 20 Hz) to the low end of the audible sound.

Comparisons show many similarities between EMG and VMG measurements. However, differences in the frequency content of these two signals suggest differences in the mechanical and the electrical responses of muscles to activation.

VMG techniques have been proposed for quantifying muscle fatigue, for automatic control of powered prostheses (Barry, 1987; Barry et al., 1992), and for diagnostic purposes for subjects with neuromuscular disorders (Rhatigan et al., 1986).

The VMG signals are affected by different variables such as the length of the muscle, the temperature, muscle properties, and possibly fibre type distribution (Barry, 1987; Frangioni, 1987; Zhang et al., 1990; Zhang et al., 1992).

CRITICAL COMMENTS

The appropriate use of accelerometers is a major concern in all applications of this measuring instrument. The author suggests that of all the measuring techniques discussed in this book, the appropriate use of accelerometers is one of the most difficult. The fact that an accelerometer provides on-line information may sometimes cloud the need to carefully study the following aspects of the situation:

- How well does the measured acceleration actually represent the acceleration in which one is interested?
- Is the measured acceleration the variable that answers the question of interest?
- What does the measured acceleration mean mechanically and biologically?

Certainly, these questions are relevant for all other applications of measuring techniques. However, it seems that they are more easily answered for other measuring techniques than for accelerometry.

3.6 OPTICAL METHODS

NIGG, B.M.
COLE, G.K.
WRIGHT, I.C.

The study of biomechanics involves analysis of movement. Sophisticated imaging devices have been developed to quantify movement of specific markers. Various possibilities for quantifying kinematics, using optical methods, are described in this section. The first part discusses various camera and marker systems currently used in biomechanical kinematic analyses. The second and third parts address three- and two-dimensional motion analysis currently used in biomechanical research.

3.6.1 CAMERAS AND MARKERS

This section discusses the different combinations of active and passive cameras and the markers most frequently used in biomechanical assessment. Cameras used in biomechanical analysis include conventional high-speed film cameras, high-speed video cameras, and cameras that sense electromagnetic wave signals. Markers, representing points on the system of interest such as prominent bony landmarks, may be passive, e.g., reflective tape, or active, e.g., electronic transmitters.

FILM CAMERAS AND MARKERS

One conventional method for quantifying movement is to use a film camera (or cine or motion picture camera). For research purposes, 16 mm or 35 mm film cameras are often used. Two basic types of film cameras are used for high-speed filming; intermittent pin registered cameras (Figure 3.6.1), and cameras that use a rotating prism (Figure 3.6.2).

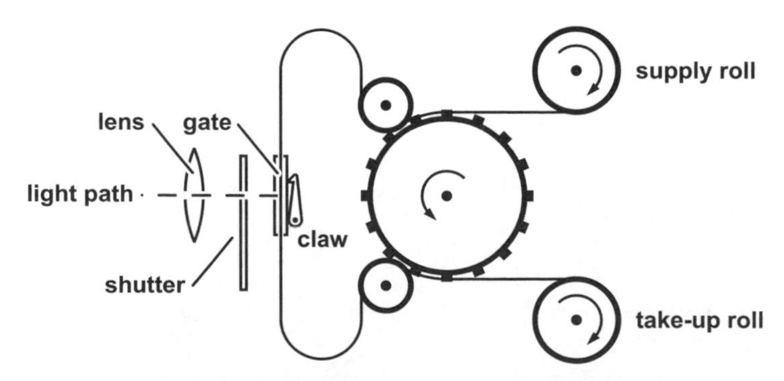

Figure 3.6.1 **Schematic illustration of an intermittent pin registered camera.**

Intermittent pin registered cameras use mechanical devices to hold the film stationary with pins, for a fraction of a second behind the lens and shutter, in the gate. During this time, the film is exposed. This process is repeated for every frame. A pin registered film camera can achieve sampling frequencies of up to 500 Hz (frames per second), which is adequate for most biomechanical applications.

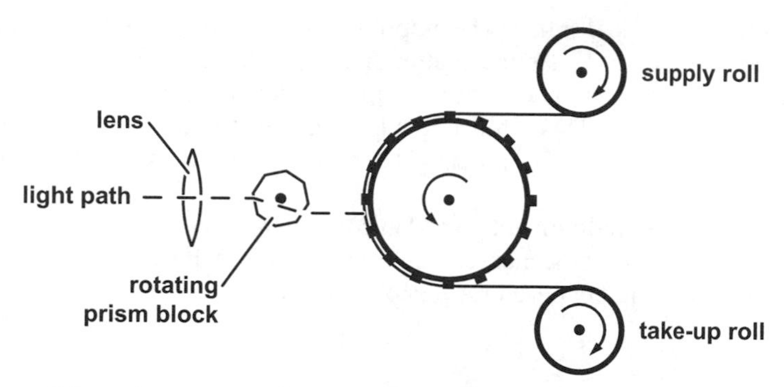

Figure 3.6.2 **Schematic representation of a rotating prism camera.**

In *rotating prism film cameras*, light passes through the lens onto a rotating prism from which the light is directed onto the film (Figure 3.6.2). Film and prism move constantly. Rotating prism-based cameras provide sampling frequencies of up to 10 kHz.

Stickers applied to points of interest or points painted onto the skin at the location of interest can act as markers for film analysis of motion. In special cases, when soft tissue movement may interfere with the movement of interest, using a stencil of the segment of interest with defined marker points can help determine the position of points of interest.

Time calibration of high-speed film cameras can be done internally or externally. Most cameras have internal time calibration systems that allow the determination of the film frequency. The application of time markers onto the film is usually done a few frames before or after the gate, a fact that has to be taken into consideration when determining the actual film frequency. The film frequency can also be determined using external clocks or light signals (Miller & Nelson, 1973) that appear in each frame. Determining the actual film frequency is important, since the film needs a certain time interval to reach the designated frame rate, and the frequency may not remain constant after this initial time interval.

High-speed film cameras are mechanical masterpieces and are, consequently, expensive. Film cameras also use expensive film material that must be developed. In the end, economical considerations concerning the costs of film and film development may limit the progress of a planned project. However, high-speed film is still one of the best methods of quantifying movement, since the human eye may be able to draw important conclusions during film analysis that may be difficult to achieve with other systems. Many biomechanics laboratories use high-speed film techniques in the initial phase of new projects.

VIDEO CAMERAS AND MARKERS

Practical problems, such as film costs, the need to change film during measurements, and time-consuming film analysis, have helped motivate the development of video techniques for quantifying movement in biomechanical applications. The film camera is replaced by a video camera. The basic principles of optics and camera function in video cameras are identical to those of film cameras. The major difference is that the film is replaced with an array of light sensitive cells (sensors). The process of data collection is discussed below for a black and white camera.

Like a film camera, a video camera has a shutter mechanism that opens and lets light enter for a fraction of a second. The light sensitive cells detect the incoming light, become excited, and emit an electrical signal. The amplitude of this signal is proportional to the intensity of the incoming light. When the shutter closes, the light sensitive cells remain excited for a limited time, and a computer-controlled multiplexer system goes through each cell and reads the electrical signal. This information is transmitted to a computer where it is stored as a set of $n \times m$ pixels. After this process is completed, the shutter opens again and the cyclic process continues.

Video cameras are predominantly used in combination with automatic or semiautomatic data analysis systems. Thus, markers are used that allow this (semi-) automatic process. The main features of markers used for this purpose are:

- Contrast,
- Shape, and
- Colour.

Markers covered with reflective material are most commonly used for black and white video cameras. Additionally, the shape of a marker may be used for identification purposes during data analysis. For colour cameras, the colour of a marker may be used as an additional feature to recognize and identify a marker during the analysis. A marker size corresponding to a field of several pixels (optimally 10 to 20) is required for optimal data analysis.

Video cameras are typically less expensive than film cameras. Video tape also costs much less than film.

PASSIVE CAMERAS USING ACTIVE MARKERS

The camera-marker systems discussed above use passive markers with the markers' spatial positions registered by the passive camera system. However, one type of optical system in biomechanical applications uses active markers. The most frequently used technology is based on markers that emit a signal in the infrared (IR) part of the spectrum.

Cameras that quantify IR signals are similar, in principle, to the cameras described for film and video. Two types of that quantify IR signals are: signal sensing and source sensing. In addition to the features described above, IR signal-sensing cameras may use filters that select a defined part, the IR part, of the frequency spectrum of the optical waves reaching the lens, allowing only this part to pass to the photo diode sensors (Woltring & Marsolais, 1980).

IR source sensor systems use active markers. Light emitting diodes (LED) are attached to points of interest of a segment. They emit signals with a distinct frequency in the IR part of the spectrum. Each LED attached to the segment of interest has its own specific frequency and can easily be identified at every point in time during a measurement. Active LED-based markers in the IR spectral wavelength are also used with video cameras for high-contrast and large distance measurements.

ACTIVE CAMERAS AND PASSIVE MARKERS

The camera systems described above are passive. Light beams from the object pass through the lens and shutter opening and produce a reaction in the picture plane, e.g., film. Another type of camera, called an active camera, sends a beam of light to an object. The camera has three scanners (rotating mirrors) that each sweep a narrow ($\approx$1 cm wide) planar beam of light through the space of interest. The three scanners emit the light sequentially (Figure 3.6.3). Photo diodes are used as markers mounted to the segment of interest. Light

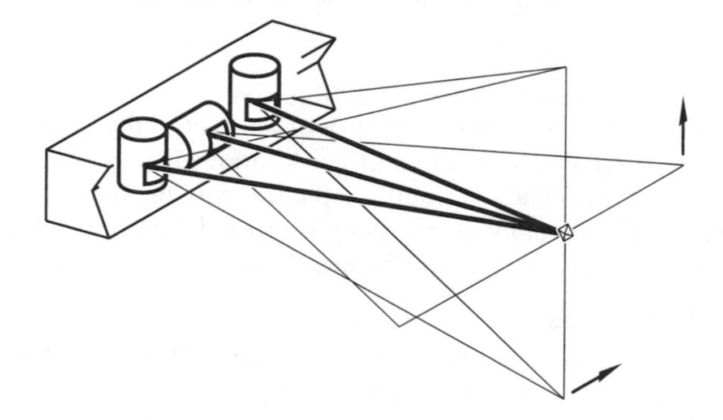

Figure 3.6.3 **Schematic illustration of an active camera system using three rotating mirrors (from Mitchelson, 1988, with permission).**

reaching a diode results in an electrical signal which, together with the positional information from the active mirror, is used to provide information for the reconstruction of positional data. The angular information from the three sequential mirror signals is used to reconstruct the three-dimensional coordinates of the diodes of interest (Mitchelson, 1988).

PANNING AND TILTING VIDEO CAMERAS

The kinematics tracking systems used in biomechanical motion analysis typically use a set of fixed cameras that can record the subjects movements in a relative small volume. To measure the subject's kinematics in a much larger volume, e.g. a ski jumper, a system of moving cameras that can follow the subject in very different natural or artificial setups was designed (Scheirman et al., 1998). This system comprises of a set of cameras that have two optical encoders in the tripod heads, so at any moment their position is recorded in both vertical (pan) and horizontal (tilt) direction. This camera angular coordinates are stored on each image, and, during post-processing, a specific software can reconstruct the three-dimensional coordinates of the markers. The calibration is done moving a few calibrated rods in the volume of interest before or after the measurements.

OTHER POSSIBILITIES

Other possibilities for quantifying movement exist besides film, video, IR source sensors, and active cameras. They include chronophotography (sometimes called chronozyklophotography), stroboscopy, and flashing light sources and are typically used in applications where a special effect should be demonstrated.

Chronophotography

Chronophotography (or chronozyklophotography) uses a conventional camera. A second shutter (a disc with an opening) is mounted in front of the camera and rotates with a given frequency. The camera shutter is open for the period of the entire movement of interest. The resulting picture has the moving object and the stationary background n times exposed.

Ideal setups for good chronocyclographic pictures require a background that is dark compared to the moving object. Chronophotography is often used to illustrate special aspects of a movement and is less used for scientific analysis in biomechanics.

Stroboscopy

Stroboscopy is based on the same idea of multiple exposure as chronophotography. The movement is executed in a dark environment (with respect to the object of interest). The camera shutter is open for the entire duration of the movement of interest. Sequences of flashes of light are then used, and the result is a picture with the moving object in multiple exposure. Like chronophotography, stroboscopy is primarily used for illustration purposes and rarely for scientific analysis in biomechanics.

Flashing light sources

Flashing light sources mounted to the different segments of a moving object may be used to quantify or illustrate movement. This technique is, in principle, identical to the IR source sensor method, except that the frequency spectrum used is in the visible range. This specific approach is not often used in biomechanical analysis.

3.6.2 DETERMINING MARKER POSITIONS

In this section, general comments are made about reconstructing marker positions in three-dimensional kinematic motion analysis.

CALCULATION OF THREE-DIMENSIONAL COORDINATES FROM SEVERAL CAMERAS

Essentially, a camera provides a two-dimensional image of a three-dimensional situation on a film medium. Determining three-dimensional spatial coordinates from several two-dimensional sets of information is commonly performed using the DLT (direct linear transformation) method (Abdel-Aziz & Karara, 1971). For m markers, the DLT method provides a (linear) relationship between the two-dimensional coordinates of a marker, i $(i = 1, \ldots, m)$, on the film and its three-dimensional location in space. For n cameras, the relationship between the coordinates of the markers on the film of camera, j $(j = 1, \ldots, n)$, and the spatial three-dimensional coordinates of this marker are determined by:

$$x_{ij} = \frac{a_{1j}X_i + a_{2j}Y_i + a_{3j}Z_i + a_{4j}}{a_{9j}X_i + a_{10j}Y_i + a_{11j}Z_i + 1}$$

$$y_{ij} = \frac{a_{5j}X_i + a_{6j}Y_i + a_{7j}Z_i + a_{8j}}{a_{9j}X_i + a_{10j}Y_i + a_{11j}Z_i + 1}$$

where for a given marker i:

x_{ij} = x coordinate of marker i on the film measured with camera j
y_{ij} = y coordinate of marker i on the film measured with camera j
X_i = x coordinate of marker i in the three-dimensional space
Y_i = y coordinate of marker i in the three-dimensional space
Z_i = z coordinate of marker i in the three-dimensional space
a_{kj} = coefficient k in the transformation formulas for marker i

A calibration of N points with known coordinates x_r, y_r, z_r, (r = 1, ..., N), is used to determine the coefficients, a_{kj}, for each camera j. At least six calibration points (corresponding to 12 equations) are required to determine the 11 coefficients. The overdetermined system of linear equations is solved using a least squares fit technique.

Two equations describe each camera. Consequently, the DLT method provides more equations than unknowns for the position of a marker i as long as at least two cameras are involved (can see the marker). The determination of the coordinates for each marker does not have a unique result, but is an approximation with errors Δ_j. The norm of residuals (NR) for one coordinate is defined as:

$$NR = \sqrt{\Delta_1^2 + \Delta_2^2 + \ldots + \Delta_n^2}$$

The DLT method, and any similar method, provide one distinct solution for x_i, y_i, and z_i by minimizing the norm of residuals, as long as the number of cameras involved in the process remains constant. However, in practical applications, especially when four to six cameras are used, markers are often lost by specific cameras for a certain time. Camera and lens systems have a limited accuracy. Every set of coordinates, x_{ij} and y_{ij}, is affected with some error. This error is implicit in the calculated set of coordinates, x_i, y_i, and z_i. As a result, there will be a difference in coordinate values if the set is calculated using two, three, four or more cameras. The difference is largest for the jump from two to three cameras and becomes smaller with each increase in number of cameras. This effect is illustrated in Figure 3.6.4.

CORRECTION FOR LENS DISTORTION

The accuracy of the reconstructed three-dimensional positions of a marker depends upon many different factors including the:

- Accuracy of the calibration frame,
- Quality of the DLT reconstruction,
- Quality of the lenses used, and
- Deformation of the film or the inaccuracy of the image plane.

Inaccuracies caused by the first two factors can be reduced by using additional control points and a more accurate calibration frame. To eliminate inaccuracies due to the last two factors, the DLT must be expanded to:

$$x_{ij} + \Delta x_{ij} = \frac{a_{1j}x_i + a_{2j}y_i + a_{3j}z_i + a_{4j}}{a_{9j}x_i + a_{10j}y_i + a_{11j}z_i + 1}$$

$$y_{ij} + \Delta y_{ij} = \frac{a_{5j}x_i + a_{6j}y_i + a_{7j}z_i + a_{8j}}{a_{9j}x_i + a_{10j}y_i + a_{11j}z_i + 1}$$

where Δx_{ij} and Δy_{ij} are terms describing errors due to lens and camera distortion. Two approaches can be used to determine these correction terms Δx_{ij} and Δy_{ij}, to improve the accuracy of marker position reconstruction.

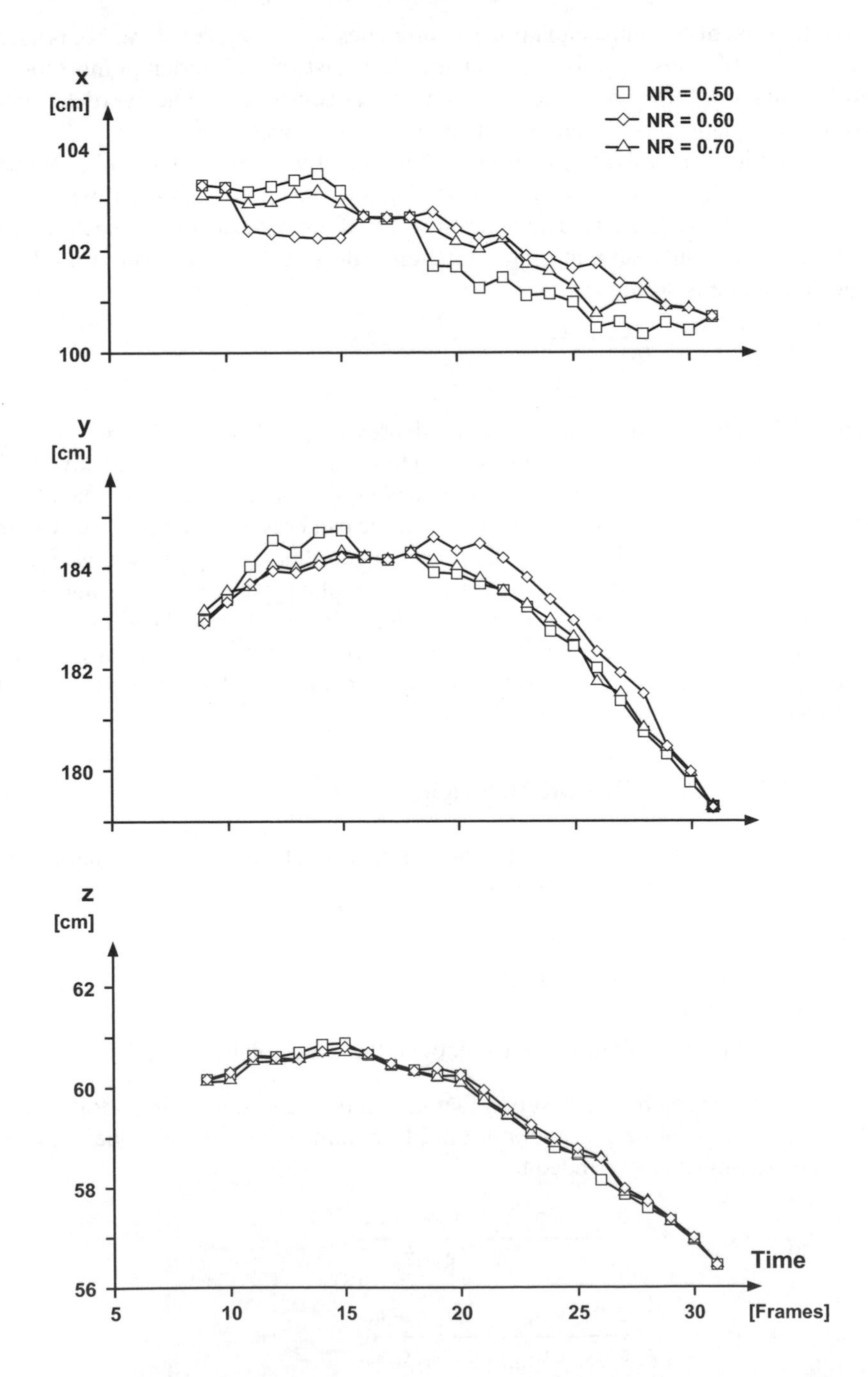

Figure 3.6.4 **Difference in coordinate calculation using a DLT method. The coordinates of a marker in the laboratory coordinate system were determined using the results for two, three, and four cameras. In this case, the inclusion/exclusion of cameras has been produced by changing the norm of residuals (NR). Note the number of cameras does not remain constant for a given NR.**

Approach 1

The terms Δx_{ij} and Δy_{ij} are determined by adding additional terms to the basic DLT. The additional terms are used to account for the various factors that can cause non-linear relationships between the marker positions and image locations. Δx_{ij} and Δy_{ij} can be expressed as:

$$\begin{aligned}
\Delta x_{ij} \quad = \quad & c_{1j} + c_{2j}x_{ij} + c_{3j}y_{ij} + c_{4j}x_{ij}^{2} + c_{5j}y_{ij}^{2} + \\
& (x_{ij} - x_{oj})(k_{1j}r_{ij}^{2} + k_{2j}r_{ij}^{3} + k_{3j}r_{ij}^{4} + k_{4j}r_{ij}^{5} + k_{5j}r_{ij}^{6}) + \\
& P_{1j}(r_{ij}^{2} + 2(x_{ij} - x_{oj})^{2}) + 2P_{2j}(x_{ij} - x_{oj})(y_{ij} - y_{oj})
\end{aligned}$$

$$\begin{aligned}
\Delta y_{ij} \quad = \quad & c_{6j} + c_{7j}x_{ij} + c_{8j}y_{ij} + c_{9j}x_{ij}^{2} + c_{10j}y_{ij}^{2} + \\
& (y_{ij} - y_{oj})(k_{1j}r_{ij}^{2} + k_{2j}r_{ij}^{3} + k_{3j}r_{ij}^{4} + k_{4j}r_{ij}^{5} + k_{5j}r_{ij}^{6}) + \\
& P_{2j}(r_{ij}^{2} + 2(x_{ij} - x_{oj})^{2}) + 2P_{1j}(x_{ij} - x_{oj})(y_{ij} - y_{oj})
\end{aligned}$$

where:

$c_{1j}, c_{2j}, c_{3j}, c_{6j}, c_{7j}, c_{8j}$	account for linear lens distortion
$k_{1j}, k_{2j}, k_{3j}, k_{4j}, k_{5j}$	account for symmetrical lens distortion
P_{1j}, P_{2j}	account for asymmetrical lens distortion or inaccuracies in the location of the point of symmetry
$c_{4j}, c_{5j}, c_{9j}, c_{10j}$	account for non-linear components of film deformation
r	distance from the point of symmetry
x_{oj}, y_{oj}	point of symmetry or the projection of the optical axis on the image plane

These lens correction parameters can be determined simultaneously with the 11 DLT parameters using a least square fit method to minimize the norm of residuals. This method is sensitive to the number of control points used in the calibration process. To successfully apply this method, additional control points are required during the calibration. The minimum number of control points for the DLT is six, but this minimum number of points increases by one for every two additional parameters added to the DLT.

The improvement in accuracy provided by these additional terms depends upon the amount of distortion present in the particular camera. If higher order terms for symmetrical lens distortion (k_{2j}, k_{3j}, k_{4j}, k_{5j}), asymmetrical lens distortion (P_{1j}, P_{2j}) or non-linear film deformation (c_{4j}, c_{5j}, c_{9j}, c_{10j}), are included, when in fact there is little or no distortion present in the system, these additional terms can introduce errors. These errors can lead to large errors in marker position reconstruction, particularly when extrapolating outside the volume defined by the control points. Furthermore, many of these additional terms provide only insignificant improvement to the accuracy of method. Therefore, often only one or two of the symmetrical lens distortion terms (k_{1j}, k_{2j}) are included.

Approach 2

Assuming that the major error components result from the lens and camera, and assuming that calibration or camera placement, or both, have only a minor influence on the magnitude of the error, the corrections for these distortions for a given camera can be determined once and used for every subsequent data collection. This can be done by doing a calibration with a fine grid of many control points filling the entire field of view of the camera. The image refinement components can be saved as a vector field of correction vectors (Figure 3.6.5) or as a set of parameters such as the terms c_{kj}, k_{kj}, and P_{kj} described in Approach 1. By using this procedure, additional control points above the minimum of six, are not required during subsequent calibration and data collection.

Figure 3.6.5 **Illustration of a camera system correction as a vector field of correction values in image coordinates (from Ladin et al., 1990, with permission).**

ERRORS IN LANDMARK COORDINATES

Errors in three-dimensional motion analysis may occur for many different reasons. Some errors are discussed in the following sections. Some of these errors may be caused by errors in determining landmark coordinates.

(i) Errors due to relative camera placement

In three-dimensional motion analysis, each camera, i, is connected by an imaginary line to each marker, k. Ideally, for n cameras, these lines intersect spatially at one point. However, in real situations, these lines do not intersect at one point, because the directions of these lines are not infinitesimally accurate. Spatial coordinates of one marker point are, therefore, determined using distances between a point on a line (connecting the camera and the marker) and the assumed actual spatial position of this marker. Usually, a least squares fit method is applied to determine the location of marker, k, under the condition that:

$$\sqrt{d_{1k}^2 + d_{2k}^2 + \ldots + d_{nk}^2} = \text{minimum}$$

where:

d_{ik}	=	distance between a point on the line connecting camera, i, and marker, k, and the estimated spatial position of marker, k
i	=	symbol for the cameras
k	=	symbol for the markers

The quality of the determined coordinates depends, although not exclusively, on the angle of intersection between the two lines. Small errors in orientation of two perpendicular lines does not significantly influence the coordinates of the lines, whereby, the distance between these two lines is minimal. However, if the angle between two intersecting lines is small, errors in the orientation of these lines may produce substantial differences in coordinates. This suggests that cameras should be placed so that the angles between two optical axes intersect at an angle that nears 90° but is not less than 60°.

Cameras are often placed in an umbrella configuration for motion analysis of human or animal locomotion (Figure 3.6.6). For reasons of convenience, the cameras are often arranged close to the ground level. As a result, the coordinate of each marker within this plan should be well-established. However, the coordinates in the two mutually perpendicular planes may be affected with considerable errors. An improvement in accuracy may be obtained by placing the cameras such that they are not all in the same plane.

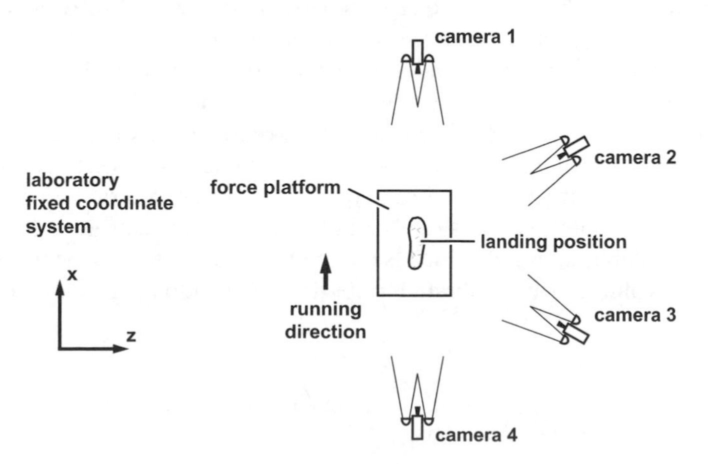

Figure 3.6.6 **Schematic illustration for a camera placement in a typical umbrella configuration, as used often in gait analysis.**

(ii) Errors due to number of cameras

Ideally, all cameras see each marker at every point in time. In real situations, this is not the case, and markers are lost from view. In film analysis, the positions of these lost markers may be determined using a best estimate based on the position of the body. In video analysis, these markers are lost until they are once again visible to the camera. The position of a marker at a given point in time is calculated using only those cameras that see the marker. This may be critical for certain camera placements and camera numbers.

As an example, an experiment is performed using three cameras: 1, 2, and 3. Assume that the optical axes of cameras 1 and 2 intersect at an angle of 90°, and the optical axes of cameras 1 and 3 intersect with an angle of 20°. At time, t_1, the position of marker, k, is determined using information from all three cameras. At time, t_2, information from camera 2 is missing. At time, t_3, information from all three cameras is again available. The results for this example are illustrated for the x and y coordinates in Figure 3.6.4.

For instance, frames 15 and 19 use (for three different norm of residuals) a different number of cameras than do frames 16, 17, and 18, (which all use the same number of cameras) to determine the position of marker, k.

Obviously, a change from six to five cameras would induce less change in the final coordinate calculation than a change from three to two cameras. In general, noise in coordinates decreases with an increasing number of cameras. However, this is not necessarily equivalent to an increase in accuracy of the determined marker positions.

(iii) Errors due to calibration

The accuracy of the measured data depends considerably on the accuracy of the calibration procedure. It is often assumed that accuracy increases the more the calibrated volume fills the field of view. However, the situation is more complex than this assumption. Lenses are not ideal structures, and have errors that influence the accuracy of the collected data. Consequently, data should be corrected for lens errors. Several software packages do this, as described above. If data is not corrected for lens errors, calibration of the central part of the field of view, where the lens usually has fewer errors than at its boundaries, may be better than using calibration frames that exploit the entire field of view. Additionally, errors in determining the position of markers depend on the accuracy of the calibration marker set-up. Errors in the location of these markers affect the accuracy of marker coordinates.

However, most current commercial camera systems offer calibration procedures and software that reduce calibration errors substantially. The most elegant and effective calibration method uses a calibration rod, which is moved in the calibration space and covers the whole calibration volume. This calibration rod provides data to achieve excellent calibration results.

(iv) Errors due to digitization

Conventional film analysis typically uses manual digitization. Results from manual digitization may be affected by:

- Random inaccuracies of the digitizing operator in positioning the digitization pen. Such errors can be reduced by employing appropriate marker shapes (spherical), markers with dots in the middle, and multiple digitization, and
- Placement of cameras (see above).

Video analysis often uses automatic digitization. The accuracy of automatic digitization depends on various factors such as:

- Number of cameras (see above),
- Placement of cameras (see above),
- Marker shape (spherical markers are typically used because they have the same shape irrespective of the direction from which they are viewed),
- Size of the markers (the centroid can be determined with better accuracy for a large than for a small marker),
- Merging of markers (the centroid of a marker cannot be correctly established),
- Partial cover of a marker (the centroid of a marker cannot be correctly established), and
- Threshold level of the video system (determines how well the video screen estimates the spherical image of the marker).

Most of these problems can be solved with appropriate preparations of the setup and adequate software.

3.6.3 DETERMINING RIGID BODY KINEMATICS

The three-dimensional marker positions are used to determine limb segment positions and orientations. This process, and associated errors, are discussed in the first two parts of this section. Position and orientation of the limb segments can be used to describe the relative orientation and movement of limb segments. Methods used to describe relative limb segment orientations commonly used in biomechanics are discussed in the balance of this section

RECONSTRUCTION OF RIGID BODY MOTION FROM MARKER POSITIONS

Marker based coordinate systems

A coordinate system is defined for each limb segment or body of interest (Figure 3.6.7). Limb segments are often assumed to be a rigid body. Thus, if the position and orientation of the segment fixed coordinate system is known, then the position of any point on the limb

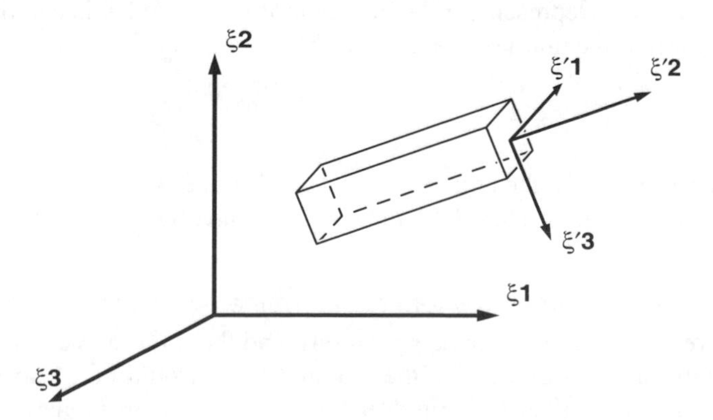

Figure 3.6.7 **Orientation of a rigid body in space with an embedded Cartesian coordinate system, ξ', relative to a reference Cartesian coordinate system, ξ.**

segment is known. Since the positions of the optical markers are measured, it is easiest to define the segment fixed coordinate system based on the markers that are attached to that segment. This can be done with three or more than three markers:

(i) Using three markers to determine a coordinate system

Three markers is the minimum number required to completely define a coordinate system. An example of how a segment coordinate system can be defined for a segment of interest is outlined below. Assume:

A	=	first marker on the segment of interest
B	=	second marker on the segment of interest
C	=	third marker on the segment of interest
$\mathbf{a}$	=	position vector of point A in the lab coordinate system
$\mathbf{b}$	=	position vector of point B in the lab coordinate system
$\mathbf{c}$	=	position vector of point C in the lab coordinate system
$\boldsymbol{\xi}'\mathbf{1}$	=	direction vector of the first axis of the segment coordinate system
$\boldsymbol{\xi}'\mathbf{2}$	=	direction vector of the second axis of the segment coordinate system
$\boldsymbol{\xi}'\mathbf{3}$	=	direction vector of the third axis of the segment coordinate system

Any point can be used as the origin, O, of the segment coordinate system. Assume that the point A has been selected to be the origin. For this assumption, the unit vector $\boldsymbol{\xi}'\mathbf{1}$ representing the direction of the first axis can be defined as the direction vector from point A to point B:

$$\boldsymbol{\xi}'\mathbf{1} = \frac{\mathbf{b}-\mathbf{a}}{|\mathbf{b}-\mathbf{a}|}$$

The unit vector $\boldsymbol{\xi}'\mathbf{3}$, representing the direction of the third axis can be defined as the cross-product between the unit vector $\boldsymbol{\xi}'\mathbf{1}$ and the unit vector pointing from A to C:

$$\boldsymbol{\xi}'\mathbf{3} = \boldsymbol{\xi}'\mathbf{1} \times \frac{\mathbf{c}-\mathbf{a}}{|\mathbf{c}-\mathbf{a}|}$$

The unit vector $\xi'2$, representing the direction of the second axis can then be defined as the cross product of the unit vectors $\boldsymbol{\xi}'\mathbf{3}$ and $\boldsymbol{\xi}'\mathbf{1}$:

$$\boldsymbol{\xi}'\mathbf{2} = \boldsymbol{\xi}'\mathbf{3} \times \boldsymbol{\xi}'\mathbf{1}$$

In this arbitrary example, the unit vectors $\boldsymbol{\xi}'\mathbf{1}$, $\boldsymbol{\xi}'\mathbf{2}$ and $\boldsymbol{\xi}'\mathbf{3}$ are the bases vectors of the limb fixed coordinate system. They determine position and orientation of the segment of interest.

(ii) Using more than three markers to determine a coordinate system

If there were no errors in the marker positions, and the limb segment was truly a rigid body, then additional markers beyond the required three would be of no use. However, since limb segments may not be truly rigid, and there may be errors in marker positions, additional markers may improve the accuracy of limb segment position and orientation reconstructions. A least squares method (Veldpaus et al., 1988) or similar method (Söderquist &

Wedin, 1993) can be used to determine the segment position and orientation from the marker positions.

These methods are relatively sophisticated algorithms, but simply described, they do the following. Given the positions of the markers fixed to some segment at two different times (or positions of the segment of interest), the displacement and rotation of that segment that occur between these two points of time are determined that minimize the sum:

$$\min \sum_{k=1}^{n} \left(\left| \mathbf{m}_{ki} - (\mathbf{d}_{ij} + R_{ij}\mathbf{m}_{kj}) \right| \right)$$

where:

$\mathbf{m}_{ki}$	=	position vector of marker k at time i
$\mathbf{m}_{kj}$	=	position vector of marker k at time j
n	=	total number of markers
R_{ij}	=	rotation that occurs from time i to time j
$\mathbf{d}_{ij}$	=	position vector of point A in the lab coordinate system

Thus, the displacement and rotation are selected that move the markers at time i so that they lie as close as possible to the markers at time j. By using these optimization methods, inaccuracies in the position description caused by changes in the relative marker positions due to deformation of the body are averaged over the whole time interval of data collection.

Anatomically significant coordinate systems

The selection or definition of a coordinate system fixed to a specific segment is an important step. Usually, the coordinate system is defined such that each of the coordinate directions has some anatomical significance. It is common, for example, for the coordinate system to be aligned with the anatomical coordinate system so that when movements or displacements are measured, they can be described in clinically relevant terms.

However, markers often cannot be located on limb segments in positions appropriate for the definition of a anatomically significant coordinate system. In such cases, it is often easiest to define the orientation of the limb segment fixed coordinate system using a neutral position.

The use of a neutral position consists of positioning the subject in a neutral or reference position (often the anatomical position). Measurements are then made to determine the marker positions relative to the selected anatomically significant coordinate system. The coordinate system fixed to the limb segment does not need to be explicitly constructed from the locations of the markers. The positions measured during the neutral trial can then be compared to the measured positions during the movement of interest to determine absolute position and orientation of the anatomically relevant limb segment fixed coordinate system.

The comparison of the measured marker positions to the neutral position marker positions can be done using the methods developed for using three or more markers to track a body described above. The least squares method (Veldpaus et al., 1988) or singular value decomposition method of Söderquist and Wedin (1993) can be used in the following way for 3 or more markers:

${}^{L}\mathbf{m}_1, {}^{L}\mathbf{m}_2, \ldots {}^{L}\mathbf{m}_n$	position vectors of markers 1 through n, expressed relative to the anatomically significant limb segment fixed coordinate system L (measured using a neutral position)
${}^{A}\mathbf{m}_{1i}, {}^{A}\mathbf{m}_{2i}, \ldots {}^{A}\mathbf{m}_{ni}$	position vectors of markers 1 through n at time i, expressed relative to the absolute coordinate system A

Output:

$\mathbf{d}_i$	the displacement of the limb segment fixed coordinate system relative to the absolute coordinate system at time i
R_i	the orientation of the limb segment fixed coordinate system relative to the absolute coordinate system at time i

This method produces similar results to the rigid body reconstruction method using an explicit definition of the coordinate system in terms of marker positions with the two following advantages:

- The limb segment fixed coordinate system can be defined in a convenient way using a neutral position, and
- Three or more markers can be used in an optimal way to minimize errors caused by errors in the marker data or non-rigidity of the limb segment.

ERRORS IN RIGID BODY MOTION USING MARKER INFORMATION

Markers attached to a segment of interest may not always represent true skeletal locations. These differences are referred to as relative and absolute errors. The placement of the markers can also substantially influence the propagation of marker coordinate errors and relative marker movement errors when calculating segment kinematics:

Relative marker error is the relative movement of two markers with respect to each other.
Comment: the relative marker error establishes how well the markers on the segment estimate the rigid body assumption. This error is caused by soft tissue movement.

Absolute marker error is the movement of one specific marker with respect to specific bony landmarks of a segment.
Comment: this error establishes the actual accuracy of the measurement.

(i) Errors due to relative marker movement

The three-dimensional motion of body segments is typically described using rigid body mechanics. The spatial coordinates of segment landmarks are used directly or indirectly to define body-fixed coordinate systems for each segment. Markers placed on the skin of a subject or animal are assumed to represent the location of bony landmarks of the segment of interest. However, skin movement and movement of underlying bony structures are not necessarily identical, and substantial errors may be introduced in the description of bone movement when using skin-mounted marker arrays (Lesh et al., 1979; Ladin et al., 1990). Three major approaches have been suggested to correct skin displacement errors: invasive marker placement, data treatment, and marker attachment systems.

Invasive methods provide the most accurate results for bone movement. Bone pins have been used to assess movement of the tibia and femur, e.g., Levens et al. (1948). Knee joint kinematics have been compared for data collected from skin-mounted markers and markers attached to bone pins (Ladin et al., 1990). The results for both methods showed differences of up to 50% for the knee angle. Results from invasive methods are, however, not applicable in most research and clinical settings for movement analysis. Therefore, other approaches are needed.

Mathematical algorithms have been proposed for error reduction in raw data, including skin displacement effects (Plagenhoef, 1968; Miller & Nelson, 1973; Winter et al., 1974; Zernicke et al., 1976; Soudan & Dierckx, 1979; Woltring, 1985; Veldpaus et al., 1988). Smoothing algorithms are based on the assumption that the noise is additive and random with a zero mean value (Woltring, 1985). However, errors due to skin displacement may not have zero means. Mathematical algorithms, as currently used, do not seem to be appropriate approaches for solving the errors due to marker movement relative to bony landmarks.

Marker attachment systems (frames) have recently been developed to reduce errors due to relative marker movement. The following discussion concentrates on the tibia. However, it may be used in analogy for other segments of human or animal bodies. Positional data from four different lightweight marker attachment systems (frames) were compared to data from skin-mounted markers. Frame 1 was constructed from thermoplastic material, frame 2 of rectangular aluminum rods, frame 3 of plastic material of medium stiffness, and frame 4 used laminated plastic strips. The fifth marker attachment system was the attachment of the markers to the skin. Skin displacement errors were analyzed in terms of relative and absolute errors.

The *relative* error was determined as the change in three-dimensional length, ΔL_{ik}, between any two markers with respect to the static length, where:

$$\Delta L_{ik} = L_{ik}^{stat} - L_{ik}^{dyn}$$

and:

$$L_{ik} = \sqrt{(x_i - x_k)^2 + (y_i - y_k)^2 + (z_i - z_k)^2}$$

where:

x_i, y_i, z_i = coordinates of marker, i, at the time, t

x_k, y_k, z_k = coordinates of marker, k, at the time, t

The results for the relative errors are illustrated in Figure 3.6.8. Marker 1 was mounted close to the distal anterior part of the tibia and marker 2 at the proximal anterior part of the tibia. The four frames substantially reduced the relative movement of marker 1 with respect to marker 2 when compared to the results for the skin-mounted markers. The biggest length difference during stance between markers 1 and 2 was more than 3 cm in this example. Such differences in length (as measured for the skin-mounted markers) of a structure, which is assumed to be rigid, interfere with the accuracy of the results. These results suggest that markers attached to frames may fulfil the requirements of a rigid body, and that markers attached directly to the skin may not fulfil these conditions satisfactorily.

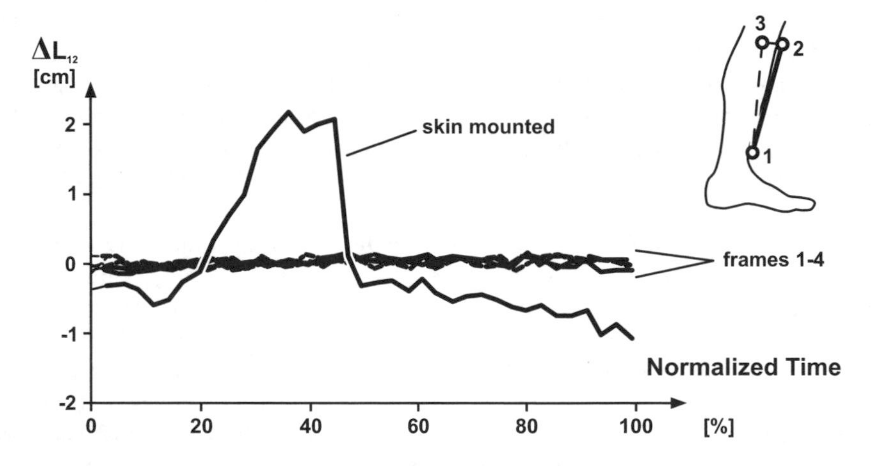

Figure 3.6.8 Actual three-dimensional length between the markers 1 and 2 during ground contact in running (time 0 corresponds to heel strike, time 100% to take off) for four marker frames and for skin-mounted markers. The graph illustrates results for one subject and is representative of the general trend.

The use of an array of more than three markers may help solve for relative marker movement. In this case, the distance between all markers is monitored and the markers with the minimal changes in distance, for instance, are used for further calculation.

(ii) Errors due to absolute marker movement

Markers used to determine segmental movement are typically attached to the skin of test subjects. However, there may be relative movement between skin and the underlying bone, especially when high inertia forces are present, e.g., during impact in running, or when high changes in muscle activity occur. Relative movement between skin and bone-mounted markers in dynamic situations has been determined in a few invasive experimental studies (Murphy, 1990; Angeloni et al., 1993; Holden et al., 1994; Lafortune et al., 1994; Andriacchi & Toney, 1995; Capozzo et al., 1996; Reinschmidt, 1996; Reinschmidt et al., 1997a). The agreement between skin and bone marker based kinematics in walking and running ranged from good to virtually no agreement (Reinschmidt et al., 1997b). The difference between the absolute skin marker position and the skeletal marker position was relatively small for the leg, but substantial for the foot/shoe and for the thigh (Figure 3.6.9). The results indicated that skin or shoe markers overestimated the actual bone movement.

(iii) Errors due to inadequate placement of the markers

The placement of the markers on the segment of interest may be a further source of error in determining marker coordinates. Possible factors include:

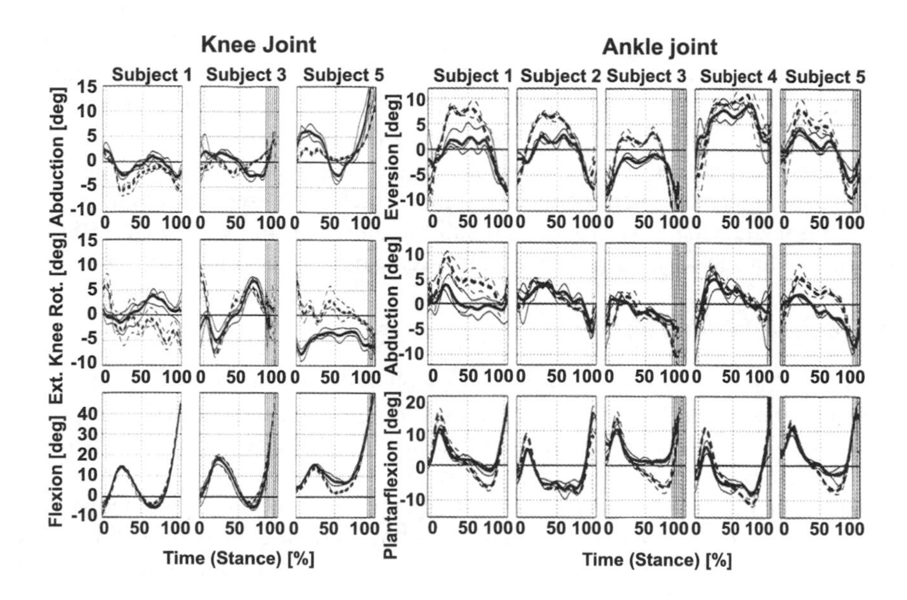

Figure 3.6.9 **Illustration of the positional differences between shin and skeletal muscles for the calcaneus (heel of the shoe), tibia, and femur. The results for the ankle joint are on the left. The results for the knee joint are on the right. Shaded areas at the beginning and end are areas associated with high measuring errors (from Reinschmidt et al., 1997b, with permission).**

- Co-linearity (the markers should define a plane and should not be on one line),
- Distance between markers (large distances between markers improve accuracy), and
- Marker distribution with respect to the axis of rotation of the segment (helical axes are most accurately determined when the helical axis passes directly through the centroid of the marker distribution).

Most of these factors can be controlled with a careful preparation of each segment of interest (Söderquist & Wedin, 1993).

NEW DEVELOPMENTS

In recent years, markerless movement analysis started to emerge. This technique does not rely on optical markers to identify motion variables. This new technique provides total independence from any sort of marking devices. There are two different approaches used in markerless analysis:

Model-based markerless capture

The model based markerless capture uses the same image capture devices as the marker-based method. However, the image processing algorithm is usually very complex, e.g., silhouette estimation and matching, and is done successive steps. First, the silhouette is extracted from the background. Second, the picture is segmented, and third, a correspondence between a chosen model of motion and the extracted silhouette is established, usually using an iterative closest point (ICP) algorithm. The method is semi-automatic and often imposes some limits on the complexity of movement that can be reliably traced.

The main disadvantage of this method is that the quality of the results depends on the right choice of a predetermined model of motion, which sometimes proves to be difficult. However, this method has proved to be as reliable as any of the marker-based methods, with the obvious advantage of not depending on markers and marker-related errors.

Model-free markerless capture

The model free markerless capture is completely independent of the subject or the type of motion under scrutiny. The method is still in its infancy and some technical aspects must still be solved. There are no differences between this method and the methods presented above at the hardware level. Furthermore, the image analysis algorithms are almost the same as described earlier. However, this method uses some additional mathematical algorithms that can automatically identify different body moving parts by fitting primitive shapes into a three-dimensional shape. The most common fitting technique uses multiple superquadric shapes to decompose three-dimensional point data into primitive sub-shapes using neural networks, fuzzi clustering cuts and other methods.

A combination of a marker-based and a markerless system has been presented recently (Mündermann et al., 2005). This method identifies body segments using a full body laser scan with markers positioned at the subject (a kind of a calibration). For the actual movement to be studied, the algorithms used to isolate the body from the background are no different from the other markerless methods. The experimental results obtained with this method were comparable with those obtained by a marker based system.

DESCRIPTION OF RIGID BODY ORIENTATION IN THREE-DIMENSIONS

Several methods for representing the orientation between two body segments in three-dimensional space have been proposed (Goldstein, 1950; Chao, 1980; Grood & Suntay, 1983; Yeadon, 1990; Woltring, 1991; Cole et al., 1993). This section discusses some of these methods (Cole et al., 1993).

The position of a particle in three-dimensional space may be represented as a vector with three components, each representing one of the three translational degrees of freedom (DOF). The description of the position of a rigid body in three-dimensional space additionally requires information that describes its orientation in space. If a Cartesian coordinate system with axes, ξ'_i (i = 1, 2, 3), is embedded in a rigid body, then the orientation of the rigid body (Figure 3.6.7) relative to a reference Cartesian coordinate system with axes, ξ_i (i = 1, 2, 3), is defined by a 3 × 3 rotation matrix, [R], with components, $R_{ij} = \cos(\xi'_i, \xi_j)$ (i, j = 1, 2, 3). This matrix is also referred to as the direction cosine matrix or attitude matrix, and it can be used as an operator to transform the components of a vector from one coordinate system to the other:

$$\{\xi'\} = [R]\{\xi\} \quad \text{and} \quad \{\xi\} = [R]^{-1}\{\xi'\}$$

where $\{\xi\}$ and $\{\xi'\}$ are a common vector represented in the two systems. The rotation matrix has the properties that its inverse is equal to its transpose, $[R]^{-1} = [R]^T$, and its determinant is equal to one, $|R| = 1$. A detailed description of a least squares method for the determination of the rotation matrix from the coordinates of landmarks fixed to a body is given in Veldpaus et al. (1988).

For the interpretation of body segment orientation in biomechanical analyses, a parametric representation of the rotation matrix is usually desired. Current parametric representations include:

- Cardan/Euler angles,
- Joint Coordinate Systems (JCS),
- Finite helical axes and rotations, and
- Helical angles.

These parametric representations are discussed in this section, with emphasis given to Cardan/Euler angles and the JCS. The discussion is based on the following assumptions:

- Cartesian coordinate systems have been embedded in each body segment of interest, with the x-axis pointing anteriorly, the y axis vertically, and the z-axis to the right when the subject is in the anatomical position,
- The orientation of a body segment relative to an inertial reference system is specified by a rotation matrix, and
- The orientation of one body segment with respect to another, i.e., joint orientation, is specified by a rotation matrix.

Cardan/Euler angles

In the Cardan/Euler angle representation of three-dimensional orientation, the rotation matrix is parameterized in terms of *three independent angles,* resulting from an ordered sequence of rotations about the axes (i, j, k), of a selected Cartesian coordinate system (x_1, y_1, z_1), to obtain the attitude of a second coordinate system (x_2, y_2, z_2), relative to the first (Figure 3.6.10). When the first and the last axes are the same, ($i = k$), the term Euler an-

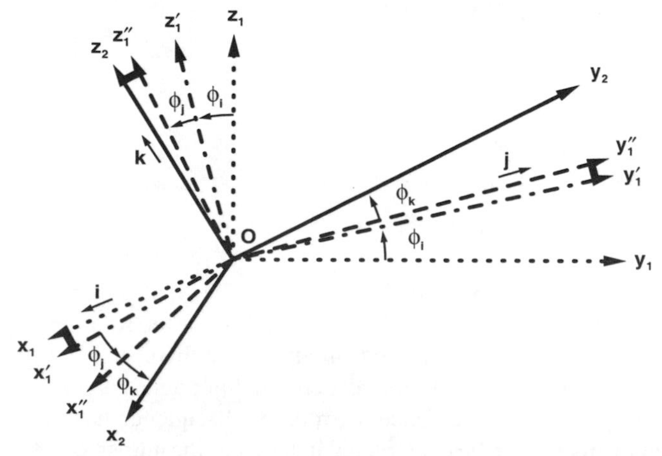

Figure 3.6.10 **Sequential rotations ϕ_i, ϕ_j, and ϕ_k about the axes i, j, and k from the starting coordinate system (x_1, y_1, z_1) to obtain the orientation of the second coordinate system (x_2, y_2, z_2). The axes i, j, and k have been arbitrarily defined as the x, y, and z axes, respectively, although they are not restricted to this definition.**

gles is used. When all three axes are different, the term Cardan angles is used, and it is this representation that is typically used in biomechanics (Woltring, 1991). When referring to other sources, caution must be used, because some investigators use the terms Cardan and Euler angles differently. The i and k axes correspond to axes in each of the two segment embedded coordinate systems, while the j axis is referred to as the nodal axis.

The parametric representation is mathematically defined by:

$$[R_{ijk}] = [R_k(\phi_k)][R_j(\phi_j)][R_i(\phi_i)]$$

where:

$[R_{ijk}]$ = parametric representation of the rotation matrix for the chosen sequence of rotations

$[R_i(\phi_i)]$ = parametric representation of the rotation matrix for a rotation, ϕ_i, around the specific axis i. An analogous definition is used for the rotations about the axes j and k

The parametric rotation matrices for rotations about the x, y, and z axes of a Cartesian coordinate system are defined as:

$$[R_x(\phi_x)] = \begin{bmatrix} 1 & 0 & 0 \\ 0 & \cos\phi_x & \sin\phi_x \\ 0 & -\sin\phi_x & \cos\phi_x \end{bmatrix}$$

$$[R_y(\phi_y)] = \begin{bmatrix} \cos\phi_y & 0 & -\sin\phi_y \\ 0 & 1 & 0 \\ \sin\phi_y & 0 & \cos\phi_y \end{bmatrix}$$

$$[R_z(\phi_z)] = \begin{bmatrix} \cos\phi_z & \sin\phi_z & 0 \\ -\sin\phi_z & \cos\phi_z & 0 \\ 0 & 0 & 1 \end{bmatrix}$$

Since matrix multiplication is not commutative, the final parametric rotation matrix, $[R_{ijk}]$, depends on which of the axes x, y or z is chosen for the rotations i, j, and k. The parameterization of the rotation matrix into Cardan angles, therefore, is *sequence dependent*. For a given segment or joint orientation, different component values can be obtained depending on the sequence selected. Once the rotational sequence has been determined, the parametric rotation matrix can be constructed in terms of the angles ϕ_x, ϕ_y, and ϕ_z. The resulting expressions can then be equated to the numerical values of the rotation matrix for the selected orientation to solve for each angle. For example, if the axes i, j, and k are arbitrarily defined as the axes z, y, and x, respectively, the parametric rotation matrix is constructed as:

$$[R_{zyx}] = [R_x(\phi_x)][R_y(\phi_y)][R_z(\phi_z)]$$

$$[R_{zyx}] = \begin{bmatrix} (C\phi_z C\phi_y) & (S\phi_z C\phi_y) & (-S\phi_y) \\ (-S\phi_z C\phi_x + C\phi_z S\phi_y S\phi_x) & (C\phi_z C\phi_x + S\phi_z S\phi_y S\phi_x) & (C\phi_y S\phi_x) \\ (S\phi_z S\phi_x + C\phi_z S\phi_y C\phi_x) & (-C\phi_z S\phi_x + S\phi_z S\phi_y C\phi_x) & (C\phi_y C\phi_x) \end{bmatrix}$$

where:

$$S\phi_k = \sin\phi_k$$

$$C\phi_k = \cos\phi_k$$

If the numerical components of the rotation matrix, [R], are labeled as, R_{mn} (m = 1, 2, 3; n = 1, 2, 3), and counter-clockwise rotations are chosen as positive, the sine and cosine of each angle can be calculated from:

$$\sin\phi_y = -R_{13} \qquad \cos\phi_y = \sqrt{1 - \sin^2\phi_y}$$

$$\sin\phi_x = \frac{R_{23}}{\cos\phi_y} \qquad \cos\phi_x = \frac{R_{33}}{\cos\phi_y}$$

$$\sin\phi_z = \frac{R_{12}}{\cos\phi_y} \qquad \cos\phi_z = \frac{R_{11}}{\cos\phi_y}$$

From these equations, each angle can be determined in the range, $-\pi < \phi < \pi$, so that the time history of the angle is continuous (Yeadon, 1990).

(i) Sequence selection in Cardan/Euler angles

The sensitivity of calculations of joint orientation to the rotational sequence is illustrated in Figure 3.6.11. The differences in component values, $\Delta\phi$, for two different sequences of ordered rotations were determined for the ankle joint complex, with the second and third rotations the reverse of those shown in the previous example. The results indicate that, theoretically, there are substantial differences in orientation components for different Cardanic sequences. However, the differences are not always relevant in practical applications to human movement. Differences in relative segmental orientations of the lower extremities in running due to different sequences of ordered rotations are minimal. However, in a side shuffle movement, with a larger in-eversion range-of-motion, the differences can be large. In range-of-motion (ROM) measurements, the differences are substantial, and results from different sequences of ordered rotations cannot be compared (Table 3.6.1). These sensitivity considerations illustrate the need for a well-defined convention for biomechanical analysis.

In biomechanics, the three Cardan angles are typically used to represent the anatomical components of joint orientation and movement, e.g., flexion-extension, adduction-abduction, and inversion-eversion. The anatomical components of movement are defined with respect to the anatomical position, as follows (Moore, 1980; Snell, 1973):

Flexion-extension is motion of a segment in a *parasagittal plane.*

Adduction-abduction is motion of a segment *away from or toward the sagittal plane.*

Axial rotation of a segment is rotation of a segment about its *longitudinal axis.* Note: the terminology of the third rotational component differs depending on the joint.

For practical use in biomechanics, the sequence of ordered rotations should be chosen so that the anatomical definitions are satisfied. Rotational sequences that do not conform to these definitions should be avoided. The following argument provides evidence that the sequence of Cardan angles that satisfies the definitions of flexion-extension, adduction-abduction, and third component rotation occurs in the following order:

Table 3.6.1 Comparative numerical results for measurements of angular positions for the ankle joint complex for running, side shuffle, and range of motion (ROM) positions (unpublished, Cole, 1992, with permission).

ACTIVITY	COMPONENT	ANGLE SEQUENCE PL-DO/AD-AB/IN-EV	ANGLE SEQUENCE PL-DO/IN-EV/AD-AB
RUNNING	DO	25	21.4
	AB	10	10.6
	EV	20	19.7
SIDE SHUFFLE	DO	20	30.3
	AB	15	18.1
	IN	35	33.6
ROM	PL	40	26.8
	AB	40	41.8
	IN	20	15.2

- Flexion-extension,
- Adduction-abduction, and
- Axial rotation.

The flexion-extension component, first in the sequence, must occur within the parasagittal plane. The second rotation must occur in a plane perpendicular to the sagittal plane since the coordinate system is orthogonal. Therefore, the second rotation satisfies the anatomical definition of adduction-abduction. Axial rotation is the third rotation in the sequence. After the first two rotations are completed, the remaining axis must correspond to the longitudinal axis of the distal segment, which again satisfies the anatomical definition for the final rotation. Consequently, the proposed sequence of (1) flexion-extension, (2) adduction-abduction, and (3) axial rotation conforms to the anatomical definitions of segment rotations.

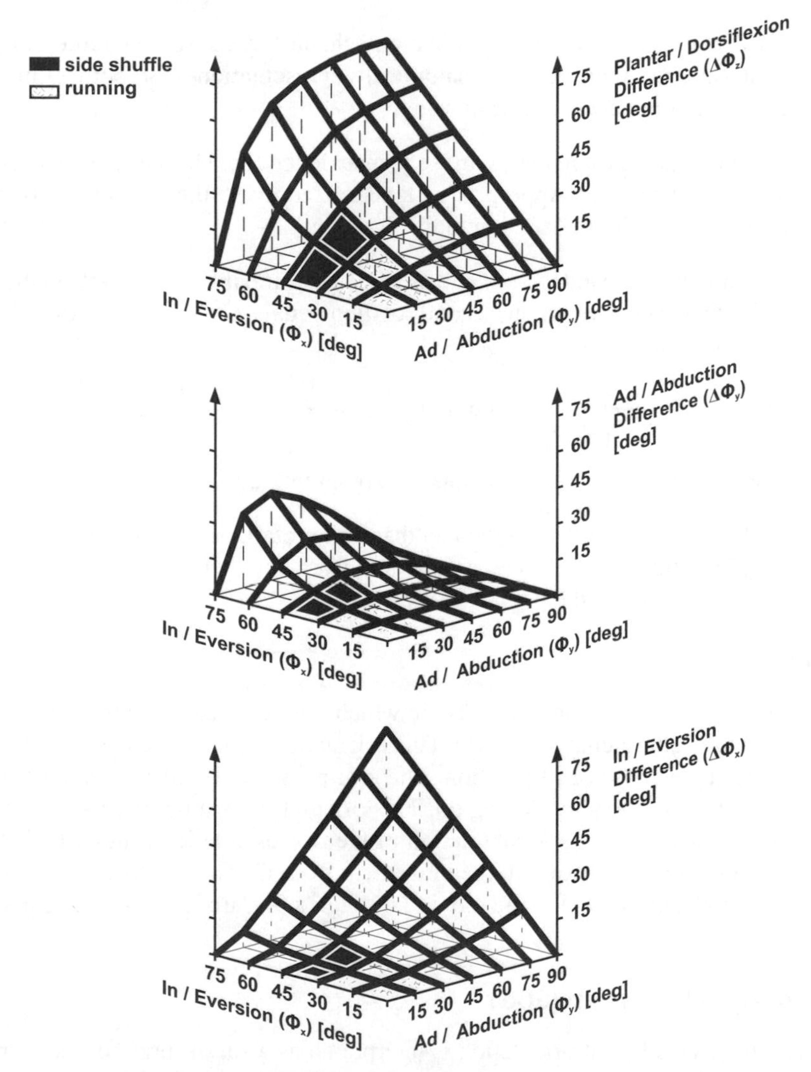

Figure 3.6.11 **Differences in the calculated values for plantar-dorsiflexion (top), ad-abduction (centre), and in-eversion for two different sequences of rotation with the second and third rotation reversed. The primary sequence that is illustrated is plantar-dorsiflexion, ad-abduction, and in-eversion. ϕ_z was arbitrarily chosen as 45°. ϕ_x and ϕ_y were varied from 0° to 90° and the rotation matrices were generated based on the primary sequence. The representations of ϕ_x, ϕ_y, and ϕ_z in the secondary sequence were calculated from the generated rotation matrices and compared to the primary values (from Cole et al., 1993, with permission).**

It can easily be demonstrated that the sequence (1) flexion-extension, (2) axial rotation, and (3) adduction-abduction does not satisfy the above anatomical definitions. If axial rotation is chosen as the second rotation, the third rotation no longer conforms to the definition of adduction-abduction, since it is not constrained to a plane perpendicular to the sagittal plane.

(ii) Advantages and disadvantages of Cardan angles

(1) Cardan angles are widely used in biomechanics, because they represent joint orientation analogous to the anatomical representation that both clinicians and researchers are accustomed to using.

(2) The fact that Cardan angles are sequence dependent has sometimes been viewed as a disadvantage. Appropriate standardization of the sequence, as proposed above, is one simple solution to this problem.

(3) One major disadvantage of Cardan angles is gimbal lock. A mathematical singularity that occurs when the second rotation equals $\pm \pi/2$. For example, ϕ_x and ϕ_z are determined from:

$$\sin\phi_x = \frac{R_{23}}{\cos\phi_y} \quad \text{and} \sin\phi_z = \frac{R_{12}}{\cos\phi_y}$$

If $\phi_y = \pm\pi/2$, then ϕ_x and ϕ_z are undefined.

(4) Problems may result from the fact that finite rotations are not vector quantities. A simple subtraction of one orientation from another one can yield erroneous results in terms of the path of motion.

Example 1

Consider an initial position of the arm in which it is outstretched laterally, at 90° to the vertical, with the palm facing anteriorly. This orientation is quantified as 90° of abduction of the arm from the anatomical position. The arm is moved in a transverse plane until it points anteriorly with the palm facing up. This orientation is quantified as 90° of flexion from the anatomical position. A subtraction of the two orientations suggests that the arm underwent adduction and flexion in the movement from the first to the second position. In reality, the arm axially rotated while moving through a 90° arc in a transverse plane of the body.

Joint Coordinate System (JCS)

Three-dimensional joint orientation is interpreted as a set of three rotations that occur about the axes of a Joint Coordinate System (JCS) (Figure 3.6.12).

(1) One axis, with unit vector, $\mathbf{e}_1$, is selected to be the medio-lateral (z) axis of the proximal segment coordinate system. This is the rotational axis for flexion-extension at the joint.

(2) Another axis, with unit vector, $\mathbf{e}_3$, is selected to be the longitudinal axis of the distal segment. The axial rotation component is measured about this longitudinal axis of the distal segment.

(3) These two segment-fixed axes define the remaining axis, with mutually-perpendicular unit vector, $\mathbf{e}_2$. This remaining axis is the cross-product of the two segment-fixed axes, and it defines the axis of rotation for adduction-abduction at the joint.

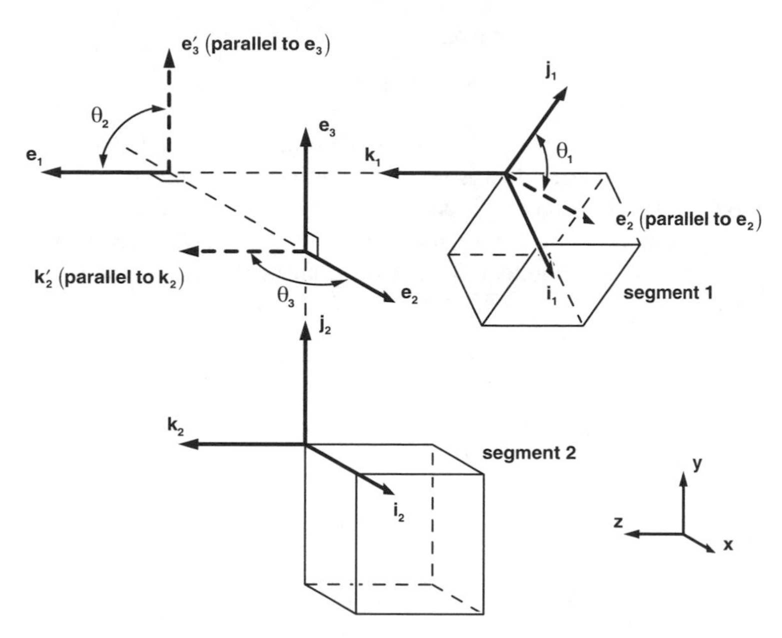

Figure 3.6.12 **Illustration for the definition of the Joint Coordinate System (JCS) (from Grood and Suntay, 1983, with permission).**

The equations for the JCS were originally presented for the specific application to the knee joint (Grood & Suntay, 1983). These equations are presented below in a modified form to make them applicable in the general case (Cole et al., 1993). The equations cannot be expressed in general terms using the coordinate axes x, y, and z. Depending on which axes are selected for the two segment-fixed axes, it is possible to obtain both a right- and left-handed Joint Coordinate System.

To ensure a right-handed Joint Coordinate System, an alternative labelling of segment-fixed coordinate axes is used:

F-axis: Flexion-extension axis. Chosen as the axis of the segment coordinate system that is oriented predominantly in the medio-lateral direction. The axis is described by the unit vector, **f**.

L-axis: Longitudinal axis. Chosen as the axis of the segment coordinate system that is oriented predominantly lengthwise along the distal segment. The axis is described by the unit vector, **l**.

T-axis: Third axis. Calculated as the cross-product of the longitudinal axis and the flexion-extension axis. The axis is defined by the unit vector, **t**, resulting from the cross-product:

$$\mathbf{t} = \mathbf{l} \times \mathbf{f}$$

where:

$\times$ = symbol for vector cross product

The unit vectors, **l**, **f**, and **t** are obtained from the rotation matrix that specifies the orientation of the segment of interest that is relative to an inertial reference system. As an illustration, if the flexion-extension axis is chosen as the z-axis, then, $\mathbf{f} = \{R_{31}, R_{32}, R_{33}\}$.

The unit vectors that describe the orientations of the axes of the JCS between a reference segment, i, and a target segment, j, relative to an inertial reference system are described as follows.

$$\mathbf{e}_{1_{ij}} = \mathbf{f}_i$$

$$\mathbf{e}_{2_{ij}} = \left(\frac{\mathbf{e}_{3_{ij}} \times \mathbf{e}_{1_{ij}}}{\left|\mathbf{e}_{3_{ij}} \times \mathbf{e}_{1_{ij}}\right|}\right) \cdot A$$

where:

$$A = \begin{cases} -1 \text{ if} & \left(\mathbf{e}_{3_{ij}} \times \mathbf{e}_{1_{ij}}\right) \bullet \mathbf{t}_j < 0 \\ & \text{and:} \\ +1 \text{ otherwise} & \left(\left(\mathbf{e}_{3_{ij}} \times \mathbf{e}_{1_{ij}}\right) \times \mathbf{e}_{3_{ij}}\right) \bullet \mathbf{f}_j > 0 \end{cases}$$

$\bullet$ = dot product

$\cdot$ = scalar multiplication

$$\mathbf{e}_{3_{ij}} = \mathbf{l}_j$$

The three angles that represent the three-dimensional orientation of the target segment, j, with respect to the reference segment, i, relative to a neutral position are calculated as follows.

For the angle of rotation about the axis for flexion-extension:

$$\phi_{1_{ij}} = \arccos(\mathbf{e}_{2_{ij}} \bullet \mathbf{t}_i) \cdot \operatorname{sign}(\mathbf{e}_{2_{ij}} \bullet \mathbf{l}_i)$$

For the angle of rotation about the axis for adduction-abduction:

$$\phi_{2_{ij}} = \arccos(\mathbf{r} \bullet \mathbf{l}_j) \cdot \operatorname{sign}(\mathbf{e}_{1_{ij}} \bullet \mathbf{e}_{3_{ij}})$$

where:

$$\mathbf{r} = \left(\frac{\mathbf{e}_{1_{ij}} \times \mathbf{e}_{2_{ij}}}{\left|\mathbf{e}_{1_{ij}} \times \mathbf{e}_{2_{ij}}\right|}\right)$$

For the angle of rotation about the axis for axial rotation:

$$\phi_{3_{ij}} = \arccos(\mathbf{e}_{2_{ij}} \bullet \mathbf{t}_j) \cdot \operatorname{sign}(\mathbf{e}_{2_{ij}} \bullet \mathbf{f}_j)$$

where:

$$\text{sign}(x) = \begin{cases} 1 & \text{if } x \geq 0 \\ -1 & \text{if } x < 0 \end{cases}$$

Counter-clockwise rotations relative to the neutral position have been defined as positive. In comparison to the equations originally presented by Grood and Suntay (1983), the scalar, A, has been added to ensure that each of the three angles is continuous in the range, $-\pi < \phi < \pi$, and the unit vector, **r**, has been added to ensure that counter-clockwise rotations about the axis, $\mathbf{e}_2$, are always positive regardless of which axes are chosen for, $\mathbf{e}_1$ and $\mathbf{e}_3$.

A comparison of the JCS to the proposed sequence of Cardan angles shows the two parametric representations to be identical:

- The first body fixed axis, $\mathbf{e}_1$, corresponds to the first rotational axis of the Cardanic sequence,
- The axis, $\mathbf{e}_2$, corresponds to the nodal axis, and
- The second body fixed axis, $\mathbf{e}_3$, corresponds to the final rotational axis of the Cardanic sequence.

The three angles are calculated differently in the two methods. However, the principle remains the same and the results are identical.

(iii) Advantages and disadvantages of the JCS

(1) The advantages and disadvantages of the JCS for representing three-dimensional joint orientation are identical to those described for Cardan angles.

(2) The major difference between the JCS and the Cardan angle representation is conceptual. For a given joint orientation, the sequential nature of the Cardan angle approach implies that a movement from the neutral position is occurring to obtain the joint orientation of interest, which is not necessarily the case. The JCS, on the other hand, is conceptually an actual representation of a specific joint orientation. This difference would suggest that the JCS approach is the preferable one.

(3) The JCS can also be used to describe displacements. The Cardan and Euler angles only describe the relative orientation of two bodies. However, when using the JCS, the displacement can be described as three displacements, in turn, along each of the axes $\mathbf{e}_1$, $\mathbf{e}_2$, and $\mathbf{e}_3$.

Finite helical axis

Any finite movement of a body from a reference position can be described as a rotation about and a translation along a line in space (Woltring, 1991). This line in space is called the *finite helical axis*. To completely describe a movement, the following finite helical axis parameters must be specified:

$\mathbf{n}$ = the unit vector describing the orientation of the finite helical axis
$\mathbf{s}$ = a point on the helical axis, locating it in space
θ = the amount of rotation about the helical axis
d = the amount of translation along the helical axis

For any movement, the finite helical axis passes through the centre of rotation. For pure rotations, the finite helical axis is simply the axis of rotation, like the hinge of a door. If there are translations in the same plane as the rotations, then the axis can lie some distance from the body. If there are translations parallel to the axis, then the motion resembles that of a screw, and $d > 0$. The helical axis description of a movement can be awkward if there are translations with very small rotations. In cases like this, the location of the axis **s** can be very far from the body.

In addition to describing simply the relative orientations of two bodies, the finite helical axis provides information about how the displacement took place. When the movement of one body is described relative to a connected body, the finite helical axis can provide information about the joint between the two bodies. For any movements of the forearm with respect to the upper arm, for example, the helical axis will pass though the centre of the elbow joint and the direction of the rotation **n** will always be in the same direction, because the elbow joint is (for the most part) a revolute joint. For spherical joints, e.g., the hip joint, the axis will always pass through the centre of the hip, but the orientation can vary. For joints that are neither spherical nor revolute, the position and orientation of the helical axes can change. Helical axes can be used for describing the movement at joints because of information they provide about how the joint works.

The helical axis orientation parameters, **n**, θ, can be calculated from the rotation matrix for a selected angular displacement from one coordinate frame to the next. The axis position **s** and the translation d, can be determined from the displacement of the origin of the coordinate frame. The methods used to calculate the helical axis parameters in biomechanical applications are described in detail elsewhere, e.g., Spoor & Veldpaus (1980) and Woltring et al. (1985). In short, the helical axis can be found because for a given rotation and displacement, all points on the helical axis remain on the helical axis.

(iv) Advantages of finite helical axes

(1) When plotted relative to anatomical landmarks, finite helical axes provide a functional representation of the movement occurring at the joint, especially when the finite increments of rotation are chosen to be small.

(2) The finite helical axis approach provides information about the actual axes of rotation in a joint and linear translations of one body with respect to another, which cannot be obtained using Cardan or Euler angles.

(3) Finite helical axes do not suffer from gimbal lock. No orientations exist that cannot be represented by finite helical axes.

(v) Disadvantages of finite helical axes

(1) The finite helical axis does not provide a representation of joint orientation in terms of three anatomically and clinically meaningful parameters, as do Cardan angles and the JCS.

(2) Helical axis parameters are sensitive to noise in the spatial landmark coordinates, and to the magnitude of the finite rotation (Spoor, 1984; Woltring et al., 1985). However, the influence of the former can be substantially reduced with appropriate smoothing of the landmark coordinates (Woltring & Huiskes, 1985; Lange et al., 1990).

Helical angles

The concept of helical angles has been proposed as an alternative to Cardan or JCS angles (Woltring, 1991; Woltring, 1992). Helical angles are a condensation of the information included in finite helical axes, and only include the components involved in the change or difference in orientation (translations are not described). The three helical angles are calculated from the components of the finite helical axis:

$$\theta = (\theta_x, \theta_y, \theta_z)^T = \theta \mathbf{n}$$

where:

$\mathbf{n}$ = unit direction vector of the finite helical axis (as above)

Helical angles have many of the same advantages (and disadvantages) as the finite helical axes. However, one advantage that they have over the helical axes is that they are as compact a description of orientation as Euler or Cardan angles (with only three parameters).

Helical angles (and finite helical axes) eliminate the problem of gimbal lock. Additionally, for rotations about a single coordinate axis, helical angles are the same as Cardan or JCS angles. As with Cardan or JCS angles, helical angles are not additive. However, they reduce the non-linear properties of the parametric representation of joint orientation (Woltring, 1991).

Final comments

The selection of the appropriate methodology for describing three-dimensional joint orientation must be determined by the application. Each approach has its own inherent advantages and disadvantages. The JCS approach, for example, produces a result that is easily understood in clinical and anatomical environments. The helical axis approach, on the other hand, provides results that better represent the functional movement that occurs at a joint.

3.7 STRAIN MEASUREMENT

SHRIVE, N.

3.7.1 MEASURING POSSIBILITIES

The measurement of strain is fundamental to understanding and defining material behaviour, and to clarifying structural behaviour and design. The measurement of strain is important for developing implants to assess how the material behaves under load and determining how the distribution of load in the structure is affected by geometry and applied loading.

Strain measuring techniques for stiff materials, such as bone, metals, and ceramics, that have developed over the years allow good sensitivity to be achieved fairly cheaply. The technology for measuring strain on more flexible materials, such as tendons and ligaments, has not followed the pace set by more standard techniques. Measuring strain on more flexible materials has faced special problems that have proven difficult to overcome. Frequently, the basic issue is: Has the strain being measured been changed by the very attempt to measure it?

GENERAL

Strain gauges measure engineering strain, which is defined as:

$$\varepsilon = \frac{\text{change in length}}{\text{original length}} = \frac{L - L_o}{L_o} = \frac{\Delta L}{L_o}$$

where:

L = current length
L_o = original length
ΔL = change in length

True strain is defined as the integral, from the original to the current length, of the increment in length divided by the immediately preceding length.

$$\varepsilon_{true} = \int_{L_o}^{L} \frac{dl}{l}$$

$$= \ln\left(\frac{L}{L_o}\right)$$

When measurements of strain are made, the overriding principle for the measuring system is that it is highly flexible. Very little load must be required to deform the measuring device. The device will then not interfere with the strain being measured, and the load that is normally carried by the material being strained, will continue to be carried by that material, rather than being transferred into the measuring device. This is in direct contrast to a load measuring device that is in series with the load being measured. Here the overriding principle is that the device is stiff. Highly stiff load cells do not interfere with the deformation characteristics of the system.

There are three basic techniques for measuring strain. The first is to measure the equivalent of strain and to convert the measurement to mechanical strain by a gauge factor. Electrical resistance strain gauges, such as foil gauges or liquid mercury strain gauges, in Wheatstone bridge configuration are based on this principle, as they measure electrical strain in the circuit.

The second technique is to measure and amplify ΔL, the change in length, and to divide by an accurate measure of L_0. Extensometers and optical interference methods are based on this principle. Methods based on these first two techniques have the potential to be both accurate and sensitive. Indeed, sensitivities in the order of one microstrain can be achieved with some semiconductor and foil gauges. While this level of accuracy and sensitivity is not normally needed in biomechanics, it indicates what can be achieved.

The third technique is to measure lengths and derive strain from length. Methods based on this technique will be inherently inaccurate at low strains. This inaccuracy occurs because the difference between two large numbers (the lengths) will be inaccurate when the difference is of the same order of magnitude as the error in the large numbers. Simply:

$$(L \pm E) - (L_0 \pm E_0) \quad = \quad (L - L_0) \pm (E + E_0) \quad = \quad \Delta L \pm (E + E_0)$$

where E and E_0 are the errors in the length measurements.

When ΔL is substantially larger than $(E + E_0)$ the strain reading will be meaningful. Thus, if large strains are measured, techniques based on length measurement may provide reasonable indications of the actual level of strain. At low strain levels, however, strain measurements from such methods will have a large error.

Strain measuring devices include:

- Electrical resistance strain gauges,
- Extensometers,
- Optical methods, and
- Other methods.

These devices are discussed in the following paragraphs.

ELECTRICAL RESISTANCE STRAIN GAUGES

General principles

The resistance of an electric conductor is a function of the dimensions of the conductor and its resistivity.

$$R = \frac{sL}{A}$$

where:

R = electrical resistance

s	=	resistivity
L	=	length
A	=	cross-sectional area

From the mechanics of deformable bodies, it is known that when such a conductor (like a wire) is stretched, L increases and A decreases. Therefore, the electrical resistance R, changes. The change in electrical resistance can be related to the change in length. With appropriate circuitry:

$$\frac{\Delta R}{R_o} = G_F \frac{\Delta L}{L_o} \quad (3.7.1)$$

where:

R_o	=	initial electrical resistance for the unstrained material
ΔR	=	change in resistance due to the strain in the material
G_F	=	gauge factor
L_o	=	initial length of the unstrained material
ΔL	=	change in length of the strained material

G_F is the gauge factor, which is typically about two for bonded wire and foil gauges in their linear range (as the strain gets higher, G_F becomes more and more non-linear with strain. Special circuitry is needed or allowances made for the non-linearity).

The relationship of equation 3.7.1, above, holds as long as the conductor in the gauge stretches with the specimen. This occurs if the gauge is firmly glued to the specimen. In a strain gauge, the conductor has a very small cross-section, requiring only small forces for extension. For example, if the conductors in a gauge (Figure 3.7.1) have a total width of 2

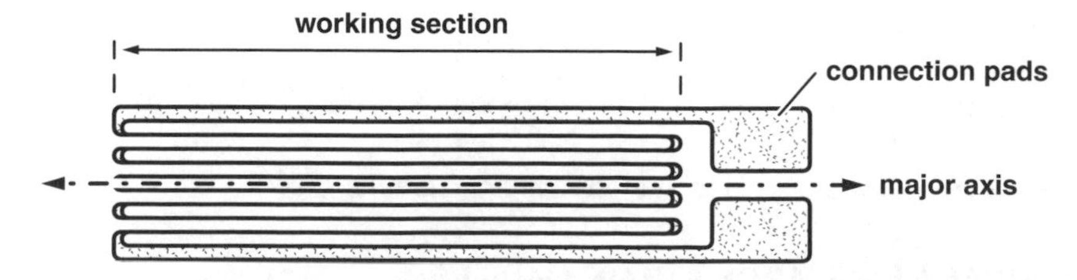

Figure 3.7.1 **Schematic of a foil strain gauge. The gauge measures extensional strain in the direction of the major axis. The strain is an average measure over the length of the working section.**

mm and thickness 5 micrometres (in some foil gauges, the conductor is only 2.5 micrometres thick), less than 1 N will cause yield of a typical conductor. This level of force can easily be transmitted through a variety of adhesives, given the area of the gauge bonded to the specimen surface.

Thus, concern about the gauge reinforcing the specimen and altering the normal strain distribution should arise with very thin structures or flexible materials, or both, e.g., cartilage, ligament or tendon in biomechanics. In such cases, consideration should be given to the level of reinforcement provided, and the alteration to the strain measured. A related issue more pertinent to biomechanics is the potential for a chemical reaction between the adhesive and the test material. Strain would be measured on this newly created material, not the specimen material, and the strain in the specimen would be altered by the presence of the new material.

Strain gauges measure extensional strain in the direction of the major axis of the gauge (Figure 3.7.1). However, both the lateral strain and the shear strain in the plane of the gauge can effect the extensional strain reading. In most cases, these effects are negligible. However, users may have to compensate for effects that are not negligible. Manufacturers normally indicate the magnitude of this transverse sensitivity through the K_t factor of the gauge. If this factor is larger than the accuracy desired or if rosette gauges are used, then the effects above may need to be considered. The extensional strain is averaged over the working area of the gauge, and the division of the conductor into a series of connected wires in the major axis direction amplifies the change in R over that which would be obtained by a single conductor covering the same working area as a single sheet. In gauge selection, the size of the working area of the gauge must be considered in light of the size and material of the specimen, the rate of local strain variation, and the desired information. For example, a large gauge does not reveal much of value where strains vary rapidly and knowledge of the variation is desired. Similarly, a small gauge is of equally little value when an average strain over a composite is required. Foil gauges range in working length from about 0.2 mm to 100 mm, thus providing considerable choice.

More than one gauge is required if knowledge of extensional strain in more than one direction on the surface is desired. Two and three gauge rosettes are available. A three gauge rosette is required for determining principal strains and their directions (unknown at the beginning of the test). The gauge directions are typically offset from each other by 45° or 60°.

Circuitry

Strain levels in stiff materials are typically measured in terms of microstrain. With a gauge factor of 2, the change in electrical resistance will be of the same order of magnitude (10^{-6}) compared to the original resistance (R_0) of the conductor. Therefore, the Wheatstone bridge (Figure 3.7.2) is used to measure the small changes.

The choice on a strain gauge conditioning amplifier will be for 1/4, 1/2 or full bridge operation. In 1/4 bridge operation, only one resistor (a gauge) will be active (change during the test). Let this be R_1 in Figure 3.7.2. R_3 and R_4, in this case, are precision resistors of equal value inside the conditioning amplifier. R_2 is also in the amplifier or could be a separate (dummy) gauge. If in the amplifier, R_2 is adjusted to have the same resistance as the active gauge R_1 in the unstrained condition. Thus, the voltage drop across the centre of the bridge V_{out}, is zero. The circuit is balanced. As soon as strain is applied, R_1 changes by ΔR and V_{out} is no longer zero. The magnitude of V_{out} depends on ΔR.

If R_2 is active, in addition to R_1, usually sensing strain of the opposite sign to R_1, then a half bridge circuit is being used. An example is simple bending where a gauge or one side of a beam would sense tension while one on the other side would sense compression. When all four gauges are active, with all resistance changes arranged to maximize the change in V_{out}, full bridge operation is being used. For example, strain gauge based extensometers use this system.

The ability to measure strain accurately with this circuitry depends on the stability of the excitation voltage and the ability to measure V_{out}. The precision of the instruments used to

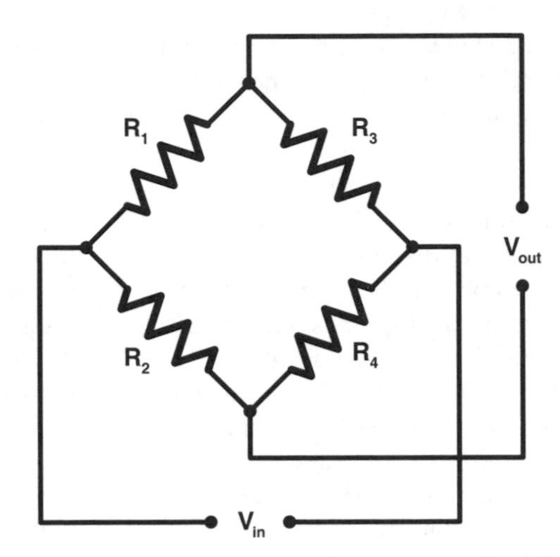

Figure 3.7.2 **Wheatstone bridge circuitry for strain gauge operation.**

perform these two tasks is paramount for the strain sensitivity of the circuit. It is also important to recognize that when the excitation voltage is turned on, current flows through the arms of the bridge, heating the resistors. The circuit must be allowed to reach thermal equilibrium before taking measurements.

In a circuit, the resistance of the lead wires to the gauges needs to be considered. If the wire is long or thin, or both, then the lead wire can have sufficient resistance to desensitize the strain gauge. Basically, R_o in equation 3.7.1 is increased without an equivalent increase in ΔR.

Temperature compensation

The resistance of a strain gauge will change with temperature as the material on which the gauge is mounted expands and contracts with the change in temperature. In situations where the temperature of the specimen will change during a test, the thermal effect must be isolated from the load effect, in order for the latter to be determined. There are two basic approaches to solve this problem.

In temperature compensating circuits, the active gauge might be R_1 in Figure 3.7.2, as before. R_2 would be a compensating dummy gauge mounted on a piece of the same material as the test specimen. This piece of material would be subject to the same thermal regime as the test specimen. Thus, the resistance of both R_1 and R_2 would change by the same amount with temperature, having a null effect on V_{out}. This technique requires that the temperature of both the test specimen and the separate piece change simultaneously. Some circuits are automatically temperature compensating, e.g., a half-bridge where the gauges are bonded on either side of a beam being bent.

The second approach is self-temperature-compensation (STC). Certain constantan or nickel chromium alloys can be heat-treated to obtain specific thermal characteristics when rolled for incorporation in foil gauges. These gauges expand sympathetically with the material on which they are mounted and, therefore, do not measure thermal strain. STC gauges of this type are designed to compensate for the thermal expansion of a particular material, typically steel. A steel compensating gauge will not compensate correctly for temperature changes in other materials, e.g., titanium. Temperature changes in the lead wires to the gauge must also be considered in STC circuits. A three-wire circuit is required in 1/4 bridge operation to eliminate the temperature effect in the lead wires (Figure 3.7.3).

Temperature rises in small components due to strain gauge current can be reduced by using a pulsed power source.

Semiconductor strain gauges have a smaller current carrying capacity, which reduces the allowable excitation voltage. Therefore, apparently high gauge factors cannot always produce as large an output signal as expected.

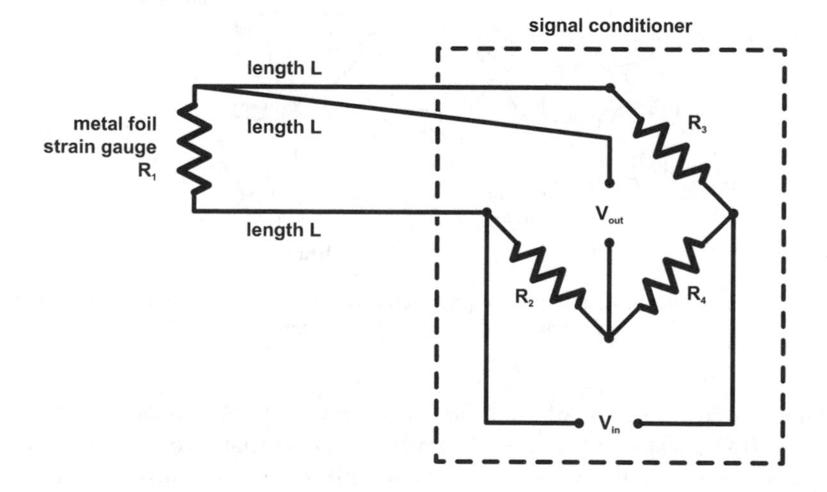

Figure 3.7.3 **Three wire arrangement for temperature compensation of the wires in quarter bridge operation with a STC gauge.**

Electric resistance strain gauges

Foil gauges

Foil strain gauges are strain measuring sensors consisting of a thin layer of a metal conductor arranged in an array of connected parallel lines (Figure 3.7.1). These gauges were described in the preceding paragraphs on the general principles of electric resistance strain gauges. These are the most commonly used form of strain gauge sensors. The expressions foil gauge and strain gauge are often used synonymously.

Semiconductor strain gauges

Semiconductor strain gauges are wafers of semiconductor doped to the form of a strain gauge. They have much higher gauge factors than the foil and wire gauges above, and, therefore, have higher output for a given strain and excitation voltage. The sensitivity with currently available instrumentation is about 1 microstrain.

Liquid mercury strain gauges

The liquid mercury strain gauge was introduced (Edwards et al., 1970) as a method for measuring strain on soft tissues. As shown in Figure 3.7.4, the conductor, which changes resistance, is usually a column of mercury, or less frequently, a gallium-indium electrolyte solution. The lead wires near the ends of the column are sewn to the tissue being studied. This causes a problem with the definition of the gauge length, L_0. If there is no slippage between the wire, the suture, and the tissue, the gauge length is the distance between the two sutures. Therefore, the gauge length is not necessarily the length of the mercury column. With this method, strain measures of up to 50% have been reported.

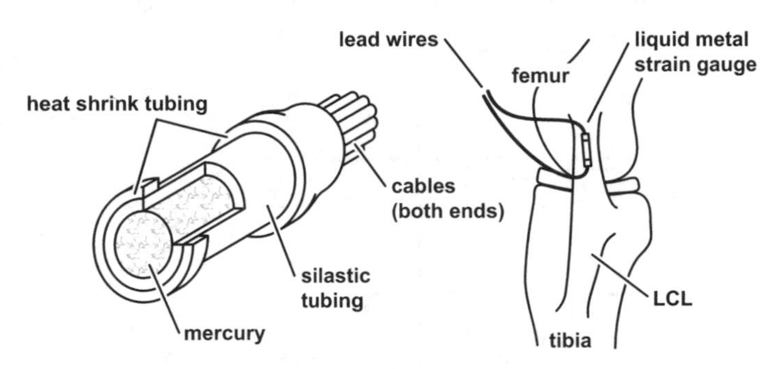

Figure 3.7.4 **Schematic illustrations of a liquid metal strain gauge and its construction (left), and a possible arrangement on a lateral collateral ligament (right).**

The gauge is used as one arm of a Wheatstone bridge, and has sometimes been used in addition to a 120 Ω resistor in the arm. However, as the resistance of a Liquid Mercury Strain Gauge (LMSG) is in the order of 1 Ω, the circuit is very insensitive. Greater sensitivity can be achieved by creating a bridge circuit with 1 Ω resistors.

LMSGs have a non-linear response to strain, and individual gauges have given reproducible results. However, there is considerable variation in response from gauge to gauge. In work on soft tissues, these gauges have been widely used.

EXTENSOMETERS

Extensometers were originally mechanical devices that relied on mechanical magnification of the increase in length of the gauge length, which was achieved by the movement of levers around pivot points. However, electromechanical extensometers were introduced with the advent of strain gauges. The basic principle used is that the highest strain that develops in a cantilever is at the root of a cantilever (Figure 3.7.5a). The displacement at the end of the cantilever is related to the strain generated. With two such arms firmly attached to a specimen, the distance between them is the gauge length, and the increase in the length of the gauge length is simply the sum of the two displacements measured (Figure 3.7.5b). This sum can easily be converted to specimen strain. Some extensometers have the gauges on the portal section of the portal frame made by the two arms and their connector.

The critical issues for an extensometer are the contact between the specimen and the extensometer arms, and the force required to open the arms. The latter needs to be small relative to the force applied to the specimen, so as not to interfere with the strain being mea-

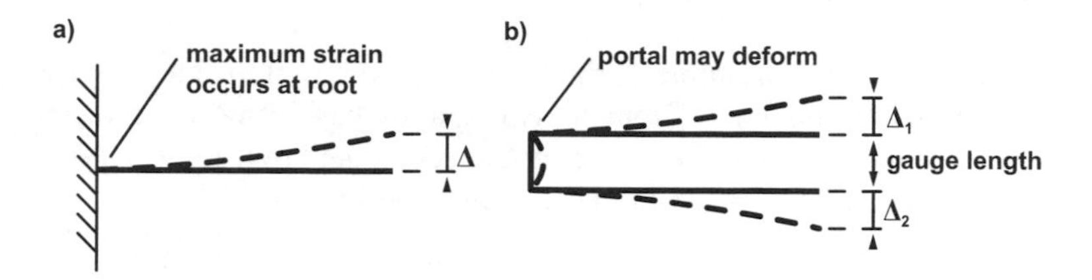

Figure 3.7.5 **(a) The maximum strain in a cantilever given a simple deflection at its end occurs at the root. With two such arms (b), an extensometer is formed. The distance between them is the gauge length. Depending on the construction of the extensometer, the portal connecting the two arms may deform and be strain gauged.**

sured. An extensometer designed for large steel specimens may have an opening force that is too big for a thin aluminium sheet. The contact needs to be firm, with no slip between arm and specimen. Thus, extensometers made for metals have sharp edges and are held tight to the specimen through spring loading. These extensometers slice through soft tissues with considerable ease.

Thus, Shrive et al. (1992 and 1993) developed a variation on the extensometer theme to accommodate soft tissues (Figure 3.7.6a). In this instance, the extensometer is not mounted on the specimen. The weight therefore does not distort the specimen, which would occur if the extensometer was suspended on the specimen as usual. Here, the weight is supported by a light flexible ring. Normal extensometers (held on the specimen) accommodate rigid body displacement of the gauge length by moving with the specimen (Figure 3.7.6b). With the new extensometer, rigid body displacement of the gauge length is accommodated by the ring rolling in shear. Low contact forces are controlled by the ring, and the cantilever tips are serrated to provide grip on soft tissue.

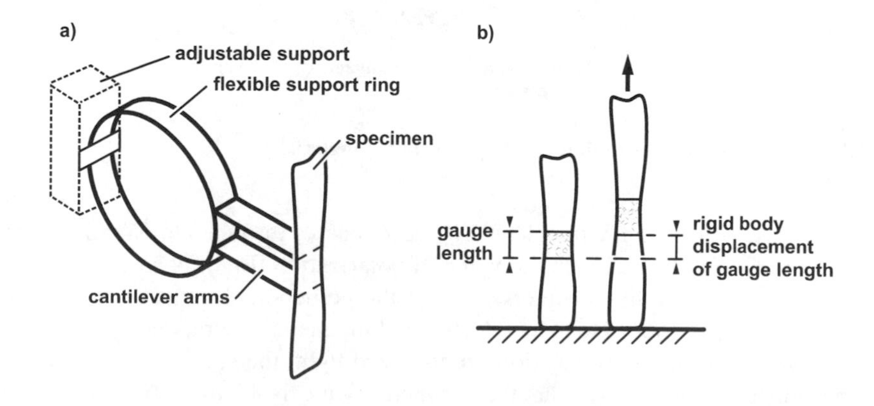

Figure 3.7.6 **Schematic illustration of a soft tissue extensometer. The soft tissue extensometer uses standard principles, but is supported externally to the specimen (a). Rigid body deformation of the gauge length when the specimen extends (b), is accommodated by the support ring deforming in shear. The ring also applies the low contact forces needed to hold the tips of the arms on the specimen.**

OPTICAL TECHNIQUES

The two major optical strain measuring methods used in biomechanics are photoelasticity and direct dimensional measurement. Holography has been used in recent years and is, therefore, discussed briefly. The Moiré fringe technique, however, has been used minimally in biomechanical applications and is not discussed.

Photoelasticity

Photoelasticity is a stress-induced response and, therefore, more properly called photoelastic stress analysis. However, since directions of principal strain are the same as those of principal stress, the effect can also be considered a strain response with results related to strains. The method is based on the phenomenon that the refractive indices of certain polymers change with the level of stress. The different velocities which result in polarized light when it is shone through a piece of stressed material, cause the light waves to interfere in relation to principal stresses. The interference results in patterns of dark and light fringes that can be analyzed to obtain both the magnitude and direction of stress in the material.

Equipment needed for photoelastic analysis is shown in Figure 3.7.7. Light from a light

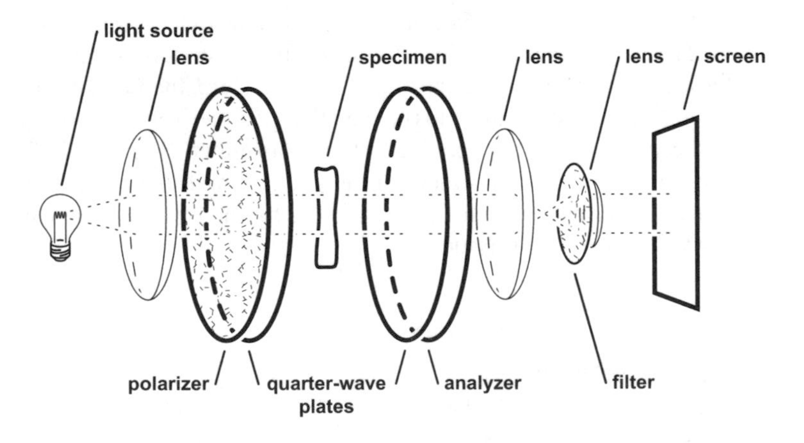

Figure 3.7.7 **Schematic of equipment for photoelastic analysis.**

source is focused into a beam with a lens. The beam passes through a polarizing plate that transmits only those light waves in the plane of polarization. Light is an electromagnetic wave travelling along the line of intersection of the perpendicular electric and magnetic waves. In normal light, the wave pairs (electric and magnetic) are randomly oriented relative to the direction of light propagation. In polarized light, the electric wave is always in one plane and one plane only, and the magnetic wave is in the perpendicular plane (Figure 3.7.8).

The quarter wave plate splits the polarized light into two perpendicular components with a phase difference of $\pi/2$. This is called circularly polarized light. In transmission photoelasticity, the circularly polarized light passes through a stressed transparent model, where different phase shifts occur because of the different refractive indices in the differently stressed regions of the model. The second quarter wave plate converts the wave back to polarized light. The analyzer combines the components of the phase shifted light into only one plane. The results are a series of dark and light fringes.

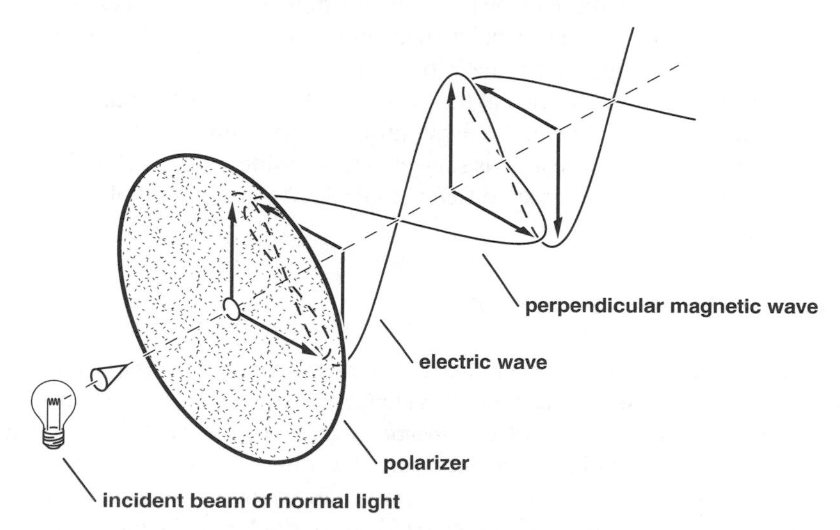

Figure 3.7.8 **Schematic illustration of electric and magnetic waves of polarized light. Normal light has electric and magnetic waves in all planes. Polarized light has an electric wave in one plane only, with the associated magnetic wave perpendicular to it.**

Two sets of fringe patterns can be obtained. With the quarter wave plates absent, isoclinic fringes can be obtained. Dark (isoclinic) fringes are seen where the axes of the polarizer and analyzer correspond to the directions of principal stress. Rotation of the polarizer and analyzer relative to the specimen reveals the directions of principal stress throughout the specimen, since the dark isoclinic will move in the specimen with the rotation of the plates. With the quarter wave plates in the sequence, isoclinic fringes are removed, and isochromatic fringes revealed. These latter fringes are dependent on the difference between the principal stresses and can be used to estimate magnitudes of stress. The zero order isochromatic fringe is black, and with white light the subsequent fringes appear as spectra of colours that make numerical analysis difficult. It is better to use a monochromatic source of light so that the fringes are distinct dark and light lines.

If the model is two-dimensional, the fringes can be projected onto a screen to be photographed directly as the model is loaded. For three-dimensional models, a more complex procedure called stress freezing is followed. In stress freezing, the model is loaded and the temperature raised (typically, in an oven). The stiffness of the model reduces dramatically and the model deforms more. The specimen is then cooled, still under load, to room temperature. The load is removed, but the effects of the load are frozen into the model. Essentially, the secondary bonds, which connected the major molecules, were released by thermal energy in the heating process (causing the reduction in stiffness and the realignment of the major molecules). New secondary bonds were formed between the major molecules on cooling. When this specimen is sliced, and each slice viewed in transmission polariscope, a picture of the three-dimensional stresses in the original model can be built up.

In models where more than one material is used, for example, a model of a hip implant in a femur, careful selection of the modelling materials is required. To obtain a meaningful stress pattern, the moduli of the materials selected to represent the bone and implant should be equally in proportion to those of the bone and the implant. If stress freezing is employed, then the ratio of the moduli must be maintained throughout the applied temperature regime.

In photoelastic coatings, the circularly polarized light is shone onto the coating, passes through the coating, and is reflected at the surface of the object under study by an aluminium-impregnated layer of cement. The light must then pass through the photoelastic layer again before analysis. Thus, the light is subjected to a double dose of refraction in the photoelastic layer, and the equations are different to those of transmission photoelasticity by a factor of two.

Direct dimensional measurement

Various techniques have been employed to measure lengths between markers optically. These include high speed cinematography, video dimensional analysis, and holography.

Noyes et al. (1984) used *high speed cinematography* to quantify strain in ligaments and tendons. Ink lines were marked on a soft tissue and filmed at 400 frames per second. Sequential frames of the developed film were then projected onto a digitizer plate, and points on the edges and in the middle of the ink bands were digitized. As the ink bands moved apart, the distance between the digitized points increased. The accuracy of the estimated distances between points. Therefore, depending on the accuracy of the digitizing equipment itself, and the ability of the user to select the same point on the ink mark consistently. The technique is labour intensive and was superseded by video dimensional analysis, which is now more automated.

Video dimensional analysis is based on a technique developed for analyzing video pictures for dimension (Yin et al., 1972). A VDA system (Figure 3.7.9) consists of a video

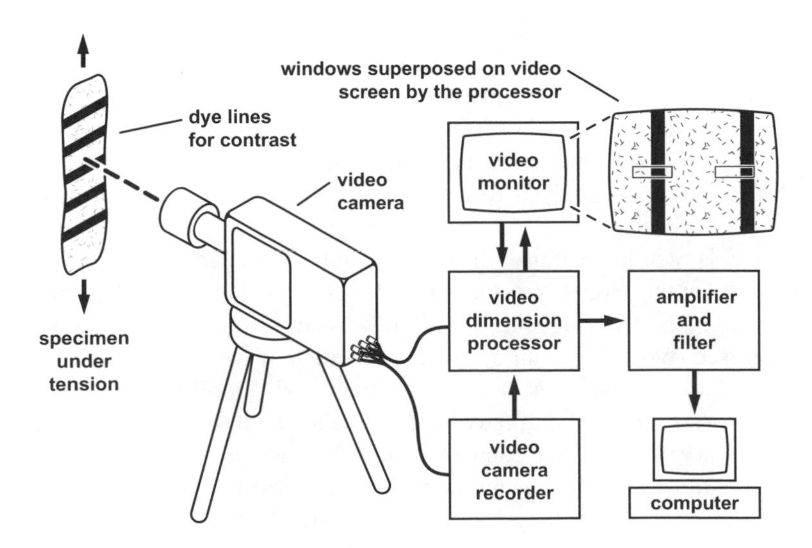

Figure 3.7.9 Schematic of a VDA System.

camera that is focused on the tissue or sample being tested. The camera must be oriented such that the direction of test in the monitor view is horizontal (along a raster line). The output voltage from the camera is sent through a video dimension analyzer micro processor and possibly a VCR. The signal, if in real time or on replay from the VCR, can have two windows added to the view in the monitor (Woo et al., 1983). The operator of the system can adjust the position and length of the two windows to cover the edges of two dye lines on the specimen, using a pair of light and dark contrast lines. As the electron beam sweeps along a raster line where the windows are superimposed, the contrast in the first window (for example, dark to light) starts a timer. When the beam passes the sharp contrast in the second window (this time light to dark), the timer is turned off. Thus, the initial or gauge length is established as the time it takes the beam to move from one contrast to the other with the specimen in the undeformed state. As the specimen is deformed and the dye lines move apart, the time between the two contrasts increases and can be related to the new length between the dye lines. The change in time over the original time can be related to the change in length divided by the original length, i.e., strain. With long enough windows, rigid body displacement of the gauge length can easily be accommodated.

A major problem with these techniques is accuracy at low strains. At low strains, the error in the measurement of the inter-marker distance is of the same order of magnitude as the difference in length being estimated.

Variations on this video analysis theme exist, for example, point markers on tissues rather than dye lines. The coordinates of the centres of the pixels darkened by the points are calculated, and distances between point centres determined. Again, these length measuring techniques are subject to the inherent inaccuracy of the approach, for low strains.

The accuracy of early VDA systems was about 0.5% strain, but this has improved with more accurate image analysis algorithms to about 0.1% strain. Such strain sensitivity is acceptable when measuring a 10% or 40% strain at failure of a soft tissue, but may be unacceptable if subtle variations in strain are needed, or cycling at low loads/deformations is of interest. A second problem with video and film methods is that a planar view is taken. If the object moves in or out of that plane by rigid body rotation, a strain that has not actually occurred will be measured by these techniques (Figure 3.7.10).

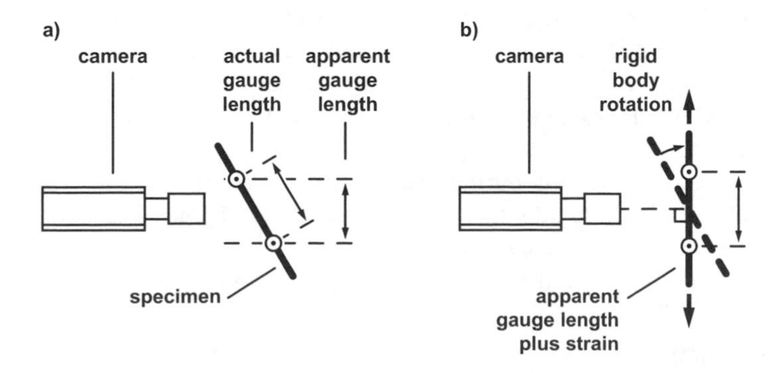

Figure 3.7.10 **A source of error for VDA or film measurements is out of plane rotation (a). The specimen, not normal to the viewing plane of the camera initially, rotates into that plane without extensional deformation. The camera now sees the actual gauge length, but the system records it as the apparent gauge length plus strain (b). For example, a 5^0 rotation would give an apparent strain of 0.38%.**

Holography

Holography uses the interference of two beams of monochromatic light that began their paths simultaneously. As a result, a laser beam is split into two sub-beams: the reference beam and the object beam. The object beam is reflected off the object and then interferes with the reference beam as both are projected onto photographic emulsion (Figure 3.7.11). The emulsion is typically coated on a glass plate. Since the emulsion retains information not just on the intensity of light from the object (as a photograph would), but also on the phase of that light with respect to the reference beam, it is possible to reconstruct a three-dimensional image of the object.

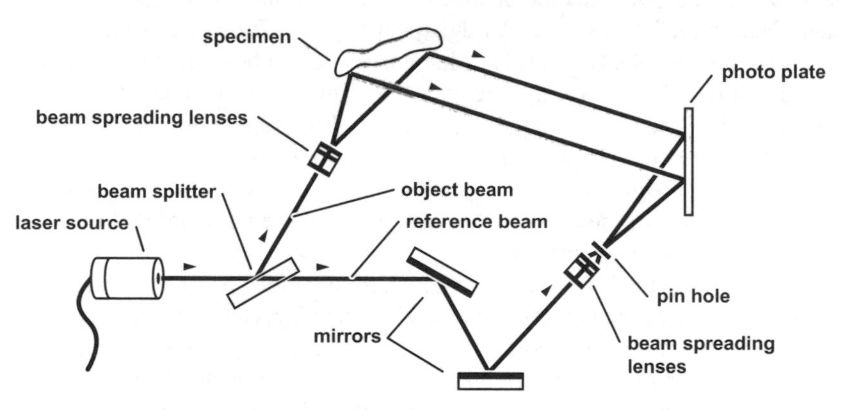

Figure 3.7.11 **Schematic of equipment for holography.**

Holographic interferometry requires the recording of two holograms. The hologram first is recorded when the specimen is unloaded and the second recorded after, or during, loading. Essentially, the two object beams interfere with each other. Where the path lengths of the light beams are changed by a multiple of half a wavelength, the beams cancel each other out. Where the path lengths differ by a multiple of a whole wavelength, the beams add to each other. Thus, a series of dark stripes appears in front of the object where the beams cancel. These can be related to the deformation of the specimen.

Double exposure of the emulsion on the glass includes the object wave interference and the hologram, while the fringe pattern can be viewed at will. However, the view is for one loading condition.

If more continuous viewing during loading is required, then a hologram must first be made of the object. The hologram then replaces the photographic plate, and must do so precisely. Subsequent loading of the specimen allows the fringes to be viewed in real time, but there is no permanent record of the result for any one loading position.

OTHER METHODS

Other non-standard techniques have been developed to overcome various difficulties with strain measurement in biomechanics. Three discussed here are: brittle lacquers, the Hall Effect displacement transducer, and Stress Pattern Analysis by Thermal Emission (SPATE).

Brittle lacquers

Brittle lacquers are materials that can be applied to the surface of a specimen and then allowed to dry. The thin layer of material that forms the coat provides negligible reinforcement to relativity large and stiff structures like bones. After drying, subsequent loading of the specimen in tension causes the lacquer to crack perpendicular to tensile strain. The threshold level of strain for cracking to begin can be as low as 500 $\mu\varepsilon$.

If compressive loading is applied, brittle lacquers will not measure compressive strain, other than indirectly. The indirect method is to apply the load, then apply the lacquer and to let it cure while the specimen remains under load. Once the lacquer is dry (some 16 hours later), the specimen can be unloaded. Removing compression allows the specimen to expand back to its normal shape and the lacquer may crack. Problems with creep and drying of the specimen need to be considered in this type of test.

It is difficult to obtain accurate quantitative values of strain with lacquers. However, they can indicate regions of high strain if the progression of cracking is observed.

Hall Effect displacement transducer

The Hall Effect displacement transducer was developed for use on soft tissues (Arms et al., 1983) and consists of a piston and a cylinder (Figure 3.7.12). A Hall Effect displacement transducer is on the cylinder, while the end of the piston is a magnet. The voltage produced by the transducer depends on the position of the magnet. In a 5 mm transducer, there is about 1 mm of piston travel where a linear relationship exists between the position of the piston and the Hall Effect voltage.

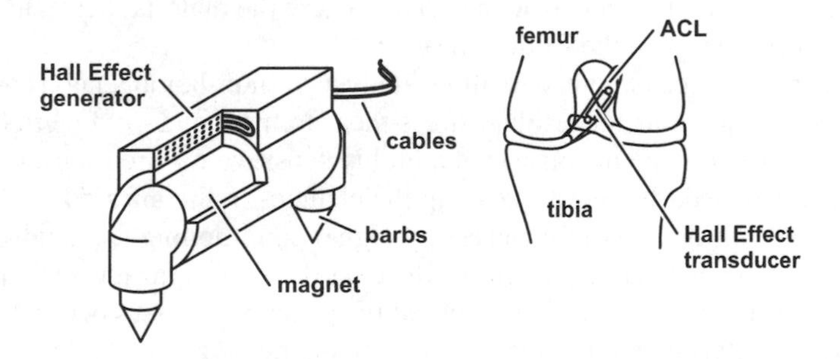

Figure 3.7.12 Schematic illustrations of a Hall Effect transducer on the left, and a possible arrangement on an ACL (right).

Both the piston and the cylinder are attached to barbs that are used to fix the transducer to soft tissues. Suturing is also required to hold the two components in place. Therefore, as with the LMSG, there is difficulty with definition of the initial gauge length. Subsequent straining may take the device out of the linear range. Practically, it requires considerable skill to suture the two components in place so that there is free movement of the piston in the cylinder. Misalignment can cause the piston to jam in the cylinder, especially during cyclic loading.

Axial rotation of the piston in the cylinder causes a change in the output voltage. Thus, situations in which there is torsional and axial displacement give inaccurate strain values.

This effect has been reduced with the more recent Differential Variable Reluctance Transducer (DVRT).

Hall Effect transducers have two immense advantages. They are small and can be implanted in vivo like the LMSG.

SPATE

Stress Pattern Analysis by Thermal Emission provides the acronym SPATE (Oliver, 1987). The method is based on the minute temperature changes that materials undergo when they are stressed. The thermoelastic equation for linear elasticity is:

$$\Delta T = -K_m T \Delta\sigma$$

where:

ΔT = peak to peak change in temperature in ^{o}K due to $\Delta\sigma$
$\Delta\sigma$ = peak to peak change in the sine wave component of the sum of the principal applied stress at the reference original frequency
T = mean temperature of the specimen in ^{o}K
K_m = thermoelastic constant

The equation is applicable to isotropic, homogeneous materials loaded adiabatically in their linear elastic range. Cyclic loading is required, and the small changes in temperature (thousandths of a degree K) are reversible. Thus, these are not the temperature changes observed in cyclic loading of viscoelastic materials, where the material heats up due to dissipation of hysteresis energy in the form of heat.

SPATE theory has been extended to allow analysis of non-homogeneous or anisotropic materials, or both. Equations for residual stress measurement and plastic effects have also been developed. The equipment consists of a highly sensitive infrared sensor. As the specimen is cyclically loaded, the sensor picks up the changes in the infrared photon emission from the surface. With the aid of the necessary signal conditioning and analysis, a picture can be built of the cyclic changes in temperature on the surface of the specimen. The hot spots are where the largest changes in the sum of the principal stresses occur. Methods have been developed to determine individual stress components.

3.7.2 APPLICATIONS

Strain gauges find wide application in force transducers (load cells) and displacement and strain transducers (extensometers). Strain gauges can be applied to metal components with relative impunity. Therefore, artificial joint or limb components can provide interesting data (Weightman, 1977; Little, 1985 and 1990; Bartel et al., 1986). Protection of the circuitry against fluid attack and corrosion is necessary. Gauges have also been embedded into the cement between prosthesis and bone (Lewis et al., 1982; Crowninshield & Tolbert, 1983; Burke et al., 1984; Miles & Dall, 1985; Little & O'Keefe, 1989). Fracture site loads have been monitored with an external fixator (Wang et al., 1997), while soft tissue loads have been estimated indirectly through the use of strain gauges bonded to metal components – devices such as the buckle transducer and the IFT (see section 3.3.2).

However, the extensive difficulties in gluing strain gauges to biological materials have led to few direct applications. Strains on bones have been measured in vitro or in vivo where the short-lived adhesion (two to three days) has not been an issue. Bonding directly to soft tissues has not been achieved, due to the high water content of these structures and the lack of knowledge of the effect of any chemical reaction of the adhesive with the tissue.

The validation of strain gauge measurements should always be a matter of concern to the user. The selection of the technique to be used should involve a careful assessment of the capabilities and limitations of the various methods available. The following points should be considered:

(1) How well are the gauges aligned with the intended direction of measurement?

(2) What is the effect of the strain gradient where the gauge is located – if there is one?

(3) What happens when the circuit is turned on and heats up? This is of particular importance with non-metallic applications, e.g., ceramics, polymers.

(4) Has the bonding or the gauge (plus bonding) affected the material being tested, either chemically or through reinforcement? Has the strain you wanted to measure been affected by your attempt to measure it?

(5) Will the temperature vary in the test from ambient, therefore, requiring temperature compensation?

(6) Does the loading procedure affect the result?

(7) How reproducible and repeatable are the results?

(8) If you are strain gauging a model of a particular structure, how good is your model, e.g., appropriate relative stiffness of components, geometric errors, scaling factors?

More detailed reviews of these factors are provided elsewhere (Little et al., 1991, Pople, 1983).

Strains in bones and bone plates have been measured by holography (Kojima et al., 1986; Shelton et al., 1990), as have strains in prostheses (Manley et al., 1987) and an external fixator (Jacquot et al., 1984). The three-dimensional deformation of the maxilla and zygomatic arch caused by forces applied through headgear has also been studied (Zentner et al., 1996), as has the effect of splinting on mandibular implants (Besimo & Kempf, 1995). Holography, however, is generally not used widely. Photoelastic models are more common, having been used for example, to examine stresses in models of hips (Haboush, 1952; Steen Jensen, 1978) and hip replacements (Orr et al., 1985a, b), the ankle (Kihara et al., 1987) and the knee (Chand et al., 1976). Photoelastic coatings on the other hand, appear to have been used mainly in pilot studies to determine where strain gauges should be placed in subsequent tests (Finlay et al., 1986 and 1989; Walker and Robertson, 1988). Brittle lacquers and SPATE have also been used to assess metallic and bony structures in the body (Gurdjian &

Lissner, 1947; Evans & Lissner, 1948; Kalen, 1961; Harwood & Cummings, 1986; Friis et al., 1989; Kohles et al., 1989).

Soft tissue strains have proven more difficult to assess. Methods have been classified as contact or non-contact, rather than by the fundamental principle of the measurement technique. Conceptually, non-contact methods do not interfere with measured strain, since the tissue is not touched. However, permanent staining of a specimen through the application of dye lines or spots suggests that the dye has interacted chemically with some component of the tissue. How this affects the strain of the tissue is not known, and leaves open the question of just how much the strain has been altered and if the method is really non-contact. Contact problems and gauge length definition have not stopped researchers attempting to determine strain magnitudes in vivo with either the Liquid Mercury Strain Gauge or Hall Effect transducer. Liquid Mercury Strain Gauges have been used in vivo on horses (Lochner et al., 1980) and ponies (Riemersma et al., 1980), and in vitro on human joints (Colville et al., 1990; Terry et al., 1991). The Hall Effect transducer has been used in vivo on the Anterior Cruciate Ligament of humans (Howe et al., 1990). In vitro measurements have typically been made with film and video systems, despite the inherent inaccuracy of most methodologies. For the large strains at soft tissue failure, inaccuracies are not of major concern. The soft tissue extensometer might provide the accuracy and sensitivity to determine more subtle strain effects in soft tissues. However, none of these devices will reveal the variation in strain across the cross-section of most ligaments, nor the level of pre-strain (the strain in the material before the measuring device was attached). Further advances are needed before we will be able to define zero strain in a soft tissue, and the variation of strain in a ligament at one particular joint angle, let alone how that strain distribution changes with joint angle even before load is applied.

3.8 EMG

VON TSCHARNER, V.
HERZOG, W.

3.8.1 INTRODUCTION

An electromyogram (EMG) is a recording of the electrical potential at some distance from the muscle fibres that was generated by multiple activated muscle fibres. The EMG measures the effect of temporal imbalances of ions around muscle fibres at the location of the electrodes. The electrical potential difference between two points indicates the energy that can be gained by moving an electric charge from one to the other point. There are entire books devoted to EMG, e.g., Basmajian and De Luca (1985), and Loeb and Gans (1986), and it would be presumptuous to try to rival them. However, most of these texts focus on the physiology, rather than the mechanics, of muscle and neglect issues that specifically relate to biomechanical problems. The relationship between electromyographic signals and muscular force represents one of the issues that has not been addressed thoroughly in these texts. This chapter attempts to address this issue from two points of view:

- Presenting results of studies employing electro-neuromuscular stimulation (ENMS) of α-motor neurons to investigate the EMG-force relationship under precisely controlled conditions, and
- Discussing the findings of studies employing direct and simultaneous measurements of EMG and force in a freely moving animal model.

The interplay of muscles and muscle fibres while moving is another current issue. This interplay strongly influences time and frequency aspects of the EMG signals, which are related to specific physiological effects. Recent developments of wavelet based time/frequency analysis have led to methods applicable to electromyograms that were recorded while moving. This yields a better understanding of time/frequency aspects, and a new area of applications of EMG intensity patterns, which is the study of movement related behaviours using pattern recognition methods and classifying these patterns by neural networks. Understanding the basic concept of the nature of the EMG signal, its recording, and processing is essential. The first sections of this chapter describe the most important aspects of the EMG signal and its analysis.

3.8.2 EMG SIGNAL

The membrane around a muscle fibre prevents a free flow of ions, and ion pumps and the neuromuscular stimulation tightly control the membrane potential by the mechanisms discussed below.

A resting potential is created by ion pumps and ion exchange processes, which use metabolic energy to generate and upkeep the separation of charges. At rest, the membrane carries a strong electrochemical potential. The Na^+ concentration is 7.7 times higher outside the cell, and the K^+ ion concentration is 30 times higher inside the cell, creating the chemical potential. As a result of the separation of charges, a resting electrical potential of about

-90 mV is established inside the cell (Wilkie, 1968). At rest, the ion channels are closed and there is little ion diffusion across the membrane.

Upon stimulation, when an action potential of a motor neuron reaches the pre-synaptic terminal (see chapter 2.7), a series of chemical reactions take place at the neuromuscular endplate, culminating in the release of acetylcholine (Ach.). Acetylcholine diffuses across the synaptic cleft, binds to receptor molecules of the muscle fibre membrane and causes a change in membrane permeability by opening ion channels. Most importantly, Na+ ions, mainly driven by the electrical potential of the membrane, depolarize the membrane. The outflow of K+ ions, mainly driven by the chemical potential (gradient of the K+ ions) against the electrical potential, repolarizes the membrane. With sufficient stimulation, the potential inside the cell rises temporarily from its resting value of -90 mV to about 30 to 40 mV. The transient change of the potential inside the muscle fibre at one specific location is called the muscle fibre action potential (MFAP) (Figure 3.8.1).

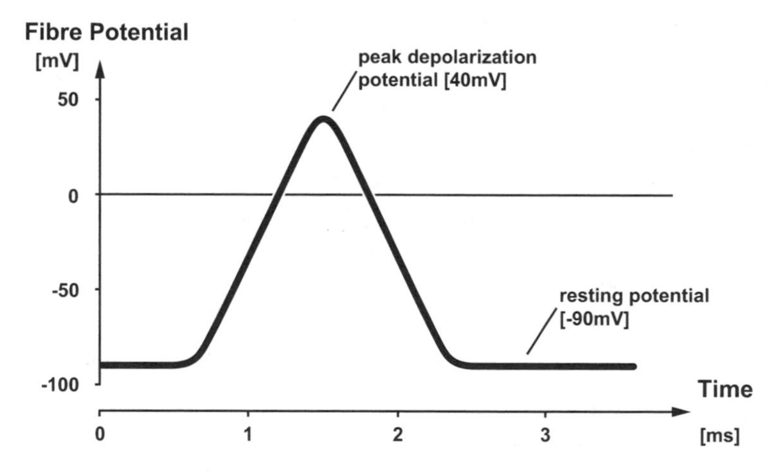

Figure 3.8.1 **Schematic representation of a muscle fibre action potential (MFAP).**

Skeletal muscles contract in response to electrochemical stimuli (see section 2.7.3). If the local MFAP following the stimulation is large enough, it elicits a chain reaction that opens the adjacent sodium and potassium channels, thus starting the flow of ions, which in turn causes a depolarization and repolarization at locations proximal and distal to the initial stimulation. The muscle fibre conduction velocity (MFCV) indicates the speed at which the chain reaction propagates along the muscle fibre. As a result of the propagation, a similar muscle fibre action potential (MFAP) can be observed at all points distal and proximal of the initial stimulation. Therefore, the muscle fibre action potential travels along the muscle fibre at the speed of the muscle fibre conduction velocity. However, muscle fibre action potential and muscle fibre conduction velocity are two separate phenomena.

A motor unit (MU) is a group of muscle fibres that are all stimulated by the same motor axon.

In an intact muscle, a fibre is never stimulated by itself but always with all other fibres that make up a motor unit. The motor endplates (see Figure 2.7.10) for each muscle fibre are not aligned but are spread over a finite distance and over a finite depth along the motor unit. The volume containing the neuromuscular junctions is called the innervation zone (IZ) (Figure 3.8.2). The depolarization of the muscle fibres, therefore, creates a drop of sodium

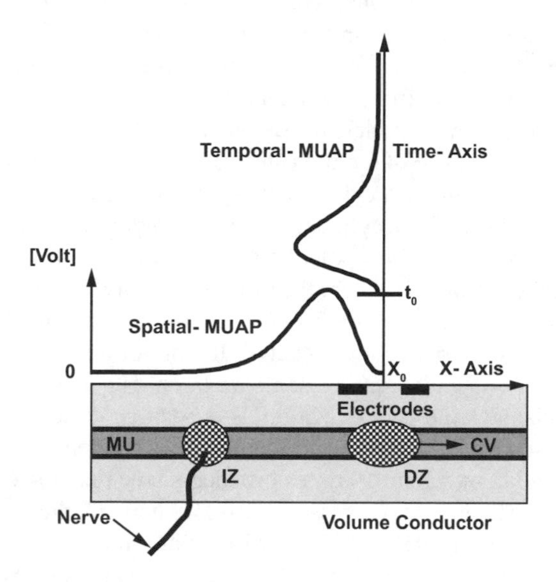

Figure 3.8.2 **Illustration of the spatial and temporal motor unit action potential (MUAP). IZ = innervation zone, DZ = depolarization zone, CV = conduction velocity, MU = motor unit.**

ions within the innervation zone. This drop of sodium ions results in an electrochemical gradient around the innervation zone, creating an electrochemical potential of the sodium ions. Because of the muscle fibre conduction, the area of depolarization moves to the right and to the left of the innervation zone. This moving depolarization zone (DZ) is, at the time of innervation (t=0), identical with the innervation zone. However, with time, the depolarization zone moves along the motor unit. Shortly after the sodium influx starts, the potassium outflux starts, counterbalancing the sodium flux by forming its electrochemical gradient. The electrical potential at some point distant to the depolarization zone is thus a result of the superposition of the two electrochemical gradients. The time dependency of this electrical potential depends on:

- Muscle fibre action potentials,
- Muscle fibre conduction velocities of the different fibres, and
- Ionic properties of the tissue in a volume surrounding the motor unit.

The electrical potential measured over time at some point in the vicinity of the motor unit is called the motor unit action potential (MUAP) (Figure 3.8.2).

In contrast to the muscle fibre action potential (MFAP), which is well-localized, the motor unit action potential is not well-localized. The motor unit action potential is not well-localized because the innervation zone is large compared to a point inside a muscle fibre and inhomogeneous, because the neuromuscular endplates are scattered all over the innervation zone. The muscle fibre action potential travels at the muscle fibre conduction velocity.

Thus, the concentrations of sodium and potassium ions also move along the muscle fibres of the motor unit. This propagation velocity represents the propagation velocity of the depolarization zone and is called the motor unit conduction velocity. This can be visualized as a sphere the size of the innervation zone, which contains an imbalance of sodium and potassium ions traveling along the motor unit. However, if the muscle fibre conduction velocity is slightly different for the different muscle fibres, then the depolarization zone gets broader as it moves farther away from the innervation zone, creating dispersion. Often, a three current model (see the review article by Stegeman, 2000), mimicking an imbalance of ions, is used to estimate the shape of a motor unit action potential. In addition, a model of a volume conductor with a string in a cylindrical volume is used to compute the motor unit action potential at the surface of this cylinder. An electromyogram is a recording of one or more superimposed motor unit action potentials through electrodes at some point sufficiently far away from the motor unit, where the in-homogeneity within the innervation zone and the depolarization zone become irrelevant (Stegeman & Linssen, 1992).

In contrast to the muscle fibre action potential, the motor unit action potential starts at ground level and becomes negative during depolarization. However, because of historical reasons one usually displays the depolarization as a positive deflection. Thus, the actual EMG signal is composed of the sum of all motor unit action potentials. The motor unit action potential has a similar, but about 10 times broader, shape than the corresponding muscle fibre action potential (Stegeman & Linssen, 1992). In general, the actual EMG obtained from voluntarily contracting muscles is a record of many motor units firing at different mean rates. Since the placement of the recording electrode on a muscle determines its geometric relation to the motor units, recordings obtained using different electrode placements on the same muscle will be different.

FACTORS INFLUENCING EMG SIGNAL FREQUENCY CONTENT

Because the motor unit action potential varies in time, the signal resulting from the superposition of motor unit action potentials shows an oscillatory pattern. The EMG reflects oscillatory properties of the EMG generated by the moving motor unit action potential. The frequency content of the EMG signal thus depends on specific properties of the motor unit action potentials.

Motor unit conduction velocity

The motor unit conduction velocity (MUCV) is the property most frequently used to explain changes in frequency content of the EMG. The most used relationship is the scaling of the geometrical motor unit action potential by the motor unit conduction velocity. If the time-dependent motor unit action potential remains unchanged, and other disturbing effects become negligible, then the geometric motor unit action potential along the motor unit becomes scaled by the motor unit conduction velocity as shown in Figure 3.8.2 (Lindstrom, 1974). An increase in motor unit conduction velocity shifts the frequency spectrum to higher frequencies. This phenomenon is often used to quantify the effect of fatigue of muscles. The motor unit conduction velocity depends on the fibre type and fibre size. However, the scaling effect is not the only characteristic that affects the frequency content of an EMG signal. Thus, the relationship between motor unit conduction velocity and frequency should be used with caution (Farina et al., 2004).

Motor unit action potential

A change of the shape of the motor unit action potential (MUAP) changes the frequency content of the EMG signal (Merletti & LoConte, 1997), however, one generally does not know whether the change in shape of the motor unit action potential or the change in conduction velocity altered the frequency content. To resolve this uncertainty, the motor unit conduction velocity should be measured in parallel with changes in frequency content. Motor unit action potentials are typically different for various muscle fibre types. Therefore, changing the fibre type selection for a specific muscular task changes the spectra during this task.

Motor unit recruitment pattern

The motor unit recruitment pattern indicates when, and how frequently, the motor unit is activated while a muscle is stimulated. It is generally believed that the motor unit recruitment pattern is random during isometric contractions, and therefore has little influence on the frequency content of the EMG. However, some contributions should be observable at low frequencies (Stegeman, 2000; Farina et al., 2004). During spontaneous or explosive movements, motor units are activated in a synchronized way (Weytjens & Van Steenberghe, 1984; Merlo et al., 2005). Synchronization results in large and broad pulses in the EMG, thus generating strong low frequency components. Corresponding power spectra show a strong additional peak at low frequencies.

Motor unit depth

The depth of the motor unit influences the spectral properties of an EMG signal because of the filtering effect of the surrounding tissue. Spectral shifts may occur if groups of motor units at different distances from the electrodes are activated, or if the electrodes move with respect to a motor unit during contraction or movement. This effect may become substantial during physical activities. The effect has been quantified by model calculations (Farina et al., 2004). The effect on the spectra is a narrowing and shifting of its shape to lower frequencies.

Distribution of motor unit conduction velocity

The distribution of motor unit conduction velocity (MUCV) causes the spectra to shift to lower frequencies as the electrodes are placed farther away from the innervation zone. This effect reflects the dispersion caused by the various conduction velocities of the different fibres. The result is a lower amplitude and a lower mean frequency for EMGs farther away from the innervation zone. The effect is quantified by reporting the transfer function (Hunter et al., 1987).

Muscle length

The length of the muscle changes the spectral content of the spectra (Doud & Walsh, 1995), shifting the spectra to higher frequencies as the muscles gets shorter. The effect has to be considered in cases where length changes are large. However, usually length changes are not the dominant effects in a study.

Muscle-tendon interface

The motor unit action potential disappears at the muscle-tendon interface. If the electrodes are placed too close to the muscle-tendon interface a distorted motor unit conduction velocity may disrupt the expected signal. There are also other effects that may happen at the end of the muscles (Stegeman, 2000; Gootzen et al., 1991). As a result, the area close to the muscle-tendon interface should be avoided for electrode placement.

Crosstalk

Crosstalk occurs if the activity of a nearby muscle contributes to the EMG of the muscle under investigation (Winter et al., 1994; Farina et al., 2004). Crosstalk is an important concern in many studies, and little is known about how to separate it from the main signal. However, crosstalk is caused by distant motor units, and therefore, primarily contributes low frequency components to the EMG. Contribution from crosstalk is most likely a non-traveling signal. If array electrodes are used, crosstalk shows up as a common signal on adjacent electrode pairs. These non-conducting signals mimic infinitely fast traveling signals and can be deleted if detected.

Movement artefacts

Movement artefacts occur especially in measurements of explosive and impact like activities, e.g. sports activities. Movement artefacts usually contribute increasingly less signal to the EMG with increasing frequency. Movement artefacts are often excluded by discarding the lowest frequency components of an EMG (Conforto et al., 1999), or by selectively dismissing the time periods where the movement artefacts occurred.

Electrochemical imbalance

Electrochemical imbalance occurs when the muscle is suddenly activated. In the inactive state, ions outside the muscle fibre are at an electrochemical equilibrium. When the muscle is activated, there is a shortage of sodium ions and a delayed excess of potassium ions propagating along the motor unit. This imbalance persists for as long as the muscle is activated. If the muscle is activated for a long enough period, a new electrochemical quasi-steady state situation builds up. The effect of the electrochemical imbalance is most likely to contribute low frequency components to the frequency spectrum at the time of the onset of activation. This effect should be considered if spectral analysis is performed at the onset of a strong muscular activation.

In conclusion, the intensity, phase and frequency of an EMG signal are affected by many different muscle characteristics. Thus, a substantial set of variables are available to asses the behaviour of muscles when using appropriate signal analysis.

3.8.3 EMG SIGNAL RECORDING

Electromyographic signals are typically recorded by electrodes measuring differences in the electrical potential between two points. EMG electrodes may be grouped broadly into:

- Indwelling electrodes (electrodes inside the muscle), and
- Surface electrodes (at the surface of the skin).

Electrodes are typically connected to the amplifiers in a monopolar or bipolar manner.

Indwelling electrodes come in different shapes and are often homemade. We describe the indwelling wire electrodes used successfully in our research centre for chronic recordings from cat hind limb muscles, e.g., Herzog et al. (1993b). For further discussion, see Loeb and Gans (1986). For our indwelling bipolar electrodes, we use Teflon-insulated, multi-stranded, stainless steel biomedical wire (Bergen BW9.48, Figure 3.8.3). For

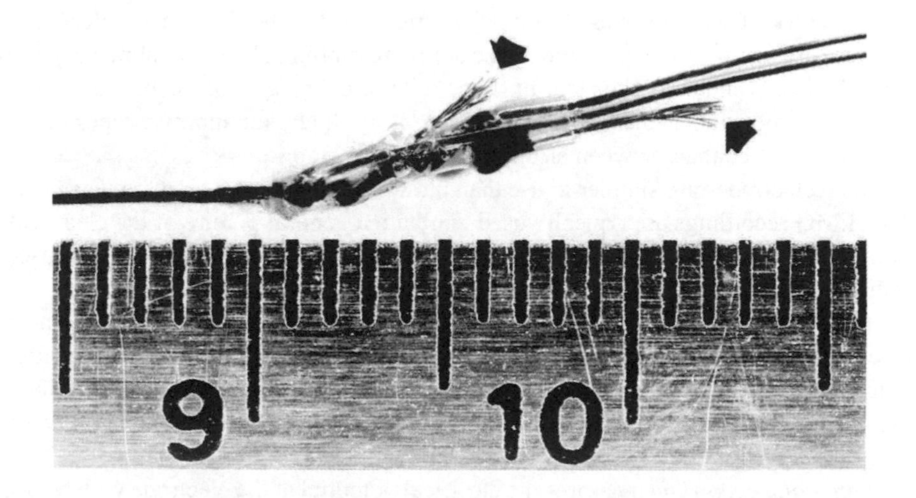

Figure 3.8.3 **Indwelling, bipolar wire electrodes used for chronic EMG recording in cat hindlimb muscles. The arrows indicate where the insulation was removed from the wires to produce the recording sites (from Herzog et al., 1993b, with permission).**

recording, two mms of wire are exposed at the tip by mechanically removing or heat-shrinking, or both, the Teflon insulation (Figure 3.8.3, arrows). About 10 mm of the wire is bent backward to produce a permanent, hook-like deformation. A silk suture is tied to the bent area of both wires so that the distance between the two exposed tips is about 5 to 7 mm. A ground electrode is constructed from the same wire as the recording electrodes by exposing about 50 mm of the insulation on one end and forming a loop held in place by Silastic tubing.

For chronic implantation, all electrodes are autoclaved and implanted under sterile conditions. Silk sutures and electrodes are attached to a small surgical needle (MIL-TEX MS-140), which is pulled through the largest part of the muscle belly in the direction of the muscle fibres. Once the recording area of the electrodes in the mid-belly region of the muscle is reached, the silk sutures emerging from the entry and exit holes are tied to the muscle fascia to prevent movement of the electrodes relative to the muscle. The ground electrode is typically placed on a bony surface such as the medial surface of the tibia.

This type of indwelling electrode allows for recording from a specific muscle. Crosstalk from other muscles may be virtually eliminated by choosing the recording site and the inter-electrode distance appropriately. Indwelling electrodes also allow measurements from small and deep-lying muscles. The problems of skin movement during locomotion, and irritation to the animal caused by fixing the electrodes on the skin, are avoided. This type of indwelling electrodes allows for chronic recordings in the freely moving animal over a period of weeks, and possibly months. This type of mounting was used for the measurements relating EMG and measurements of force.

Surface electrodes are placed on the skin overlying the muscle of interest. Like indwelling electrodes, surface electrodes come in a variety of types. Probably the most frequently used surface electrodes are commercially available silver-silver chloride electrodes. Before recording with surface electrodes, the electrical impedance of the skin must be decreased by shaving the area of electrode placement, and applying rubbing alcohol or abrasive pastes to remove dead cells and oils. The recording electrodes are then attached using commercially available electrode gel. Typically, slight pressure is applied with adhesive tapes or elastic bands to improve contact between electrode and skin.

Surface electrodes are simpler to use than indwelling electrodes, and are non-invasive. Surface EMG recordings are typically used to obtain a general picture of the electrical activity of an entire muscle or muscle group. However, surface electrodes can only be used for recording from relatively large superficial, muscles.

EMG recordings can be done in a monopolar or a bipolar configuration. Both methods require a ground electrode that yields a common reference potential for comparison and electrical stability. It is common practice to use a bony area not too distant from the measuring area to place the ground electrode and to connect this electrode to the ground of the measuring instrument.

A *monopolar recording* measures the electrical potential at the electrode with respect to the ground electrode. The advantage of the monopolar configuration is that the resulting signal reflects the actual shape of the motor unit action potential at the location of the electrode. It is, therefore, the measurement with the least distortion of the signal. However, it also records all potential fluctuations that are induced in the tissue between the measuring electrode and the ground. These fluctuations may be much bigger than the motor unit action potential. Therefore, the monopolar configuration is only appropriate in a well-controlled and shielded environment.

A *bipolar configuration* uses two electrical contacts to measure a potential difference. The two electrodes are connected to the positive and negative input of a differential amplifier. If the negative input of the differential amplifier is connected to the electrode located closer to the innervation zone the signal shows a positive deflection due to sodium influx. The two electrodes are used to determine the potential of one electrode with respect to the other electrode. For two electrodes placed along the motor unit one measures the difference between two points along the motor unit. The bipolar configuration is most frequently used in EMG quantification. EMG measurements are reported in units of volts measured at the position of the electrodes.

In bipolar measurements, the ground electrode provides a common reference point, which is allowed to alter, within limits, over time. The fluctuation of the potential at the ground electrode represents a common signal to both entries of the differential amplifier. The common mode rejection factor of an amplifier indicates how well this common mode signal is suppressed.

The shape of the signal measured in a bipolar configuration does not reflect the shape of the actual motor unit action potential. If the electrodes are small in diameter, compared to the width of the motor unit action potential (Figure 3.8.2) and the inter-electrode distance is short, compared to the geometrical motor unit action potential, the recorded signal represents the differential of the motor unit action potential. This type of detection is called Single Differential Detection (SDD). In practical work, the electrodes are large and the inter-electrode distance is not small. The signal in this case represents the convolution of the electrode shape with the geometrical motor action potential.

When forming the differential of a signal, any constant additional factor does not contribute to the resulting signal. Additionally, by adding a high pass filter to reduce drifting of the SDD signal loses the original base line and an offset is added to the differential. Therefore, the original motor unit action potential cannot be obtained by integrating the SDD, since part of the information was lost. The advantage of bipolar EMG measurements is that a signal with optimal common mode rejection can be obtained.

Arrays of electrodes are increasingly common. The potential of each electrode is recorded in a monopolar or bipolar amplifier setting. Occasionally, SDDs are again differentiated to obtain double differential detection (DDD). The main advantage of an array configuration is that the location of the innervation zone and the motor unit conduction velocity can easily be determined. The array configuration requires many amplifiers, making signal processing an elaborate task. From the authors' point of view, a tripolar electrode configuration, with monopolar or bipolar signal detection, is the best compromise, as it allows a detailed signal analysis. This configuration will probably be used more frequently in the future.

The *positioning of the electrodes* is critical for EMG recording. Bipolar electrodes must be placed on one side of the innervation zone. For big surface muscles, the locations of the innervation zone are well known. The SENIAM website (www.seniam.org) lists innervation zones for many muscles. EMG recordings close to the tendon-muscle interface distort the signals. Optimal EMG signals are measured in the middle third of the distance between the innervation zone and the muscle-tendon interface.

The *inter-electrode distance* should be such that a slope of a geometrical motor unit action potential is detectable. For large surface muscles, an inter-electrode distance of about 20 mm is suitable. For small muscles, this inter-electrode distance is too large. However, selecting an inter-electrode distance that is too short results in a low signal amplitude and poor EMG detection. A suitable compromise must be used for each muscle.

The *electrode shape* must be smaller than the inter-electrode distance to avoid a short between the electrodes. Most frequently, round electrodes are used for EMG measurements. The electrode shape is not of primary concern in current measurements.

3.8.4 EMG SIGNAL PROCESSING

EMG signals are analyzed in many different ways. Examples of raw and processed EMG signal are illustrated in Figure 3.8.4. The main variables of interest that can be determined from such a signal are discussed in the next few paragraphs.

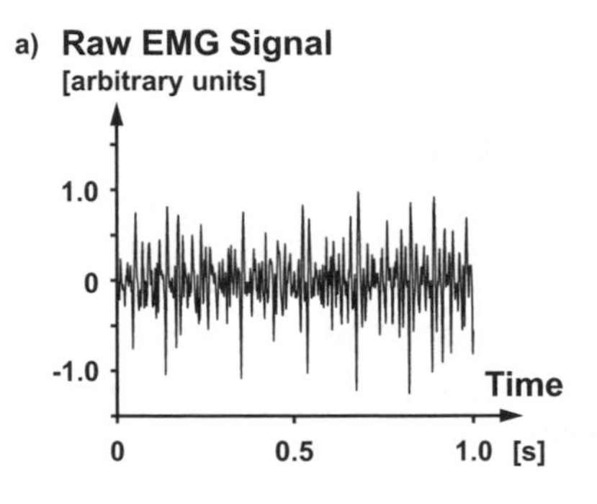

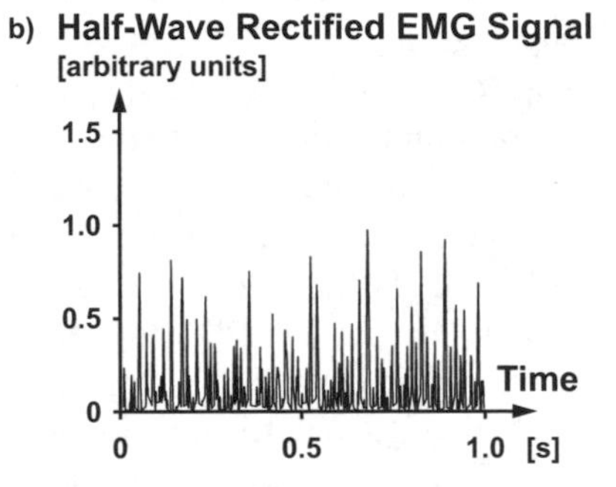

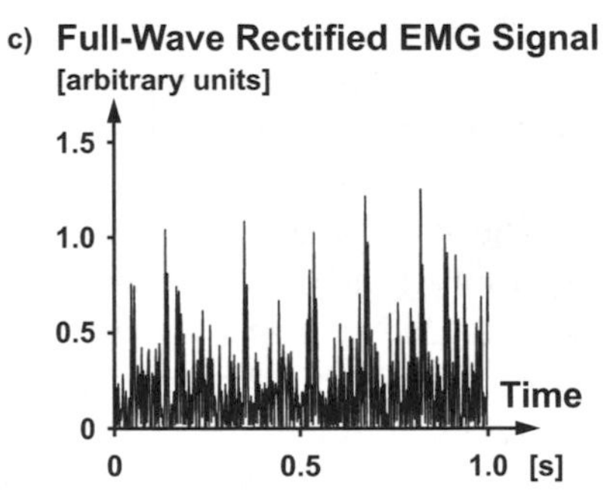

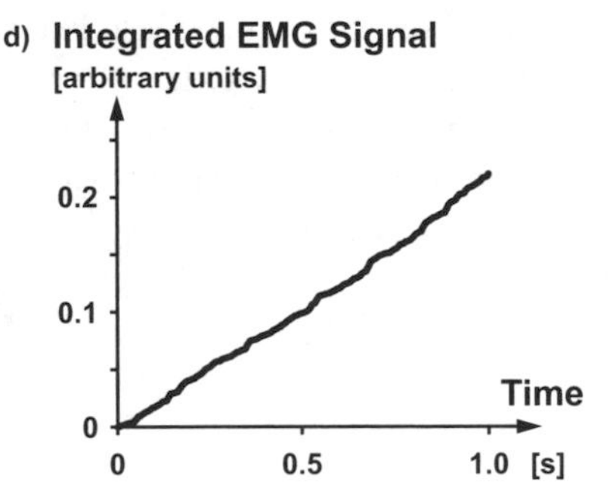

Figure 3.8.4 **(a) Raw EMG signal from a human rectus femoris muscle during an isometric knee extensor contraction at about 70% of maximal voluntary contraction. (b) The corresponding half wave rectified signal. (c) The corresponding full wave rectified signal. (d) The corresponding integrated signal.**

EMG SIGNAL ANALYSIS IN TIME DOMAIN

Onset

The onset of a muscular activity occurs when the EMG amplitude exceeds a certain threshold level. Onset of activity is best determined from the raw EMG signal. Typically, a threshold value is set and the instant when the signal passes this threshold value is defined as the onset of muscle activity (Figure 3.8.4). Onset of EMG activity determined from a filtered signal may include a time delay and thus may not be accurate.

Magnitude

The magnitude of EMG activity may be obtained using different methods. Some scientists use the rectified (|EMG (t)|) and smoothed EMG signal to represent magnitude (Figure 3.8.4). However, when smoothing the EMG signal, some frequency components are lost. For example, when using a low-pass filter, selective high frequency components are lost, and this must be kept in mind as it affects the magnitude of the signal. The most commonly used approaches of magnitude measurements are described below.

Envelope of rectified signal: The signal is first rectified and various methods are used to create an envelope around the peaks of the signal. The peak heights become a dominant component of the extracted magnitude.

Integrated EMG: The integrated EMG (IEMG) is the mathematical integration of the rectified EMG signal with respect to time (Figure 3.8.4). Since integration of the EMG signal gives steadily increasing values over time, integrations are typically performed either over a sufficiently small time period, T, or the integrator is reset to zero when the integrated value reaches a specified limit. The integrated EMG as measure of magnitude has often been related to muscular force, e.g., Bigland and Lippold (1954), Bouisset and Goubel (1971), and Thorstensson et al. (1976).

The RMS: The magnitude of the EMG signal can be represented by the root mean square, RMS, value. RMS values are calculated by summing the squared values of the raw EMG signal, determining the mean of the sum, and taking the square root of the mean. The RMS is frequently used in studying muscular fatigue. The square of the RMS is equal to the power and is the appropriate measure of magnitude reflecting the electrical activity of the muscle (see below).

Power: The square of the RMS, $(RMS)^2$, is equal to the power of the EMG signal for the same time interval. in the paragraph Fourier transforms and model calculations of randomly superimposed identical motor unit action potentials recorded in a bipolar manner show that the power, in this case, is directly proportional to the number of motor unit action potentials, and that feature is not shared by other measures of magnitude (amplitude cancellation: Farina, 1996). Power is also additive. Thus, the sum of the power obtained from repeated trials represents the power of the sum of the EMGs of the individual trials. This means power can be averaged for a repeated set of trials. However, power is not a visually appealing magnitude, when plotted. Large activities dominate low activities.

Motor unit conduction velocity

The motor unit conduction velocity (MUCV) is the velocity with which the depolarization signal travels along the muscle fibre. The motor unit conduction velocity can be quantified by using tripolar electrodes and bipolar recordings. There is a time delay (phase shift) between the signal quantified using electrodes 1 and 2 and the signal using electrodes 2 and 3 (Figure 3.8.5). This time delay can be used to determine the conduction velocity. To measure conduction velocity sampling rates of at least 10 kHz, or higher, are required.

The motor unit conduction velocity is important because it decreases when a muscle is fatigued. Faster motor unit conduction velocities yield higher frequency components in EMG recordings. Therefore, many researchers preferred measuring changes in frequency content to asses the effect of altered conduction velocities because this can be done with much lower sampling rates (1 kHz).

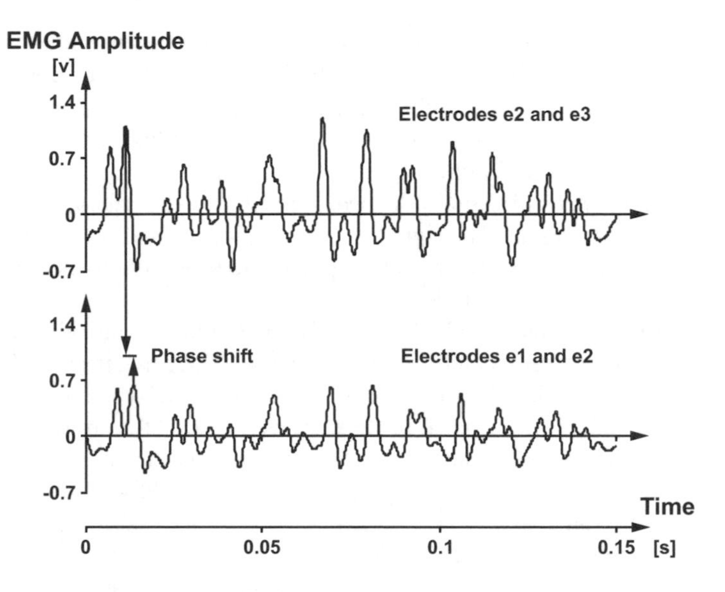

Figure 3.8.5 **Bipolar recordings of an EMG signal of the abductor pollicis brevis muscle using an array of 3 electrodes. Electrode e1 is closest to the tendon, e2 is in the middle, and e3 is closest to the innervation zone. The inter-electrode distance was 6 mm and the electrode diameter was 2 mm. The signal farther away from the innervation zone has a phase shift with respect to the signal recorded closer to the innervation zone. The inter-electrode distance divided by the phase shift represents the average motor unit conduction velocity.**

EMG SIGNAL ANALYSIS IN THE TIME/FREQUENCY DOMAIN

Two forms of analyses of EMG signals are discussed in this section: the Fourier transform and the wavelet transform. This section assumes that the reader is familiar with the basics of Fourier transforms (Mallat, 1998). However, let's recapitulate some practical aspects of the Fourier transform of real valued signals first and get familiar with the notion of time space and frequency space. The word space refers to the fact that the signal is represented in a vector space. Depending on the coordinates used to describe this space, we call it time space, frequency space, or wavelet space.

Fourier transforms

Let's assume we have a digitized signal, s, of length N that has been sampled at a rate ν (Hz); therefore, the time interval per sample is $\Delta t = 1/\nu$. Time, t, at sample n is $t = n \cdot \Delta t$ with n covering the range from 0 to N-1 and the duration of the entire signal is $T = N \cdot \Delta t$. The signal can be viewed as a vector in an Euclidian vector space with the base vectors $a_n = (0\,0\,0....1....0\,0)^T$, where n indicates the locations of 1 in these vectors. The base vectors a_n are orthogonal to each another, thus $a_n^T \cdot a_m = 0$ for n not equal to m, and 1 otherwise. If the base vectors, a_n, are arranged as columns of the matrix A, then A is the identity matrix, and the base vectors represent the axes in this space. The magnitudes of these axes are obtained as the projections of s onto these vectors.

$$s_n = a_n^T \cdot s$$

or, using A we get the identity

$$s = A^T \cdot s$$

Thus the amplitude at time $t = n \cdot \Delta t$ is equal to the magnitude along a_n. The space spanned by the base vectors, a_n, is called time space of dimension N.

Let's assume that there is another set of base vectors consisting of sine and cosine waves. Remember, $e^{-i\omega t} = \cos(\omega t) - i \cdot \sin(\omega t)$. We use an exponential notation for these waves, thus always using the cosine and sine wave of the same frequency.

$$(b_k)_n = \frac{1}{\sqrt{N}} \qquad \text{for k=0}$$

$$(b_k)_n = \frac{\sqrt{2}}{\sqrt{N}}^{i \cdot 2\pi \cdot k \cdot \frac{1}{T} \cdot n \cdot \Delta t} \qquad \text{with k covering the range 1 to N/2-1}$$

$$(b_k)_n = \frac{1}{\sqrt{N}}^{i \cdot 2\pi \cdot k \cdot \frac{1}{T} \cdot n \cdot \Delta t} \qquad \text{for k=N/2}$$

The frequency for each K is $f_k = \Delta f \cdot k$, with $\Delta f = 1/T$. For k=0 and k=N/2, b_k has no complex part.

b_k are base vectors of length N that form the Euclidian vector space as described above. If we view the real and the complex part as separate base vectors, we obtain N orthogonal base vectors. The magnitudes of the projections onto these axes are $c_k = b_k^T \cdot s$. If B represents the matrix with columns b_k, the Fourier transform can be written as

$$c = B^T \cdot x$$

The magnitudes c_k are called the Fourier coefficients. The space spanned by vectors b_k is called frequency space.

The energy, p, of the signal in T is the sum of the signal squared, thus the sum of $p_n = s_n \cdot s_n$ for all n. This is equal to the length squared of the vector of the signal in time space. The length squared can also be expressed in the frequency space as the sum of $p_{nk} = c_k^T \cdot \text{conjugate}(c_k)$ for all k. A plot of p_k vs. f_k is called the power spectrum of the signal.

The signal in the frequency space can be recovered in the time space using the inverse Fourier transform. If conjugate(B) represents the matrix with columns conjugate(b_k), the reconstructed time series becomes

$$s = \mathrm{real}(\mathrm{conjugate}(B) \cdot c)$$

Performing a Fourier transform is computationally demanding. If N is a power of 2, then numerical methods are available that significantly reduce the computational effort. A Fourier transform using these methods is called a fast Fourier transform.

Now, let's apply the Fourier transform to an EMG signal. Let's assume that we have a number of randomly occurring motor unit action potentials, MUAP, indexed by m, during a time period T. The Fourier transform of the sum of all MUAPs is equal to the Fourier transform of the individual MUAPs.

$$c = B^T \cdot (MUAP_1 + MUAP_2 + ...) = B^T \cdot MUAP_1 + B^T \cdot MUAP_2 = \sum(c_m)$$

Thus, the power is

$$p = \sum(c_m) \cdot \mathrm{conjugate}(\sum(c_m)) = \sum(c_m \cdot \mathrm{conjugate}(c_m)) + \mathrm{crossproducts}$$

Cross products represent the correlation between the different c_m. If the different MUAPs are independent of each another, the average cross product goes to 0. Thus, the total power is equal to the sum of the power of the individual MUAPs with added noise represented by the cross products. In conclusion, the power spectrum of randomly superimposed MUAPs is equal to the sum of the individual power spectra plus some noise. However, if MUAPs do not occur randomly, e.g., if they are synchronized, the cross products do not average out to be 0, and thus they contribute to the power spectrum.

Fourier transforms are used to quantify the frequency spectrum of a signal and to detect frequency changes between contractile conditions. The power density spectra, or simply power spectrum, of an EMG signal indicates how the power of the EMG signal is distributed among the various frequencies.

The median or mean frequency of power spectra are often used as indicators of fatigue. Increasing fatigue is associated with a loss of high frequency content in the EMG, e.g., Lindström et al. (1977), Komi and Tesch (1979), Bigland-Ritchie et al. (1981), Hagberg (1981), and Petrofsky et al. (1982). If it is assumed that the spatial motor unit action potentials do not change with time, then their frequency is linearly related to the conduction velocity. However, motor unit action potentials have been shown to change shape under fatiguing conditions (Merletti & LoConte, 1997). Furthermore, synchronization of MUAPs (Merlo et al., 2005) tends to decrease the frequency content of the signal. Fatigue is also associated with an increased repolarization time causing the motor unit action potentials to become more asymmetric. Thus, a change of the median frequency is not necessarily a good indicator of changes in motor unit conduction velocity.

Wavelet transform (general comments)

The general wavelet transform has been developed as an extension of the windowed Fourier transform and Gabor transform (Mallat, 1998). Initially, the wavelet transform was used for data compression and image analysis. Currently, wavelet based methods are used in data analysis for many biological and medical data sets (Aldroubi & Unser, 1996).

For the wavelet transform, the period T of a FFT is shortened to T1 so the time interval becomes just a short sine wave containing a few oscillations. The new window has the values 1 within the period T1 and 0 elsewhere in the period, T. This is a first approximation of a short wave or a wavelet. A signal in T1 will have the same projection as the projection on a full sine wave, whereas, a signal outside T1 will have a value of 0 when projected onto this short wave. The sine waves in T1 can be of low or high frequency. For high frequency sine waves, the window T1 can be short. For low frequency sin waves the window must be large enough to encompass at least one period of the sine wave. Instead of using sine waves a set of short waves can be used that are shifted along the time axis and scaled to shorter windows for higher frequencies. These short waves do not have to be sinusoidal but have to fulfill a minimal set of rules to be called wavelets. Usually one specific wavelet is taken and scaled in time to become shorter wavelets with the same basic shape. The resolution changes, depending on the wavelet scaling.

A wavelet transform is a projection of the signal on a set of base vectors represented by the scaled and shifted wavelets.

There are infinite possibilities for creating wavelets. Some wavelets are optimized for specific applications. Morlet, Paul and Daubechies wavelets are most commonly used. Ideal wavelets are orthogonal to one another (Mallat, 1998, pp. 249 to 254). Ideal wavelets allow the reconstruction of the original signal by a linear combination with the least amounts of weights. However, orthogonal wavelets do not provide much visually comprehensive information about the frequency in an EMG signal. For the analysis of frequencies, symmetrical wavelets with an oscillatory pattern should be used, e.g., Morlet and Cauchy wavelets. In this text, the Cauchy wavelets are used and modified. Currently, continuous wavelet transforms are used for time information and continuous or discrete wavelet transforms for frequency information, e.g., von Tscharner (2000), Karlsson et al. (2000), and Karlsson and Gerdle (2001).

For EMG analyses with wavelets, similar variables to the conventional methods mentioned above are of interest, but with an optimized time or frequency resolution. EMG wavelet analysis provides information about intensity at a certain time or frequency, or both. A short burst of EMG intensity in a wavelet analysis is called a muscular event. Furthermore, the frequency distribution and the conduction velocity of this event can be determined. These variables quantify the interplay of muscles during a movement.

Wavelet transform (mathematical description)

The wavelet used in this text is of the Cauchy type. It is a real wavelet, in the mathematical sense, and has a symmetric oscillatory pattern in time space.

Equation of wavelet.

$$\Psi(f_c, m, t) = \Gamma(m+1) \cdot e^m \cdot \frac{f_c}{(m - i \cdot 2\pi f_c t)^{m+1}} \tag{3.8.1}$$

where:

Ψ = wavelet function
f = frequency
f_c = centre frequency
m = mode
t = time
Γ = gamma function

Each wavelet has a real and a complex part, indicated by the symbol, i (Figure 3.8.6, bottom). The power spectrums of the real and complex part are identical. The equation of the wavelet in frequency space (Figure 3.8.6, top) is

$$\mathrm{FFT}\Psi(f, f_c, m) = \left(\frac{f}{f_c}\right)^m \cdot e^{\left(\frac{-f}{f_c}+1\right)\cdot m} \tag{3.8.2}$$

where:

$\mathrm{FFT}\Psi$ = fast Fourier transform of Ψ

Heisenberg's uncertainty principle indicates that high frequency and a high time resolution cannot occur simultaneously. To achieve a physiologically meaningful frequency and time resolution, wavelets in this text are scaled non-linearly.

Equations for scaling

$$cf = \frac{1}{0.3}\cdot(0.45+j)^{1.959} \tag{3.8.3}$$

$$m_j = cf_j \cdot 0.3 \tag{3.8.4}$$

where:

j = wavelet number (j = 1…11)

In this text, a set of 11 wavelets is typically used. This number of wavelets seems sufficient for most EMG analysis applications. However, if more or less frequency components are expected, higher or lower numbers of wavelets can be used with the same scaling equation.

The above Cauchy wavelets are ideally suited as filters. However, when analyzing the EMG signals, it is often desirable to determine the power. The power of the EMG is obtained by the Fourier transform and is then distributed among the wavelets by multiplication by the wavelets of the filter bank. Thus, the power of the EMG at frequency f is distributed into those filters that are not 0 at f. Thus, if the above wavelets are divided by the square root of the sum of squares of the individual wavelets at each frequency, then the actual power can be determined directly (von Tscharner, 2000). These modified wavelets have a slightly different shape of the filters in the filter bank. In time space these modified wavelets are represented by wavelets of a slightly different shape than the Cauchy wavelets. We called these wavelets power-wavelets because they are ideal for the determination of power from the EMG signal. The filter bank of the power-wavelets is shown in Figure 3.8.6 and the corresponding centre frequencies, in Hz, are:

f(1)	6.902
f(2)	19.287
f(3)	37.711
f(4)	62.089
f(5)	92.359
f(6)	128.471
f(7)	170.386
f(8)	218.068
f(9)	271.487
f(10)	330.619
f(11)	395.438

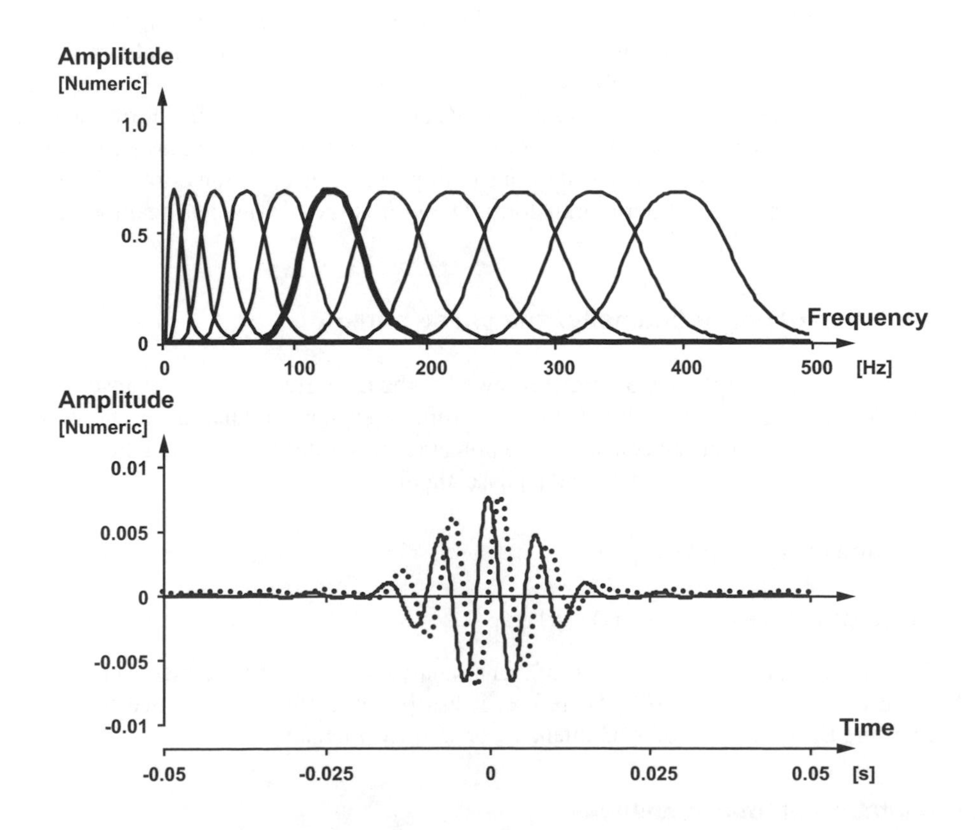

Figure 3.8.6 **Illustration of the power wavelet filter bank in frequency space (top) and one specific wavelet in time space obtained from the Fourier transform (bottom). The specific wavelet from the filter bank (top) corresponding to the bottom curve is highlighted. The solid line represents the real part and the dotted line represents the imaginary part of the wavelet.**

Computation of wavelet transform

The computation of the wavelet transform can be done either in time space or in frequency space.

Computation of wavelet transform in time space

The wavelet transform is a projection of the signal onto real and imaginary wavelets. Mathematically, these projections can be computed as convolutions of the signal with the wavelets. The result of the wavelet transforms are weight factors obtained for each wavelet index (j) at every time point. Thus, one weight factor for the real part, $r_j(t)$, and one for the imaginary part, $i_j(t)$, of each wavelet is obtained.

Computation of wavelet transform in frequency space

Mathematically, a convolution of two functions is equivalent to a multiplication of the two functions in frequency space. Because multiplications are much easier to perform than convolutions, working in frequency space is often preferred. Thus, the wavelet transform with one wavelet can be done in frequency space as a multiplication of the Fourier transformed wavelet with the Fourier transformed EMG signal. The wavelet transform separates frequencies of the signal that are not under the wavelet curve from those that are under the curve and thus act as a filter. The result of this multiplication is then transformed back into time space using an inverse Fourier transform. Doing this for all wavelets again yields the weight factors $r_j(t)$ and $i_j(t)$.

Computation of power (intensity) and phase angle

For a given j and t, $r_j(t)$ and $i_j(t)$ can be viewed as the magnitudes along the abscissa and ordinate of an orthogonal two-dimensional coordinate system. In analogy to the Fourier transform, $r_j(t)$ and $i_j(t)$ can be converted to a power representing the square of the length of the vector in this coordinate system and a phase angel.

$$\text{power}_j(t) = r_j^2(t) + i_j^2(t) \tag{3.8.5}$$

$$\text{phase–angle}_j(t) = \text{arctg}(i_j(t)/r_j(t))$$

The power is defined in the sense of signal analysis, but not in the mechanical sense where power can produce work. Therefore, it has become common practice to call the power of the EMG signal the EMG intensity, or, simply, intensity.

Computation of total intensity

Total intensity can be computed for each time point by summing the intensity over all wavelets. The total intensity and the RMS squared are similar quantities, as both relate to power. However, total intensity has a higher time resolution. Thus, wavelet analysis and classical analysis are the same when considering total intensity.

Graphical representation of wavelet results

Wavelet results can be depicted as conventional line graphs or as grey-scale graphs. An example of both possibilities is illustrated in Figure 3.8.7.

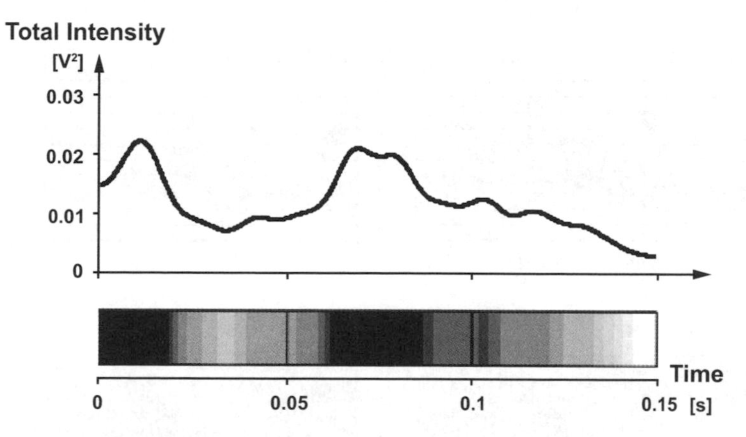

Figure 3.8.7 **Total intensity computed from the wavelet transformed signal shown in Figure 3.8.5. The intensity can either be represented as a line diagram (top) or in a grey-scale diagram (bottom). The grey-scale diagram shows the largest intensities in black, and low intensities in different shades of grey.**

Scalograms are excellent ways to illustrate EMG results from wavelet analysis for the whole frequency spectrum of interest. The single wavelet results are arranged in a column arrangement with the wavelet corresponding to the lowest frequencies at the bottom and the wavelet corresponding to the highest frequencies at the top (Figure 3.8.8).

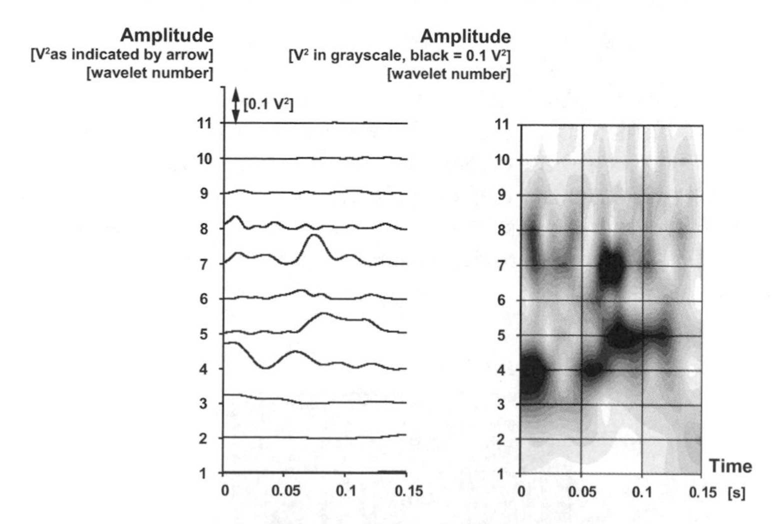

Figure 3.8.8 **Illustration of a scalogram including 11 non-linearly scaled wavelets for a line graph and for a grey-scale graph. Grey-scale values between the wavelets are obtained by interpolation. The lowest frequency wavelet is at the bottom and the highest frequency wavelet at the top of the graph. The illustration is from measurements using a bipolar electrode for the abductor pollicis brevis muscle, as shown in Figure 3.8.5**

Scalograms can also be used to compare two different situations. The two illustrated scalograms in Figure 3.8.9 correspond to the EMG signal collected from electrodes 1 and 2 (right) and 2 and 3 (left). This comparison can be done for the intensity scalograms (top) and for the phase scalograms (bottom).

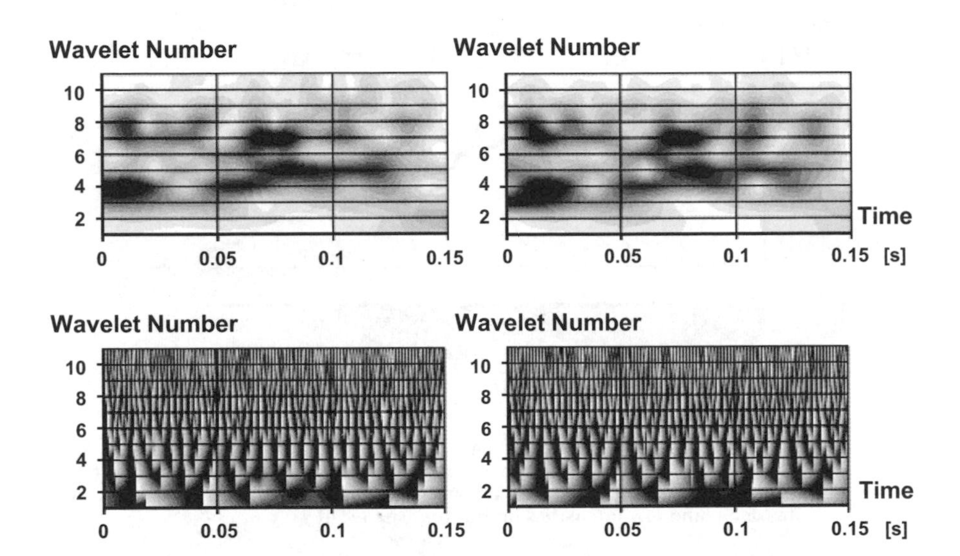

Figure 3.8.9 **Illustration of scalograms for intensity (top), phase-angle (bottom) for EMG measurements using a tri-array electrode. Results for electrodes 1 and 2 are on the right and results for electrodes 2 and 3 are on the left.**

Phase shift

The subtraction of two phase scalograms shown in Figure 3.8.9 using the grey-scale depiction is powerful in detecting phase shifts between signals from two specific conditions (Figure 3.8.10). The phase shift is represented by the width of the white areas in the lower frequency wavelets of this phase diagram (Figure 3.8.10). At higher frequencies the phase shift becomes longer than one oscillatory cycle and thus lines split up as can be seen at wavelet 6. Dividing the inter-electrode distance by the phase shift yields the motor unit conduction velocity.

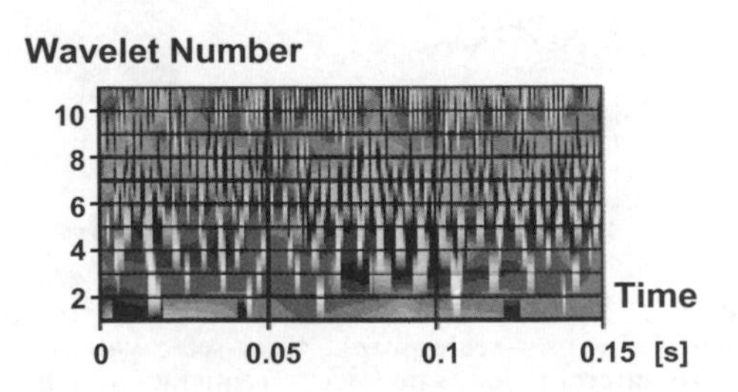

Figure 3.8.10 **Illustration of the phase shift using the subtraction of two grey-scale phase scalograms for the two different situations mentioned in Figure 3.8.9.**

Wavelet spectrum

Plotting the centre frequencies on the abscissa and the intensity on the ordinate of a two-dimensional figure, one obtains a wavelet spectrum. A wavelet spectrum is similar, although not identical to a power spectral density representation of the signal. In the power spectral density representation, each frequency point represents power of one single frequency, whereas, in the wavelet spectrum, the intensity reflects the power in a whole frequency band and the position of the band is indicated by the centre frequency.

Because of the stochastic nature of an EMG signal a wavelet spectrum extracted at a single time point has large fluctuations. Averaging is helpful in obtaining meaningful wavelet spectra. Averaging can be done by either averaging the scalograms of some trials and then extract the spectra, or averaging the spectra within a reasonably long time period (Figure 3.8.11 and Figure 3.8.12).

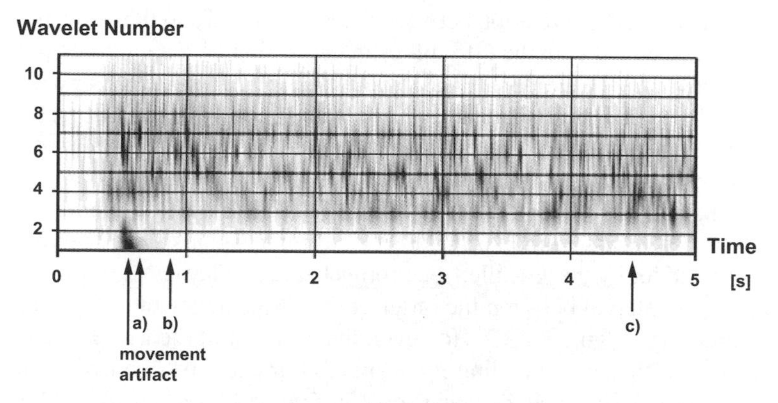

Figure 3.8.11 **EMG scalogram for a fatiguing exercise of the abductor pollicis brevis during a maximal voluntary contraction.**

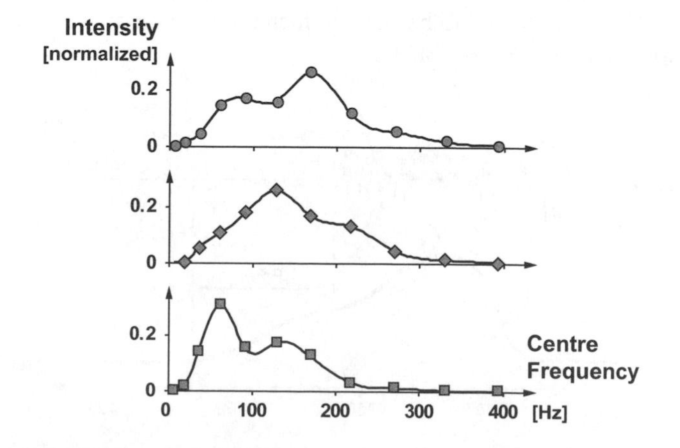

Figure 3.8.12 **Wavelet spectra of EMG measurements as described in Figure 3.8.11 normalized to total intensity = 1 and averaged for 150 ms. The top spectrum starts at arrow a in Figure 3.8.11, thus just after the movement artifact. The middle and bottom spectrat start at arrow b and c.**

3.8.5 EMG-FORCE RELATION

We are not attempting to duplicate information that is presented in several excellent textbooks dealing exclusively with EMG. Rather, we would like to address one aspect of electromyography that has played a major role in the field of biomechanics: the relation that may exist in a muscle between EMG and force.

The idea that EMG should relate in some way to muscular force is appealing. After all, an increase in the firing rates of motor units is known to increase force and the integrated form of the EMG (Basmajian & De Luca, 1985), and recruitment of an increasing number of motor units also increases muscular force and the corresponding IEMG (Guimaraes et al., 1994a). Thus, there must be at least a qualitative relation between the EMG signal and the corresponding force of a muscle. The existence of such a relation is not questioned in the scientific community, but its precise, quantitative nature is hotly debated.

The difficulty in finding a relation between the EMG signal and the corresponding force is at least partly associated with the difficulties in measuring EMG and force properly from individual muscles, and the temporal disassociation of the two signals. For example, when measuring EMG signals from a particular muscle in an intact system, it is possible that signals from surrounding muscles may also be recorded (crosstalk). Furthermore, force recordings from individual muscles may be readily obtained, but only in animal models, because of the legal and ethical constraints on taking such invasive measurements in humans. Finally, the temporal disassociation of the EMG and force signals is well recognized and is typically accounted for by the so-called electromechanical delay (EMD). EMD is normally defined as the time interval between the onset of the EMG signal and the onset of the corresponding force, e.g., Figure 3.8.13. However, the duration of the EMG signal may differ from the duration of the corresponding force signal. For example, the EMG and force signals in the cat soleus during a single step shown in Figure 3.8.13 have an EMD of approximately 72 ms (as defined above), but the EMG signal terminates approximately 109 ms before the force signal. The question that arises in this situation is whether the EMD is constant and may be defined by a single value or whether it varies continuously throughout the step cycle. If the latter is assumed to be correct, then the relation between the continuously varying EMD and time must be established.

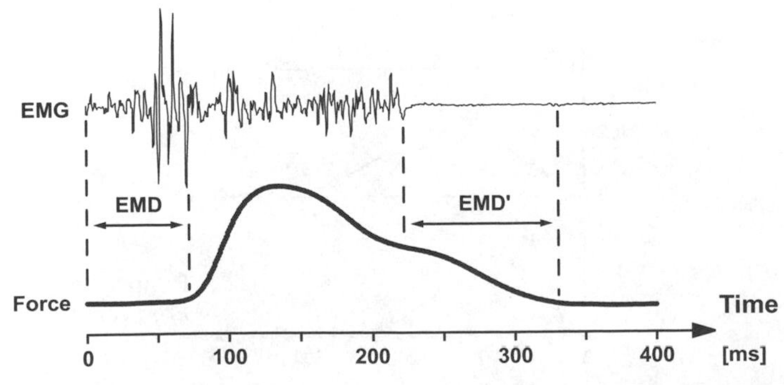

Figure 3.8.13 **Force and EMG signals obtained from cat soleus muscle during walking at a nominal speed of 1.2 m/s on a motor driven treadmill. The signals represent a normal step from the onset of EMG to paw off. The electromechanical delay for this particular step was, EMD = 72 ms, and EMG finished, EMD = 109 ms, before soleus force became zero.**

EMG-force relations of skeletal muscles have often been described for is… tractions, e.g., Lippold (1952), Milner-Brown and Stein (1975), and Moritani and de V… (1978). Isometric contractions under steady state conditions, i.e., constant force, allow for the simplest comparisons between force and EMG, because (a) one value of force may be compared to one value of the processed EMG, (b) the temporal disassociation between EMG and force may be neglected in a steady state contraction lasting for a long period of time, i.e., a second or more, and (c) the instantaneous contractile conditions of the muscle may be determined easily. Despite the apparent simplicity of isometric contractions, many different relations between EMG and force have been suggested for these contractile conditions, possibly because different muscles may have different relations between EMG and force (Bigland-Ritchie et al., 1980), and possibly because forces in muscles have been estimated indirectly from the forces or moments produced by an entire group of agonistic muscles; as well, EMG signals may have been recorded, processed, and analyzed in many different ways.

As a result of the difficulty in quantifying EMG-force relations in isometrically contracting muscle, descriptions of EMG-force relations for dynamically contracting muscle are rare and controversial. Most of the dynamic experiments have been performed using isokinetic contractions on strength dynamometers. These dynamometers typically enforce a relatively constant angular velocity of joint movement. Needless to say, a constant angular velocity of joint movement cannot necessarily be associated with a constant rate of change in length of the muscle-tendon complex, because of the possible variations of the muscular moment arm as a function of the joint angle, and the varying velocity (vector) of the attachment site of the muscle relative to the direction of shortening of the muscle as a function of the joint angle. Furthermore, a constant contractile speed of the muscle-tendon complex may not translate into a constant speed of shortening of the contractile elements (fibres and sarcomeres) because of the varying force in the muscle and the associated changes in length of the elastic elements arranged in series with the contractile elements. Only a few studies have attempted to relate EMG and force (moments) during normal, unrestrained movements (Hof & van den Berg, 1981a; Olney & Winter, 1985; Ruijven & Weijs, 1990).

To study the basic relation that may exist between the EMG signal and the force of a muscle, we felt it necessary not only to measure the force of the muscle and the uncontaminated EMG, but also to control precisely the activation to the muscle and to activate the muscle in the most physiological way possible in an attempt to produce an EMG signal indistinguishable from a signal obtained during voluntary muscle contraction. Our approach to this problem and the corresponding results are presented below.

EMG-FORCE RELATION OF CAT SOLEUS USING CONTROLLED STIMULATION

In studying the EMG-force relation of mammalian skeletal muscle, we used the cat soleus as the experimental model. Myotomy of the back muscles and laminectomy were performed, exposing the nerve roots of the seventh lumbar (L7) and the first sacral (S1) vertebrae as described by Rack and Westbury (1969). The ventral roots of L7 and S1 containing the α-motor neurons of the soleus were separated into ten filaments and placed over hook-like electrodes for stimulation. The electromyographical activity of the soleus during ventral root stimulation was measured using indwelling, bi-polar wire electrodes inserted in the mid-belly region along the axis of the soleus fibres. Soleus forces were measured by attaching the distal insertion of the soleus tendon containing a remnant piece of bone to a force transducer (Guimaraes et al., 1994a).

Each of the prepared ventral root filaments could be stimulated independently of the remaining filaments. Stimulations were performed using pseudo-random interstimulus intervals (with a coefficient of variation of 12.5%) based on a Gaussian distribution (Zhang et al., 1992). The independent stimulation of the ten ventral root filaments at different mean frequencies using pseudo-random interstimulus intervals allowed for the creation of EMG signals that were similar (in the time and frequency domain) to EMG signals measured during voluntary contractions of the same muscle(Figure 3.8.14).

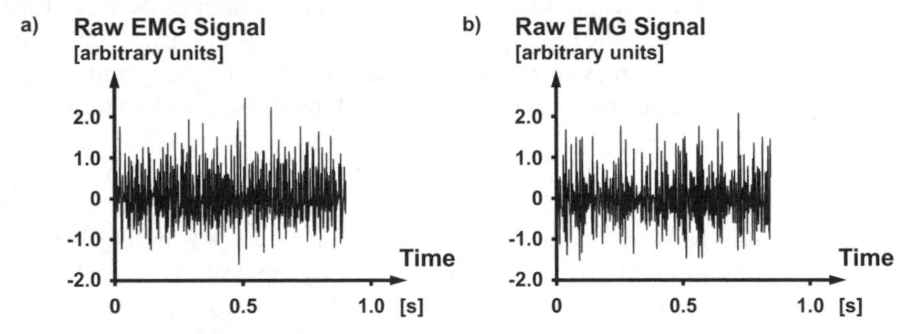

Figure 3.8.14 **One of the EMG signals shown in this figure was obtained from cat soleus muscle during walking, the other signal was obtained using pseudo-random stimulation of ten ventral root filaments that contained α-motor neurons to the soleus muscle (we leave it up to the reader to decide which is which) (from Zhang et al., 1992, with permission).**

A single ventral root filament prepared as described above contained an estimated 15 to 25 individual motor units. Applying a single stimulus to a single ventral root filament produced a compound motor unit action potential (CMUAP) of particular shape and duration for isometric contractions at a particular muscle length. When stimulating a single ventral root filament repeatedly and at different frequencies, the shape and duration of CMUAP was preserved, so the integration of a single full-wave rectified CMUAP remained approximately constant for rates of stimulation ranging from 5 to 50 Hz, and consequently, the corresponding IEMG was linearly related to the mean stimulation rate (Figure 3.8.15a).

The force response of soleus to increasing mean rates of stimulation of a single ventral root filament is shown in Figure 3.8.15b. By contrast to the behaviour of the IEMG, mean forces were non-linearly related to the mean stimulation rate and tended to saturate beyond frequencies of approximately 25 Hz in the cat soleus. The resulting IEMG-mean force relation for stimulation of a single ventral root filament is shown in Figure 3.8.15c.

A graded increase in the stimulation of the isometric soleus muscle was achieved by first increasing the number of stimulated ventral root filaments and simultaneously increasing the mean stimulation rate of already active ventral root units (Table 3.8.1, trials 1-10), and then by increasing the mean stimulation rates of all ventral root filaments progressively (Table 3.8.1, trials 11-27). The response of an isometrically held soleus muscle to such a stimulation protocol was a gradual increase in the IEMG (Figure 3.8.16a) and an "S"-shaped increase in the corresponding force (Figure 3.8.16b). For trials 1-5 (Figure 3.8.16b), the force response remained almost zero, perhaps because the frequencies of stimulation used in these first few trials were very low, in fact, lower than the threshold frequencies observed during recruitment of motor units in skeletal muscles during voluntary movements

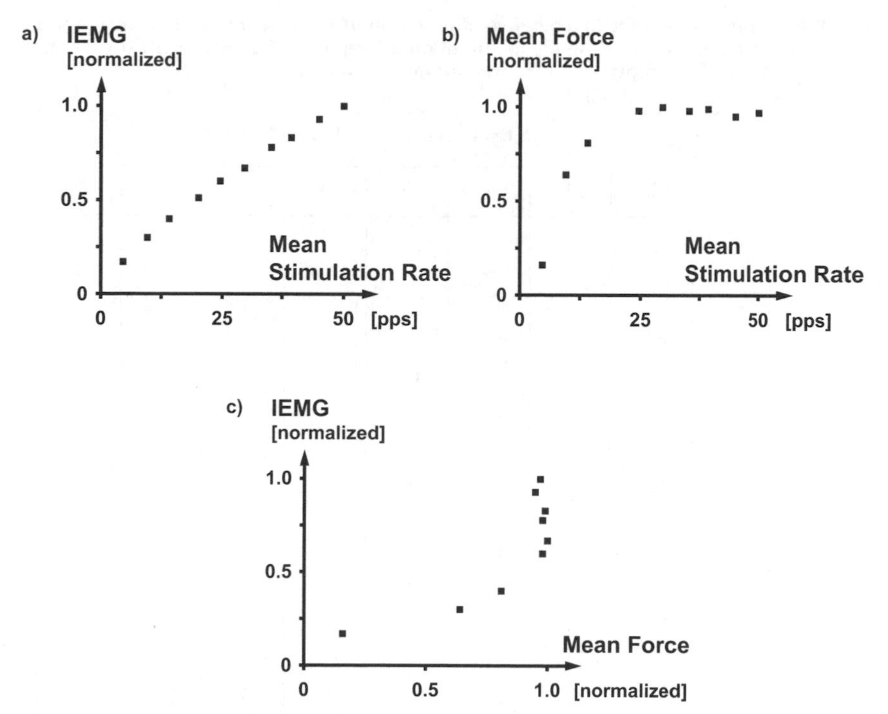

Figure 3.8.15 **IEMG (a) and mean force (b) as a function of mean stimulation rate to a single ventral root filament of cat soleus muscle, and the corresponding IEMG-mean force relation (c) (from Guimaraes et al., 1994a, with permission).**

(Burke, 1981). Therefore, trials 1-5 in Table 3.8.1 may be considered non-physiologic. In the corresponding relation between IEMG and mean force (Figure 3.8.16c), these trials are labeled L, for Low stimulation region.

Beyond trial 17, the increase in force with increasing stimulation starts to level off (Figure 3.8.16b). Trial 17 is the first trial in which the majority of the ventral root filaments were stimulated at mean frequencies of 25 Hz or more. Therefore, it appears that the leveling off of the force response beyond trial 17 is associated with the saturation in muscular force observed in single ventral root filaments at stimulation frequencies of approximately 25 Hz (Figure 3.8.15b). This particular region of force saturation is labeled H, for High stimulation region, in Figure 3.8.16c showing the IEMG-mean force relation. The forces in soleus muscle that were reached using trials 17-27 (Table 3.8.1) are not reached typically during normal voluntary movements in the cat (Walmsley et al., 1978; Hodgson, 1983; Herzog et al., 1993a). Therefore, it has been argued that the H region shown in Figure 3.8.16c, like the L region, is not relevant for the IEMG-force relation of cat soleus muscle during voluntary movements (Guimaraes et al., 1994a).

The IEMG-mean force relation for trials 6 to 16 (Table 3.8.1) could be approximated well with a straight line for all animals tested ($r^2 \geq 0.97$ in all cases). It appears, therefore, that the relation between mean force and IEMG is linear for isometrically contracting cat soleus muscle within the physiologically relevant region, i.e., the intermediate or I region (Figure 3.8.15c), and non-linear, i.e., S-shaped, for the whole range of stimulations used in the protocol shown in Table 3.8.1.

Table 3.8.1 Protocol used for independent stimulation of 10 ventral root filaments. The numbers in the table show the mean stimulation frequency for each ventral root filament in each trial. Empty spaces represent no stimulation.

TRIAL #	VENTRAL ROOT FILAMENT #									
	1	2	3	4	5	6	7	8	9	10
1	3									
2	3	3								
3	5	5	3							
4	5	5	3	3						
5	7	7	5	5	3					
6	9	9	7	7	5	3				
7	11	11	9	9	7	5	3			
8	13	13	11	11	9	7	5	3		
9	15	15	13	13	11	9	7	5	3	
10	17	17	15	15	13	11	9	7	5	3
11	19	19	17	17	15	13	11	9	7	5
12	21	21	19	19	17	15	13	11	9	7
13	23	23	21	21	19	17	15	13	11	9
14	25	25	23	23	21	19	17	15	13	11
15	27	27	25	25	23	21	19	17	15	13
16	29	29	27	27	25	23	21	19	17	15
17	31	31	29	29	27	25	23	21	19	17
18	33	33	31	31	29	27	25	23	21	19
19	35	35	33	33	31	29	27	25	23	21
20	37	37	35	35	33	31	29	27	25	23
21	39	39	37	37	35	33	31	29	27	25
22	41	41	39	39	37	35	33	31	29	27
23	43	43	41	41	39	37	35	33	31	29
24	45	45	43	43	41	39	37	35	33	31
25	47	47	45	45	43	41	39	37	35	33
26	49	49	47	47	45	43	41	39	37	35
27	51	51	49	49	47	45	43	41	39	37

The results presented in Figure 3.8.15 and Figure 3.8.16 were all obtained for isometrically contracting soleus muscle at a given length. It is well accepted that the EMG-force relation of skeletal muscles under dynamic conditions needs to account for the instantaneous contractile conditions of the target muscle: the instantaneous length, rate of change in length, and activation (Hof and van den Berg, 1981a; Olney and Winter, 1985). The effect of muscle length on the EMG, force, and the corresponding IEMG versus mean force relation is shown in Figure 3.8.17a-c for isometric contractions at three lengths of cat soleus muscle corresponding to ankle joint angles of 80°, 105°, and 130° (180° = full extension of the ankle joint, therefore, the ankle joint angle of 80° corresponds to the longest, and the angle of 130° to the shortest muscle length).

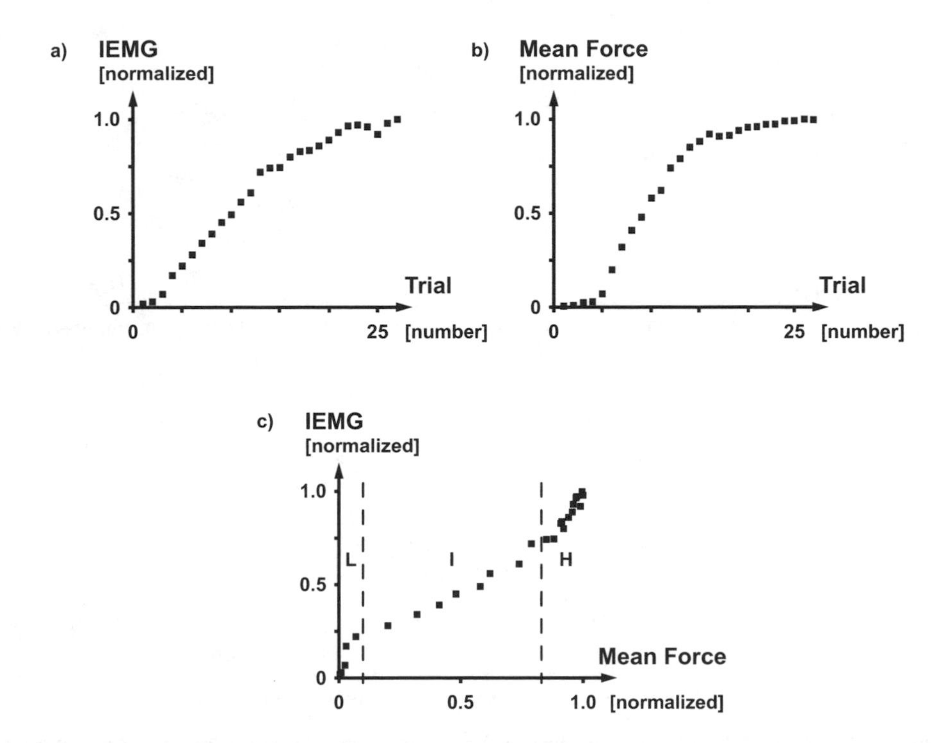

Figure 3.8.16 IEMG (a) and mean force (b) response of cat soleus muscle subjected to the stimulation protocol described in Figure 3.8.1, and the corresponding IEMG-mean force relation (c). L is the low stimulation region, unphysiologic; H is the high stimulation and high force region, unphysiologic; I is the intermediate stimulation region, physiologic (from Guimaraes et al., 1994a, with permission).

The IEMG results of the experiment shown in Figure 3.8.17a were independent of muscle length. We feel that this independence can be obtained only if the location of the recording electrodes relative to the muscle does not change considerably during shortening and lengthening of the muscle, and if the interelectrode distance is fixed. Mean force values clearly depended on the muscle length. In this particular case, higher forces at all levels of stimulation were obtained for longer muscle lengths (Figure 3.8.17b). This result corresponds to the observation that cat soleus muscle operates on the ascending limb of the force-length relation for the lengths tested here (Rack and Westbury, 1969; Herzog et al., 1992).

The IEMG-mean force relations of cat soleus muscle clearly depend on muscle length (Figure 3.8.17c). They are S-shaped for each length, as described above for the single length, with an intermediate region that may be approximated adequately with a straight line. As discussed above, it is this intermediate region that appears to be most relevant for voluntary muscular contractions.

The straight-line approximations of the IEMG-mean force relations obtained in the intermediate regions for the three different muscles lengths have different slopes. The slopes become progressively steeper with decreasing muscle length, indicating that more IEMG is required to produce a given force at short, compared to long, muscle lengths in the cat soleus muscle. Similar findings have been reported during voluntary contractions of human soleus (Close et al., 1960).

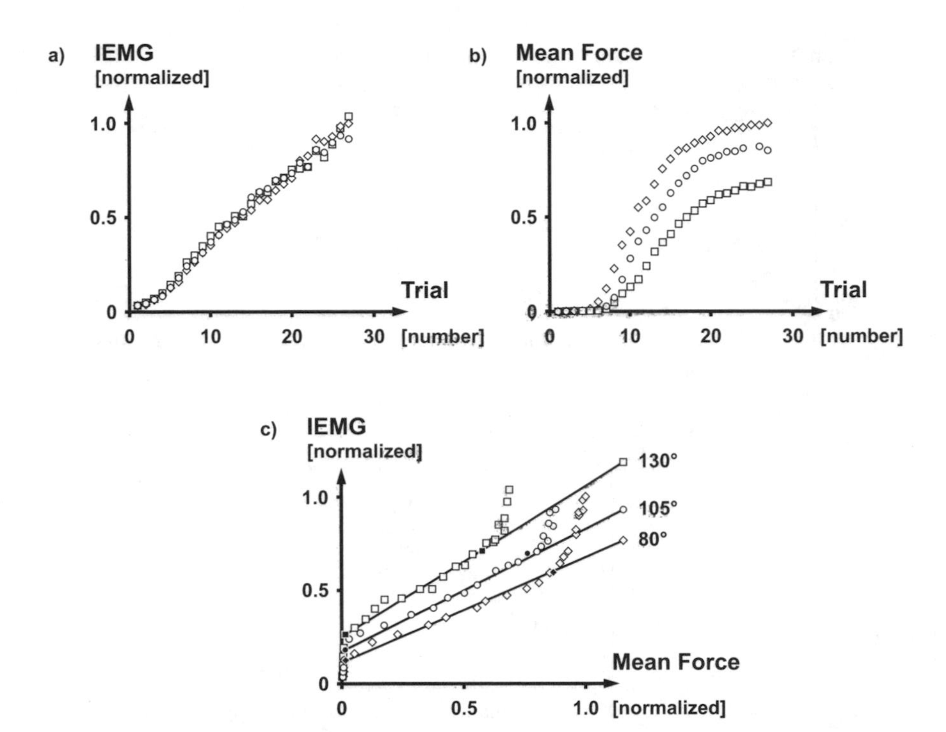

Figure 3.8.17 **IIEMG (a) and mean force (b) response of cat soleus subjected to the stimulation protocol described in Figure 3.8.1 for three different muscle lengths, and the corresponding IEMG-mean force relation (c). The intermediate region of the IEMG-mean force relations were approximated by a straight line (by eye). The slopes of the straight lines become progressively steeper with decreasing muscle length (180° = full extension of the ankle joint) (from Guimaraes et al., 1994b, with permission).**

To assess whether the differences in the forces produced at different muscle lengths depend only on muscle length or whether they depend on muscle length and the level of stimulation, the force obtained at a given level of stimulation for one muscle length was divided by the corresponding force at another muscle length. If forces were independent of the level of stimulation (and so, depended only on muscle length), one would expect the resultant force ratio to be a constant value; however, this is not the case. Figure 3.8.18 shows the ratio

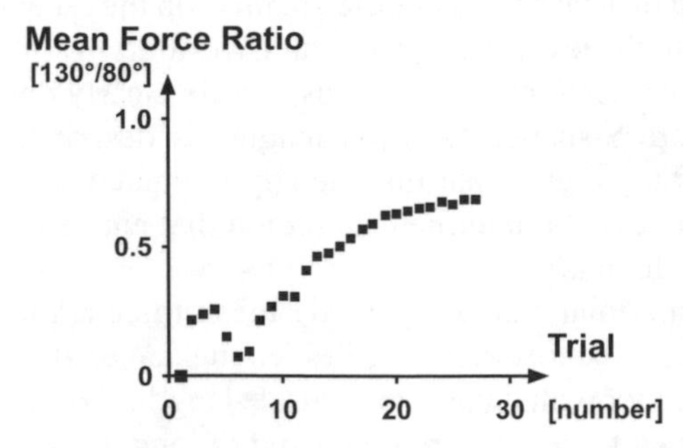

Figure 3.8.18 **Ratio of the forces obtained from the shortest (ankle angle = 130°) and the longest muscle length (ankle angle = 80°) for all 27 levels of stimulations shown in Figure 3.8.1 (from Guimaraes et al., 1994b, with permission).**

that was found by dividing the forces obtained at the shortest muscle length (130°) by the corresponding forces obtained at the longest muscle length (80°) for all 27 levels of stimulation used in this protocol (Table 3.8.1). The resulting ratio was not constant, indicating that there is a non-linear scaling of the force-length property of cat soleus muscle for submaximal levels of stimulation.

EMG-FORCE RELATION OF CAT SOLEUS DURING UNRESTRAINED LOCOMOTION

To assess the EMG-force relation of skeletal muscle during voluntary, dynamic contractions, it is necessary to measure the forces and EMGs directly from the target muscle. Simultaneous measurements of force and EMG have been performed quite frequently in cat hindlimb muscles during locomotion, e.g., Walmsley et al. (1978), Hodgson (1983), and Abraham and Loeb (1985). However, researchers have not used such data systematically to relate the instantaneous muscular forces to the corresponding EMG records, with the exception of Sherif et al. (1983), who described qualitatively the relation between EMG and force in the cat medial gastrocnemius muscle.

When attempting to assess the EMG-force relation in muscle during locomotion in the cat, two major difficulties are added to the interpretation of the data. First, the stimulation received by the muscle at any instant in time is not known during voluntary contractions, and second, the variable contractile conditions of the muscle fibres during locomotion must be determined or estimated. It has been suggested that these contractile conditions, i.e., the length and rate of change in length of the contractile elements of the muscle, affect the EMG-force relation substantially and thus must be known (Inman et al., 1952; Close et al., 1960; Heckathorne & Childress, 1981; Hof & van den Berg, 1981a, b, c, d; Basmajian & De Luca, 1985; Olney & Winter, 1985; van den Bogert et al., 1988; Ruijven & Weijs, 1990). Variable muscle fibre lengths (and corresponding rates of change in fibre length) have been estimated for cat hindlimb muscles during locomotion (Hoffer et al., 1989; Griffiths, 1991; Allinger & Herzog, 1992; Weytjens, 1992), However, the validity of these estimates must still be established thoroughly in further experiments.

At present, there is no theoretical paradigm that allows for the accurate prediction of muscular force from EMG records during dynamic, voluntary muscular contractions. In what follows, we present an approach that has been selected to shed some light on the EMG-force relation of skeletal muscle during voluntary movement. The main goal of this approach was to describe the EMG-force relation of cat soleus during locomotion, and to assess qualitatively the association that may exist between the EMG signals and the corresponding forces.

Force and EMG signals were obtained from soleus of three cats walking at speeds ranging from 0.4 to 1.2 ms, and trotting at speeds varying from 1.5 to 1.8 ms, on a motor-driven treadmill. The EMG data were full-wave rectified and integrated over time periods that varied from step to step. The time periods chosen for integration of the EMG signals were such that the integrated EMG (IEMG) would contain a frequency spectrum like the force signals for the corresponding steps (Guimaraes et al., 1995). The IEMG signals calculated for each time period were then approximated using cubic interpolation splines to give a continuous IEMG-time history for each step cycle.

IEMG-time histories calculated in this way were plotted against the corresponding force-time histories. Figure 3.8.19 shows such IEMG-force relations for three steps of three

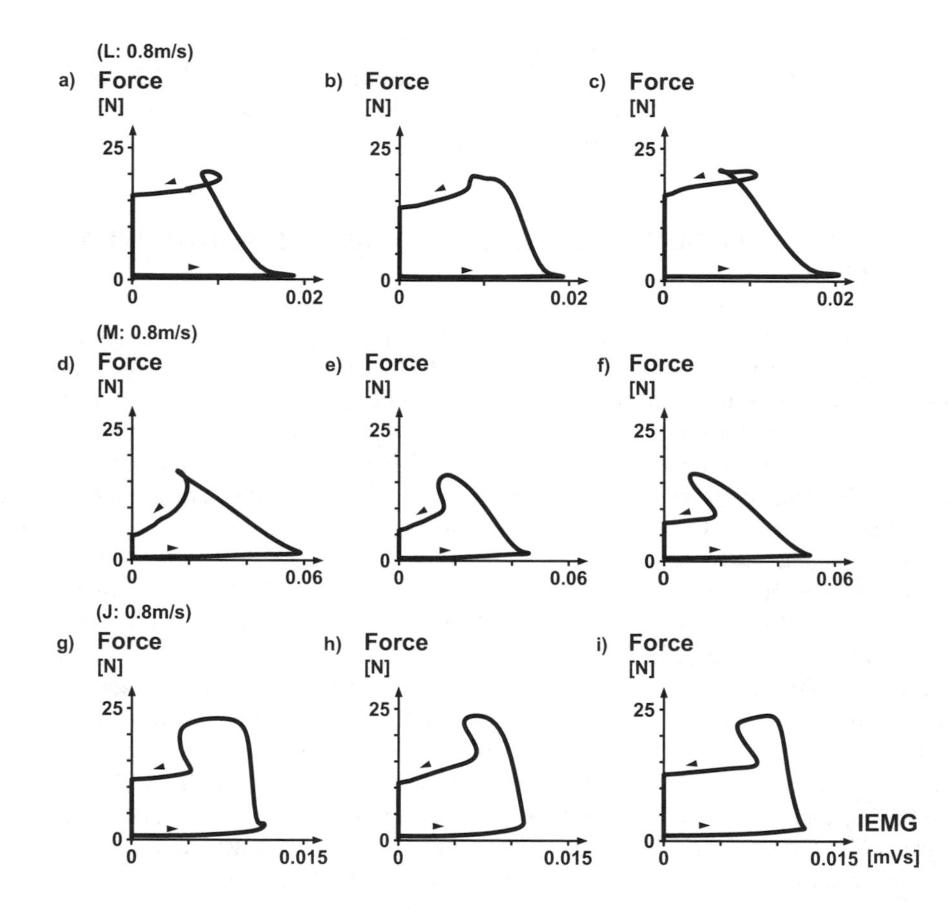

Figure 3.8.19 **IEMG-force relations obtained from three animals (L, M, J) for three non-consecutive steps of walking at 0.8 m/s on a motor-driven treadmill. Data points were obtained at the same instance in absolute time thus these graphs do not take electromechanical delay into account. Steps of the same animal are aligned horizontally in rows (from Guimaraes et al., 1995, with permission).**

animals, all walking at a nominal speed of 0.8 ms. Each graph starts at the onset of EMG, which corresponds to the data points aligned close and parallel to the horizontal, IEMG axis, follows the loops in the direction indicated by the arrows, and finishes at the end of soleus force production, which is indicated by the data points on the vertical axis going towards the origin of the graph.

EMG always precedes soleus force production, and the IEMG values reach peak levels before any appreciable force is developed (Figure 3.8.19). This fast build-up of EMG activity with little or no soleus force precedes paw contact by about 60 to 90 ms at a speed of walking of 0.8 m/s. Activation in this phase of the step cycle may be associated with preparing the contractile machinery of the muscle for force production at paw impact; that is, Ca^{2+} is released from the sarcoplasmic reticulum, attaches to troponin C, and so enables actomyosin interaction through cross-bridge formation. Cross-bridges attached in this phase, preceding paw contact, provide stiffness to the muscle, and may act as spring-like elements during and following paw impact at the beginning of the stance phase (Walmsley and Proske, 1981).

Following this initial phase, IEMGs tend to decrease, while force first increases and then decreases until the time when EMG activity becomes zero. At the end of EMG activity, soleus forces are typically still about 25 to 75% of the maximal force achieved during the step cycle (Figure 3.8.19). In the last phase of stance, soleus forces steadily decline to zero in the absence of any EMG activity.

At a walking speed of 0.8 m/s, EMG activity typically terminates about 120 to 150 ms before soleus forces become zero. Since EMG activity finishes 120 to 150 ms before soleus force becomes zero, but only starts about 60 to 90 ms before the onset of soleus force, it is obvious that the EMG signal has not only shifted relative to the force records, but is also shorter in duration than the corresponding force signal. Therefore, two of the difficulties of deriving an instantaneous relation between force and EMG during dynamic muscular contractions are associated with the relative shift of the EMG and the force signals, and the difference in length of the two signals.

To further explore the EMG-force relation of the cat soleus during locomotion, the delay between the onset of the EMG and the onset of force was accounted for by shifting the onset of the EMG signal in time to the onset of force, i.e., the electromechanical delay was offset. Furthermore, the EMG signal was linearly scaled along the time axis until the end of the EMG activity coincided with the end of soleus force production in each step cycle. In this way, a given IEMG could always be associated with a given, non-zero force of soleus during the stance phase of locomotion. Relating the IEMG obtained in this way to the corresponding soleus forces, produced the relations shown in Figure 3.8.20.

It has been suggested that any explanation of the relation between EMG and force in dynamic situations must consider the instantaneous contractile properties of the muscle, i.e., the instantaneous length and rate of change in length of the muscle or the contractile elements (Hof & van den Berg, 1981a,b,c,d; Basmajian & De Luca, 1985; Olney & Winter, 1985). We have taken this idea a step further and have calculated a theoretical activation (A) of the muscle. This theoretical activation (Allinger & Herzog, 1992) was defined as:

$$A = F_i / F_{imax} \tag{3.8.6}$$

where F_i is the force required by (or measured in) the muscle, and F_{imax} is the maximal force that the muscle can produce for the given instantaneous contractile conditions. We then compared the activation calculated theoretically with the IEMG obtained experimentally for the cat soleus during locomotion. A perfect match between the theoretically calculated activation and the IEMG would indicate that the amount of IEMG signal depends precisely on the ratio of the force produced by a muscle and the maximal possible force. Figure 3.8.21a-c shows the mean soleus forces and the corresponding theoretically determined maximal soleus forces for three animals walking at a speed of 1.2 m/s. For all animals, the measured soleus force is much smaller than the theoretically determined maximal force for the first 20 to 40% of the stance phase; however, beyond the initial 40% of the stance phase, the two forces become similar in magnitude. Sometimes the measured force even exceeds the theoretically determined maximal force.

Calculating the theoretical activation, defined in equation (3.8.6), from the results shown in Figure 3.8.21a-c, and comparing the theoretical activation to the experimentally determined IEMG of cat soleus at a walking speed of 1.2 m/s, gave the results shown in Figure 3.8.21d-f. The IEMG values were always larger in the initial part of the stance phase (≈0 to 20%) and tended to be lower in the final part of the stance phase (≈80 to 100%) than

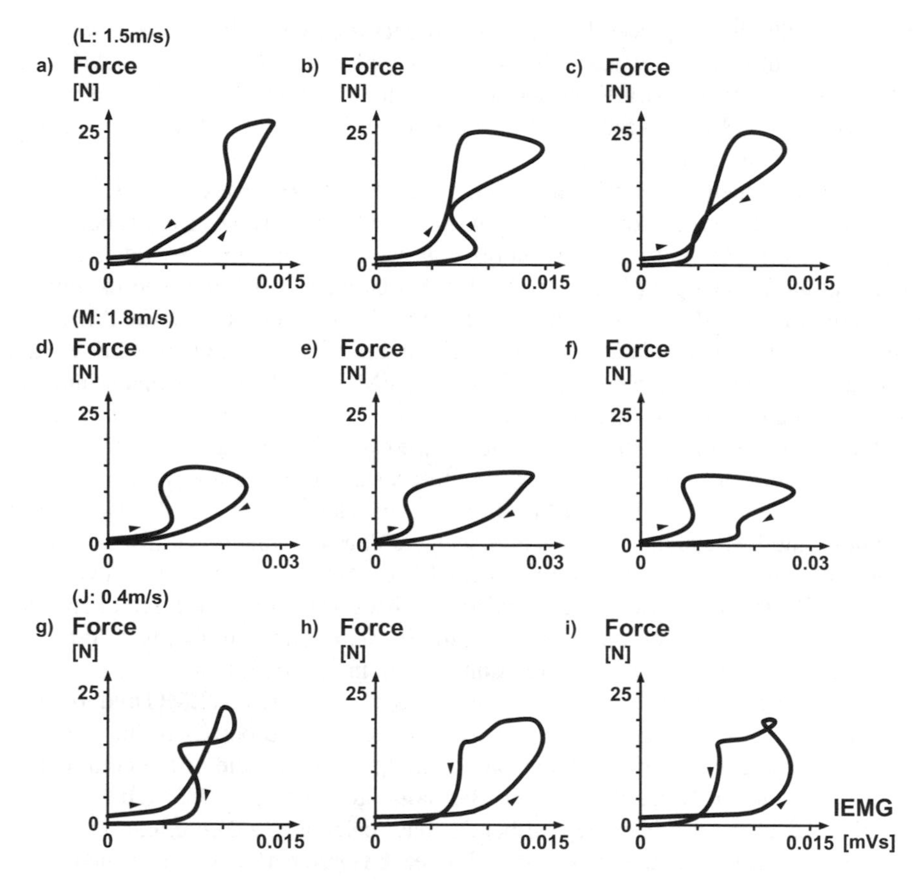

Figure 3.8.20 **IEMG-force relations obtained during three consecutive steps of trotting (animals L and M) and walking (animal J) at the speeds indicated on the graphs. The electromechanical delay between the EMG and force signals was taken into account in these graphs, as described in the text (from Guimaraes et al., 1995, with permission).**

the corresponding activation predicted theoretically. It appears that the actual activation required in cat soleus is much higher in the initial part of the stance phase than one would expect from the maximal available force, and the forces required during this phase. This high activation is most likely associated with the initial priming of the muscle for force production; the activation of the filaments and the cross-bridges, and the corresponding stiffening of the muscle (Walmsley & Proske, 1981).

Similarly, in the final 20% of the stance phase, IEMG values are typically lower than expected theoretically. In this final phase, soleus forces are still relatively high compared to the maximal available soleus force. Nevertheless, the IEMG values are typically less than 50% of the maximum, and decrease rapidly to zero (Figure 3.8.21d-f). As for the initial part of the stance phase, the systematic deviations of the IEMG from the theoretically calculated activation are most likely associated with the relatively slow changes in force following activation or deactivation of the muscle. Contrary to the initial part of the stance phase, in this final part of stance, the sharp decline of the IEMG is followed slowly by a decrease in muscular force, resulting in the smaller values of the IEMG compared to the corresponding val-

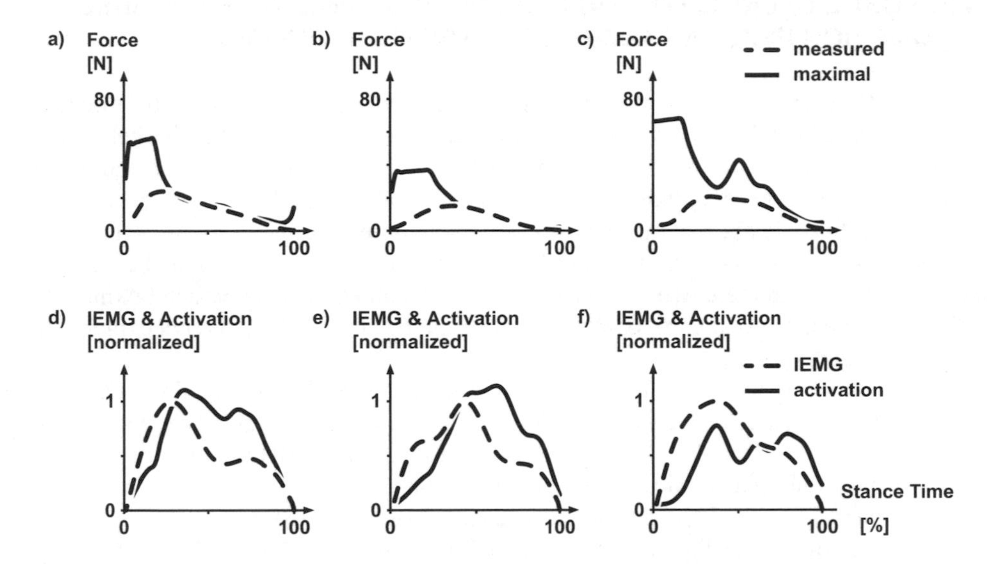

Figure 3.8.21 **Force-time histories and maximal possible force-time histories of cat soleus muscle during the stance phase of walking at 1.2 m/s for three different animals (a-c), and the corresponding IEMG-time and theoretically calculated activation-time histories for the same steps as shown above (d-f) (from Guimaraes et al., 1995, with permission).**

ues of the theoretical activation for the last 20% of the stance phase. Also, calculations of the theoretical activation (equation (3.8.6)) do not account for potentiation of force of a shortening muscle that was stretched just before the shortening phase, e.g., Cavagna et al. (1968). Since the soleus undergoes a stretch-shortening cycle during the stance phase of walking, force potentiation may occur during the latter stages of stance, and so, large forces may be produced with relatively little activation.

In the middle part of the stance phase (between approximately 20 to 80% of the stance time, Figure 3.8.21d-f), the IEMG values form a bi-modal curve. The first peak of this bi-modal curve is reached between about 25 to 45% of the stance time, followed by the second, smaller peak (or sometimes just a shoulder rather than a peak) at approximately 70 to 80% of the stance time. The first peak of the IEMG appears to coincide with the end of the initial priming of the muscle for the following force production (Figure 3.8.21). The second peak appears to be associated with a burst of EMG required near the end of the step phase to maintain soleus forces at a given level, while the contractile conditions become highly unfavorable for force production (Figure 3.8.21a-c).

The activation-time histories predicted theoretically are typically bi-modal or multi-modal, like the experimentally determined IEMG-time histories in the middle part of the stance phase (Figure 3.8.21d-f). Despite this similarity between the two curves, it is apparent that the theoretical activations do not match the IEMG-time curves in detail. We speculate from these results that IEMG, and the activation calculated theoretically based on the instantaneous contractile conditions of the muscle, represent the same phenomenon, and that the theoretically predicted activation may be a possible predictor of IEMG in the middle part of the stance phase, if the instantaneous contractile conditions of the muscle can be obtained accurately.

EMG-FORCE RELATION OF CAT PLANTARIS DURING UNRESTRAINED LOCOMOTION USING AN ADAPTIVE FILTERING APPROACH

Up to this point, we have always attempted to describe or explain the EMG-force relation using basic knowledge of the physiology of force and EMG production in skeletal muscle. This approach provides insight into the behaviour of muscle, and in the end may be used to formulate a general paradigm that contains testable hypotheses. However, this approach is not the only one that may be chosen. For example, if the goal of a project is to give the best possible force predictions from EMG, without any real concern about the biological relationship between these two signals, then a purely mathematical approach linking EMG to force may be the best and the simplest method (Zhang et al., 1993). Having explored this route, selected results are presented below.

Before choosing a theoretical approach to predict dynamic forces from EMG exclusively, some characteristics of the force and EMG signal have to be established. Most importantly, it must be realized that the characteristics of force and EMG are time-dependent or non-stationary, during dynamic contractions. Therefore, conventional filtering and smoothing techniques, based on a fixed (stationary) design, were rejected.

Adaptive filtering techniques can account for non-stationarities and have been used successfully in the analysis of a variety of biological signals, for example, electrocardiograms (Ferrara & Widrow, 1982; Yelderman et al., 1983), electrogastric signals (Kentie et al., 1981; Chen et al., 1990), and vibroarthrographic signals (Zhang et al., 1991). The adaptive filtering system used for plantaris force estimations from the corresponding EMG of normally walking and running cats (Herzog et al., 1993a) is shown in Figure 3.8.22. In the figure, z^{-1} represents a one-sample delay and j stands for the time instant. The symbol d_j stands for the primary input which contains the force signal, s_{jo}, plus additive noise, n_{jo}. The reference input, x_j, which is the output of a demodulator consisting of a full-wave rectifier and a low-pass filter, contains noise, n_{j1}, and the signal s_{j1}, which is related to s_{jo}, but does not necessarily have the same waveform as the signal s_{jo}, at theprimary input. The

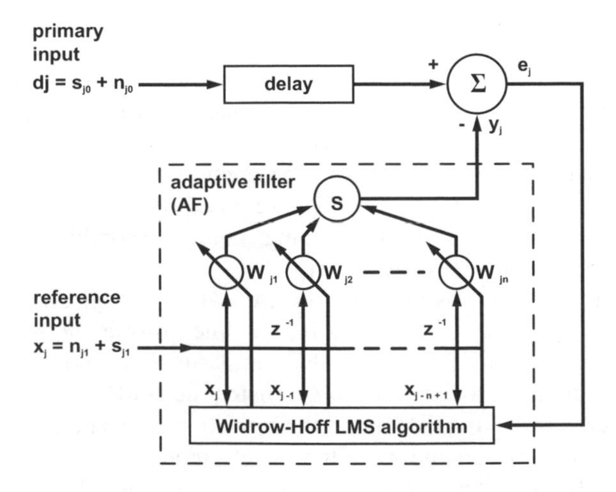

Figure 3.8.22 **An adaptive filtering system with the tapped delay line. z^{-1} represents a one-sample delay and j stands for the time instant.**

noise component, n_{jo}, is assumed to be uncorrelated with n_{j1}, s_{j1} or s_{jo}. The output signal, y_j, is formed as the weighted sum of a set of input signal samples:

$$x_j, x_j, \ldots, x_j - N + 1$$

Mathematically, the output, y_j, is equal to the inner product of the input vector, X_j, and the weight vector, W_j:

$$y_j = X_j^T W_j \tag{3.8.7}$$

where the input signal vector is defined as:

$$X_j = \left[x_j, x_{j-1}, \ldots, x_{j-N+1}\right]^T \tag{3.8.8}$$

and the filter weight vector is:

$$W_j = \left[w_{ji}, w_{j2}, \ldots, w_{jN}\right]^T$$

During the adaptation process, the weights are adjusted according to the least mean squared (LMS) algorithm (Zhang et al., 1994; Widrow, et al., 1975). Let P represent the cross-correlation vector defined by:

$$P = E\{d_j X_j\} \tag{3.8.9}$$

and let R represent the input correlation matrix defined by:

$$R = E\{X_j X^T{}_j\} \tag{3.8.10}$$

where E {} stands for the expectation operator. Thus, a general expression of the mean-squared error (MSE) as a function of the weight vector can be obtained as follows:

$$e_j = d_j - y_j = d_j - X^T{}_j W_j \tag{3.8.11}$$

$$E\{e^2{}_j\} = E\{d^2{}_j\} - 2P^T W_j + W^T{}_j R W_j \tag{3.8.12}$$

Using the steepest-descent method and approximating the gradient MSE by gradient squared error, we obtain:

$$W_{j+1} = W_j - \mu\Delta_j \tag{3.8.13}$$

$$\Delta_j = \frac{de^2{}_j}{dW_j} \tag{3.8.14}$$

and the well-known Widrow-Hoff LMS algorithm is derived as (Widrow et al., 1975):

$$W_{j+1} = W_j + 2\mu e_j X_j \tag{3.8.15}$$

where μ is the step size which controls the convergence rate of the adaptation and the stability of the system. The larger the value of μ, the larger is the gradient noise that is introduced, but the faster the algorithm converges, and vice versa. Therefore, in the non-stationary environment, step-size optimization becomes a critical issue for successful adaptive signal processing using the LMS algorithm. Generally, a compromise has to be made between a fast convergence, which is necessary to track the statistical changes in the input signal, and the slow adaptation which is needed to reduce the gradient noise (or misadjustment). A careful selection of performance measurements is important for step-size optimization (Zhang et al., 1991; Zhang et al., 1994).

After the convergence of the system, the adaptive filter with the Widrow-Hoff LMS algorithm creates a least mean squares estimate of the force signal, n_{jo}, in the primary input by filtering the reference input, i.e., the rectified EMG signals. According to the adaptive theory (Widrow et al., 1975), the filter output is the best estimate of the signal, n_{jo}, in the sense of the mean square error. A critical requirement for the successful operation of this scheme is that the primary and the reference signal components must be uncorrelated with the primary noise components, but correlated with each other. This statement is explained using the following demonstration (for convenience, the subscript j is omitted). From Figure 3.8.22, the system output is the error, given by:

$$e = s_o + n_o - y \tag{3.8.16}$$

where s_o represents the primary signal component. Then, the MSE can be obtained as:

$$E\{e^2\} = E\{n_o^2\} + E\{(s_o - y)^2\} + 2E\{n_o(s_o - y)\} \tag{3.8.17}$$

Since n_o and $(s_o - y)$ are uncorrelated and have a mean of zero, $E\{n_o(s_o - y)\} = 0$, the equation (3.8.17) becomes:

$$E\{e^2\} = E\{n_o^2\} + E\{(s_o - y)^2\} \tag{3.8.18}$$

Thus, minimizing the MSE for a given input can be achieved by minimizing $E\{(s_o - y)^2\}$. That is:

$$\min E\{e^2\} = \min E\{(s_o - y)^2\} + E\{n_o^2\} \tag{3.8.19}$$

The MSE minimization yields the least mean square estimate of the signal, s_o. This result is obtained by the adaptive filtering process governed by the LMS algorithm without requiring any prior knowledge of the statistics of the signal and the noise, except the requirement that the signal components in the primary input and the reference input are correlated with each other, but are uncorrelated with the noise. According to the adaptive filtering principle, the output of the system, y_j, is the best estimate of the muscle force, s_o, in the primary input.

In Figure 3.8.23, the actual forces recorded in the cat plantaris muscle are shown for eight walking steps on a 10° downhill slope at a speed of 0.4 m/s; in Figure 3.8.24, the corresponding EMG signals are shown. The actual plantaris forces were used as the desired signals of the primary input to the adaptive filter, and the demodulated plantaris EMG signal was used as the reference input. The estimated force using the adaptive filter with $\mu = 0.01$ is given in Figure 3.8.25(dashed line), together with the corresponding actual force (solid line). As can be seen, cat plantaris forces estimated exclusively from rectified and low-pass filtered EMG signals were adequate after the convergence (about a quarter of the first step cycle) of the adaptive filtering process.

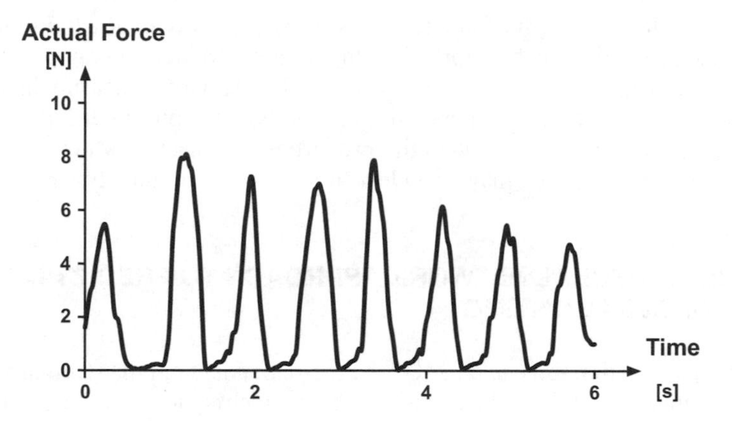

Figure 3.8.23 **The actual forces recorded in the cat plantaris muscle during eight walking steps on a 10° downhill slope at a speed of 0.4 m/s.**

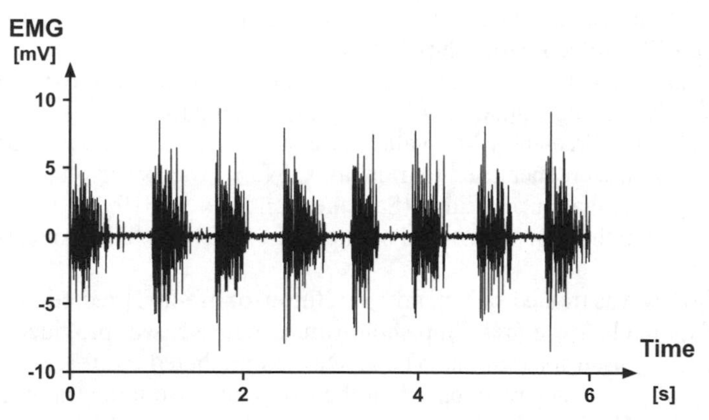

Figure 3.8.24 **The EMG signals of cat plantaris corresponding to the force results of Figure 3.8.23.**

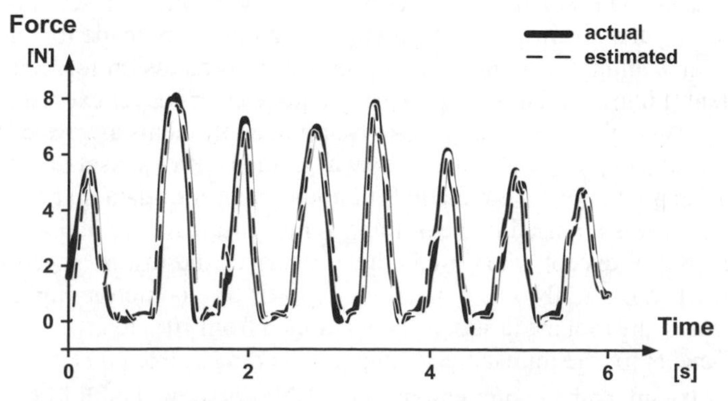

Figure 3.8.25 **Estimated plantaris forces using adaptive filtering with the LMS algorithm (dashed line), and the actual plantaris forces (solid line) observed during normal cat locomotion.**

It has been stated that EMG-force relations of dynamically contracting muscles must contain muscle properties, such as force-length and force-velocity relations. However, instantaneous force-length-velocity properties of muscles are hard to measure in vivo. The results of this study indicate that adequate estimates of dynamic muscular forces are possible without using the instantaneous contractile properties of muscles, which suggests, therefore, that an adaptive approach may provide a simple way to predict dynamic muscle forces.

AN ARTIFICIAL NEURAL NETWORK APPROACH TO PREDICT DYNAMIC MUSCLE FORCES FROM EMG

As pointed out earlier, force and EMG are time-dependent or non-stationary, during dynamic contractions. Aside from the adaptive filtering techniques described in the previous section, Artificial Neural Network (ANN) techniques provide excellent properties to relate non-stationary biological signals.

As for the adaptive filtering techniques, ANN approaches are not suited to provide insight into the biological relationship between EMG and force signals; rather ANN techniques represent a powerful numerical set of tools that might be able to capture the essential features of the EMG-force relationship.

To test the predictive ability of ANNs, soleus and gastrocnemius forces and EMGs were measured in freely moving animals at 4 to 6 speeds, ranging from 0.4 m/s walking to 2.4 m/s trotting. The actual force and EMG values of a subset of animals, locomotion speeds or steps of a given trial, were then used to train an ANN that was based on a feedforward network with a backpropagation algorithm (Savelberg and Herzog, 1997; Liu, 1997). The specific network used in this study consisted of one input layer, two hidden layers, and a single cell, output layer.

Once the ANN was trained sufficiently, i.e., the error of force predictions from EMG of the training data fell below a preset threshold value, the ANN was provided with EMG input of data sets not used for training. The ANN then predicted the force-time histories of these new sets of data exclusively based on the experience from the training sets, and the newly provided EMG-time histories. The following three conceptual types of force-time history predictions were made, (1) intrasession predictions, (2) intrasubject predictions, and (3) intersubject predictions.

For the intrasession predictions, the ANN was trained with a subset of data from a single animal walking at a given speed. The force predictions were made for another subset of data from the same animal walking at the same speed. Intrasession force predictions were always excellent (Figure 3.8.26) with root mean square errors never exceeding 7% between the predicted and actual forces, and cross-correlation coefficients always exceeding 0.98.

For the intrasubject predictions, the ANN was trained with a subset of data from a single animal, and force predictions were made for another subset of data that contained a speed of locomotion that was not used in the training set. Intrasubject predictions were generally good (Figure 3.8.27), except when predicting forces for the lowest speed of locomotion (0.4 m/s), and when the ANN was trained with speeds of locomotion higher than 0.4 m/s. Including all cases, the root mean square errors ranged from 10% to 28%, and the cross-correlation coefficients for the intrasubject comparisons ranged from 0.66 to 0.94.

Finally, for the intersubject predictions, the ANN was trained with force and EMG data of a subset of animals, and force predictions from EMG were made from data of one animal that was not used in the training set. To our surprise, the intersubject force predictions were

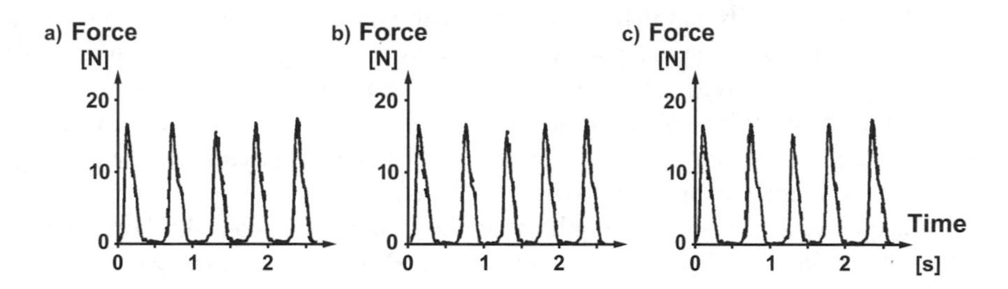

Figure 3.8.26 **Dynamic predictions of force from EMG signals using an ANN approach. Cat soleus forces were predicted for the intrasession protocol (see text) with the cat walking at 0.8 m/s. Solid line = actual forces; dashed line = predicted forces.**

generally very good (Figure 3.8.28) with root mean square errors always below 13%, and cross-correlation coefficients between the predicted and actual forces always exceeding 0.91.

The results of this study point out the tremendous ability of ANNs to capture the essential features of the EMG-force relationship of skeletal muscle during dynamic, non-stationary tasks. Many of the predictions made using the ANN approach for dynamic tasks exceeded the quality of the predictions of other approaches during static tasks, which, in contrast to the ANN approach, contained information on the structure, size, physiology, and contractile conditions of the muscle. Although from a purely practical point of view, the predictions of dynamic muscle force obtained here were excellent, and required no input about the muscle's contractile conditions there are obvious drawbacks of the ANN approach: no biological insight about EMG or in vivo muscle force production is obtained; force and EMG measurements must be performed at one point from the target muscle to allow for training of the ANN; and the enormousness of the numerical procedures underlying ANNs prevents for an easy reformulation of the model in simple or analytical terms. However, if the sole purpose of an application is to provide accurate force predictions of in vivo skeletal muscles under dynamic conditions from the corresponding EMG signals, then the ANN approach can give excellent results, even for intersubject predictions (Figure 3.8.28).

As mentioned in the Introduction, the purpose of this chapter was not to copy information on EMG that is readily available in textbooks devoted exclusively to EMG. Rather, we attempted to present information on a specific topic of EMG, the EMG-force relation of skeletal muscles. The results of three very different and new approaches were discussed. In the first approach, we pointed out the possibility of how a controlled stimulation technique on the ventral root level could be used to create a "physiologic"-looking EMG, and how, under these controlled conditions, EMG and force of a muscle could be measured for any contractile conditions desired. The second and third approaches dealt with the relations between EMG and force of dynamically contracting muscles during locomotion. Forces and EMGs were measured directly from the target muscles in both cases. In the second approach, the EMG-force relations were described, and then qualitatively analyzed, using the known physiological properties of muscle as guidelines. The third approach was based exclusively on a theoretical input-output-type system. Physiological (or other) properties of muscle were disregarded. The origin of the input signal (EMG) and the estimated output variable (force) were of no importance. Interestingly, though this last approach often gave excellent estimates of muscular force, no insight into the system's behaviour could be gained from it.

Book chapters typically summarize the well-established facts and condense the wealth of knowledge and information in a particular area into easily digestible portions. Not so in this chapter. We have, on purpose, chosen to focus on an area of research that is hotly debated and full of controversy. Many of the results presented here have not been published elsewhere at the time of writing this chapter. The results shown are correct, however, many of the ideas are new and may turn out to be, in the words of A.V. Hill, "false trails". As long as the results and discussions were stimulating to the student who reads this book and to the researcher who is interested in the relation between EMG and force in dynamically contracting skeletal muscles, then the chapter will have served its purpose.

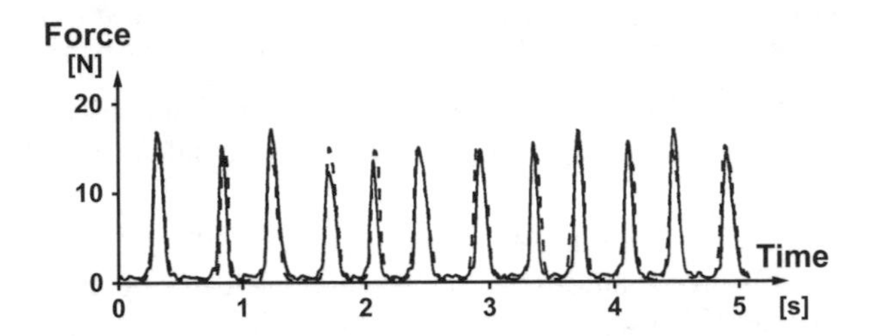

Figure 3.8.27 **Dynamic predictions of force from EMG signals using an ANN approach. Cat soleus forces were predicted for the intrasubject protocol (see text) with the cat walking at 1.2 m/s. Solid line = actual forces; dashed line = predicted forces.**

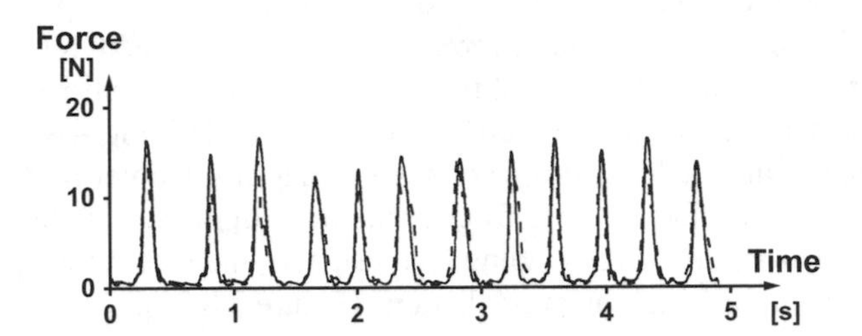

Figure 3.8.28 **Dynamic predictions of force from EMG signals using an ANN approach. Cat soleus forces were predicted for the intersubject protocol (see text) with the cat walking at 1.2 m/s. Solid line = actual forces; dashed line = predicted forces.**

3.8.6 EXAMPLES OF WAVELET EMG ANALYSIS

Movement artefacts and fatigue

A subject was asked to apply maximum voluntary isometric contraction (MVC), while pressing with the right thumb against a force transducer. During this action, the EMG of the muscle abductor pollicis brevis was measured for a period of five seconds (Figure 3.8.11, page 429).

Movement artefacts

Wavelet analysis is a possible tool for identifying movement artefacts in EMG signals (Conforto, 1999). A movement artefact, e.g. Figure 3.8.11, page 429, typically occurs at low frequencies (corresponding to wavelet 0). Movement artefacts often have a triangular shape, and some artefacts last a long time, but with decreasing intensity for higher frequencies. When analyzing the EMG signal, the triangular time frequency area of the movement artefact should be discarded.

Muscle fatigue

Fatigue is defined as the inability to maintain a certain force level. Fatigue is frequently studied using a time/frequency analysis of the EMG signal. A fatiguing effect can be identified by a gradual decay of the intensity towards lower frequencies. An EMG scalogram for such a fatigue effect is shown in Figure 3.8.11 (page 429). The corresponding wavelet spectra (Figure 3.8.12, page 429) shows that the frequency content does change towards lower frequency with increasing time.

In conventional fatigue analysis from EMG signals, it is common practice to measure the mean or the median frequency. However, the results in Figure 3.8.11 (page 429) and Figure 3.8.12 (page 429) suggest that some caution is appropriate. In the initial phase of muscle activation there is a strong, low frequency peak around 60 to 90 Hz followed by some higher frequency components (Figure 3.8.12, page 429). The low frequency peak disappears as time progresses. At the end of about 4.5 seconds, the general spectrum has shifted to lower frequencies and has changed its shape. Thus, the wavelet-based analysis confirms the general trend of fatigue shifting the spectra to lower frequencies (Karlsson et al., 2000). However, considering the large change in spectral shape, it can be questioned if the median or mean frequencies are really the appropriate variables to describe fatigue related phenomena.

EMG during cycling

Power-wavelets are able to detect frequency shifts of a muscle during selected dynamic activities (von Tscharner, 2002). The example of cycling (Figure 3.8.29) for the gastrocnemius muscle shows that the muscular activity during a cycling movement occurs at certain time points, thus reflecting muscular events. A muscular event uses a specific set or a specific distribution of motor units and shows the spectrum resulting from this combination. The results in this experiment clearly show that the intensity reflecting muscle activity starts at low frequencies (low wavelets) and gradually changes to higher frequencies between about 160 and 300 degrees.

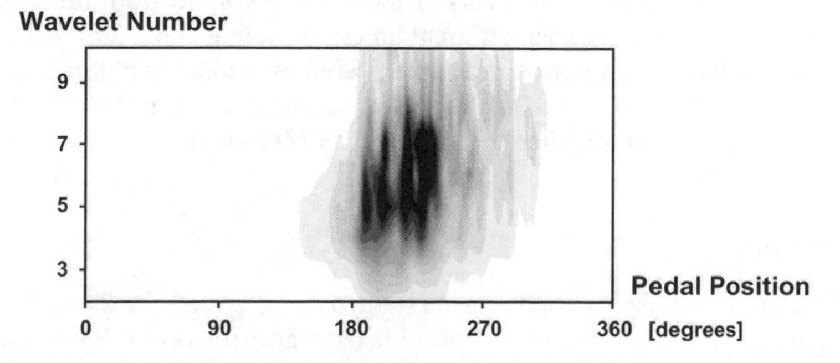

Figure 3.8.29 **Illustration of the wavelet scalogram for the muscle activity of the gastrocnemius medialis muscle during one pedal revolution, while cycling on an ergometer. (0 degrees represents the rear horizontal position).**

Frequency spectrum and muscle fibre activation

Morlet type wavelets were used to measure short time mean frequency or instantaneous mean frequency (Karlsson & Gerdle, 2001). Linear increasing instantaneous mean frequencies with increasing force levels were observed and the slope and the intercept at 0 force level were determined. A detailed knowledge of the fibre type composition of the vastus lateralis was available from biopsies. The presence of type 2a and 2c muscle fibres correlated positively with the instantaneous mean frequencies. A principal component analysis revealed that the intercept of the regression line correlated especially well with the area of the type 2 fibres. This indicated that the type of activated fibres significantly defined the spectra of the EMG. It became, therefore, tempting to study if frequency changes are a result of the composition of activated fibre types.

Experiments on rats revealed that it was possible to generate intensity patterns that revealed motor unit activity at high frequencies and at low frequencies at different times indicating that the two types of motor units created separate spectra (Wakeling & Syme, 2002). Additionally, experiments with fish, who have geometrically separated fast and slow twitch fibres, showed that these two fibre types have spectra with significantly different shapes, and that the frequency of the maxima of the spectrum of the slow twitch fibres was about a third of the frequency of the maximum of the spectrum of the fast twitch fibres (Wakeling et al., 2002). In the same study, the authors showed that the intensity spectra of cats were strongly dependent on whether a fast or a slow movement was required. Thus, the animal studies confirmed that spectral differences observed in intensity patterns can result from task dependent recruitment of different fibre types.

Human muscles typically contain a combination of fast and slow twitch fibres. It is, therefore, extremely difficult if not impossible to assess these groups of fibres separately. An indirect possibility to estimate how much EMG intensity is caused by each of the two main fibre types derives from analysis of the EMG spectra (von Tscharner & Goepfert, 2006). The method demands that effects of fatigue and changes in muscle length can be neglected. For this condition, large spectral changes are used to assign different groups of motor units to groups containing different fibre type compositions. This method is not well-developed and needs further work.

EMG during running

Low and high frequency components of the EMG recorded from the biceps femoris muscle while running were found at different times (Wakeling et al., 2001). The frequency content of the tibialis anterior muscle is relatively high immediately before heel-strike (von Tscharner et al., 2003) (Figure 3.8.30), suggesting that more fast twitch muscle fibres are involved in the pre-activation. After heel-strike, the EMG intensity is at a lower frequency than before heel-strike.

Muscle atrophy

Details relating to muscle atrophy can be studied using wavelet analysis. Muscles of the lower leg get atrophic when a subject has ankle osteoarthritis (OA). EMG analysis using wavelet methodology showed substantial frequency shifts in some of these muscles. It was speculated that a likely reason for these frequency shifts was the predominant atrophy of fast twitch muscle fibres.

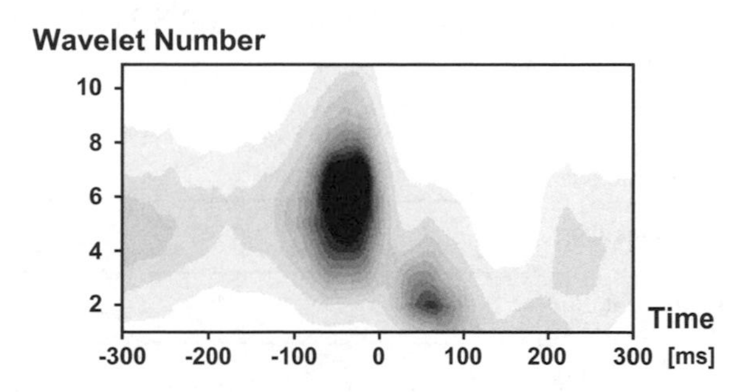

Figure 3.8.30 **EMG scalograms (averaged over 40 runners) for the tibialis anterior. Heel-strike is at the time 0. The period from 0 to 190 ms represents stance phase.**

Motor unit conduction velocity

Among other factors, motor unit conduction velocity (MUCV) determines the shape of frequency spectra. It is, therefore, interesting to separate spectral changes caused by motor unit conduction velocity from spectral change caused by other factors. The wavelet methodology allows filtering the EMG at various frequencies. The EMG signals recorded with a tri-electrode setup from the abductor pollicis brevis (Figure 3.8.9) illustrates the potential of the wavelet methodology. There is a strong correlation between the electrode pair (2, 3) closer to the innervation zone, the source signal, and the electrode pair (1, 2) that was farther away from the innervation zone, the response signal. A combination of the intensity and the phase aspect was used to extract conduction velocities at different frequencies. This, and other examples, illustrate that wavelet methodology reveals many details about spectral aspects of the EMG.

Pattern recognition of EMG

Often, there is an interest in gaining information about movement pattern. Pattern recognition of EMGs becomes possible by using intensity pattern as input for a principle component analysis and applying pattern recognition methods and neural networks (von Tscharner & Goepfert, 2003). A first attempt for this approach was made when changes in whole intensity patterns while cycling on a cycle ergometer were studied (von Tscharner, 2002). The results showed changes in timing of the muscular events. Furthermore, changes in intensity and frequency of muscular events progressed in a systematic way with time. This approach reveals detailed information for changing from a rested into an increasingly fatigued physical condition.

Multi-muscle Intensity pattern

When studying complex movement behaviour, multi-muscle intensity pattern appear to be appropriate. For running, such a multi-muscle intensity pattern has been quantified for the gastrocnemius medialis, tibialis anterior, hamstring, rectus femoris, and vastus lateralis (Figure 3.8.31) (von Tscharner & Goepfert, 2003). Figure 3.8.31 allows for a global interpretation of a multi-muscle problem.

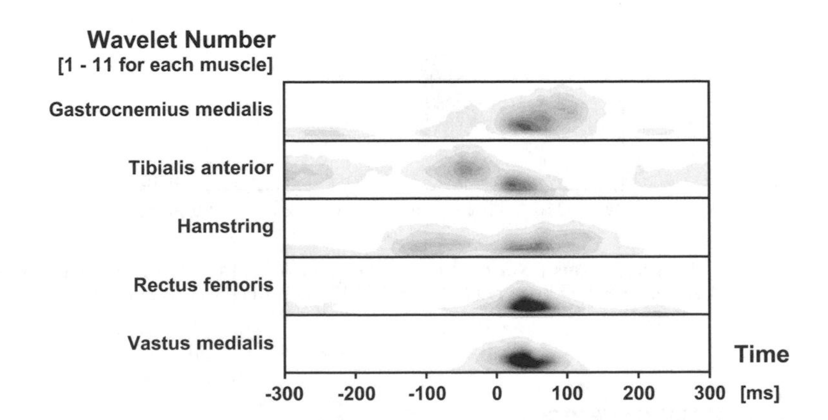

Figure 3.8.31 **Illustration of a multi-muscle intensity pattern obtained as an average over 40 subjects while heel-toe running (barefoot). The pattern includes the scalograms for the gastrocnemius medialis, tibialis anterior, hamstring, rectus femoris, and vastus lateralis.**

An analysis of such multi-muscle intensity patterns revealed differences in timing and frequency content between male and female runners. Pattern recognition methods were applied and it could be determined if a runner was female or male. In turn, this indicated that the differences of the intensity were global over the whole pattern and highly significant. The example showed that wavelet transformed EMGs of multiple muscles yield a new tool in biomechanics to study and discriminate movements involving many muscles simultaneously.

At the time this chapter was written, the examples given herein did not represent generally applied methods. The examples show a concept of EMG analysis based on wavelets that the authors believe might significantly improve future research of human movements. If this chapter stimulates students to progress with new methods and improve current methods, it has served its purpose.

3.9 INERTIAL PROPERTIES OF THE HUMAN OR ANIMAL BODY

NIGG, B.M.

3.9.1 INERTIAL PROPERTIES

The inertial properties of the human body, such as mass, centre of gravity, and mass moments of inertia, are often required for quantitative analyses of human motion. This section discusses techniques developed for obtaining data for calculating inertial properties. The discussion concentrates on theoretical considerations, experimental methods, and theoretical methods.

THEORETICAL CONSIDERATIONS

Inertial properties are defined below for two- and three-dimensional cases. Consider first a general case (Figure 3.9.1), in which the origin of a rectangular coordinate system (x, y, z) is located at the centre of mass, CM, of a rigid body.

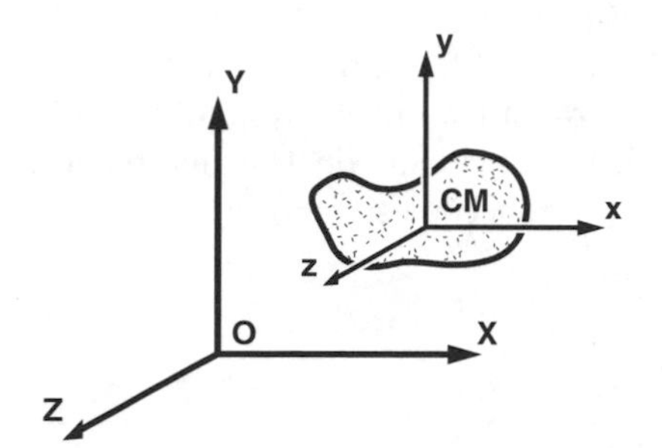

Figure 3.9.1 **A rigid body with a coordinate system with its origin in CM = (x_{cm}, y_{cm}, z_{cm}), and a general coordinate system with its origin in O = (0, 0, 0).**

In the three-dimensional case, where an object moves in all three dimensions, rotation can occur about any axis in space. In this situation, the mass moment of inertia about the centroid of the total mass is given by the matrix:

$$I^{CM} = \begin{bmatrix} I_{xx}^{CM} & -I_{xy}^{CM} & -I_{xz}^{CM} \\ -I_{yx}^{CM} & I_{yy}^{CM} & -I_{yz}^{CM} \\ -I_{zx}^{CM} & -I_{zy}^{CM} & I_{zz}^{CM} \end{bmatrix} \qquad (3.9.1)$$

I_{xx}^{CM}, I_{yy}^{CM}, I_{zz}^{CM} are the moments of inertia about the x, y, and z axes, respectively, through the CM.

where:

$$I_{xx}^{CM} = \int (y^2 + z^2)dm$$

$$I_{yy}^{CM} = \int (x^2 + z^2)dm$$

$$I_{zz}^{CM} = \int (x^2 + y^2)dm \tag{3.9.2}$$

I_{xy}^{CM}, I_{yx}^{CM}, I_{xz}^{CM}, I_{zx}^{CM}, I_{yz}^{CM} and I_{zy}^{CM} are the products of inertia.

where:

$$I_{xy}^{CM} = I_{yx}^{CM} = \int xy\,dm$$

$$I_{xz}^{CM} = I_{zx}^{CM} = \int xz\,dm$$

$$I_{yz}^{CM} = I_{zy}^{CM} = \int yz\,dm$$

For a general moment of inertia for a point A with A ≠ CM, the moment of inertia can be determined using the parallel axis theorem.

For practical purposes, chose a coordinate system with the axes coinciding with the principle axes. In this simplification, the products of inertia become zero and the moment of inertia about the centroid (CM) becomes:

$$I^{CM} = \begin{bmatrix} I_{xx} & 0 & 0 \\ 0 & I_{yy} & 0 \\ 0 & 0 & I_{zz} \end{bmatrix} \tag{3.9.3}$$

To simplify, reduce the space of interest to two dimensions, assuming:

- A two-dimensional problem in the x, y plane,
- A symmetrical object,
- That the moment of inertia with respect to CM is of interest, and
- That the axis of rotation is perpendicular to x, y plane.

Equation (3.9.1) can then be written as:

$$I^{CM} = \int r^2\,dm = \int (x^2 + y^2)\,dm \tag{3.9.4}$$

The determination of the total moment of inertia about the centroid requires detailed information on the location and mass of each differential mass element, dm, under consideration. Usually, information of this detail is unavailable. The procedure can often be further simplified by assuming that the body is made of a homogeneous material of uniform density, ρ, which simplifies equation (3.9.4) to:

$$I^{CM} = \rho \int r^2 \, dV \qquad (3.9.5)$$

where:

dV = differential volume element

The moment of inertia can be determined by direct measurement or through calculation. The various inertial parameters that may need to be determined, depending upon the calculation method used are:

m = total mass of the object of interest
dm = differential mass element
r = perpendicular distance of a mass element from the axis under consideration
ρ = density of the material
dV = volume element corresponding to the mass elements

x_{CM}, y_{CM}, z_{CM} } coordinates of the CM of the object of interest

x_A, y_A, z_A } coordinates of an arbitrary point A

$x_{CA} = x_{CM} - x_A$
$y_{CA} = y_{CM} - y_A$
$z_{CA} = z_{CM} - z_A$

The next sections discuss various experimental and theoretical methods of determining inertial parameters.

EXPERIMENTAL METHODS

This section discusses experimental methods used to determine mass, centre of mass, centre of gravity, and moments of inertia. Historically, the first experimental attempts to determine these values were made on cadavers. Braune and Fischer (1889), Dempster (1955), Clauser et al. (1969), and Chandler et al. (1975) used cadaver studies to determine mass, volume, density, and centre of mass of the total human body and selected segments.

Mass and density

The determination of the *mass* is defined as:

$$m = \frac{W}{g}$$

where:

m = mass of object of interest
g = acceleration due to gravity
W = weight of object of interest

The definition of the object of interest is a problem in biomechanical applications. For example, because the human thigh is not well defined researchers must answer many questions arbitrarily. For example: How much of the muscles crossing the knee or the hip should be included? Should the cuts be made in an extended or a flexed position? Should the cuts be perpendicular or inclined? Similar questions have been answered arbitrarily in the past.

The *average density*, ρ, of an object is defined as:

$$\rho = \frac{m}{V} \qquad (3.9.6)$$

where:

ρ = average density of the object of interest
m = mass of the object of interest
V = volume of the object of interest

Two following two methods are commonly used to determine the average density of an object:

- In the first method, mass is determined by weighing the object in air. By lowering the object into water and weighing the displaced water, the object's volume can be determined (after correcting for temperature differences). Mass and volume are used to determine the density.
- In the second method, the object is weighed first in air and then in water. The density can then be determined using:

$$\rho = \frac{W_{air} - W_{water}}{V \cdot g} \qquad (3.9.7)$$

where:

ρ = density of the object of interest
W_{air} = weight of the object in air
W_{water} = weight of the object in water
V = volume of the object of interest
g = acceleration due to gravity

The volume displacement method was used to determine the volumes of segments of living people (Dempster, 1955; Clauser et al., 1969). The subjects lowered the limb into the measuring tank to a given depth, and the displaced water was measured. The volume displacement method proved less reliable for living people than for cadavers, with repeat measurement variations of 3 to 5%. These variations were attributed to difficulty keeping the subject's limb stationary while the displaced volume was drained.

Densities determined with the above methods give an average density for the entire segment. However, the individual tissues that make up a human segment have different densities. Cadaver segments have been dissected to determine bone, muscle, and fat densities. Volumes of the individual tissues can be determined through the use of methods such as computerized axial tomography or magnetic resonance imaging. These volumes can then be used to determine overall segment masses (indicated later in Table 3.9.2).

The physical properties of muscle, skin, adipose tissue (fat), and bone are different. Experimental data relative mass and density for various body segments from six embalmed cadavers (Clarys & Marfell-Jones, 1986) is summarized in Table 3.9.1. The density of skin is relatively constant for the four locations tested, and differences are smaller than 1%. Adipose density varies by about 4%, with the lowest values measured for the arm and the highest values for the foot. Muscle density varies by about 2%, with the lowest values measured for the foot and the highest for the forearm. Bone density varies by 14%, with the highest values measured for the forearm and the lowest for the foot. The high variation of bone density suggests that these differences should be included in precise calculations of inertia parameters.

Table 3.9.1 Relative mass and density for arm, forearm, leg, and foot based on data from six embalmed cadavers (from Clarys & Marfell-Jones, 1986, with permission).

BODY SEGMENT	TISSUE	RELATIVE MASS [%]	DENSITY [gcm^{-3}]
ARM	skin	6.90	1.050
	adipose	36.45	0.954
	muscle	41.85	1.049
	bone	14.80	1.224
FOREARM	skin	8.35	1.051
	adipose	23.55	0.961
	muscle	51.50	1.054
	bone	16.60	1.308
LEG	skin	6.00	1.055
	adipose	28.95	0.958
	muscle	42.70	1.042
	bone	22.35	1.2075
FOOT	skin	13.25	1.0586
	adipose	30.95	0.992
	muscle	24.70	1.037
	bone	31.10	1.1525

Centre of gravity and centre of mass

The centre of gravity of an object is the location of the centroid of all gravitational forces, G, of its mass elements, dm:

$$\mathbf{r}_{CG} = \frac{\int \mathbf{r}\, dG}{\int dG} \tag{3.9.8}$$

The centre of mass of an object is the location of the centroid of all mass elements, dm:

$$\mathbf{r}_{CM} = \frac{\int \mathbf{r}\, dm}{\int dm} \tag{3.9.9}$$

The locations of the centre of gravity and the centre of mass are the same in a first approximation. However, a more detailed look at the centre of gravity and the centre of mass of a skyscraper illustrates the difference between these two properties (Figure 3.9.2).

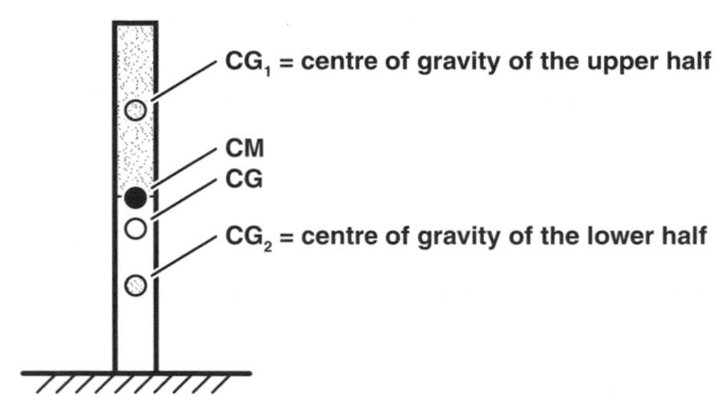

Figure 3.9.2 **Illustration of the difference between centre of gravity, CG, and centre of mass, CM, of a long object, e.g., a skyscraper, exposed to gravity.**

The centre of mass of a skyscraper is at the centre of the structure. Consider a skyscraper's upper and lower half separately. Let the centre of gravity of the upper half be at CG_1 and the centre of gravity of the lower half at CG_2. Since CG_2 is closer to the earth's centre than CG_1, the force acting at the lower half is greater than the force acting at the upper half. Consequently, the CG of the whole structure is closer to the ground than the CM. However, in general, these differences are small and can be neglected.

Several methods have been used to determine the centre of gravity (Table 3.9.2). Braune and Fischer (1889) used thin metal rods driven into frozen cadaver segments at right angles to the three cardinal planes. The segments were suspended from each rod and the rod's planes of intersection with the segment were marked. The intersection of the three planes defined the centre of gravity.

Dempster (1955) used a balance plate to locate the plane of the centre of gravity along the longitudinal axis relative to the ends of the segments. This plate consisted of a plate positioned so that it could pivot around the turned down-ends of one of the diagonals. Similar balance plate configurations can be used on living people to determine the whole body centre of mass, but not segmental centroids.

The moment table technique is more commonly used than the balance plate. A table (or plate) is supported at one end by a knife edge on a metal stand, and at the other end by a knife edge placed on a scale (Figure 3.9.3). If the weight of the object under consideration is known, the location of the centre of gravity can be found by determining moments about the metal stand end, using the force reading from the scale and the distance between the knife edges.

Table 3.9.2 Relative segment mass and location of centre of gravity from selected cadaver studies. Relative segment masses are given as a percentage of total body mass. Location of the centre of gravity is given as a percentage of the distance along the longitudinal axis from the proximal joint.

SEGMENT	RELATIVE MASS $\left(\frac{m_i}{m}\right)$ RELATIVE DISTANCE $\left(\frac{L_i}{L}\right)$									
	HARLESS (1806)		BRAUNE & FISCHER (1889)		FISCHER (1906)		DEMPSTER (1955)		CLAUSER (1969)	
	$\frac{m_i}{m}$ [%]	$\frac{L_i}{L}$ [%]	$\frac{m_i}{m}$ [%]	$\frac{L_i}{L}$ [%]	$\frac{m_i}{m}$ [%]	$\frac{L_i}{L}$ [%]	$\frac{m_i}{m}$ [%]	$\frac{L_i}{L}$ [%]	$\frac{m_i}{m}$ [%]	$\frac{L_i}{L}$ [%]
HEAD	7.6	36.2	7.0		8.8		7.9		7.3	
TRUNK	44.2	44.8	46.1		45.2		46.9		50.7	
ARM	5.7		6.2	52.6	5.4	44.6	4.9	51.2	4.9	41.3
UPPER ARM	3.2		3.3	47.0	2.8	45.0	2.7	43.6	2.6	51.3
FOREARM & HAND	2.6		3.0	47.2	2.6	46.2	2.2	67.7	2.3	62.6
FOREARM	1.7	42.0	2.1	42.1			1.6	43.0	1.6	39.0
HAND	0.9	39.7	0.8				0.6	50.6	0.7	
LEG	18.4		17.3	40.7	17.6	41.2	15.7	43.4	16.1	38.2
UPPER LEG	11.9	48.9	10.7	43.9	11.0	43.6	9.6	43.3	10.3	37.2
LOWER LEG & FOOT	6.6		6.5	51.9	6.6	53.7	5.9	43.4	5.8	47.5
LOWER LEG	4.6	43.3	4.7	42.0	4.5	43.3	4.5	43.3	4.4	37.1
FOOT	2.0	44.4	1.7	43.4	2.1		1.4	42.9	1.5	44.9

$$x = \frac{L \cdot W_{scale}}{W_{object}}$$

where:

x	=	distance between the location of the centre of gravity of the object and the joint pivot axis A
L	=	distance between the knife edges
W_{scale}	=	reading of the scale
W_{object}	=	weight of the object of interest

Note, the scale is set to zero for the initial setup where the table is unloaded.

Van den Bogert (1989) used the photographic suspension method, which was initially developed by Braune and Fischer (1889), to determine the centre of gravity. The segment is hung from a rope attached to a point on the segment. The segment and rope are then photographed and the whole process is repeated using a second suspension point on the segment. The two photographs are overlaid and the centre of gravity located at the point of intersection of the extrapolations of the ropes.

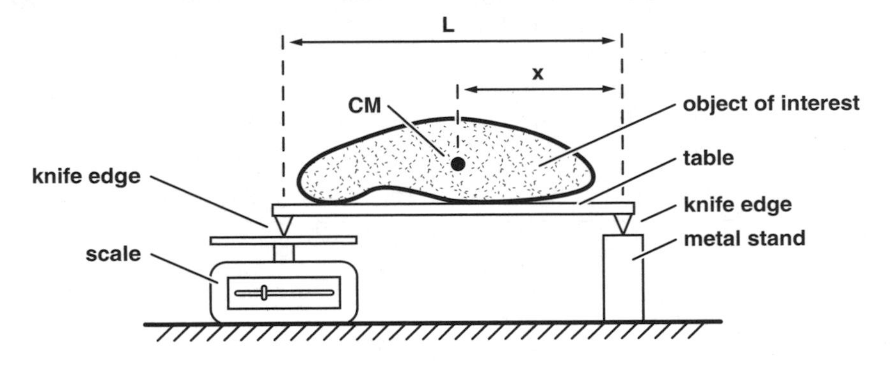

Figure 3.9.3 **Illustration of a moment table used to determine the centre of gravity.**

Values (as found by Harless, 1860; Braune & Fischer, 1889; Fischer, 1906; Dempster, 1955; and Clauser et al., 1969) for the related masses and for the location of the centre of gravity of a segment, as a percentage of the distance along the longitudinal axis of the segment relative to the proximal end, are given in Table 3.9.2.

Moments of inertia

Several methods have been proposed to experimentally determine moments of inertia. The methods use a pendulum or a torsional pendulum approach.

The pendulum method consists of suspending the object of interest from a fixed point, setting it in motion by shifting it a few degrees from its equilibrium position, and measuring the time it takes to swing for one period of oscillation. The moment of inertia I_o about an axis through the point of suspension I_o, is given by:

$$I_o = \frac{W_{air} \cdot h \cdot T^2}{4\pi^2}$$

where:

I_o = moment of inertia about an axis through the point of suspension
W_{air} = weight of the object in air
h = distance from the centre of gravity of the object to the point of suspension
T = period of one oscillation

The moment of inertia for an axis through the centre of mass, I_{CM}, can be determined using the parallel axis theorem:

$$I_{CM} = I_o - mh^2 \qquad (3.9.10)$$

where:

m = mass of the object of interest

Several authors have used pendulum methods to determine the moment of inertia of the whole body and of body segments.

Dempster (1955) screwed the proximal joint centres of a frozen limb segment onto a bar with knife edges on both ends. The segment was supported in a metal frame by the bar while the knife edges sat upon the frame. The segment was allowed to swing freely through an arc of approximately 5° while pivoting on the knife edges. The period of oscillation was measured, and the moment of inertia about the centre of gravity calculated from equations (3.9.10) and (3.9.11). The trunk, with the head and neck attached, was frozen and suspended by placing the bar with the knife edges through a horizontal bore hole directed from the centre of one acetabulum to the other. The moment of inertia was determined for the segments. The shoulder portions were held in their natural configurations using plaster of paris moulds while being frozen. The frozen shoulder was then screwed to the bar at the medial end of the clavicle and the moment of inertia determined as above. Table 3.9.3, provides a summary of the moments of inertia determined by Dempster, along with a comparison to values determined by Hatze (1980).

Chandler et al. (1975), determined the three principle moments of inertia and the products of inertia of various body segments using a specimen holder box. The segment to be measured was frozen and strapped into the box. The moment of inertia and products of inertia were determined using the pendulum technique. Inertia was determined by swinging the box around the coordinate system's x, y, and z axes, around an xy axis (located in the xy plane), a yz axis (located in the yz plane), and around an xz axis (located in the xz plane). The distance from the centre of mass of the segment to each of the swing axes was measured using a photographic suspension method. From these measurements, and knowledge of the orientation of the box coordinate system relative to the segment coordinate system, the three principle moments of inertia and direction angles were calculated. This procedure provided an inertial tensor for the centre of gravity.

Later Chandler and colleagues evaluated the procedure using a solid aluminum bar whose inertial properties were known (Chandler et al., 1975). Six trials produced deviations from the theoretical moment of inertia values that ranged from −10.4 to +3.8%. The deviations clustered in the range −3 to +3%, indicating a reasonable degree of accuracy for the method. Their results, however, are not included in Table 3.9.3, because they do not relate directly to the coordinate system used by other authors.

Lephart (1984) refined the method developed by Chandler et al. (1975) by correcting for a systematic error, which over-predicted the oscillation time, and therefore overestimated the moment of inertia. With this correction, the absolute error for the principle moments of inertia was less than 5%.

Santschi (1963) determined the moments of inertia for 66 living male subjects in the following eight body positions: standing, standing with arms over head, spread eagle, sitting, sitting with forearms down, sitting with thighs elevated, mercury configuration, and relaxed. The accuracy for the moment of inertia values ranged from 2 to 8%.

Table 3.9.3 Experimentally determined values for segmental transverse moment of inertia (based on data from Dempster, 1955; Hatze, 1980, with permission).

SEGMENT	DEMPSTER (1955)	HATZE (1980)
	TRANSVERSE MOMENT OF INERTIA [kgm^2]	TRANSVERSE MOMENT OF INERTIA [kgm^2]
HEAD - NECK	0.0310	0.0337
LEFT ARM	0.0222	0.0203
RIGHT ARM	0.0220	0.0229
LEFT FOREARM	0.0055	0.0086
RIGHT FOREARM	0.0072	0.0093
LEFT HAND	0.0009	0.0010
RIGHT HAND	0.0011	0.0010
LEFT LEG	0.0650	0.0798
RIGHT LEG	0.0620	0.0747
LEFT FOOT	0.0037	0.0051
RIGHT FOOT	0.0040	0.0051

Van den Bogert (1989) introduced the *torsional pendulum method*, which used a vertically oscillating turntable to measure moments of inertia for horse segments. The turntable consisted of a vertical shaft supported by bearings that held cross members holding the segment of interest. Springs attached to a fulcrum on the shaft maintained the oscillation. A segment was placed on the cross members such that the plane in which it normally moved was horizontal (sagittal plane). Photographs taken from above were used to determine the distance from the centre of gravity to the axis of rotation. The centre of gravity was determined using the photographic suspension method. The moment of inertia, I, in the sagittal plane was calculated from:

$$I_{CG} = k(T^2 - T_o^2) - md^2 \quad (3.9.11)$$

where:

I_{CG} = moment of inertia for the segment for an axis through the CG of the segment perpendicular to the sagittal plane
k = rotational spring constant divided by $4\pi r^2$
T = measured period of oscillation for the table and segment
T_o = measured period of oscillation for the unloaded table
m = mass of the segment
d = distance between the axis of the rotating table and the CG

The rotational spring constant, k, was determined by measuring the oscillation time of eight steel cylinders of known moments of inertia, and calculating the best fit to the measured T and known I values. Using the calibrated spring constant, k, moments of inertia of the calibrated objects could be determined with an accuracy of 0.8%.

Moments of inertia are often important for human movement analysis, especially during sport. Consequently, moments of inertia have been estimated for various body positions and body axes. Examples of such estimates (Donskoi, 1975) are illustrated in Figure 3.9.4 for body positions and rotation axes of the human body that are typical in gymnastics.

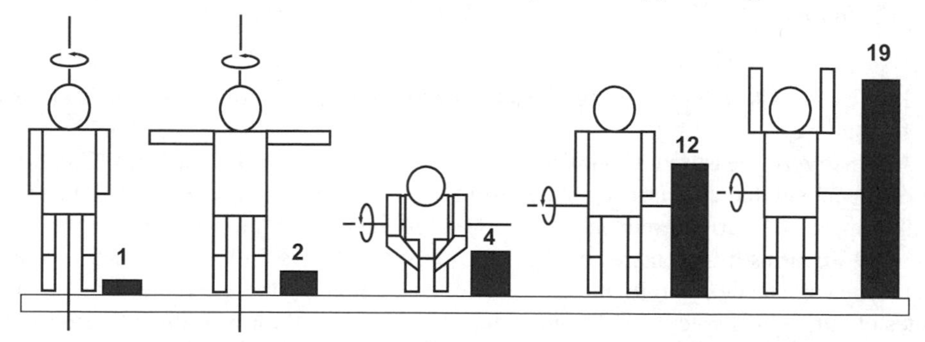

Figure 3.9.4 **Moments of inertia (in kgm^2) for the human body in selected positions for different axes (from Donskoi, 1975, with permission).**

Comprehensive methods

Computerized axial tomography

Computerized axial tomography (CAT) has been used to determining inertial properties of humans or animals (Rodrigue et al., 1983; Huang & Saurez, 1983). CAT takes a three-dimensional x-ray of a small section of the object of interest. Scans are taken at small intervals along the segment, generating a series of crosssections. These crosssections can be digitized, and the various tissue areas (generally broken down into bone, muscle, and fat) can be calculated. The crosssectional area multiplied by section width gives the section volume.

Summations of section volumes yield the segment volume. Densities for each tissue type can be taken from dissection measurements. For living people, densities can be taken from published values or calculated directly from the CAT number (Huang & Saurez, 1983) using the equation:

$$\rho = A \cdot N_{CAT} + B \tag{3.9.12}$$

where:

ρ = density of the segment of interest
N_{CAT} = CAT number of the tissue, as indicated by Huang and Saurez (1983)
A, B = coefficients

A and B can be calculated during a calibration procedure (Huang & Saurez, 1983) from:

$$A = \frac{(\rho_2 - \rho_1)}{(n_2 - n_1)}$$

$$B = \rho_1 + n_1 \frac{(\rho_2 - \rho_1)}{(n_2 - n_1)}$$

where:

ρ_1, ρ_2 = mass densities of the two known materials used for the calibration process
n_1, n_2 = N_{CAT} for the same materials, respectively

The densities within the crosssections can be calculated on a point-by-point basis using this method.

Alternatively, measured or published values can be used for the densities. Combining these values with the volumes calculated from the digitizing process can determine the inertial properties. Rodrigue and Gagnon (1983) used the CAT method to estimate the densities of 24 forearms from 6 male and 4 female human cadavers. Volume estimates for each tissue type were made by digitizing the CAT scans. Combining these estimates with density values obtained by dissection determined the overall segment mass, segment volume, and average segment density. Estimates of segment density obtained using the CAT method were within 2% of measured values.

Quick release method

Fenn (1938) developed the quick release method, which is an angular motion method, to determine the moment of inertia of the forearm. This method requires the subject's limb to exert a constant force against a device that is then suddenly released, causing the limb to accelerate. The angular acceleration of the limb is measured and the moment of inertia determined from:

$$I = \frac{F \cdot d}{\alpha}$$

where:

F = applied force
d = distance between the point of application of the force and the pivot point
α = measured angular acceleration of the limb

The underlying assumption is that the movement of the limb is purely inertial and that the viscoelastic properties of the limb can be neglected. The mean moment of inertia (in the sagittal plane) of the forearm, as measured by Bouisett and Pertuzon (1967) on 11 subjects, was 0.0599 kgm^2, which is comparable to other published results using different methods (Table 3.9.4). Therefore, the underlying premise is reasonably valid. The variation in results between consecutive tests was only 3%.

Relaxed oscillation method

The basic concept underlying the relaxed oscillation method is applying a single or oscillating force to a resting limb, and determining the moment of inertia from the resulting response of the limb. The method has been applied to the forearm and is generally done on an apparatus that (for the forearm) consists of a beam. A handle is placed at one end of the beam, and an elbow cup is placed at the other end of the beam. The test subject sits beside the apparatus, with the upper arm loosely hanging down and the elbow flexed at 90°. If a force is applied to the arm and beam, they start to oscillate (Peyton, 1986). A small aluminum strip attached to the beam acts as a spring element. A strain gauge mounted on the aluminum strip can be used to measure the movement and the oscillation time, T, which can be used to determine the moment of inertia, I, as:

$$I = \frac{k \cdot T^2}{4\pi^2} \tag{3.9.13}$$

where:

k = relaxation constant determined by statically loading the device and measuring the deflection
T = oscillation time (period)

The moment of inertia of the arm is determined by repeating the procedure with the apparatus empty to determine its moment of inertia, and then subtracting this value from the one determined in equation (3.9.14). Assumptions are: the device and forearm are rigidly coupled, almost all of the elastic deformation is occurring at the spring element, and the damping of the system is small.

Errors are introduced into the method because:

- The axis of rotation of the forearm at the elbow may not coincide with the axis of rotation of the device (up to 5% error),
- Forearm oscillations will be transmitted through the upper arm to the shoulder joint, resulting in a slight underestimation of the moment of inertia, and
- The forearm is not a rigid structure, but this potential source of error can be compensated for by having the subject tense the forearm muscles.

The form of the output signal indicates if the subject is preparing his or her muscles properly by relaxing the joint and tensing the forearm. The mean and range of forearm values determined with this method compare well to others found in literature on this subject (Table 3.9.4).

Table 3.9.4 Experimentally determined moments of inertia for the forearm and hand segment about the humeral axis from selected authors using different methods.

REFERENCES	NUMBER OF SUBJECTS	MOMENT OF INERTIA OF THE FOREARM [kgm^2]	
		MEAN	RANGE
BRAUNE AND FISCHER (1889)	2	0.0505	n. a.
FENN (1938)	1	0.0590	n. a.
HILL (1940)	1	0.0277	n. a.
WILKE (1950)	1	0.0530	n. a.
DEMPSTER (1955)	8	0.0577	0.0397 - 0.0852
BOUISSET AND PERTUZON (1968)	11	0.0599	0.0430 - 0.080
KWEE (1971)	4	n.a.	0.03 - 0.051
ALLUM AND YOUNG (1976)	4	0.0740	0.0690 - 0.0820
PEYTON (1986)	8	0.0646	0.0476 - 0.0862

These values represent the moment of inertia of the entire limb about the humeral axis.

The method was also applied with the limb oscillating continuously, driven by a torque motor (Allum & Young, 1976). The lateral force at the control stick was measured with a strain gauge, linear (tangential) acceleration was measured with a linear accelerometer, and angular position of the beam was measured with a potentiometer. The frequency of oscillation varied from 1.4 to 9.0 Hz, and the amplitudes of acceleration were chosen to increase linearly with the frequency. During testing, subjects were asked to relax their shoulder musculature and maintain a firm wrist. Five cycles of oscillation in this position were recorded, and the average peak-to-peak accelerations for the five cycles were plotted on the graph. The slope of the graph was used as a measure of the subject's forelimb moment of inertia about the humeral axis. Results are compared for forearm plus hand moments of inertia in Table 3.9.4.

Gamma-scanner method

The intensity of a gamma ray beam decreases as it passes through a substance. This intensity reduction can be measured and used to determine the density of a substance. Zatziorsky and Seluyanov (1983) measured 100 subjects using the gamma ray scanner technique. The subject was placed in a reclined position on a couch, and their entire body scanned with a gamma ray emitter moving above the body. A collimated detector, moving underneath the body directly under the emitter measured the reduced gamma ray radiation. A surface density profile for the entire body was produced in this way. The resulting segment mass is comparable to data presented by Dempster (1955), except for the thigh, which is considerably higher for results from Zatziorsky and Seluyanov's (1983) method (Table 3.9.5).

Table 3.9.5 Relative segment mass comparisons for different selected experimental methods. Relative segment masses are given as a percentage of the total body mass. The values for Hatze were calculated based on volumes from Hatze (1980) and average segment densities from Dempster (1955). The difference in the results between Dempster and Zatziorsky & Seluyanov may be explained by the different subjects studied. Zatziorsky & Seluyanov studied athletes.

SEGMENT	RELATIVE SEGMENT MASS [%]					
	DEMPSTER (1955)	CLAUSER ET AL. (1969)	HATZE (1980)	ZATZIORSKY & SELUYANOV (1983)	JENSEN (1986)	JENSEN (1986)
	ADULTS	ADULTS	ADULTS	ADULTS	12 YEARS OLD	15 YEARS OLD
HEAD	7.9	7.3	6.5	6.9	10.1	6.7
TRUNK	49.6	50.7	45.1	43.5	41.7	41.6
THIGH	9.7	10.3	11.9	14.2	11.0	12.1
SHANK	4.5	4.0	5.5	4.3	5.3	5.6
FOOT	1.4	1.5	1.5	1.4	2.1	2.1
UPPER ARM	2.7	2.6	2.9	2.7	3.2	3.5
LOWER ARM	1.6	1.6	1.8	1.6	1.7	1.7
HAND	0.6	0.7	0.6	0.6	0.9	0.8

Magnetic resonance imaging (MRI)

Magnetic resonance imaging (MRI) is a non-invasive, non-radioactive technique for obtaining crosssectional structural images. MRI uses the change in orientation of the magnetic moment of hydrogen nuclei for a particular tissue that is generated when the tissue is placed in a magnetic field and stimulated with a radio frequency wave. The decay signal is measured by a receiver coil. The received signal is processed and generates an image. The image shows the crosssection of the scanned object.

Martin et al. (1989) applied MRI to determine inertial properties for baboon segments. Eight embalmed baboon segments were procured (four forearms, two upper arms, and two lower legs), mounted in a jig parallel to the scanning direction, and longitudinally scanned in 15 mm increments. The segments were weighed in water and in air, and volume and average density were calculated. The centres of gravity were determined using the reaction board method, and the moment of inertia about the traverse axis was measured using the pendulum method. Coordinate data describing the perimeter of the crosssectional areas of fat, muscle, and bone were determined by digitizing their respective boundaries from the MRI image. The crosssectional area was converted to real area using a scaling factor obtained from an image of an object with a known size.

The muscle's crosssectional area was determined by subtracting the fat and bone areas from the segmental area of the particular image. Image sections were represented as frustra of right circular cones to determine slice volume. Tissue densities were determined by dissection for muscle and bone, while fat densities were taken from Clauser et al. (1969). MRI tissue volumes multiplied by tissue densities yielded image section masses, which were summed to determine segment masses. The location of the centre of gravity, with respect to the proximal end of the segment, was determined by multiplying each individual section mass by the distance from the proximal end to the section's centre of gravity. The moment of inertia of the segment about the traverse axis through the segment's centre of gravity, was determined by the sum of all the moments of inertia of the image sections about the segment's centre of gravity, using the parallel axis theorem.

MRI predicted the location of the centre of gravity relative to the proximal end well (43.3% versus 44.6% measured), and the mean segment density exactly (1.124 kg/m^3) relative to measured values. MRI overestimated segment volumes (633 cm^3 versus 595 cm^3), segment masses (770 g versus 720 g), and moments of inertia (0.00333 kgm^2 versus 0.00321 kgm^2). A thin layer of embalming fluid between the segment and the plastic bag containing the segment may have introduced volumetric errors of +7%, because the fluid looked exactly like skin on the MRI image. When this overestimation was taken into account, MRI tended to slightly underestimate the moment of inertia. Additional errors could occur because of the difficulty in distinguishing between cancellous and cortical bone (with largely differing densities) near segment end points, and from tissue movement during moment of inertia measurements.

Overall, the accuracy of MRI is better than or similar to other methods discussed so far. It appears to be an accurate, non-invasive technique for determining inertial properties and will likely be used often in the future.

Volume contour mapping

Weinbach (1938) developed volume contour maps from front and right side view photographs of an individual. The authors assumed that a horizontal section through the body is elliptical in shape. A contour map was prepared by plotting the crosssectional area of a given section through the body on the ordinate of a graph and plotting the distance from the soles of the feet to the given section on the abscissa of the graph. The body's volume of the body was given by the area under the curve. Further graphical manipulation allowed the calculation of the centre of gravity and moment of inertia. See Weinbach (1938) for a detailed description of this method.

THEORETICAL METHODS

The experimental methods discussed in the last few pages seem appropriate to determine relative masses, location of segmental centres of mass, and volumes. However, Hatze (1986) has suggested that theoretical approaches and anthropomorphic models are better at experimentally determining principal moments of inertia and the inclination of the principal axes.

Instead of experimentally determining the inertial properties, the human body can be represented as a mathematical model, allowing the inertial properties to be determined mathematically. Five theoretical approaches are discussed as examples of alternative modelling methods:

- Hanavan model (Hanavan, 1964),
- Photogrammetric method (Jensen, 1978),
- Hatze model (Hatze, 1980),
- Yeadon model (Yeadon, 1989a, b), and
- Regression equations.

Mathematical models allow the calculation of inertial properties of body segments and of the whole body in many positions. Therefore, mathematical models can be used to simulate human movement. In general, anthropometric measurements are taken of a subject directly, or from photographs that are later digitized, to determine dimensions used as input into the model.

The Hanavan model

The Hanavan model (1964) is made of fifteen simple geometric solids (Figure 3.9.5).

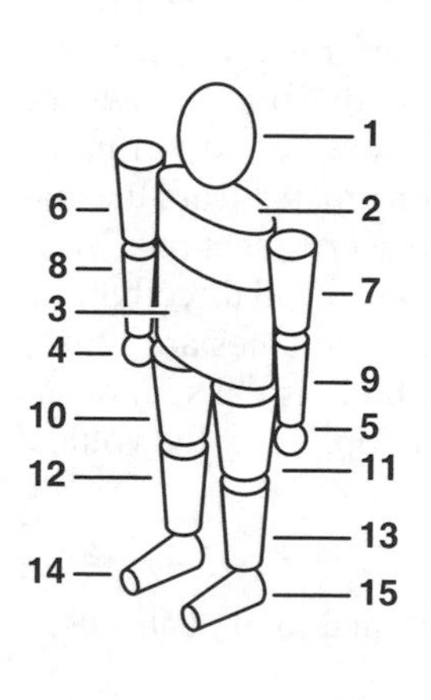

SEGMENT	NUMBER(S)	CONSTRUCTION
HEAD	1	right circular ellipsoid
UPPER TORSO	2	right elliptical cylinder
LOWER TORSO	3	right elliptical cylinder
HAND	4, 5	sphere
UPPER ARM	6, 7	frustum of right circular cone
LOWER ARM	8, 9	frustum of right circular cone
UPPER LEG	10, 11	frustum of right circular cone
LOWER LEG	12, 13	frustum of right circular cone
FOOT	14, 15	frustum of right circular cone

Figure 3.9.5 **Schematic illustration of the Hanavan model with description of specific segments. (from Hanavan, 1964, with permission).**

Twenty-five anthropometric measurements were taken from each individual subject, and used to tailor the geometric solids. The model was based on several assumptions:

- The human body can be represented by a set of rigid bodies of simple geometric shapes and uniform density,
- The regression equations used for the segment weights are representative of the spectrum of body weights of interest,
- The limbs move about fixed points when the body changes position, and
- The limbs are connected by massless hinge joints.

Centre of gravity, mass, and moments of inertia of each segment were derived from the geometry of each of the simple solids. The Barter regression equations (Barter, 1957) were used to calculate the segment weights. If the calculated total body weight did not equal the measured body weight, the difference was distributed proportionally over the segments. Cardan angles were used to describe the position of each of the moveable segments of the body with respect to the body coordinate system.

The model was tested using data provided by Santschi et al. (1963). Anthropometric measurements from 66 subjects were used. Centre of gravity location and principle moments of inertia were calculated for eight body positions and compared to measured data from Santschi et al. (1963). The mathematical model predicted the centre of gravity of the total body within 1.8 cm, and the moments of inertia within 10% of measured values, which may be affected with errors.

Photogrammetric method

The human body may be modelled using elliptical zones (Jensen 1976, 1978, 1986, 1989). In the photogrammetric method, the entire body is sectioned (mainly in the traverse plane) into zones 2 cm wide, and sixteen segments: head, neck, upper trunk, lower trunk, upper arm, lower arm, hand, upper leg, lower leg, and foot. Segment division follows the basic procedures described in Dempster (1955), but with sections modified to fall in the transverse plane. Each zone is represented by an ellipse, which is constructed from the major and minor axes of the two neighbouring planes, and is, therefore, a crosssectional average area of the zone. The centroid and volume of each zone can be calculated directly from the geometry of an elliptical plate. Segment volumes and whole body volumes are simply the sum of the zone volumes. The mass of the segment and whole body is the sum of the mass of each zone (determined by the average density of the zone multiplied by the volume of the zone).
Two assumptions are made:

- The zone centroids lie on the link connecting the distal and proximal joint centroids, and
- The reference orientations of the segment axes are parallel to the body axes.

Because the ellipses are symmetrical, segment axes and the body axes become principle axes. The mass moment of inertia can be calculated from the moment of inertia of an elliptical plate, and the body mass moments of inertia can be determined using the parallel axis theorem to sum all the segment moments of inertia.

When the method is applied, a subject lies prone upon blocks in a manner that allows all segments to be viewed from the top and side, i.e., the subject hyperextendd the neck, planter flexes the foot, and extends the fingers. The segments are positioned parallel to the body axis. Horizontal and vertical grids are marked on surfaces adjacent to the subject and photographed with the subject. The photographs are digitized and segmented. Whole body and segment inertial parameters are calculated from the digitized records. Segment average densities are taken from published sources.

The body mass calculated using the photogrammetric method is generally within 2% of the measured body mass, which indicates a high degree of model accuracy. The results of a comparison of segment mass calculations and other studies are given in Table 3.9.5. According to Jensen, application of the method is quick, requiring about ten minutes per subject to mark reference points on the body and photograph the individual. Manual digitization requires about two hours.

The Hatze model

Hatze (1980) developed a 17 segment model offering the following improvements over past models. The Hatze model:

- Includes the shoulders as separate entities,
- Differentiates between male and female subjects,
- Considers actual shape fluctuations of each individual segment,
- Accounts for varying densities across the crosssection and along the longitudinal axis of the segment,
- Adjusts the density of certain segments according to the value of a special subcutaneous fat indicator,
- Does not assume that segments are symmetrical,
- Accounts for body morphological changes such as obesity and pregnancy,
- Is valid for children, and
- Models the lungs at a lower density.

Representation of a subject with the model requires 242 separate anthropometric measurements. The greatest simplifying assumption used in the model is the assumption that the segments are rigid. See Hatze (1980), and other publications by Hatze referred to therein, for a detailed description of how segments are built up from various geometric elements, and a sample of anthropometric measurements and corresponding calculations.

Hatze used four subjects to test the model. The mean error for total body mass predication was 0.26% with a maximum error of 0.52%. This indicates an accurate estimate of body mass. The values for moment of inertia about the transverse plane of one subject in the test group were compared to those for a similarly structured cadaver from the Dempster (1955) study. The results are comparable and appear in Table 3.9.3.

Direct comparisons made against various segment inertia properties measured directly from the subject indicated that the overall agreement of the model with experimental data was about 3%, with a maximum error of about 5%.

The Yeadon model

Yeadon (1989a, b) described an 11 segment model (segmental inertial parameters can be calculated for 20 separate body segments) using 40 separate solids. Ninety-five anthropometric measurements were taken from an individual and used to define the shape of the model. In developing his model, Yeadon assumed:

- That the segments are rigid bodies,
- That no movement occurs at the neck, wrists or ankles,

- That the solids comprising a segment have coinciding longitudinal axes, and
- That density values are uniform across each solid.

Segment masses, locations of centroids, and principle moments of inertia about the centroids are calculated based on the geometry of the solid.

The Yeadon model predicted the total body masses of three subjects within 2.3%. If a correction is made for air contained in the lungs (instead of uniform thoracic density) the total body mass error is reduced to approximately 1%. Yeadon (1989a, b) did not provide information about the accuracy of other inertial properties.

Regression equations

Regression equations can provide a quick and easy way to determine inertial parameters for an individual. Regression equations are developed from a group of baseline measurements on cadavers or living people, and used to extrapolate internal properties to individuals outside the baseline group. Therefore, regression equations avoid some time-consuming procedures. Regression equations developed by Barter (1957); Clauser et al. (1969); Hinrichs (1985); Yokoi et al. (1985); Zatziorsky and Seluyanov (1985); Ackland et al. (1988); and Yeadon and Morlock (1989) are discussed. Actual equations and anthropometric measurements (where applicable) can be found in the relevant publications.

Barter (1957), attempting to overcome the limitations imposed by small sample sizes of the Braune and Fischer (1889) and Dempster (1955) studies, combined the results of all available studies. Barter took the masses of selected segments, applied statistical regression analysis, and derived regression equations that can be used to determine the various segment masses as a function of total body mass. The standard error in his equations ranged from a low of 0.3 kg for both feet to a high of 2.9 kg for the head, neck, and trunk.

Clauser et al. (1969) developed regression equations to determine the mass and location of the centre of gravity for a segment from various anthropometric measurements. The standard error for segment mass ranged from a low of 0.002 kg for the hand to a high of 0.93 kg for the head and trunk. The standard error for the centre of gravity location ranged from a low of 0.16 cm for the forearm to a high of 1.5 cm for the total leg.

Yokoi et al. (1985) applied the photogrammetric method of Jensen (1976) to 184 subjects (93 boys and 91 girls ranging from 5 to 15 years of age) to determine inertial properties. Instead of developing regression equations, the Kaups index:

$$k = \frac{BW}{H} \cdot 1000$$

where:

k = Kaups index
BW = body weight in kg
H = height in cm

along with age, gender, and body type (lean, normal or overweight) was used to classify the calculated segment inertial properties (segment mass, centre of gravity, and radius of gyration about the centre of gravity) into 16 groups. The authors only comment on the method's accuracy was that using the above classifications should provide more accurate estimates for children's segmental inertial properties than using adult properties.

Ackland et al. (1988) used the photogrammetric method to determine inertial properties. The 5-year study used 13 adolescent male subjects. Prediction equations were developed from the results of the photogrammetric method, allowing the determination of segmental inertial properties (mass, centre of gravity location, and moments of inertia) from anthropometric measurements. Direct comparison between predicted and actual measured values were not made. However, centre of gravity locations of the thigh (43.6% from proximal end of segment) and leg (41.8%) were compared to published data.

Hinrichs (1985) developed a set of regression equations for inertial parameters based on data presented by Chandler et al. (1975). The equations predict moments of inertia about the transverse and longitudinal axes passing through the centre of mass, using various anthropometric measurements. These equations are based upon a small sample size (six cadavers) and, therefore, should be treated with caution.

Yeadon and Morlock (1989) developed regression equations to determine segmental moments of inertia based on the Chandler data using linear and non-linear regression equations. Equations developed using right limb cadaver data were tested on left limb data. In addition, anthropometric measurements taken from a 10-year-old boy were input into the Yeadon mathematical model and used to determine segmental inertial properties. The regression equations predicted the left arm properties with a standard error of 21% for linear approaches, and 13% for non-linear approaches. The standard error for modelled inertial properties for the 10-year-old boy was 28.6% for the linear approaches and 20% and for the non-linear approaches. The error for the non-linear approach appears reasonable considering the degree a 10-year-old boy differs from the sample group of six cadaver adults.

Zatziorsky and Seluyanov (1985) combined anthropometric measurements with the gamma ray scanner technique and developed a series of regression equations for mass, centre of gravity, and principle moment of inertia for various body segments. Equations are based on data from 100 subjects and yield a standard error of generally less than 10%.

To summarize, regression equations can provide quick values for inertial parameters using only a few anthropometric measurements. However, these values should only be used as first approximations until they can be more thoroughly evaluated on large sample sizes of actual measurements.

A comparison of the theoretical models developed by Hanavan (1964), Jensen (1978), Hatze (1980), and Yeadon (1990) is summarized for selected variables in Table 3.9.6.

Table 3.9.6 Comparison of number of segments, number of anthropometric measurements, errors, and number of subjects used in original studies by Hanavan (1964), Jensen (1978), Hatze (1980), and Yeadon (1990).

VARIABLE	HANAVAN (1964)	JENSEN (1978)	HATZE (1980)	YEADON (1990)
NUMBER OF SEGMENTS	15	16	17	11
NUMBER OF ANTRHO-POM. MEASUREMENTS	25	--	242	95
NUMBER OF SUBJECTS	66	3	4	3
ERROR MASS	N/A	< 2%	0.26% 0.52%*	2.3% 1.0%**
CM **I^{CM}**	1.8 cm 10%	N/A N/A	N/A N/A	N/A N/A

* maximal error
** if correction made for air contained in lungs

DISCUSSION

An assumption of uniform density within the segment and throughout the body is often made while estimating inertial properties using experimental methods or mathematical models. It is important to understand the effect this assumption has on the final results. Ackland et al. (1988) investigated the effect a uniform density assumption has on the calculation of inertial properties. In this study, a right leg segment of a cadaver was measured using computerized tomography, and subsequently dissected to measure mass and density properties. Tomography and dissection results showed that the leg density was not uniform in either crosssection or longitudinally. To determine the effect non-uniform leg density had on inertial parameters, inertial properties, centre of mass, and moment of inertia about the transverse axis were measured using a Mettler balance and the pendulum method before dissection. The photogrammetric method developed by Jensen was used to model the leg with five elliptical zones and inertial properties were calculated.

Using a uniform density obtained from Dempster (1955) the photogrammetric method overestimated measured cadaver segment mass by 8.2% (due to an overestimation in volume that likely occurred as a result of using only five zones). However, segment mass was underestimated by 3.5% using the computerized tomography data. The use of a variable density value (determined from dissection) with computerized tomography data resulted in a mass underestimation of 2.8% relative to the measured data.

With computerized tomography, the centre of gravity locations of the cadaver leg were 41.7% of the segment length, measured from the proximal end of the segment using variable measured densities, and 41.2% using constant densities (Dempster data). These compared favorably to the measured location of 41.2%. All locations were closer to the proximal end of the segment than those suggested by Dempster (43.3%) and Braune and Fischer (44.2%). Computer tomography scans conducted on a living person indicated a centre of gravity location of 42.2% using variable density, versus 41.9% using a constant density taken from Dempster. The centre of gravity location determined by the photogrammetric method for the live person was at 40.0% of the total length.

The transverse moment of inertia of the cadaver leg using uniform density and computerized tomography data was within 3% of measured values, while the photogrammetric method overestimated the moment of inertia by 13.2% (again, possibly due to overestimating volume). Variable density coupled with the computerized tomography data resulted in an overestimate of 4%. The uniform density assumption, as applied to a leg in conjunction with computerized tomography, predicts mass within 1% of the variable density value, predicts centre of mass locations within 0.5% of segment length, and predicts moment of inertia just as well as the variable density calculations. Values determined using the photogrammetric technique were generally not as accurate in this example, likely due to the large zones chosen. The uniform density assumption does not appear to radically affect the determination of inertial properties, although this needs to be verified for other parts of the body. Errors introduced in determining the volume of the segment appear to have a much greater effect on the results than do the uniform density assumptions.

Many different methods are available for determining inertial properties of human or animal bodies in vivo or in vitro. Determining in vivo values is more difficult and associated with errors. The choice of method to be used depends on the accuracy required. Accuracy of inertial properties is, for instance, less important if a comparison measurement for one subject in two comparative situations is made, e.g., subject running in two different shoes. However, the accuracy of inertial properties is important if resultant joint or muscle-tendon forces in a subject during running must be determined.

The various methods for determining inertial properties discussed above are grouped according to their accuracy in Table 3.9.7. The most accurate methods are also the most time demanding methods.

Table 3.9.7 Summary of approaches for different levels of accuracy and degrees of complexity.

INERTIAL PROPERTY	USE	ACCURACY	APPROACH
MASS	easy	low	average values from literature
	↓	↓	regression equations from literature
			volume from displacement techniques average density from literature
			volume and density from CAT or MRI
	difficult	high	mathematical models with uniform density
			mathematical models with non-uniform density determination
CENTRE OF GRAVITY	easy	low	relative segmental position from literature
	↓	↓	regression equations from literature
			from CAT or MRI using uniform or variable densities
	difficult	high	mathematical models with non-uniform density determination
MOMENT OF INERTIA	easy	low	from literature
	↓	↓	regression equations from literature
			from CAT or MRI using uniform or variable densities
	difficult	high	quick release or relaxed oscillation technique (for selected segments) or pendulum (for total body)
			mathematical models with non-uniform density determination

3.10 ADDITIONAL EXAMPLES

This section has been divided into basic and advanced examples. In general, basic questions can be answered by studying the text or by performing calculations based on information provided by the text. The advanced questions go beyond what can be found in the text.

BASIC QUESTIONS

1. (a) Write the general form of a matrix of moment of inertia about the centre of mass of a rigid body in three-dimensional space. Provide the mathematical definition of each of the items in the matrix.

   (b) Write the general form of a matrix of moment of inertia if the local coordinate system is chosen to be the principle axes of the rigid body.

2. A shank and a leg are schematically illustrated in the planar Figure 3.10.1. The moment of inertia of the shank about its centre of mass is $\mathbf{I}_s$, and the moment of inertia of the thigh about its centre of mass is $\mathbf{I}_t$. Determine the moment of inertia of the entire leg about the hip joint.

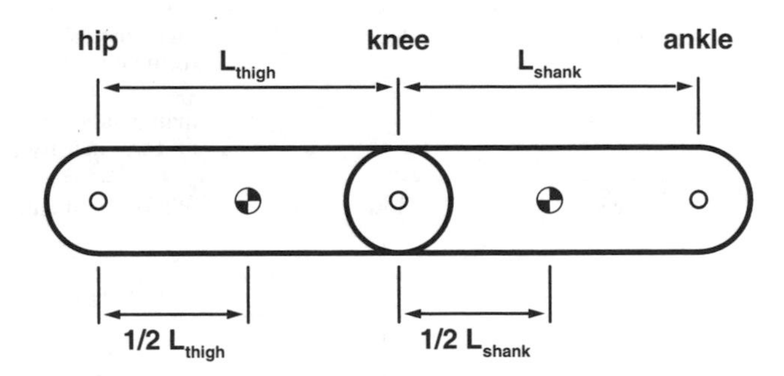

Figure 3.10.1 **Schematic illustration of a shank and a leg.**

3. Using the shank and the leg (with $\mathbf{I}_s$, $\mathbf{I}_t$, m_s, and m_t known and a knee angle of 90×) determine:

   (a) The location of the centre of mass, CM.

   (b) The moment of inertia about the centre of mass, CM.

   (c) The moment of inertia about the ankle joint.

4. A rigid body with the mass, m, is connected through a joint to a fixed mass (Figure 3.10.2). The rigid body can rotate about this joint and behaves like a pendulum. Derive the equation for the moment of inertia if the frequency of the pendulum system, f, can be measured.

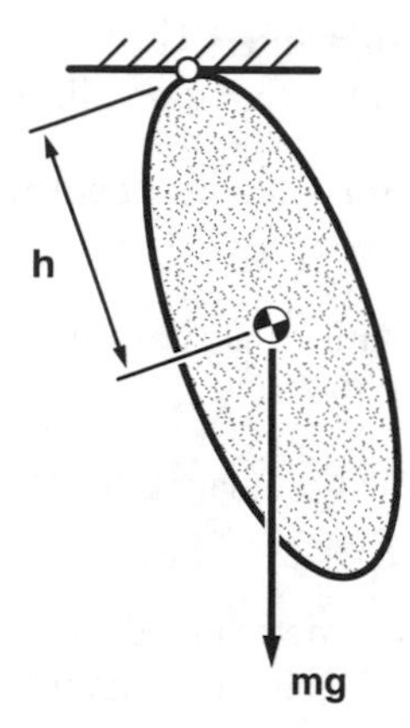

Figure 3.10.2 **Schematic illustration of a rigid body connected through a point to a fixed mass that can behave like a pendulum.**

5. A steel rod is mounted to a rigid mass (Figure 3.10.3). The rod can oscillate, and has length, L, stiffness, k, and the natural frequency, f. Determine the moment of inertia of the steel rod.

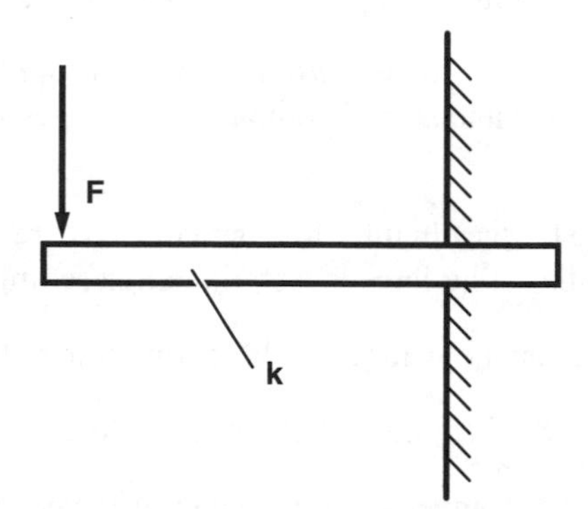

Figure 3.10.3 **Schematic illustration of a steel rod mounted to a rigid mass.**

6. Discuss the difference between Cardan and Euler angles.

7. The calculated results of angular displacements depend, among other factors, on the chosen sequence of axes of rotation. List the sequence of rotation axes that allows the use of the definitions for flexion-extension, ab-adduction, and in-eversion for the ankle joint, and provide the justification for the selected sequence from the anatomical definitions.

8. The calculated results of angular displacements depend, among other factors, on the chosen sequence of axes of rotation. List the sequence of rotation axes that allows the use of the definitions for flexion-extension, ab-adduction, and internal-external rotation for the knee joint, and provide the justification for the selected sequence from the anatomical definitions.

9. Movement determined from skin markers is often different to movement determined from bone-fixed markers. List and describe possibilities to minimize this difference.

10. (a) Write the rotation matrices for rotations about the x, y, and z axes of a Cartesian coordinate system.

    (b) Determine the parametric rotation matrix for Euler angles in the sequence z, x, and z.

    (c) Determine the parametric rotation matrix for Cardan angles in the sequence z, x, and y.

11. The velocity of a marker has been determined from film and video information. Estimate the error of the calculated velocity. Use the following assumptions:

    - The marker position was digitized with an accuracy of 0.5 mm,
    - The distance between two positions of the marker is 1 cm,
    - The time interval between two frames is accurate to 0.001 s, and
    - The time interval between two frames is 0.01 s.

12. Name at least five different measuring possibilities for force measurement.

13. List and describe two possible force transducers to measure the force of Achilles tendon in vitro in human specimens, and compare advantages and disadvantages of the transducers.

14. (a) Draw a Wheatstone bridge circuitry for a situation where two force transducers are used to determine the acting force in a strain gauge setting.

    (b) Write and solve the equations for the Wheatstone bridge for this example.

    (c) Explain how results can be used to determine the force in a structure.

15. Discuss the possibilities to compensate for changes effected by changes in temperature in strain gauge measurements.

16. Name at least three methods for measuring pressure distribution in biomechanics, discuss their principal functioning, and the advantages and disadvantages of each method.

17. Discuss the advantages and disadvantages of the three basic force measuring possibilities, (piezo-electric, strain gauge, and capacitance) with respect to frequency response, quasi-static measurements, linearity, crosstalk, and three-dimensional measurements.

18. (a) Draw the force-time curves for heel-toe running in the three components: anterio-posterior, medio-lateral, and vertical. Include units and order of magnitude for both force and time.

    (b) Show how the information from part (a) would be used to determine if the subject was running at a constant speed.

19. Explain the information content of force-deformation and stress-strain diagrams.

20. Describe at least one method to quantify the energy loss of a bouncing mass, assuming that you have information from a force plate or an accelerometer, but no kinematic information.

21. Acceleration measurements from bone and skin-mounted acceleration sensors typically provide different results.

    (a) Discuss the differences between acceleration results determined from bone or skin-mounted accelerometers.

    (b) Discuss the major reasons for these differences.

    (c) Which acceleration is needed to determine the ground reaction force with the help of inverse dynamics (top to bottom approach) – skin-mounted, bone-mounted, or other acceleration? Explain and provide the reason.

22. Discuss how the following parameters influence experimentally determined acceleration results when quantifying the acceleration of the tibia during locomotion: mass, location, fixation, resonance frequency, damping of the mounted accelerometer, and frequency content of the input signal

23. A steel shot with an attached accelerometer is to be used in a surface drop test experiment. Draw the acceleration-time diagrams seen by the accelerometers, with the upward acceleration being positive for the following two cases:

    (a) The shot lands and does not bounce off the surface.

    (b) The shot bounces off the surface (draw the acceleration curve until the time the shot contacts the surface the second time).

24. A person is standing still on a force plate with dimensions as illustrated in Figure 3.10.4. The distances between the centres of the force sensors are 0.6 m and 0.9 m. The forces measured in the four corners are $\mathbf{F}_1 = 100$ N, $\mathbf{F}_2 = 250$ N, $\mathbf{F}_3 = 350$ N, and $\mathbf{F}_4 = 200$ N.

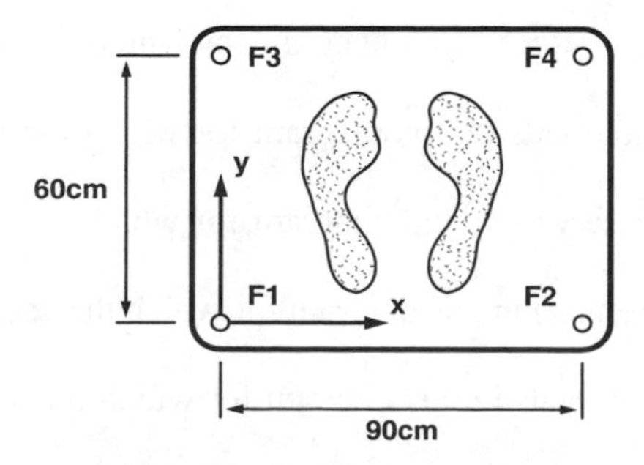

Figure 3.10.4 **Schematic illustration of a person standing on a force plate.**

(a) Determine the centre of pressure.
(b) Comment on the force distribution on the two illustrated feet, and explain where the forces are high and where they are low.

25. The movement of a segment (Figure 3.10.5) is confined to the x-y plane. Two accelerometers are mounted firmly to the segment. The accelerometer on the left end of the segment (point a) measures acceleration in the direction indicated with an amplitude of
$a_1 = 10 \cdot \sqrt{2}\ m \cdot s^2$. The accelerometer at the right end of the segment (point b) measures an acceleration of $a_2 = 0\ m \cdot s^2$, determine:

   (a) The linear acceleration of the centre of mass, CM.
   (b) The angular velocity of the CM at the time t = 1 s.
   (c) The angular acceleration of the CM.

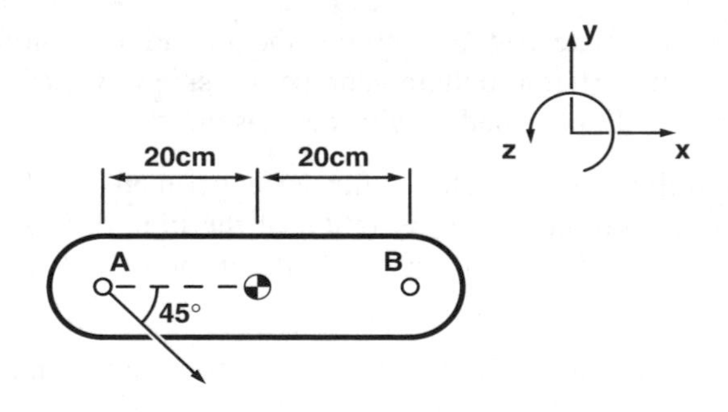

Figure 3.10.5 Schematic illustration of a limb segment.

Use the following assumptions:

- Gravity can be neglected,
- The limb segment is rigid, and
- The connection between the accelerometers and limb segment is rigid.

26. A vertical ground reaction force has been measured as illustrated in Figure 3.10.6.

   (a) Determine the time point(s) in this diagram for which the velocity of the CM is zero.
   (b) Determine the time intervals in this diagram for which the velocity of the CM is negative.
   (c) Determine the time intervals in this diagram for which the velocity of the CM is positive.
   (d) Determine the time point(s) in this diagram for which the acceleration of the CM is zero.
   (e) Determine the time intervals in this diagram for which the acceleration of the CM is negative.
   (f) Determine the time intervals in this diagram for which the acceleration of the CM is positive.
   (g) Determine the time point(s) in this diagram for which the CM reaches its minimal height.
   (h) Describe the movement executed by the subject.

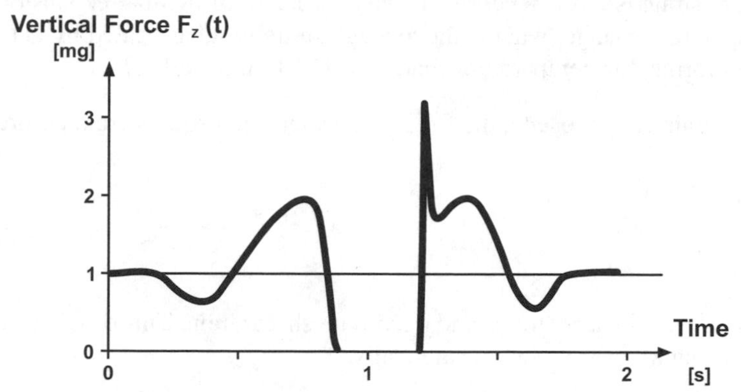

Figure 3.10.6 **Schematic illustration of the vertical ground reaction force, F_z (t), as a function of time for a jump from standing still with a wind-up movement.**

27. Discuss some of the advantages and disadvantages of surface EMG recordings versus indwelling EMG recordings.

28. Discuss some of the advantages and disadvantages of bi-polar (or multi-polar) and mono-polar EMG recordings.

29. Explain and discuss the factors contributing to electromechanical delay.

30. A given amount of electromyographical signal in a muscle can be associated with a very high or a very low force. Explain why this statement is correct. Give examples.

31. Compare the magnitude of engineering and true strains for a specimen that was extended by:

    (a) 1%.
    (b) 5%.
    (c) 10% of its original length.

32. If a length measuring technique is used to measure strain, calculate the range of possible strain measurements for nominal 1% and 10% strains if the error in the length measurement is:

    (a) 0.5%.

    (b) 0.05%.

33. Poisson's ratio is the ratio of the lateral strain to the longitudinal strain when a material is loaded uniaxially.

    $$\varepsilon_{lat} = -\upsilon\varepsilon_{long}$$

    If the Elastic (Young's) modulus relating uniaxial stress to uniaxial strain is $E(\sigma = E\varepsilon)$, show that the bulk modulus, relating volumetric strain to hydrostatic stress, $K = \dfrac{E}{3(1-2v)}$.

34. If thermal strain is $\alpha\Delta T$ where α is the coefficient of thermal expansion and ΔT is the temperature change, what is the average strain in a bar heated by ΔT at one end, rising uniformly to a temperature change of $11\Delta T$ at the other?

35. In a 45× strain gauge rosette, the following extensional strains are measured:

$$\varepsilon_0 = 475\ \mu\varepsilon$$
$$\varepsilon_{45} = -225\ \mu\varepsilon$$
$$\varepsilon_{90} = -325\ \mu\varepsilon$$

Determine the principal strains and maximum shear strain. This problem can be solved graphically using Mohr's circle or analytically.

36. In a 60× strain gauge rosette, the following extensional strains are measured:

$$\varepsilon_0 = +600\ \mu\varepsilon$$
$$\varepsilon_{60} = -360\ \mu\varepsilon$$
$$\varepsilon_{120} = +300\ \mu\varepsilon$$

Determine the principal strains and maximum shear strain. This problem can be solved graphically using Mohr's circle or analytically.

ADVANCED QUESTIONS

1. When determining the centre of force with force plate measurements, substantial noise can be observed in the raw data during initial ground contact and takeoff.

   (a) Discuss the reason for this excessive noise.

   (b) Propose methods to determine the actual centre of force for these situations.

2. Axial acceleration of the tibia has been quantified during heel-toe running for two sport shoe conditions (soft and hard). The axial acceleration amplitudes for the soft sole condition were 15 g, and axial acceleration amplitudes for the hard soled shoe were 20 g. Discuss the possible interpretation of these results with respect to internal loading at the ankle joint. Use the following assumptions:

   - The accelerometers were strapped tightly to the tibia, and
   - The maximal amplitudes were recorded in the first 15 milliseconds after heel strike.

3. When discussing acceleration results, two variables are often used (the amplitude of the acceleration and the peak-to-peak amplitude of the acceleration). Discuss these two variables with respect to their mechanical and/or biological appropriateness. Specifically, do both variables have a sound mechanical and/or biological and physical meaning?

4. Explain from the point of view of the property of a signal why it is infinitely harder to obtain accurate muscle force predictions from EMG during dynamic than during static tasks.

5. How would you design an experiment to clarify and explain the exact relationship of muscle force and EMG? Assume that technical assistance and financial support are available, as required.

6. Assume that you have measured acceleration at the tibia with the help of skin-mounted accelerometers. The result of one trial is illustrated in Figure 3.10.7 (accelerometer 1). Discuss this result critically, identify possible problems, and accept or reject the derived results for further use.

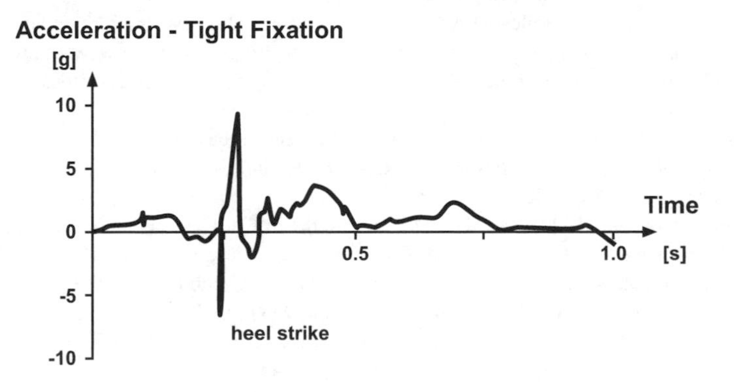

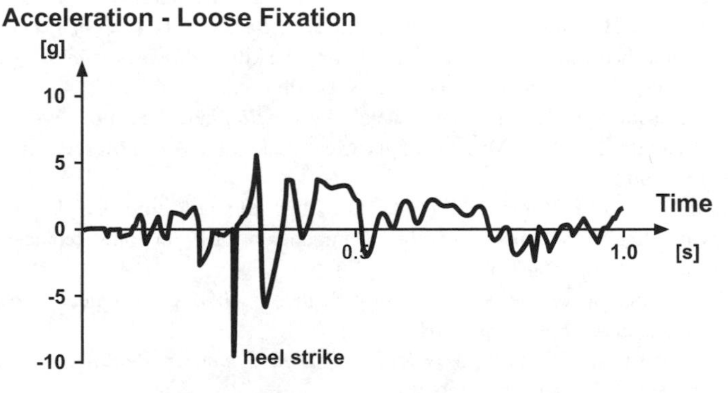

Figure 3.10.7 **Examples of acceleration-time-signals for two mounting models: a tight fixation (top) and a loose fixation (bottom).**

3.11 REFERENCES

Abdel-Aziz, Y.I. and Karara, H.M. (1971) Direct Linear Transformation from Comparator Co-ordinates Into Object Space Co-ordinates. *Proc. ASP/UI Symposium on Close-range Photogrammetry.* Am. Soc. of Photogrammetry, Falls Church, VA. pp. 1-18.

Abraham, L.D. and Loeb, G.E. (1985) The Distal Hindlimb Musculature of the Cat. *Exp. Brain Res.* **58,** pp. 580-593.

Ackland, R., Blanksby, B.A., and Bloomfield, J. (1988) Inertial Characteristics of Adolescent Male Body Segments. *J. Biomech.* **21 (4),** pp. 319-327.

Ackland T.R., Hensen P.W., and Bailey, D.A. (1988) The Uniform Density Assumption and its Effect Upon the Estimation of Body Segment Inertial Parameters. *Int. J. Sport Biomech.* **4,** pp. 146-155.

Aldroubi, A. and Unser, M. (1996) *Wavelets in Medicine and Biology.* CRC Press, Boca Raton, New York, London, Tokyo.

Allard, P., Blanchi, J., and Aissaoui, R. (1995) Bases of Three-dimensional Reconstruction. *Three-dimensional Analysis of Human Movement* (eds. Allard, P., Stokes, I., and Blanchi, J.). Human Kinetics, Champaign, IL. pp. 19-40.

Allinger, T. and Herzog, W. (1992) Calculated Fibre Lengths in Cat Gastrocnemius During Walking. *Proceedings of NACOB II.* Amer. Soc. of Biomechanics, Chicago, IL. pp. 81-82.

Allum, J.H.J. and Young, L.R. (1976) The Relaxed Oscillation Technique for the Determination of the Moment of Inertia of Limb Segments. *J. Biomech.* **9 (1),** pp. 21-26.

Arcan, M. and Brull, M.A. (1976) A Fundamental Characteristic of the Human Body and Foot, the Foot-Ground Pressure Pattern. *J. Biomech.* **9 (7),** pp. 435-457.

Aritomi, H., Morita, M., and Yonemoto, K. (1983) A Simple Method of Measuring the Footsole Pressure of Normal Subjects Using Prescale Pressure-detecting Sheets. *J. Biomech.* **16 (2),** pp. 157-165.

Arms, S.W., Boyle, J., Johnson, R.J., and Pope, M.H. (1983) Strain Measurements in the Medial Collateral Ligament of the Human Knee: An Autopsy Study. *J. Biomech.* **16 (7),** pp. 491-496.

Asang, E. (1974) Biomechanics of the Human Leg in Alpine Skiing. *Biomechanics IV* (eds. Nelson, R.C. and Morehouse, C.A.). University Park Press, Baltimore. pp. 236-242.

Barry, D.T. (1987) Acoustic Signals from Frog Skeletal Muscle. *Biophys. J.* **51,** pp. 769-773.

Barry, D.T., Hill, T., and Im, D. (1992) Muscle Fatigue Measured with Evoked Muscle Vibrations. *Muscle & Nerve.* **15,** p. 303.

Bartel, D.L., Bicknall, V.L., and Wright, T.M. (1986) The Effect of Conformity, Thickness, and Material on Stresses in Ultra High Molecular Weight Components for Total Joint Replacement. *J. Bone Jt. Surgery.* **68 (A-7),** pp. 1041-1051.

Barter, J.T. (1957) Estimation of the Mass of Body Segments. *WADC Technical Report (TR-57-260).* Wright-Patterson Air Force Base, OH.

Basmajian, J.V. and De Luca, C.J. (1985) *Muscles Alive. Their Functions Revealed by Electromyography* (5ed). Williams & Wilkins, Baltimore.

Bauman, J.H. and Barnd, P.W. (1963) Measurement of Pressure Between Foot and Shoe. *The Lancet.* **1,** pp. 629-633.

Beckwith, T.G., Buck, N.L., and Marangoni, R.D. (1982) *Mechanical Measurements* (3ed). Addison-Wesley, Reading, MS.

Bergmann, G., Graichen, F., and Rohlmann, A. (1993) Hip Joint Loading During Walking and Running, Measured in Two Patients. *J. Biomech.* **26 (8),** pp. 969-990.

Bernstein, N.A. (1947) *On the Construction of Movements.* Moscow, Medgiz. (In Russian).

Bernstein, N.A. (1967) *The Co-ordination and Regulation of Movements.* Pergamon Press, Oxford.

Bernstein, N.A., Mogilanskaia, Z., and Popova, T. (1934) *Technics of Motion-study.* Moscow, Leningrad. (In Russian).

Besimo, C. and Kempf, B. (1995) In-vitro Investigation of Various Attachments for Overdentures on Osseointegrated Implants. *J. Oral Rehab.* **22 (9),** pp. 691-698.

Bigland, B. and Lippold, O.C.J. (1954) Motor Unit Activity in the Voluntary Contraction of Human Muscle. *J. Physiol.* **125,** pp. 322-335.

Bigland-Ritchie, B., Kukulka, C.G., and Woods, J.J. (1980) Surface EMG/force Relations in Human Muscle of Different Fibre Composition. *J. Physiol.* **308**, pp. 103-104.

Bigland-Ritchie, B., Donovan, E.F., and Roussos, C.S. (1981) Conduction Velocity and EMG Power Spectrum Changes in Fatigue of Sustained Maximal Efforts. *J. Appl. Physiol.* **51,** pp. 1300-1305.

Bishop, P.J., Norman, R.W., and Kozey, J.W. (1984) An Evaluation of Football Helmets Under Impact Conditions. *Am. J. of Sports Med.* **12 (3)**, pp. 233-236.

Bishop, P.J. and Briard, B.D. (1984) Impact Performance of Bicycle Helmets. *Can. J. of Appl. Sports Sci.* **9 (2)**, pp. 94-101.

Bloach, A. (1935) New Methods for Measuring Mechanical Stress at Higher Frequencies. *Nature.* **136,** pp. 223-224.

Bobbert, M.F. and Schamhardt, H.C. (1990) Accuracy of Determining the Point of Force Application with Piezoelectric Force Plates. *J. Biomech.* **23 (7)**, pp. 705-710.

Bouche, R.R. (1974) Vibration Measurements. *Process Instruments and Controls Handbook* (2ed) (ed. Considine, D.M.). McGraw-Hill, New York. **9,** pp. 68-74.

Bouisset, S. and Pertuzon, E. (1967) Experimental Determination of the Moment of Inertia of Limb Segments. *Biomech. I.* Karger, Basel. pp. 106-109.

Bouisset, S. and Goubel, F. (1971) Interdependence of Relations Between Integrated EMG and Diverse Biomechanical Quantities in Normal Voluntary Movements. *Activitas Nervosa Superior.* **13,** pp. 23-31.

Boyer, K.A. and Nigg, B.M. (2004) Muscle Activity in the Leg is Tuned in Response to Impact Force Characteristics. *J. Biomech.* **37,** 1583-1588.

Boyer, K.A. and Nigg, B.M. (2006) Soft Tissue Vibrations within One Soft Tissue Package. *Journal of Biomechanics* **39**, 645-651.

Boyer, K.A. and Nigg, B.M. (in press) Muscle Tuning during Running: Implications of an Un-tuned Landing. *Journal of Biomechanical Engineering.*

Braune, W. and Fischer, O. (1889) Über den Schwerpunkt des menschlichen Körpers, mit Rücksicht auf die Ausrüstung des deutschen Infanteristen. *Abhandlung der Königl. Sächsischen Gesellschaft der Wissenschaften.* **26,** pp. 561-672.

Bregler, C. and Malik, J. (1998) Tracking people with twists and exponential maps. *IEEE Conference on Computer Vision and Pattern Recoginition*, Santa Barbara, CA. pp. 8-15.

Breyer, H.G., Enes-Gaiao, F., and Kuhn, H.J. (1979) Biegebeanspruchung des Femurkopfprosthesenschaftes bei hypertrochanter Osteotomie. *Proceedings: Pauwels Symposium.* Free University Berlin, Berlin. pp. 136-143.

Brown, T.D., Sigal, L., Njus, G.O., Njus, N.M., Singerman, R.J., and Brand, R.A. (1986) Dynamic Performance Characteristics of the Liquid Metal Strain Gage. *J. Biomech.* **19 (2),** pp. 165-173.

Burke, D.W., Davies, J.P., O'Connor, D.O., and Harris, W.H. (1984) Experimental Strain Analysis in the Femoral Cement Mantle of Simulated Total Hip Arthroplasties. *Trans. Orthop. Res. Soc.* **9**, p. 296.

Burke, R.E. (1981) Motor Units: Anatomy, Physiology, and Functional Organization. *Handbook of Physiology-The Nervous System II* (ed. Brooks, V.). Williams & Wilkins, Baltimore. pp. 345-422.

Cady, W.G. (1964) *Piezoelectricity: An Introduction to the Theory and Applications of Electromechanical Phenomena in Crystals* (2ed). Dover Publications, New York.

Calderale, P.M. (1983) The Use of Strain Gauges for Biometrics in Europe. *Eng. in Med.* **12 (3)**, pp. 117-133.

Calderale, P.M., Gola, M.M., and Gugliotta, A. (1979) New Theoretical and Experimental Developments in the Mechanical Design of Implant Stems of Stem-femur Coupling: New Improved Procedures. *Proceedings: Pauwels Symposium.* Free University Berlin, Berlin. pp. 199-208.

Carlet, G. (1872) Sur la Locomotion Humaine. *Ann. Sci. Naturelles.* **5,** pp. 1-92.

Cavagna, G.A. (1964) Mechanical Work in Running. *J. Appl. Physiol.* **19 (2),** pp. 249-256.

Cavagna, G.A., Dusman, B., and Margaria, R. (1968) Positive Work Done by a Previously Stretched Muscle. *J. Appl. Physiol.* **24 (1)**, pp. 21-32.

Cavanagh, P.R. (1990) The Mechanics of Distance Running: A Historical Perspective. *Biomechanics of Distance Running* (ed. Cavanagh, P.R.). Human Kinetics, Champaign, IL. pp. 1-34.

Cavanagh, P.R. and Ae, M. (1978) A Technique for the Display of Pressure Paths from a Force Platform. *J. Biomech.* **11,** pp. 487-491.

Cavanagh, P.R. and Lafortune, M.A. (1980) Ground Reaction Forces in Distance Running. *J. Biomech.* **13,** pp. 397-406.

Cavanagh, P.R., Williams, K.R., and Clarke, T.E. (1981) A Comparison of Ground Reaction Forces During Walking Barefoot and in Shoes. *Biomechanics VII-B* (eds. Morecki, A., Fidelus, K., Kedzior, K., and Wit, A.). University Park Press, Baltimore. pp. 151-156.

Cavanagh, P.R., Hennig, E.M., Bunch, R.P., and Macmillan, N.H. (1983) A New Device for the Measurement of Pressure Distribution Inside the Shoe. *Biomechanics VIII-B* (eds. Matsui, H. and Kobayashi, K.). Human Kinetics, Champaign, IL. pp. 1089-1096.

Cavanagh, P.R., Ulbrecht, J.S. and Caputo G.M. (2000) New Developments in the Biomechanics of the Diabetic Foot. *Diabetes Metab. Res. Rev.* **16 (Suppl 1),** pp. S6-S10.

Centre Georges Pompidou, Musee National d'Art Modern (1977) *E.J. Marey, 1830/1904, La Photographie du Mouvement.*

Chand, R., Haug, E., and Rim, K. (1976) Stresses in the Human Knee Joint. *J. Biomech.* **9 (6)**, pp. 417-422.

Chandler, R.F., Clauser, C.E., McConville, J.T., Reynolds, H.M., and Young, J.W. (1975) Investigation of the Inertial Properties of the Human Body. *Technical Report (DOT HS-801-430).* Wright-Patterson Air Force Base, OH.

Chao, E.Y.S. (1978) Experimental Methods for Biomechanical Measurements of Joint Kinematics. *CRC Handbook of Engineering in Medicine and Biology.* CTC Press, West Palm Beach, FL. pp. 385-409.

Chao, E.Y.S. (1980) Justification of Triaxial Goniometer for the Measurement of Joint Rotation. *J. Biomech. Eng.* **13,** pp. 989-1006.

Chen, J., Vandewalle, J., Sansen, W., Vantrappen, G., and Janssens, J. (1990) Multichannel Adaptive Enhancement of the Electrogastrogram. *IEEE Trans. on Biomedical Engineering.* BME **37 (3)**, pp. 285-294.

Cheung, G., B. S., and Kanade, T. (2003) Shape-from-silhouette of articulated objects and its use for human body kinematics estiamtion and motion capture. *Proc. of CVPR ,* pp. 55-59.

Cheung, K. M., Kanade, T., Bouguet, J.-Y., and Holler, M. (2000) A real time system for robust 3D voxel reconstruction of human motions. *Proceedings of the 2000 IEEE Conference on Computer Vision and Pattern Recognition*, **2**, pp. 714-720.

Chodera, J.D. and Lord, M. (1979) Pedobarographic Foot Pressure Measurement and the Applications. *Disability* (eds. Kenedi, R.M., Paul, J.P. and Hughes, J.). MacMillan, London. pp. 173-181.

Chu, M.L., Yazadani-Ardakani, S., Gradisar, I.A., and Askew, M.J. (1986) An In-vitro Simulation Study of Impulsive Force Transmission Along the Lower Skeletal Extremity. *J. Biomech.* **19,** pp. 979-987.

Clarke, T.E. (1980) The Pressure Distribution Under the Foot During Barefoot Walking. Unpublished doctoral Dissertation, Pennsylvania State University, University Park, PA.

Clarys, J.P. and Marfell-Jones, M.J. (1986) Anthropometric Prediction of Component Tissue Masses in the Minor Limb Segments of the Human Body. *Human Biol.* **58 (5)**, pp. 761-769.

Clauser, C.E., McConville, J.T., and Young J.W. (1969) Weight, Volume, and Centre of Mass of Segments of the Human Body. *AMRL Technical Report (TR-69-70).* Wright-Patterson Air Force Base, OH.

Close, J.R., Nickel, E.D., and Todd, F.N. (1960) Motor-unit Action Potential Counts. Their Significance in Isotonic and Isometric Contractions. *J. Bone Jt. Surg.* **42 (A-7),** pp. 1207-1222.

Cole, G.K., Nigg, B.M., Ronsky, J.L., and Yeadon, M.R. (1993) Application of the Joint Coordinate System to 3-D Joint Attitude and Movement Representation: A Standardization Proposal. *J. Biomech. Eng.* **115**, pp. 344-349.

Colville, M.R., Marder, R.A., Boyle, J.J., and Zarins, B. (1990) Strain Measurement in Lateral Ankle Ligaments. *Am. J. Sports Med.* **18 (2)**, pp. 196-200.

Conforto, S., Alessio, T.D., Pignatelli, S. (1999) Optimal Rejection of Movement Artefacts from Myoelectric Signals by Means of a Wavelet Filtering Procedure. *J. Electromyogr. Kinesiol.* **9**, pp. 47-57.

Croone, W. (1664) *De Ratione Motus Musculorum.* London, England.

Crowninshield, J. and Tolbert, J. (1983) Cement Strain Measurement Surrounding Loose and Well Fixed Femoral Component Stems. *J. Biomed. Matls. Res.* **17**, pp. 819-828.

Cunningham, D.M. and Brown, G.W. (1952) Two Devices for Measuring the Forces Acting on the Human Body During Walking. *Proceedings of the Society for Experimental Stress Analysis IX-2* (eds. Mahlmann, C.V. and Murray, W.M.). Massachusetts Institute of Technology.

Davis, B.L., Perry, J.E., Neth, D.C., and Waters, K.C. (1998) A Device for Simultaneous Measurement of Pressure and Shear Force Distribution on the Plantar Surface of the Foot. *J. Appl. Biomech.* **14**, pp. 93-104.

de Lange, A., Huiskes, R., and Kauer, J.M. (1990) Effects of Data Smoothing on the Reconstruction of Helical Axis Parameters in Human Joint Kinematics. *J. Biomech. Eng.* **112**, pp. 107-113.

Dempster, W.T. (1955) Space Requirements of the Seated Operator. *WADC Technical Report (TR-55-159)*. Wright-Patterson Air Force Base, OH.

Denoth, J. (1980) Ein mechanisches Modell zur Beschreibung von passiven Belastungen. *Sportplatzbeläge* (eds. Nigg, B.M. and Denoth J.). Juris Verlag, Zürich. pp. 45-67.

Denoth, J. (1980) Materialeigenschaften. *Sportplatzbeläge* (eds. Nigg, B.M. and Denoth, J.). Juris Verlag, Zürich. pp. 54-66.

Donskoi, D.D. (1975) *Grundlagen der Biomechanik*. Verlag Bartels 4 Wernitz KG, Berlin.

Doud, J.R., Walsh, J.M. (1995) Muscle Fatigue and Muscle Length Interaction: Effect on the EMG Frequency Components. *Electromyogr. Clin. Neurophysiol.* **35(6)**, pp. 331-339.

DuBois-Reymond, E. (1849) *Untersuchungen ueber thierische elektricitaet*. Reimer Verlag von G, Berlin. **2 (2)**.

Duchenne, G.B.A. (1959) *Physiologie des Mouvements* (trans. Kaplan, E.B.). W.B. Saunders, London. (Original 1867).

Eaton, E.C. (1931) Resistance Strain Gauge Measures Stresses in Concrete. *Eng. News Records.* **107,** pp. 615-616.

Edwards, R.G., Lafferty, J.F., and Lange, K.O. (1970) Ligament Strain in the Human Knee Joint. *J. Basic Eng., Trans. ASME.* pp. 131-136.

Elftman, H. (1934) A Cinematic Study of the Distribution of Pressure in the Human Foot. *Anat. Rec.* **59 (4),** pp. 481-487.

Elftman, H. (1938) The Force Exerted by the Ground in Walking. *Arbeitsphysiologie.* **10,** pp. 485-491.

Elliott, B.C., Blanksby, B.A., and Ellis, R. (1980) Vibration and Rebound Characteristics of Conventional Oversized Tennis Rackets. *Research Quarterly for Exercise and Sport.* **51 (4)**, pp. 608-615.

English, T.A. and Kilvington, M. (1979) In-vivo Records of Hip Load Using a Femoral Implant with Telemetric Output (A Preliminary Report). *J. Biomed. Eng.* **1,** pp. 111-115.

Evans, F.G. and Lissner, H.R. (1948) Stresscoat Deformation Studies of the Femur Under Static Vertical Loading. *Anat. Rec.* **100**, pp. 159-190.

Farina, D., Merletti, R., and Enoka, R. M. (2004) The Extraction of Neural Strategies from the Surface EMG. *J. Appl. Physiol.* **96**, pp. 1486-1495.

Fenn, W.O. (1938) The Mechanics of Muscular Contraction in Man. *J. Appl. Physics.* **9 (3),** pp. 65-177.

Ferrara, E.R. and Widrow, B. (1982) Fetal Electrocardiogram Enhancement by Time-sequenced Adaptive Filtering. *IEEE Trans. on Biomedical Engineering.* BME **29 (6),** pp. 458-460.

Finlay, J.B., Bourne, R.B., Landsberg, R.P.D., and Andreae, P. (1986) Pelvic Stresses In-vitro - I: Malsizing of Endoprostheses. *J. Biomech.* **19 (9)**, pp. 703-714.

Finlay, J.B., Rorabeck, C.H., Bourne, R.B., and Tew, W.M. (1989) In-vitro Analysis of Proximal Femoral Strains using PCA Femoral Implants and a Hip-abductor Muscle Simulator. *J. Arthro.* **4 (4)**, pp. 335-345.

Fischer, O. (1906) *Theoretische Grundlagen für eine Mechanik der lebenden Körper*. Teubner, Berlin.

Frangioni, J.V., Kwan-Gett, T.S., Dobrunz, L.E., and McMahon, T.A. (1987) The Mechanism of Low-frequency Sound Production in Muscle. *Biophys. J.* **51**, pp. 775-783.

Frederick, E.C., Hagy, J.L., and Mann, R.A. (1981) Prediction of Vertical Impact Forces During Running. *J. Biomech.* **14 (7),** p. 498.

Friis, E.A., Cooke, F.W., Henning, C.E., and Samani, D.L. (1989) Effect of Bone Block Shape on Patellar Stress in ACL Reconstruction: An Evaluation Using SPATE. *15th Annual Meeting, Society for Biomaterials*.

Gage, H. (1967) *Accelerographic Analysis of Human Gait* (eds. Byars, E.E., Contini, R., and Roberts, V.L.). Biomechanics Monograph. ASME, New York.

Galvani, L. (1953) *DeViribus Electricitatis* (trans. Green, R.). University Press, Cambridge. (Original 1791).

Garg, D.P. and Ross, M.A. (1976) Vertical Mode Human Body Vibration Transmissibility. *IEEE transactions of Systems, Man, and Cybernetics* **SMC-6,** 102-112.

Gohlke, W. (1959) *Einfüehrung in die piezoelektrische Messtechnik.* **2**. Auflage, Akad. Verlagsges, Leipzig.

Goldstein, H. (1950) *Classical Mechanics.* Addison-Wesley, London.

Griffiths, R.J. (1991) Shortening of Muscle Fibres During Stretch of the Active Cat Medial Gastrocnemius Muscle: The Role of Tendon Compliance. *J. Physiol.* London. **436,** pp. 219-236.

Grimaldi, F.M. (1665) *Physicomathesis de Lumine.* Bologna, Italy.

Grood, E.S. and Suntay, W.J. (1983) A Joint Coordinate System for the Clinical Description of Three-dimensional Motions: Application to the Knee. *J. Biomech. Eng.* **105 (2),** pp. 136-144.

Gootzen, T.H.J.M., Stegeman, D.F., Van Oosterom, A. (1991) Finite Limb Dimensions and Finite Muscle Length in a Model for the Generation of Electromyographic Signals. *Electroencephalogr. Clin. Neurophysiol.* **81**, pp. 152-162.

Gross, T.S. and Nelson, R.C. (1988) The Shock Attenuation Role of the Ankle During Landing from a Vertical Jump. *Med. and Science in Sports and Exercise.* **20 (5),** pp. 506-514.

Grundy, M., Blackburn, P.A., Tosh, P.A., McLeish, R.D., and Smith, L. (1975) An Investigation of the Centres of Pressure Under the Foot While Walking. *J. Bone Jt. Surg.* **57 (B-1),** pp. 98-103.

Guimaraes, A.C.S., Herzog, W., Hulliger, M., Zhang, Y.T., and Day, S. (1994a) EMG-force Relation of the Cat Soleus Muscle: Experimental Simulation of Recruitment and Rate Modulation Using Stimulation of Ventral Root Filaments. *J. Exp. Biol.* **186,** pp. 77-93.

Guimaraes, A.C.S., Herzog, W., Allinger, T.L., and Zhang, Y.T. (1994b) Effects of Muscle Length on the EMG-force Relation of the Cat Soleus Muscle Using Non-periodic Stimulation of Ventral Root Filaments. *J. Exp. Biol.* **193,** pp. 49-64.

Guimaraes, A.C.S., Herzog, W., Hulliger, M., Zhang, Y.T., and Day, S. (1995) EMG-force Relation of the Cat Soleus Muscle During Locomotion, and its Association with Contractile Conditions. *J. Exp. Biol.* **198,** pp. 975-987.

Gurdjian, E.S. and Lissner, H.R. (1947) Deformations of the Skull in Head Injury as Studied by the Stresscoat Technique. *Am. J. Surg.* **73**, pp. 269-281.

Gurfinkel, V.S. and Safornov, V.A. (1969) Apparatus for Measurements of Moments of Inertia of Different Human Body Segments. *Author's Certificate N. 255483.* Moscow. State Committee on Invention and Discoveries of U.S.S.R. (In Russian).

Haboush, E.J. (1952) Photoelastic Stress and Strain Analysis in Cervical Fractures of the Femur. *Bulletin Hospital Joint Diseases.* **13**, pp. 252-258.

Hagberg, M. (1981) Muscular Endurance and Surface Electromyogram in Isometric and Dynamic Exercise. *J. Appl. Physiol: Resp., Env., & Ex. Phys.* **51,** pp. 1-7.

Hanavan, E.P. (1964) A Mathematical Model of the Human Body. *AMRL Technical Report (TR-64-102).* Wright-Patterson Air Force Base, OH.

Harris, C.M. (1995) *Shock and Vibration Handbook.* (4th ed.). McGraw-Hill, New York, USA.

Harless, E. (1860) Die statischen Momente der menschlichen Gliedmassen. *Abhandlung der math.-phys. Cl. der Königlich Bayerischen Akademie der Wissenschaften.* **8**, pp. 69-96.

Harwood, N. and Cummings, W.M. (1986) Applications of Thermoelastic Stress Analysis. *Strain.* **22 (1)**, pp. 7-12.

Hatze, H. (1980) A Mathematical Model for the Computational Determination of Parameter Values of Anthropometric Segments. *J. Biomech.* **13,** pp. 833-843.

Hatze, H. (1986) *Methoden biomechanischer Bewegungsanalyse.* Öesterreichischer Bundesverlag, Wien.

Hauser, W. and Schaff, P. (1987) Ski Boots: Biomechanical Issues Regarding Skiing Safety and Performance. *Int. J. of Sport Biomech.* **3,** pp. 326-344.

Heckathorne, C.W. and Childress, D.S. (1981) Relationships of the Surface Electromyogram to the Force, Length, Velocity, and Contraction Rate of the Cineplastic Human Biceps. *Am. J. Phys. Med.* **60 (1),** pp. 1-19.

Hehne, H.J. (1983) *Das Patellofemoralgelenk.* Enke Verlag, Stuttgart.

Hennig, E.M. (2002) Invited Wei Lun Public Lecture: The Human Foot During Locomotion - Applied Research for Footwear. Hong Kong: The Chinese University of Hong Kong.

Hennig, E.M., and Milani, T.L. (1995) In-shoe Pressure Distribution for Running in Various Types of Footwear. *J. Appl. Biomech.* **11 (3)**, pp. 299-310.

Hendry, A.W. (1948) *An Introduction to Photoelastic Analysis*. Blackie & Sons Ltd., Glasgow.

Hennig, E.M., Cavanagh, P.R., Albert, H.T., and Macmillan, N.H. (1981) A Piezoelectric Method of Measuring the Vertical Contact Stress Beneath the Human Foot. *J. Biomed. Eng.* **4,** pp. 213-222.

Hennig, E.M., Cavanagh, P.R., and Macmillan, N.H. (1983) Pressure Distribution by High Precision Piezoelectric Ceramic Force Transducers. *Biomechanics VIII-B* (eds. Matsui, H. and Kobayashi, K.). Human Kinetics, Champaign, IL. pp. 1081-1088.

Hertzberg, H.T., Dupertius, C.V., and Emanuil, J. (1957) *J. Photogrammetric Engineering*. **23**, pp. 942-947.

Herzog, W., Leonard, T.R., Renaud, J.M., Wallace, J., Chaki, G., and Bornemisza, S. (1992) Force-length Properties and Functional Demands of Cat Gastrocnemius, Soleus, and Plantaris Muscles. *J. Biomech.* **25 (11),** pp. 1329-1335.

Herzog, W., Stano, A., and Leonard, T.R. (1993) A Telemetry System to Record Force and EMG Recording from Cat Ankle Extensor and Tibialis Anterior Muscles. *J. Biomech.* **26**, pp. 1463-1471.

Herzog, W., Leonard, T.R., and Guimaraes, A.C.S. (1993a) Forces in Gastrocnemius, Soleus, and Plantaris Tendons of the Freely Moving Cat. *J. Biomech.* **26 (8),** pp. 945-953.

Herzog, W., Stano, A., and Leonard, T.R. (1993b) Telemetry System to Record Force and EMG from Cat Ankle Extensor and Tibialis Anterior Muscles. *J. Biomech.* **26**, pp. 1463-1471.

Herzog, W., Archambault, J.M., Leonard, T.R., and Nguyen, H.K. (1996) Evaluation of the Implantable Force Transducer for Chronic Tendon-force Recordings. *J. Biomech.* **29 (1),** pp. 103-109.

Heywood, E.J., Jeutter, D.C. and Harris, G.F. (2004) Tri-axial Plantar Pressure Sensor: Design, Calibration and Characterization. *Proceedings of the 26th Annual International Conference of the IEEE EMBS* (eds.). San Francisco, pp. 2010-2013.

Hill, A.V. (1940) The Dynamic Constants of Human Muscle. *Proc. Roy. Soc.* **128 (B),** pp. 263-274.

Hinrichs, R.N. (1985) Regression Equations to Predict Segmental Moments of Inertia from Anthropometric Measurements: An Extension of the Data of Chandler et al., 1975. *J. Biomech.* **18 (18),** pp. 621-624.

Hodgson, J.A. (1983) The Relationship Between Soleus and Gastrocnemius Muscle Activity in Conscious Cats - A Model for Motor Unit Recruitment? *J. Physiol.* **337,** pp. 553-562.

Hof, A.L. and van den Berg, J. (1981a) EMG to Force Processing I: An Electrical Analog of the Hill Muscle Model. *J. Biomech.* **14 (11),** pp. 747-758.

Hof, A.L. and van den Berg, J. (1981b) EMG to Force Processing II: Estimation of Parameters of the Hill Muscle Model for the Human Triceps Surae by Means of a Calf-ergometer. *J. Biomech.* **14 (11),** pp. 759-770.

Hof, A.L. and van den Berg, J. (1981c) EMG to Force Processing III: Estimation of Model Parameters for the Human Triceps Surae Muscle and Assessment of the Accuracy by Means of a Torque Plate. *J. Biomech.* **14,** pp. 771-785.

Hof, A.L. and van den Berg, J. (1981d) EMG to Force Processing IV: Eccentri-concentric Contractions on a Spring-flywheel Set Up. *J. Biomech.* **14 (11),** pp. 787-792.

Hoffer, J.A., Caputi, A.A., Pose, I.E., and Griffiths, R.I. (1989) Roles of Muscle Activity and Load on the Relationship Between Muscle Spindle Length and Whole Muscle Length in the Freely Walking Cat. *Prog. Brain Res.* **80,** pp. 75-85.

Holden, J.P., Grood, E.S., and Cummings, J.F. (1991) The Effect of Flexion Angle on Measurement of Anteromedial Band Force in the Goat ACL. *Proc. 37th Meeting of the Orthopedic Res. Soc.* p. 588.

Hosein, R. and Lord, M. (2000) A Study of In-shoe Plantar Shear in Normals. *Clin. Biomech.* **15 (1),** pp. 46-53.

Howe, J.G., Wertheimer, C., Johnson, R.J., Nichols, C.E., Pope, M.H., and Beynnon, B. (1990) Arthroscopic Strain Gauge Measurement of the Normal Anterior Cruciate Ligament. *Arthroscopy: The Journal of Arthroscopic and Related Surgery.* **6 (3)**, pp. 198-204.

Hsiao, H., Guan, J. and Weatherly, M. (2002) Accuracy and Precision of Two In-shoe Pressure Measurement Systems. *Ergonomics.* **45 (8),** pp. 537-555.

Huang, H.K. and Saurez, F.R. (1983) Evaluation of Cross-sectional Geometry and Mass Density Distribution of Humans and Laboratory Animals Using Computerized Tomography. *J. Biomech.* **16 (10),** pp. 821-832.

Hughes, R., Rowlands, H. and McMeekin, S. (2000) A Laser Plantar Pressure Sensor for the Diabetic Foot. *Medical Engineering & Physics.* **22 (2),** pp. 149-154.

Huiskes, R. (1980) Some Fundamental Aspects of Joint Replacement. *Acta. Orthop. Scand.* Suppl. p. 185.

Huiskes, R. and Sloof, T.J. (1979) Experimentelle und rechnerische Spannungsanalyse der Hüftgelenkverankerung. *Proceedings: Pauwels Symposium.* Free University Berlin, Berlin. pp. 173-198.

Hunter, I.W., Kearney, R.E, Jones, L.A. (1987) Estimation of the Conduction Velocity of Muscle Action Potentials using Phase and Impulse Response Function Techniques. *Med. Biol. Eng. Comput.* **25(2),** pp. 121-126.

Hutton, W.C. and Dhanandran, M. (1981) The Mechanics of Normal and Halux Valgus Feet: A Quantitative Study. *Clin. Orthop.* **157,** pp. 7-13.

Inman, V.T., Ralston, H.J., Saunders, J.B. de C.M., Feinstein, B., and Wright Jr., E.W. (1952) Relation of Human Electromyogram to Muscular Tension. *Electroenceph. Clin. Neurophysiol.* **4,** pp. 187-194.

Jacquot, P., Rastogi, P.K., and Pflug, L. (1984) Mechanical Testing of the External Fixator by Holographic Interferometry. *Orthopaedics.* **7 (3),** pp. 513-523.

Jacob, H.A.C. and Huggler, A.H. (1979) Spannungsanalysen an Kunststoffmodellen des menschlichen Beckens sowie des proximalen Femurendes mit und ohne Prothese. *Proceedings: Pauwels Symposium.* Free University Berlin, Berlin. pp. 118-135.

Jensen, R.K. (1976) Model for Body Segment Parameters. *Biomechanics V-B* (ed. Komi, P.V.). University Park Press, Baltimore. pp. 380-386.

Jensen, R.K. (1978) Estimation of the Biomechanical Properties of Three Body Types Using a Photogrammetric Method. *J. Biomech.* **11,** pp. 349-358.

Jensen, R.K. (1986) Body Segment Mass, Radius, and Radius of Gyration Proportions of Children. *J. Biomech.* **19 (5),** pp. 359-368.

Jensen, R.K. (1989) Changes in Segment Inertial Proportions Between 4 and 20 Years. *J. Biomech.* **22 (6-7),** pp. 529-536.

Kalen, R. (1961) Strains and Stresses in the Upper Femur Studied by the Stresscoat Method. *Acta. Orthop. Scand.* **31 (2),** pp. 103-113.

Karara, H. (1980) Non-Metric Cameras. *Developments in Close Range Photogrammetry* (ed. Atkinson, K.). Applied Science Publishers Ltd., London, England. **1,** pp 65-70.

Karlsson, S., Yu, J. and Akay, M. (2000) Time-Frequency Analysis of myoelectric Signals During Dynamic Contractions: A Comparative Study. IEEE Trans. *Biomed. Eng.* **47(2),** pp. 228-238.

Karlsson, S., Gerdle, B., (2001) Mean Frequency and Signal Amplitude of the Surface EMG of the Quadriceps Muscles Increase with Increasing Torque—A Study using the Continuous Wavelet Transform. *J. Electromyogr. Kinesiol.* **11 (2),** pp. 131-140.

Kentie, M.A., van der Schee, E.J., Grashuis, J.L., and Smout, A.J.P.M. (1981) Adaptive Filtering of Canine Electrogastrographic Signals. Part 2: Filter Performance. *Med. & Biol. Eng. & Comput.* **19,** pp. 765-769.

Kihara, T., Unno, M., Kitada, C., Kubo, H., and Nagata, R. (1987) Three-dimensional Stress Distribution Measurement in a Model of the Human Ankle Joint by Scattered-light Polarizer Photoelasticity: Part 2. *Appl. Optics.* **26,** pp. 643-649.

King, A.I., Ruan, J.S., Zhou, C., Hardy, W.N., and Khalil, T.B. (1995) Recent Advances in Biomechanics of Brain Injury Research: A Review. *J. Neurotrauma.* **12 (4),** pp. 651-658.

Kohles, S.S., Vanderby Jr., R., Belloli, D.M., Thielke, R.J., Bowers, J.R., and Sandor, B.I. (1989) Differential Infrared Thermography: A Correlation with Stress and Strain in Cortical Bone. *AMD - Vol 98, Biomechanics Symposium* (eds. Torzilli, P.A. and Friedman, M.H.). **ASME**. pp. 81-84.

Kojima A., Ogawa, R., Izuchi, N., Matsumoto, T., Iwata, I., and Nagata, R. (1986) Holographic Investigation of the Mechanical Properties of Tibia Fixed with Internal Fixation Plates. *Biomechanics: Basic and Applied Research* (eds. Bergmann, A., Kolbel, R., and Rohlmann, A.). Martinius Nijoff. pp. 243-248.

Komi, P.V. (1990) Relevance of In-vivo Force Measurements to Human Biomechanics. *J. Biomech.* **23 (S-1)**, pp. 23-24.

Komi, P.V. and Tesch, P. (1979) EMG Frequency Spectrum, Muscle Structure, and Fatigue During Dynamic Contractions in Man. *Eur. J. Appl. Physiol. Occup. Physiol.* **32,** pp. 41-50.

Komi, P.V., Salonen, M., Jarvinen, N., and Kokko, O. (1987) In-vivo Registration of Achilles Tendon Forces in Man. *Int. J. Sports Med.* **8,** pp. 3-8.

Koulaxouzidis, A.V., Holmes, M.J., Roberts, C.V. and Handerek, V.A. (2000) A Shear and Vertical Stress Sensor for Physiological Measurements using Fibre Bragg Gratings. *Proceedings of the 22nd Annual International Conference of the IEEE EMBS.* Volume 1 (ed. Enderle, J.D.). Chicago, pp. 55-58.

Kotzar, G.M., Davy, D.T., Goldberg, V.M., Heiple, K.G., Berilla, J., Heiple Jr., K.G., Brown, R.H., and Burstein, A.H. (1991) Telemeterized In-vivo Hip Joint Force Data: A Report on Two Patients After Total Hip Surgery. *J. Ortho. Res.* **9 (15),** pp. 621-633.

Kwee, H.H. (1971) *Neuromuscular Control of Human Forearm Movements with Active Dynamic Loading.* Ph.D. Thesis, McGill University, Montreal. Nat'l Library of Canada, Ottawa.

Ladin, Z., Mansfield, P.K., Murphy, M.C., and Mann, R.W. (1990) Segmental Analysis in Kinesiological Measurements. *Image Based Motion Measurement, SPIE* (ed. Walton, J.S.). **1356,** pp. 110-120.

Lafortune, M.A. and Cavanagh, P.R. (1987) Three-dimensional Kinematics of the Patella During Walking. *Biomechanics X-A* (ed. Jonsson, B.). Human Kinetics, Champaign, IL. pp. 337-341.

Lafortune, M.A. and Hennig, E.M. (1989) Contribution of Angular Motion and Gravity to Tibial Acceleration. *Med. and Science in Sports and Exercise.* **23 (3)**, pp. 360-363.

Laing, P., Deogan, H., Cogley, D., Crerand, S., Hammond, P., and Klenerman, L. (1992) The Development of the Low Profile Liverpool Shear Transducer. *Clinical Physics and Physiological Measurement.* **13**, pp. 115-124.

Lanyon, L.E. (1972) In-vivo Bone Strain Recorded from Thoracic Vertebrae of Sheep. *J. Biomech.* **12,** pp. 593-600.

Laughlin, D.R. (1989) A Magneto-hydrodynamic Angular Motion Sensor for Anthropomorphic Test Device Instrumentation. *Proc. 33rd Stapp Car Crash Conf.* Appl. Tech. Assoc., Albequerque, New Mexico. Paper 12SC31.

Lee, V.S., Solomonidis, S.E., and Spence, W.D. (1997) Stump-socket Interface Pressure as an Aid to Socket Design in Prostheses for Trans-femoral Amputees-A Preliminary Study. *J. Eng. in Medicine, Part H.* **211 (2),** pp. 167-180.

Lephart, S.A. (1984) Measuring the Inertial Properties of Cadaver Segments. *J. Biomech.* **17 (7),** pp. 537-543.

Lesh, M.D., Mansour, J.M., and Simon, S.R. (1979) A Gait Analysis Subsystem for Smoothing and Differentiation of Human Motion Data. *J. Biomech. Eng.* **101 (3),** pp. 205-212.

Levens, A.S., Inman, V.T., and Blosser, J.A. (1948) Transverse Rotation of the Segments of the Lower Extremity in Locomotion. *J. Bone Jt. Surg.* **30 (A-4),** pp. 859-872.

Lewis, J., Askey, M., and Jaycox, D. (1982) A Comparative Evaluation of Tibial Component Designs of Total Knee Prostheses. *J. Bone Jt. Surgery.* **64 (A-1)**, pp. 129-135.

Light, L.H., McLellan, G.E., and Klenerman, L. (1980) Skeletal Transients on Heel Strike in Normal Walking with Different Footwear. *J. Biomech.* **13,** pp. 477-480.

Lindholm, L.E. and Oeberg, K.E. (1974) An Opto-electric Instrument for Remote On-line Movement Monitoring. *Biomechanics IV* (eds. Nelson, R.C. and Morehouse, C.A.). University Park Press, Baltimore. pp. 510-512.

Lindstrom, L. (1974) Contributions to the Interpretation of Myoelectric Power Spectra. *Ph.D. thesis*, Chalmers University, Goteborg.

Lindström, L., Kadefors, R., and Petersén, I. (1977) An Electromyographic Index for Localized Muscle Fatigue. *J. Appl. Physiol.* **43,** pp. 750-754.

Lippold, O.C.J. (1952) The Relation Between Integrated Action Potential in Human Muscle and its Isometric Tension. *J. Physiol.* **117 (4),** pp. 492-499.

Little, E.G. (1985) A Static Experimental Stress Analysis of the Geomedic Knee Joint Using Embedded Strain Gauges. *Eng. Medicine.* **14 (2)**, pp. 69-74.

Little, E.G. (1990) Three-dimensional Strain Rosettes Applied to an Analysis of the Geomedic Knee Prosthesis and Cement Fixation. *Appl. Stress Analysis*. Elsevier, New York.

Little, E.G., Tocher, D., and McTague, D. (1980) The Validation of Strain Gauge Measurements. *Strain Measurement in Biomechanics, Canadian Medical and Biological Engineering Society.* pp. 73-86.

Little, E.G. and O'Keefe, D. (1989) An Experimental Technique for the Investigation of Three-dimensional Stress in Bone Cement Underlying a Tibial Plateau. *Proc. Inst. Mech. Eng.* **203 (1)**, pp. 35-41.

Liu, M.M. (1997) *Dynamic Muscle Force Predictions From EMG Signals Using Artificial Neural Networks.* M.Sc. Thesis, Department of Electrical Engineering. University of Calgary.

Liu, Z.J. and Herring, S.W. (2000) Bone Surface Strains and Internal Bony Pressures at the Jaw Joint of the Miniature Pig during Masticatory Muscle Contraction. *Arch. Oral. Biol.* **45 (2),** pp. 95-112.

Liu, W. and Nigg, B.M. (2000) A Mechanical Model to determine the Influence of Masses and Mass Distribution on the Impact Force during Running. *J. Biomech.* **33,** 219-224.

Lochner, F.K., Milne, D.W., Mills, E.J., and Groom, J.J. (1980) In-vivo and In-vitro Measurement of Tendon Strain in the Horse. *Am. J. Vet. Res.* **41 (12)**, pp. 1929-1937.

Loeb, G.E. and Gans, C. (1986) *Electromyography for Experimentalists.* The University of Chicago Press, Chicago, IL.

Lord, M. (1981) Foot Pressure Measurement: A Review of Methodology. *J. Biomed. Eng.* **3,** pp. 91-99.

Lord, M. and Hosein, R. (2000) A Study of In-shoe Plantar Shear in Patients with Diabetic Neuropathy. *Clin. Biomech.* **15 (4),** pp. 278-283.

Lord, M., Hosein, R., and Williams, R.B. (1992) Method for In-shoe Shear Stress Measurement. *J. Biomed. Eng.* **14**, pp. 181-186.

Luck, J., Small, D. and Little, C. Q. (2001) Real-time tracking of articulated human models using a 3D shape-from-silhouette method. *Robot Vision, International Workshop Rob.Vis.* **98**, pp. 19-26.

Mak, A.F.T., Zhang, M. and Boone, D.A. (2001) State-of-the-art Research in Lower-limb Prosthetic Biomechanics-Socket Interface. *Journal of Rehab. Res. & Dev.* **38 (2),** pp. 161-174.

Mallat, S.G. (1998) *A Wavelet Tour Of Signal Processing.* Academic Press, San Diego, Lon-don, Boston, New York, Sydney, Tokyo, Toronto.

Maness, L.W., Benjamin, M., Podoloff, R., Bobick, A., and Golden, R.F. (1987) Computerized Occlusal Analysis: A New Technology. *Quintessence International.* **18 (4)**.

Manley, M.T., Ovryn, B., and Stern, L.S. (1987) Evaluation of Double-exposure Holographic Interferometry for Biomechanical Measurements In-vitro. *J. Ortho. Res.* **5 (1)**, pp. 144-159.

Marey, E.J. (1972) *Movement.* Arno, New York. (Original 1895).

Marey, M. (1873) De la Locomotion Terrestre chez les Bipedes et les Quadrupedes. *J. de l'Anat. et de la Physiol.* **9,** p. 42.

Martin, A.D. (1991) *Variability in the Measures of Body Fat.* Sport and Exercise Sciences Institute, University of Manitoba, Winnipeg.

Martin, P.E., Mungoile, M., Marzke M.W., and Longhill J.M. (1989) The Use of Magnetic Resonance Imaging for Measuring Segment Inertial Properties. *J. Biomech.* **22 (4),** pp. 367-376.

Martini, K.H. (1983) Multicomponent Dynamometers Using Quartz Crystals as Sensing Elements. *ISA Transactions.* **22 (1)**, pp. 35-46.

Merletti, R., Lo Conte, L.R. (1997) Surface EMG Signal Processing During Isometric Contractions. *J. Electromyogr. Kinesiol.* **7(4)**, pp. 241 .250.

Merlo, E., Pozzo, M., Antonutto, G., di Prampero, P. E., Merletti, R., Farina, D. (2005) Time-Frequency Analysis and Estimation of Muscle Fiber Conduction Velocity from Surface EMG Signals During Explosive Dynamic Contractions. *J. Neurosci. Meth.* **142**, pp. 267-274

Mesnager, M. (1930) Sur la Détermination Optique des Tensions Intérieures dans les Solides Trois-dimensions. *Compt. Rendu.* **190 (22),** pp. 1249-1250.

Miles, A.W. and Dall, D.M. (1985) An Experimental Study of Femoral Cement Stress in Total Hip Replacement: Influence of Structural Stiffness of the Femoral Stem. *Eng. Medicine.* **14 (3)**, pp. 133-135.

Miller, D.I. and Nelson, R.C. (1973) *Biomechanics of Sport*. Lea & Febiger, Philadelphia.

Milner-Brown, H.S. and Stein, R.B. (1975) The Relation Between the Surface Electromyogram and Muscular Force. *J. Physiol.* **246,** pp. 549-569.

Mitchelson, D.L. (1988) Automated Three-dimensional Movement Analysis Using the CODA-3 System. *Biomedizinische Technik.* **33 (7-8)**, pp. 179-182.

Miura, M., Miashita, M., Matsui, H., and Sodeyama, H. (1974) Photographic Method of Analyzing the Pressure Distribution of the Foot Against the Ground. *Biomechanics IV* (eds. Nelson, R.C. and Morehouse, C.A.). University Park Press, Baltimore. pp. 482-487.

Moeslund, T., et al. (2001) A Survey of computer vision-based human motion capture. *Computer Vision and Image Understanding* **81**, pp. 231-268

Moore, K.L. (1980) *Clinically Oriented Anatomy*. Williams & Wilkins, Baltimore.

Morice, P.B. and Base, G.D. (1953) The Design and Use of a Demountable Mechanical Strain Gauge for Concrete Structures. *Magazine of Concrete Res.* **5 (13)**, pp. 37-42.

Moritani, T. and deVries, H.A. (1978) Re-examination of the Relationship Between the Surface Integrated Electromyogram and Force of Isometric Contraction. *Am. J. Phys. Med.* **57 (6),** pp. 263-277.

Morlock, M. and Nigg, B.M. (1991) Theoretical Considerations and Practical Results on the Influence of the Representation of the Foot for the Estimation of Internal Forces with Models. *Clin. Biomech.* **6,** pp. 3-13.

Morris, J.R. (1973) Accelerometry: A Technique for the Measurement of Human Body Movements. *J. Biomech.* **6 (6),** pp. 729-736.

Morton, D.J. (1930) Structural Factors in Static Disorders of the Foot. *Amer. J. of Surgery.* **9,** pp. 315-326.

Mündermann L., et al. (2005) *Videometrics VIII IS&T/SPIE*, San Jose, CA. pp. 268-287.

Mündermann, L., Anguelov, D., Corazza, S., Chaudhari, A.M., and Andriacchi, T.P. (2005) Validation of a Marker-less Motion Capture System for the Calculation of Lower Extremity Kinematics. *ISB XXth Congress*, Cleveland, Ohio.

Mündermann, A., Wakeling, J.M., Nigg, B.M., Humble, R.N., and Stefanyshyn, D. J. (2005) Foot orthoses affect frequency components of muscle activity in the lower extremity. *Gait. Posture. 8 June 2005, e-publication ahead of print.*

Neukomm, P.A. and Denoth, J. (1978) Messmethoden. *Biomechanische Aspekte zu Sportplatzbelägen* (ed. Nigg, B.M.). Juris Verlag, Zürich. pp. 28-45.

Nicol, K. and Hennig, E.M. (1976) Time Dependent Method for Measuring Force Distribution Using a Flexible Mat as a Capacitor. *Biomechanics V-B* (ed. Komi, P.V.). University Park Press, Baltimore. pp. 433-440.

Nicol, K. and Rusteberg, D. (1991) Measurement of Pressure Distribution on Curved and Soft Surfaces: Applications in Medicine. Institut für Sport und Sportwissenschatt der Johann Wolfgang Goethe-Universität.

Nigg, B.M. and Neukomm P.A. (1973) Erschütterungsmessungen beim Skifahren. *Med. Welt.* **24 (48)**, pp. 1883-1885.

Nigg, B.M., Neukomm, P.A., and Unold, E. (1974) Biomechanik und Sport: Über Beschleunigungen die am menschlichen Körper bei verschiedenen Bewegungen auf verschiedenen Unterlagen auftreten. *Orthopäde.* **3,** pp. 140-147.

Nigg, B.M. (1977) *Menschliche Microvibrationen*. Birkhäuser Verlag, Basel.

Nigg, B.M and Denoth, J. (1980) *Sportplatzbeläge*. Juris Verlag, Zürich.

Nigg, B.M., Denoth, J., and Neukomm, P.A. (1981) Quantifying the Load on the Human Body: Problems and Some Possible Solutions. *Biomechanics VII-B* (eds. Morecki, A., Fidelus, K., Kedzior, K., and Wit, A.). University Park Press, Baltimore. pp. 88-99.

Nigg, B.M. (1985) Loads in Selected Sports Activities: An Overview. *Biomechanics IX-B* (eds. Winter, D.A., Norman, R.W., Wells, R.P., Hayes, K.C., and Patla, A.E.). Human Kinetics Pub., Champaign, IL. pp. 91-96.

Nigg, B.M., Bahlsen, H.A., Lüthi, S.M., and Stokes, S. (1987) The Influence of Running Velocity and Midsole Hardness on External Impact Forces in Heel-toe Running. *J. Biomech.* **20 (10)**, pp. 951-959.

Nigg, B.M. (1988) The Assessment of Loads Acting on the Locomotor System in Running and Other Sport Activities. *Seminars in Orthopaedics Vol. 3.* **4 (Dec.),** pp. 197-206.

Nigg, B.M. (1997) Impact Forces in Running. *Curr. Opin. Orthopaed.* **8,** 43-47.

Nigg, B.M. and Liu, W. (1999) The Effect of Muscle Stiffness and Damping on Simulated Impact Force Peaks during Running. *J. Biomech.* **32,** 849-856.

Nigg, B.M. and Wakeling, J.M. (2001) Impact Forces and Muscle Tuning: A New Paradigm. *Exerc. Sport Sci. Rev.* **29,** 37-41.

Nolte, V. (1979) *Die Handschrift des Ruderers*. Messtechnische Briefe. **15,** pp. 49-53.

Norman, R.W., Bishop, P.J., Pierrynowski, M.R., and Pezzack, J.C. (1979) Aircrew Helmet Protection Against Potential Cerebral Concussion in Low Magnitude Impacts. *Aviat., Space, and Environ. Med.* **50 (6),** pp. 553-561.

Novotny, M. and Kilpi, J. 2.2 Pressure Sensors. Available from http://www.ad.tut.fi/aci/courses/7606010/pdf/PressureSensors.pdf, 25/02/2005.

Noyes, R.F., Butler, D.L., Grood, E.S., Zernicke, R.F., and Hefzy, M.S. (1984) Biomechanical Analysis of Human Ligament Grafts Used in Knee Ligament Repairs and Reconstructions. *J. Bone Jt. Surg.* **66 (A)**, pp. 344-352.

Oliver, D.E. (1987) Stress Pattern Analysis by Thermal Emission. *Handbook on Experimental Mechanics* (ed. Kobayashi, A.S.). pp. 610-620.

Olney, S.J. and Winter, D.A. (1985) Predictions of Knee and Ankle Moments of Force in Walking from EMG and Kinematic Data. *J. Biomech.* **18 (1),** pp. 9-20.

Orizio, C. (1993) Muscle Sound: Bases for the Introduction of a Mechanomyographic Signal in Muscle Studies. *CRC Reviews in Biomedical Engineering.* **21 (3)**, pp. 201-243.

Orr, J.F., James, W.V., and Bahrani, A.S. (1985) The Effects of Hip Prosthesis Stem Cross-sectional Profile on the Stresses Induced in Bone Cement. *Eng. Medicine.* **15**, pp. 13-18.

Orr, J.F., James, W.V., and Bahrani, A.S. (1985) A Preliminary Study of the Effects of Medio-lateral Rotation on Stresses in an Artificial Hip Joint. *Eng. Medicine.* **14**, pp. 39-42.

Oster, G. and Jaffe, J.S. (1980) Low Frequency Sounds from Sustained Contraction of Human Skeletal Muscle. *Biophysical. J.* **30**, pp. 119-128.

Perry, J.E., Davis, B.L., and Hall, J.O. (1995) Profiles of Shear Loading in the Diabetic Foot. *XVth Congress of the International Society of Biomechanics: Book of Abstracts* (eds. Hakkinen, K., Keskinen, K.L., Komi, P.V., and Mero, A.). Jyväskylä, Finland, Gummerus. pp. 722-723.

Petrofsky, J.S., Glaser, R.M., and Phillips, C.A. (1982) Evaluation of the Amplitude and Frequency Components of the Surface EMG as an Index of Muscle Fatigue. *Ergonomics.* **25,** pp. 213-223.

Peyton, A.J. (1986) Determination of the Moment of Inertia of Limb Segments by a Simple Method. *J. Biomech.* **19 (5),** pp. 405-410.

Plagenhoef, S.C. (1968) Computer Program for Obtaining Kinetic Data on Human Movement. *J. Biomech.* **1,** pp. 221-234.

Platt, D.M., Wilson, A.M., and Goodship, A.E. (1992) Techniques for Tendon Force Measurement. *Experimental Mechanics: Technology Transfer Between High Tech Engineering and Biomechanics* (ed. Little, E.G.). Elsevier, New York. pp. 389-398.

Podoloff, R.M. and Benjamin, M. (1989) A Tactile Sensor for Analyzing Dental Occlusion. *Sensors.* **6,** pp. 41-47.

Polliack, A.A., Sieh, R.C., Craig, D.D., Landsberger, S., McNeil, D.R. and Ayyappa, E. (2000) Scientific Validation of Two Commercial Pressure Sensor Systems for Prosthetic Fit. *Prosthet. Orthot. Int.* **24 (1),** pp. 63-73 (Abstract).

Pople, J. (1983) Errors and Uncertainties in Strain Measurement. *Strain Gauge Technology.* Applied Science Publishers, New York. pp. 209-264.

Prokop, L. (1972) Die Auswirkungen von Kunststoffbahnen auf den Bewegungsapparat. *Oesterr. J. für Sportmedizin.* **2,** pp. 3-19.

Quesada, P.M. and Rash, G.S. (2000) Quantitative Assessment of Simultaneous Capacitive and Resistive Plantar Pressure Measurements During Walking. *Foot & Ankle International.* **21 (11),** pp. 928-934.

Quesada, P.M., Rash, G.S. and Jarboe, N. (1996) Assessment of Pedar and F-Scan revisited. *Clin. Biomech.* **12 (3),** pp. 15.

Rack, P.M.H. and Westbury, D.R. (1969) The Effects of Length and Stimulus Rate on Tension in the Isometric Cat Soleus Muscle. *J. Physiol.* **204,** pp. 443-460.

Razian, M.A. and Pepper, M.G. (2003) Design, Development and Characteristics of an In-shoe Triaxial Pressure Measurement Transducer Utilizing a Single Element of Piezoelectric Copolymer Film. *IEEE Transactions on Neural Systems and Rehabilitation Engineering* **11 (3),** pp. 288-293.

Redner, S. (1968) Photoelasticity. *Encyclopedia of Polymer Science and Technology.* **9**, pp. 590-610.

Reinschmidt, C., van den Bogert, A.J., Lundberg, A., Nigg, B.M., Murphy, N., Stacoff, A., and Stano, A. (1997a) Tibiofemoral and Tibiocalcaneal Motion During Walking: External Versus Skeletal Markers. *Gait and Posture.* **6 (1997),** pp. 98-109.

Reinschmidt, C., van den Bogert, A.J., Nigg, B.M., Lundberg, A., and Murphy, N. (1997b) Effect of Skin Movement on the Analysis of Skeletal Knee Joint Motion During Running. *J. Biomech.* **30,** pp. 729-732.

Rhatigan, B.A., Mylrea, K.C., Lonsdale, E., and Stern, L.Z. (1986) Investigation of Sounds Produced by Healthy and Diseased Human Muscular Contraction. *IEEE Trans. Biomed. Eng.* **BME-33**, pp. 967-971.

Rhodes, A., Sherk, H.H., Black, J., and Margulies, C. (1988) High Resolution Analysis of Ground Foot Reaction Forces. *Foot & Ankle.* **9 (3),** pp. 135-138.

Riemersma, D.J., van den Bogert, A.J., Schamhardt, H.C., and Hartman, W. (1988) Kinetics and Kinematics in the Equine Hind Limbs: In-vivo Tendon Strain and Joint Kinematics. *Am. J. Vet. Res.* **49 (8),** pp. 1353-1359.

Rodrigue, D. and Gagnon, M. (1983) The Evaluation of Forearm Density with Axial Tomography. *J. Biomech.* **16 (11),** pp. 907-913.

Rosenhahn, B., Kersting, U. G., Smith, A. W., Gurney J. K., Brox, T. and Klette, R. (2005) A system for marker-less human motion estimation. (eds W. Kropatsch, R. Sablatnig, and A. Hanbury), *Pattern Recognition*, Springer LNCS 3663, Vienna, Austria. pp. 230-237.

Saha, S. and Lakes, R.S. (1979) The Effect of Soft Tissue on Wave Propagation and Vibration Tests for Determining the In-vivo Properties of Bone. *J. Biomech.* **10 (7),** pp. 393-401.

Salmons, S. (1969) The 8th Annual International Conference on Medical and Biological Engineering: Meeting Report. *Biomed. Eng.* **4,** pp. 467-474.

Sanders J.E., Bell, D.M., Okumura, R.M. and Dralle, A.J. (1998) Effects of Alignment Changes on Stance Phase Pressures ad Shear Stresses on Transtibial Amputees: Measurements from 13 Transducer Sites. *IEEE Trans. Rehab. Eng.* **6 (1),** pp. 21-31.

Sanders, J.E., Zachariah, S.G., Jacobsen, A.K. and Fergason, J.R. (2005) Changes in Interface Pressures and Shear Stresses over Time on Trans-tibial Smputee Subjects Ambulating with Prosthetic Limbs: Comparison of Diurnal and Six-Month Differences. *J. Biomech.*

Sanderson, D.J., Hennig, E.M. and Black, A.H. (2000) The Influence of Cadence and Power Output on Force Application and In-shoe Pressure Distribution during Cycling by Competitive and Recreational Cyclists. *Journal of Sports Sciences.* **18**, 173-181.

Santschi, W.R., Du Bois, J., and Omoto, C. (1963) *Moments of Inertia and Centres of Gravity of the Living Human Body.* Aerospace Medical Research Laboratory (Report No. TDR-63-36). Wright-Patterson Air Force Base, OH.

Sauerwald, F. and Wieland, H. (1925) Über die Kerbschlagprobe nach Schule-Moser. *Z. Metalkunde.* **17**, pp. 358-364, 392-399.

Savelberg, H.H.C.M. and Herzog, W. (1997) Prediction of Dynamic Forces From Electromyographic Signals: An Artifical Neural Network Approach. *J. Neurosci. Methods.* **78,** pp. 65-74.

Scheirman, G., Porter, J., Leigh, M., and Musick, D. (1998) An integrated method to obtain three-dimensional coordinates using panning and tilting video cameras. *Proc. of the 16th annual symposium of the International Society of Biomechanics in Sports*, Konstanz, Germany.

Schneider, E. (1980) *Leistungsanalyse bei Rudermannschaften.* Limpert Verlag, Bad Homburg.

Schwartz, R.P. and Heath, A.L. (1937) Some Factors Which Influence the Balance of the Foot in Walking: The Stance Phase of Gait. *J. Bone Jt. Surg.* **19 (2),** pp. 431-442.

Shelton, J.C., Gorman D., and Bonfield W. (1990) Application of Holographic Interferometry to Investigate Internal Fracture Fixations Plates. *J. Mater. Sci. Mater. in Medicine.* **1**, pp. 146-153.

Shem, K.L., Breakey, J.W. and Werner, P.C. (1998) Pressures at the Residual Limb-Socket Interface in Transtibial Amputees with Thigh Lacer-Slide Joints. *Journal of Prosthetics and Orthotics.* **10 (3),** pp. 51-55.

Shereff, M.J., Bregman, A.M., and Kummer, F.J. (1985) The Effect of Immobilization Devices on the Load Distribution Under the Foot. *Clin. Orthop.* **192,** pp. 260-267.

Sherif, M.H., Gregor, R.J., Liu, L.M., Roy, R.R., and Hager, C.L. (1983) Correlation of Myoelectric Activity and Muscle Force During Selected Cat Treadmill Locomotion. *J. Biomech.* **16 (9),** pp. 691-701.

Shrive, N.G., Damson, E.L., and Frank, C.B. (1992) Technology Transfer Regarding the Measurement of Strain on Flexible Materials with Special Reference to Soft Tissues. *Experimental Mechanics: Technology Transfer Between High Tech Engineering and Biomechanics* (ed. Little, E.G.). Elsevier, New York. pp. 121-130.

Shrive, N.G., Damson, E.L., Meyer, R.A., and Iverslie, S.P. (1993) Soft Tissue Extensometer. U.S. Patent Application 08/112,841.

Snell, R.S. (1973) *Clinical Anatomy for Medical Students.* Little Brown & Company, Boston. pp. 1-5.

Soames, R.W. (1985) Foot Pressure Patterns During Gait. *J. Biomech. Eng.* **7,** pp. 120-126.

Söderquist, I. and Wedin, P. (1993) Determining the Movements of the Skeleton using Well-configured Markers. *J. Biomech.* **26,** pp. 1473-1477.

Solakian, A.G. (1935) A New Photoelastic Method. *Mech. Eng.* **57**, pp. 767-771.

Soudan, K. and Dierckx, P. (1979) Calculation of Derivatives and Fourier Co-efficients of Human Motion Data, while using Spline Functions. *J. Biomech.* **12 (1),** pp. 21-26.

Spoor, C.W. (1984) Explanation, Verification, and Application of Helical-axis Error Propagation Formulas. *Human Movement Sci.* **3,** pp. 95-117.

Spoor, C.W. and Veldpaus, F.E. (1980) Rigid Body Motion Calculated from Spatial Coordinates of Markers. *J. Biomech.* **13,** pp. 391-393.

Steen Jensen, J. (1978) A Photoelastic Study of a Model of the Proximal Femur. *Acta. Orthop. Scandinavia.* **49**, pp. 54-59.

Stegeman, D. F. and Linssen, W. H. J. P. (1992) Muscle Fiber Action Potential Changes and Surface EMG: A Simulation Study. *J. Electromyogr. Kinesiol.* **2(3**), pp. 130-140.

Stegeman, D. F., Blok, J.H., Hermens, H. J. and Roeleveld, K. (2000) Surface EMG Models: Properties and Applications. *J. Electromyogr. Kinesiol.* **10(5)**, pp. 313-326.

Stein, P.K. (1992) The Lack of Technology Transfer Within Low-tech Engineering Disciplines: How it has Affected Experimental Stress Analysis History. *Experimental Mechanics* (ed. Little, E.G.). Elsevier, New York. pp. 201-213.

Sterzing, T.F. and Hennig, E.M. (1999) Measurement of Plantar Pressures, Rearfoot Motion and Tibial Shock During Running 10 km on a 400 m Track. *Proceedings of Fourth Symposium on Footwear Biomechanics.* (eds. Hennig, E.M., Stefanyshyn, D.J.). International Society of Biomechanics, Canmore, Canada, pp. 88-89.

Stokes, I.A.F., Faris, I.B., and Hutton, W.C. (1975) The Neuropathic Ulcer and Loads on the Foot in Diabetic Patients. *Acta. Orthop. Scand.* **46,** pp. 839-847.

Stokes, M.J. and Dalton, P.A. (1991) Acoustic Myographic Activity Increases Linearly up to Maximal Voluntary Isometric Force in the Human Quadriceps Muscle. *J. Neurol. Sci.* **101**, pp. 163-167.

Terry, G.C., Hammon, D., France, P., and Norwood, L.A. (1991) The Stabilizing Function of Passive Shoulder Restraints. *Am. J. Sports Med.* **19 (1)**, pp. 26-34.

Theyson, H., Güth, V., and Abbink, F. (1979) Fotogrammetrische Methoden zur 3-dimensionalen Vermessung der Fuβsohle sowie eine Methode zur exakten Bestimmung des Druckablaufs unter den verschiedenen Bereichen der Fuβsohle bein Gehen. *Funktionelle Diagnostik in der Orthopädie.* Stuttgart. pp. 87-90.

Thorstensson, A., Karlsson, J., Viitasalo, J.H.T., Luhtanen, P., and Komi, P.V. (1976) Effect of Strength Training on EMG of Human Skeletal Muscle. *Acta. Physiol. Scand.* **98,** pp. 232-236.

Tichy, J. and Gautschi, G. (1980) *Piezoelektrische Messtechnik.* Springer Verlag, Berlin.

Totsu, K., Haga, Y. and Esashi, M. (2005) Ultra-miniature Fiber-optic Pressure Sensor using White Light Interferometry. *J. Micromech. Microeng.* **15 (1),** pp. 71-75.

Unold, E. (1974) Erschütterungsmessungen beim Gehen und Laufen auf verschiedenen Unterlagen mit verschiedenem Schuhwerk. *Jugend und Sport.* **8**, pp. 289-292.

Urry, S. (1999) Plantar Pressure Measurements Sensors. *Meas. Sci. Technol.* **10 (1),** pp. R16-R32.

Valiant, G.A., McMahon, T.A., and Frederick, E.C. (1987) A New Test to Evaluate the Cushioning Properties of Athletic Shoes. *Biomechanics X-B* (ed. Jonsson, B.). Human Kinetics, Champaign, IL. pp. 937-941.

van Campen, D.H., Croon, H.W., and Lindwer, J. (1979) Mechanical Loosening of Knee-endoprostheses Intermedullary Stems: Influence of Dynamic Loading. *Proceedings 25th Annual ORS.* **98,** pp. 20-22.

van den Bogert, A.J. (1989) *Computer Simulation Locomotion in the Horse.* Ph.D. Thesis, University of Utrecht, Netherlands.

van den Bogert, A.J., Hartman, W., Schamhardt, H.C., and Sauren, A.A.H.J. (1988) In-vivo Relationship Between Force, EMG, and Length Change in Deep Digital Flexor Muscle of the Horse. *Biomechanics XI-A* (eds. Hollander, A.P., Huijing, P.A., and van Ingen Schenau, G.J.). Free University, Amsterdam. pp. 68-74.

van Ruijven, L.J. and Weijs, W.A. (1990) A New Model for Calculating Muscle Forces from Electromyograms. *Eur. J. Appl. Physiol.* **61,** pp. 479-485.

Veldpaus, F.E., Woltring, H.J., and Dortmans, L.J. (1988) A Least-squares Algorithm for the Equiform Transformation from Spatial Marker Co-ordinates. *J. Biomech.* **21 (1),** pp. 45-54.

Voloshin, A.S. and Wosk, J. (1981) Influence of Artificial Shock Absorbers on Human Gait. *Clin. Ortho.* **160,** pp. 52-56.

Voloshin, A.S. and Wosk, J. (1982) An In-vivo Study of Low Back Pain and Shock Absorption in the Human Locomotor System. *J. Biomech.* **15 (1),** pp. 21-27.

Voloshin, A.S., Burger, C.P., Wosk, J., and Arcan, M. (1985) An In-vivo Evaluation of the Leg's Shock Absorbing Capacity. *Biomechanics IX-B* (eds. Winter, D.A., Norman, R.W., Wells, R.P., Hayes, K.C., and Patla, A.E.). Human Kinetics, Champaign, IL. pp. 112-116.

von Gierke, H.E. (1971) Biodynamic Models and their Applications. *J. Acoust. Soc. Am.* **50,** 1397-1413.

von Tscharner, V. (2000) Intensity Analysis in Time-Frequency Space of Surface Myoelectric Signals by Wavelets of Specified Resolution. *J. Electromyogr. Kinesiol.* **10(6)**, pp. 433-45.

von Tscharner, V. (2002) Time-Frequency and Principal-Component Methods for the Analysis of EMGs Recorded during a Mildly Fatiguing Exercise on a Cycle Ergometer. *J. Electromyogr. Kinesiol.* **12(6)**, pp. 479-492.

von Tscharner, V., Goepfert, B., Nigg, B.M. (2003) Changes in EMG Signals for the Muscle Tibialis Anterior while Running Barefoot or with Shoes Resolved by Non-Linearly Scaled Wavelets. *J. Biomech.* **36(8)**, pp.1169-1176.

von Tscharner, V. and Goepfert, B. (2003) Gender-dependent EMGs of Runners Resolved by Time/Frequency and Principal Pattern Analysis. *J. Electromyogr .Kinesiol.* **13(3)**, pp. 253-272.

von Tscharner, V. and Goepfert, B. (2006) Estimation of the Interplay between Groups of Fast and Slow Muscle Fibers of the Tibialis Anterior and Gastrocnemius Muscle While Running. *J. Electromyogr. Kinesiol.* **16(2)**, pp. 188-197.

Yue, Z. and Mester, J. (2002) A Model Analysis of Internal Loads, Energetics, and Efects of Wobbling Mass during the Whole-body Vibration. *J. Biomech.* **35,** 639-647.

Wakeling,J.M. and Nigg, B.M. (2001) Modification of Soft Tissue Vibrations in the Leg by Muscular Activity. *J Appl.Physiol* **90,** 412-420.

Wakeling, J.M., Pascual, S.A., Nigg, B.M., von Tscharner, V. (2001) Surface EMG Shows Distinct Populations of Muscle Activity When Measured during Sustained Sub-Maximal Exercise. *Eur. J. Appl. Physiol.* **86(1)**, pp. 40-47.

Wakeling, J.M., Kaya, M., Temple, G.K., Johnston, I.A., Herzog, W. (2002) Determining Patterns of Motor Recruitment during Locomotion. *J. Exp. Biol.* **205(3)**, pp. 359-369.

Wakeling, J.M., Nigg, B.M., and Rozitis, A.I. (2002) Muscle Activity Damps the Soft Tissue Resonance that Occurs in Response to Pulsed and Continuous Vibrations. *J. Appl. Physiol.* **93,** 1093-1103.

Wakeling, J.M., Syme, D.A. (2002) Wave Properties of Action Potentials from Fast and Slow Motor Units of Rats. *Muscle Nerve.* **26(5)**, pp. 659-668.

Wakeling, J.M., Liphardt, A.M., and Nigg, B.M. (2003) Muscle Activity Reduces Soft-tissue Resonance at Heel-strike during Walking. *J. Biomech.* **36,** 1761-1769.

Wakeling, J.M. and Liphardt, A.M. (2005) Task-specific Recruitment of Motor Units for Vibration Damping. *J. Biomech.* 13 May 2005, *e-publication ahead of print.*

Walker, P.S. and Robertson, D.D. (1988) Design and Fabrication of Cementless Hip Stems. *Clin. Orthop.* **235**, pp. 25-34.

Walmsley, B., Hodgson, J.A., and Burke, R.E. (1978) Forces Produced by Medial Gastrocnemius and Soleus Muscles During Locomotion in Freely Moving Cats. *J. Neurophysiol.* **41 (5),** pp. 1203-1216.

Walmsley, B. and Proske, U. (1981) Comparison of Stiffness of Soleus and Medial Gastrocnemius Muscles in Cats. *J. Neurophysiol.* **46,** pp. 250-259.

Wang, Z.G., Peng, C.L., Zheng, X.L., Wang, P., and Wang, G.R. (1997) Force Measurement on Fracture Site with External Fixation. *Medical and Biological Engineering and Computing.* **35 (3),** pp. 289-290.

Weightman, B. (1976) The Stress in Total Hip Prosthesis Femoral Stems: A Comparative Experimental Study. *Engineering in Medicine.* **2,** pp. 138-147.

Weightman, B. (1977) Stress Analysis. *The Scientific Basis of Joint Replacement* (eds. Swanson, S.A.V. and Freeman, M.A.R.). Pitman, Kent, England. pp. 18-45.

Weinbach, A.P. (1938) Contour Maps, Centre of Gravity, Moment of Inertia, and Surface are of Human Body. *Hum. Biol.* **10,** pp. 356-371.

Weytjens, J.L.F. (1992) *Determinants of Cat Medial Gastrocnemius Muscle Force During Simulated Locomotion.* Thesis, Department of Clinical Neuroscience. University of Calgary. Nat'l Library of Canada, Ottawa.

Weytjens, J.L.F., Van Steenberghe, D. (1984) The Effects of Motor Unit Synchronisation on the Power Spectrum of the Electromyogram. *Biol. Cybern.* **51**, pp. 71-77.

White, A.A. and Raphael, I.G (1972) The Effect of Quadriceps Loads and Knee Position on Strain Measurements of the Tibial Collateral Ligament. *Acta. Orthop. Scand.* **43,** pp. 176-187.

Widrow, B., Glover, J.R., McCool, J.M., Kaunitz, J., Williams, C.S., Hearn, R.H., Zeidler, J.R., and Goodlin, R.C. (1975) Adaptive Noise Cancelling: Principles and Applications. *Proc. IEEE.* **63**, pp. 1692-1716.

Wilke, D.R. (1950) The Relation Between Force and Velocity in Human Muscle. *J. Physiol. Lond.* **110,** pp. 249-280.

Wilke, D.R. (1968) *Studies in Biology, No. 11, Muscle.* Edward Arnold Publishers Ltd., London.

Winter, D.A. (1979) *Biomechanics of Human Movement.* John Wiley & Sons, New York.

Winter, D.A., Fuglevand, A.J., Archer, S. (1994) Crosstalk in Surface Electromyography: Theoretical and Practical Estimates. *J. Electromyogr. Kinesiol,* **4,** pp.15-26.

Winter, D.A., Sidwall, H.G., and Hobson, D.A. (1974) Measurement and Reduction of Noise in Kinematics of Locomotion. *J. Biomech.* **7,** pp. 157-159.

Wollaston, W.H. (1810) On the Duration of Muscle Action. *Philos. Trans. Roy. Soc.* pp. 1-5.

Woltring, H.J. (1985) On Optimal Smoothing and Derivative Estimation from Noisy Displacement Data in Biomechanics. *Human Movement Science.* **4 (3),** pp. 229-245.

Woltring, H.J. (1991) Representation and Calculation of 3-D Joint Movement. *Human Movement Science.* **10 (5),** pp. 603-616.

Woltring, H.J. (1992) 3-D Attitude Representation: A Standardization Proposal. *J. Biomech.*

Woltring, H.J. and Marsolais, E.B. (1980) Optoelectric (Selspot) Gait Measurement in Two- and Three-dimensional Space: A Preliminary Report. *Bulletin of Prosthetics Research.* **17 (2),** pp. 46-52.

Woltring, H.J. and Huiskes, R. (1985) A Statistically Motivated Approach to Instantaneous Helical Axis Estimation from Noisy, Sampled Landmark Coordinates. *Biomechanics IX-B* (eds. Winter, D.A., Norman, R.M., Wells, R.P., Hayes, K.C., and Patla, A.E.). Human Kinetics, Champaign, IL. pp. 274-279.

Woltring, H.J., Huiskes, R., de Lange, A., and Veldpaus, F.E. (1985) Finite Centroid and Helical Axis Estimation from Noisy Landmark Measurements in the Study of Human Joint Kinematics. *J. Biomech.* **18 (5),** pp. 379-389.

Woo, S.L.-Y., Gomez, M.A., Seguchi, Y., Endo, C.M., and Akeson, W.H. (1983) Measurement of Mechanical Properties of Ligament Substance from a Bone-ligament Preparation. *J. Ortho. Res.* **1 (1)**, pp. 22-29.

Xu, W.S., Butler, D.L., Stouffer, D.C., Grood, E.S., and Glos, D.L. (1992) Theoretical Analysis of an Implantable Force Transducer for Tendon and Ligament Structures. *J. Biomech. Eng.* **114 (2),** pp. 171-177.

Yeadon, M.R. (1989a) The Simulation of Aerial Movement - II: A Mathematical Model of the Human Body. *J. Biomech.* **23 (1),** pp. 67-74.

Yeadon, M.R. (1989b) The Simulation of Aerial Movement - III: The Determination of the Angular Moment of the Human Body. *J. Biomech.* **23 (1),** pp. 75-83.

Yeadon, M.R. (1990) The Simulation of Aerial Movement - I: The Determination of Attitude Angles from Film Data. *J. Biomech.* **23 (1),** pp. 59-66.

Yeadon, M.R. and Morlock, M. (1989) The Appropriate Use of Regression Equations for the Estimation of Segmental Inertial Parameters. *J. Biomech.* **22 (6-7),** pp. 683-689.

Yelderman, M., Widrow, B., Cioffi, J.M., Hesler, E., and Leddy, J.A. (1983) ECG Enhancement by Adaptive Cancellation of Electrosurgical Interference. *IEEE Trans. on Biomed. Eng.* EMB **30 (7),** pp. 392-398.

Yin, F.C.P., Tompkins, W.R., Peterson, K.L., and Intaglietta, M. (1972) A Video-dimension Analyser. *IEEE Trans. Biomed. Eng.* **BME (19-5)**, pp. 376-381.

Yokoi, T., Shibukawa, K., Ae, M., Ishijima, S., and Hashihara, Y. (1985) Body Segment Parameters of Japanese Children. *Biomechanics IX-B* (eds. Winter, D.A. et al.). Human Kinetics, Champaign, IL. pp. 227-232.

Zatziorsky, V. and Seluyanov, V. (1983) The Mass and Inertia Characteristics of the Main Segments of the Human Body. *Biomechanics VIII-B* (eds. Matsui, H. and Kabayashi, K.). Human Kinetics, Champaign, IL. pp. 1152-1159.

Zatziorsky, V. and Seluyanov, V. (1985) Estimation of the Mass and Inertial Characteristics of the Human Body by Means of the Best Predictive Equations. *Biomechanics IX-B* (eds. Winter D.A., Norman, R., Wells, R.P., Hayes, K.C., and Patla, A.E.). Human Kinetics, Champaign, IL. pp. 233-239.

Zentner, A., Sergl, M.G., and Filippidis, G. (1996) A Holographic Study of Variations in Bone Deformations Resulting from Different Headgear Forces in a Macerated Human Skull. *Angle Orthodontist.* **66 (6),** pp. 463-472.

Zernicke, R., Caldwell, G., and Roberts, E.M. (1976) Fitting Biomechanical Data with Cubic Spline Functions. *Research Quarterly.* **47 (1),** pp. 9-19.

Zhang, M., Turner-Smith, A.R., Tanner, A. and Roberts, V.C. (1998) Clinical Investigation of the Pressure and Shear Stress on the Trans-tibial Stump with a Prosthesis. *Medical Engineering & Physics.* **20 (3),** pp. 188-198.

Zhang, Y.T., Ladly, K.O., Liu, Z.Q., Tavathia, S., Rangayyan, R.M., Frank, C.B., and Bell, G.D. (1990) Interference in Displacement Vibroarthrography and its Adaptive Cancellation. *Proc. 16th Can. Med. and Biol. Eng. Conf.* pp. 107-108.

Zhang, Y.T., Ladly, K.O., Rangayyan, R.M., Frank, C.B., Bell, G.D., and Liu, Z.Q. (1990) Muscle Contraction Interference in Acceleration Vibroarthrography. *Proc. the IEEE/EMBS 12th Annual International Conference.* pp. 2150-2151.

Zhang, Y.T., Frank, C.B., Rangayyan, R.M., Bell, G.D., and Ladly, K.O. (1991) Step Size Optimization of Non-stationary Adaptive Filtering for Knee Sound Analysis. *Med. & Biol. Eng. & Comput.* **29** (Supplement, Part 2), p. 836.

Zhang, Y.T., Frank, C.B., Rangayyan, R.M., and Bell, G.D. (1992) A Comparative Study of Simultaneous Vibromyography and Electromyography with Active Human Quadriceps. *IEEE Trans. Biomed. Eng.* **39 (10)**, pp. 1045-1052.

Zhang, Y.T., Herzog, W., Parker, P.A., Hulliger, M., and Guimaraes, A. (1992) Distributed Random Electrical Neuromuscular Stimulation: Dependence of EMG Median Frequency on Stimulation Statistics and Motor Unit Action Potentials. *NACOB II* (eds. Draganich, L., Wells, R., and Bechtold, J.). American Society of Biomechanics, Chicago, IL. pp. 185-186.

Zhang, Y.T., Herzog, W., Sokolosky, J., and Guimaraes, A.C.S. (1993) Adaptive Estimation of Muscular Forces from the Myoelectric Signal Obtained During Dynamically Contracting Muscle. *Proceedings, Canadian Conference on Electrical and Computer Engineering.* pp. 1305-1307.

Zhang, Y.T., Rangayyan, R.M., Frank, C.B., and Bell, G.D. (1994) Adaptive Cancellation of Muscle Contraction Interference in Vibroarthrographic Signals. *IEEE Trans. on Biomed. Eng.* **41,** pp. 181-191.

Ziegert, J.C. and Lewis, J.L. (1979) The Effect of Soft Tissue on Measurements of Vibrational Bone Motion by Skin-mounted Accelerometers. *Transactions of the ASME.* **101,** pp. 218-220.

4 MODELLING

Modelling, an attempt to represent reality, is often used when understanding phenomena becomes difficult. A model can be a powerful tool for increasing the understanding of mechanisms. Therefore, modelling has been applied frequently in many daily or research situations, or both. The power of modelling is increasingly recognized in biomechanical research. Modelling, discussed in this chapter, becomes a powerful scientific tool when combined with experimental data.

4.1 DEFINITIONS AND COMMENTS

Absolute temperature: θ: a positive temperature measure consistent with the ideal gas law.

Analytical solution: General solution of a mathematical problem using symbolic manipulation, e.g., paper and pencil, as opposed to *numerical solution.*

Body weight: Body weight is the resultant force for all the segmental weights of a human or animal body.

Clausius-Duhem inequality: The form of the second law of thermodynamics for a continuous non-homogeneous system.

Conservative force field: A force field derived from a potential, V.

Deduction: Logical reasoning from a known principle to an unknown; from the general to the specific (Webster).

Degrees of freedom (DOF): A minimum number of kinematic variables (coordinates) required to specify all positions and orientations of the body segments in the system.

Direct dynamic solution: See *forward dynamic solution.*

Distribution problem: Calculation of internal forces acting on the musculoskeletal system using the known resultant joint forces and moments. The distribution problem is usually underdetermined.

Endergonic reaction: A chemical reaction for which the Gibbs free-energy of the reactants is smaller than that of the products.

Endothermic reaction: A chemical reaction for which the enthalpy of the reactants is smaller than that of the products.

Enthalpy: $H = U - pV$
A function of state for systems that can be characterized by pressure and volume alone.

Entropy: S: a thermodynamical function of state postulated in the second law of thermodynamics.

Exergonic reaction: A chemical reaction for which the Gibbs free-energy of the reactants is larger than that of the products.

Exothermic reaction: A chemical reaction for which the enthalpy of the reactants is larger than that of the products.

First law of thermodynamics: The time rate of change of the sum of kinetic plus internal energy is equal to the power of the external forces plus the heating input.

Force: Force cannot be defined. However, effects produced by forces can be described.
Comment: mechanical effects of forces are:

A force can accelerate a mass:
$\mathbf{F} = m \cdot \mathbf{a}$

A force can deform a spring:
$\mathbf{F} = k \cdot \Delta\mathbf{x}$

- Active force: Active forces in human locomotion are forces generated by movement entirely controlled by muscular activity.
- Contact force: Force resulting from physical contact between two objects.
Comment: an example of a contact force is the force between tibia and femur in the knee joint.
- External/internal forces: Force acting externally or internally, or both, to the system of interest.
- Ground reaction force: A ground reaction force (GRF) is a force that acts from the ground on the object that is in contact with the ground.
- Impact force: Impact forces in human locomotion are forces that result from a collision of two objects and that reach their maximum earlier than 50 milliseconds after the first contact of the two objects.
Comment: the most often used application of impact forces is the landing of the heel of the foot on the ground.
- Remote force: Force not resulting from physical contact that one object exerts on another.
Comment: an example of a remote force is the gravitational force.
- Resultant joint force: Vector sum of all forces transmitted by muscle-tendon units, ligaments, bones, and soft tissue across the intersegmental force.
Comment: the resultant joint force is also called *resultant intersegmental joint force.* Mathematically, the resultant joint force is defined as:

$$\mathbf{F} = \underbrace{\sum_{i=1}^{m} \mathbf{F}_i}_{\text{(muscle)}} + \underbrace{\sum_{j=1}^{l} \mathbf{F}_j}_{\text{(lig)}} + \underbrace{\sum_{k=1}^{n} \mathbf{F}_k}_{\text{(bone)}} + \underbrace{\sum_{p=1}^{s} \mathbf{F}_p}_{\text{(other)}}$$

where:

$\mathbf{F}$ = resultant intersegmental joint force
$\mathbf{F}_i$ = forces transmitted by muscles
$\mathbf{F}_j$ = forces transmitted by ligaments
$\mathbf{F}_k$ = bone-to-bone contact forces
$\mathbf{F}_p$ = forces transmitted by soft tissue, etc.

Note that the resultant joint force is an abstract quantity that is not related to the actual force measured between two bones in a joint (the bone-to-bone force).

Forward dynamic solution: Also referred to as *direct dynamic solution.* Solution of the equations of motion for a multibody system, where the input is forces and the output is movement. The equations are second order ODEs.

Free body diagram (FBD): A free body diagram consists of a sketch of the system of interest, representing all the external forces and moments acting on the system of interest, and a reference frame (coordinate system).
Comment: a properly drawn free body diagram allows the immediate application of Newton's second law.

Frictional joint moment: Moment in a joint due to friction in the joint, including both surface roughness (dry) and viscous (fluid) phenomena.
Comment: the frictional joint moment is often neglected or, if not, included in the resultant joint moment.

Gibbs free-energy: $G = H - \theta S$
A thermodynamical function of state.

Helmholtz free-energy: $F = U - \theta S$
A thermodynamical function of state.

Homogeneous process: A thermodynamical process for which all points of a system are at the same state for any given instant.

Induction: Logical reasoning from particular facts or individual cases to a general conclusion (Webster).

Inertial frame of reference: Frame of reference in which Newton's laws are valid. Relative to an inertial frame of reference, a body remains at rest or in a state of rectilinear translation with constant speed when no resultant force or moment act on it.

Initial condition: Values of all state variables at the start of a simulation. Must be supplied before a simulation can be done.

Internal energy: U: a thermodynamical function of state postulated in the first law of thermodynamics.

Inverse dynamic solution: Solution of the equations of motion for a multibody system, where the input is movement and known forces, and the output are unknown forces. The equations are algebraic equations.

Kinetic energy of a particle: $E_{kin} = \frac{1}{2}mv^2$

Lagrange multipliers: λ: additional variables used to introduce geometrical constraints in the field equations.

Mathematical model: A set of equations that describe the behaviour of a system.

Mathematical system: A set of mathematical equations and inequality relationships.

- Determinate system: A mathematical system is determinate if the number of system equations equals the number of unknowns.
- Overdetermined system: A mathematical system is overdetermined if it has more system equations than unknowns.
- Underdetermined system: A mathematical system is underdetermined if it has fewer system equations than unknowns.

Mechanical power: $P = \mathbf{F} \cdot \mathbf{v}$

Metabolic cycle: The process of dissociation of high energy compounds, such as glucose, to produce the free-energy needed to sustain animal life.

Model: An attempt to represent reality.

- Model parameter: A numerical value that represents a certain mechanical property of the system.
  Comment: a model parameter is constant in time, e.g., spring constant.
- Model variable: A numerical value that represents a certain aspect of the mechanical state of the system.
  Comment: a model variable changes its value with time, e.g., the bone-to-bone force in a joint.

Muscle model: Mathematical model with muscle length and activation as input variable, and muscle force as output variable.

Musculo-skeletal model: Mathematical model that includes the multibody dynamics of the skeletal system, and forces generated by muscles and other tissue producing forces which act on bones.

Numerical model: Also referred to as *computer model.* Implementation of a mathematical model where the variables are represented by numbers.

Numerical solution: Solution of a mathematical problem for specific values of model parameters and inputs, e.g., using a computer.

ODE: Ordinary differential equation. Equation that includes derivatives of the unknown quantities with respect to only one independent variable, usually time.

- **First order ODE:** Ordinary differential equation involving no higher derivatives of the unknowns than the first.
- **Second order ODE:** Ordinary differential equation involving no higher derivatives of the unknowns than the second.

Partial differential equation: Also known as PDE. Equation that includes derivatives of the unknown quantities with respect to more than one independent variable, and usually two or three spatial coordinates.

Particle: Matter assumed to occupy a single point in space.
Comment: the volume of the body of interest is small compared to the space in which its behaviour is of interest.

Potential energy: V : a scalar time-independent field such that the force field is (minus) its spatial gradient.

Principle of conservation of mechanical energy: When all forces acting on a system of particles derive from a potential, the sum of kinetic plus potential energy is a constant of the motion of the system.

Principle of virtual work: The equilibrium of a mechanical system is equivalent to the identical vanishing of the total virtual work performed by all external and internal forces through all virtual displacement of the system.

Resultant joint moment: Net moment produced by all intersegmental forces with lines of actions that cross the joint.
Comment: the resultant joint moments are also called *resultant intersegmental joint moments.* Resultant joint moments include resultant muscle joint moments, resultant ligament joint moments, and resultant bony contact force moments. The resultant joint moment only reflects the net effect of agonist and antagonist muscles. For that reason, the resultant joint moment is also called *net joint moment.* Mathematically, the resultant joint moment is defined as:

$$\mathbf{M} = \underbrace{\sum_{i=1}^{m} \mathbf{M}_i}_{\text{(muscle)}} + \underbrace{\sum_{j=1}^{l} \mathbf{M}_j}_{\text{(lig)}} + \underbrace{\sum_{k=1}^{n} \mathbf{M}_k}_{\text{(bone)}} + \underbrace{\sum_{p=1}^{s} \mathbf{M}_p}_{\text{(other)}}$$

where:

M = resultant intersegmental joint moment
$\mathbf{M}_i$ = moments transmitted by muscles
$\mathbf{M}_j$ = moments transmitted by ligaments
$\mathbf{M}_k$ = bone-to-bone contact moments
$\mathbf{M}_p$ = moments transmitted by soft tissue, etc.

Rigid body: Matter assumed to occupy a finite volume in space and that does not deform if subjected to external forces.

Scientific research: Investigation in some field of knowledge undertaken to discover or establish facts or principles (Webster).

Simulation: Experimentation using a model.
Comment: simulation can be done with paper and pencil. However, the repeated execution of a simple task is best done with computer program. Thus, in practice, the expression simulation is commonly understood as a synonym for computer simulation.

State variables: The variables required to describe the state of a system. For a mechanical system these are all degrees of freedom and corresponding velocities.

Steady-state solution: Solution in which all derivatives are zero.

Stiff system of equations: System of ODEs in which the ratio between smallest and largest time constant is much smaller than one.

System of interest: Sum of mass points or segments that are considered as a basic unit to discuss and solve a problem.
Comment: in biomechanical analysis, the system of interest is typically considered as being rigid. All forces acting from the outside on this system are called external forces for this system. All forces acting within the system of interest are called internal forces for this system.

Thermoelastic effect: The emission (or absorption) of heat by a thermoelastic material subjected to an isothermal contraction (or expansion).

Unique solution: One and only one solution.

Validation of a model: Providing evidence that a model is strong and powerful.
Comment: it is recommended that this term be replaced by *evaluation of a model.*

Virtual displacement: A set of small changes in the values of the degrees of freedom of a system.

4.2 SELECTED HISTORICAL HIGHLIGHTS

Selected historical highlights relating to biomechanical modelling are summarized below.

Fourth century B.C.	Aristotle used the term energy to denote something in action.
1695 Leibniz	Explicitly formulated the principle of conservation of mechanical energy for a particle, in terms of live force (kinetic energy) and dead force (potential energy).
1744 Euler	Following a suggestion of his teacher, D. Bernoulli (1700-1782), successfully used a definition of the potential force (strain energy) of an elastic bar to solve the buckling problem by means of variational calculus.
1788 Lagrange	Adopted the principle of virtual work as the basis for his analytical mechanics.
1798-1799 Rumford & Davy	Count Rumford (Benjamin Thompson) and Sir Humphry Davy proved experimentally that heat is a form of energy, not a substance.
1824 Carnot	Published his *Réflexions sur la Puissance Motrice du feu*, in which he laid down the foundation for the second law of thermodynamics.
1850 Clausius	Introduced the term entropy to describe what was essentially Carnot's caloric.
1851 Kelvin	Reconciled the work of Carnot and Joule, to give a precise formulation of the first and second laws of thermodynamics.
1857 Pasteur	Established the catalytic role of micro-organisms in the process of fermentation.
1858 Rankine	A Scottish engineer and physicist who used a graphical technique for analyzing problems of applied mechanics.
1864 Heidenhain	Established that the work potential and the heat production of a muscle increased with externally applied loads, up to a maximum.
1864 Maxwell	Showed that reciprocal figures and force polygons are related and may be used in mechanical analyses.

1875	Culmann	A Swiss scientist who published a book on graphical statics and later became the leader in the application of graphical methods to engineering problems.
1875	Weisbach	A German engineer who used diagrams of bodies with forces acting on them (forces indicated by arrows). Some of these diagrams were identical to current free body diagrams. Other diagrams only included selected forces.
1878	Gibbs	Developed the fundamentals of chemical thermodynamics.
1881	Gibbs	Developed the mathematics of vector analysis, which is used to solve the vector equations that result from free body diagrams.
1910	Smith & Longley	Outlined procedures for handling mechanical problems by using drawings and vectors.
1913	Hill	Started his famous experiments on heat measurements of contracting frog skeletal muscle.
1923	Fenn	While working in Hill's laboratory, demonstrated that energy output increased with work done against an opposing force in stimulated frog muscle, thus ruling out the viscoelastic theory of Hill and others.
1928	Reynolds	Used the term free body diagram and defined it as "the body isolated from all other bodies with arrows representing the forces acting upon it".
1938	Elftman	Estimated internal forces in the lower extremities and energy changes during walking. Used a mechanical model to make these estimations.
1938	Hill	Found an experimental relation between muscular force and velocity, and attempted a thermodynamic interpretation.
1941	Lipmann	Discovered the role of ATP as a kind of common currency of metabolic energy.
1955	Lissner	Made an attempt to interpret the mechanics of the musculo-skeletal system for physiotherapists.
1957	Huxley, A.F.	Proposed the Cross-bridge theory of muscular contraction, which allowed a consistent view of mechanical and energetic properties of skeletal muscle.
1961	Dempster	Published an article on free body diagrams as an approach to the mechanics of human posture and motion.

1965	Paul	Developed a deterministic model to estimate forces in the hip joint region.
1966	Drillis & Contini	Compared the range of measurements on a living subject using the torsional pendulum.
1967	Furnee	Presented a paper on a television and computer system for gait analysis (this subsequently became available as VICON after developments of Furnee's system at the University of Strathclyde).
1968	Riddle & Kane	Used a three-segment model to simulate the effect of arm movement for reorienting astronauts in space.
1970	Miller	Simulated somersaults with a four-segment model.
1971	Passerello & Huston	Developed a 10 segment model of the human body to simulate movement in space. At the same time, they presented a simulation model for astronaut reorientation.
1973	Seireg & Arvikar	The first to use mathematical optimization to solve the distribution problem. Published a mathematical model to estimate forces in the lower extremities using the minimization of several objective functions.
1976	Hatze	Presented a simulation model for optimizing human movement.
1978	Pedotti et al.	Published a non-linear model of the human body with biological input. The first to use a non-linear optimization approach to solve the distribution problem.
1979	van Gheluwe	Simulated back somersaults with full twists.
1981	Crowninshield & Brand	The first to use a non-linear optimization approach based on a physiological criterion to solve the distribution problem. They then presented a possible mathematical solution to the indeterminacy problem of load sharing in muscles crossing a joint, by using optimization procedures.
1981	Hatze	Presented a comprehensive model for simulating human motion during the take off phase in the long jump that relied on mechanical, muscle-physiological, and neuro-physiological knowledge.
1984	Dul et al.	Presented a possible mathematical solution to the indeterminacy problem using sequential optimization procedures. The first to attempt a validation of the results of their distribution problems using direct muscle force measurements.

1984	Hubbard	Simulated javelin flight and determined optimal release techniques.
1984	Yeadon	Presented a simulation model for aerial movement and applied it to twisting somersaults.
1987	Warsaw, Poland	First International Symposium on Computer Simulation in biomechanics.
1987	Herzog	Included contractile properties of muscles in estimating internal forces.
1987	Winters & Stark	Discussed the advantages and disadvantages of changing the complexity of muscle models.
1989	Haug	Described software for computer simulation of mechanical problems (DADS).
1990	Pandy	Simulated vertical jumping using a musculo-skeletal model and determined optimal muscle coordination strategy.
1991	Pandy & Zajac	Determined optimal muscle coordination for vertical jumping using a four-segment musculo-skeletal model.
1995	Delp	Developed interactive software for musculo-skeletal modelling.
1998	Neptune	Applied simulated annealing method to optimization of movement.

4.3 A NEARLY POSSIBLE STORY

NIGG, B.M.

On his way to class, Expe meets his colleague, Theo. While walking together, the two discuss a project Expe is currently studying. The project concerns the vertical velocity of a dropping mass before touching the ground. Researcher Expe says:

> "We finished our experiments and obtained very interesting results. For all masses used in the study, we found a non-linear relation between the dropping height, H, and the vertical velocity of the centre of mass, v, just before reaching the ground surface. The mathematical relation between these two variables is:
>
> $$v = \text{const} \,.\, H^{0.498}$$
>
> Isn't that interesting?".

Researcher Theo takes a piece of paper and says (fatherly):

> "Dear colleague Expe, we know that the total mechanical energy of a falling object remains constant (if we neglect air resistance). Furthermore, the total translational kinetic energy of the mass immediately before reaching the ground is $\frac{1}{2} \cdot m \cdot v^2$. The rotational energy of the body is assumed to be zero initially and remains so because of the absence of external moments. The loss of potential energy from the highest point to the position immediately before contacting the ground is $m \cdot g \cdot H$. Consequently, if you use the law of conservation of energy you get:
>
> $$m \cdot g \cdot H = \frac{1}{2} \cdot m \cdot v^2 \qquad \text{or:} \qquad v = \sqrt{2g} \cdot H^{0.5}$$
>
> This means that your constant is the square root of twice the gravitational acceleration, 2 g, which is about 4.43".

Researcher Expe is surprised and interested because, in fact, he found almost exactly this value for his constant. Patronizingly, his colleague, Theo, adds:

> "My dear friend, it may be helpful for you to contact me when planning your next project, so we can discuss your problem for five minutes. I would be happy to help you".

In this nearly possible story, which is partly translated and adapted from Hatze (1978), Theo represents a theoretical researcher, while Expe represents an experimental researcher. The story may be exaggerated. However, it contains several messages. The most important of these messages is that in some research projects a (theoretical) model may shed light on a problem, or help in the prediction of numerical values for variables of interest. Specifically, the use of a mechanics-based model may be helpful in biomechanical studies. The following paragraphs discuss selected examples of biomechanical modelling.

4.4 GENERAL COMMENTS ABOUT MODELLING

NIGG, B.M.

4.4.1 GENERAL CONSIDERATIONS

The term model is used in many different ways in daily life. In this book, a model is defined as follows:

A model is an attempt to represent reality.

Usually, a model is a simplified representation of reality. However, some models seem more complicated than reality (see section 4.4.4).

Models are constantly developed in daily life and research, and it may part of human nature to try to overcome our inability to cope with complex situations by using simplified models. Examples of models can be taken from diverse fields, including:

Daily life	- he is a bad boy
	- she is an angel
Religion	- theism
	- deism
Physics	- Coulomb friction
	- Bohr's model of the atom
Biomechanics	- effective mass model (Denoth, 1980)
	- 160 element model (Seireg & Arvikar, 1973)
	- simple models of walking and jumping (Alexander, 1992)

4.4.2 INFORMATION USED TO CONSTRUCT A MODEL

Model construction relies on two types of information: *knowledge* of the system being modelled, and *experimental data* that constitutes system inputs or outputs, or both. A deductive process uses knowledge of the system being modelled, and moves from general principles to specifics. An inductive process uses experimental data to attempt to arrive at a general conclusion that explains the data. A deductive solution process for a given problem, with clearly defined assumptions, yields a logically derived result, which is typically a unique solution. An inductive solution process for a given problem yields a probable result, which is typically not a unique solution, because an infinite number of models satisfy the observed input-output relationship (Figure 4.4.1).

At first glance, Figure 4.4.1 seems to imply that deduction should be used as much as possible, since deduction typically provides a unique solution. Be aware, though, that the knowledge on which a deductive approach is built may merely amount to preliminary assumptions. Changing from inductive to deductive methods may simply move the uncertainty from the number of possible answers to the number of possible assumptions.

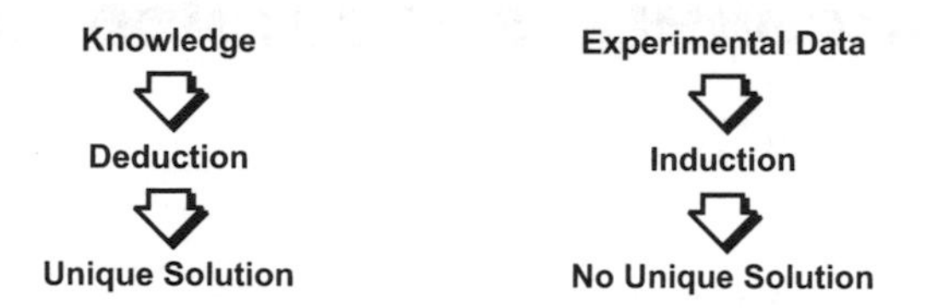

Figure 4.4.1 **Schematic illustration of the connection between types of basic information, method used, and expected results.**

4.4.3 SIMPLIFICATION

One important aspect of developing a model is to decide what should be neglected and what should be included. It seems obvious that all important aspects should be included, and that all unimportant aspects may be neglected. However, attempting to list rules and guidelines is not easy. In general, simpler is better. However, there is always the danger that a simple model may not agree with reality, because some aspects have been omitted. Developing a model and making choices about including or excluding certain aspects may be more of an art than a science.

The following example that demonstrates simplification. Let us assume that the purpose of a project is to develop a model to estimate the magnitude of the impact forces at the tibio-femoral joint during landing in heel-toe running. Obviously, eye colour is not important and can be neglected. It may even be assumed that the size and mass of the head and the mass of the foot have little influence on the magnitude of the impact forces and impact moments at the tibio-femoral joint. Finally, assume that the muscles can be neglected when estimating the magnitude of the forces and moments at the tibio-femoral joint (Nigg, 1986). These assumptions and simplifications lead to a simple model of the human body; a three segment model comprising a lower leg, a thigh, and a third segment, the rest of the body (Figure 4.4.2), which are connected by hinge joints (Denoth, 1980).

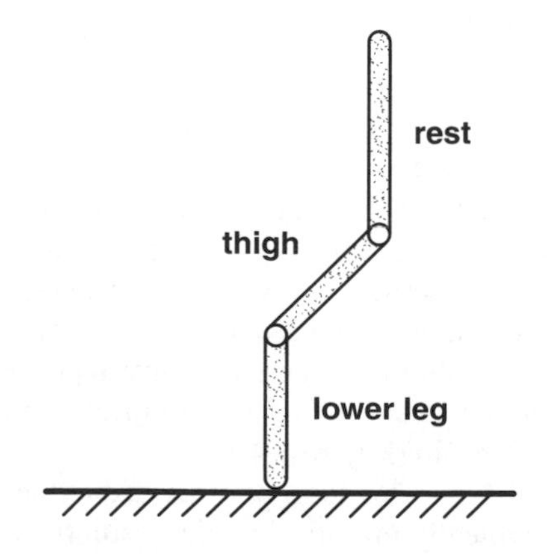

Figure 4.4.2 **Illustration of a simple model for the determination of the magnitude of impact forces and moments at the tibio-femoral joint (from Denoth, 1980, with permission).**

4.4.4 THE PURPOSE OF A MODEL

Models are used for actual or theoretical (simulated) situations. Results and conclusions from models, for both applications, can be used for two specific purposes:

Purpose a: **to increase knowledge and insight about reality, and**

Purpose b: **to estimate or predict variables of interest.**

A model may provide insight into the true nature of the system of interest, and increase our understanding of the interaction of the variables important for the situation described by the model. A model may also provide an accurate estimation of the variable(s) of interest. The two purposes of a model seem clear. However, there are at least three comments to be made in this context:

(1) One purpose of a model is to increase knowledge and insight about reality, suggesting that results and conclusions may be used to improve knowledge of, or insight into, a complex situation. However, knowledge and insight are essential for the development of a model. The fact that insight and knowledge are prerequisites for the development of a model, but are also the purpose of using a model, seems contradictory. Specifically, a model provides insight into the relationships among variables, and may indicate how these relationships are governed.

(2) A model can only provide information that has been implicitly implemented into the model. A model cannot suggest, for instance, that a set of variables (x_i,...., $x_{n'}$) should be added to the original design of the model if these variables were not included in the model. However, results and conclusions from a model can be used to study how the variables considered important for the question of interest are sensitive to changes in internal or external conditions, i.e., increase general understanding or knowledge and insight about reality.

(3) Naively, a model can be considered good if the results are accurate. An historical example illustrates the problem associated with such a view:

> Theories concerning the solar system have changed with time. About 300 years before Christ, Aristarchos of Samos proposed that the sun was fixed at the centre of the universe and the earth revolved around the sun in a circular orbit. He also suggested that the stars appeared fixed in position because their distances from the sun were tremendous compared to the distance from the earth to the sun. Some of the early astronomers accepted this heliocentric theory of the universe. In about 150 A.D., Claudius Ptolemeus proposed a theory, called the Geocentric theory, which suggested that the earth was at the centre of the universe and that the sun moved around the earth in a circle. This theory (model) succeeded in explaining and predicting planetary motion with the degree of measurement accuracy then possible. Ptolemeus' theory was taught and used extensively

from the second to the sixteenth century. Copernicus (1472-1543) revived and extended the heliocentric theory of Aristarchos, starting a revolution in scientific thought that was carried forward by Kepler, Galileo, and Newton. However, for a substantial period of time it was more accurate to calculate the path of planets as viewed from the earth or one's own position in navigation by using the old geocentric system than the new heliocentric system.

This example illustrates that it may be possible to calculate or estimate variables of interest sufficiently accurately by using a model whose conceptual construction does not correspond to reality. In addition, the example illustrates that quality criteria for a model depend on the model's actual purpose.

4.4.5 THE VALIDATION OF A MODEL

Validation of a model is commonly required in scientific work. The validation process is discussed in the following section.

The word *valid* derives from the Latin word *validus,* which translates as strong or powerful.

Validating a model can be defined as providing evidence that the model is strong and powerful for the task for which it has been designed.

Validation of a model consists of providing a set of cases for which the results of the model correspond to reality. Such cases support the statement that the model is strong and powerful for the task for which it has been designed. Evaluating the power of a model can be through direct measurements, indirect measurements, or trend measurements.

DIRECT MEASUREMENTS

In some cases, it is possible to measure the estimated variable in a limited set of experiments, e.g., force measurements in the Achilles tendon (Komi et al., 1987). Consequently, the estimated force can be compared to the measured force. If the experimental and the estimated results agree within an acceptable range, confidence in the model increases. Agreement between estimated and measured results, with changed boundary or initial conditions, may further increase confidence in a model. However, these agreements are not a guarantee that the structure of the model is similar to the real world, even if the model is commonly used in many applications.

INDIRECT MEASUREMENTS

The comparison between estimated and measured results can often not be made because it is often impossible to make experimental measurements of internal variables, e.g., force distribution in the patella-femoral joint. Therefore, direct validation cannot be performed. However, in some cases, measurements of another (measurable but not needed) variable may be made, and compared with the value predicted for this variable by the model. As an example, the forces in the tibio-femoral joint could be calculated by using the gravitational and the inertial forces produced by the segments above the knee. Additionally, the gravitational and inertial forces could be used to calculate the ground reaction force. This calculated ground reaction force could be compared with the measured ground reaction force. If the two ground reaction forces were close, one could feel comfortable about the model's potential to estimate forces in the tibio-femoral joint. However, it would still be possible that the estimated forces in the tibio-femoral joint were substantially wrong. Another example of indirect measurements includes EMG measurements for the estimation of muscle forces.

The limitations of such indirect validation procedures is illustrated with the following example. Several models estimating forces in human skeletal muscles (Pedotti et al., 1978; Crowninshield & Brand, 1981; Dul et al., 1984; Herzog, 1987) were indirectly validated or evaluated when they were first presented. However, when applied to the same situation, these models predicted substantially different muscle forces (Herzog & Leonard, 1991). Most, if not all, of these models estimated muscle forces incorrectly. The example suggests that the importance of indirect validation should not be overestimated. The potential of a model to do the task for which it was developed is often a weak evaluation of the model.

TREND MEASUREMENTS

An agreement between facts and results from a model is not necessarily an agreement in the values of the variables. One purpose of a model is to describe the general behaviour of the system of interest. In this case, agreement may consist of similar trends and developments. For example, a model might say that if x increases linearly, y will increase quadratically. If the purpose of a model is to improve our understanding of how variables in the model interact, the quality of the model depends on how well the trends that are predicted agree with the trends that are measured.

FINAL COMMENTS

Validating a model means obtaining evidence that the proposed model is strong and powerful for the purpose for which it has been developed. Validation may lead to increased confidence in a model, but it never confirms that the model corresponds to reality. A difference between the results of an estimation given by a model and an experimental measurement may indicate that the conceptual structure or the detailed composition of the model are inadequate.

The word validation is not well-defined and has developed different meanings over time. It may be appropriate to replace the term validation of a model with *evaluation of a model.* Each model should be evaluated by providing direct or indirect evidence that the model is strong and powerful for the purpose for which it was developed. Such an evaluation indicates if the results of a model are strongly supported by evidence or whether the supporting evidence is weak. In later chapters, the term evaluation is used whenever the strength and power of a model is assessed for the task it has been designed.

4.4.6 TYPES OF MODELS

A model provides information about a relationship between cause and effect. A model can be used in one of two ways: direct or inverse. In direct use, the model proceeds from cause to effect and, typically, yields a unique solution. In inverse use, a model attempts to move from the effect to the cause(s) and, typically, yields several possible solutions (not unique).

Models can be classified in various ways. One way is to call some models intuitive, and others abstract (Fischbein, 1987). Another is to recognize inductive and deductive models (Kemeny, 1959). The taxonomy of biomechanical models used in this book has four groups: analytical, semi-analytical, black box, and conceptual.

ANALYTICAL (DEDUCTIVE) MODELS

Analytical models are developed on the basis of (real or speculated) knowledge and insight. Using our understanding of the human musculo-skeletal system, its physiology and physics (mechanics), a structure can be developed that can be described with a mathematically deterministic model. The advantage of such a model is that it has a unique solution that is independent of the selected mathematical procedure. The critical point in developing such a model is the selection of the assumptions and simplifications.

Selected examples for analytical models in biomechanical applications include the model for running to determine optimal surface compliance (McMahon & Greene, 1979), the three segment model for the determination of the impact forces in the tibio-femoral joint (Denoth, 1980), the wobbling mass model used for the estimation of the actual bone-to-bone forces and joint moments of the lower extremities (Gruber et al., 1987), and the series of models for sequential joint extension in jumping (Alexander, 1989).

SEMI-ANALYTICAL MODELS

Semi-analytical models are based on knowledge and insight. However, the system of interest is too complicated to make it mathematically deterministic with the available basic information. The mathematical description of the system of interest has more unknowns than equations. Consequently, more assumptions are added.

Biomechanical examples for semi-analytical models include most models for estimating muscle forces and bone-to-bone forces from external kinematics and kinetics (Pedotti et al., 1978; Crowninshield & Brand, 1981; Winter, 1983; Dul et al., 1984; Herzog, 1987; Morlock & Nigg, 1991).

BLACK BOX MODELS

In black box models (also called regression models), a set of mathematical functions are used to determine the input-output relation. In the first step, an appropriate (but perhaps arbitrary) mathematical function is determined that best describes the relationship between known input and output pairs. In a second step, the determined mathematical function is used to predict output values for a given set of input values.

The black box approach has two applications. First, it can be used to estimate quantities that cannot be measured in certain situations. The aspect of interest is the accuracy of the output if compared to the correct value. The way the output is determined, the structure of the mathematical formula, and the functional relationship between input and output are not of interest. Second, the black box approach can be used to provide insight into possible functional relationships between input and output. A mathematical function is used to describe the input-output relationship. A subsequent analysis of the mathematical components that determine the input-output relationship may provide insight into the general structure of the relationship studied.

A biomechanical example for the black box model, in the sense of the first application, could be the prediction of the length of a long jump based on variables such as approach velocity and take off angle (Ballreich & Brüggemann, 1986). Biomechanical examples of using the black box model to explore functional relationships between input and output could be the determination of the dominant material properties, e.g. elasticity, and viscoelasticity, of the soft tissue under the human heel performed by Denoth (1980). He used a complex non-linear mathematical function to approximate the experimentally measured force-deformation curves, and then used the dominant terms of this function to describe the material.

CONCEPTUAL MODELS

Conceptual models consist of a hypothesis (which is based on insight or speculation), and procedures capable of supporting or disproving the tested hypothesis. These procedures can be experimental or theoretical. In biological research, experimental procedures for disproving or supporting a hypothesis are frequently used. The advantage of the conceptual approach is that larger concepts can be subdivided into smaller steps that can be treated individually. The disadvantage of the conceptual approach is that a hypotheses can never be proven, only disproven. Consequently, several pieces of evidence must be accumulated to provide enough support for a concept. Although this approach is common in the biological sciences, biomechanical examples for this approach are not easy to find.

A classification of tools (a model is a kind of a tool) is associated with difficulties since many appropriate possibilities exist for an organization of these tools into groups with similar characteristics. Grouping of tools typically depends deciding which of the different characteristics to selects as relevant for the subdivision. In this text, characteristics describing the mathematical solution are arbitrarily chosen as criterion. However, other approaches may be as appropriate as the one presented here.

4.4.7 DESCRIPTIVE, EXPERIMENTAL, AND/OR ANALYTICAL RESEARCH AND MODELLING

Science must start with facts and end with facts, no matter what theoretical structures it builds in between.

This statement by Einstein underlines the fact that scientific activities include a sequence of steps (Kemeny, 1959). First, a scientist is an observer. Second, a scientist describes what is observed. Third, a scientist develops theories that allow predictions. Fourth, a scientist compares the predictions with the facts. This process is cyclic in its structure (Figure 4.4.3).

The main characteristic of scientific activities is the cyclic interaction between facts and theory. In the world of biomechanics, researchers with instruments such as EMG sensors, force plates, and video cameras observe and describe the situation of interest as accurately as possible or as accurately needed, or both. Based on these observations, theories (models) are developed using induction. Such theories may be a set of mathematical formulas that are, in turn, used deductively to predict the outcome of a theoretically possible future experiment. The prediction will then be compared with the actual results (facts) of an experiment. This process continues because new facts and information are the starting point for new cycles of induction and deduction.

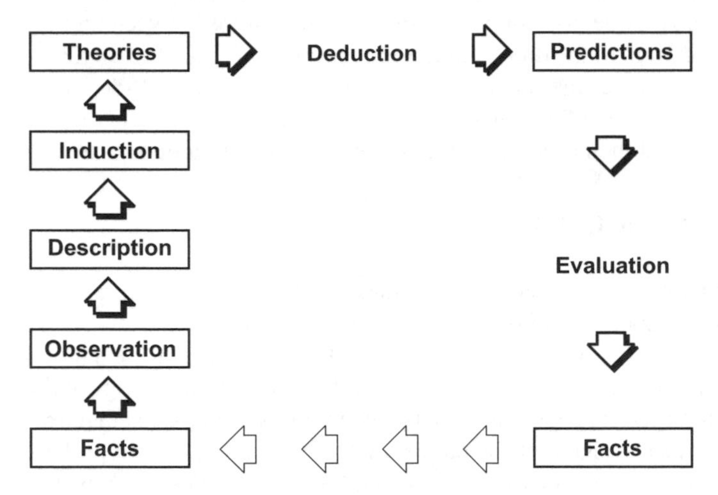

Figure 4.4.3 Schematic illustration of the cyclic interaction between facts and theory in scientific activities (from Kemeny, 1959, with permission).

Several comments may be of interest in this context:

(1) It may be wise to start a research project by describing accurately the actual situation (in biomechanics, this includes the actual kinetics and kinematics). Early developments in established sciences such as physics, chemistry or biology have included long periods of observations and description of these observations. Sometimes errors have been made because observations were not made well and not accurately documented.

(2) Quantum leaps in sciences were often a result of brilliant theories or models (Newton, Einstein, Bohr) some of which, of course, were based on a wealth of observation.

(3) The actual (mathematical) development of a model in itself is not research, but corresponds to the development of a tool. Measuring devices such as force platforms, pressure distribution insoles, accelerometers, indwelling electrodes for EMG measurement, and video cameras are models as well as tools, and their development is technology.

4.4.8 GENERAL PROCEDURES IN MODELLING

A set of general steps are common for developing a biomechanical model. These steps are:

(1) Definition of the question(s) to be answered.
(2) Definition of the system of interest.
(3) Review of existing knowledge.
(4) Selection of procedure (model) to be applied.
(5) Simplifications and assumptions.
(6) Mathematical formulation.
(7) Mathematical solution (using appropriate input data).
(8) Evaluation of the model.
(9) Discussion, interpretation, and application of the results.
(10) Conclusions.

This procedure is general and can be applied for any form of models, including inverse dynamic models, black box models, and simulation models. The specific steps are discussed further in the following paragraphs.

STEPS (1), (2), (3), (9), AND (10)

These steps are common to every research project and are not specific to developing a model. The definition of the question to be answered is the cornerstone of any reasonable research project. All the following steps are designed to answer this question. The definition of the body of interest is the next step in narrowing the field of future work.

The review of existing knowledge, which usually includes a review of the relevant literature, helps a researcher to decide whether or not further steps have to be taken to answer the question of interest, or whether the question or parts of it have already been answered. A literature review is often considered the most important initial step in a research project. Many researchers suggest that a thorough study of the available literature is the *conditio sine qua non*, the necessary condition for a successful start of a research project. However, other researchers argue that a thorough review of the literature should only be started when your own approach to the problem, i.e., project design, has been developed. This latter strategy, they argue, has the advantage that innovative thinking is not inhibited by the study of other solutions, and, that this approach is more likely to produce new and different approaches. It is important, however, to ascertain that no one has already solved the initial problem. This is increasingly difficult with apparently definitive reviews covering only a restricted region backwards in time, and restricted international citation of work between North America, Europe, and the Far East.

Discussing and interpreting results and forming conclusions are, of course, the most exciting and interesting part of any research project, independent of the applied methodology.

STEP (4)

Based on the first three steps, a decision must be made about what methodology to apply to solve the question of interest. The decision relevant for our purposes is to develop a model. If this approach is taken, the type of model developed must be specified. There are at least two philosophies about how to proceed. One is to start with a model as simple as possible and to make it more complex when needed, i.e., if the simple model was not able to provide the answer of interest. The other is to start with a complex model and to sequentially exclude its superfluous part(s).

Models range from simple, one-particle models to complex, multi-segment ones with physiological, anatomical, and neurological components. The selection process is crucial, since the task is not to find the most elaborate model, but the most appropriate one.

STEP (5)

Making the simplifications and assumptions is a challenging and difficult step in developing a model. They incorporate the modeller's philosophy and understanding of the situation. Two steps are required. The first step is deciding what to include and what to neglect, and what general assumptions to be made must be decided. The second step requires providing evidence and reasons why these assumptions are reasonable must be supplied. Note that evidence and support may range from experimental data to blunt statements such as "to simplify the mathematical calculations".

STEPS (6) AND (7)

The mathematical formulation and solution are usually simple, straightforward steps. A few standard methods are available that basically serve the same functions. Some are more elegant than others. However, the different mathematical methods should not usually influence the outcome of the results.

To make use of the model, it is necessary to have appropriate input data. This could be based upon previously collected data or may be completely hypothetical. In the latter case, some caution should be exercised to ensure that the proposed input data is reasonable. Using appropriate input data, the model is applied to produce the output data. This may be done for a large number of simulations to generate a general picture of how the system responds.

STEP (8)

Mathematical formulations of modelled situations can provide every possible result. It is important to provide evidence for the appropriateness of the findings derived from the model. This is not proof as discussed earlier, but rather some support for the confidence level in the output of the model. Note that output is a general term. It may be a number, a set of time histories, or a trend.

4.5 FORCE SYSTEM ANALYSIS

NIGG, B.M.

4.5.1 INTRODUCTION

A system of biological structures can react to external force stimuli *biologically* or *mechanically*, or both. Biological reactions include strengthening or weakening of tissues (bone, cartilage, ligament, tendon, and muscle), depending on the characteristics of the acting force inputs (see Chapter 2.8). Mechanical reactions include deformation or acceleration, or both, of the total body or of segments of it. This section concentrates on describing the mechanical behaviour of a system of rigid bodies that represent a biological system.

The body of interest (the body of a human or animal) is often represented as a system of n rigid bodies connected by idealized joints. Obviously, there are many different ways of solving a mechanical problem. This section presents and discusses one specific approach of analyzing the mechanical behaviour of a system of rigid bodies (force system analysis). Understanding the definitions presented and the steps involved in the presented force system analysis provides the tools to solve mechanical questions in a systematic way. The approach presented in this chapter should help to understand and perform the biomechanical analysis of the human or animal body.

The analysis of a force system includes the following components:

- Mechanical system of interest,
- Assumptions,
- Free body diagram,
- Equations of motion, and
- Mathematical solution.

The following paragraphs discuss the listed steps individually.

4.5.2 THE MECHANICAL SYSTEM OF INTEREST

A mechanical system is a specific collection of particles or rigid bodies. The system of interest for a specific question is the sum of the main points or segments, which are considered a basic unit needed to discuss and solve a problem. For example, when analyzing the flight path of a javelin in the air, the system of interest is the javelin. When determining the forces between the foot and the ground of a runner (the ground reaction forces), the system of interest could be the whole runner, the foot of the runner, or the ground. When determining the muscular forces across the elbow joint during a tennis stroke, the system of interest could be the lower arm or the upper arm (to list just a few possibilities). In general, to determine forces acting on a body for a specific question, several different systems of interest can be used.

Conceptually, the system of interest must be a structure on which the force of interest acts as an external force (external to this structure). The system of interest can be determined using several steps:

Step 1: A sketch of the actual situation is drawn that includes the location for which the force should be determined.
Example: when interested in the contact force between tibia and femur, draw the total body of an athlete, or the support leg for which the knee joint force should be determined.

Step 2: The total body is subdivided into two parts. The separation is made at the location where the force should be determined.
Example: when determining the force between tibia and femur, the human body or the leg is divided into two parts; the lower leg and the rest of the body.

Step 3: The system of interest is the whole or part of one of the two body parts that have been separated from each other.
Example: the system of interest in the above example could be the tibia, the lower leg, the femur, or the rest of the body, among others.

For this text, the following convention is applied: the system of interest is drawn with solid lines. Other parts of the total body, e.g., the human body, may be drawn with dotted lines, in a shaded form, or not be drawn at all.

Selecting and defining the system of interest is an important step in developing a free body diagram and for force system analysis. Many errors in the solution of a mechanical problem begin because the system of interest has not been defined properly.

EXAMPLE 1

Suppose you want to determine the force in the hip joint of the supporting leg during running for a human being. First, a schematic sketch of the runner is made. Second, the human body is separated at the hip joint. Third, the system of interest is determined. Possible systems of interest include (a) the supporting leg, (b) the supporting femur, (c) the pelvis, and (d) the rest of the body, as illustrated (Figure 4.5.1). Some of these possible systems of interest may be more appropriate than others, depending on the assumptions and the possible measurements that can be made.

This example shows that different segments of the human or animal body may be selected appropriately as the system of interest for the same question, e.g., determining the forces of the hip joint. The selection depends on knowledge of the total system, mathematical skills available to the scientist, experimental setup, and variables, which can be determined experimentally, to be used as input into the mathematical equations describing the problem. Selecting the system of interest often influences the elegance and ease of the mathematical solution.

4.5.3 ASSUMPTIONS

Once the system of interest is defined, the researcher must make assumptions defining aspects that are important, and aspects that are not. Specifically, the following assumptions must be made:

- Whether the problem is one-, two- or three-dimensional,
- Which external forces (due to tendons or ligaments, etc.) to include, and which to exclude, from further consideration,
- The magnitude and direction of possibly known forces,
- Material and structural characteristics of the system of interest, and
- Many other aspects of importance.

The rules that govern selecting appropriate assumptions are difficult to summarize. The development of a good set of assumptions depends on the experience and the *feeling* of the researcher. As a simple general rule:

- Define what is really important mechanically, and make the assumption that only forces due to those structures are acting.

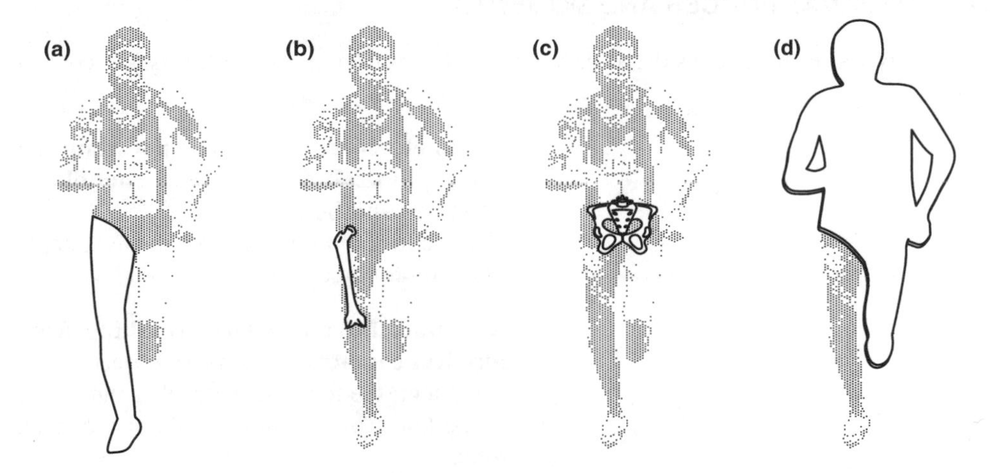

Figure 4.5.1 **Examples for possible systems of interest for estimating bone-to-bone forces in the human hip joint. The possible systems of interest are drawn in solid lines. The rest of the body is indicated in a shaded form.**

4.5.4 FREE BODY DIAGRAM

Developing a free body diagram (FBD) for the analysis of mechanical and biomechanical problems has taken decades. The most important steps in developing a free body diagram have been described earlier (Dempster, 1961).

A free body diagram consists of a:

- **Sketch of the system of interest,**
- **Representation of all the external forces and moments acting on the system of interest, and**
- **Reference frame (coordinate system).**

THE SKETCH OF INTEREST

The sketch of the system of interest includes only the system of interest, and nothing else. If, for instance, the system of interest is the leg (as illustrated in Figure 4.5.1, left drawing), only the leg is drawn. The ground and the hip, for example, are not drawn. It is important to draw the sketch so that the important geometrical aspects of the system of interest are maintained. For example, using the leg as the system of interest in Figure 4.5.1, the foot should be positioned so that it is approximately below the center of mass of the whole body. If, the foot during heel landing in heel-toe running is the system of interest, it is important to draw the heel lower than the toe. With respect to shape and rigidity, it may be acceptable to simplify the foot as a triangle and the leg as a parallelogram, and to assume that these systems are rigid bodies.

THE EXTERNAL FORCES AND MOMENTS

Two types of forces considered in free body diagrams are: remote forces and contact forces.

A remote force is a force not resulting from physical contact that one object exerts on another.
Comment: gravitational force is an example of a remote force.

A contact force is a force resulting from physical contact between two objects.
Comment: the force between tibia and femur in the knee joint is an example of a contact force.

The most important (and typically the only) remote force used in biomechanical force system analyses is the weight of the system of interest. Note that the weight drawn in the free body diagram is the weight of the system of interest and not the weight of the total body! If, for instance, the foot has been selected as the system of interest, the weight of the rest of the body is included implicitly in the forces and moments drawn at the ankle joint, and in the ground reaction force.

Contact forces (moments) are forces (moments) caused through the direct contact of the system of interest with other bodies. Contact forces include:

- Joint contact forces,
- Muscle-tendon forces,
- Ligamentous forces,
- Forces between the hand and an object,
- Forces between the foot and the ground, and
- Air or water resistance forces, etc.

Often, forces are summarized into *resultant forces*. The weight of the different parts of the hand is typically summarized into the resultant weight of the hand. The various local forces acting between the ground and the foot during walking are typically summarized into the ground reaction force. Additionally, one resultant joint force, and one resultant joint moment, can represent any loading situation with respect to a specific point of a joint, e.g., a joint center.

The resultant joint force is the vector sum of all forces transmitted by muscle-tendon units, ligaments, bones, and soft tissue across the joint.

Comment: the resultant joint force is also called *resultant intersegmental joint force.* Mathematically, the resultant joint force is defined as:

$$\mathbf{F} = \underbrace{\sum_{i=1}^{m} \mathbf{F}_i}_{\text{(muscle)}} + \underbrace{\sum_{j=1}^{l} \mathbf{F}_j}_{\text{(lig)}} + \underbrace{\sum_{k=1}^{n} \mathbf{F}_k}_{\text{(bone)}} + \underbrace{\sum_{p=1}^{s} \mathbf{F}_p}_{\text{(other)}}$$

where:

$\mathbf{F}$ = resultant intersegmental joint force
$\mathbf{F}_i$ = forces transmitted by muscles
$\mathbf{F}_j$ = forces transmitted by ligaments
$\mathbf{F}_k$ = bone-to-bone contact forces
$\mathbf{F}_p$ = forces transmitted by soft tissue, etc.

Note that the resultant joint force is an abstract quantity that is not related to the actual force measured between two bones in a joint (the bone-to-bone force).

The resultant joint moment is the net moment produced by all intersegmental forces with lines of actions that cross the joint.

Comment: the resultant joint moments are also called *resultant intersegmental joint moments.* Resultant joint moments may include resultant muscle joint moments, resultant ligament joint moments, and resultant bony contact force moments. The resultant joint moment only reflects the net effect of agonist and antagonist muscles. For that reason, the resultant joint moment is also called *net joint moment.*

$$\mathbf{M} = \underbrace{\sum_{i=1}^{m} \mathbf{M}_i}_{\text{(muscle)}} + \underbrace{\sum_{j=1}^{l} \mathbf{M}_j}_{\text{(lig)}} + \underbrace{\sum_{k=1}^{n} \mathbf{M}_k}_{\text{(bone)}} + \underbrace{\sum_{p=1}^{s} \mathbf{M}_p}_{\text{(other)}}$$

where:

$\mathbf{M}$ = resultant intersegmental joint moment
$\mathbf{M}_i$ = moments transmitted by muscles

$\mathbf{M}_j$ = moments transmitted by ligaments
$\mathbf{M}_k$ = bone-to-bone contact moments
$\mathbf{M}_p$ = moments transmitted by soft tissue, etc.

Free body diagrams allow the researcher to work with actual forces or resultant forces. Both of these approaches are discussed briefly in the following section.

Actual forces and moments approach

This approach uses forces and moments that could be actually measured if appropriate transducers were to be used. The *actual forces and moments approach* starts with defining the force-transmitting elements included in the analysis. Subsequently, all actual forces and moments are drawn into the free body diagram. They include:

(1) Remote forces
- Weight of the system of interest.

(2) Contact forces
- Forces due to adjacent bones.
- Forces due to adjacent ligaments.
- Forces due to adjacent muscle-tendon units.
- Reaction forces due to contact with other external bodies, e.g., ground.

(3) Frictional joint moments
- Moments due to friction in the joints (these are usually assumed to be small and are, therefore, neglected).

EXAMPLE 2

Question

Draw the free body diagram (FBD) for the foot of a person standing on the forefoot of one leg.

System of interest

Foot.

Assumptions

(1) The foot is a rigid structure.
(2) The foot has one idealized joint, the ankle joint, which is responsible for plantar- and dorsi-flexion between foot and leg.
(3) The structures responsible for contact forces are the Achilles tendon, the tibia at the ankle joint, and the ground.
(4) There is no friction in the ankle joint.
(5) The weight of the foot can be neglected.
(6) The problem can be solved two dimensionally.

The *actual forces and moments approach* is advantageous for simple applications, as illustrated in Figure 4.5.2. This approach is also used for simple estimations of internal forces or to determine the order of magnitude of internal forces. It has the advantage of showing the actual forces in the drawing, and that it is possible to relate the drawing to the real situation, recognizing forces acting in selected structures. However, in applications with more complex force distributions, this approach may not be appropriate and can be replaced by another approach, the *resultant forces and moments approach*.

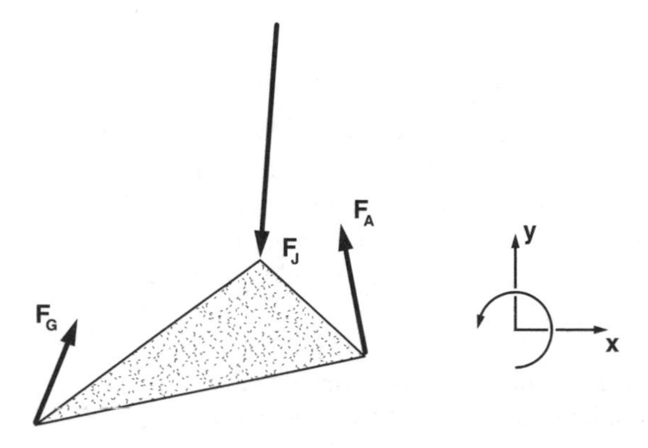

Figure 4.5.2 **Example for a free body diagram, estimating the force in the Achilles tendon when standing on the forefoot of one leg using the *actual forces and moments approach*.**

Resultant forces and moments approach

This approach uses equipollent resultant forces and moments (Andrews, 1974). The free body diagram includes:

(1) Remote forces

- Weight of the system of interest.

(2) Contact forces and contact moments

- Resultant joint forces and moments: the complicated force distributions at the distal and proximal joints are replaced by the equipollent resultant joint forces and moments.
- Resultant surface forces and moments acting on the system of interest: the sometimes complicated force distributions acting on the segment surface are replaced by an equipollent resultant surface force and moment acting at some arbitrarily, but appropriately, located point.

The calculated resultant joint forces and moments do not correspond to actual forces and moments, and cannot be measured with appropriate transducers, as in the previous approach. They are abstract quantities that are often not used as final results, but they are used as input into a second step, distributing these forces and moments to specific structures, e.g., ligaments, tendon, and bone.

EXAMPLE 3

Question

Draw the FBD for the foot of a person standing on the forefoot of one leg.

System of interest

Foot.

Assumptions

(1) The foot is a rigid structure.
(2) The foot has one idealized joint, the ankle joint, which is responsible for plantar- and dorsi-flexion between foot and leg. The foot does not perform any in-eversion or ab-adduction movements, or both.
(3) The structures responsible for contact forces in this specific example are the Achilles tendon, the tibia at the ankle joint, the rigid foot at the ankle joint, and the ground.
(4) There is no friction in the ankle joint.
(5) The weight of the foot can be neglected.
(6) The problem can be solved two dimensionally.

FBD:

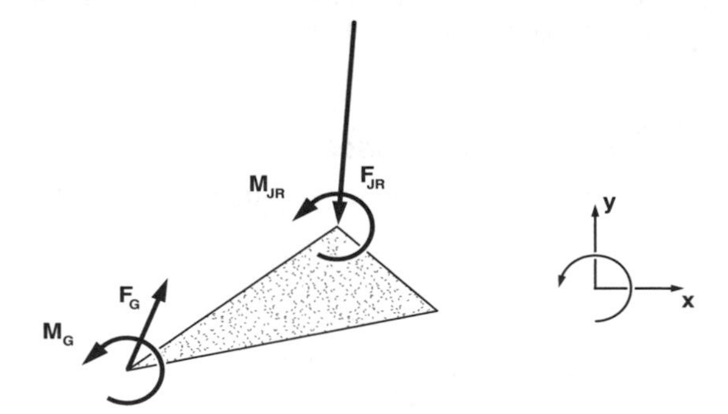

Figure 4.5.3 **Example of a free body diagram, estimating the force in the Achilles tendon when standing on the forefoot of one leg using the *resultant joint forces and moments approach*.**

The *resultant forces and moments approach* is appropriate for situations in which the complexity of the acting forces prohibits drawing a simple diagram, and in which the subsequent calculations favor a two-step approach. In the two-step approach, step 1 concerns the resultant joint moments and forces, and step 2 concerns the distribution of resultant joint moments and forces).

The inertia forces approach

Free body diagrams should, in addition to the above mentioned forces, include inertia forces (Roth, 1989). Using this approach, d'Alembert's principle is applied. This approach is not discussed further. This text will concentrate on the application of Newton's laws. See textbooks for more information on d'Alembert's principle.

THE REFERENCE FRAME (COORDINATE SYSTEM)

Each free body diagram must include a reference frame (coordinate system) that defines the direction of the positive axes for translation and rotation. Often, the following convention for a Cartesian coordinate system is used:

y = vertical direction
x = a-p direction
z = medio-lateral direction

In the following chapters, the *actual forces and moments approach* is applied for the simple applications, and the *resultant forces and moments approach* is applied for more complex applications. In selected cases, the same problem is solved twice, once using each approach.

4.5.5 EQUATIONS OF MOTION

The construction of a free body diagram, the formulation of the corresponding equations of motion, and the solution and interpretation of them, may be complicated for biological/biomechanical applications. Consequently, it may be advantageous to use a systematic convention for the formulation of the equations of motion. Such a convention may reduce several errors that occur as the equations of motion are produced. Here, a convention is outlined for the *resultant joint forces and moments approach* for a planar case. It can be adapted for the three-dimensional case.

Assumptions

(1) A right handed Cartesian coordinate system is defined, with x, y, and z axes perpendicular.
(2) Longitudinal axis of the segment remains in the x-y plane.

Proposed convention

(1) Define the system of interest.
(2) Use solid lines to represent the system of interest.
(3) Use dotted lines (if preferred) to represent other (adjacent) systems to illustrate a particular situation.
(4) Number all systems. If appropriate, number the ground 0 and use a sequence from there.
(5) Draw all force components in positive axis directions in the FBD.
(6) Draw moments in positive x axis direction in the FBD.
(7) F_{ijx} denotes a force component in the x direction exerted by system j on system i.
(8) M_{ijx} denotes a moment component about the x axis exerted by system j on system i.
(9) $C_i = (x_i, y_i)$ is the centre of mass of segment i.
(10) $J_{ik} = (x_{ik}, y_{ik})$ is the joint centre between the segment of interest i and the neighbouring segment k.

(11) m_i is the mass of segment i.
(12) g is the acceleration due to gravity in m/s^2.
(13) Use a superscript dot for the first time derivative, and a superscript double dot for the second time derivative.
(14) I_{iz} is the moment of inertia of segment i about the z-axis through its centre of mass.
(15) φ_{iz} is the angle the segment's longitudinal axis between the two joint centres, $J_{i(i-1)}$ and $J_{i(i+1)}$, makes with the positive x axis.

The corresponding free body diagram, FBD, is Figure 4.5.4:

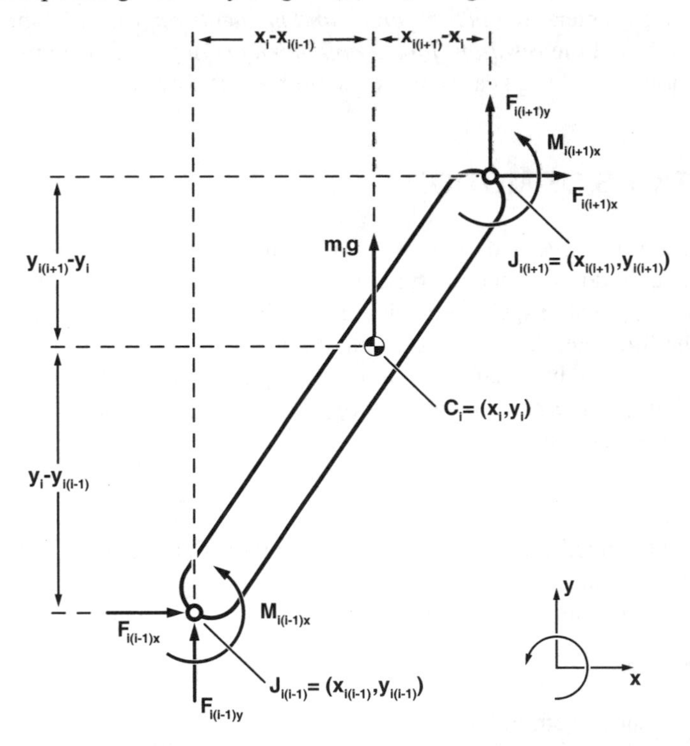

Figure 4.5.4 **Standardized general free body diagram for a body i with two joint centres.**

The convention proposed in this section is one of several appropriate possibilities. The proposed convention that all forces and moments should be drawn in positive axis direction may deserve some further explanation.

There are at least three possibilities for drawing forces and moments in a FBD:

(1) Draw all forces and moments the way they act.
(2) Draw the forces and moments in an arbitrary way.
(3) Draw all forces and moments in a positive axis direction.

For all three approaches, the calculated forces and moments follow the same rule: the forces and moments act as drawn in the FBD if the calculated forces and moments are positive. The forces and moments act in opposite direction to the forces and moments drawn in the FBD if the calculated forces and moments are negative. For the three possibilities to draw forces and moments that means:

(1) If a calculated force or moment has a negative sign, the force or moment was incorrectly drawn in the FBD, and acts in opposite direction than the one drawn.
(2) If a calculated force or moment has a negative sign the force or moment was incorrectly drawn in the FBD and acts in opposite direction than the one drawn. This is, of course, the same statement as the statement shown in (1) above. The difference is that for the second approach more forces and moments will (most likely) be negative.
(3) If a calculated force or moment has a positive sign, the force or moment acts in a positive axis direction. If a calculated force or moment has a negative sign, the force or moment acts in a negative axis direction. The result for this third approach, therefore, relates to the selected coordinate system, and not to the drawing. It is suggested that this is a more general approach. However, it may be counterintuitive to draw specific forces in a positive axis direction, e.g., weight.

Too much emphasis should not be put into this aspect. All approaches are appropriate if properly used and understood.

Using these conventions, the two-dimensional general equations of motion for segment i with two joints are:

Translation:

$$m_i \ddot{x}_i = F_{i(i+1)x} + F_{i(i-1)x}$$

$$m_i \ddot{y}_i = F_{i(i+1)y} + F_{i(i-1)y} + m_i g$$

Rotation:

$$\begin{aligned} I_{iz}\ddot{\varphi}_{iz} = {} & M_{i(i+1)z} + M_{i(i-1)z} \\ & -(y_{i(i+1)} - y_i)F_{i(i+1)x} + (y_i - y_{i(i-1)})F_{i(i-1)x} \\ & +(x_{i(i+1)} - x_i)F_{i(i+1)y} - (x_i - x_{i(i-1)})F_{i(i-1)y} \end{aligned}$$

The first two equations are merely statements of the principle of linear momentum (Euler's 1st law or principle of the motion of the mass centre, applied to bodies of finite extent), which requires that an inertial reference frame, R, be used. The third equation, the moment equation, uses the net moment, the moment of inertia, and the angular acceleration of the body relative to the inertial reference frame, R.

The proposed convention has the advantage that the equations of motion are independent of the particular (second approach) free body diagram drawn. Such a procedure is advantageous when analyzing a complex multi-link system in combination with a computer program. The next stage (after calculating the resultant joint forces and moments) is to distribute the resultant joint forces and moments to the structures that cross a particular joint.

4.5.6 MATHEMATICAL SOLUTION

The equations of motion have n unknowns. The number of equations m can be equal, smaller or larger than the number of equations. These special cases are:

equal	$m = n$	determined
smaller	$m < n$	underdetermined
larger	$m > n$	overdetermined

If the number of unknowns and the number of equations are equal, the system has a unique solution. If not, appropriate mathematical procedures must be applied to find a solution.

4.6 MATHEMATICALLY DETERMINATE SYSTEMS

NIGG, B.M.

4.6.1 INTRODUCTION

This section attempts to show how simple models using particles may be, and how models have been used, to discuss biomechanical situations. The discussion of what level of complexity of a biomechanical model is appropriate will probably never end. Complex models, like the one used by Yeadon (1984) to analyze airborne movements, or the one by Seireg and Arvikar (1973) that contained 29 muscles per leg, are certainly impressive from a mathematical point of view alone. However, one can disagree with the view that because the human body is complex, biomechanists should always reproduce as much as possible of its complexity in their models. This section attempts to illustrate McNeill Alexander's (1992) proposition that, in certain cases, simple models may be appropriate. We start with the simplest possible model, a particle with no force acting on it, and progress step-wise to more complicated models that use one or two segments. Examples have been selected to provide an insight into important aspects of modelling.

We start with the simplest possible case, which is a particle with no force acting on it, and then proceed to more complex examples. Some readers may disregard the rather trivial initial examples and jump to the more demanding ones. For others, these simple examples may provide an adequate tool to refresh aspects of mathematics. First, they attempt to illustrate different possibilities to model determinate systems of particles with simple mathematical tools, and second, they attempt to discuss the physical and biological interpretation that these models may include.

Every example uses the same setup and starts with explanatory comments that provide insight into the problem. The comments may explain why the model was developed at all and what may be concluded from the results. The following headings are used:

Question — Question to be answered.

Assumptions — Assumptions used in this example.

FBD — Free body diagram (FBD).
Illustrations: in some cases, it may be advantageous to add a drawing to illustrate the actual situation at different time points. Such drawings may or may not include forces. However, adding a real free body diagram is recommended, in addition to such drawings. Free body diagrams may also be added to the text during the solution of the problem if this sheds light on developing additional steps in the process.

Equations of motion — Equations of motion used in this example.
EM for translation of particles, and EM for translation and rotation of rigid bodies.

Initial conditions — Initial conditions for this example.

Final conditions	Final conditions for this example.
Solution	Solution of the problem.

At the end of each example, its results are discussed and its biomechanical importance is addressed.

4.6.2 MECHANICAL MODELS USING PARTICLES

EXAMPLE 1 (the simplest case - no force)

The simplest possible model in mechanics is a particle in space with no force acting on it. The setup and the mathematics for this example are trivial, and the result is obvious. However, the example may serve as a simple start to a journey that leads to more complex models.

Question

Describe the one-dimensional movement of a particle upon which no force is acting.

Assumptions

(1) The question can be treated as a one-dimensional problem.

FBD:

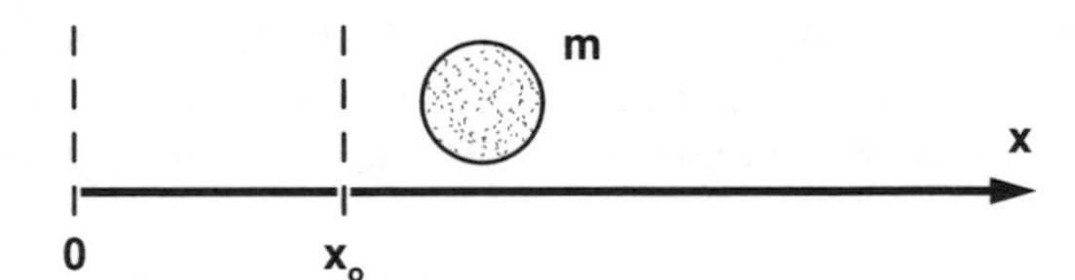

Figure 4.6.1 **Free body diagram, FBD, for one particle with no force acting on the particle.**

Equations of motion

$$m\ddot{x}(t) = 0$$

Initial conditions

$$x(0) = x_o$$

$$\dot{x}(0) = v_o$$

Solution

$$\ddot{x}(t) = 0$$

Integration of this equation provides:

$$\dot{x}(t) = const = c_1$$

Using the initial condition:

$$\dot{x}(0) = v_o$$

The constant, c_1, can be determined as v_o:

$$\dot{x}(t) = v_o$$

A second integration provides:

$$x(t) = v_o t + c_2$$

Using the initial condition:

$$x(0) = x_o$$

provides the general equation for the position of the particle of interest as a function of time:

$$x(t) = x_o + v_o t \tag{4.6.1}$$

This is the well-known equation for uniform rectilinear motion. The position, $x(t)$, depends on the initial position, the initial velocity, v_o, and the time. Of course, it may be difficult to find a biomechanical application for this model.

EXAMPLE 2 (one particle with a constant force acting on it)

Another simple case is a particle in space with a constant force acting on it. One practical example from physics is a mass in the air with only the gravitational force acting on it. Again, although we know the result of this example in advance, i.e., the movement of a particle with constant acceleration, this example leads the way to a thorough understanding of the possibilities of modelling biomechanical problems.

Question

Describe the one-dimensional movement of a particle upon which a constant force is acting.

Assumptions

(1) A constant force, F, is acting on the particle in vertical (y) direction:

$$F = mg$$

where:

$$g = 9.81 \text{ m/s}^2$$

(2) The mass can be considered as a particle.
(3) Air resistance can be neglected.
(4) All forces are drawn in a positive axis direction.

FBD:

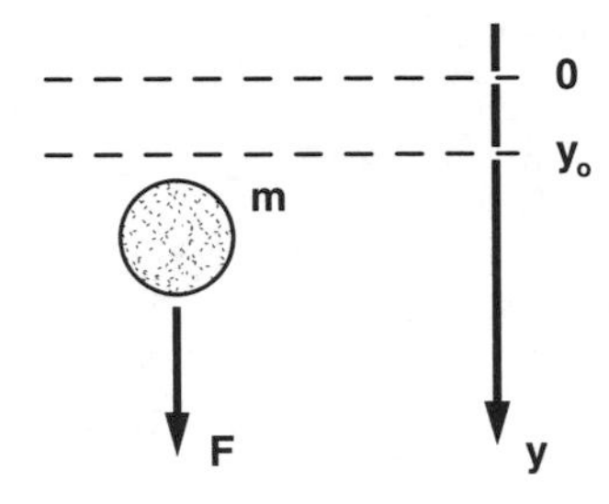

Figure 4.6.2 **FBD for a particle with a constant gravitational force, F = - mg, acting on it.**

Equations of motion

$$m\ddot{y}(t) = mg$$

Initial conditions

$$y(0) = y_o$$

$$\dot{y}(0) = v_o$$

Solution

Integration of the acceleration provides the velocity:

$$\dot{y}(t) = g\,t + \text{const}$$

Using the initial condition provides the equation for the velocity:

$$\dot{y}(t) = g\,t + v_o$$

Integration of the velocity provides the position:

$$y(t) = 1/2\,g\,t^2 + v_o\,t + \text{const}$$

Using the initial condition provides the equation for the position:

$$y(t) = y_o + v_o\,t + 1/2\,g\,t^2 \tag{4.6.2}$$

This is the equation of uniformly accelerated rectilinear motion. The equation can be used for the determination of the vertical motion of a projectile with no air resistance.

EXAMPLE 3 (force of a linear spring acting on a particle)

This example corresponds in a simplistic way to the real-life situation of a gymnast taking off from a trampoline or a diver landing on a diving board (with gravity neglected). The particle would correspond to the athlete and the spring to the trampoline or the diving board. The model permits the calculation of where the particle is at a certain point in time.

Question

Describe the movement of a particle upon which the force of a linear spring is acting.

Assumptions

(1) The force of the linear spring is:

$$F = -k \cdot x$$

where:

k = spring constant
x = spring deformation

(2) The mass of the spring can be neglected.
(3) The question can be treated as a one-dimensional problem.

FBD:

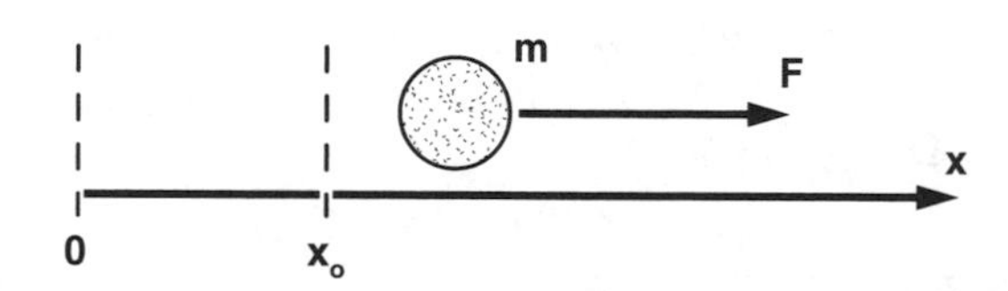

Figure 4.6.3 **Free body diagram, FBD, for a particle in contact with an ideal linear spring.**

Equations of motion

$$m\ddot{x}(t) = F = -kx(t)$$

$$\ddot{x}(t) = -\frac{k}{m}x(t)$$

and from spring theory:

$$\frac{k}{m} = \Omega^2 \qquad (\Omega = \text{circular frequency})$$

consequently:

$$\ddot{x}(t) = -\Omega^2 x(t)$$

which is a second order linear differential equation.

Initial conditions

$$x(0) = x_o$$

$$\dot{x}(0) = v_o$$

Solution

General solution:

$$x(t) = A\sin\omega t + B\cos\omega t$$

$$\dot{x}(t) = A\omega \cdot \cos\omega t - B\omega \cdot \sin\omega t$$

$$\ddot{x}(t) = -A\omega^2 \cdot \sin\omega t - B\omega^2 \cdot \cos\omega t$$

$$= -\omega^2(A \cdot \sin\omega t + B \cdot \cos\omega t)$$

$$\ddot{x}(t) = -\omega^2 \cdot x(t)$$

consequently:

$$\omega^2 = \Omega^2$$

and with the initial conditions:

$$x(0) = B = x_o$$

$$\dot{x}(0) = A \cdot \omega = v_o$$

$$A = \frac{v_o}{\omega} = \frac{v_o}{\Omega}$$

$$x(t) = \frac{v_o}{\Omega} \cdot \sin\Omega t + x_o \cos\Omega t \qquad (4.6.3)$$

Equation (4.6.3) describes the position of a particle during contact with an ideal linear spring. The maximal position (deformation of the spring), velocity, and acceleration can be determined. For the sake of simplicity, the initial condition for the position is simplified to:

Initial conditions

$$x(0) = x_0 = 0$$

This provides for the position the general equation:

$$x(t) = v_o \cdot \sqrt{\frac{m}{k}} \cdot \sin\sqrt{\frac{k}{m}} \cdot t$$

and for the minimal position (maximal deformation) the equation:

$$x_{max} = v_o \cdot \sqrt{\frac{m}{k}}$$

The equation for the velocity, using the same initial conditions, is:

$$\dot{x}(t) = v_o \cdot \cos\sqrt{\frac{k}{m}} \cdot t$$

which provides for the maximal velocity (at the beginning and at the end of contact):

$$\dot{x}_{max} = v_o$$

The equation for the acceleration, using the same initial conditions, is:

$$\ddot{x}(t) = -v_o\sqrt{\frac{k}{m}} \cdot \sin\sqrt{\frac{k}{m}} \cdot t$$

which provides the maximal acceleration at the time of the maximal deformation:

$$\ddot{x}_{max} = -v_o\sqrt{\frac{k}{m}}$$

Figure 4.6.4 illustrates the movement characteristics for position, velocity, and acceleration of the mass point. Note the phase shift between position, velocity, and acceleration for

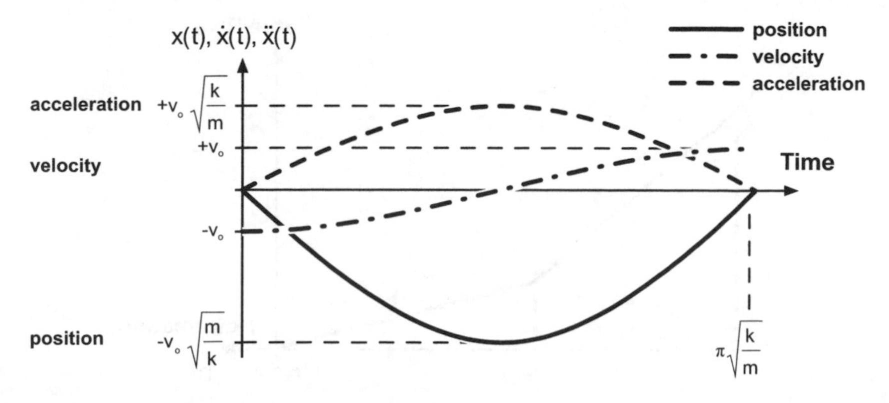

Figure 4.6.4 **Position, velocity, and acceleration as functions of time for a particle dropping onto an ideal linear spring.**

the given assumptions. These considerations could be applied to the movement of a diver during the last contact with the diving board. The velocity would be zero at the lowest position of the board, and the acceleration would be maximal at this point in time. The actual situation in diving corresponds reasonably well to reality. The main difference is that the position is not a clean sinusoidal curve, but a slightly asymmetrical sinusoidal curve.

EXAMPLE 4 (force of two idealized springs acting on a particle)

One could imagine a foot landing on a surface and describe the foot as a particle, with the heel pad and the surface as two springs in series. This conception of the situation is rather simplistic, but it may provide some insight into the problem of interest.

Question

Describe the movement of a particle as it drops onto a system of two ideal linear springs. Specifically, answer the following questions:

(1) Determine the velocity at contact, v_{crit}, for which the first spring bottoms out.
(2) Determine the position, $z(t)$, of the particle for the two springs.
(3) Determine the maximal acceleration (deceleration) of the particle.
(4) Draw and discuss the graphs for $z(t)$, $\dot{z}(t)$, and $\ddot{z}(t)$.

Assumptions

(1) Spring 1 is in contact with the mass. Spring 2 is in contact with spring 1.
(2) Spring 1 is much softer than spring 2, i.e., $k_1 << k_2$.
(3) Spring 1 compresses first completely to z_b before spring 2 starts to compress.
(4) Spring 1 remains compressed at position z_b while spring 2 is compressed. Consequently, the movement can be subdivided into two parts, a first part for which spring 1 changes length while spring 2 remains at its original length, and a second part for which spring 1 remains compressed at z_b while spring 2 changes length. The idealized force-deformation diagram for this case is illustrated in Figure 4.6.5.

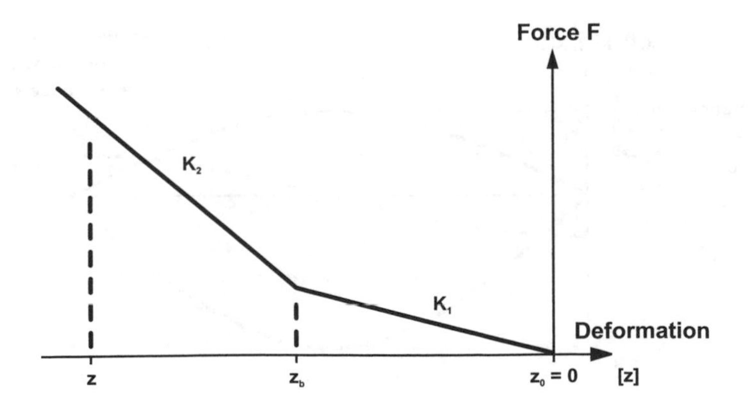

Figure 4.6.5 **Assumed force deformation diagram of the two spring systems.**

(5) The forces are drawn in positive axis direction.

(6) The two springs are ideal. The force is independent of the velocity, v_o. There is no loss of energy.

$$F_1 = -k_1 \cdot z(t)$$

$$F_2 = -k_2 \cdot z(t)$$

(7) The masses of the springs can be neglected.

(8) The numerical values for the example are:

m	=	4 kg	=	dropping mass
z_b	=	0.04 m	=	total compression of spring 1
k_1	=	10^4 N/m	=	spring constant 1
k_2	=	10^6 N/m	=	spring constant 2
v_o	=	4 m/s	=	initial velocity of mass (for question 4)

(9) Gravity can be neglected.

(10) The problem can be solved one dimensionally.

Answer to question 1 - determination of critical velocity

FBD:

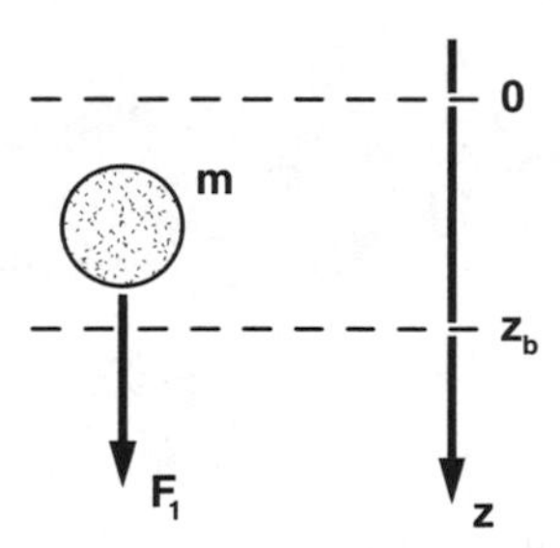

Figure 4.6.6 **Free body diagram, FBD, for the first part of the problem (question 1), in which the second spring is assumed to be infinitely stiff. This FBD corresponds to Figure 4.6.3.**

Equations of motion

Assuming that only spring 1 is compressed:

$$m\ddot{z}(t) = F_1 = -k_1 \cdot z(t)$$

or:

$$\ddot{z}(t) = -\Omega_1^2 \cdot z(t)$$

Initial conditions

$$z(0) = 0$$

$$\dot{z}(0) = v_o$$

Solution

$$z(t) = A \cdot \sin\omega_1 t + B \cdot \cos\omega_1 t$$

$$\dot{z}(t) = A \cdot \omega_1 \cdot \cos\omega_1 t - B\omega_1 \cdot \sin\omega_1 t$$

$$\ddot{z}(t) = -\omega_1^2 \cdot z(t)$$

where:

$$\omega_1^2 = \Omega_1^2$$

Using the initial condition:

$$z(0) = 0$$

provides:

$$B = 0$$

and the initial condition:

$$\dot{z}(0) = v_o$$

provides:

$$A = \frac{v_o}{\Omega_1} = \frac{v_o}{\sqrt{k_1}} \cdot \sqrt{m}$$

consequently:

$$z(t) = \frac{v_o}{\Omega_1} \cdot \sin\Omega_1 t$$

Calculation of the critical velocity, v_c, at which spring 1 is totally compressed (bottoms out):

$$v_c = z_b \cdot \Omega_1$$

Spring 1, the softer spring in this example, does not compress completely if the landing velocity, v_o, is smaller than the critical velocity, ($v_o < v_c$). However, spring 1 compresses completely if $v_o \geq v_c$.

Result for the selected numerical values:

$$v_c = 0.04\text{m}\sqrt{\frac{100\text{m}}{0.04\text{ms}^2}}$$

$$v_c = 2 \text{ m/s}$$

The critical velocity for this particle is 2 m/s, which corresponds to a drop from about 0.2 m (20 cm). If the mass is dropped from a height of about 0.2 m, the material of the soft spring compresses completely (bottoms out). The values used correspond to material properties of the human heel as follows: the mass of a leg is about 4 to 6 kg, which is the mass used. The spring constant of the material of the human heel was reported in the order of magnitude of about $2 \cdot 10^4$ N/m (Nigg & Denoth, 1980). Consequently, the numbers used in the numerical calculations and the subsequent results correspond in their order of magnitude to the landing of a heel on the ground. A landing velocity of the heel of about 2 m/s in barefoot running may produce complete compression of the heel pad. Note that barefoot runners typically do not land on their heel, but land flat footed. The human heel is, of course, a viscoelastic material, rather than an elastic material. The numerical results and their translation into the practical application must, therefore, be taken with a grain of salt. More accurate and appropriate results might be determined with a more complex model that contained viscous elements (dampers), and considered the effect produced by the increasing contact area between heel and ground.

Answer to question 2 - determination of the general position of the particle

Figure 4.6.7 shows an illustration of the situation.

FBD:

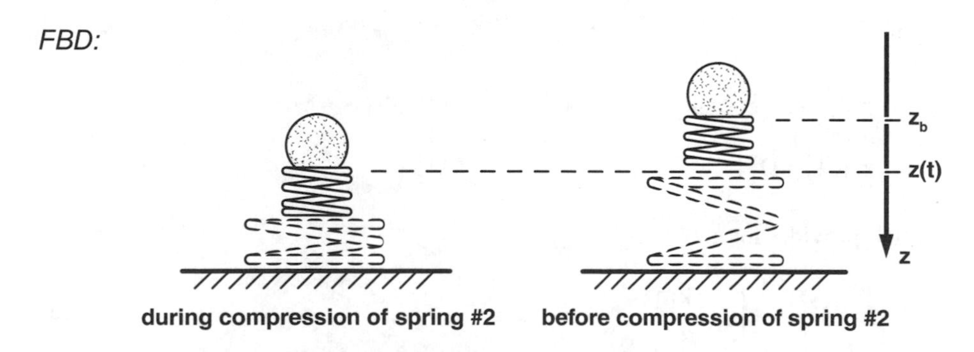

Figure 4.6.7 **Illustration of the situation for the point at which spring 1 is totally compressed. The system of interest for this example is the mass of the particle and the totally compressed massless spring 1.**

Equations of motion

The force acting on the particle can be composed from the force needed to completely compress the first spring, and the force needed to compress the second spring partially.

$$m\ddot{z}(t) = \text{force (spring 2)} + \text{force (spring 1)}$$

$$m\ddot{z}(t) = -k_2[z(t) - z_b] - k_1 z_b$$

Initial conditions

$$z(t_b) = z_b$$

$$\dot{z}(t_b) = v_b$$

Solution

$$z(t) = C \cdot \sin\omega_2 t + D \cdot \cos\omega_2 t + a \tag{4.6.4}$$

$$\dot{z}(t) = C\omega_2 \cdot \cos\omega_2 t - D\omega_2 \cdot \sin\omega_2 t \tag{4.6.5}$$

$$\ddot{z}(t) = -\omega_2^2[z(t) - a] \tag{4.6.6}$$

Determination of ω_2, a, and D:

$$-\Omega_2^2[z(t) - z_b] - \Omega_1^2 \cdot z_b = -\omega_2^2[z(t) - a]$$

$$-\Omega_2^2 \cdot z(t) + \Omega_2^2 \cdot z_b - \Omega_1^2 \cdot z_b = -\omega_2^2 \cdot z(t) + \omega_2^2 \cdot a$$

consequently:

$$\omega_2 = \Omega_2 \tag{4.6.7}$$

and:

$$-\Omega_2^2 \cdot z_b - \Omega_1^2 \cdot a = \omega_2^2 \cdot a$$

$$z_b(\Omega_2^2 - \Omega_1^2) = \Omega_2^2 \cdot a$$

which provides for a:

$$a = z_b\left[1 - \frac{k_1}{k_2}\right] \tag{4.6.8}$$

$$z_b = C \cdot \sin\Omega_2 t + D \cdot \cos\Omega_2 t + z_b - z_b \cdot \frac{k_1}{k_2}$$

For the time $t_b = 0$:

$$D = z_b \cdot \frac{k_1}{k_2} \tag{4.6.9}$$

Determination of C:

$$\text{for } \dot{z}\ (t = t_b = 0) = v_b$$

$$v_b = C \cdot \Omega_2 \cdot \cos\Omega_2 t - D \cdot \Omega_2 \cdot \sin\Omega_2 t \quad (4.6.10)$$

$$C = \frac{v_b}{\Omega_2}$$

v_b can be replaced using the conservation of energy:

$$\frac{1}{2} m v_o^2 = \frac{1}{2} k_1 \cdot z_b^2 + \frac{1}{2} m v_b^2 \quad (4.6.11)$$

$$v_b^2 = v_o^2 - \Omega_1^2 \cdot z_b^2 \quad (4.6.12)$$

Substituting equation (4.6.12) in equation (4.6.11) provides for C:

$$C = \sqrt{\frac{v_o^2 \cdot m}{k_2} - \frac{k_1}{k_2} \cdot z_b^2} \quad (4.6.13)$$

Substituting equations (4.6.7), (4.6.8), (4.6.9), and (4.6.13) in equation (4.6.6) provides the position of the mass as a function of time:

$$z(t) = \sqrt{\frac{v_o^2 \cdot m}{k_2} - z_b^2 \cdot \frac{k_1}{k_2}} \cdot \sin\left[\sqrt{\frac{k_2}{m}} \cdot t\right] + z_b \cdot \frac{k_1}{k_2} \cdot \cos\left[\sqrt{\frac{k_2}{m}} \cdot t\right] + z_b\left[1 - \frac{k_1}{k_2}\right]$$

z(t) is the answer to question 2. It provides the general position of the particle as a function of time, for case 2.

Answer to question 3 - determination of the maximal acceleration

Using equation (4.6.6):

$$\ddot{z}(t) = -\Omega_2^2 (C \cdot \sin\Omega_2 t + D \cdot \cos\Omega_2 t)$$

which can be written as:

$$\ddot{z}(t) = -\Omega_2^2 \cdot \sqrt{C^2 + D^2} \cdot \sin(\Omega_2 t + \delta)$$

For the maximal acceleration:

$$\ddot{z}_{max} = -\Omega_2^2 \cdot \sqrt{C^2 + D^2}$$

$$\ddot{z}_{max} = -\Omega_2^2 \sqrt{v_o^2 \cdot \frac{m}{k_2} - \frac{k_1}{k_2} \cdot z_b^2 + z_b^2 \cdot \frac{k_1^2}{k_2^2}}$$

$$\ddot{z}_{max} = -\sqrt{v_o^2 \cdot \frac{k_2}{m} - z_b^2 \cdot k_1 \cdot \frac{k_2}{m^2} + z_b^2 \cdot \frac{k_1^2}{m^2}} \tag{4.6.14}$$

The maximal acceleration (deceleration) depends on the landing velocity of the particle, the mass, and the two material constants of the two ideal springs. The result for the numerical values of this example (answer to question (4)) is illustrated in Figure 4.6.8.

The setup and the results of this example need a few additional comments with respect to their methodology and interpretation:

(1) In real life, a situation like the one described in this example rarely occurs. Usually the two materials, e.g., heel, shoe sole, sport surface, landing mats, do not behave like linear springs, and the second material starts to deform at the same time as the first one. A real force-deformation diagram for the loading part would realistically be curved, not composed of two straight lines. As well, a loss of energy for the above-mentioned materials can usually be expected.

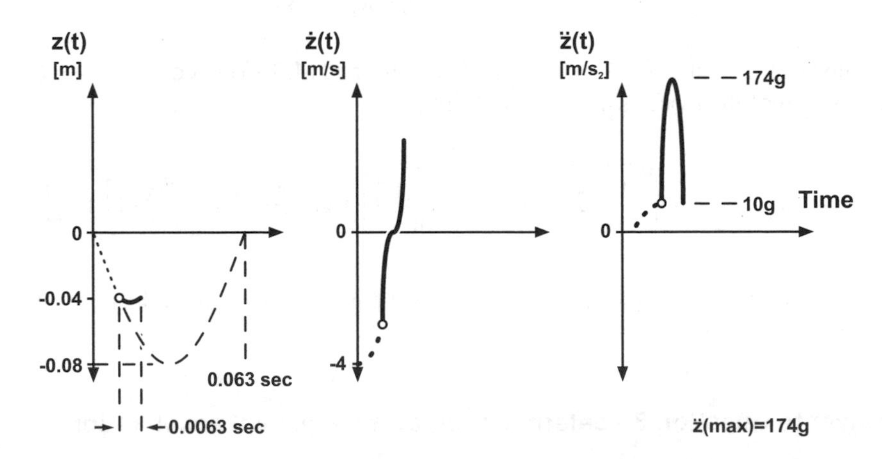

Figure 4.6.8 **Graphical illustration of position, $z(t)$, velocity, $\dot{z}(t)$, and acceleration, $\ddot{z}(t)$, for a particle dropping on two springs in series.**

(2) The mass was assumed to be a particle with no volume. In reality, masses have a volume and a geometrical shape. The force-deformation diagrams of two spheres with diameters of 5 and 10 cm (if everything else is kept constant) are different. This fact is not reflected in this model.

(3) This model should not be used to estimate accurate decelerations, since the assumptions are general. The value of this model lies in explaining some general findings. It can, for instance, be used to explain the bottoming out phenomenon. Corners in the position-, velocity-, and acceleration-time diagram (or in the force-time diagram) indicate that a material bottoms out.

(4) The results of this model illustrate that sport shoes or playing surfaces designed to protect the human heel should have material properties (material stiffness) in the same order of magnitude as the material constants of the human heel. If their stiffness is much higher than the stiffness of the human heel, they will only be effective after the heel pad is mostly compressed. Artificial track surfaces are generally much stiffer than the human heel. One should not expect that these surfaces would affect the impact forces significantly.

EXAMPLE 5 (force depending on deformation and velocity)

Experimental measurements from drop tests with a human heel (Nigg & Denoth, 1980; Misevich & Cavanagh, 1984) suggest that the force acting on the heel depends on the deformation $x(t)$ and on the velocity of deformation $\dot{x}(t)$. The heel pad, therefore, has viscoelastic properties.

Question

Determine the movement of a mass (particle) dropping onto a system with viscoelastic properties.

Assumptions

(1) The human heel has some spring-like behaviour.

(2) Friction is involved and, therefore, the loss of energy is not zero. The material exhibits viscoelastic behaviour.

(3) Force $= F(t) = F_{damp} + F_{spring} = -k \cdot x(t) - r \cdot \dot{x}(t)$

This mathematical description of human tissue is simplistic. A more sophisticated approach proposed by Nigg and Denoth (1980) was:

$$\sigma(x,\dot{x}) = a \cdot x^2 + b \cdot x \cdot \dot{x}$$

(4) The problem can be treated one dimensionally.

(5) The mass is considered as a particle.

(6) The spring and damping elements are arranged in parallel.

(7) The forces are drawn in the positive axis direction.

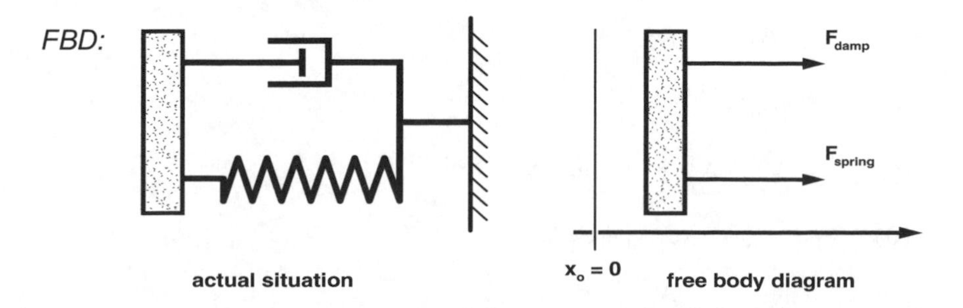

Figure 4.6.9 **Illustration of the setup (left) and FBD of a particle acted upon by a spring and a damper (right).**

Equations of motion

$$m\ddot{x}(t) = -kx(t) - r\dot{x}(t)$$

or:

$$m\ddot{x}(t) + r \cdot \dot{x}(t) + k \cdot x(t) = 0$$

Initial conditions

$$x(0) = 0$$

$$x(0) = v_o$$

Solution

$$x(t) = A \cdot e^{-ct}$$

$$\dot{x}(t) = -c \cdot A \cdot e^{-ct}$$

$$\ddot{x}(t) = c^2 \cdot A \cdot e^{-ct}$$

$$m \cdot c^2 \cdot A \cdot e^{-ct} - r \cdot c \cdot A \cdot e^{-ct} + k \cdot A \cdot e^{-ct} = 0$$

$$mc^2 - rc + k = 0$$

$$c_{1,2} = \frac{r \pm \sqrt{r^2 - 4mk}}{2m}$$

or:

$$c_{1,2} = \frac{r}{2m} \pm \sqrt{\frac{r^2}{4m^2} - \frac{k}{m}} \tag{4.6.15}$$

If $r^2 \neq 4mk$ the result indicates two values for c, c_1, and c_2. Therefore:

$$x(t) = A_1 \cdot e^{-c_1 t} + A_2 \cdot e^{-c_2 t}$$

$$x(0) = 0 = A_1 + A_2$$

$$A_1 = -A_2$$

$$x(t) = A_1(e^{-c_1 t} - e^{-c_2 t})$$

$$\dot{x}(t) = -c_1 \cdot A_1 \cdot e^{-c_1 t} + c_2 \cdot A_1 \cdot e^{-c_2 t}$$

$$\dot{x}(0) = v_o = -c_1 A_1 + c_2 A_1$$

$$A_1 = \frac{v_o}{c_2 - c_1}$$

Which provides equation (4.6.16):

$$x(t) = \frac{v_o}{c_2 - c_1}(e^{-c_1 t} - e^{-c_2 t}) \tag{4.6.16}$$

where:

$$c_1 = \frac{r}{2m} + \sqrt{\frac{r^2}{4m^2} - \frac{k}{m}}$$

$$c_2 = \frac{r}{2m} - \sqrt{\frac{r^2}{4m^2} - \frac{k}{m}}$$

$$c_2 - c_1 = -\frac{1}{m}\sqrt{r^2 - 4mk}$$

If $r^2 = 4mk$ the general form may be shown to be:

$$x(t) = A_1 e^{-ct} + A_2 t e^{-ct}$$

where:

$$c = \frac{r}{2m}$$

$$x(0) = 0 = A_1$$

$$x(t) = A_2 t e^{-ct}$$

$$\dot{x}(t) = -cA_2 t e^{-ct} + A_2 e^{-ct}$$

$$\dot{x}(0) = v_o = A_2$$

which provides equation (4.6.17):

$$x(t) = v_o t e^{-ct} \tag{4.6.17}$$

If the landing velocity, v_o, the mass, m, the damping coefficient, r, and the spring constant, k, are known, one can predict the movement, $x(t)$, of the particle (shot) during the contact with the heel. The same result can also be obtained experimentally, which may provide an opportunity to compare theory and experimental results.

The outlined procedure allows us to vary systematically one variable, e.g., the mass, for given additional parameter settings and, consequently, to understand the influence of this variable on the deformation in a functional sense. This procedure is called *simulation*. Simulation can, for instance, be used if the maximal compression of a material is known. If the material (k and r) and the velocity of touch down are known, the mass for which the material bottoms out can be determined. If the mass and the material are known, the velocity of touch down for which the material bottoms out can be determined. For a further discussion of simulation see Chapter 4.8.

General comments on damping

Damping is a well known concept in mechanics. Expressions such as light, heavy, and critical damping are frequently used in mechanics and have their importance in biomechanical analysis. For the simplest case of a spring - damper system in parallel, critical damping is defined as:

$$r_{crit} = \sqrt{4mk} = 2m\Omega$$

where:

Ω = circular frequency

The three distinguished cases mentioned previously are defined as:

Heavy damping:	$r > r_{crit}$	Since c_1 and c_2 are negative, the position (x) approaches zero as t increases. The mass returns to its equilibrium position without any oscillations.
Critical damping:	$r = r_{crit}$	A mass in a critically damped system regains its equilibrium position in the shortest possible time without any oscillations.
Light damping:	$r < r_{crit}$	A mass oscillates around an equilibrium position with diminishing amplitude.

EXAMPLE 6(a simple muscle model)

The ideas for this section derive from some considerations by Hörler (1972a). They deal with a simple muscle model that in the simplest possible (reasonable) case, describes a muscle with the help of a contractile element, a spring, and a damper. One may wonder if such a simple mathematical model might predict the force-time characteristics of an isometric muscle contraction and, if yes, what mechanical characteristics of the system would best match the theoretical predictions of such a force-time curve.

Question

Determine the mechanical characteristics of a simplistic muscle model, including a contractile element, a spring, and a damper.

Assumptions

(1) The muscle mass can be concentrated in one particle.

(2) The mechanical properties of a muscle, which are typically distributed over the whole muscle, can be separated and modelled independently.

(3) The elasticity of the considered muscle is assumed to be constant. Consequently, a simple linear spring can be used.

(4) The damping behaviour of the considered muscle is assumed to be constant. Consequently, a simple damper can be used.

(5) The muscle can be represented with a spring on one side of the mass and a damper in parallel with a contractile element on the other side of the mass.

(6) The internal force produced by the contractile element will change from zero to its maximal value immediately at the time t = 0.

This means:

$$F(t) = 0 \quad \text{for} \quad t < 0$$
$$F(t) = F_{max} \quad \text{for} \quad t \geq 0$$

(7) The symbols used in this example are:

t	=	time
m	=	muscle mass
k	=	spring constant
r	=	coefficient of viscous damping
$x(t)$	=	position of muscle mass (particle)
$F(t)$	=	internal force produced by the contractile element
$K(t)$	=	external force at muscle insertion, P (A)

(8) The forces are drawn in the positive axis direction.

The FBD and an illustration of the situation are shown in Figure 4.6.10.

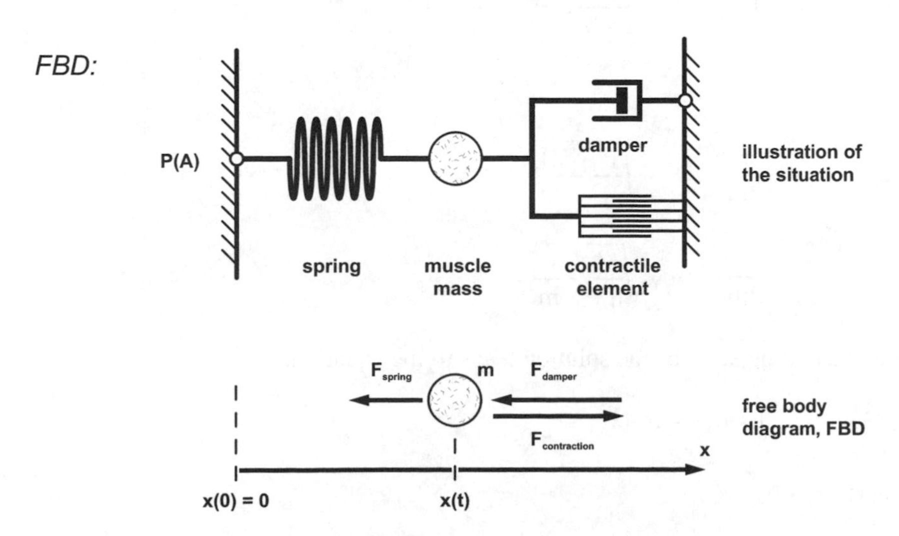

Figure 4.6.10 **Illustration of the simplistic muscle model (top) and the corresponding FBD (bottom) (from Hörler, 1972a, with permission).**

Equations of motion

The equation of motion for the particle is:

$$m\ddot{x}(t) = F(t) - kx(t) - r\dot{x}(t)$$

Initial conditions

$$x(0) = 0$$

$$\dot{x}(0) = 0$$

Final conditions

$$x(\infty) = x_{max}$$

$$\dot{x}(\infty) = 0$$

Solution

The position before the time $t = 0$ is used as the zero point of the x-axis. The displacement of the mass, $[x(t) - x(0)]$ which corresponds to $x(t)$, also indicates the elongation of the spring. For the force, $K(t)$, acting at point $P(A)$ one can write:

$$K(t) = kx(t)$$

Because the problem is similar to the problem discussed in the previous chapter, the various steps are not specifically presented. For non-critical damping, $a \neq b$, the solution leads to the equation:

$$\frac{K(t)}{K_{max}} = 1 + \frac{b}{a-b} \cdot e^{-at} - \frac{a}{a-b} \cdot e^{-bt}$$

where:

$$a = \frac{r}{2m} + \sqrt{\frac{r^2}{m^2} - \frac{4k}{m}}$$

$$b = \frac{r}{2m} - \sqrt{\frac{r^2}{m^2} - \frac{4k}{m}}$$

For critical damping, $a = b$, the solution leads to the equation:

$$\frac{K(t)}{K_{max}} = 1 - e^{-at} - te^{-at}$$

where:

$$a = b = \frac{r}{2m}$$

The result is a two parametric equation for the parameters a and b. The values of a and b can be determined with the help of an experiment. The experiment to determine a and b was performed for the biceps brachialis muscle group. The subject was strapped to a force measuring device that measured the force exerted by the lower arm in the direction of the line of action of the biceps (Figure 4.6.11). At the time $t(0) = 0$, the subject had to perform

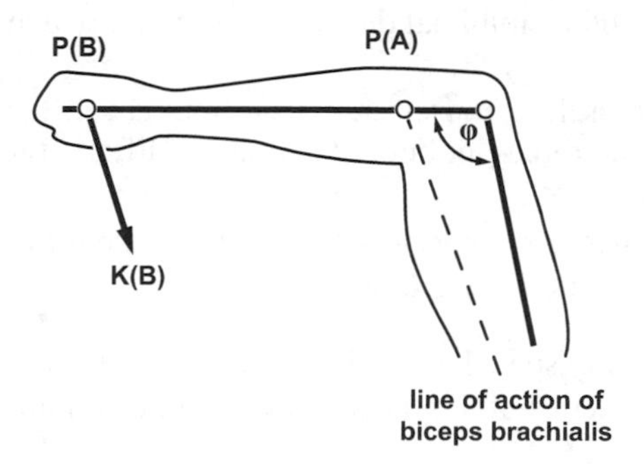

Figure 4.6.11 **Schematic illustration of the experimental setup for determining the coefficients a and b in the relative force equation.**

a maximal isometric contraction, and the corresponding force, $K(t)$, was measured. The experiment was performed for four different angles between upper and lower arm (elbow angles φ). The experimentally determined result for $K(t)$ was simulated with the simple mechanical muscle model using a least square fit method. This procedure provided the values for parameters a and b that are listed in Table 4.6.1.

Table 4.6.1 **Experimentally determined values for the parameters a and b for which the experimental and theoretical results for K(t) fit best.**

φ [°]	a [s^{-1}]	b [s^{-1}]
60	10.6	10.6
90	11.8	11.8
125	14.7	14.7
137	19.9	19.9

The results indicate that the experimental and theoretical $K(t)$ results fit best when the two parameters a and b are equal, a = b. Based on the results of the previous section, this corresponds to critical damping. The dimensions of the forces in the mechanical elements (spring and damper) are such that the displacement of the muscle mass occurs without any vibrations in the shortest possible time. The results from this experiment, therefore, suggest that (a) muscle properties are tuned for a certain outcome, and (b) the tuning of the muscle corresponds to the tuning of a critically damped movement. It can be argued that this is the case in reality since muscle vibrations cannot be seen in such isometric contractions, at least not externally.

The calculations have also been performed with the assumption that the muscle force, F(t), does not change immediately but changes as a function of time.

$$F(t) \quad = \quad F_{max}(1 - e^{-ct}) \qquad \text{for} \quad t \geq 0$$

The results for this more realistic approach support the finding that muscle movement is tuned.

The idea of *muscle tuning* may, if further supported, be important for several reasons:

- Muscle wobbling should be minimal during locomotion, which can be confirmed experimentally.
- Muscle tuning (and, therefore, muscle activity) should be influenced by the magnitude or frequency of external forces, or both. It should be different for different surface and shoe combinations.
- Muscle tuning (and, therefore, muscle activity) should be influenced by actual muscle strength and, consequently, muscle fatigue.

These, of course, are only speculations. However, they illustrate how a simple one particle model may prompt a series of thoughts and sometimes findings.

EXAMPLE 7 (a model for jumping ability)

The jumping ability of animals is rather surprising. One often tends to relate jumping ability to body size. However, this association is not without problems. First, size is not well-defined. It is not clear whether size corresponds to height, to weight, or to something else. Second, animals of similar size have completely different jumping abilities. A horse can jump rather high, while a cow can barely jump, yet they are of similar size. Additionally, some small animals, such as fleas, are able to jump fantastic heights. It has been suggested that the study of geometrically similar creatures may provide the answer to the question of how body size of body dimensions influences the jumping ability. The following model has been proposed to solve this question (Hörler, 1973).

Question

Develop a model to explain the relationship between body dimensions and the ability to jump.

Assumptions

(1) Only geometrically similar creatures will be considered.
(2) The model is restricted to the vertical jump from standing position without wind-up movement.
(3) The joint angles of the starting position are the same for all similar creatures.
(4) The muscle force responsible for the movement of the centre of mass can be described by a constant force, F.
(5) The human body considered in this context can be described by a particle with the mass, m.

The symbols used in the following are:

t	=	time
a	=	distance which the centre of mass moves upwards during contact with the ground

F	=	average muscle force responsible for movement
W	=	body weight = mg
H	=	height of flight
L	=	length (= height) of the body (body size)
λ	=	similarity quotient

(6) The comparison of a test body with a general body provides the similarity relations.

definition: $L = \lambda \cdot L_T$

length: $a = \lambda \cdot a_T$

because of the definition of the similarity quotient.

force: $F = \lambda^2 \cdot F_T$

because of the general rule that the muscular force increases proportionally with the physiological cross-sectional area of the muscle (which corresponds to the second power of the length).

weight: $W = \lambda^3 \cdot W_T$

because the weight increases proportionally with the volume (which corresponds to the third power of the length).

(7) For the numerical calculations for a test subject (measured in an experiment):

$a_T = 0.2$ m

$L_T = 1.6$ m

$H_T = 0.3$ m

(8) The problem can be solved using a one-dimensional approach.

An illustration of the situation and a free body diagram are shown in Figure 4.6.12. The top part illustrates the position of the particle representing the jumper at different times; the bottom part shows the free body diagram.

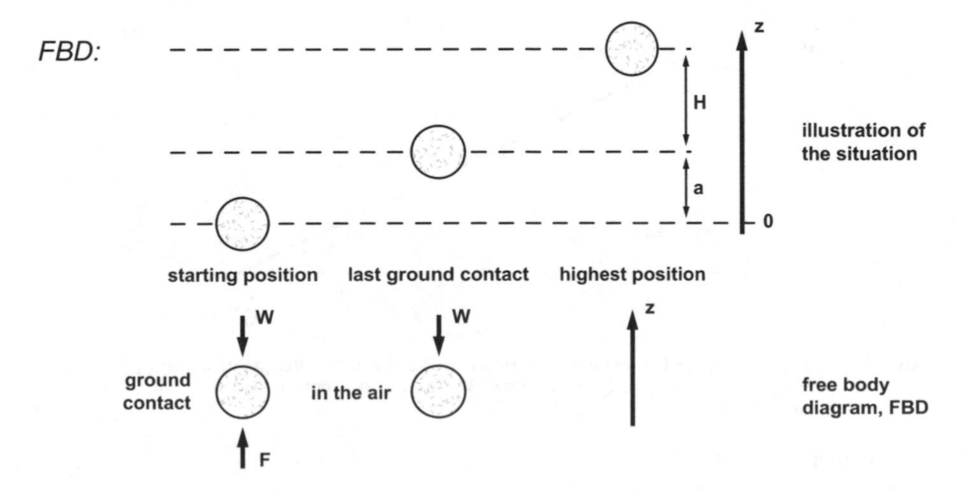

Figure 4.6.12 **Illustration of the general situation (top) and the free body diagram for the particle while in contact with the ground (bottom left) and while in the air (bottom right) (from Hörler, 1973, with permission).**

Solution

This problem uses energy conservation as one possible approach to reaching a solution. The reader may try to solve the problem in a different way.

The energy at take off corresponds to the work produced by the resultant force over the distance, a, that the particle is travelling vertically during push-off.

$$(F - W) \cdot a = m \cdot g \cdot H$$

$$H = a\left[\frac{F}{W} - 1\right]$$

Using the test body:

$$H = \lambda \cdot a_T\left[\frac{\lambda^2 \cdot F_T}{\lambda^3 \cdot W_T} - 1\right]$$

$$H = a_T\left[\frac{F_T}{W_T} - \lambda\right]$$

$$H = a_T\left[\frac{F_T}{W_T} - \frac{L}{L_T}\right] \tag{4.6.18}$$

Equation (4.6.18) describes the general relationship between the height of a jump, (H), and the body size, (L), for a test subject, T, (Figure 4.6.13).

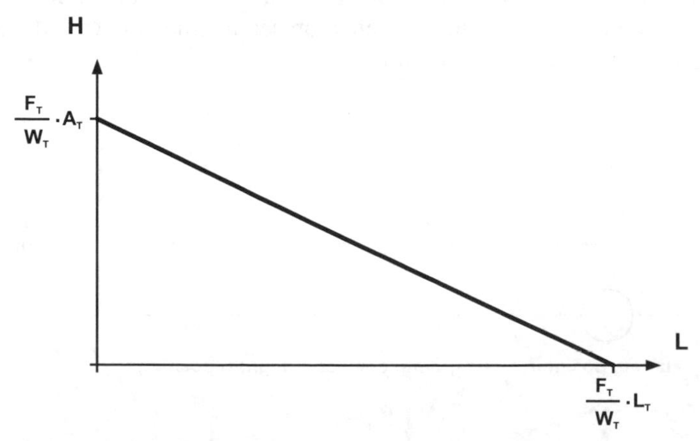

Figure 4.6.13 **Illustration of the general relationship between the height of a jump, H, and the length, L, of the body (body size) for a test subject, T (from Hörler, 1973, with permission).**

This result shows that:

- The bigger the body size, the less the height of jumping.
- There is a size limit. If the body exceeds this limit the subject cannot jump, and once fallen, would not be able to raise to its legs.

Numerical considerations when using equation (4.6.18) for a test subject are:

$$H_T = a_T\left[\frac{F_T}{W_T} - \frac{L_T}{L_T}\right]$$

$$\frac{F_T}{W_T} = \frac{H_T}{a_T} + 1 = \frac{0.3}{0.2} + 1 = 2.5$$

Generally for a subject similar to the test subject:

$$H = 0.2 \cdot \left[2.5 - \frac{L}{1.6}\right]$$

$$H = \frac{1}{2} - \frac{L}{8}$$

where L and H are measured in metres.

The relationship between body size, L, and jumping height, H, is illustrated in Figure 4.6.14. The graphical illustration for our test subject runs into problems for smaller sizes. The model, therefore, should be improved, which we will do in the next step.

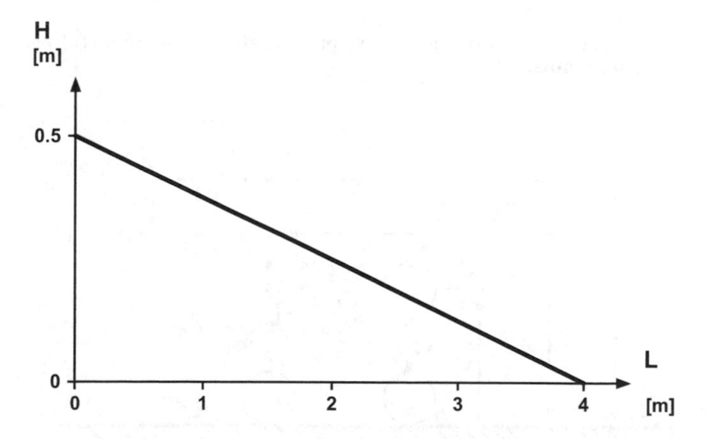

Figure 4.6.14 **Illustration of the relationship between body size and jumping height (from Hörler, 1973, with permission).**

EXAMPLE 8 (simple example for a system of particles)

This example is a continuation of the model for clumsiness (Example 7). It is an attempt to improve the initial model so that the results become more relevant for actual high jumping.

The relevance of this second step in this model is illustrated by a hypothetical discussion between athletes. The small athlete says to a tall one athlete, “It is easy for you to jump high, because your centre of mass is already high at take off". The tall athlete answers, "No, it is easy for you to jump high, because you need much less force to move your body mass, which is smaller than mine". Who is right?

Assumptions

(1) The model of the human body consists of two particles. Half of the body mass is concentrated in the upper particle and the other half in the lower one (Figure 4.6.15 and Figure 4.6.16).

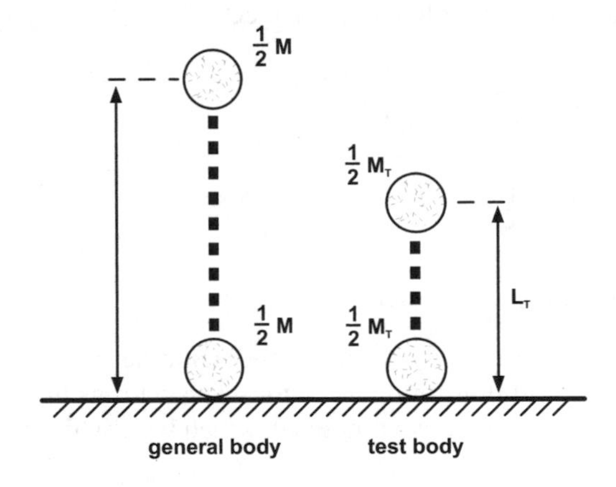

Figure 4.6.15 **Schematic illustration of the two particle simulation (model) (from Hörler, 1972b, with permission).**

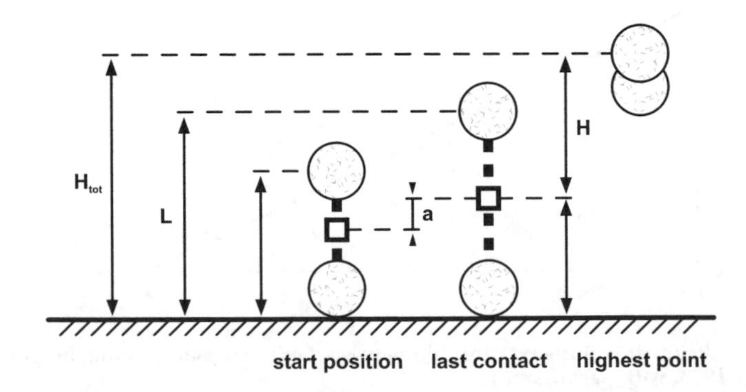

Figure 4.6.16 **Schematic illustration of the two body model at lowest position, at take off, and at the highest point. The two masses are only connected by forces acting between them (from Hörler, 1972b, with permission).**

(2) The movement analyzed is the vertical jump without a wind-up movement.

(3) The body has the ability to pull together its two masses. This assumption is subdivided into two parts:

- One where the time for contraction is not a problem.
- One where the time for contraction is a problem.

(4) The length of the body (size) in its starting position is 75% of its standing length.

(5) All the other assumptions that were made for the one particle model are used analogously in this second step.

Solution

The solution is determined in two steps: the first step where the subject has enough time to pull together, and the second step where the subject does not have enough time to pull together.

The subject has enough time to pull together

The following calculations are made under the assumption that the subject has ample time to pull the two masses together. This may, of course, not correspond to reality.

Figure 4.6.16 provides the equation:

$$H_{tot} = \frac{L}{2} + H$$

and using H from the previous section:

$$H = \frac{1}{2} - \frac{L}{8} \qquad \text{(in metres)}$$

provides for H_{tot}:

$$H_{tot} = \frac{1}{2} + \frac{3}{8}L \qquad (4.6.19)$$

This means that for the specific test-subject (L = 1.6 m) the total height of the jump would be 1.10 metres. The centre of mass would have its highest point 1.10 metres above ground (Figure 4.6.17).

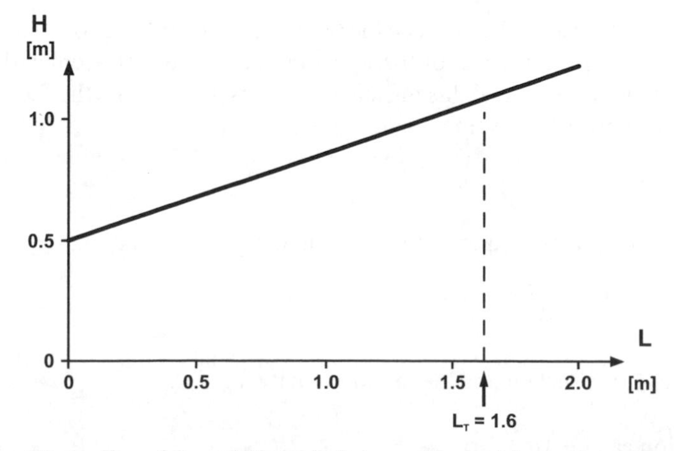

Figure 4.6.17 **Illustration of the relationship between H_{tot} and body size determined from the two mass model, if the subject has enough time to pull the two masses together (from Hörler, 1972b, with permission).**

However, the human body usually does not have enough time to pull together during a jump. Consequently, the model will be developed in a further step with the attempt to incorporate this aspect.

Calculations and considerations if there is not enough time to pull together

The assumption that the subject has enough time to pull together is not realistic. The pulling together process requires time, and it is more realistic to assume that this process depends on the acceleration of the masses and the available muscle forces. A possible model for this situation is illustrated in Figure 4.6.18.

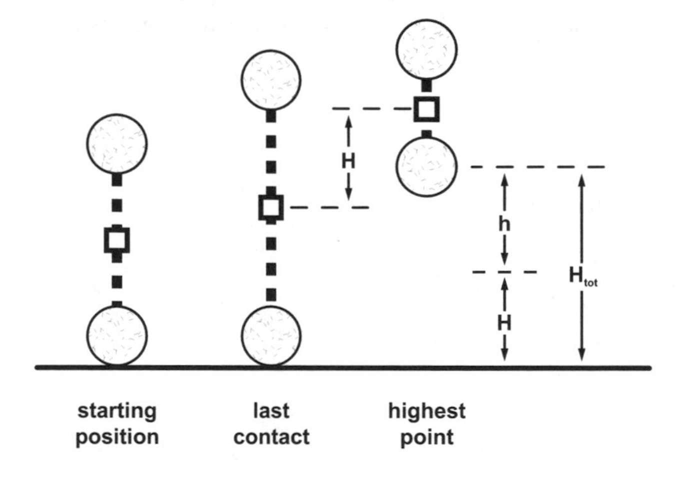

Figure 4.6.18 **Same situation as in 4.6.16, except that the body does not have enough time to contract (from Hörler, 1972b, with permission).**

The symbol h represents the distance the body is able to pull the lower mass towards the centre of mass during the time Δt of the upwards movement. This upwards movement is produced by a muscle force, and this muscle force is assumed to be the force F that was responsible for the upwards movement during ground contact.

$$H_{tot} = H + h$$

For the (vertical) displacement one can write the general relation:

$$x = \frac{1}{2}\ddot{x} \cdot \Delta t^2$$

if the force, F, is constant and the initial velocity $v_o = 0$.

This relation can be used for:

(i) The movement of the centre of mass during flight.

(ii) The movement of the lower mass during flight.

$$\text{(i)} \quad H = \frac{1}{2} g \cdot \Delta t^2 \tag{4.6.20}$$

because the only force acting during the flight is the gravitational force.

$$\text{(ii)} \quad h = \frac{1}{2} \ddot{x} \cdot \Delta t^2$$

$$\frac{1}{2} m \ddot{x}(t) = F - \frac{1}{2} W$$

$$\ddot{x}(t) = \frac{2}{m}\left[F - \frac{1}{2} W\right]$$

$$h = \frac{1}{2} \cdot \frac{2}{m}\left[F - \frac{1}{2} W\right] \Delta t^2$$

$$h = \frac{1}{m}\left[F - \frac{1}{2} W\right] \cdot \Delta t^2 \tag{4.6.21}$$

Elimination of Δt from equations (4.6.20) and (4.6.21) yields:

$$h = \frac{1}{m}\left[F - \frac{1}{2} W\right] \cdot 2\frac{H}{g} = 2H\left[\frac{F}{W} - \frac{1}{2}\right]$$

$$H_{tot} = H + 2H \cdot \frac{F}{W} - H = 2H \cdot \frac{F}{W}$$

H from equation (4.6.22) can be used:

$$\frac{F}{W} = \frac{F_T}{W_T} \cdot \frac{L_T}{L}$$

$$H_{tot} = 2 \cdot a_T \cdot \frac{F_T}{W_T}\left[\frac{F_T}{W_T} \cdot \frac{L_T}{L} - 1\right] \tag{4.6.22}$$

For the numerical example the total jumping height is determined by:

$$H_{tot} = \frac{4}{L} - 1 \tag{4.6.23}$$

The combination of equations (4.6.19) and (4.6.23) provides the relationship between the body size, L, and the height of the jump, H_{tot}, that is illustrated in Figure 4.6.19.

The maximum of the curve corresponds to the ideal length of the body for this test subject if everything is changed to be geometrically similar.

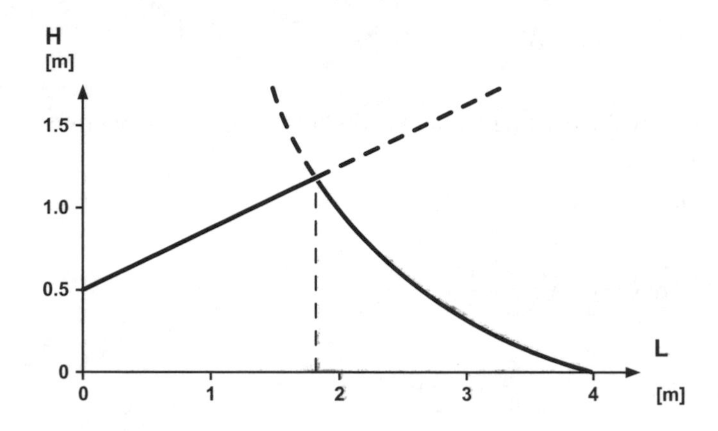

Figure 4.6.19 **Relationship between body size and jumping height if the subject lacks enough time to pull the two masses together (from Hörler, 1972b, with permission).**

Summary

The results of the two models using one and two particles for the description of the relationship between body size and jumping height can be summarized as follows:

(1) Assuming geometrically similar body constructions, there would be an optimal body length (size) for the vertical jump studied. In other words, the optical illusion in the Tarzan movies that shows Tarzan as three metres tall, but with the movement (jumping) abilities that he would have if his stature were normal, is not realistic. Tarzan should be much clumsier.

(2) This result is based on several assumptions. The most important is that the body weight increases faster than the muscle force for increasing body size (length).

(3) The numerical considerations were made with a test-subject (a semi-sportive young woman) of 1.6 m body length. They showed that optimal body length would be 1.83 m. The number, 1.83 m, is not important. This result suggests in a general sense that good high jumpers will be tall but not extremely tall.

(4) The idea that the performance of a human or an animal is limited by mechanical constraints has been further studied for birds by Pennycuick (1993). He discussed the mechanical limitations on evolution and diversity for birds, and showed that there is a mass limit above which birds would not be able to fly. He found that there is no living species weighing more than 14 kg that flies, and that the diversity of birds dwindles as mass increases.

EXAMPLE 9 (elastic and viscoelastic shoe soles and surfaces)

The following model illustrates the influence of the elastic and viscous properties of heel pads, shoe soles or playing surfaces, or all, on the energy demands during locomotion

(Anton & Nigg, 1990), and follows the publication of Nigg and Anton (1995) with the permission of the publisher.

Introductory comments

Three materials are used to cushion the landing of the heel during running: the running surface, the midsole material of the shoe, and the soft tissue of the heel. These materials have elastic and viscous components. To show the effects of these materials, a simple thought experiment is discussed (Alexander et al., 1986). It is illustrated in Figure 4.6.20.

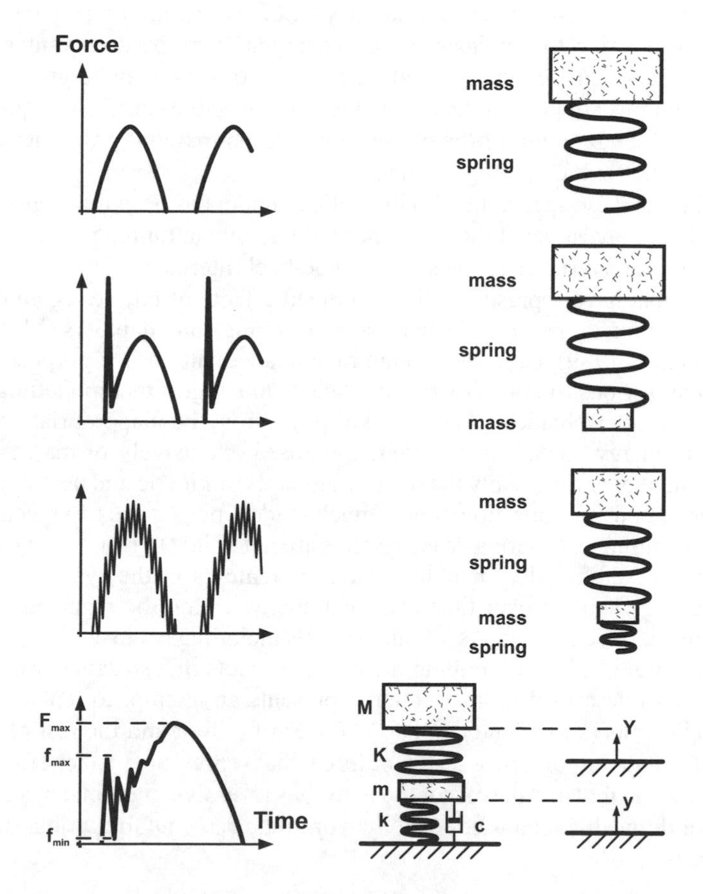

Figure 4.6.20 **Schematic illustration of the effects of elastic or viscoelastic elements, or both, on the ground reaction forces during impact and active phases (from Alexander et al., 1986, with permission).**

The first step in this thought experiment involves a mass and a spring that drops onto a rigid surface. The resultant force-time diagram (top of Figure 4.6.20) shows the corresponding ground reaction force. The second step involves a second somewhat smaller mass, m, which is added at the lower end of the dropping structure. For this drop test, the force-time diagram looks different. It has a first force peak, followed by a second force peak, with a slower loading rate than the first peak (second group from the top in Figure 4.6.20). This force-time curve resembles a running force-time curve. In the third step of the thought experiment, another spring is added to the dropping system. This spring could represent an additional elastic element, for instance, a shoe sole. For this drop test, the force-time diagram resembles the first force-time diagram with the addition of a high frequency modulation of the original signal. In the last step (bottom of Figure 4.6.20) the lower spring is replaced by a spring-dashpot combination. The force-time curve for this drop test shows an initial impact force peak with some subsequent dampened vibrations, followed by an active force peak.

The thought experiment may suggest that the combination of elastic and viscous elements in the surface-shoe-heel material may be of critical importance for economical running. Vibrations due to dominant elastic behaviour may be disadvantageous to running economy. It is speculated that muscular activity and work may be greater when high frequency vibrations are present than when there are no vibrations. Consequently, it is speculated that the combination of the elastic and viscous material properties of surface-shoe-heel may influence the running economy.

To examine these speculations, a theoretical model has been developed. The purpose of this model is to investigate how the work requirements in running are affected by different viscoelastic characteristics of the surface-shoe-heel interface.

Models that have represented the combined effects of all the leg and hip muscles involved in running by passive elements such as springs and dampers (McMahon & Green, 1979; Blickhan, 1989), have been found to be inadequate for the purpose of this research. The fact that serious running is strenuous seems to suggest that modelling muscles by energy-conserving mechanical elements like springs may be inappropriate, because the total mechanical energy content in a system composed exclusively of masses and springs remains constant over time. Only the relative amounts of kinetic and potential energy change. In the human body, the question of how much work is performed in a mechanical system is, therefore, meaningless. Spring-mass systems are not suited to respond to the purpose of this investigation. Additionally, including damper elements in the system decreases the mechanical energy content over time. The lost energy cannot be regained for lack of active components in these models. As a result, models including exclusively masses, springs, and dampers are not ideal for describing the energy aspects of a sustained running motion.

The model described in this section represents an attempt to replace the passive mechanical elements (spring and dampers) between the foot and the rest of the body, with a strategic formulation of how a resultant force, that represents the net effect of all the muscles between the foot and the rest of the body, has to evolve over time in a running situation. The model derived is then used to study work requirements for various surface-shoe-heel characteristics.

The model

Question

Determine the influence of the elastic and viscous elements that cushion the ground contact on the energy demands of a runner.

Assumptions

(1) The human body is subdivided into two masses. One mass, m_1, represents the foot of the support leg. The other mass, m, represents the rest of the body (Figure 4.6.21).

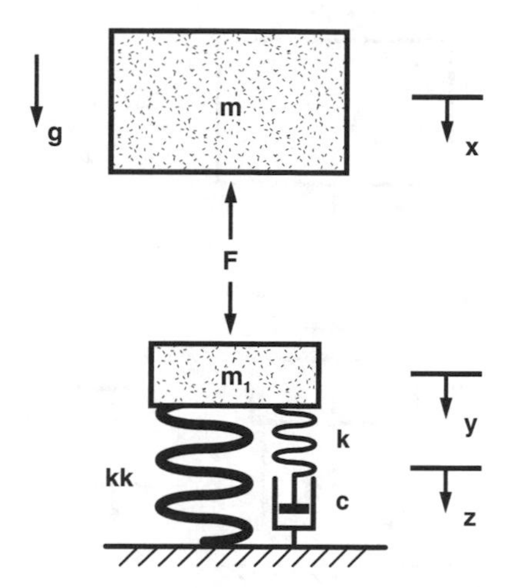

Figure 4.6.21 **Illustration of the mechanical model used in this example (from Nigg & Anton, 1995, with permission).**

This second mass, m, is called the upper body in the following.

(2) The effects of all the extremities except the support leg on the upper body are neglected.

(3) The horizontal velocity of the upper body is assumed to be constant. The model neglects its movement in the horizontal direction, considering only its vertical movement. The model is, therefore, one-dimensional.

(4) A spring-damper combination, k, kk, and c, represents the combined material properties of the surface, the shoe midsole, and the human heel.

(5) The surface is assumed to be rigid.

(6) A force, F, acts between the upper body and the foot.

(7) A force, F_G (ground reaction force), acts between the rigid ground and the lower end of the spring-damper combination.

(8) During the flight phase, the runner's body moves freely in the conservative gravitational force field. For that time interval, the total mechanical energy (sum of kinetic and potential energy) is assumed (as a first approximation) to be constant.

(9) The mathematical analysis is limited to the stance phase.

(10) During stance phase, the upper body loses height and is simultaneously slowed down. The total mechanical energy of the body decreases from an initial value when it touches down to a minimal value, and increases again towards take off.

(11) The force, F, between the foot and the upper body, develops so as to minimize the work it performs in bringing the upper body from touch down to take off. In its mathematical expression, this assumption leads to an open loop optimal control problem.

(12) The muscles that generate the force, F, are assumed to be incapable of energy storage. Therefore, the work performed is counted positive for the upper body moving up and down.

(13) The problem can be solved one dimensionally.

The free body diagram for this example is illustrated in Figure 4.6.22.

FBD:

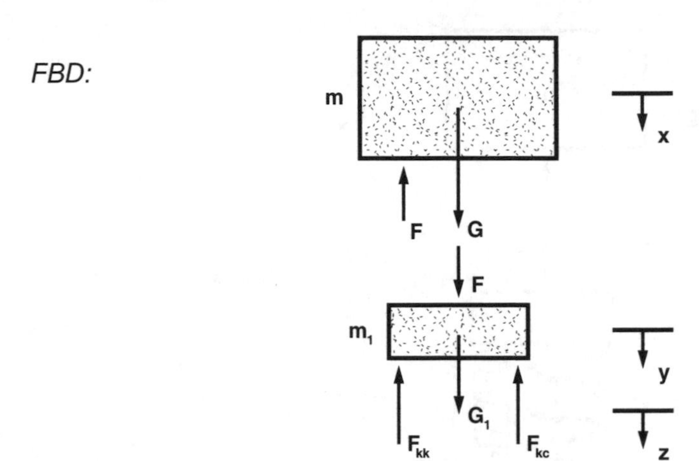

Figure 4.6.22 **FBD of the two particle model for the energy demand considerations during running.**

Equations of motion

The equations of motion for the illustrated setups are:

$$m\ddot{x} = m \cdot g - F \tag{4.6.24}$$

$$m_1\ddot{y} = m_1 \cdot g + F - (kk + k) \cdot y + k \cdot z \tag{4.6.25}$$

$$\dot{z} = \frac{k}{c}(y - z) \tag{4.6.26}$$

where:

g = earth acceleration
x, y, z = coordinates

Solution

The question will be solved as an optimization problem. In this process, the equations of motion act as constraints for the optimization problem.

$$\int_o^t [|F(\dot{x} - \dot{y})| + a \cdot F^2 + b \cdot f(\dot{F})]dt \rightarrow \min \tag{4.6.27}$$

where:

$$f(\dot{F}) = (\dot{F} - \dot{F}_{max})^2 \quad \text{if} \quad \dot{F} > \dot{F}_{max}$$

$$f(\dot{F}) = 0 \quad \text{if} \quad \dot{F}_{min} \le \dot{F} \le \dot{F}_{max}$$

$$f(\dot{F}) = (\dot{F} - \dot{F}_{min})^2 \quad \text{if} \quad \dot{F} < \dot{F}_{min}$$

The first term under the integral represents the work performed by the force, F. The second term under the integral represents the fact that the physiological cost to the system is higher at high force levels than at lower ones. The third term under the integral permits the limitation of the maximal rate of force increase and decrease, dF/dt. The second time derivative of F is the unknown function that will be determined by the optimization process. The second derivative of F was chosen as the unknown so that boundary conditions for F and dF/dt could be specified. The factors a and b permit the adjustment of the relative importance of the second and third terms under the integral with respect to the first term.

The ground reaction force, F_G, is given by the equation:

$$F_G = (kk + k) \cdot y - k \cdot z$$

Pontryagin's maximum principle is applied to solve the optimization problem. The first term under the integral supplies the work required for a one step cycle once the solution to the optimization process is substituted.

The model was initially tested using the following inputs:

$$\begin{aligned}
\Delta t_1 &= 0.1\ \text{s} \quad \text{corresponding to running} \\
\Delta t_2 &= 0.6\ \text{s} \quad \text{corresponding to walking} \\
m &= 70\ \text{kg} \\
m_1 &= 7.5\ \text{kg} \\
kk &= 2.5 \cdot 10^5\ \text{N/m} \\
k &= 2.5 \cdot 10^6\ \text{N/m} \\
c &= 8.4 \cdot 10^3\ \text{kg/s} \\
\dot{F}_{max} &= +7.5 \cdot 10^4\ \text{N/s} \\
\dot{F}_{min} &= -7.5 \cdot 10^4\ \text{N/s}
\end{aligned}$$

The stiffness and damping values chosen provide for a stiff and critically dampened foot-surface interface.

The initial and terminal boundary conditions chosen were:

$$\begin{aligned}
\text{for } t &= 0\ \text{s} \\
x &= 0\ \text{m} \\
y &= 0\ \text{m} \\
z &= 0\ \text{m} \\
\dot{x} &= +0.6\ \text{m/s} \\
\dot{y} &= 0\ \text{m/s} \\
\dot{z} &= 0\ \text{m/s} \\
F &= 0\ \text{N} \\
F_G &= 0\ \text{N}
\end{aligned}$$

$$\begin{aligned}
\text{for } t &= \Delta t_i \quad (0.1\ \text{s or } 0.6\ \text{s}) \\
\dot{x} &= -0.6\ \text{m/s} \\
x - y &= 0\ \text{m} \\
F_G &= 0\ \text{N}
\end{aligned}$$

The second terminal boundary condition, x - y = 0 m, stipulates that the length of the leg at take off is the same as at touch down. However, take off may occur with the masses m and m1 located at different heights than at touch down.

Results of the model and discussion

The output of the model consists of the ground reaction force, F_G, and the force, F, which acts between the upper body and the foot. The estimation of the ground reaction force is not needed since it is possible to measure the ground reaction force experimentally. However, the ground reaction force can be used to evaluate (validate) the model. If the ground reaction force predicted by the model and the actual ground reaction force are similar, confidence in the other results increases. Consequently, a first step compares the predicted ground reaction forces with the experimentally determined ground reaction forces. Figure 4.6.23 shows the predicted ground reaction force for running (contact time = 0.1 s).

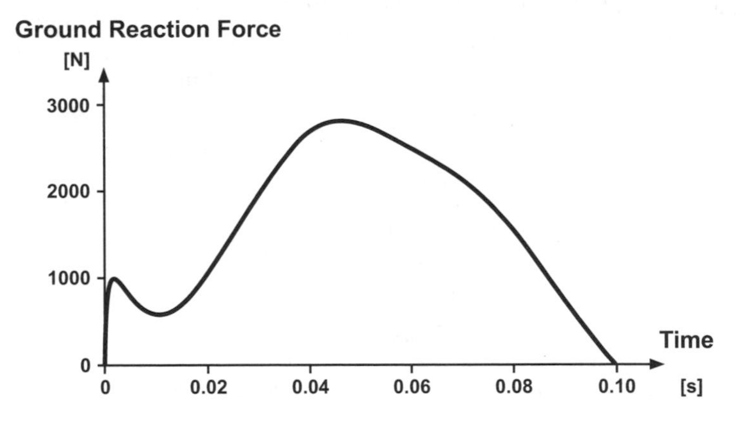

Figure 4.6.23 **Predicted ground reaction force, F_G, for a contact time of 0.1 s (running) (from Nigg & Anton, 1995, with permission).**

The estimated ground reaction force shows some similarity to experimentally determined ground reaction forces (Cavanagh & Lafortune, 1980; Nigg & Lüthi, 1980). The estimate shows the initial impact force peak and the subsequent active force peak.

Figure 4.6.24 shows the predicted ground reaction force for walking. The general characteristics for ground reaction forces for walking are present in this predicted force-time curve. The curve has a camel-like shape like typical walking curves.

The two comparisons between predicted and experimentally determined ground reaction forces show good agreement in shape as well as in magnitude. Based on this agreement, the credibility of the values of other variables determined by this model may increase. Note that only general assumptions were made for the calculation with this model. In addition to some geometrical assumptions, e.g., that landing and take off speed are the same, the main assumption was that F is selected so that the mechanical work performed by F is minimal. This simple mechanical system with basic assumptions produces ground reaction forces that correspond in magnitude and shape to the actual ground reaction forces measured experimentally. Furthermore, the shape of the ground reaction force changes from short to long contact times, the same way it changes from running to walking.

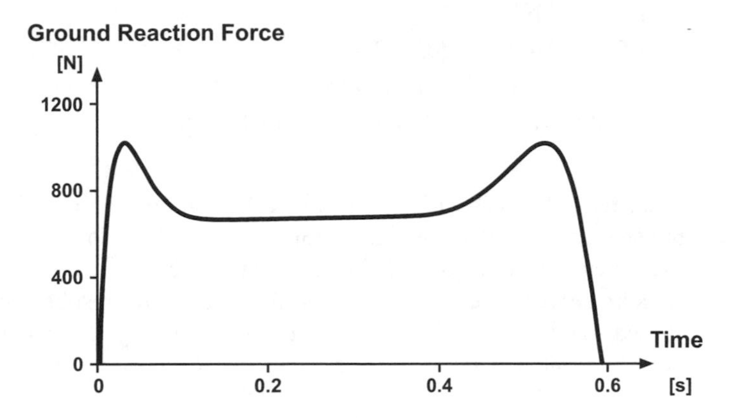

Figure 4.6.24 **Predicted ground reaction force, F_G, for a contact time of 0.6 s (walking) (from Nigg & Anton, 1995, with permission).**

The second step in the model calculations estimated the forces, F, between foot and upper body. In a first approximation, these forces can be considered as the forces in the ankle joint. Figure 4.6.25 illustrates the force-time diagram for the force F for the contact time of

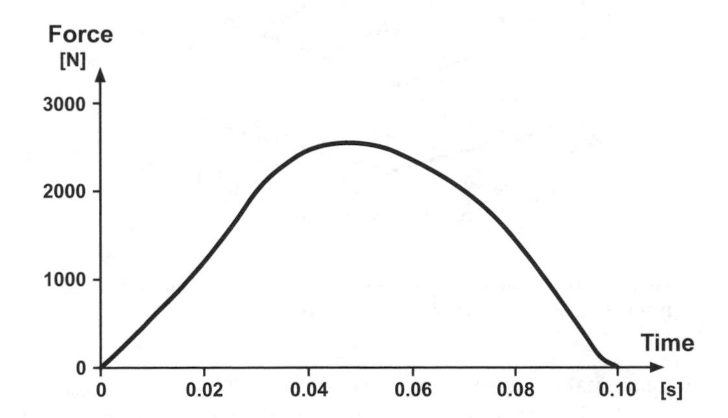

Figure 4.6.25 **Illustration of the internal force, F, as a function of time for the short contact time, 0.1 s, corresponding to the movement running (from Nigg & Anton, 1995, with permission).**

0.1 s. Notice that the peak value of F is about 10% smaller than the peak value of the ground reaction force, F_G, for the same movement (F_{max} = 2500 N and F_{Gmax} = 2800 N) due to dynamic effects at the foot level. Additionally, the impact peak in the ground reaction force, which is solely due to dynamic effects at the foot level, is absent in the force curve for F at the ankle joint level. The force-time curve, F(t), for the longer contact time of 0.6 s is nearly identical to the ground reaction force curve for the same contact time. Dynamic effects do not have a noticeable effect and no diagram for them is, therefore, shown.

In the third step, the work performed by F was calculated. This calculation has been done for the following material constants of the elements between the foot and the surface:

kk = $1.25 \cdot 10^5$ N/m (case 1)
kk = $2.5 \cdot 10^5$ N/m (case 2)
k = variable from $2.5 \cdot 10^5$ N/m to $6.25 \cdot 10^5$ N/m
c = variable from $2.5 \cdot 10^3$ kg/s to $17.5 \cdot 10^3$ kg/s

The material constant, kk, has been chosen so that the maximal static deflection of the foot mass, m_1, under a load of 2500 N, was 2 cm for case 1 and 1 cm for case 2. The values used correspond reasonably well to the actual forces and deflections in human movement.

The ranges for k and c have been chosen so that they extend from subcritical damping of a system composed solely of m_1, k, kk, and c, to critical damping of a system including m, m_1, k, kk, and c. This range was assumed to cover the actual range of possibilities for running.

The estimations of the performed work during a step cycle (Figure 4.6.26 and

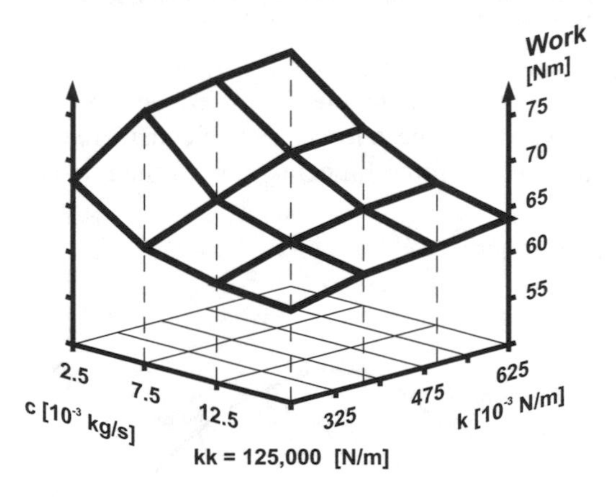

Figure 4.6.26 **Work required per step cycle for kk = $1.25 \cdot 10^5$ N/m and variable k and c (case 1) (from Nigg & Anton, 1995, with permission).**

Figure 4.6.27) indicate that the amount of work required is generally higher for case 1 (the softer spring constant for kk) than for case 2 (the harder spring constant for kk). For case 1, with the softer spring, kk, the work performed decreases steadily with increasing c and decreasing k. Higher values for c, the damper, make the foot-surface interface dynamically stiffer, which results in a lower damper deflection and, consequently, in lower damping. Increasing values of k communicate more force to the damper, which results in higher damper deflections and higher damping.

Case 2, in which the spring stiffness of kk is higher, presents a completely different relationship between the material properties of k and c, and the work performed. The influence of the damping coefficient, c, is quite small, but the influence of k becomes more interesting. There is a critical range in k values over which the work requirements change rapidly, whereas, they remain fairly constant over the remaining intervals of k and c. From a practical standpoint, this means that there is a critical combination of material properties at which relatively small changes in material properties are associated with work increases of about 10%.

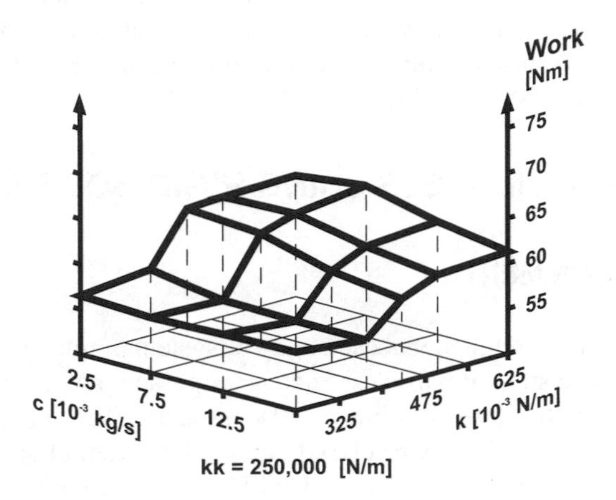

Figure 4.6.27 **Work required per step cycle for kk = 2.50 · 105 N/m and variable k and c (case 2) (from Nigg & Anton, 1995, with permission).**

This example is only a first step in attempting to determine the effect of various material properties and movement changes on the work requirements during running. The model does not allow us to provide specific details about the critical material properties for reduced or increased work requirements. As a matter of fact, the material constants k and c cannot really be separated. The required work depends on the combined effect of k and c. However, the model illustrates that specific combinations of material properties may be advantageous or disadvantageous from an energy point of view.

This simple model also shows that the work requirement is not exclusively dependent on how much energy is lost in the damper. Work is performed by the muscles in slowing down the upper body after touch down, and in accelerating it in anticipation of take off. The amount of work performed is the sum of force multiplied by displacement increments. The functional dependencies of force and displacement of time are interdependent for the problem under investigation, and they depend on the viscoelastic foot-shoe-surface interface. They must assume a form that fulfils the task of bringing the mass m from touch down to lift-off under the given boundary conditions. Note that even if c were set to zero, the resulting work requirement would still be unequal to zero and would assume different values for different spring constants kk.

There is evidence that part (but only part) of the kinetic energy at touch down is stored in elastic tissue during the stance phase (Alexander & Vernon, 1975). This is not inconsistent with the approach taken here. Even if the muscles are required to perform only part of the work given by the first term in the integral equation, it would still make sense for this portion to be minimal. However, the muscle tendon units have been assumed in this approach to be unable to store energy. The fact that the theoretically estimated ground reaction force is close to the experimentally determined ground reaction force, may suggest that the idea of storage of elastic energy in the muscle-tendon units during running should be carefully reconsidered.

The modelling approach used here determined an integral force, F, which could be considered the result of all forces produced due to muscle activity, gravity, and inertia at the ankle joint level. The initial and terminal boundary conditions chosen were realistic for running. The resultant F_G-time curves as estimated from the model, and the F_G-time curves as determined from experiments, show good agreement in magnitude and shape. This may suggest that the presented optimal control model for running and its underlying postulate of minimum performed work is acceptable.

4.6.3 MECHANICAL MODELS USING RIGID BODIES

INTRODUCTORY COMMENTS

The preceding sections assumed that each body or system of interest could be treated as a single particle or as a system of particles. Such an approach, however, is not always appropriate. In this section, the systems of interest discussed are considered to be rigid bodies.

Forces acting on a rigid body may be classified as either external or internal forces. External forces result from the action of other bodies. They include contact and remote forces and are responsible for the kinematic behaviour of the rigid body of interest.

The possible movements of a rigid body are translation and rotation.

The degree of freedom (DOF) of a rigid body is the number of independent variables necessary to describe its position in space.

The maximal number of variables necessary to describe translation is three. The maximal number of variables necessary to describe rotation is three. Consequently, in a general case, a rigid body has three translational, and three rotational, degrees of freedom.

Kinetics and kinematics of a rigid body may be described by using the two fundamental principles of linear and angular momentum, and the conservation of mechanical energy theorem:

- Conservation of linear momentum.
- Conservation of angular momentum.
- Conservation of energy.

These three sets of equations describe the movement of a rigid body. In most applications, however, not all of them are needed. Other ways to discuss movements of a rigid body have been discussed in Chapter 1. However, the three laws of conservation are used in the following examples. In a majority of such analyses in biomechanics, the laws of conservation of linear and angular momentum are used.

The conservation of linear momentum can be written in its components as:

$$m\ddot{x} = F_{x1} + F_{x2} + \ldots + F_{xn}$$

$$m\ddot{y} = F_{y1} + F_{y2} + \ldots + F_{yn}$$

$$m\ddot{z} = F_{z1} + F_{z2} + \ldots + F_{zn}$$

where:

m	=	mass of the rigid body of interest
x, y, z	=	coordinates of the centre of mass of the rigid body
F_{ji}	=	force component of force i in j direction

The conservation of the angular momentum in the most general case can be written as:

$$\{M\} = [I]\,\{\alpha\} + [\omega]\,[I]\,\{\omega\} \qquad (4.6.28)$$

where:

{}	=	vector symbol
[]	=	tensor symbol
{M}	=	moment vector
[I]	=	inertia tensor
{α}	=	angular acceleration vector
[ω]	=	angular velocity vector expressed as a skew symmetric second order tensor
{ω}	=	angular velocity vector

or in the form of equation (4.6.29):

$$\begin{bmatrix} M_x \\ M_y \\ M_z \end{bmatrix} = \begin{bmatrix} I_{xx} & -I_{xy} & -I_{xz} \\ -I_{yx} & I_{yy} & -I_{yz} \\ -I_{zx} & -I_{zy} & I_{zz} \end{bmatrix} \begin{bmatrix} \alpha_x \\ \alpha_y \\ \alpha_z \end{bmatrix} + \begin{bmatrix} 0 & -\omega_z & \omega_y \\ \omega_z & 0 & -\omega_x \\ -\omega_y & \omega_x & 0 \end{bmatrix} \begin{bmatrix} I_{xx} & -I_{xy} & -I_{yz} \\ -I_{yx} & I_{yy} & -I_{yz} \\ -I_{zx} & -I_{zy} & I_{zz} \end{bmatrix} \begin{bmatrix} \omega_x \\ \omega_y \\ \omega_z \end{bmatrix}$$

I_{xx}, I_{yy}, I_{zz}	=	moments of inertia with respect to, CM
I_{ij}	=	product of inertia with respect to the axes i and j
α_x, α_y, α_z	=	angular acceleration components
M_x, M_y, M_z	=	moment components of moment, **M**

Consequently, a maximum of six equations are available (three for translation and three for rotation) for the mathematical description of the kinematics or kinetics, or both, of one rigid body. A two-dimensional example for one rigid body with four external forces acting on it is discussed below. Also given are the free body diagram (Figure 4.6.28) and the corresponding equations of motion.

$$m\ddot{x} = -F_{1x} + F_{2x} + F_{3x} - F_{4x}$$

$$m\ddot{y} = -F_{1y} - F_{2y} + F_{3y} + F_{4y} - mg$$

$$I_{zz}\ddot{\phi}_z = [(\mathbf{r}_1 \times \mathbf{F}_1) + (\mathbf{r}_2 \times \mathbf{F}_2) + (\mathbf{r}_3 \times \mathbf{F}_3) + (\mathbf{r}_4 \times \mathbf{F}_4)]_z$$

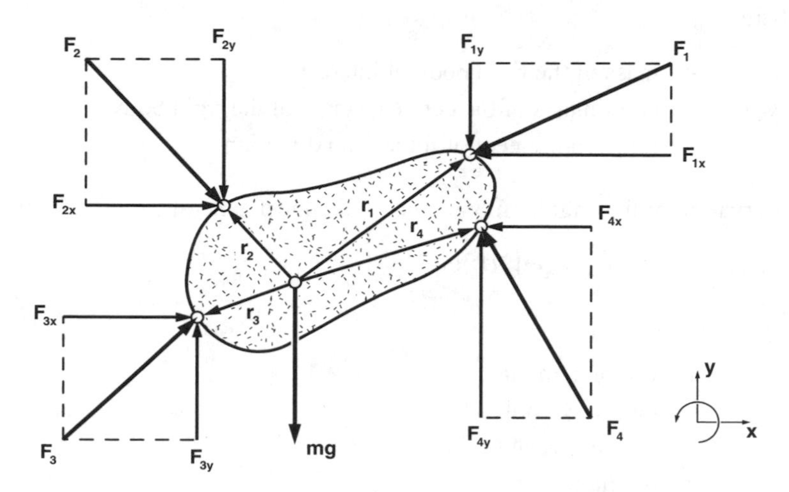

Figure 4.6.28 **Rigid body with four forces acting externally.**

where:

$\times$ = sign for vector multiplication

$\mathbf{r}_i$ = vector from the origin to the point of application of the force $\mathbf{F}_i$

Note that the right hand side of the moment equation is the component of the moment in z-axis direction, which is, of course, a scalar quantity. The subscript z beside the bracket indicates this fact.

Section 4.6.3 is again a section with many examples. The examples carry two messages. First, they attempt to illustrate different possibilities to model determinate systems of rigid bodies with simple mathematical tools. Second, they attempt to discuss the physical and biological interpretation which simple models of rigid bodies may include.

EXAMPLE 10(stability in somersaulting)

Somersaults are performed in many sport activities, including gymnastics, diving, and ski acrobatics. Coaches and athletes know, based on practical experience, that some rotations are stable and others unstable. In other words, in some types of somersaults it is difficult to maintain the initial rotation around the same axis, while in others it is easy. Lay-out somersaults, for instance, are difficult, but tucked somersaults are easy to balance. Furthermore, the magnitude of the principal moments of inertia play an important role in the stability of a specific rotation. This section provides two model considerations to shed light on somersaults.

Plausibility considerations

Consider a rectangular block of a homogeneous material with sides a, b, and c (Figure 4.6.29), and principal axes 1, 2, and 3. This block is thrown into the air, rotating

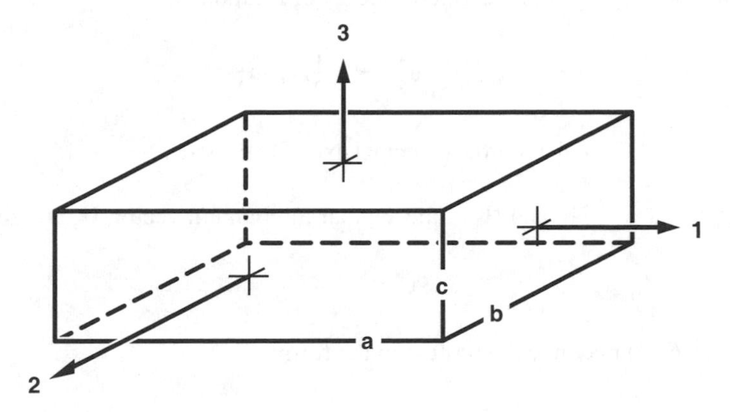

Figure 4.6.29 **Rigid body, with three principal axes. Axis 3 having the maximal and axis 1 having the minimal moment of inertia.**

around one of the three principle axes. During the airborne phase only gravity acts on the block, so the angular momentum, H, remains constant. One would, therefore, expect that the rotation, which at the beginning is around one particular principal axis of the rectangular block, should continue around the same principal axis during the whole airborne phase. These expectations can be experimentally verified for rotations around axes 1 and 3, the axes with the maximal (3) and minimal (1) values for the principal moments of inertia. However, rotations that start around axis 2, the axis with the median value for the principal moment of inertia, are unstable, and the block starts to twist and tilt (Nigg, 1974).

Question

Illustrate with the conservation of energy the fact that some rotations are stable and others unstable.

Assumptions

(1) Homogeneous rigid rectangular block.
(2) No air resistance.
(3) $a > b > c$.
(4) $I_{11} < I_{22} < I_{33}$.

Solution

In this example, the conservation of energy is used.

Symbols:

I_{ii} = moment of inertia for a principal axis (central principal moments of inertia where central is referring to the centre of mass)

I_a = moment of inertia of the rectangular block with respect to a momentary axis of rotation, a

E_{kr} = kinetic energy of rotation

$\mathbf{H}$ = angular momentum vector with the components (H_1, H_2, H_3)

ω = angular velocity about the principle axis i

For a principal-axis-system the kinetic energy of rotation is:

$$E_{kr} = \frac{1}{2} I_{11} \cdot \omega_1^2 + \frac{1}{2} I_{22} \cdot \omega_2^2 + \frac{1}{2} I_{33} \cdot \omega_3^2 \qquad (4.6.29)$$

where the axes 1, 2, and 3 are the principal axes of the body.

The components H_1, H_2, and H_3 of the angular momentum vector, **H**, are given by:

$$H_i = I_{ii} \cdot \omega_i \quad (i = 1, 2, 3) \qquad (4.6.30)$$

Equation (4.6.29) may now be written in the form:

$$1 = \frac{H_1^2}{2E_{kr} \cdot I_{11}} + \frac{H_2^2}{2E_{kr} \cdot I_{22}} + \frac{H_3^2}{2E_{kr} \cdot I_{33}} \qquad (4.6.31)$$

Equation (4.6.31) represents the surface of an ellipsoid, aligned with the body axes in the angular momentum space on which the tip of the angular momentum vector must lie.

The magnitude of the angular momentum vector is given by:

$$H^2 = H_1^2 + H_2^2 + H_3^2 \qquad (4.6.32)$$

This may be written in the form:

$$1 = \frac{H_1^2}{H^2} + \frac{H_2^2}{H^2} + \frac{H_3^2}{H^2} \qquad (4.6.33)$$

Equation (4.6.33) represents the surface of a sphere in the angular momentum space on which the tip of the angular momentum vector must lie.

Both equations (4.6.31) and (4.6.33) must be satisfied throughout a given motion. This means that the angular momentum vector will travel along the intersection of the surface of the ellipsoid, and the surface of the sphere. The results will be discussed for special cases.

Special case 1

The momentary axis of rotation coincides with principal axis 1.

$$I_{11} = I_a$$

Since I_{11} is the smallest moment of inertia, the length of the radius of the sphere is the same as the length of the smallest semi-axis (in direction 1) of the ellipsoid. Therefore, the ellipsoid and the sphere have two common points, or the sphere contacts the surface of the ellipsoid at the two points where they intersect with axis 1 (Figure 4.6.30 left). A small variation of the momentary angular velocity direction corresponds to a small variation of a stable state of equilibrium. No neighbouring point is a possible solution. Consequently, the rotation around the principal axis with the minimal value of moment of inertia is stable.

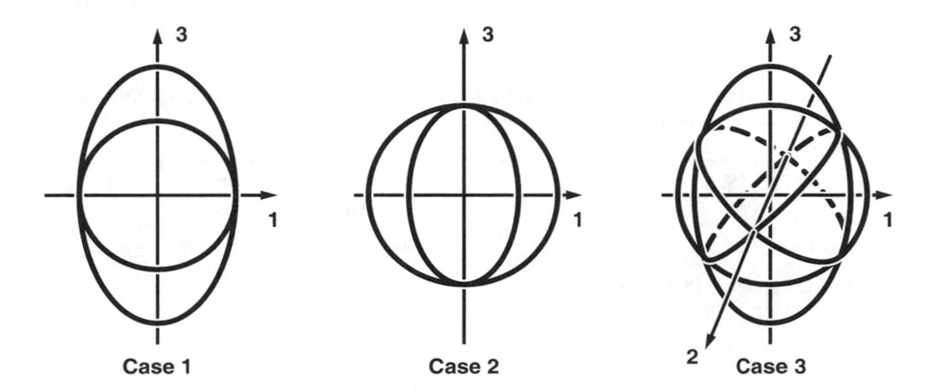

Figure 4.6.30 **Illustration of stable and unstable mathematical solutions in the angular momentum space for the selected special cases 1, 2, and 3, which correspond to initial rotations around each of the principal axes of rotation.**

Special case 2

The momentary axis of rotation coincides with principal axis 3.

$$I_{33} = I_a$$

Since $I_{33} > I_{22} > I_{11}$, the ellipsoid in this special case has two common points of contact with the sphere (Figure 4.6.30 centre). All other points of the sphere surface fall outside the ellipsoid surface. A small variation of the momentary angular velocity direction signifies a small variation of a stable state of equilibrium. However, all neighbouring points of the solution point are not a possible solution. Consequently, the rotation around the principal axis with the maximal value of moment of inertia is a stable rotation.

Special case 3

The momentary axis of rotation coincides with principal axis 2.

$$I_{22} = I_a$$

Since $I_{11} < I_{22} < I_{33}$, the sphere surface and the ellipsoid surface have two intersecting curves (Figure 4.6.30 right side). All points of these two curves are possible solutions for the angular momentum vector $\mathbf{A}_a$. The neighbouring points to the initial solution (axis 2 intersection) are possible solutions. A small variation of the momentary angular velocity direction signifies, therefore, a small variation in an unstable state of equilibrium, since all the neighbouring points of the initial solution are also possible solutions. Axis 2, therefore, may move away from the initial direction of the angular momentum. Consequently, rotation around the principal axis with the median value of moment of inertia is unstable.

Application to human movement

In the standing position, the axis with the median value of moment of inertia is the transverse axis. The axis with the smallest moment of inertia is the longitudinal axis. The axis with the largest moment of inertia is the antero-posterior or sagittal axis. Estimated moments of inertia for the three principal axes of the human body in the standing position are summarized in Table 4.6.2.

Table 4.6.2 Estimated moments of inertia for the three principal axes: longitudinal, transverse, and antero-posterior.

AXIS	ESTIMATED MOMENT OF INERTIA [kgm^2]
LONGITUDINAL	$\approx$ 1
TRANSVERSE	$\approx$ 12
ANTERO-POSTERIOR	$\approx$ 13

Rotation around the transverse axis occurs in the layout somersault. From the considerations above, we know that this rotation is unstable. The layout somersault is, therefore, not as simple as commonly assumed. On the other hand, initiating twisting during a layout somersault would seem to be much less difficult than generally assumed. To initiate a twist during the airborne somersault phase:

- There has to be an initial rotation around the axis with the median moment of inertia (for the layout somersault, the transverse axis).
- The body has to be in the layout position (note in the tucked position the moment of inertia around the transverse axis is the greatest).

The initially symmetrical body must be made asymmetrical.

For the somersault in the tucked position, the moments of inertia may be estimated using an ellipsoid approach with axis a as the transverse axis, axis b as the sagittal axis, and axis c as the longitudinal axis:

$$I_a = \frac{m}{5}(b^2 + c^2)$$

$$I_b = \frac{m}{5}(a^2 + c^2)$$

$$I_c = \frac{m}{5}(a^2 + b^2)$$

With the assumptions:

$$m = 60 \text{ kg}$$
$$a = 0.2 \text{ m}$$
$$b = 0.3 \text{ m}$$
$$c = 0.5 \text{ m}$$

the moments of inertia in the tucked position are estimated to be:

transverse axis	I_a	= 4.1 kgm^2
sagittal axis	I_b	= 3.5 kgm^2
longitudinal axis	I_c	= 1.6 kgm^2

Consequently, a rotation around axis b, the sagittal axis, is unstable, while rotations around axis a, the transverse axis, and axis c, the longitudinal axis, are stable for somersaults in the tucked position. It is, therefore, more difficult to initiate a twisting movement from the tucked position.

The moments of inertia in the layout position for the transverse and sagittal axes are relatively close. A change in body configuration, e.g., backwards arching, may change which axis has the median moment of inertia, stablilizing a previously unstable rotation.

The mechanics of twisting somersaults

The rotations of a rigid body during airborne movements in general, and the mechanics of twisting somersaults in particular, have been discussed extensively by Yeadon (1993a, b, c, d, and 1984). Unstable rotations during somersaults have been discussed extensively in these publications, and the following paragraphs summarize their findings.

Rotations of a rigid body are typically described using three rotational angles, as illustrated in Figure 4.6.31:

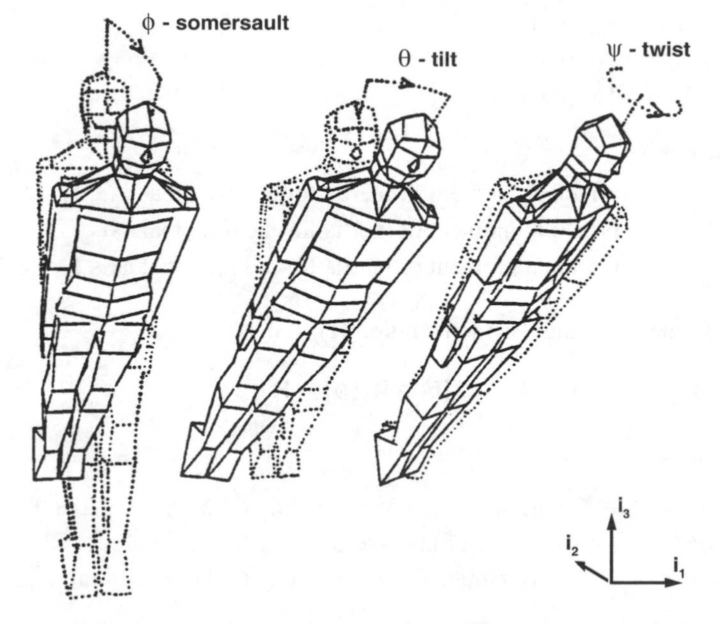

Figure 4.6.31 **Illustration of the three rotational angles used to describe rotations of a rigid body during airborne motions (from Yeadon, 1984, with permission of Pergamon Pres Ltd., Headington Hill Hall, Oxford 0X3 0BW, UK).**

Φ	SOMERSAULT	Rotation about an axis fixed in space.
Ψ	TWIST	Rotation about a longitudinal body axis.
Θ	TILT	Rotation away from the somersault plane.

Airborne motions of a rigid body fall into the twisting and wobbling modes and the singular solution separating them. In the singular mode, the rotation occurs about the axis with the medial moment of inertia. The rotation is unstable. In the twisting mode, the twist increases monotonically. In the wobbling mode, the twist oscillates about a mean value. The objective of this section is to derive the two modes and the singular solution, and to discuss their implications for the airborne movement of humans.

Derivation of wobbling and twisting modes

A rigid body with zero net torque conserves angular momentum, H:

$$\mathbf{H} = I_{ff} \cdot \omega_{fi} \tag{4.6.34}$$

where:

I_{ff} = whole body inertia tensor
ω_{fi} = angular velocity of the body relative to the inertial frame i

The conservation of angular momentum can be expressed in terms of a local frame, f. For this purpose, I_{ff}, may be oriented with the principal axes of the body so that $(I_{ff})_f$, is a diagonal matrix:

$$(I_{ff})_f = \begin{bmatrix} A & 0 & 0 \\ 0 & B & 0 \\ 0 & 0 & C \end{bmatrix}$$

where:

A = principal moment of inertia for the principal axis $\mathbf{f}_1$
B = principal moment of inertia for the principal axis $\mathbf{f}_2$
C = principal moment of inertia for the principal axis $\mathbf{f}_3$

In the frame f, the angular momentum, $(\mathbf{H})_f$, is:

$$(\mathbf{H})_f = R_3(\Psi) \cdot R_2(\Theta) \cdot R_1(\phi) \cdot (\mathbf{H})_i \tag{4.6.35}$$

where:

$R(\Psi)$ = rotation matrix corresponding to the twist angle, Ψ
$R(\Theta)$ = rotation matrix corresponding to the tilt angle, Θ
$R(\Phi)$ = rotation matrix corresponding to the somersault angle, Φ

The angular momentum, $(\mathbf{H})_i$, is set such that the tumbling motion takes place in a specific plane:

$$(\mathbf{H})_i = \begin{bmatrix} H \\ 0 \\ 0 \end{bmatrix}$$

which leads to:

$$(\mathbf{H})_f = \begin{bmatrix} H\cos\Theta\cos\phi \\ -H\cos\Theta\sin\phi \\ H\sin\Theta \end{bmatrix} \quad (4.6.36)$$

In the frame f, the angular velocity, ω_{fi}, is:

$$(\omega_{fi})_f = \begin{bmatrix} \cos\Theta\cos\Psi & \sin\Psi & 0 \\ -\cos\Theta\sin\Psi & \cos\Psi & 0 \\ \sin\Theta & 0 & 1 \end{bmatrix} \cdot \begin{bmatrix} \dot{\phi} \\ \dot{\Theta} \\ \dot{\Psi} \end{bmatrix} \quad (4.6.37)$$

The initial equation then takes the form:

$$\begin{bmatrix} H\cos\Theta\cos\Psi \\ -H\cos\Theta\sin\Psi \\ H\sin\Theta \end{bmatrix} = \begin{bmatrix} A & 0 & 0 \\ 0 & B & 0 \\ 0 & 0 & C \end{bmatrix} \cdot \begin{bmatrix} \dot{\phi}\cos\Theta\cos\Psi + \dot{\Theta}\sin\Psi \\ -\dot{\phi}\cos\Theta\sin\Psi + \dot{\Theta}\cos\Psi \\ \dot{\phi}\sin\Theta + \dot{\Psi} \end{bmatrix} \quad (4.6.38)$$

In the absence of external torque, the conservation of rotational energy yields:

$$E_{rot} = \frac{1}{2} \cdot (A\omega_1^2 + B\omega_2^2 + C\omega_3^2) \quad (4.6.39)$$

where:

E_{rot} = (constant) rotational energy
$\omega_1, \omega_2, \omega_3$ = components of the angular velocity, $(\omega_{fi})_f$

or in terms of Θ and Ψ:

$$E_{rot} = \frac{1}{2} \cdot \left(\frac{H^2\cos^2\Theta\cos^2\Psi}{A} + \frac{H^2\cos^2\Theta\sin^2\Psi}{B} + \frac{H^2\sin^2\Theta}{C} \right) \quad (4.6.40)$$

This may alternately be written in either of the following two forms:

$$\sin^2\Psi = (1 - c_\alpha^2 \sec^2\Theta) \cdot \frac{B(A-C)}{C(A-B)} \quad (4.6.41)$$

or:

$$\cos^2\Psi = (c_\alpha^2 \sec^2\Theta - 1) \cdot \frac{A(B-C)}{C(A-B)} \quad (4.6.42)$$

where the coefficients c_α and c_β are defined as:

$$c_\alpha^2 = \frac{\frac{H^2}{C} - 2E_{rot}}{\frac{H^2}{C} - \frac{H^2}{A}} \qquad c_\beta^2 = \frac{\frac{H^2}{C} - 2E_{rot}}{\frac{H^2}{C} - \frac{H^2}{B}}$$

The *somersault rate* (angular velocity for somersault) may be expressed using the above equations as:

$$\dot{\phi} = H\left[\frac{1}{C} - \left(\frac{1}{C} - \frac{1}{A}\right)c_\alpha^2 \sec^2\Theta\right] \qquad (4.6.43)$$

The *twist rate* (angular velocity for twist) may be expressed using the above equations as:

$$\Psi = H\left[\frac{1}{C} - \frac{1}{A}\right]c_\alpha^2 \sec^2\Theta \sin\Theta \qquad (4.6.44)$$

The constants c_α^2 and c_β^2 are positive if C is the minimal principal moment of inertia. For the following considerations we assume that A>B>C (note that the previous discussions are correct for A>B>C and for B>A>C).

The constant c_α^2 attains its minimal value of zero if:

$$E_{rot1} = \frac{H^2}{2C}$$

For a given angular momentum, the energy, E_{rot1}, corresponds to the largest possible energy of rotation, since $A > B > C$. In other words, if $c_\alpha^2 = 0$, the rotational energy of the rigid body is maximal.

The constant c_α^2 attains its maximal value of 1 if:

$$E_{rot2} = \frac{H^2}{2A}$$

For a given angular momentum, the energy, E_{rot2}, corresponds to the smallest possible energy of rotation, since A>B>C. In other words, if $c_\alpha^2 = 1$, the rotational energy of the rigid body is minimal.

Consequently, the coefficient c_α^2 is positive and between 0 and 1:

$$0 \le c_\alpha^2 \le 1$$

One may define a real value, α, by the relationship:

$$\cos^2\alpha = c_\alpha^2$$

The previous equations can be used to show that, for a twist angle $\Psi = 0$, the tilt angle, Θ, is equal to α:

$$\sin^2\Psi = (1 - c_\alpha^2 \sec^2\Theta) \cdot \frac{B(A - C)}{C(A - B)}$$

For $\Psi = 0$:

$$0 = (1 - c_\alpha^2 \sec^2\Theta)$$

$$c_\alpha^2 = \cos^2\alpha = \cos^2\Theta$$

Therefore:

$$\Theta = \alpha \quad \text{for} \quad \Psi = 0 \qquad \text{(no twist)}$$

The same procedure will now be applied with c_α. The complication arises when $(c_\beta)^2$ is not restricted to be less than or equal to one. This case yields then, to the twist and wobble modes.

Using the expressions for c_α and c_β, we may write:

$$\cos^2\alpha = \cos^2\alpha_o \cdot c_\beta^2$$

where:

$$\cos^2\alpha_o = \frac{A(B - C)}{B(A - C)}$$

This leads to:

$$\cos^2\alpha = \cos^2\Theta(1 - \sin^2\Psi \sin^2\alpha_o)$$

The case in which $\Psi = 90°$ corresponds to the case in which:

$$\cos^2\Theta = c_\beta^2$$

Now, there are three cases to consider: the case when $\alpha > \alpha_o$, the case when $\alpha < \alpha_o$, and the case when $\alpha = \alpha_o$.

Case 1 (the twisting mode)

This mode is also called high energy mode. Let us consider a rod with A, B, C such that $A = B > C$. For this case, $\alpha_o = 0$ and, consequently:

$$a \geq \alpha_o$$

In general, a rotating rigid body will be said to be in the rod mode for $\alpha > \alpha_o$.

For the rod mode:

$$0 \quad < \quad c_\beta^2 \quad < \quad 1$$

By analogy to the previous considerations, there exists an angle β for which:

$$\cos^2\beta \quad = \quad c_\beta^2$$

α and β are related by:

$$\left[\frac{1}{C} - \frac{1}{A}\right]\cos^2\alpha \quad = \quad \left[\frac{1}{C} - \frac{1}{A}\right]\cos^2\beta$$

which relates the values α and β, of the tilt angle Θ, to the values 0 and 90°, for the twist angle Ψ.

The procedures of Whittaker (1937) and Whittaker and Watson (1962) may be used to obtain a general solution of the form:

$$y \quad = \quad \mathrm{dn}(pt)$$

where:

$$y \quad = \quad \frac{\sin\Theta}{\sin\alpha}$$

dn = a Jacobian elliptic function with modulus k

$$p^2 \quad = \quad \frac{H^2\sin^2\alpha(A - C)(B - C)}{ABC^2}$$

t = time

The function $\mathrm{dn}(pt)$, and the tilt angle Θ, are found to be periodic with the time period $2K/p$, where K is the complete elliptic integral of the first kind (Bowman, 1953). This oscillation of the tilt angle is called nutation.

The equation for the twist rate shows that the twist rate has the sign of $H \cdot \sin\Theta$. The twist rate is monotonic, since $\sin\Theta$ is bounded between $\sin\alpha$ and $\sin\beta$.

$$\mathrm{sign}(\Psi) \quad = \quad \mathrm{sign}(H\sin\Theta)$$

The equations describing the connection between the tilt angle and the twist angle show that the situation $\Theta = \alpha$ corresponds to the zero and half twist, and, therefore, the time taken for a full twist is twice the period of the tilt angle, Θ, and the average twist rate is:

$$\Psi_{av} \quad = \quad \frac{\pi p}{2K}$$

The somersault rate varies between H/A and H/B, and the time period of the oscillations is the same as the time period of the tilt angle, Θ, namely $2K/p$. The average somersault rate is:

$$\dot{\phi}_{av} = H\left[\frac{1}{C} - \left(\frac{1}{C} - \frac{1}{A}\right)V(\alpha)\right]$$

where:

$$V(\alpha) = \left[\frac{\cos^2\alpha}{\cos^2\Theta}\right]_{av}$$

In summary, both the somersault angle, Φ, and the twist angle, Ψ, steadily increase, while the tilt angle, Θ, oscillates between α and β (assuming that H, α, β > 0). The results for the twisting mode are summarized in Table 4.6.3 (Yeadon, 1984).

The rod movement may be described as a twisting somersault. This mode of motion will, henceforth, be referred to as the *twisting mode.*

Table 4.6.3 Summary of selected values for case 1, the twisting model.

TIME [TWIST PERIODS]	TWIST ANGLE [RAD]	TILT ANGLE	TILT RATE [RAD/S]	SOMERSAULT RATE	TWIST RATE
QUARTER	$\frac{\pi}{2}$	b (min)	0	$\frac{H}{B}$ (max)	$\left[\frac{H}{C} - \frac{H}{B}\right]\sin\beta$
HALF	p	a (max)	0	$\frac{H}{A}$	$\left[\frac{H}{C} - \frac{H}{A}\right]\sin\alpha$ (max)
THREE QUARTER	$\frac{3\pi}{2}$	b (min)	0	$\frac{H}{B}$ (max)	$\left[\frac{H}{C} - \frac{H}{B}\right]\sin\beta$
FULL	2p	a (max)	0	$\frac{H}{A}$	$\left[\frac{H}{C} - \frac{H}{A}\right]\sin\alpha$ (max)

Case 2 (the wobbling mode)

This case is also called low energy mode, and is analogous to the movement mode of a disc with the principal moments of inertia A, B, and C, where A>B=C. The derivations are not discussed in this section. They have been published by Yeadon (1993a, b, c, d).

The results can be summarized as:

- The somersault angle Φ steadily increases while,
- The tilt angle Θ oscillates between α and -α, and
- The twist angle Ψ oscillates between Ψ_o and $-\Psi_o$.

These results are summarized in Table 4.6.4 (Yeadon, 1984).

Table 4.6.4 Summary of the values for the twist and tilt angles as functions of the period and of the tilt angle rate.

TIME	TILT ANGLE	TWIST ANGLE	SOMERSAULT RATE
0	α	0	minimum
K/q	0	Ψ_o	maximum
2K/q	$-\alpha$	0	minimum
3K/q	0	$-\Psi_o$	maximum

The discussed movement of a disc may be described as a wobbling somersault. This mode will, henceforth, be referred to as the *wobbling mode.*

Case 3 (the unstable singular solution)

The section on case 1 explored the motions associated with $\alpha > \alpha_o$, and the section on case 2 explored the motions associated with $\alpha < \alpha_o$. This section discusses the singular solution that corresponds to $\alpha = \alpha_o$. For a rigid body with the principal moments of inertia A > B > C the condition:

$$E_{rot} = \frac{1}{2} \cdot \frac{H^2}{B}$$

is satisfied by steady rotations about the axis corresponding to the intermediate moment of inertia, B. This solution is obvious, however, the condition is also satisfied by a motion in which the rotation about the intermediate axis in one direction, leads to a half twist followed by a motion in the opposite direction. This latter set of solutions may be derived using the procedure of Whittaker (1937), and the fact that the condition:

$$E_{rot} = \frac{1}{2} \cdot \frac{H^2}{B}$$

implies both:

$$\sin^2 \alpha_o = \frac{C(A-B)}{B(A-C)}$$

$$\beta_o = 0$$

where:

α_o is the angle of tilt when the twist angle Ψ is zero, and
β_o is the angle of tilt at the quarter twist position.

The motion may now be quantitatively described (Table 4.6.5) (Yeadon, 1984). For ease of presentation, the time is defined as zero when the twist angle $\Psi = 0$.

Table 4.6.5 Description of the motion for the unstable rotation case.

TIME	TWIST ANGLE [DEG]	SOMERSAULT RATE	TWIST RATE
$-\infty$	−90	$\frac{H}{B}$	0
0	0	$\frac{H}{A}$	$H\left[\frac{1}{C}-\frac{1}{A}\right]\sin\alpha$
$+\infty$	−90	$\frac{H}{B}$	0

The average twist rate is zero. The average somersault rate is H/B. This, of course is to be expected, since the motion has the same energy as the rotation (with rate of H/B) around the intermediate axis.

Both the motions that satisfy:

$$E_{rot} = \frac{1}{2}\cdot\frac{H^2}{B}$$

are unstable with respect to perturbations that are not in the rotation plane. The result of such a perturbation would be degeneration of the movement into either the twisting or the wobbling mode solution space.

- If the perturbation nudges the movement into the twisting mode, successive twists of 180° will appear at periodic intervals, and the twists will be in the same direction.
- If the perturbation nudges the movement into the wobbling mode, the successive twists of nearly 180° will be in opposite directions. In between the 180° twists, little twisting action will be observed.

An arbitrarily small perturbation of the system can give rise to qualitatively disparate motions. The rigid body movement may, therefore, be described as an unstable somersault. This mode of motion will, henceforth, be described as an *unstable mode.*

Discussion

Rotations about the axis with the minimal principal moment of inertia are stable because such motions are central in the twisting mode solution space, and a small perturbation merely moves the trajectory into a similar twisting movement.

Rotations about the axis with the maximal principal moment of inertia are similarly stable because they are central in the wobbling mode solution space, and a small perturbation moves the trajectory into a similar wobbling movement.

Rotations about the axis with the intermediate principal moment of inertia are unstable because they are an asymptotic limit of both the twisting and the wobbling mode solution spaces. A small perturbation applied outside the plane of rotation results in the adoption of either a twisting or a wobbling movement.

Torque-free rotational motions of a rigid body fall into two general classes (Figure 4.6.32):

- The *twisting mode.*
- The *wobbling mode.*

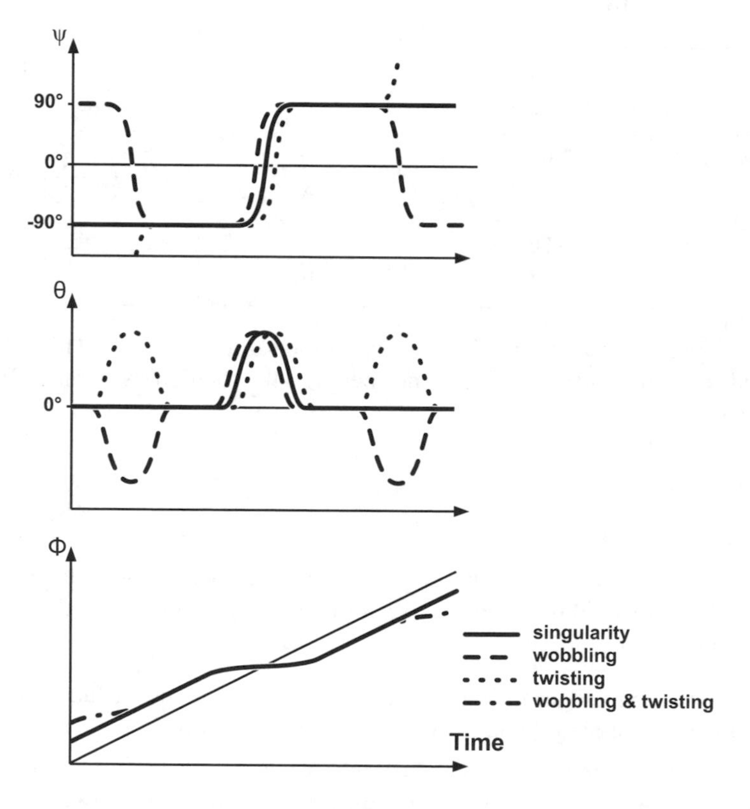

Figure 4.6.32 **Illustration for the twisting and the wobbling mode of rotations of a rigid body (from Yeadon, 1984, with permission).**

In the twisting mode, the twist angle steadily increases, whereas, in the wobbling mode, the twist angle oscillates around a mean value. Which mode a body gets into depends on the relative magnitudes of the principal moments of inertia. This suggests that these movement modes change with changes in the body configuration that alter the relative magnitudes of the principal moments of inertia. Such changes in body configuration can be used to change a non-twisting into a twisting somersault, and vice-versa. Changes in body configurations can be used to initiate or terminate twisting movements in somersaults. As well, these considerations can be taken into account when determining the optimal time and position for someone to accelerate or decelerate the twisting motion (Yeadon, 1993a, b, c, d).

EXAMPLE 11 (arm movement with no muscles)

One simple approach to discussing movement of an extremity and joint forces is to use rigid body dynamics, and not to consider any muscle involvement. For example, the movement of the upper and lower arm with respect to the shoulder joint in a horizontal plane, may be described as the movement of a double pendulum. Certainly, such an approach can only approximate a specific real-life situation. Discussing this most simple example of multiple rigid body dynamics may, however, be helpful in showing a possible procedure for modelling human movement.

Question

Determine the movement (position, velocity, and acceleration) of an idealized arm with no muscle influence in the horizontal plane.

Assumptions

(1) The elbow and shoulder are ideal hinge joints with no friction.
(2) The shoulder joint is fixed in space.
(3) The muscular influence is neglected.
(4) The anthropometrical information with respect to length, moments of inertia, and masses are known.
(5) The initial conditions for the angular position, $\varphi_i(0)$, and the angular velocity, $\dot{\varphi}_i(0)$, are known.
(6) The problem can be solved two-dimensionally.

The illustration of the situation and the free body diagrams are given in Figure 4.6.33.

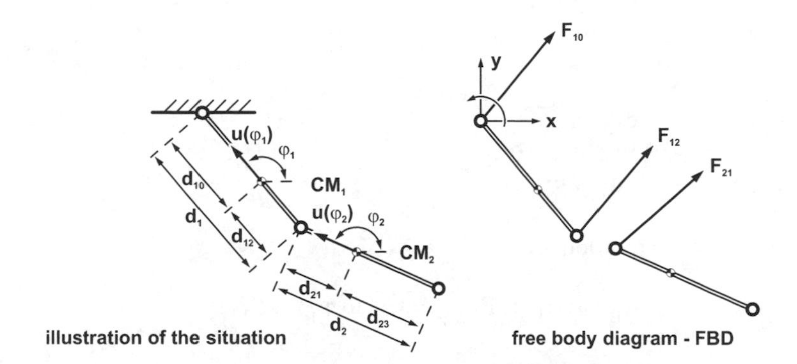

Figure 4.6.33 **Illustration of an idealized arm with no muscles, moving in a horizontal plane (left). Corresponding free body diagram (right).**

Solution

The origin of the coordinate system is positioned in the shoulder joint.

For the following calculation the symbols used are:

$\mathbf{F}_{ik}$ = force acting on body i from body k
$\mathbf{r}_i$ = vector from the origin of the coordinate system to the centre of mass (CM) of segment i
$\mathbf{r}_{ik}$ = vector from the CM of segment i to the joint between segments i and k

$\mathbf{u}(\varphi_i)$ = unit vector from the CM of segment i to the joint between segment i and segment (i-1)
d_i = distance between joint (i -1) and joint i which corresponds to the length of segment i
d_{ik} = distance between the CM of segment i and joint k
I_i = moment of inertia of segment i with respect to the centre of mass of segment i

The system illustrated in Figure 4.6.33 is a system with two DOF and with no friction in the joints. The corresponding equations of motion are:

Equations of motion

Translation:

$$m_1 \ddot{\mathbf{r}}_1 = \mathbf{F}_{10} + \mathbf{F}_{12} \tag{4.6.45}$$

$$m_2 \ddot{\mathbf{r}}_2 = \mathbf{F}_{21} \tag{4.6.46}$$

Rotation:

$$I_1 \ddot{\varphi}_1 = [\mathbf{r}_{10} \times \mathbf{F}_{10}]_z + [\mathbf{r}_{12} \times \mathbf{F}_{12}]_z$$

$$I_2 \ddot{\varphi}_2 = [\mathbf{r}_{21} \times \mathbf{F}_{21}]_z$$

and with:

$$\mathbf{r}_{10} = d_{10} \cdot \mathbf{u}(\varphi_1)$$

$$\mathbf{r}_{12} = -d_{12} \cdot \mathbf{u}(\varphi_1)$$

The equations for rotation are:

$$I_1 \ddot{\varphi}_1 = d_{10}[\mathbf{u}(\varphi_1) \times \mathbf{F}_{10}]_z - d_{12}[\mathbf{u}(\varphi_1) \times \mathbf{F}_{12}]_z \tag{4.6.47}$$

$$I_2 \ddot{\varphi}_2 = d_{21}[\mathbf{u}(\varphi_2) \times \mathbf{F}_{21}]_z \tag{4.6.48}$$

Constraints:

$$\mathbf{r}_1 = -d_{10} \cdot \mathbf{u}(\varphi_1) \tag{4.6.49}$$

$$\mathbf{r}_2 = -d_1 \cdot \mathbf{u}(\varphi_1) - d_{21} \cdot \mathbf{u}(\varphi_2) \tag{4.6.50}$$

where:

$$\mathbf{u}(\varphi_i) = (\cos\varphi_i,\ \sin\varphi_i,\ 0)$$

The general procedure for the solution has two steps:

- Step 1 - find $\mathbf{F}_{ik}$ in terms of $\mathbf{r}_i$ and their time derivatives, using the equations (4.6.45) and (4.6.46).
- Step 2 - find $\mathbf{r}_i$ and their time derivatives in terms of the angular coordinates, φ_i and their time derivatives, using the equations (4.6.49) and (4.6.50).

These steps will now be applied to this example:

Step 1

$$-\mathbf{F}_{12} = m_2\ddot{\mathbf{r}}_2 = +\mathbf{F}_{21} \tag{4.6.51}$$

$$\mathbf{F}_{10} = m_1\ddot{\mathbf{r}}_1 + m_2\ddot{\mathbf{r}}_2 \tag{4.6.52}$$

Equations (4.6.51) and (4.6.52) in equations (4.6.47) and (4.6.48):

$$I_1\ddot{\varphi}_1 = d_{10}[\mathbf{u}(\varphi_1)\times(m_1\ddot{\mathbf{r}}_1 + m_2\ddot{\mathbf{r}}_2)]_z + d_{12}[\mathbf{u}(\varphi_1)\times m_2\ddot{\mathbf{r}}_2]_z$$

$$I_1\ddot{\varphi}_1 = d_{10}[\mathbf{u}(\varphi_1)\times m_1\ddot{\mathbf{r}}_1]_z + d_1[\mathbf{u}(\varphi_1)\times m_2\ddot{\mathbf{r}}_2]_z$$

$$I_2\ddot{\varphi}_2 = d_{21}[\mathbf{u}(\varphi_2)\times m_2\ddot{\mathbf{r}}_2]_z$$

$$I_1\ddot{\varphi}_1 = m_1 d_{10}[\mathbf{u}(\varphi_1)\times\ddot{\mathbf{r}}_1]_z + m_2 d_1[\mathbf{u}(\varphi_1)\times\ddot{\mathbf{r}}_2]_z \tag{4.6.53}$$

$$I_2\ddot{\varphi}_2 = m_2 d_{21}[\mathbf{u}(\varphi_2)\times\ddot{\mathbf{r}}_2]_z \tag{4.6.54}$$

Step 2 using equation (4.6.49) and equation (4.6.50)

$$\dot{\mathbf{r}}_1 = -d_{10}\left[\frac{d\mathbf{u}(\varphi_1)}{d\varphi_1}\cdot\frac{d\varphi_1}{dt}\right] = -d_{10}\cdot\mathbf{u}'(\varphi_1)\cdot\dot{\varphi}_1$$

$$\ddot{\mathbf{r}}_1 = -d_{10}[\mathbf{u}''(\varphi_1)\cdot\dot{\varphi}_1^2 + \mathbf{u}'(\varphi_1)\cdot\ddot{\varphi}_1] \tag{4.6.55}$$

Similar calculations lead to equation (4.6.56) for segment 2:

$$\ddot{\mathbf{r}}_2 = -d_1[\mathbf{u}''(\varphi_1)\cdot\dot{\varphi}_1^2 + \mathbf{u}'(\varphi_1)\cdot\ddot{\varphi}_1] - d_{21}[\mathbf{u}''(\varphi_2)\cdot\dot{\varphi}_2^2 + \mathbf{u}'(\varphi_2)\cdot\ddot{\varphi}_2] \tag{4.6.56}$$

Rules for calculation:

$$\begin{aligned}\mathbf{u}(\varphi_i)\times\mathbf{u}'(\varphi_i) &= (\cos\varphi_i,\ \sin\varphi_i,\ 0)\times(-\sin\varphi_i,\ \cos\varphi_i,\ 0)\\ &= (0,\ 0,\ \cos^2\varphi_i + \sin^2\varphi_i)\\ &= (0,\ 0,\ 1)\end{aligned}$$

$$\mathbf{u}(\varphi_i) \times \mathbf{u}''(\varphi_i) = (\cos\varphi_i, \sin\varphi_i, 0) \times (-\cos\varphi_i, -\sin\varphi_i, 0)$$

$$= (0, 0, 0)$$

$$\mathbf{u}'(\varphi_i) \times \mathbf{u}''(\varphi_i) = (-\sin\varphi_i, \cos\varphi_i, 0) \times (-\cos\varphi_i, -\sin\varphi_i, 0)$$

$$= (0, 0, 1)$$

$$\mathbf{u}(\varphi_1) \times \mathbf{u}''(\varphi_2) = (0, 0, \sin(\varphi_1 - \varphi_2))$$

$$\mathbf{u}(\varphi_1) \times \mathbf{u}'(\varphi_2) = (0, 0, \cos(\varphi_1 - \varphi_2))$$

In the next step the equations (4.6.55) and (4.6.56) are substituted in equations (4.6.53) and (4.6.54) using the rules for calculating cross products:

$$I_1 \cdot \ddot{\varphi}_1 = m_1 \cdot d_{10}\{\mathbf{u}(\varphi_1) \times [-d_{10}[\mathbf{u}''(\varphi_1) \cdot \dot{\varphi}_1^2 + \mathbf{u}'(\varphi_1) \cdot (\ddot{\varphi}_1)]]\}$$

$$+m_2 \cdot d_1\{\mathbf{u}(\varphi_1) \times [-d_1[\mathbf{u}''(\varphi_1) \cdot \dot{\varphi}_1^2 + \mathbf{u}'(\varphi_1) \cdot (\ddot{\varphi}_1)]]\}$$

$$+m_2 \cdot d_1\{\mathbf{u}(\varphi_1) \times [-d_{21}[\mathbf{u}''(\varphi_2) \cdot \dot{\varphi}_2^2 + \mathbf{u}'(\varphi_2) \cdot (\ddot{\varphi}_2)]]\}$$

$$= m_1 \cdot d_{10}(-d_{10} \cdot \ddot{\varphi}_1) + m_2 \cdot d_1(-d_1 \cdot \ddot{\varphi}_1)$$

$$-m_2 \cdot d_1 \cdot d_{21}[\sin(\varphi_1 - \varphi_2) \cdot \dot{\varphi}_2^2 + \cos(\varphi_1 - \varphi_2) \cdot \ddot{\varphi}_2]$$

Which leads to equation (4.6.57):

$$[I_1 + m_1 d_{10}^2 + m_2 \cdot d_1^2] \cdot \ddot{\varphi}_1 + [m_2 \cdot d_1 \cdot d_{21} \cdot \cos(\varphi_1 - \varphi_2)] \cdot \ddot{\varphi}_2$$

$$= -[m_2 d_1 d_{21} \cdot \sin(\varphi_1 - \varphi_2)] \cdot \dot{\varphi}_2^2 \qquad (4.6.57)$$

Equation (4.6.54) is treated the same way, which leads to equation (4.6.58):

$$m_2 d_1 \cdot d_{21} \cdot \cos(\varphi_2 - \varphi_1)]\ddot{\varphi}_1 + [I_2 + m_2 \cdot d_{21}^2]\ddot{\varphi}_2$$

$$= -[m_2 d_1 d_{21} \sin(\varphi_2 - \varphi_1)] \cdot \dot{\varphi}_1^2 \qquad (4.6.58)$$

or, in a more general form:

$$A \cdot \ddot{\varphi}_1 + B \cdot \ddot{\varphi}_2 = E \cdot \dot{\varphi}_2^2$$

$$C \cdot \ddot{\varphi}_1 + D \cdot \ddot{\varphi}_2 = F \cdot \dot{\varphi}_1^2$$

This is a system of non-linear, second order differential equations. Such a system is usually solved using numerical methods. However, numerical solutions may well have more than one solution. Caution and intuition are important for an iteration approach.

Here, the calculations for an example with the following numerical values are presented:

$$
\begin{aligned}
m_1 &= 3 \text{ kg} \\
m_2 &= 2 \text{ kg} \\
d_{10} &= d_{12} = d_{21} = 0.15 \text{ m} \\
d_1 &= d_2 = 0.3 \text{ m} \\
I_1 &= I_2 = 0.01 \text{ kg m}^2
\end{aligned}
$$

and for the initial conditions, IC:

$$
\begin{aligned}
\varphi_1(0) &= 0 \\
\varphi_2(0) &= 0 \\
\dot{\varphi}_1(0) &= 2\pi/\text{sec} \\
\dot{\varphi}_2(0) &= \pi/\text{sec}
\end{aligned}
$$

The basic procedure for this numerical solution is to use the initial conditions for the time t to determine the angular position and velocity for the time $t + \Delta t$. Starting with the initial conditions at the time 0, which are known, a possible example for such an approach could be:

$$
\begin{aligned}
\varphi(\Delta t) &= \varphi(0) + \Delta t \cdot \dot{\varphi}(0) \\
\dot{\varphi}(\Delta t) &= \dot{\varphi}(0) + \Delta t \cdot \ddot{\varphi}(0)
\end{aligned}
$$

or in a more general form:

$$
\begin{aligned}
\varphi(t + \Delta t) &= \varphi(t) + \Delta t \cdot \dot{\varphi}(t) \\
\dot{\varphi}(t + \Delta t) &= \dot{\varphi}(t) + \Delta t \cdot \ddot{\varphi}(t)
\end{aligned}
$$

The angular accelerations are determined for each point in time using the equations:

$$
\begin{aligned}
A \cdot \ddot{\varphi}_1 + B \cdot \ddot{\varphi}_2 &= E\dot{\varphi}_2^2 \\
C \cdot \ddot{\varphi}_1 + D \cdot \ddot{\varphi}_2 &= F\dot{\varphi}_2^1
\end{aligned}
$$

which provide for the accelerations:

$$
\ddot{\varphi}_1 = \frac{DE\dot{\varphi}_2^2 - BF\dot{\varphi}_1^2}{AD - BC}
$$

$$
\ddot{\varphi}_2 = \frac{AF\dot{\varphi}_1^2 - CE\dot{\varphi}_2^2}{AD - BC}
$$

Using this approach, the angular position, velocity, and accelerations can easily be determined in steps of, for instance, 1 ms:

for t = o
$$\ddot{\varphi}_1 = 0$$
$$\ddot{\varphi}_2 = 0$$
$$\dot{\varphi}_1 = 2\pi$$
$$\dot{\varphi}_2 = \pi$$
$$\varphi_1 = 0$$
$$\varphi_2 = 0$$

for t = 1 ms
$$\ddot{\varphi}_1(1) = -0.190919$$
$$\ddot{\varphi}_2(1) = 0.51543$$
$$\dot{\varphi}_1(1) = 6.28$$
$$\dot{\varphi}_2(1) = 3.14$$
$$\varphi_1(1) = 0.00628$$
$$\varphi_2(1) = 0.00314$$

for t = 2 ms
$$\ddot{\varphi}_1(2) = -0.382016$$
$$\ddot{\varphi}_2(2) = 1.030977$$
$$\dot{\varphi}_1(2) = 6.283$$
$$\dot{\varphi}_2(2) = 3.142$$
$$\varphi_1(2) = 0.01256$$
$$\varphi_2(2) = 0.00628$$

This iteration procedure can be executed for the time interval of interest. The results for this iteration procedure for the first 1000 ms (1 s) are illustrated in Figure 4.6.34.

EXAMPLE 12 (a model for impact forces)

Impact forces have been defined (Chapter 3.1) as forces that result from a collision of two objects that have a maximum force (impact force peak) earlier than 50 milliseconds after the first contact of the two objects. Impact forces occur during daily activities such as walking, but are generally of more interest in activities with higher forces, such as running, jumping, and boxing.

The most commonly discussed and studied impact force variable is the maximal vertical impact force, F_{zi}, during landing in running. A simple mechanical model using two rigid bodies and two springs connecting these rigid bodies on each side was used (Nigg, 1986) to understand factors important in the development of an appropriate model for impact force analysis. The experiment was performed by dropping this mechanical structure from different heights onto a force plate that measured the vertical ground reaction force. Additionally, the vertical impact forces were simulated with a mathematical model that described this situation (Figure 4.6.35).

The results of the experiment and the simulation (Nigg, 1986) indicate that the angle between the two rigid segments (knee angle), and the landing velocity (of the heel), are important factors influencing the magnitude of the measured or calculated maximal vertical impact forces, F_{zi}. However, the results suggest that the stiffness of, and the force in the two springs, do not influence the maximal vertical impact forces, F_{zi}, for impact forces

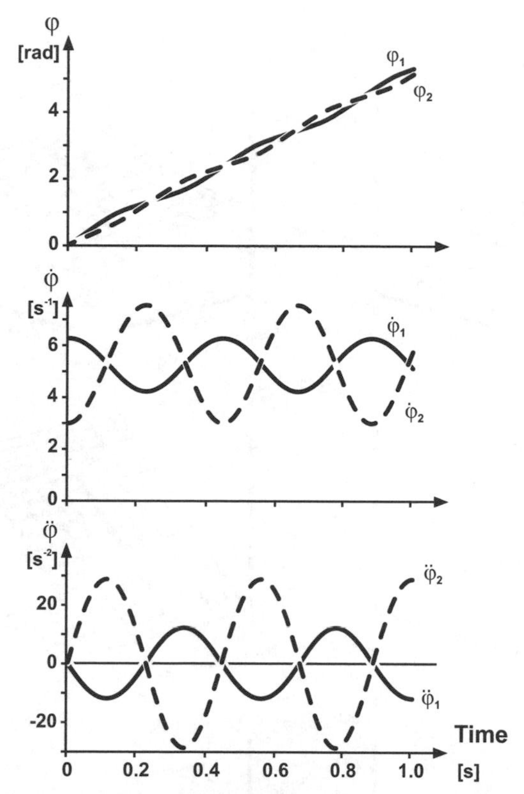

Figure 4.6.34 **Graphical illustration of the numerical solution. Segment 1 and 2 pull and push each other. The amplitudes are, however, higher for the movement of segment 2 than for the movement of segment 1.**

with a maximum earlier than 10 milliseconds after first ground contact. This result may be used to establish the assumptions for a model of the human body that can estimate impact forces. The results of this experiment suggest that, in a first approximation, muscle forces and muscle stiffness can be neglected in a model used to estimate impact forces in landing. This finding has been supported in an impact simulation study that included muscles with actual muscle properties (Gerritsen and van den Bogert, 1993). Note that the strength and stiffness of the springs do, however, influence the magnitude of the measured or calculated result for impact forces that reach their maximum later (see section 4.9).

Based on these considerations the following model was developed (Lemm, 1978; Denoth, 1986).

Question

Develop a model that allows for the determination of factors important for impact forces.

Assumptions

(1) The problem can be solved two dimensionally.

(2) The model consists of three rigid segments. Segment 1 corresponds to the leg. Segment 2 corresponds to the thigh. Segment 3 corresponds to the rest of the body.

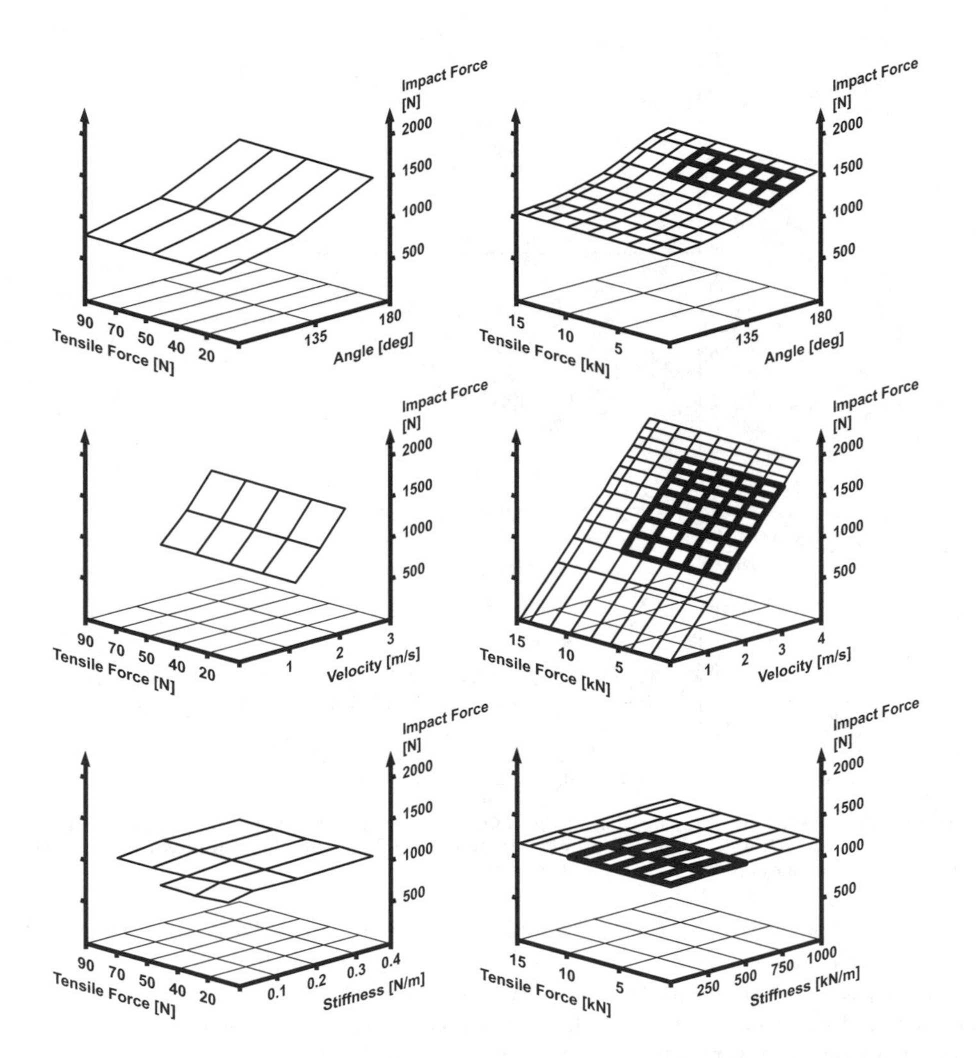

Figure 4.6.35 **Maximum impact forces resulting from a drop of a two segment structure (left) and from a computer simulation of this same experiment (right). Top results: velocity = 2 m/s, stiffness of surface = 4000 N/m. Middle results: knee angle = 135°, stiffness of surface = 4000 N/m. Bottom results: velocity = 2 m/s, knee angle = 135°. The thicker lines indicate results for this physiological range (from Nigg, 1986, with permission).**

(3) The segments are assumed to have a length, L_i. The width and depth are assumed to be very small compared to the other geometrical measures.
(4) The muscle influence is neglected.
(5) The movement of interest is heel-toe running.
(6) Running surface, heel pad, and ankle joint are considered one system with a linear spring constant. The spring constant is known.
(7) The masses, m_i, consist of a rigid and a non-rigid part. For this model the rigid masses are assumed to be:

m_1 (rigid)	=	0.5 m (leg)	=	50% of the total leg mass
m_2 (rigid)	=	0.5 m (thigh)	=	50% of the total thigh mass
m_3 (rigid)	=	0.2 m (rest)	=	20% of the total rest mass

(8) The moments of inertia, I_i, are known.
(9) The joints are assumed to be frictionless.
(10) The symbols used in this example are:

φ_i angle between the horizontal axis (x-axis) and segment i (long segment axis) measured at the posterior side

d_i length of segment i

$\mathbf{r}_i$ vector from the origin of the coordinate system to the centre of mass, CM_i, of segment i

where:

$$\mathbf{r}_1 = (x_1, y_1, m_1)$$
$$\mathbf{r}_2 = (x_2, y_2, m_1)$$
$$\mathbf{r}_3 = (x_3, y_3, m_1)$$

m_i mass of the element i

d_{ik} distance between the centre of mass of segment i, CM_i, and the end of segment i towards segment k

$m_i\mathbf{g}$ ravitational forces acting on segment i

where:

$$\mathbf{g} = (0, -g)$$

$\mathbf{F}_{ik}$ (actual) joint force in the joint between the segments i and k

where:

$$\mathbf{F}_{ik} = -\mathbf{F}_{ki}$$

$\mathbf{F}_{10}$ ground reaction force

$\mathbf{u}(\varphi_i)$ unit vector from the centre of mass of segment i, CM_i, directed towards segment (i + 1).

Equations of motion

Translation:

$$m_1\ddot{\mathbf{r}}_1 = m_1\mathbf{g} + \mathbf{F}_{10} + \mathbf{F}_{12}$$

$$m_2\ddot{\mathbf{r}}_2 = m_2\mathbf{g} + \mathbf{F}_{21} + \mathbf{F}_{23}$$

$$m_3\ddot{\mathbf{r}}_3 = m_3\mathbf{g} + \mathbf{F}_{32}$$

Rotation:

Since the problem is two-dimensional, the moment equations are written with respect to the centre of mass for the third component:

$$I_1\ddot{\varphi}_1 = [\mathbf{M}(F_{10}) + \mathbf{M}(F_{12})]_z$$

$$= [(\mathbf{r}_{10} \times \mathbf{F}_{10}) + (\mathbf{r}_{12} \times \mathbf{F}_{12})]_z$$

$$I_1\ddot{\varphi}_1 = [-d_{10}\mathbf{u}(\varphi_1) \times \mathbf{F}_{10} + d_{12}\mathbf{u}(\varphi_1) \times \mathbf{F}_{12}]_z$$

$$I_2\ddot{\varphi}_2 = [-d_{21}\mathbf{u}(\varphi_2) \times \mathbf{F}_{21} + d_{23}\mathbf{u}(\varphi_2) \times \mathbf{F}_{23}]_z$$

$$I_3\ddot{\varphi}_3 = [-d_{32}\mathbf{u}(\varphi_3) \mathrm{x} \mathbf{F}_{32}]_z$$

Constraints:

$$\mathbf{r}_2 = \mathbf{r}_1 + d_{12}\mathbf{u}(\varphi_1) + d_{21}\mathbf{u}(\varphi_2)$$

$$\mathbf{r}_3 = \mathbf{r}_2 + d_{23}\mathbf{u}(\varphi_2) + d_{32}\mathbf{u}(\varphi_3)$$

$$= \mathbf{r}_1 + d_{12}\mathbf{u}(\varphi_1) + d_2\mathbf{u}(\varphi_2) + d_{32}\mathbf{u}(\varphi_3)$$

and for the time derivatives of the constraint functions:

$$\dot{\mathbf{r}}_2 = \dot{\mathbf{r}}_1 + d_{12}\mathbf{u}'(\varphi_1)\dot{\varphi}_1 + d_2\mathbf{u}'(\varphi_2)\dot{\varphi}_2$$

$$\ddot{\mathbf{r}}_2 = \ddot{\mathbf{r}}_1 + d_{12}[\mathbf{u}''(\varphi_1)\dot{\varphi}_1^2 + \mathbf{u}'(\varphi_1)\ddot{\varphi}_1]$$

$$+d_{21}[\mathbf{u}''(\varphi_2)\dot{\varphi}_2^2 + \mathbf{u}'(\varphi_2)\ddot{\varphi}_2]$$

$$\ddot{\mathbf{r}}_3 = \ddot{\mathbf{r}}_1 + d_{12}[\mathbf{u}''(\varphi_1)\dot{\varphi}_1^2 + \mathbf{u}'(\varphi_1)\ddot{\varphi}_1]$$

$$+d_2[\mathbf{u}''(\varphi_2)\dot{\varphi}_2^2 + \mathbf{u}'(\varphi_2)\ddot{\varphi}_2]$$

$$+d_{32}[\mathbf{u}''(\varphi_3)\dot{\varphi}_3^2 + \mathbf{u}'(\varphi_3)\ddot{\varphi}_3]$$

Where $\mathbf{u}'(\varphi_i)$ and $\mathbf{u}''(\varphi_i)$ are the first and second derivatives of the unit vectors with respect to φ_i, and where:

$$\mathbf{u}(\varphi_i) = (\cos\varphi_i, \sin\varphi_i, 0)$$

$$\mathbf{u}'(\varphi_i) = (-\sin\varphi_i, \cos\varphi_i, 0)$$

$$\mathbf{u}''(\varphi_i) = (-\cos\varphi_i, -\sin\varphi_i, 0) = -\mathbf{u}(\varphi_i)$$

Elimination of $\mathbf{r}_2$, $\mathbf{r}_3$, $\mathbf{F}_{12}$, $\mathbf{F}_{23}$, and $\mathbf{F}_{10}$ provides:

$$c_{11}\ddot{x}_1 + c_{12}\ddot{y}_1 + c_{13}\ddot{\varphi}_1 + c_{14}\ddot{\varphi}_2 + c_{15}\ddot{\varphi}_3 = D_1$$

$$c_{21}\ddot{x}_1 + c_{22}\ddot{y}_1 + c_{23}\ddot{\varphi}_1 + c_{24}\ddot{\varphi}_2 + c_{25}\ddot{\varphi}_3 = D_2$$

$$c_{31}\ddot{x}_1 + c_{32}\ddot{y}_1 + c_{33}\ddot{\varphi}_1 + c_{34}\ddot{\varphi}_2 + c_{35}\ddot{\varphi}_3 = D_3$$

$$c_{41}\ddot{x}_1 + c_{42}\ddot{y}_1 + c_{43}\ddot{\varphi}_1 + c_{44}\ddot{\varphi}_2 + c_{45}\ddot{\varphi}_3 = D_4$$

$$c_{51}\ddot{x}_1 + c_{52}\ddot{y}_1 + c_{53}\ddot{\varphi}_1 + c_{54}\ddot{\varphi}_2 + c_{55}\ddot{\varphi}_3 = D_5$$

or, in a more general form:

$$C\ddot{\mathbf{X}} = \mathbf{D}$$

where:

$$\ddot{\mathbf{X}} = (\dot{x}_1, \ddot{y}_1, \ddot{\varphi}_1, \ddot{\varphi}_2, \ddot{\varphi}_3)$$

and where the matrix C is symmetric with $c_{ik} = c_{ki}$. The corresponding coefficients are described by Lemm (1978).

The solution of such a system of differential equations can be performed using numerical techniques. In the case of this specified example, it was solved using the 4th order Runge-Kutta approach. Lemm (1978) and Denoth (1986) used this model to develop the concept of an effective mass.

EXAMPLE 13 (a wobbling mass model)

Rigid body models, together with inverse dynamics methods, have typically been used to estimate the internal forces acting on the musculo-skeletal system during human movement. However, in reality, the body is not composed of a set of linked rigid bodies. Rather, each body segment consists of a rigid part (bone), and a non-rigid part (skin, muscle, ligament, tendon, connective tissue, and other soft tissue structures). During an impact, such as heel-strike in running or landing from a vertical jump, the skeletal structures of the body experience sudden, high accelerations, whereas, the soft tissue's movement is delayed, initiating damped vibrations of the soft tissue relative to the bone. In estimating internal forces proximal to the point of contact using inverse dynamics, the externally applied force and the acceleration of the segments of interest are typically measured. The measured ground reaction force (assuming impact with the ground) contains inertial components of both rigid and soft tissue structures, and is not influenced by the rigid body assumption. However, acceleration measurements may introduce errors in the estimation of internal forces because accelerations of the bone lead to overestimation, and accelerations of the soft tissue to underestimation of the internal forces.

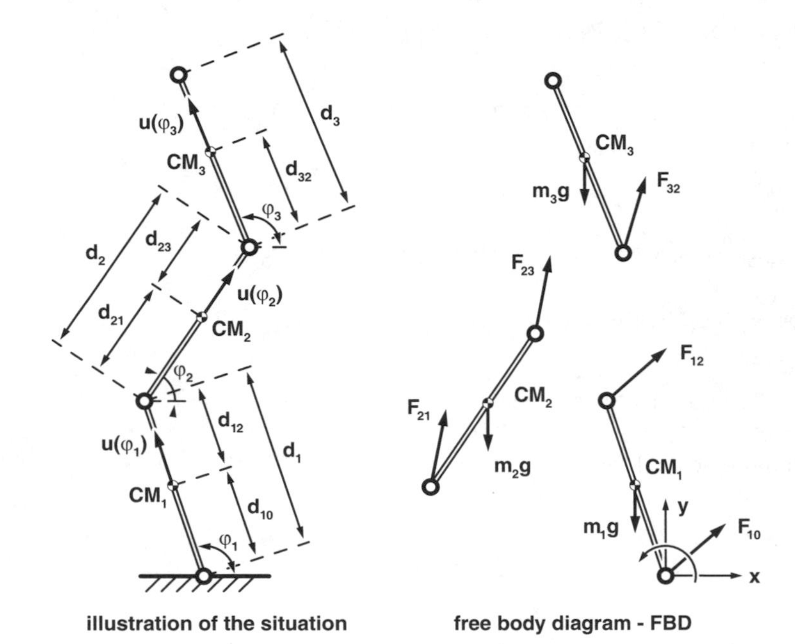

Figure 4.6.36 **Free body diagram (right) and descriptive diagram (left) for the three segment model.**

The potential errors associated with rigid body models led to the conclusion that the approximation of the human body with rigid segments is justified only for movements that are not too rapid, and is irrelevant for high impacts (Denoth et al., 1984). To account for the effect of relative displacements between the rigid and soft tissues on the joint forces during impact of the foot with the ground, a two-dimensional model of the body was developed, in which each segment was a combination of a skeletal part and a soft tissue part (Denoth et al., 1984). This model was further refined by Gruber (1987), who used it to illustrate the potential errors in estimating the forces and moments at the knee and hip joints using a rigid body model.

Question

Determine the resultant forces and moments at the knee and hip joints when landing on the heel of one foot from a vertical jump (Figure 4.6.37).

Assumptions

(1) All motion occurs in the sagittal plane.
(2) The human body is composed of three segments: the shank, which is numbered segment 1, the thigh, which is numbered segment 2, and the trunk, which is numbered segment 3.
(3) The segments are joined together by frictionless hinge joints.
(4) Movements of the upper extremities are neglected, and their masses are included in the mass of the trunk.
(5) All skeletal structures within a segment are modelled as a single rigid body with the geometry of a cylinder with length L_{ir}, diameter d_{ir}, and mass m_{ir}, for $i=1,2,3$.

(6) The soft tissues or wobbling mass of each segment are modelled as a rigid body with a geometry of a cylinder of length L_{iw}, diameter d_{iw}, and mass m_{iw}, for i=1,2,3.
(7) The moment of inertia of each skeletal mass and each soft tissue mass in the plane of motion can be approximated from the length and radius of each cylinder.
(8) Relative motion between the soft tissues and the skeletal mass of each segment has three degrees of freedom, two translational and one rotational.
(9) Quasi-elastic, strongly damped forces and moments are associated with relative movement between the soft tissue and skeletal mass of each segment. These forces and moments act at the centres of mass of the soft and skeletal tissues.

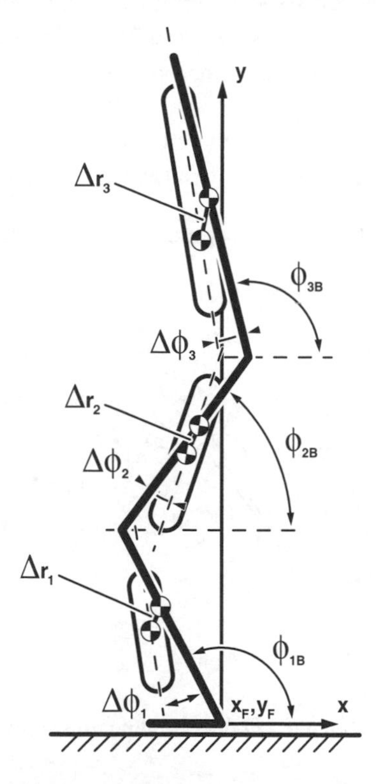

Figure 4.6.37 Illustration of the three-link model with skeletal and soft tissue masses and a mass-less foot (from Gruber, 1987, with permission).

Symbols

The symbols used for this example are (all vectors are defined in an inertial reference frame):

m_{ir} = mass of the rigid part of segment i (corresponds to mass of the bone)

m_{iw} = mass of the soft tissue of segment i (corresponds to the wobbling mass). Note that the soft tissue will be called wobbling mass in the following

$\mathbf{g}$ = acceleration due to gravity

$\mathbf{r}_{ik}$ = position vector from the centre of mass of the rigid part of segment i to the joint between segments i and k

$\mathbf{r}_{ir}$ = position of the centre of mass of the rigid part of segment i in the inertial reference frame

$\mathbf{r}_{iw}$ = position of the centre of mass of the wobbling mass of segment i in the inertial reference frame

φ_{ir} = angle that each rigid segment makes with the horizontal axis of the inertial reference frame in the sagittal plane

φ_{iw} = angle that each wobbling mass segment makes with the horizontal axis of the inertial reference frame in the sagittal plane

$\mathbf{F}_{ik}$ = force acting on segment i from segment k

$\mathbf{F}_{ie}$ = elastic force associated with the translation of the wobbling mass

$\mathbf{F}_{id}$ = damping force associated with the translation of the wobbling mass

$\mathbf{M}_{ik}$ = resultant moment acting on segment i at the joint between segments i and k

M_{ie} = elastic moment associated with the rotation of the wobbling mass

$\mathbf{M}_{id}$ = damping moment associated with the rotation of the wobbling mass

The free body diagram for the skeletal segments of the model is shown in Figure 4.6.38.

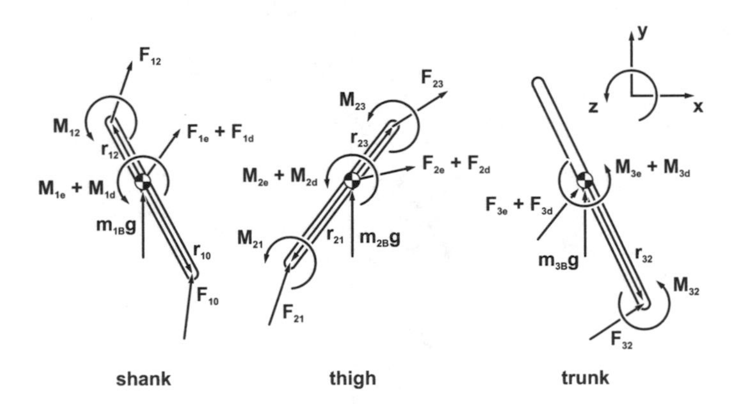

Figure 4.6.38 **FBD of the skeletal parts of the wobbling mass model.**

The free body diagram for the soft tissue segments of the model is shown in Figure 4.6.39.

Equations of motion

The equations of motion for each of the rigid segments of the model are:

Translation:

$$m_{1r}\ddot{\mathbf{r}}_{1r} = \mathbf{F}_{10} + \mathbf{F}_{12} + \mathbf{F}_{1e} + \mathbf{F}_{1d} + m_{1r}\mathbf{g}$$

$$m_{2r}\ddot{\mathbf{r}}_{2r} = \mathbf{F}_{21} + \mathbf{F}_{23} + \mathbf{F}_{2e} + \mathbf{F}_{2d} + m_{2r}\mathbf{g}$$

$$m_{3r}\ddot{\mathbf{r}}_{3r} = \mathbf{F}_{32} + \mathbf{F}_{3e} + \mathbf{F}_{3d} + m_{3r}\mathbf{g}$$

Rotation:

$$I_{1r}\ddot{\varphi}_{1r} = \left[(\mathbf{r}_{10} \times \mathbf{F}_{10}) + (\mathbf{r}_{12} \times \mathbf{F}_{12}) + \mathbf{M}_{12} + \mathbf{M}_{1e} + \mathbf{M}_{1d}\right]_z$$

$$I_{2r}\ddot{\varphi}_{2r} = \left[(\mathbf{r}_{21} \times \mathbf{F}_{21}) + (\mathbf{r}_{23} \times \mathbf{F}_{23}) + \mathbf{M}_{21} + \mathbf{M}_{23} + \mathbf{M}_{2e} + \mathbf{M}_{2d}\right]_z$$

$$I_{3r}\ddot{\varphi}_{3r} = \left[(\mathbf{r}_{32} \times \mathbf{F}_{32}) + \mathbf{M}_{32} + \mathbf{M}_{3e} + \mathbf{M}_{3d}\right]_z$$

The equations of motion for each of the soft tissue segments (wobbling masses) are:

$$m_{iw} \cdot \ddot{\mathbf{r}}_{iw} = -\mathbf{F}_{1e} - \mathbf{F}_{1d} + m_{iw} \cdot \mathbf{g}$$

$$m_{2w} \cdot \ddot{\mathbf{r}}_{2w} = -\mathbf{F}_{2e} - \mathbf{F}_{2d} + m_{2w} \cdot \mathbf{g}$$

$$m_{3w} \cdot \ddot{\mathbf{r}}_{3w} = -\mathbf{F}_{3e} - \mathbf{F}_{3d} + m_{3w} \cdot \mathbf{g}$$

Rotation:

$$I_{1w} \cdot \ddot{\varphi}_{1w} = \left[-\mathbf{M}_{1e} - \mathbf{M}_{1d}\right]_z$$

$$I_{2w} \cdot \ddot{\varphi}_{2w} = \left[-\mathbf{M}_{2e} - \mathbf{M}_{2d}\right]_z$$

$$I_{3w} \cdot \ddot{\varphi}_{3w} = \left[-\mathbf{M}_{3e} - \mathbf{M}_{3d}\right]_z$$

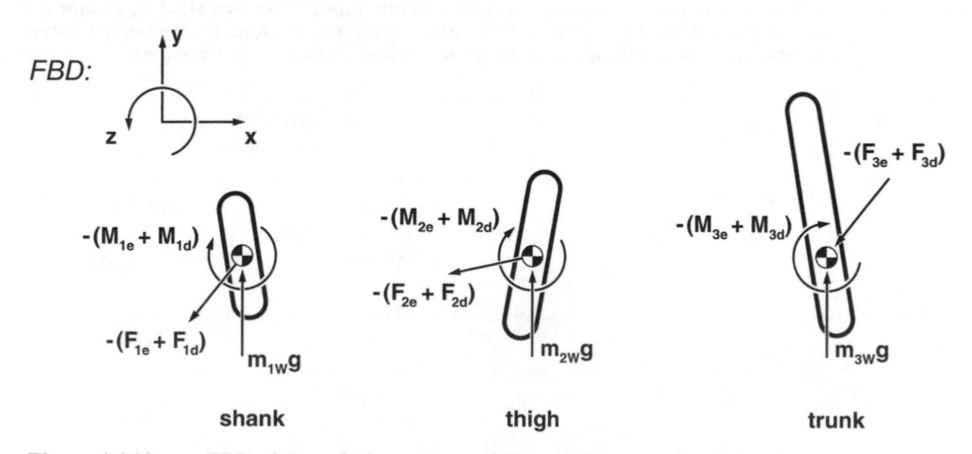

Figure 4.6.39 FBD of the soft tissue parts of the wobbling mass model.

The model has 14 degrees of freedom with the following inputs:

- The initial conditions for segmental orientations, φ_{1r} (0), were determined from high speed film of a subject performing the required movement.
- Foot acceleration was measured as a function of time using an accelerometer placed on the lateral malleolus of the subject.

- The vertical and antero-posterior components of the ground reaction force were measured throughout the movement.
- The length of each segment was determined based on measurements taken of the subject.
- The distribution of the subject's mass between the three segments was determined based on the published cadaver data of Clauser et al. (1969).
- The distribution of the segmental masses to the skeletal and soft tissue masses was done based on empirical measurements, and by trial and error.
- Moments of inertia were estimated based on the length of each segment, and an assumption for the radius of each cylindrical segment.
- The stiffness and damping coefficients associated with relative translations between skeletal and soft tissue masses were determined experimentally.
- The stiffness and damping coefficients associated with the relative rotation between skeletal and soft tissue masses were determined by trial and error.

Solution

- The set of coupled equations of motion were solved using standard integration techniques. The initial conditions for the rigid segments were determined from film analysis. The initial conditions for each wobbling segment were assumed to be identical to the initial conditions of the corresponding rigid segment 1.

The differences between the results predicted for the wobbling mass model and the rigid body mass model are summarized in Table 4.6.6. The values were estimated from the force-time and moment-time graphs of Gruber's (1987) publication.

Table 4.6.6 Comparison of selected maximal results during impact from the rigid body and the wobbling mass model. Approximate values were taken from the presented force/moment-time curves (based on data from Gruber, 1987, with permission).

SEGMENT	VARIABLE	UNIT	PREDICTED PEAK VALUES	
			WOBBLING MASS MODEL	RIGID BODY MODEL
KNEE	F_{ant}	N	200	800
	F_{post}	N	no peak	1000
	M_{12} (flexion)	Nm	no peak	200
HIP	F_{ant}	N	200	1100
	F_{post}	N	300	2500
	M_{23} (flexion)	Nm	no peak	1200
	M_{23} (extension)	Nm	no peak	800

Furthermore, the graphs for the anterior-posterior forces in the knee and hip joint (Figure 4.6.40) as well as the flexion/extension moments of these two joints (Figure 4.6.41) are shown as a function of time.

The results can be summarized as follows:

(1) The anterior-posterior forces in the knee and hip joint were predicted differently by the rigid body and wobbling mass models. The rigid body model predicted high impact peaks in the anterior-posterior direction, while the wobbling mass model predicted only small anterior-posterior forces during the impact phase.

(2) The flexion/extension moments at the knee and hip joint were predicted differently by the rigid body and wobbling mass models. The rigid body model predicted high impact peaks, while the wobbling mass model predicted no impact peaks.

(3) In general, the wobbling mass model predicted smaller internal forces and moments than did the rigid body model.

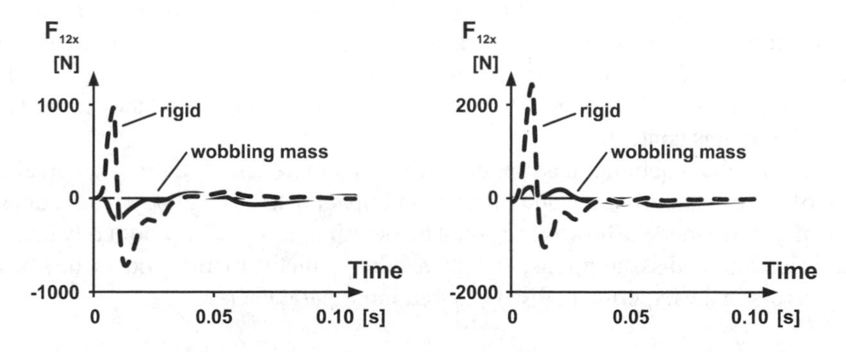

Figure 4.6.40 **Horizontal components of the resultant forces at the knee joint (left), and at the hip joint (right), estimated by the rigid body and wobbling mass models for a drop jump movement with landing at the heel (from Gruber, 1987, with permission).**

Comments

This example brings to light two important considerations: the magnitude of the absolute error for the results predicted with the two models, and errors that may have contributed to the differences in the results of the wobbling mass and rigid body models.

Errors in the results of both the wobbling mass model and the rigid body model arise from several sources, including the various assumptions that have been made and the experimentally measured inputs that have been used. The accelerations measured using the skin-mounted accelerometer at the malleolus seemed unusually high for the move being performed. LaFortune (1991) has shown that skin-mounted accelerometers can overestimate peak accelerations of the tibia by as much as 50% in comparison to bone- mounted accelerometers.

The differences in the results between the two models are of greater interest, however, in the context of this book. One would like to know whether the wobbling mass model ex-

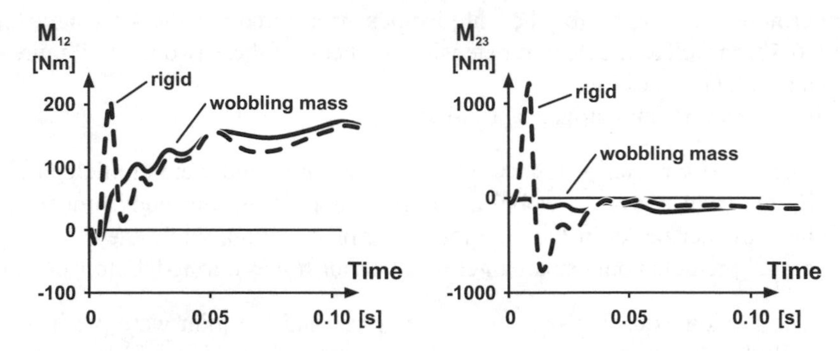

Figure 4.6.41 **Resultant moments at the knee joint (left), and at the hip joint (right), estimated by the rigid body and wobbling mass models for a drop jump movement with landing at the heel (from Gruber, 1987, with permission).**

erts a corrective influence over the results or, in other words, whether or not the reduction of internal loading in the knee and hip joint predicted by the wobbling mass model corresponds to reality. Possible errors that may affect the results are that muscles are attached to one specific segment (for instance, gastrocnemius), and that the soft tissue is allowed to rotate. However, Gruber (1987) does not elaborate on these questions, and they cannot be discussed further in this context.

In summary, the wobbling mass model provides a more appropriate conceptual representation of the mechanics of the body during an impact than a rigid body model does. The accuracy of the two models, however, cannot be determined since they both rely on numerous simplifications and assumptions, and the results predicted by the models may be influenced by errors in the experimentally measured input parameters.

4.6.4 COMMENTS FOR SECTION 4.6

Chapter 4.6 discussed various mathematical models that were determinate, which means that the number of unknowns and equations were equal. This is a specific situation for biomechanical modelling and, as discussed in the next section, is uncommon. However, in summary:

(1) It may be possible in biomechanics to discuss specific questions using simple deterministic models. The complexity of a model is not a priority related to the power and potential of a model to answer a question or to increase our understanding of a situation.

(2) Even the simple two-dimensional models that use two or three rigid bodies and no muscles provide results that must be solved with the help of numerical techniques. These numerical techniques, however, usually give more than one solution, so caution is required in using them.

4.7 MATHEMATICALLY INDETERMINATE SYSTEMS

HERZOG, W.

4.7.1 INTRODUCTION

If a mathematical system contains more equations than unknowns, it is said to be overdetermined, and in general, there is no solution. For example, the system shown below contains three equations (4.7.1) to (4.7.3), and two unknowns, x and y.

$$x + y = 7 \tag{4.7.1}$$

$$x - y = 3 \tag{4.7.2}$$

$$x + 2y = 12 \tag{4.7.3}$$

From equations (4.7.1) and (4.7.2) one obtains a unique solution for x (x=5) and y (y=2). However, this solution does not satisfy equation (4.7.3). If, for example, equations (4.7.1) and (4.7.3) are used to solve for x and y, the solution will be different (x=2, y=5), and it will not satisfy equation (4.7.2). Solutions exist if the right hand sides obey special conditions, e.g., if 12 is replaced by 9 in equation (4.7.3).

If a system contains more unknowns than equations, it is said to be underdetermined, and in general, there are an infinite number of possible solutions. For example, the system shown below contains one equation, (4.7.4), and two unknowns, x and y.

$$x \cdot y = 12 \tag{4.7.4}$$

Possible solutions for this equation include:

(x=1, y=12), or

(x=2, y=6), or

(x=3, y=4), or

(x=120, y=0.1), etc.

In biomechanics, there is a specific underdetermined problem, the so-called distribution problem (Crowninshield & Brand, 1981a), which has received overwhelming attention in the past two decades. The distribution problem is used to solve for internal forces acting on the musculo-skeletal system by using the known resultant joint forces and moments. In the calculation of internal forces, the number of unknowns typically exceeds the number of available equations describing the mechanics of the system. To find a unique solution for the distribution problem in biomechanics, the underdetermined system may be made determinate either by increasing the number of system equations, e.g., Pierrynowski and Morrison (1985), or by decreasing the number of unknowns, e.g., Paul (1965), until the number of equations and unknowns is the same. However, when increasing the number of system equations, some non-trivial assumptions must be made, and when decreasing the number of unknowns, some relevant information about the specific system behaviour may get lost. For these and some well-supported physiological reasons, researchers have attempted to solve the distribution problem in biomechanics by using mathematical optimization approaches. In this chapter, we focus on solving the distribution problem in biomechanics using optimization theory.

4.7.2 BASIC CONCEPTS

When attempting to determine internal (ligamentous, muscular, bony contact) forces in and around a biological joint from the known resultant joint forces and moments, certain modelling assumptions must be made. These assumptions include how a joint is defined and how forces are transmitted across the joint by internal structures. A brief discussion of these assumptions follows.

A biological joint is typically defined as a point that may be associated with an anatomical landmark, e.g., the lateral malleolus for the ankle joint, or that may be defined mathematically and may move relative to the bones that make up the joint, e.g., the instantaneous centre of zero velocity concept. In mechanics, we typically think of a joint as a point that is contained in both segments that make up the joint. When performing an analysis of internal forces, a fictitious surface, which is not necessarily planar, is passed through the anatomical joint space. This surface severs all tissues that transverse the joint. It is typically assumed that bony contact regions, muscles, and ligaments are the only structures that transmit non-negligible forces across a joint. For each structure that transmits force across a joint, the point of application of that force, Q, is chosen so that the moment produced by that structure about point Q is zero (Figure 4.7.1). Of course, each structure produces a moment about points other than point Q; in particular, about the joint centre, O.

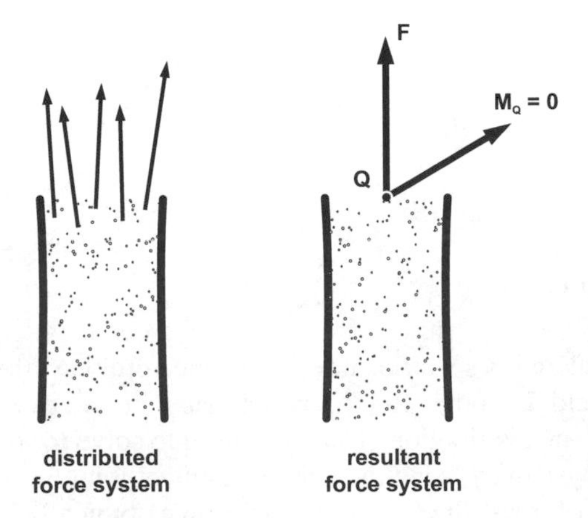

Figure 4.7.1 **Equipollent replacement of a distributed force system, e.g., in a ligament attaching to a bone, by a resultant force and moment. In biomechanics, we tend to associate the point of application of the resultant force with a point, Q, where the resultant moment of the distributed force system is zero.**

4.7.3 JOINT EQUIPOLLENCE EQUATIONS

The joint equipollence equations relate the muscular, ligamentous, and bony contact forces to the resultant joint force and moment. Using the assumptions made above, the force and moment equipollence equations are as follows (Figure 4.7.2):

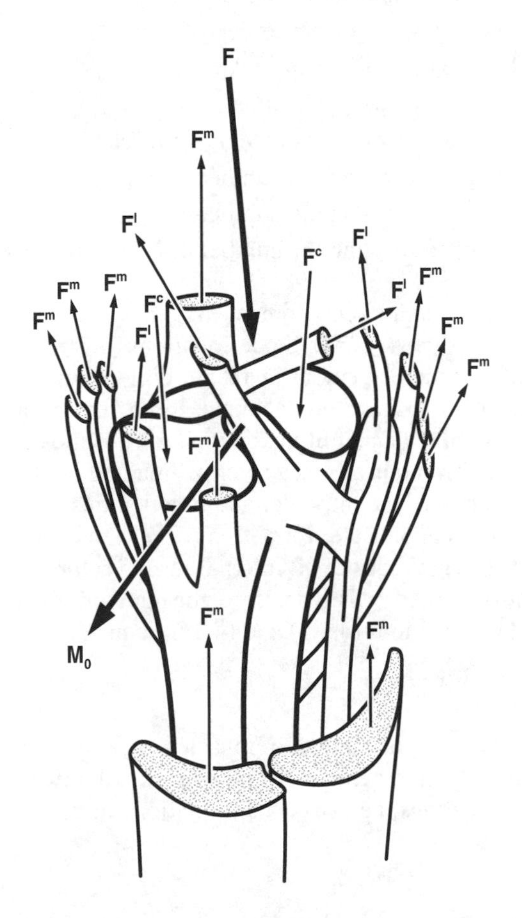

Figure 4.7.2 **Force distribution in a joint and its equipollent replacement by a resultant external force (F) and a resultant external joint moment (M_0) (from Crowninshield & Brand, 1981a, with permission).**

$$\mathbf{F} = \sum_{i=1}^{N} (\mathbf{F}_i^m) + \sum_{j=1}^{P} (\mathbf{F}_j^l) + \sum_{k=1}^{Q} (\mathbf{F}_k^c) \tag{4.7.5}$$

$$\mathbf{M}_o = \sum_{i=1}^{N} (\mathbf{r}_{i/0} \times \mathbf{F}_i^m) + \sum_{j=1}^{P} (\mathbf{r}_{j/0} \times \mathbf{F}_j^l) + \sum_{k=1}^{Q} (\mathbf{r}_{k/0} \times \mathbf{F}_k^c) \tag{4.7.6}$$

where:

$\mathbf{F}$	=	variable resultant external joint force
$\mathbf{M}_o$	=	variable resultant external joint moment
$\mathbf{F}^m$	=	internal muscular forces
$\mathbf{F}^l$	=	internal ligamentous forces
$\mathbf{F}^c$	=	internal bony contact forces
$\mathbf{r}_{i/0}$	=	location vector for muscular force i
$\mathbf{r}_{j/0}$	=	location vector for ligamentous force j
$\mathbf{r}_{k/0}$	=	location vector for bony contact force k
N	=	integer indicating the number of muscular forces
P	=	integer indicating the number of ligamentous forces
Q	=	integer indicating the number of bony contact forces

The resultant joint force, **F**, and joint moment, $\mathbf{M}_o$, may be obtained using the inverse dynamics approach, e.g., Andrews (1974). Since equations (4.7.5) and (4.7.6) are two vector equations, they yield six scalar equations in a three-dimensional system. The unknowns include all muscular, ligamentous, and bony contact force vectors, and the corresponding location vectors. Therefore, the number of unknowns exceeds the number of system equations. Using anatomical information from cadaver or imaging studies, the unknown location vectors, and the direction of the muscular and ligamentous force vectors, may be determined or estimated. This anatomical information reduces the number of unknowns substantially, leaving just the magnitudes of all internal force vectors and the direction of the bony contact force vectors as unknowns. Therefore, the number of scalar unknowns (SU) in the system, represented by equations (4.7.5) and (4.7.6), equals:

$$SU = N + P + 3Q \tag{4.7.7}$$

Where N, P, and Q represent the number of muscles, ligaments, and bony contact areas of the joint under consideration. In general, the number of unknowns will exceed the number of available system equations, i.e., six scalar equations in the three-dimensional case, or:

$$N + P + 3Q > 6 \tag{4.7.8}$$

Therefore, the problem is mathematically underdetermined.

4.7.4 SOLVING MATHEMATICALLY UNDERDETERMINED SYSTEMS USING OPTIMIZATION THEORY

Any mathematically underdetermined system may be made determinate by decreasing the number of unknowns or increasing the number of system equations, or both, until the number of unknowns and system equations match. These approaches have been used to solve the underdetermined distribution problem in biomechanics (Paul, 1965; Morrison, 1968; Pierrynowski & Morrison, 1985). However, the approach used most often to solve the distribution problem is mathematical optimization. Optimization procedures are an elegant way of solving this type of mathematical problem, and are also believed to be good indicators of the physiology underlying force-sharing among internal structures. This belief goes as far back as Weber and Weber (1836), who stated that locomotion is performed in such a way as to optimize, i.e., minimize, metabolic cost.

Optimization problems, in general, are defined by three quantities: the cost function, the design variables, and the constraint functions. The cost function is the function to be optimized. For the distribution problem in biomechanics, cost functions have been defined as:

Minimize ϕ where:

$$\phi = \sum_{i=1}^{N} F_i^m \qquad \text{(e.g., Seireg \& Arkivar, 1973)} \tag{4.7.9}$$

or:

$$\phi = \sum_{i=1}^{N} (F_i^m / pcsa_i)^3 \qquad \text{(e.g., Crowninshield \& Brand, 1981b)} \tag{4.7.10}$$

or:

$$\phi = \sum_{i=1}^{N} (F_i^m / M_{maxi})^3 \qquad \text{(e.g., Herzog, 1987)} \tag{4.7.11}$$

where:

F_i^m = force magnitude of the ith-muscle

$pcsa_i$ = physiological cross-sectional area of the ith-muscle

M_{maxi} = variable maximal moment that the ith-muscle can produce as a function of its instantaneous contractile conditions

N = total number of muscles considered

Design variables are the variables that are systematically changed until the cost function is optimized and all constraint functions are satisfied. The design variables must be contained in the cost function, and for the distribution problem, they typically are the magnitudes of the individual (muscle) forces.

The constraint functions restrict the solution of the optimization approach to certain boundary conditions. For example, in the distribution problem, typical inequality constraints are:

$$F_i^m \geq 0, \quad \text{for } i = 1, \ldots, N \tag{4.7.12}$$

and typical equality constraints are:

$$\mathbf{M}_o = \sum_{i=1}^{N} (\mathbf{r}_{i/o} \times \mathbf{F}_i^m) \tag{4.7.13}$$

which indicate that muscular forces must always be zero or positive (tensile), and resultant joint moments, $\mathbf{M}_o$, are assumed to be satisfied by the vector sum of all moments produced by the muscular forces. An optimization problem without constraint functions is called an unconstrained problem.

UNCONSTRAINED PROBLEM:ONE DESIGN VARIABLE

When trying to find optimal solutions to a problem, the difficulty of solving the problem analytically is directly related to the number of design variables and constraints. An unconstrained problem with one design variable can typically be solved in a straightforward way.

Let us assume that we are trying to find local minima of the following cost function:

$$f(x) = x^2 + 2x + 5 \tag{4.7.14}$$

The design variable in this case is x. The necessary condition for local minima of equation (4.7.14) is given by equation (4.7.15), and the sufficient condition for local minima of equation (4.7.14) is given by (4.7.16):

$$\frac{df(x)}{dx} = 0 \tag{4.7.15}$$

$$\frac{d^2f(x)}{dx^2} > 0 \tag{4.7.16}$$

Solving our problem for the necessary condition gives:

$$\frac{df(x)}{dx} = 0 = 2x + 2 \tag{4.7.17}$$

or:

$$x = -1$$

The point $x = -1$ is called a stationary point. If $d^2f(x)/dx^2$ is positive for $x = -1$, then the stationary point represents a minimum; if it is negative, the stationary point is a maximum

$$\frac{d^2f(x)}{dx^2} = 2 \tag{4.7.18}$$

(for any value of x).

Therefore, the solution $x = -1$ is a minimum. Since the solution, $x = -1$, is the only stationary point, we also know that it is a global minimum (Figure 4.7.3).

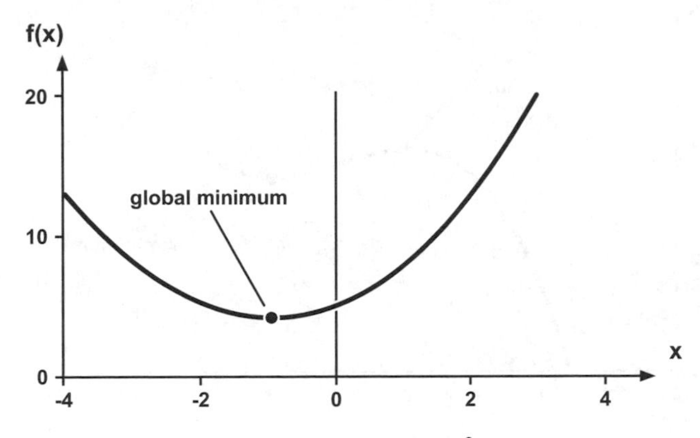

Figure 4.7.3 **Illustration of the equation** $f(x) = x^2 + 2x + 5$.

As a second example, let us assume that we would like to find local minima of the function:

$$f(x) = x^3 - 7.5x^2 + 18x - 10 \tag{4.7.19}$$

Taking first and second derivatives with respect to the design variable (x), and testing the necessary and sufficient conditions (equations (4.7.15) and (4.7.16)) gives:

$$\frac{df(x)}{dx} = 0 = 3x^2 - 15x + 18 \tag{4.7.20}$$

$$\frac{d^2f(x)}{dx^2} = 6x - 15 \tag{4.7.21}$$

Stationary points may be obtained from equation (4.7.20). They are $x_1 = 2$ and $x_2 = 3$. Replacing the two solutions in equation (4.7.21), we realize that solution one, x_1, yields -3, and solution two, x_2, yields +3. Therefore, solution one represents a a local maximum ($x_1 = 2$), and solution two represents a local minimum ($x_2 = 3$) (Figure 4.7.4).

UNCONSTRAINED PROBLEM:MORE THAN ONE DESIGN VARIABLE

When trying to find a solution to an unconstrained, multi-design variable problem, a necessary condition for a minimum solution is that the first partial derivatives of the cost function, ($f(x)$), with respect to each design variable, x_i, are equal to zero:

$$\frac{\partial f(x)}{\partial x_i} = 0 \quad \text{for } i = 1,\ldots, N \tag{4.7.22}$$

where N is the total number of design variables. A sufficient condition for x to give a local minimum of this type of problem is that the Hessian matrix, $H(x)$, given by:

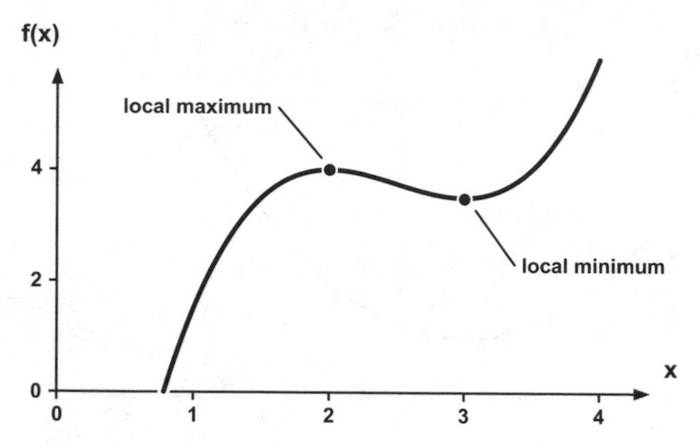

Figure 4.7.4 **Illustration of the equation** $f(x) = x^3 - 7.5x^2 + 18x - 10$.

$$H(x) = \begin{bmatrix} \frac{\partial^2 f(x)}{\partial x_1^2} & \cdots & \frac{\partial^2 f(x)}{\partial x_1 \partial x_n} \\ \cdots & \cdots & \cdots \\ \frac{\partial^2 f(x)}{\partial x_n \partial x_1} & \cdots & \frac{\partial^2 f(x)}{\partial x_n^2} \end{bmatrix} \tag{4.7.23}$$

is positive definite (if -H(x) is positive definite, then x gives a local maximum). Various ways are available to test this definiteness, e.g., all eigenvalues of H must be positive. We illustrate an easily computed test in the following.

EXAMPLE 1

Question

Find a minimum solution of the function $f(x_1, x_2)$, where:

$$f(x_1, x_2) = 5x_1 - \frac{x_1^2 \cdot x_2}{16} + \frac{x_2^2}{4x_1} \tag{4.7.24}$$

The necessary conditions for stationary points, i.e., equation (4.7.22), for this problem become:

$$\frac{\partial f(x_1,x_2)}{\partial x_1} = 0 = 5 - \frac{2x_1x_2}{16} - \frac{x_2^2}{4x_1^2} \tag{4.7.25}$$

$$\frac{\partial f(x_1,x_2)}{\partial x_2} = 0 = -\frac{x_1^2}{16} + \frac{2x_2}{4x_1} \tag{4.7.26}$$

Solving equations (4.7.25) and (4.7.26) gives two stationary points, P_1 and P_2, where:

$P_1(x_1 = 4, x_2 = 8)$

and:

$$P_2(x_1 = -4, x_2 = -8)$$

It suffices, for a minimum solution, that the Hessian matrix be positive definite. The second partial derivatives of equation (4.7.24) are:

$$\frac{\partial^2 f(x_1,x_2)}{\partial x_1^2} = -\frac{x_2}{8} + \frac{x_2^2}{2x_1^3} \tag{4.7.27}$$

$$\frac{\partial^2 f(x_1,x_2)}{\partial x_2^2} = \frac{1}{2x_1} \tag{4.7.28}$$

$$\frac{\partial^2 f(x_1,x_2)}{\partial x_1 \partial x_2} = -\frac{x_1}{8} - \frac{x_2}{2x_1^2} \tag{4.7.29}$$

Therefore, the Hessian matrix for the first stationary point, P_1 $(x_1 = 4, x_2 = 8)$, becomes:

$$H(4,8) = \begin{bmatrix} -\frac{1}{2} & -\frac{3}{4} \\ -\frac{3}{4} & \frac{1}{8} \end{bmatrix} \tag{4.7.30}$$

We calculate the two principal determinants $d = \det\left[-\frac{1}{2}\right]$ and $D = \det H(4,8)$. Then, H(4,8) is positive definite if d and D are positive.

Since $d = \left[-\frac{1}{2}\right]$ and $D = \left[-\frac{5}{8}\right]$, H(4,8) is not positive definite. In fact, since d and D are negative, the stationary point P_1 does not represent a local minimum (P_1 is actually a saddle point.) The above procedure may be repeated with the second stationary point, P_2 $(x_1 = -4, x_2 = -8)$. Again, P_2 does not represent a local minimum, but another saddle point.

There is an important class of problems for which equation (4.7.22) is sufficient on its own for x to give a minimum, which is even global. This is the class of convex problems, i.e., problems for which f is a convex function. Several tests for convexity exist. For example, it suffices if H(y) is non-negative definite for all $y = (y_1, y_2)$.

GENERAL CONSTRAINED PROBLEM

The distribution problem in biomechanics is typically solved by minimizing the cost function of a general constrained problem. Specifically, suppose we want to minimize f(x) subject to the constraints h_i (x) = 0 (i = 1, 2,..., s) where f and h_i are differentiable and $x = (x_1, \ldots, x_N)$ is a vector of design variables which are assumed to be non-negative. These non-negativity constraints $x_j \geq 0 (j = 1, 2, \ldots, N)$ make the problem of the mathematical programming type, and in certain cases, it can be solved via the Karush-KuhnTucker (KKT) conditions.

One such case, which we shall consider, is where f is convex (see above) and each h_i is affine, i.e., $h_i(x) = a_i^T x + b_i$ for some (row) vector a_i and constant b_i. In this case, the KKT conditions are necessary and sufficient for a global minimum, and they admit the following interpretation.

First, select the design variables that are zero, and label them x_k (the other design variables are positive). Then solve the *equality* constrained problem of minimizing f(x) subject to:

$$\text{all} \quad h_i(x) = 0 \quad \text{and all} \quad x_k = 0 \tag{4.7.31}$$

This problem can be attacked by Lagrange multipliers λ_i and μ_k. Let:

$$L(x) = f(x) + \sum_{i=1}^{s} \lambda_i h_i(x) + \sum_{k} \mu_k x_k$$

and solve the mathematical system consisting of the stationary point condition:

$$0 = \frac{\partial L}{\partial x_j}(x) \quad (j = 1, 2, \ldots, N) \tag{4.7.32}$$

together with the equality constraints (4.7.31).

If N is large, the task of selecting which design variables should be zero can be formidable. For small N, however, the problem may be tackled as follows: first try the case where no x_k is zero (so all x_j are positive and the final summation is absent from $L(x)$). If the system (4.7.31 and 4.7.32) is soluble with all $x_j \geq 0$, then we have found the global minimum. If not, then we try the case where just one x_k is zero. N such possibilities exist, and if any of the resulting systems has a solution with all $x_j \geq 0$, then again we have found the minimum. If not, then we try those cases where exactly two of the x_k are zero, and so on.

In the following subsection, we illustrate this procedure for a case with N = 2.

EXAMPLE 2

The following example is based on the work by Crowninshield and Brand (1981b). These authors derived their particular optimization approach based on experimentally determined stress versus endurance-time relations of skeletal muscles. The intent of their approach was to solve the distribution problem by assigning forces to the muscles involved in a particular task, such that endurance time of the task, i.e., the time the task can be maintained, was maximized. It can be shown that this intent was not achieved by Crowninshield and Brand (1981b). However, a formal proof of this statement goes beyond the scope of this book.

In the approach proposed by Crowninshield and Brand (1981b) to solve the distribution problem, i.e., equations (4.7.5) and (4.7.6), the force equation (4.7.5) was not considered. However, the moment equation (4.7.6) was used, assuming that ligamentous and bony contact forces do not contribute significantly to the resultant external joint moment. The cost function (equation (4.7.10)) requires the minimization of the sum of the cubed muscular stresses, which was said to be equivalent to maximizing muscular endurance. The optimization approach of Crowninshield and Brand (1981b) may thus be formulated as:

Minimize the cost function ϕ, where:

$$\phi = \sum_{i=1}^{N} (F_i^m / pcsa_i)^3 \tag{4.7.33}$$

and:

F_i^m, for i = 1,..., N are the design variables

subject to the constraints:

$$F_i^m \geq 0, \qquad i = 1,\ldots, N \tag{4.7.34}$$

$$\mathbf{M}_o = \sum_{i=1}^{N} (\mathbf{r}_{i/0} \times \mathbf{F}_i^m) \tag{4.7.35}$$

where all symbols have been defined before.

Let us assume, for the sake of simplicity, that the system of interest is a one-joint, planar system in which the joint of interest is crossed by two agonistic muscles. Let us further assume that we are interested in calculating the force-sharing between these two muscles; i.e., we would like to express the force in muscle 1, F_1^m, as a function of the force in muscle 2, F_2^m. Therefore, we minimize:

$$\phi = \sum_{i=1}^{2} (F_i^m / pcsa_i)^3 \tag{4.7.36}$$

subject to:

$$M_o = \sum_{i=1}^{2} (r_{i/0} \cdot F_i^m) \tag{4.7.37}$$

or:

$$0 = \sum_{i=1}^{2} (r_{i/0} \cdot F_i^m) - M_o \tag{4.7.38}$$

and we admit only solutions for muscular forces that satisfy $F_i^m \geq 0$. Strictly speaking, ϕ is not convex for all values of F_1^m . Nevertheless, this problem can be converted to a convex one of the type considered in section 4.7.6 (Herzog & Binding, 1993), and, as a result, the necessary and sufficient conditions for a minimum solution are:

$$\frac{\partial L(F)}{\partial F_1^m} = 0 \tag{4.7.39}$$

where L depends on which F_i^m are selected to be zero. As in section 4.7.4, we first try the case where both F_i^m are positive. Then:

$$L = \left(\sum_{i=1}^{2} (F_i^m/pcsa_i)^3\right) + \lambda\left[\sum_{i=1}^{2} (r_{i/0} \cdot F_i^m) - M_o\right] \tag{4.7.40}$$

The design variables are F_i^m; therefore:

$$\frac{\partial L(F)}{\partial F_1^m} = [3(F_1^m)^2/pcsa_1^3] + \lambda r_{1/0} = 0 \tag{4.7.41}$$

$$\frac{\partial L(F)}{\partial F_2^m} = [3(F_2^m)^2/pcsa_2^3] + \lambda r_{2/0} = 0 \tag{4.7.42}$$

Solving equations (4.7.41) and (4.7.42) for λ, and setting them equal to one another, yields:

$$\frac{3(F_1^m)^2}{(pcsa_1^3 \cdot r_{1/0})} = \frac{3(F_2^m)^2}{(pcsa_2^3 \cdot r_{2/0})} \tag{4.7.43}$$

and simplifying equation (4.7.43), and solving it for F_1^m, results in:

$$F_1^m = \left(\frac{r_{1/0}}{r_{2/0}}\right)^{1/2} \cdot \left(\frac{pcsa_1}{pcsa_2}\right)^{3/2} \cdot F_2^m \tag{4.7.44}$$

Equation (4.7.44) expresses the force in muscle 1, F_1^m, as a function of the force in muscle 2, F_2^m. The relation of force-sharing between the two muscles is influenced by the ratio of their moment arms $(r_{1/0}/r_{2/0})^{1/2}$ and their physiological cross-sectional areas $(pcsa_1/pcsa_2)^{3/2}$. The ratio of the physiological cross-sectional areas is a constant, because each physiological cross-sectional area is constant for a given muscle. Moment arms of muscles typically change as a function of joint angle, but on many occasions, the moment arms of two agonistic muscles change in a similar way thus giving an approximate constant ratio. Therefore, in reality, the force-sharing equation (4.7.44) derived from the optimization approach of Crowninshield and Brand (1981b) often gives a linear relationship between muscle forces of two agonistic muscles.

Let us now solve the force-sharing between these two muscles using a numerical example, where

$$M_o = 10$$

$$pcsa_1 = pcsa_2 = 1$$

$$r_{1/0} = 1 \quad \text{and} \quad r_{2/0} = 2$$

All values are in arbitrary (but, we assume, consistent) units. Therefore, the force-sharing equation for this problem (4.7.44) becomes:

$$F_1^m = \left(\frac{1}{2}\right)^{1/2} \cdot \left(\frac{1}{1}\right)^{3/2} \cdot F_2^m \tag{4.7.45}$$

or:

$$F_1^m = 0.71 F_2^m \quad \text{and} \quad F_2^m = 1.41 F_1^m \tag{4.7.46}$$

and the constraint equation (4.7.37) becomes:

$$10 = 1 \cdot F_1^m + 2 \cdot F_2^m \tag{4.7.47}$$

Solving equations (4.7.46) and (4.7.47) for F_1^m and F_2^m gives $F_1 = 2.61$ and $F_2 = 3.69$ with an associated cost (equation (4.7.36)) of 68.0. Both F_i^m are indeed positive, so we have found the global minimum.

The graphical solution of this problem is illustrated in Figure 4.7.5. It has F_1^m on the horizontal axis and F_2^m on the vertical axis.

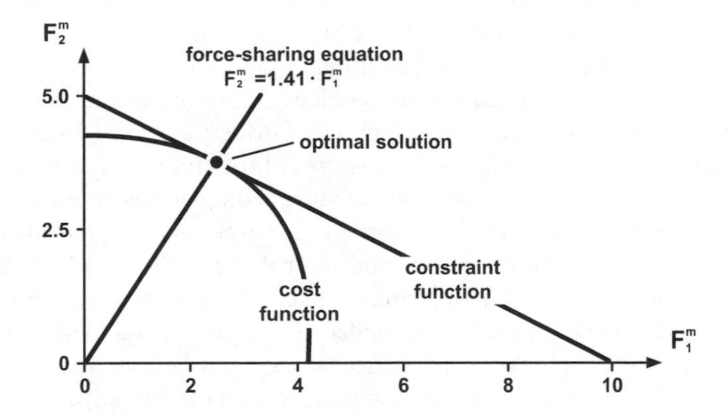

Figure 4.7.5 **Graphic illustration of the optimal solution of force sharing between two muscles in a one degree of freedom system (for details see text).**

4.8 SIMULATION

VAN DEN BOGERT, A.J.
NIGG, B.M.

4.8.1 INTRODUCTION

Research on human movement does not always use human subjects. In most areas of biomedical research, animal models are used when rigorous control of experimental conditions is required, measurements require invasive techniques, or when the use of human subjects is not ethically acceptable. In the context of biomechanics of human movement, animal models are typically used to study basic biological mechanisms. A good example is research on muscle mechanics (chapter 2.7), where full experimental control over muscle length and activation was required, and direct measurements of variables such as muscle force, muscle fibre length, and even heat production are done. This approach would not be possible in human subjects.

For more applied questions, related to specific human injuries or specific human movement tasks, animal models are not appropriate because the dynamics and anatomy are not comparable. For instance, the study of knee injuries in distance runners has a basic biology component and a mechanical component. Basic biology, or in this case the reaction of cartilage to mechanical loads, can be studied in animal models. It is, however, obviously impossible to use animal models to learn how to reduce forces in the knee joint with appropriate orthotics or other mechanical interventions. Furthermore, in human subjects, the force in the joint cannot be measured, other factors contributing to injury can not be controlled, and ethical problems arise in the study of more acute, serious injuries. Analogous to the use of animal models, it is, therefore, desirable to study such movement-related questions with methods that allow full experimental control and access to all mechanical variables for measurement. In the case of human movement, one possible option currently available is to build a model of the system under investigation, and then perform experiments on the model. Although physical and mechanical models have been used, *computer models* (also *numerical models*) are almost exclusively used for this purpose. Simulation is, therefore, defined as follows:

> ***Simulation* is the process of performing experiments on a numerical model.**

The term *experiment* implies that it is crucially important that the model reflects the causality of the system that is investigated. The input of the model should precede the output in a causative sense (see section 4.8.9).

Once such a numerical model is developed, experiments can be performed quickly and at low cost. It should, however, be kept in mind that the experiment is always performed on the model only, and conclusions may not apply to the human system when the model is not valid. This is a central problem in modeling and simulation. It is important that a model replicates those features of the system that are essential for the question that is studied. This requires careful design of the model and appropriate validation tests once the model is finished.

In biomechanics, simulation is applied in four different areas: dynamics of movement, tissue mechanics, fluid flow, and measuring techniques. In dynamics of movement, simulations allow the experimenter to apply controlled perturbations to the anatomy, muscle coordination, and control system, and observe the resulting changes in movement and forces. Tissue mechanics can be studied using finite element models (FEM), which are the numerical equivalent of partial differential equations (PDE), which describe the local relationships between stress and strain in tissue (chapter 3.7). These methods have been applied to bone, cartilage, tendon, ligament, muscle, and even brain tissue during impact. Fluid flow models, used in cardiovascular mechanics, use finite element techniques to solve hydrodynamics equations, with or without including interactions with vascular tissue mechanics. Biomechanical measuring techniques (chapter 3) often suffer from errors, and simulation may be used to determine how errors in the raw data are propagated to the final result of the calculation. The term sensitivity analysis is sometimes used, but when correctly done with realistically simulated statistical error distributions, the technique is known as *Monte-Carlo simulation.*

This chapter focuses on the simulation of movement.

4.8.2 DIFFERENTIAL EQUATIONS

A mathematical model for the simulation of movement always takes the form of a second-order system of ordinary differential equations (ODE). This term indicates that the equation contains not only the unknown movement variables, e.g., joint angles, but also their time-derivatives up to the second order, i.e., accelerations. Such equations have been derived for simple systems in sections 4.6.2 and 4.6.3. A one-dimensional mass-spring damper system (chapter 4.6, Fig. 4.6.9) was described by an equation, which is a single second-order linear ODE, with the unknown $x(t)$. A two-segment planar arm (chapter 4.6, Figure 4.6.33) was described by two coupled second-order non-linear ODEs [chapter 4.6, equations (4.6.43) and (4.6.44)]. These equations are non-linear because they involve non-linear functions (sin and cos) of the unknowns $\varphi_1(t)$ and $\varphi_2(t)$. The only reason for making a distinction between linear and non-linear ODEs is that linear equations can be solved *analytically*, which means that general solutions can be found in the form of equations giving movement explicitly as a mathematical function of time. The solutions for second-order linear ODEs take the form of combinations of exponential or harmonic (sine and cosine) functions, or both. However, even moderately complex models of the human musculo-skeletal system lead to essentially non-linear equations.

For non-linear equations, analytical solutions can only be found in special cases. But even without solving the equation completely, analytical methods can be used to study certain aspects of the behaviour of the system. As an example, consider a cyclist going downhill without pedaling. The cyclist will be considered as a particle and rolling friction will be ignored. A free body diagram (Figure 4.8.1) reveals that there are three forces acting on the cyclist: gravity, ground reaction force, and air resistance. Considering only the component of force and movement parallel to the inclined plane, the equation of motion can be formulated as:

$$m\ddot{x} = mg\sin\alpha - F_{air} \tag{4.8.1}$$

From wind tunnel tests, it is known that the air drag F_{air} is proportional to the square of the velocity:

$$F_{air} = kv^2 = k \cdot \dot{x}^2 \tag{4.8.2}$$

A typical value for k is 0.15 Ns m^{-1} (de Groot, 1995). Combining the two equations, and substituting the derivative of velocity for acceleration, we obtain:

$$\dot{v} = g\sin\alpha - \frac{k}{m}v^2 \tag{4.8.3}$$

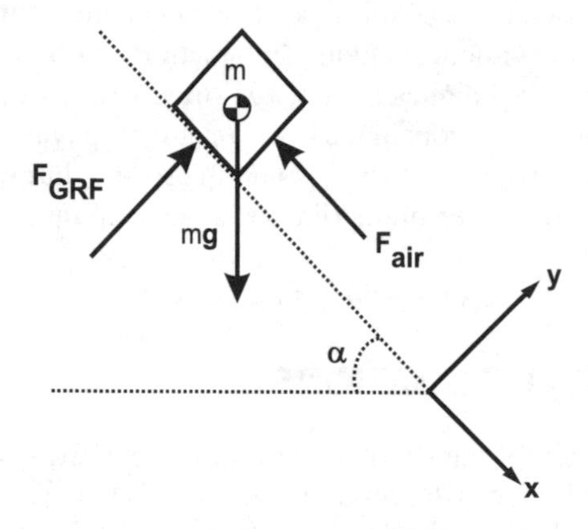

Figure 4.8.1 **Free body diagram of a cyclist coasting downhill.**

Note that this is a first-order ODE, where the velocity v is the unknown. Although we will not attempt an analytical solution, it is easy to determine a *steady-state* solution, if there is one. In a steady state, nothing changes and all derivatives are zero. Setting $\dot{v} = 0$ in equation (4.8.3) and solving for v gives the steady-state velocity:

$$v_{SS} = \sqrt{\frac{mg\sin\alpha}{k}} \tag{4.8.4}$$

For a typical cyclist, with $m = 75$ kg, on a slope of 10°, the steady-state velocity will, therefore, be just over 29 m s^{-1}.

Stability is another aspect of system behaviour that can be studied analytically. Stability can be studied by looking at small perturbations from equilibrium and determining if forces respond to push the system back to, or away from, equilibrium. A major advantage of analytical methods is they allow a general understanding of how the behaviour of the system depends on system parameters, such as mass or stiffness. However, during movement, we are usually interested in simulating the entire time course of the movement. Only in rare cases can this be done by analytical methods. Usually, a numerical solution of the ODE is required. This is where mathematical modelling becomes computer simulation.

4.8.3 NUMERICAL SOLUTION METHODS

Every first order ODE can be formulated as:

$$\dot{x}(t) = f(x, t) \tag{4.8.5}$$

The function f may be a simple mathematical function, as in equation (4.8.3) or a function that requires a substantial computation, as we see later for biomechanical systems. The variable x is the *state variable* of the model.

A large class of numerical solution methods for ordinary differential equations is based on Taylor's theorem, which relates derivatives of a function to a change in function value over a finite time interval. If $x(t)$ is a sufficiently smooth function, Taylor's theorem states:

$$x(t+h) = x(t) + h\dot{x}(t) + \frac{h^2}{2}x^{(2)}(t) + \ldots + \frac{h^n}{n!}x^{(n)}(t) + O(h^{n+1}) \tag{4.8.6}$$

where $\dot{x}$ is the first derivative of x, $x^{(n)}$ is the nth derivative, and $O(h^{n+1})$ is an error term proportional to h^{n+1}. A nth order ODE solver is obtained by truncating the right-hand side (the Taylor series) after the nth term, and choosing h small enough that the error term is small enough to be neglected. The simplest method is the forward Euler method, which is a first order method:

$$\begin{aligned} x(t+h) &= x(t) + h\dot{x}(t) \\ &= x(t) + h\, f(x(t), t) \end{aligned} \tag{4.8.7}$$

If the state, e.g., position or velocity, $x(t)$ of the system at time t is known, the right-hand side f of the differential equation (4.8.5) can be calculated, multiplied by a time step h, and added to the previous state $x(t)$ to obtain the new state $x(t+h)$ at time $t+h$. This is repeated again and again until t reaches the end of the time interval that we are interested in. This solution method can also be applied to a system of simultaneous differential equations, as shown later. Note that this solution process requires that the initial state x at $t=0$ is known. Results of a simulation always depend on this *initial condition.* The steady-state solution, however, does not depend on the initial condition, because damping causes the system to gradually "forget" its initial state.

The Euler method is the simplest method, but has a larger error for the same step size than higher order methods, where the error is proportional to a higher power of the step size h (remember that h is very small). This means that to obtain the same accuracy, the Euler method must use a smaller step size than some other methods, and the function f needs to be calculated more frequently. However, the forward Euler method is adequate for work where computer speed is not a limiting factor and will be used in the examples in this chapter. Details of other methods may be found in textbooks on numerical methods, e.g., Shampine and Gordon (1975) and Press et al. (1992). Matlab[1] provides a convenient programming environment for mathematical modelling and a student version is available at low cost. Matlab has very good built-in ODE solvers, which can be found by typing "help funfun". However, for the examples in this chapter we use the forward Euler method to facilitate translation into other programming languages.

As an example, the Matlab program in Listing 1 applies the forward Euler method to the differential equation (4.8.3) describing the downhill coasting cyclist. The output is a graph of velocity versus time (Figure 4.8.2). As an example of an experiment, a second simulation is done to simulate the effect of a 1000 m altitude where air density is approximately 10% lower.

---

1. MathWorks Inc., Natick MA http://www.mathworks.com.

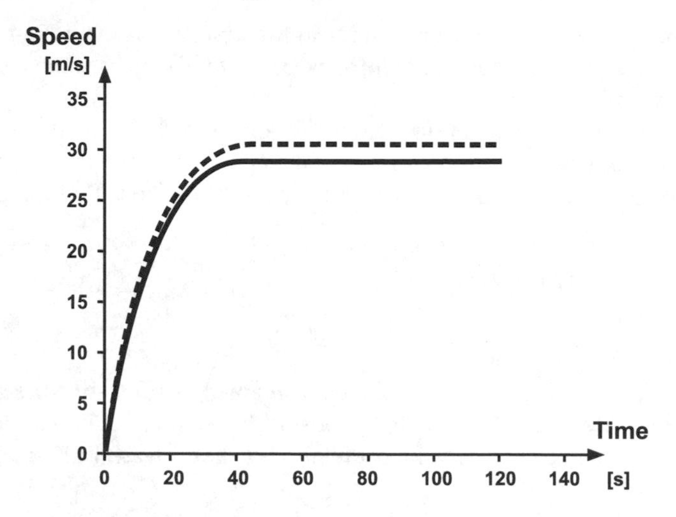

Figure 4.8.2 **Simulation of a cyclist coasting downhill. Solid curve:** k **=0.15 Ns/m, dashed curve:** k **=0.135 Ns/m.**

The choice of step size has an effect on accuracy, as can be seen from Taylor's theorem. This is demonstrated in Figure 4.8.3, showing the effect of step size on the results. Accuracy obviously suffers with increasing step size. The solution at h =20 s no longer converges to the steady state, which indicates that the numerical method has become unstable. This occurs for the forward Euler method when h becomes larger than twice the smallest time-constant of the system. A time-constant can be roughly defined, for a single ODE, as the ra-

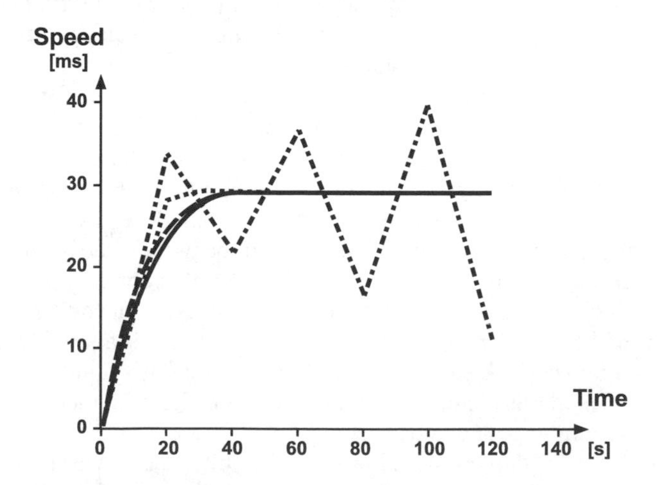

Figure 4.8.3 **Effect of step size on the simulation program in Listing 1. Solid line:** h **=0.01 s, dashed line:** h **=2 s, dotted line:** h **=10 s, dash-dotted line:** h **=20 s.**

tio between the simulated variable x and its derivative. When the solution *explodes*, as in the example with h =20 s, this is a clear indication that h should be reduced. An even lower value of h is necessary to achieve a good accuracy. Commercial software and some of the computer implementations in Shampine and Gordon (1975) automatically adjust the step size to achieve a certain, user-specified accuracy. When using a method with fixed step size, as in Listing 1, one can test the accuracy of the solution simply be running the simulation again with half the step size. If the difference between the two sets of results is smaller than the desired accuracy, the step size was sufficiently small. Otherwise, reduce h again by a factor two and repeat the test. For this model, the step size of 0.01 s, which was chosen initially, is clearly small enough to guarantee sufficient accuracy.

Complex systems typically require a set of n simultaneous ODEs:

$$\begin{aligned} \dot{x}_1 &= f_1(x_1, x_2, \ldots, x_n, t) \\ \dot{x}_2 &= f_2(x_1, x_2, \ldots, x_n, t) \\ &\ldots \\ \dot{x}_n &= f_n(x_1, x_2, \ldots, x_n, t) \end{aligned} \quad (4.8.8)$$

Once the state, which now consists of all *state variables* $x_1 \ldots x_n$, is known, all right-hand sides can be computed, and a forward time step can be made by applying a numerical solution method sequentially to each variable.

Equations of motion for mechanical systems generally lead to a set of n *second-order* ODEs, but can be transformed into a set of $2n$ first-order ODEs so that standard numerical methods can be applied. If the original set of equations is:

$$\begin{aligned} \ddot{x}_1 &= f_1(x_1, x_2, \ldots, x_n, \dot{x}_1, \dot{x}_2, \ldots, \dot{x}_n, t) \\ \ddot{x}_2 &= f_2(x_1, x_2, \ldots, x_n, \dot{x}_1, \dot{x}_2, \ldots, \dot{x}_n, t) \\ &\ldots \\ \ddot{x}_3 &= f_3(x_1, x_2, \ldots, x_n, \dot{x}_1, \dot{x}_2, \ldots, \dot{x}_n, t) \end{aligned} \quad (4.8.9)$$

we can transform this to a new set of variables $y_1 \ldots y_{2n}$, with:

$$\begin{aligned} y_k &= x_k \qquad (k = 1 \ldots n) \\ y_k &= \dot{x}_{k-n} \qquad (k = n + 1 \ldots 2n) \end{aligned} \quad (4.8.10)$$

which transforms equation (4.8.9) to:

$$\begin{aligned} \dot{y}_1 &= y_{n+1} \\ \dot{y}_n &= y_{2n} \\ \dot{y}_{n+1} &= f_1(y_1, y_2, \ldots, y_n, t) \\ \dot{y}_{2n} &= f_n(y_1, y_2, \ldots, y_n, t) \end{aligned} \quad (4.8.11)$$

This is an equation of the form (4.8.8) with 2n state variables that can be solved using the existing numerical methods.

In systems of simultaneous ODEs, there are as many time constants as state variables, and it is the smallest time constant that determines whether the numerical method is stable. The largest time constant, however, indicates what the duration of the simulation should be to examine the complete system dynamics. When the ratio between the smallest and largest time constant is very small, we consider the ODE to be *stiff*. An example of a stiff system is a simulation of a walking horse, with a largest time constant representing the stride cycle of about 1 s, while the smallest time constant is about 0.1 ms, which is due to the interaction between hoof and ground (van den Bogert et al., 1989). Using a simple numerical method, such as described above, time steps would have to be extremely small compared to the duration of the simulation, which leads to long computation times. Unfortunately, this problem becomes worse with higher order methods. A special class of methods has, therefore, been developed for stiff problems (Gear, 1971). The simplest of these is the *backward* Euler method. To make one time step in solving equation (4.8.5), $x(t+h)$ is solved from:

$$x(t+h) = x(t) + h\ f(x(t+h), t) \tag{4.8.12}$$

Since the function f is usually a complex non-linear function, this requires iterative solution, making one time step costly in terms of computation time. However, the method is stable for step sizes much larger than the smallest time constant thus resulting in better overall performance for stiff systems.

4.8.4 EQUATIONS OF MOTION FOR MECHANICAL SYSTEMS

In section 4.6.3, a straightforward method for formulation of equations of motion for multi-segment systems was introduced. This method is based on the Newton-Euler equations for each body segment. In two dimensions, these equations are:

$$\ddot{\mathbf{r}} = \frac{1}{m}\sum_k \mathbf{F}_k \tag{4.8.13a}$$

$$\ddot{\varphi} = \frac{1}{I}\sum_k M_k \tag{4.8.13b}$$

where $\mathbf{r}$ is the position vector of the center of mass of the segment in the global reference frame, φ is the orientation of the body, m is the mass, and I is the moment of inertia of the segment about the centre of mass.

The forces $\{\mathbf{F}_k\}$ and moments $\{M_k\}$ acting on the body, are identified using a free body diagram. Note that these equations are the same equations that can be used to solve unknown forces and moments from measured movements. That type of analysis, *inverse dynamic analysis*, is not considered to be a simulation, since the input (movement and ground reaction force data) is not causative to the output (internal forces and moments). Proper simulation is only possible using *forward dynamic analysis*, where the inputs are forces that cause the movement.

From equations (4.8.13) we see that a system with n body segments has $3n$ degrees of freedom, since each body segment has a two-dimensional position vector $\mathbf{r} = (x, y)$, and one orientation angle φ. In three dimensions, there would be 6n degrees of freedom, and equation (4.8.13b) would be modified [see section 1.3, equation (1.3.6)].The next step is to take into account the kinematic connections (joints) between the segments, which reduce the number of degrees of freedom. Identify a suitable set of $m \leq n$ variables (with m = degrees of freedom, also referred to as *generalized coordinates* of the system), and eliminate all other kinematic variables from the equations of motion. In example 11 of chapter 4.6, this procedure was followed to derive equations of motion for a two-segment model of the arm. Although straightforward, the method is tedious and leads to large algebraic equations even for systems of moderate complexity. Once derived, the equations of motion can be simulated using a standard numerical method, as shown in Listing 2 for the two-segment system. Note that this system has two degrees of freedom, and that the equations of motion have been formulated so that the two segment orientations can be solved as a function of time. With two degrees of freedom, we obtain four first-order ODEs. Hence, the forward Euler step appears four times in the program, once for each angle and once for each angular velocity; the state variables of the system. Results of this simulation are shown in Figure 4.8.4. The two segments had initial angular velocities of zero and 10 rad/s. The simulation

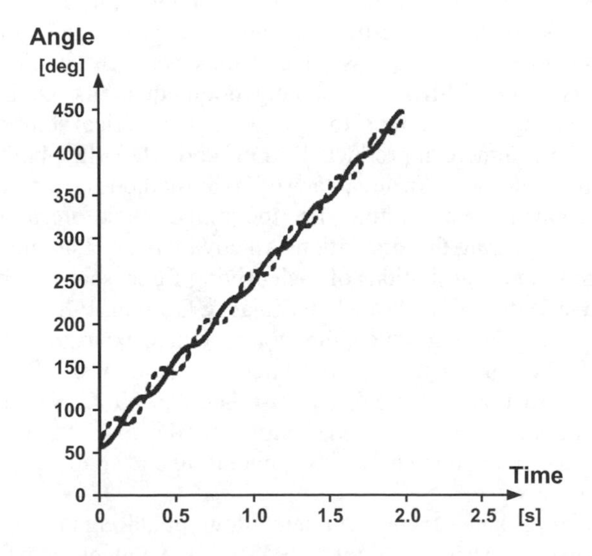

Figure 4.8.4 **Arm movement simulated using program in Listing 2. Solid line: segment 1, dashed line: segment 2.**

shows that the angular momentum is transferred back and forth between the two segments, in a 300 ms cycle. Note that segment 2 moves briefly in the opposite direction during each cycle. The whole system also rotates around the shoulder joint at approximately one revolution (360°) every two seconds.

Since the Newton-Euler method is cumbersome and difficult to perform systematically, alternative methods are often used. The oldest of these is Lagrange's equations. After identifying a set of generalized coordinates $\mathbf{q} = (q_1 \ldots q_n)$, one uses the kinematic connections to formulate the potential energy V of the system (which includes gravity and elastic forces) as a function of $\mathbf{q}$, and the kinetic energy T as a function of $\mathbf{q}$ and $\dot{\mathbf{q}}$. The Lagrangian L is defined as the difference T – V and Lagrange's equations are:

$$\frac{d}{dt}\left(\frac{\partial L}{\partial \dot{q}_k}\right) - \frac{\partial L}{\partial q_k} = Q_k \tag{4.8.14}$$

from which the generalized accelerations $\ddot{q}_k$ are easily solved to obtain a system of simultaneous second order ODEs, which are identical to what would have been obtained using the Newton-Euler formulation. The generalized forces Q_k are obtained from the original forces and moments by applying the principle of virtual work (Arnold, 1978). This procedure requires considerable algebraic manipulation, but the procedure is more systematic than derivations from the Newton-Euler equations. The algrabraic manipulations can even be automated using symbolic algebra software such as Mathematica or Maple. Even more suited for automated derivation of equations of motion are Kane's equations (Kane & Levinson, 1985). This method is used in the commercial software SD/FAST[1] and AUTOLEV[2]. SD/FAST is also used as the *dynamics engine* for SIMM[3] and Pro/Mechanica[4]. Both of these packages provide a comprehensive user-interface to help create models, numerical solutions, and visualizing results of simulations.

Another class of methods is based on the use of Newton-Euler equations with additional constraint equations. Since the constraint equations, e.g., equations (4.4.49) and (4.4.50), are algebraic equations which express relationships between positions of two segments, this leads to a mixed set of differential and algebraic equations (DAE; Haug, 1989). This method for formulating the equations, together with a numerical solution method, has been implemented in the commercial products DADS[5] and ADAMS[6], both of which include a comprehensive user interface. An advantage of DAE methods is that the reaction forces associated with the kinematic constraints, i.e., the joint contact forces in a musculo-skeletal model, are computed during the simulation. An advantage of Lagrange's equations is that it is easy to quantify the contributions of each driving force, such as a muscle, to the accelerations induced in each degree of freedom (Zajac & Gordon, 1989). For instance, it can be seen that the hamstrings during stance, function as a knee extensor because the large upper body mass effectively couples knee and hip extension (Andrews, 1985). It seems, therefore, that DAE methods are more suitable for injury-related questions, while methods using generalized coordinates more readily provide insight into the control aspects of movement. All commercial packages mentioned here allow modelling and simulation in three dimensions and in two dimensions.

All commercial packages mentioned here allow modelling and simulation in three dimensions and in two dimensions. Currently SIMM is the only product that includes the possibility of including muscles in the model, all others are mainly intended for robotics and machine design. However, muscles and other force-generating elements can be added to any mechanical model, as shown in the next sections.

---

1. Symbolic Dynamics, Inc., Mountain View, CA.
2. OnLine Dynamics, Inc., Sunnyvale, CA.
3. *Software for Interactive Musculo-skeletal Modelling*, Musculo-graphics Inc., Evanston, IL.
4. Parametric Technology Corp., Waltham, MA.
5. Dynamic Analysis and Design System, CADSI, Oakdale, IA.
6. Mechanical Dynamics, Inc., Ann Arbor, MI.

4.8.5 MUSCLE MODELS

Movement simulations can be driven using a variety of methods. In simulations of human movement during free fall or space flight, movement of the body can be produced by well-coordinated limb movements (Pasarello & Huston, 1971). Since the required muscle forces are extremely small, it is appropriate to consider joint rotations as the cause of the movement (Yeadon, 1993c). The required muscle forces may be computed, but are not relevant to the problem as long as they do not exceed the force-generating capacity of the muscles. Limb orientations or joint movements are also more suitable for communication with non-experts than muscle forces or moments. When gravitational forces and ground contact are both present, joints can no longer be moved arbitrarily and muscle forces are a more suitable input for movement simulation. These can be simply represented by joint torques as a function of time, e.g., Hubbard (1980). Although simple and effective, this approach has limitations. First, when experimenting with joint torque inputs, it is never certain that the value used is within the capability of human muscles. This is especially problematic when optimizing performance, requiring addition of constraints to prevent unphysiological solutions (see also the later section on optimization). Joint torque input also ignores certain mechanisms which are important for control of movement, especially movements of longer duration. When a muscle is stretched, its force output increases due to the mechanical properties. This provides shock absorption, stabilization, and damping of movement. Also, muscles often span two joints (biarticular muscles). This provides a mechanical coupling between joints that may facilitate control and make movement more efficient (van Ingen Schenau & Bobbert, 1993) and is absent in torque-driven models.

When muscle forces are included in a simulation of movement, it is usually desirable to include several well-known mechanical properties of muscle: the force-length relationship, the force-velocity relationship, and the activation dynamics. A muscle model which includes these properties will allow the muscle forces to respond realistically to changes in length and neural stimulation. Muscle models exist in various levels of complexity, but for movement simulation, the three-component Hill model has been used almost exclusively (Winters, 1990; Zajac, 1989). The model consists of a contractile element (CE) and two non-linear elastic elements, the parallel elastic element (PEE) and the series elastic element (SEE), as illustrated in Figure 4.8.5. The CE represents the muscle fibers, the SEE represents tendon and other elastic tissue in series with the fibers, and the PEE represents the passive properties of the fibers and elastic tissue surrounding the muscle fibers. The elastic elements are described by a mathematical force-length relationship which has the property that no force is generated below a certain length (the slack length), and stiffness increases as force increases. A simple, two-parameter model for such an element is:

$$F(L) = \begin{cases} 0 & (\text{if } L < L_{slack}) \\ k(L - L_{slack})^2 & (\text{if } L \geq L_{slack}) \end{cases} \tag{4.8.15}$$

The stiffness parameter k can be scaled such that the tendon strain at maximal muscle force is about 3 to 5% or about half the level where damage occurs. This is supported by literature data (Winters, 1990), but it should be noted that functional demands on different tendons are different. For instance, finger tendons are very long and accurate control of movement is only possible when strains remain low, compared to, for instance, the tendons in the lower extremity where elastic energy storage occurs.

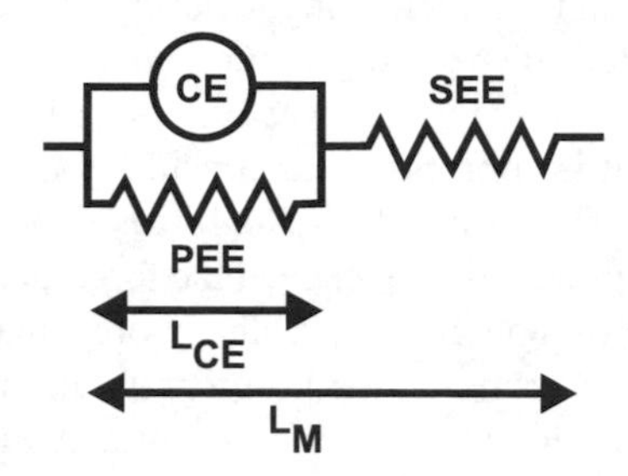

Figure 4.8.5 **The three-component Hill muscle model. CE: contractile element, PEE: parallel elastic element, SEE: series elastic element.**

The contractile element is described by Hill's force-velocity relationship:

$$F_{CE} = \frac{bF_o + a\dot{L}_{CE}}{b - \dot{L}_{CE}} \quad (\text{for } \dot{L}_{CE} < 0) \tag{4.8.16}$$

Note that this is equivalent to the traditional formulation [equation (2.7.26), chapter 2.7], since the contraction velocity v is equal to the negative of the time-derivative of contractile element length L_{CE}. The length-dependency is usually implemented by making the isometric force F_o dependent on the length L_{CE}. Theoretically, the function $F_o(L_{CE})$ has a shape as in Fig. 2.7.24, chapter 2.7, based on the properties of sliding filaments. However, the function is smoother in real muscle and can be parameterized by a simple mathematical function, for instance:

$$F_o(L_{CE}) = F_{max}\left[1 - \left(\frac{L_{CE} - L_{CEopt}}{W}\right)^2\right] \tag{4.8.17}$$

L_{CEopt} is the muscle fibre length where force is maximal. According to the sliding filament theory, the width parameter W, the maximum amount of shortening or lengthening in the CE, should be equal to $0.56L_{CEopt}$ for human muscle (Walker & Schrodt, 1973). Due to the effects of pennation and series arrangement of fibres, W tends to be higher for the complete CE of real muscles (van den Bogert et al., 1998). Equation (4.8.17) is substituted in equation (4.8.16) to obtain the force-length-velocity relationship, and finally multiplied by a scaling factor A, with a value between 0 and 1, which represents the activation or *active state* of the muscle. This results in a force-length-velocity-activation relationship for the CE:

$$F_{CE} = f_{CE}(L_{CE}, \dot{L}_{CE}, A) \tag{4.8.18}$$

where the right-hand side is a known mathematical function. This model for the contractile element is simple, but not entirely consistent with certain observations in isolated muscle preparations. For instance, equation (4.8.17) predicts that the maximal shortening velocity of the CE is bF_o/a, and, therefore, depends on length but not on active state. A more complete muscle model, also including the eccentric force-velocity relationship (for $\dot{L}_{CE} > 0$), is described by van Soest and Bobbert (1993), and more recently by McLean et al. (2003). For a thorough review of muscle modelling, see Winters (1990).

From Figure 4.8.5 it can be deduced that the force in the SEE is equal to the sum of forces in the CE and PEE. Also, the length of the SEE is the length L_M of the whole muscle-tendon complex minus L_{CE}. Therefore, the total state equation for the muscle is obtained as follows:

$$f_{SEE}(L_M - L_{CE}) = f_{CE}(L_{CE}, \dot{L}_{CE}, A) + f_{PEE}(L_{CE}) \tag{4.8.19}$$

All functions f are known, so the lengthening velocity of the CE can be solved from equation (4.8.19):

$$\dot{L}_{CE} = f(L_{CE}, L_M, A) \tag{4.8.20}$$

This is a first-order ODE, with L_{CE} as state variable. In experiments on isolated muscle, as described in section 2.7.5, activation A and length L_M are controlled by the experimenter. In that case, L_{CE} is the only unknown, and equation (4.8.20) alone is sufficient to simulate such experiments, as is shown in the example in Listing 3. Activation is assumed maximal ($A=1$) and the imposed length change $L_M(t)$ is at first isometric, followed by an isokinetic (constant velocity) ramp shortening, and finally isometric again. Note that the PEE is neglected, since the muscle length is assumed to be shorter than the slack length of the PEE. The simulated force (Figure 4.8.5) shows first the build-up of isometric force during lengthening of the SEE, then at $t=1$ s, the force drops suddenly because the initial shortening occurs only in the SEE. The CE can not shorten as quickly, due to the force-velocity relationship. At $t=2$ s, the shortening stops and the force stabilizes at the isometric force for the final muscle length. Simulations such as these are helpful when designing experiments in muscle mechanics, and to interpret results of such experiments.

Muscle properties for human muscles, such as F_{max}, L_{CEopt}, W, a, b, and L_{slack} cannot be measured directly in human muscles. Animal data may be used, after appropriate scaling to muscle size or values derived from theoretical considerations (section 2.7.5). It is important, however, to compare the performance of the model to macroscopic human data such as isometric strength data, and adjust the muscle properties where needed (Pandy & Zajac, 1990; Gerritsen et al., 1996).

The activation level A can be given as input for the muscle model, but it is more realistic to consider the *activation dynamics* of muscle. Activation of a muscle requires a polarization wave to reach all parts of the muscle, which eventually results in a concentration of Ca^{2+} ions which leads to the generation of force. The active state A can be identified as the concentration of calcium ions bound to troponin, relative to the maximum. This concentration cannot change instantaneously, since it is the result of electrical and chemical processes. To simulate this, use a model that describes how active state depends on the neural stimulation signal $s(t)$. A simple linear model for activation dynamics is:

$$\dot{A} = \frac{1}{\tau}[s(t) - A] \tag{4.8.21}$$

which is a first order ODE with state variable A. When $s(t)$ is larger than A, the active state rises and approaches the value $s(t)$ asymptotically with a time constant τ. The opposite happens when $s(t)$ drops below A. In a slightly more complex version (He et al., 1991), the model takes into account that activation occurs at a higher speed than deactivation:

$$\dot{A} = (c_1 s(t) + c_2)(s(t) - A) \tag{4.8.22}$$

where the activation time constant $\tau_a = (c_1 + c_2)^{-1}$ is usually assumed to be shorter than the deactivation time constant $\tau_d = c_2^{-1}$ (Winters & Stark, 1987). Other, more complex, activation models have been used, ranging from linear second-order models, e.g., Winters and Stark (1987), to a non-linear model that includes separate differential equations for motor unit recruitment and stimulation and the effect of fibre shortening on the activation process (Hatze, 1981). If a model of activation dynamics is not included, optimizations of a musculo-skeletal model may produce unrealistic results, as discussed in section 4.8.8.

It should be kept in mind that the three-component Hill model is based on a limited set of experimental observations, such as isometric, isotonic, and isokinetic contractions, usually performed at maximal activation. Interactions between the effects of length change and activation may, therefore, be poorly predicted. Also, history-dependent effects such as stretch potentiation are not included in the model. The Hill model is half mechanistic (PEE and SEE), and half empirical (CE) so it may not be valid when the CE is in a state that did not occur in the experiments on which the model was based. More complex models, based on the cross-bridge mechanism of muscle contraction have been formulated as a set of coupled ODE's (Zahalak, 1986). Such models should be valid for a wider range of conditions but require more complex programming, more computation time, and most importantly, also seem to produce certain results that disagree with certain observations (Cole et al., 1996). Cross-bridge models have hardly been applied for movement simulation and the Hill model, in spite of its limitations, is presently considered the most practical muscle model for simulation of human movement.

When muscles are connected to a skeleton with many degrees of freedom, the muscle model needs to be coupled to the equations of motion, resulting in a combined system of equations representing the musculo-skeletal system. This is discussed in the next section.

4.8.6 SIMULATION USING MUSCULO-SKELETAL MODELS

The output of a muscle model is the magnitude of muscle force, a scalar F. The equations of motion (4.8.13a) require a force vector (magnitude and direction) and also a point of application, so that the moment can be computed. Also, the muscle model requires the instantaneous length of the muscle L_M as input. All of these require a model of the musculoskeletal anatomy. One method for doing this is to model the muscle as a straight line of action between origin and insertion (Figure 4.8.6). If the kinematic state variables are known, the position vectors of origin, $\mathbf{r}_O$, and insertion, $\mathbf{r}_I$, can be calculated from the position and orientation of the two body segments. Then, muscle length is simply:

$$L_M = |\mathbf{r}_O - \mathbf{r}_I| \tag{4.8.23}$$

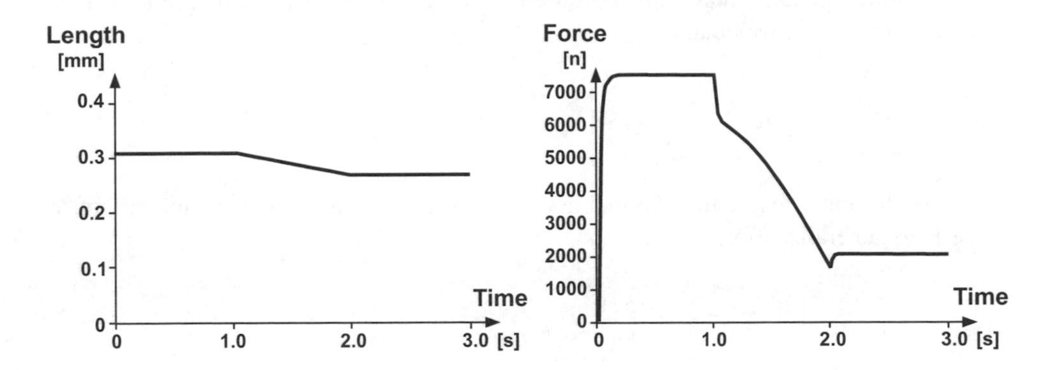

Figure 4.8.6 **Simulation of a muscle undergoing isokinetic ramp shortening between t=1 and t=2 s. Left: imposed length change, right: force generated by the muscle. See text and Listing 3 for details of the model.**

and the force vector applied to the insertion point $\mathbf{r}_I$ is:

$$\mathbf{F} = F\frac{\mathbf{r}_I - \mathbf{r}_O}{|\mathbf{r}_I - \mathbf{r}_O|} \tag{4.8.24}$$

The same vector, but in opposite direction, is applied to the origin point $\mathbf{r}_O$. Straight-line muscle models have limitations, and usually require additional *wrapping* points to model muscles and tendons that follow a curved path over underlying tissue (Delp & Loan, 1995). Failure to properly model the path of the muscle may result in underestimation of moment arms and hence, overestimation of muscle forces required for movement. This was a problem in early estimates of articular contact forces in the hip. Model calculations were as much as 100% higher than direct measurements using an instrumented prosthesis (Brand et al., 1994).

For some applications, it may be more practical at this point to include only the effect of muscle forces on movement, i.e., the moments exerted with respect to the joint centres. In the Newton-Euler formulation, this removes muscle forces entirely from equation (4.8.13a) so that other variables in that equation (the joint reaction forces) will no longer be correct. For questions related to control of movement, this is acceptable. For injury-related questions, however, accurate joint reaction forces are usually required. In the Lagrange formulation of the equations of motion, the generalized forces only contain the effect of muscles on movement. The contribution of a muscle force, F, to a generalized force, Q_k, can be determined using the principle of virtual work. When the system undergoes a hypothetical small change $\delta\mathbf{q}$ in the generalized coordinates $\mathbf{q}$, the muscle will undergo a virtual length change δL_M. The principle of virtual work states that the work done by the muscle should be equal to the work done by the generalized forces, i.e., the dot product of the generalized force vector $\mathbf{Q}$ and the generalized displacement $\delta\mathbf{q}$. Since muscle does positive work when shortening (negative δL_M), this can be expressed as:

$$-F\delta L_M = \mathbf{Q}\cdot\delta\mathbf{q} \tag{4.8.25}$$

Expanding the dot product, and considering that L_M may be a function of all generalized coordinates q_k, we obtain:

$$-F\sum_k \frac{\partial L_M}{\partial q_k}\delta q_k = \sum_k Q_k \delta q_k \qquad (4.8.26)$$

Since this must be true for all possible virtual displacements, the contribution of muscle force F to generalized force Q_k is:

$$Q_k = -F\frac{\partial L_M}{\partial q_k} \qquad (4.8.27)$$

If a generalized coordinate is a joint angle, the corresponding generalized force will be the moment of the muscles at that joint. If the generalized coordinates are segment orientations, the muscle moment will contribute to the generalized forces for both segments. The partial derivatives $\frac{\partial L_M}{\partial q_k}$, which are the moment arms, may be derived from an anatomical model, for instance, using straight lines, or measured directly in cadaver specimens or in vivo using three-dimensional imaging techniques. Moment arms can be simply measured as the distance from the muscle's line of action to the joint centre, requiring assuming a fixed joint centre. A more elegant method is to collect data of muscle length at various joint angles (or other generalized coordinates), and use regression analysis to find a mathematical relationship describing L_M as a function of the generalized coordinates. Partial derivatives may then be computed analytically. This method was first used by Grieve et al. (1978) to determine moment arms of the ankle plantarflexors, and applied later to determine moment arms for hip and ankle muscles (Spoor et al., 1990; Visser et al., 1991). The partial derivative is a moment arm in meters if L_M is measured in meters, and q_k is measured in radians. It is important that such regression models not be extrapolated beyond the range of joint angles in the experimental protocol. Extrapolation may result in unreliable moment arms and, assuming a constant moment arm, result in a better model.

Summarizing, the coupling between skeleton and muscles requires three sets of calculation at each time step of the simulation:

(1) Compute muscle lengths from the bone positions in the skeleton.
(2) Compute muscle force from the state equations of the muscles.
(3) Apply the muscle forces to the skeleton.

For a complete musculo-skeletal model, passive forces should also be considered. Ligaments, when their function is to guide joint kinematics, need not be included, since their function has been incorporated in the joint model. For instance, the cruciate ligaments in the knee can be assumed to be inextensible during normal movements, resulting in a moving centre of rotation at the intersection of the cruciates. These *guiding* ligaments do not perform mechanical work, and can be omitted from the equations of motion. They will, however, have an effect on the articular contact force. To model this, joints can no longer be considered to be kinematic mechanisms and deformation needs to be modelled. This can be done, e.g., Blankevoort and Huiskes (1996), but is outside the scope of this chapter. Internal passive force also occurs at the end of the range of motion of a joint. This may be result from stretching ligaments, compressing muscles, or a combination of both. The net effect on the movement can be modelled as an extra joint moment, which is a function of one or more joint angles. Suitable models can be found in the literature, e.g., Riener and Edrich (1999). Passive tension in ligaments spanning several joints, which occurs in some animals (van Ingen Schenau & Bobbert, 1993), can also be modelled as a simple passive muscle model with only a SEE and coupled to the skeleton in the same way as muscles.

As an example of a combined muscle and skeletal model (a musculo-skeletal model), Listing 4 presents a simulation of a maximal effort kicking motion while seated (Figure 4.8.7). The muscle model from Listing 3 is used, the moment with respect to the knee joint

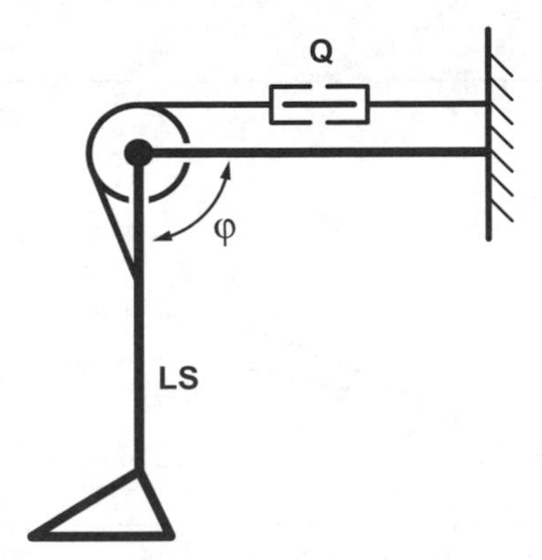

Figure 4.8.7 **Illustration of the example used for illustration of the simulation process. Q =quadriceps muscle; LS = leg segment.**

is computed assuming a constant moment arm, and the moment is applied to the equation of motion of a single rotating body segment (the shank-foot combination). Segment properties (length, mass, moment of inertia, centre of mass) were obtained from Winter (1979). The results (Figure 4.8.8) show that the force increases at first, but then quickly drops off as the muscle shortens due to the effects of the force-length and the force-velocity relationships. Such interactions between skeletal dynamics and muscle dynamics can only be observed in a model in which both are represented and coupled. As an example of an experiment, the dashed line shows what happens with an increased moment arm. Although the initial acceleration is larger, as expected, the joint does not even reach full extension because the force drops off too quickly. This is an example of how simulation can be used to look at the effect of design on performance.

4.8.7 MODELLING OF EXTERNAL FORCES

External forces are forces acting between the system of interest and the environment. These forces must be included in the right-hand side of the equations of motion (4.8.13a) for the system. Common external forces are: gravity, contact forces, and aerodynamic forces. *Gravity* is simply modeled as a constant force vector of magnitude mg and downward direction, applied to the centre of mass of each rigid body segment in the model, where m is the mass of the body segment and g is the acceleration of gravity (9.81 m/s^2). *Contact forces* can often be included by modelling the contact as a kinematic connection like a joint. This was, for example, appropriate for the foot-ground contact forces in a simulation of vertical jumping (Pandy & Zajac, 1990), provided that the simulation is terminated when tensile forces in the connection begin to occur. A similar model for walking could, therefore, only

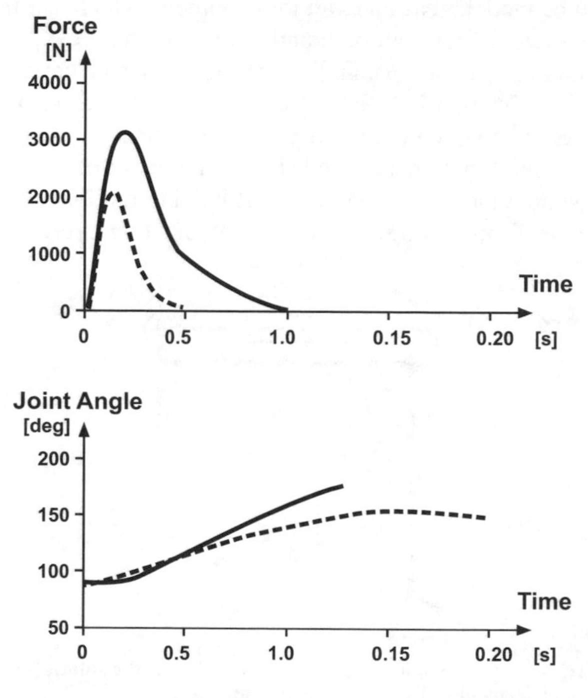

Figure 4.8.8 **Simulation of a maximal effort ballistic knee extension (Listing 4). Solid line: with muscle moment arm r_f=33 mm. Dashed line: r_f=66 mm.**

be used for approximately 48% of the stride cycle, which is the time that one toe remains connected to the ground (Yamaguchi & Zajac, 1990).

A model of cycling on an ergometer can use a similar method to include external forces: hinge joints between pelvis and ground, and between crank and ground, effectively generate the external forces required to keep the pelvis attached to the seat and to prevent the crankshaft from moving (Neptune & van den Bogert, 1998). This technique, sometimes referred to as the hard constraint method, becomes difficult to implement when a connection is not permanent, for instance the contact between foot and ground during simulation of a complete walking movement. At heel strike, a joint is suddenly added which requires the heel velocity to decrease instantaneously from some value to zero. This discontinuity in velocity is transmitted through the entire system and is accompanied by an infinitely high force with a finite impulse (time integral). These discontinuities and impulses can be calculated, as shown by Hatze and Venter (1981) but the equations are complex. It may also be more *realistic* to model the contact by a force-deformation model, especially when simulation results are compared to experimental force measurements which do not show such impulsive forces.

Implementation of a force-deformation model, sometimes referred to as a soft constraint, requires three components:

(1) A calculation of element deformation x d from the generalized coordinates of the system. When forces are velocity-dependent, the deformation velocity $\dot{x}$ is also required, which depends on the generalized velocities $\dot{\mathbf{q}}$.

(2) A constitutive equation that calculates the force $F(t)$ as a function of deformation time history $x(t < t)$.

(3) Equations that convert the external force F into generalized forces Q_i (or forces and moments) that can be applied to the equations of motion.

A disadvantage of force-deformation models is that they may make the system of differential equations stiff, especially when the contact stiffness is high and the mass of contacting body segments is small relative to the total mass of the system (van den Bogert et al., 1989). Steps (1) and (3) are very straightforward to program when the Newton-Euler formulation is used for the equations of motion, and more complicated for the Lagrange formulation. The procedures are like the implementation of muscle forces shown in the previous section. The following constitutive equation has been used for foot-ground contact:

$$F = ax^b + cx^d\dot{x}^e \tag{4.8.28}$$

Suitable values for the parameters a-e can be determined by fitting this equation to measured force-deformation data (Wright et al., 1998; Hardin et al., 2004). Contact forces are often distributed forces, so it may be necessary to use many such contact elements, distributed over the entire contact surface. Care must be taken that the stiffness parameters a are scaled to be proportional to the contact area represented by one contact element, as was done for a three-dimensional model of an athletic shoe (Wright et al., 1998). The resulting model is a discrete element model (DEM) rather than a finite element model (FEM), since there is no interaction between the elements. A DEM is considerably faster to compute than a FEM and usually adequate for incorporating contact in human movement simulations, provided accurate predictions of stresses in the contact material are not of interest.

Aerodynamic forces are important during movements at high speed. A good example is the simulation of ski jumping (Müller, 1996), which would be completely unrealistic without aerodynamic forces. In fact, the entire purpose of such simulations is to determine how athletes can make the best use of aerodynamic forces. Practically, this is done by collecting wind tunnel data and developing regression models that predict the total aerodynamic force on the body as a function of the generalized coordinates $\mathbf{q}$. The force is typically represented as a lift and a drag component, each proportional to the square of the velocity by a factor which is a function of the orientation of the body segments relative to the airflow. Such equations are adequate to model the effect of aerodynamics on the flight trajectory, but it should be noted that this does not provide information about the distribution of these forces over the different body segments. This would require finite element models of air flow, which are extremely computationally expensive and has only recently been applied to this problem (Asai et al., 1997). Another application where aerodynamic forces are important is the simulation of javelin flight to find optimal release conditions (Hubbard & Rust, 1984).

4.8.8 OPTIMIZATION STUDIES

As defined in section 4.8.1, simulation consists of performing experiments on a numerical model. Since it is easy to perform hundreds, sometimes thousands of experiments on a numerical model, a special type of experimentation, *optimization*, becomes feasible. The first optimization studies using a musculo-skeletal model were done by Hatze (1976) to find the optimal combination of muscle activation patterns $s(t)$ in five muscles that produced the fastest kicking movement in a two-segment model of the lower extremity. In a more recent simulation of vertical jumping (Pandy & Zajac, 1990), the goal was to find the optimal muscle activation patterns for a maximal-height jump. Solutions to such problems can be found using iterative methods, which require the execution of many simulations, while the unknowns are automatically varied by a suitable search algorithm. Traditionally, optimal control methods have been used which solve for arbitrary functions $s(t)$, but more recently such problems have been formulated as parameter optimizations (Pandy et al., 1992; van Soest et al., 1993). The unknown functions are described using equations with N unknown parameters, for instance switching times (van Soest et al., 1993) or function values sampled at certain time intervals (Pandy et al., 1992). The problem is then simply to find the parameter vector $\mathbf{p} = (p_1, p_2 \ldots p_N)$ which results in the best performance. Performance is a scalar function f of $\mathbf{p}$, which can be calculated in two steps: (1) perform a simulation with $\mathbf{p}$ as input, and (2) determine the performance, e.g., jumping height, from the results of the simulation. Performance measures are easily defined for maximal-effort tasks such as mentioned above. In submaximal movements, finding a strategy that minimizes energy consumption, muscle forces, or another optimization criterion may be useful. This is an elegant method to investigate which criteria govern control of movement. If the result of the optimization is like observed human behaviour, support is obtained for the validity of the hypothesized control principle. A good example is the notion that movements are executed with minimal jerk or rate of acceleration (Flash & Hogan, 1985).

Although muscle coordination is most frequently optimized, other model parameters can be optimized using the same methods. A good example is the problem of designing an optimal bicycle with respect to a certain performance criterion (van den Bogert, 1994). When optimizing design of equipment, care must be taken to optimize the human's muscle coordination and equipment at the same time. Humans can and will adapt to changes in equipment and this learning effect may be very significant. Such design problems, therefore, result in optimizations of a large number of simultaneous unknowns.

Performance optimization is not always the objective of optimization. In some cases we merely wish to create a model which replicates observations on human subjects. This requires solution of the so-called *tracking problem*. The inputs are parameterized as before, but the function f now quantifies the difference between observed and simulated variables. In humans, we can typically observe only external forces and kinematics. If $v_i(t)$ is the simulated result for variable i, and $v_i^o(t)$ is the corresponding observation, a suitable weighted least-squares formulation for the objective function f is (Neptune and van den Bogert, 1998):

$$f(\mathbf{p}) = -\sum_i \frac{1}{\sigma_i^2} \int_0^T [v_i(t) - v_i^o(t)]^2 dt \qquad (4.8.29)$$

where σ_i^2 is the variance between multiple observations of variable i, averaged over the duration T of the movement. Thus, more reproducible observations are weighted more heavily and the final result after optimization will be that tracking errors in each variable are proportional to the experimental variability of that variable. One reason for solving the tracking problem for a musculo-skeletal model is to obtain information about internal muscle forces during a specific movement. This problem can also be solved using inverse dynamics and applying a static optimization to solve the distribution problem, as shown earlier in this chapter, but there are two important differences. First, when optimizing a forward dynamics model it is automatically guaranteed that the solutions are consistent with properties of the system. Specifically, this procedure will only produce muscle forces that do not exceed the known capabilities of the muscle, i.e., maximal force as a function of length and velocity and maximal rate of force development. Secondly, if the number of unknown parameters $\mathbf{p}$ is small enough, a mathematically unique solution is obtained without requiring additional optimization criteria to solve the distribution problem. Effectively, the distribution problem is avoided by assuming simple activation functions $s(t)$ of the muscles, since these were parameterized using a small number of parameters. This can be justified by assuming minimal complexity of the control system that generates the signals $s(t)$. Solving the tracking problem requires orders of magnitude more computation time than solving the inverse problem, and is just becoming feasible for complex musculo-skeletal models (Neptune and Hull, 1998). Simulations developed with optimization of the tracking criterion, once validated, can be used for experiments that are not possible on human subjects (Gerritsen et al., 1996)

Many numerical methods for optimization of non-linear functions have been developed (Press et al., 1992). All of these start from an initial guess of $\mathbf{p}$, and then iteratively improve the solution by repeatedly calculating f while varying the inputs $\mathbf{p}$. The most efficient algorithms are based on gradient information, i.e., the derivatives of f with respect to each of the parameters p_i. Gradient information speeds up the search because it indicates in which direction each parameter needs to be changed to maximize the function $\dot{f}$. However, gradient-based algorithms have not performed well for optimization of complex movements, for several reasons. First, the gradients may be unreliable. If a muscle activation parameter includes switching times, gradients can be anywhere between zero and infinity due to the division of time into steps of size h when performing the simulations, e.g., van Soest et al. (1993). The simulation will then remain unchanged if a switching time is changed by a sufficiently small amount. In other words, the function f is piecewise constant. This problem may be overcome by using optimization methods that do not require gradients, such as the *simplex method* (Press et al., 1992; Matlab function fminsearch).

A more serious problem of most optimization methods is the tendency to converge on a local, rather than the global, optimum (Figure 4.8.9). To a certain extent, this can be prevented by solving the optimization problem many times with different initial guesses. However, some models have many local optima and the global solution may never be found. The *simulated annealing* (SA) method (Press et al., 1992) is an elegant and powerful method for global optimization, based on the analogy of minimization of potential energy in a crystal lattice, which is a function of the positional coordinates of all nuclei. At high temperature, large random changes in position occur. Lattice configurations with low energy will be encountered from time to time and the system likely to stay longer in such a state. As the temperature decreases, the system will freeze in the state of lowest energy. It has been proved analytically that with a sufficiently slow decrease in temperature, the SA method will converge to the global optimum. Unfortunately, this theoretically derived temperature reduction scheme is too slow for practical applications, and quenching (a faster temperature reduction) must be used. Nevertheless, the performance of the SA method for human movement optimization has been shown to be faster and to result in a better solution than other methods in a practical application (Neptune, 1999).

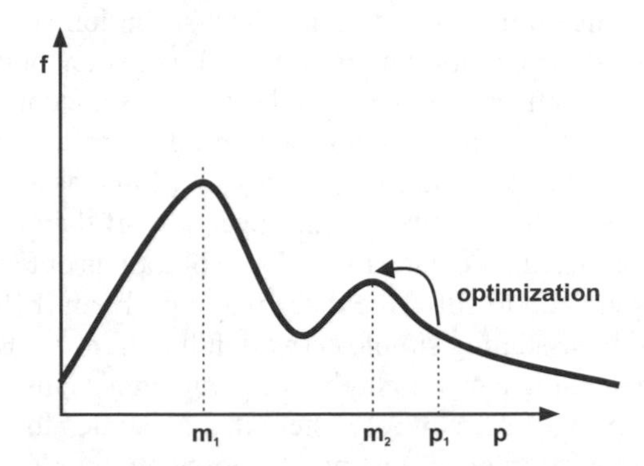

Figure 4.8.9 **Hypothetical performance function f as a function of a single parameter p. Optimization of f from the initial guess p_1 may result in finding the local maximum m_2, while the more important global maximum is located at m_1.**

As an example, we will determine the optimal release conditions for the throw of a javelin, based on a simplification of the model by Hubbard and Rust (1984). The model is asingle rigid body, with three degrees of freedom in the plane of motion. Sideways movements will be neglected. We will assume that the javelin is always released at a height of 1.5 meters and with a velocity of 30 m/s. The athlete can choose the three remaining release parameters: orientation of the javelin, angular velocity of the javelin, and direction of the throw (Figure 4.8.10). The Matlab program in Listing 5 simulates a single throw. It is written as a Matlab function which takes the three release parameters as input and returns the negative of the distance as output. Type **javelin([0.5 0.0 0.1])** to see the result of a throw with initial orientation of 0.5 radians, zero angular velocity, and an angle of attack of 0.1 radians. Now type **fminsearch('javelin',[0.5 0.0 0.1])** which performs an optimization of the distance using the simplex method, with those same release parameters as an initial guess. For this model, the following optimal release conditions were found: $\alpha = 33.4°$, $\dot{\alpha} = 0.16°/s$, and $\alpha_a = -3.7°$.

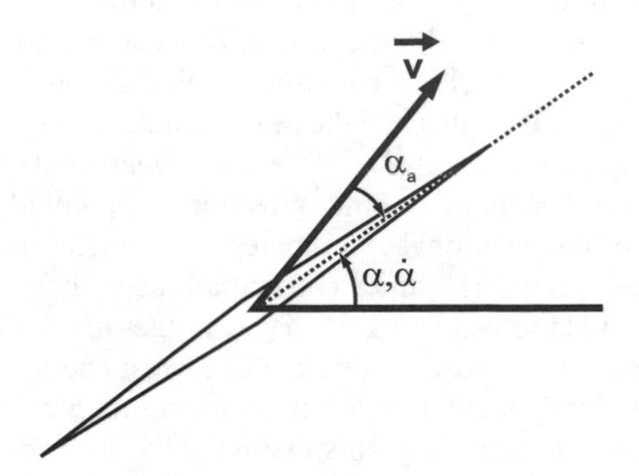

Figure 4.8.10 **Definition of release conditions of the javelin: orientation α, angular velocity $\dot{\alpha}$, and angle of attack α_a. In this example, the angle of attack is negative, indicating that the javelin is oriented clockwise relative to the velocity vector $\vec{v}$.**

4.8.9 SIMULATION AS A SCIENTIFIC TOOL

When simulation is used as a scientific tool, experiments are performed on a numerical model. Direct validation of a numerical model is usually difficult because the same experiments can not be done on human subjects. This could be because the experiments are related to severe injury, because human subjects are not sufficiently reproducible, because humans are too fatigable or because the outcome is a variable which can not be measured. How then do we ensure the validity of scientific studies using computer models? Only indirect methods are available. For this reason, the term *validation* may be too strong and *evaluation* should be used instead.

First of all, a model should be *consistent with observations* that can be made on humans. When optimization of performance is carried out, and a realistic movement is obtained, the model is generally considered to be valid because no movement data were used to develop the simulation. When solving the tracking problem [minimization of equation (4.8.29)], it is expected that after optimization all variables are within two standard deviations of the mean. If this is not the case, the model is unable to perform the movement in a sufficiently realistic manner. This should be reason to closely examine the model and make improvements where necessary. Passing this test, however, does not guarantee a valid model. Due to the redundancy of the locomotor system, the model could have found a different solution than the human to achieve the same external movement and force variables. In that case, additional predictions must be elicited from the model which can then be compared to measurements that were *not* used for development of the model.

This is especially important when the tracking problem is solved. We recommend testing the response of the model to controlled interventions and compare that response to results of the same experiments on humans. Even when the final application is a study on severe injuries, it is often still possible to evaluate the model dynamics using non-destructive *perturbation tests*. Care must be taken that these experiments test those aspects of the model that are important for the final application. For development of a model for knee ligament injuries, perturbations of initial conditions were used to evaluate the validity of the model (McLean, 2003).

Finally, a model should not be overly sensitive to errors in model parameters. Critical model parameters can be identified by *sensitivity analysis*: each parameter is adjusted by a small amount and the change in the results of the simulation is examined. In some cases, this will show that certain model parameters are too critical and the results of a simulation study would depend entirely on a random error in such a parameter. In certain cases, optimization methods are helpful. In a quasi-static model of the knee joint, it was found that the behaviour was sensitive to the lengths of the ligaments (Blankevoort & Huiskes, 1996). These unknown parameters were then eliminated by solving the tracking problem. Ligament lengths which minimized the difference between simulated and measured movements were found and these values were used for subsequent applications of the model.

Another powerful safeguard against overly sensitive models is *statistical analysis*. Experiments on humans or animals are never performed on a single individual because one individual may not be representative of the population. Statistical analysis is performed to ensure valid generalizations. When using complex musculo-skeletal models, the same principle should apply. These models have many degrees of freedom, many natural frequencies, and are often unstable and chaotic. Results from a single model could well be completely irrelevant. In a simulation study on the effect of shoe hardness on impact forces in running, both positive and negative responses were found in a group of models (Wright et al., 1998). By examining the model, this could be attributed to two mechanisms which worked in opposite directions. Impact forces tend to increase initially with harder material. This then increases the rate of knee flexion, resulting in a better shock absorption by the body. In subjects with a certain movement style, the latter mechanism resulted in an overcompensation. Statistical analysis showed no significant effect for the group of models as a whole, confirming earlier results on human subjects. Thus, statistical analysis prevented incorrect generalizations which could have been made when just one model had been used. In this case, the model was sensitive to movement style. Sensitivities to other variations within the human population, such as anatomical variations, may be detected or eliminated similarly, by creating a population of models with the appropriate range of parameter values. Note, however, that modeling and simulation are sometimes used specifically to determine the influence of inter-subject variations, and in such case the statistical approach is not appropriate.

In summary, simulation experiments only tell the truth about the model that was used. Generalization to the human population is always hazardous and requires extensive validation and careful examination of the results.

4.8.10 APPLICATIONS

In recent years, the field of biomechanical movement simulation has matured sufficiently to allow its use in answering certain basic and applied questions on human movement. The best-developed areas of application are gait and sports injuries. In gait, the functional role of muscles has been identified by solving the tracking problem in a forward dynamic model, followed by an induced acceleration analysis (Neptune et al., 2001). Minimum energy optimizations have produced realistic movements, suggesting that minimal energy is the governing principle of human gait (Anderson & Pandy, 2001). In sports medicine, simulation has shown how the effect of foot orthoses on knee joint mechanics can differ between subjects (Neptune et al., 2000). Simulation has perhaps its greatest impact in the area of acute injuries, where no human experimentation is possible. There is a long history of increasingly realistic passive human movement simulations in vehicle collisions. With active muscle models and optimization in realistic musculoskeletal models, these techniques are now becoming feasible for studies on knee and ankle ligament injuries during sports (Wright et al., 2000; McLean et al., 2004). Although the basic methodologies, as described in these pages, are quite straightforward, the complexity of modeling required for these applications is still beyond the capabilities of most laboratories. We expect, however, that commercial and user-friendly software will become available to make the technology more accessible.

4.9 ADDITIONAL EXAMPLES

This section has been divided into basic and advanced examples. In general, the basic questions can be answered by studying the text or by performing some calculations based on information provided by the text. The advanced questions go beyond what can be found in the text.

BASIC QUESTIONS:

1. A mechanical system with two masses, M and m, is connected by springs and dampers as illustrated in Figure 4.9.1. Derive the equations of motion of the system, and list the required known model parameters and the required initial conditions for determining the movement of the two masses.

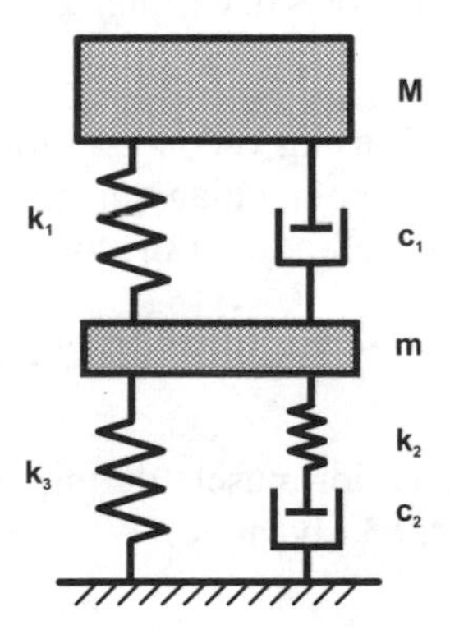

Figure 4.9.1 A mechanical system with two masses and spring-damper combinations.

2. A particle lands with the contact velocity, v_0, onto the centre of a massless board which is supported by two parallel ideal linear springs, k_1 and k_2 (Figure 4.9.2). Determine the position, z(t), of the particle for any time after contacting with the massless board.

3. Assume that the agreement between theoretically predicted results from a specific model and experimentally determined results is in the range of ±5 %. Comment on (a) the appropriateness, and (b) the validity of the model to predict these results if you are interested in predicting a result with an accuracy of 10%.

4. Discuss the advantages and disadvantages of forward dynamics compared to inverse dynamics when trying to determine muscle forces during a specific activity.

5. Discuss the advantages and disadvantages of using Newton-Euler equations of motion versus using Lagrange's equations when determining motion of a segmental model.

6. Discuss the reasons why the internal forces determined from a rigid body model are larger than the internal forces determined from a wobbling mass model.

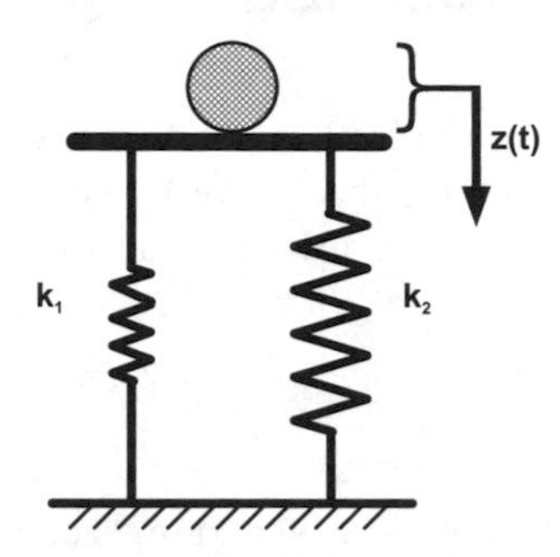

Figure 4.9.2 **Illustration of a mass on a massless board, supported by two parallel linear springs.**

7. Draw a free body diagram that allows determining the forces, applied to the spine segment L5 when shovelling snow.

8. Perform a force system analysis using vector notation, including definition of the system of interest, assumptions, free body diagram, equations of motion, and solution to determine the forces in the human body for the following conditions:

   (a) Force in the Achilles tendon for a person (m = 70 kg) standing on the toes of one leg (static).

   (b) Force in the tibialis anterior muscle during landing in heel-toe running using the data from Figure 4.9.3 (dynamic).

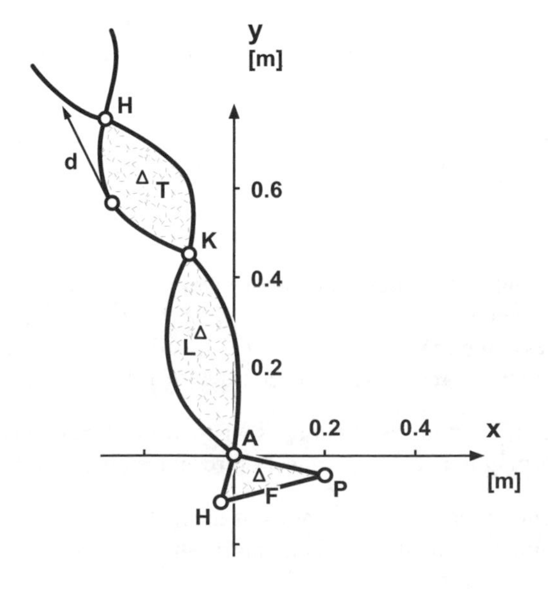

Figure 4.9.3 **Schematic illustration of a human leg during landing on the heel.**

(c) Force in the elbow joint for a person holding a mass of 10 kg in the hand assuming that the force of the biceps acts vertically and for the geometrical values indicated in Figure 4.9.4 (static).

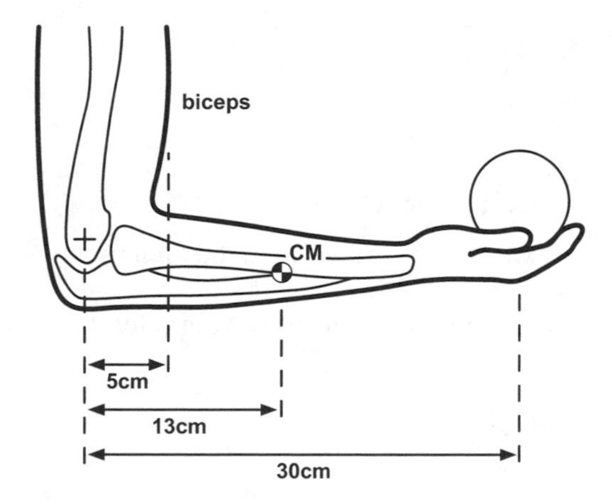

Figure 4.9.4 **Schematic illustration of the arm of a subject holding a mass of 10 kg in the hand.**

9. A runner is contacting the ground with the heel. At the time, 20 ms after first ground contact, the ground reaction force is **F** = (-100 N, 600 N, 0 N). Determine the force at the hip joint at this time point. Use the following assumptions:

- The hip joint is a hinge joint in the sagittal plane,
- Only one muscle at the posterior side of the hip joint is activated at this time point,
- The masses of the foot, leg, and thigh are:

$$m_{foot} = 2\text{ kg}, \quad m_{leg} = 4\text{ kg}, \quad m_{thigh} = 5\text{kg},$$

- The coordinates of specific landmarks are:

Ankle joint	A	=	(0.00 m, 0.00 m, 0 m)
Knee joint	K	=	(-0.10 m, 0.45 m, 0 m)
Hip joint	H	=	(-0.30 m, 0.90 m, 0 m)
Contact point with ground (heel)	C	=	(-0.02 m, -0.12 m, 0 m)
CM (foot)	F	=	(0.05 m, -0.06 m, 0 m)
CM (leg)	L	=	(-0.06 m, 0.25 m, 0 m)
CM (thigh)	T	=	(-0.20 m, 0.70 m, 0 m)
Muscle attachment (hip)	M	=	(-0.30 m, 0.70 m, 0 m)
Toe of foot	P	=	(0.20 m, -0.05 m, 0 m)

- The active muscle crossing the hip joint is attached at point, M, and has a direction which is determined by the vector **d** = (-2 m, 5 m, 0 m), and
- The velocities and accelerations at this time are:

Velocity of H, F, L, T	v_0	=	(1 m/s, -2 m/s, 0 m/s)
Acceleration of H, F, L, T	a_0	=	(0 m/s^2, 0 m/s^2, 0 m/s^2)

10. Calculate analytically the force-sharing between two agonistic muscles crossing a single joint using the cost function proposed by Pedotti et al. (1978), that is:
    Minimize ϕ, where:

$$\phi = \sum_{i=1}^{2} (F_i^m / F_{i\,max}^m)^2$$

where:

F_i^m is the force in the ith muscle, and

$F_{i\,max}^m$ is the maximal, isometric force in the ith muscle.

11. Use the following example to derive numerical values for the force-sharing problem in question 1:

$F_{1\,max}^m$ = 1000 N

$F_{2\,max}^m$ = 500 N

r_1 = 50 mm

r_2 = 40 mm

M = 25 Nm

(where r_1 is the moment arm of muscle 1 and r_2 is the moment arm of muscle 2 about the joint of interest, and M is the joint moment).

12. A mass, m, attached to a fixed point, A, by an inextensible string, is acted upon by its own weight. If the mass moves on a horizontal circle, show its speed cannot change.

13. Show that the field of gravity near the earth's surface derives from a potential.

14. Show that the gravitational force field of a heavy body derives from a potential.

15. Construct a force field that does not derive from a potential.

16. Show that the ordinary frictionless pendulum satisfies the principle of conservation of mechanical energy.

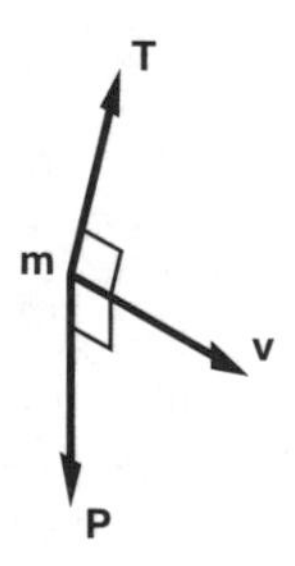

Figure 4.9.5 **Forces and velocity vectors for a frictionless pendulum.**

17. Using the principle of virtual work, derive the equations of equilibrium of a particle in space.

18. Derive the law of equilibrium of a simple lever.

19. In the simulation of the downhill coasting cyclist (Listing 1), determine which value of the step size h is required to get at most a 5% error in the speed of the cyclist at any given time.

20. Modify the simulation program in Listing 1 to add the propulsive force due to pedaling. Consider different ways to model this:

    (a) Constant force.
    (b) Constant power.

21. Modify the program in Listing 1 to compute x, the distance traveled by the cyclist, by including the ODE $\dot{x} = v$ in the model. Then include the effect of change of altitude on the air drag coefficient k.

22. Modify the equations of motion for arm movement, and the simulation model (Listing 2) to include constant joint torques at both joints.

23. Modify the equations of motion for arm movement, and the simulation model (Listing 2) to simulate movement in a vertical plane by including gravitational forces. Simulate this two-segment pendulum. What is different, compared to the simulated movement in the horizontal plane?

24. Modify the program in Listing 3 to include the eccentric part of the force-velocity relationship, i.e., $F > F_0$. Develop a suitable mathematical model from the following information:

    (a) The force-velocity relationship must be continuous at zero velocity.
    (b) At high eccentric velocities, the muscle force approaches $1.5F_0$ asymptotically. Use the modified program to simulate isokinetic lengthening protocols.

25. How many state variables are there in the model of the ballistic knee extension (Listing 4)?

26. Imagine an unloaded knee extension task with the object of hitting a target with maximal foot velocity. Assume that the target is always reached at full knee extension, and use a series of simulations to determine the optimal quadriceps moment arm for this task. How does this optimal moment arm compare to actual moment arms reported in the literature (Visser et al., 1991)?

27. Modify the simulation of knee extension to include a model for activation dynamics. Assume an initial condition $A=0$ at $t=0$. How important is activation dynamics for this movement?

28. In a real javelin, the pitching moment is a complex function of the angle of attack (Hubbard & Rust, 1984). Incorporate such a function in the simulation and examine the behaviour of the model.

29. Modify the program in Listing 5 by adding commands to draw the javelin as a line element on the screen after each forward Euler step. Examine the difference between optimal and non-optimal flight trajectories. After the simulation, examine the orientation of the javelin to see if the throw was legal (tip-first landing).

30. Add the effect of wind to the simulation of javelin flight. How should the release strategy be modified by the athlete in the presence of wind?

ADVANCED QUESTIONS:

1. Develop a relatively simple two-dimensional model of the foot to discuss the loading of an idealized lateral ligament (representing all lateral ligaments of the ankle joint complex) during a lateral side-shuffle movement. Determine the force in this idealized lateral ligament as a function of time during ground contact using experimental data from motion analysis, ground reaction forces, and appropriate assumptions.

2. Develop a relatively simple model that allows determining the force in the Achilles tendon during the take off in running using a three-dimensional vector notation. Make appropriate assumptions and use experimental data from motion analysis, and ground reaction forces.

3. Calculate analytically the force-sharing expression between three muscles in a two-joint system. Muscles 1 and 2 cross both joints and muscle 3 crosses only one joint. Use the cost function of Crowninshield and Brand (1981b), and assume that muscle forces are equal or greater than zero, that is:

   Cost function:
   Minimize ϕ, where: $\phi = \sum_{i=1}^{3} (F_i^m / pcsa_i)^3$

   Subject to the constraints:

   $h_1: \; F_i^m \geq 0$ for $i = 1, 2, 3$

   $h_2: \; M_1 = r_1 \cdot F_1^m + r_2 \cdot F_2^m + r_3 \cdot F_3^m$

   $h_3: \; M_2 = r_4 \cdot F_1^m + r_5 \cdot F_2^m$

   where:

   F_i^m is the force in the ith muscle, and

   $pcsa_i$ is the ith physiological cross-sectional area, and

   r_i $(i = 1,5)$ are the moment arms of the three muscles about the two joints.

4. Use the planar, three degrees of freedom model with six one-joint and four two-joint muscles as shown below (Figure 4.9.6), and show analytically that there are static situations for which one-joint antagonistic muscle activity is required to produce a minimum stress solution as formulated by Crowninshield and Brand (1981b), that is,

$$\text{Minimize } \phi \text{, where:} \phi = \sum_{i=1}^{10} \left(\frac{F_i}{pcsa_i} \right)^3$$

where:

F_i is the force in the ith muscle, and

$pcsa_i$ is the ith physiological cross-sectional area.

The constraints of the problem are that all three joint moments are satisfied by the moments produced by the muscle forces and that all muscle forces are tensile, i.e., $F_i \geq 0$.

Note, one possible way of solving this problem is given by Herzog and Binding (1992).

Antagonistic muscular activity is defined by $\overline{m}_i \circ \overline{M}_j < 0$, where m_i is the moment produced by the ith muscle, M_j is the resultant moment at the jth joint, and $\circ$ represents the vector dot product.

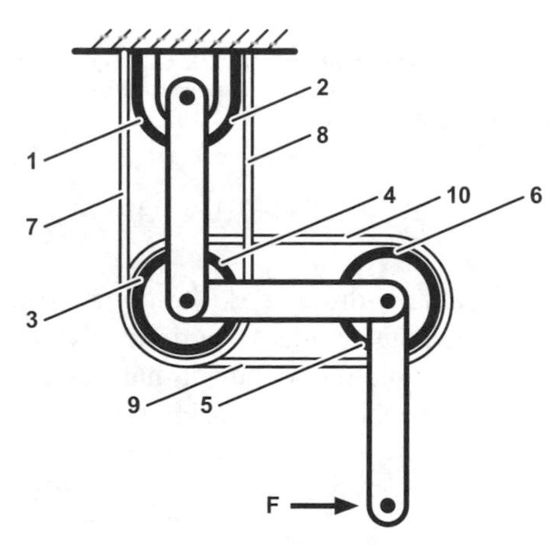

Figure 4.9.6 **Theoretical musculo-skeletal model with three rigid segments and three joints. The joints are planar hinge joints that are crossed by a pair of one-joint muscles on either side, i.e., a one-joint flexor and extensor muscle. The joints are also crossed by two (distal and proximal), and four (middle) two-joint muscles, as shown.**

5. Use the planar, two degrees of freedom model with four one-joint and two two-joint muscles as shown below (Figure 4.9.7). Show analytically that there are static situations for which co-contraction of the two two-joint muscles (muscles 5 and 6, Figure 4.9.7) is required to obtain a minimum stress solution as formulated by Crowninshield and Brand (1981b) (see advanced question 2).

   Note that one possible way of solving this problem is given by Herzog and Binding (1993).

6. Resolve the steady-state of the following differential equation and analyze the stability of the steady-state solution:

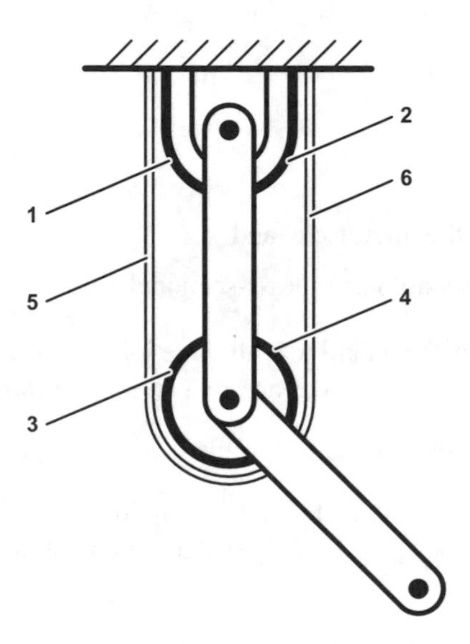

Figure 4.9.7 **Theoretical musculo-skeletal model with two rigid segments and two joints. Both joints are planar hinge joints that are crossed by a pair of one-joint and a pair of two-joint muscles, as shown.**

$$\dot{v} = \mathbf{F}_0 - bv^2 \quad (\mathbf{F}_0 > 0, b > 0)$$

7. Consider the problem of maximal-distance ski jumping. Assume that the athlete can influence lift, drag, and pitching moment continuously by changing joint angles. Choose a suitable parameterization to find the optimal strategy during flight, assuming constant initial conditions. Solve the problem using typical aerodynamic data from the literature.

```
% bicycle.m
% Program to simulate a downhill coasting cyclist

% system parameters
g = 9.81;              % acceleration of gravity (m s-2)
m = 75;                % mass of the cyclist (kg)
k = 0.15;              % drag coefficient (N s m-1)
slope = 10;            % downhill slope (degrees)
slope = slopepi/180;% same slope, in radians

% set initial conditions
t = 0;
tend = 120;            % duration of simulation in seconds
h = 20;                % integration step size in seconds
i = 1;                 % step counter
v = 0.0;               % initial velocity of cyclist
data = zeros(tend/h,2); % make space to store t,v (two columns)
data(1,:) = [t v];

% start simulation loop
while (t < tend)

% Calculate acceleration a (=dv/dt) using equation of motion
  a = gsin(slope) - (k/m)v^2;

% do one forward Euler integration step to calculate new velocity and
% store data
  v = v + ah;
  t = t + h;
  i = i + 1;
  data(i,:) = [t v];

end

% plot the results
plot(data(:,1),data(:,2));
xlabel('Time (s)');ylabel('Speed (m/s)');
```

Listing 1: **Matlab program for simulation of a cyclist coasting downhill. The program can be downloaded from http://www.bme.ccf.org/isb/tgcs/software/bogert.**

```
% arm.m
% Program to simulate a two-segment arm model in the horizontal plane

% body segment parameters
m1 = 3; m2 = 20;
I1 = 0.01; I2 = 0.01;
d1 = 0.3; d2 = 0.3;
cm1 = 0.15; cm2 = 0.15;
A = I1 + m1cm1^2 + m2d1^2; D = I2 + m2cm2^2;

% set simulation parameters and initial conditions
t = 0;                  % start at time=0
tend = 2;               % duration of simulation in seconds
h = 0.002;              % integration step size in seconds
i = 1;                  % step counter
ph1 = 1.0; ph2 = 1.0;% initial segment orientations (rad)
ph1d = 0.0; ph2d = 10.0; %initial segment angular velocitied (rad/s)
data = zeros(tend/h,3); % make space to store three columns of data

% start simulation
while (t < tend)        % continue until t reaches tend
% equations of motion
  B = m2d1cm2cos(ph1-ph2);
  C = m2d1cm2cos(ph1-ph2);
  E = -m2d1cm2sin(ph1-ph2);
  F = -m2d1cm2sin(ph2-ph1);
  ph1dd = (DEph2d^2 - BFph1d^2)/(AD - BC);
  ph2dd = (AFph1d^2 - CEph2d^2)/(AD - BC);
% do one integration step for all four state variables and store result
  ph1d = ph1d + ph1ddh;
  ph2d = ph2d + ph2ddh;
  ph1 = ph1 + ph1dh;
  ph2 = ph2 + ph2dh;
  t = t + h; i = i+1;% increment time and step counter
  data(i,:) = [t 180ph1/pi 180ph2/pi];
end
% plot the results as angles vs. time
plot(data(:,1),data(:,2),data(:,1),data(:,3),'--');
xlabel('Time [s]');ylabel('Angle [deg]');
```

Listing 2: **Matlab program for simulation of arm movement in the horizontal plane. The program can be downloaded from http://www.bme.ccf.org/isb/tgcs/software/bogert.**

```
% muscle.m
% Program to simulate muscle contractions using 3-component Hill
% model

% muscle properties for human Vastus group
Lslack = 0.223;         % slack length of SEE
Umax = 0.04;            % strain in SEE is 4% at Fmax
Lceopt = 0.093;         % optimal length of CE
width  = 0.63Lceopt;% maximum length change of CE
Fmax = 7400;            % maximal isometric force
a = 0.25Fmax;           % force-velocity parameter a
b = 0.2510Lceopt;       % f-v parameter b (Nigg and Herzog, p. 174-175)

% set initial condition for state variable Lce and initialize ODE solver
Lce = 0.087;            % this makes SEE just slack at t=0
t = 0; tend = 2.99;     % start time and duration of experiment
h = 0.001;              % integration step size in seconds
i = 1;                  % step counter
data = zeros(tend/h,2);% space to store time and force results

% start simulation
while (t < tend)
% prescribed ramp shortening profile
  if (t<=1) Lm = 0.31; end           % initial muscle+tendon length
  if (t>1 & t<2) Lm = 0.31-0.04(t-1); end
% ramp shortening at 4 cm/s
% Calculate force in SEE from current SEE length
  Lsee = Lm - Lce;      % length of SEE is total length minus CE length
  if (Lsee < Lslack)
    F = 0;              % SEE is slack, so no force
  else
    F = Fmax((Lsee-Lslack)/(UmaxLslack))^2;
% SEE force-length rel'ship
  end
% calculate isometric force at this Lce from CE force-length relationship
  F0 = max([0 Fmax(1-((Lce-Lceopt)/width)^2)]);
% calculate CE velocity from Hill's equation
  if (F > F0) disp('Error: program cannot do eccentric contractions');
          return; end
  Lcedot = -b(F0-F)/(F+a);
% note: velocity is negative for shortening!
% do one forward Euler integration step
  Lce = Lce + hLcedot;
  t = t + h; i = i + 1;
  data(i,:) = [t F];
end
plot(data(:,1),data(:,2));
xlabel('Time [s]');ylabel('Force [N]');
```

Listing 3: Matlab program for simulation of a maximally activated muscle undergoing a ramp shortening protocol. The program can be downloaded from: http://www.bme.ccf.org/isb/tgcs/software/bogert.

```
% knee.m
% program to simulate ballistic knee extensions using a
% 3-component Hill model

% muscle properties for human Vastus group
Lslack = 0.223;Umax = 0.04;   % slack length of SEE, SEE strain at Fmax
Lceopt = 0.093;               % optimal length of CE
width  = 0.63Lceopt;          % maximum length change of CE
Fmax = 7400;                  % maximal isometric force
a = 0.25Fmax;b = 0.2510Lceopt;% force-velocity parameters

% parameters for the equation of motion (Nigg and Herzog, p. 562)
m = 10; g = 9.81;             % segment mass, acceleration of gravity
Rcm = 0.264;                  % distance knee joint to centre of mass
I = 0.1832;                   % moment of inertia
Rf = 0.033;                   % moment arm of quadriceps

% initial conditions
phi = pi/2; phid = 0.0;       % start at 90 deg flexion and zero velocity
Lce = 0.31 - Lslack;          % make sure that SEE is just slack at t=0
t = 0; tend = 3.0;            % duration of experiment is three seconds
h = 0.0005; i=0;              % integration step size and step counter
data = zeros(tend/h,7);       % space to store results

% start simulation
while (t < tend & phi < pi)
% Calculate force in SEE from current SEE length
  Lm = 0.31 - (phi-pi/2)Rf;% total length from joint angle
  Lsee = Lm - Lce;            % length of SEE is total length minus CE
  if (Lsee < Lslack) F = 0;% SEE is slack, so no force
   else F = Fmax((Lsee-Lslack)/(UmaxLslack))^2;% SEE force-lengt
  end
% compute isometric force at this Lce, using force-length curve
  F0 = max([0 Fmax(1-((Lce-Lceopt)/width)^2)]);
% calculate CE velocity from Hill's equation
  if (F > F0) disp('Error: cant do eccentric contractions'); break; end
  Vce = -b(F0-F)/(F+a);
% note: velocity is negative for shortening!
% apply this muscle force, and gravity, to equation of motion
  M = RfF - mgRcmsin(phi-pi/2);% total moment at knee joint
  phidd = M/I;                % angular acceleration
% do one forward Euler integration step
  Lce = Lce + Vceh;
  phid = phid + phiddh;
  phi  = phi  + phidh;
  t = t + h; i = i+1;
  data(i,:) = [t Lce F Lm Vce phi phid];
end
subplot(221); plot(data(1:i,1),data(1:i,3));
xlabel('Time (s)');ylabel('Force (N)');
subplot(223); plot(data(1:i,1),data(1:i,6)180/pi);
xlabel('Time (s)');ylabel('Joint angle (deg)');
```

Listing 4: Matlab program for simulation of a maximal effort ballistic knee extension. The program can be downloaded from: http://www.bme.ccf.org/isb/tgcs/software/bogert.

```
function [negdistance] = javelin(p)
% Simulates javelin flight with release conditions p
% p(1) = initial angle relative to horizontal (rad)
% p(2) = initial angular velocity (rad/s)
% p(3) = initial angle of attack (rad)
% Constant model parameters:
m = 0.8;                           % mass of the javelin in kg
I = 0.4;                           % moment of inertia
v0 = 30;                           % release velocity in m/s
g = 9.81;                          % gravity
h = 0.01;                          % step size of ODE solver
% The degrees of freedom are (x,y,a): position and orientation in 2D
x = 0; y = 1.5;                    % position at release is 1.5 m above ground
a = p(1);                          % angle at release
xd = v0cos(a - p(3));              % horizontal velocity at release
yd = v0sin(a - p(3));              % vertical velocity at release
ad = p(2);                         % angular velocity at release
% Do the simulation
while (y > 0)                      % stop when it hits the ground...
  v = sqrt(xd^2 + yd^2);           % magnitude of velocity
  av = atan2(yd,xd);               % orientation of velocity vector
  aa = a - av;                     % angle of attack
  aa = rem(aa+pi,2pi)-pi;          % should be within range  pi to + pi
% Simplification of Fig. 2 in Hubbard & Rust, J. Biomech. 17:769-776
  lift = 0.15sin(2aa)v^2;          % lift force (perpendicular to velocity)
  drag = 0.5(sin(aa))^2v^2;        % drag force (parallel to velocity vector)
  pmom = -0.01aav^2;               % pitching moment
% Convert lift and drag forces to global reference frame, add gravity
  Fx = -dragcos(av) - liftsin(av);
  Fy = -dragsin(av) + liftcos(av) - mg;
% Perform forward Euler step
  xd = xd + hFx/m; x = x + hxd;
  yd = yd + hFy/m; y = y + hyd;
  ad = ad + hpmom/I; a = a + had;
end
negdistance = -x;
fprintf('Parameters: %8.3f %8.3f %8.3f --> Distance: %8.3f m\n',p,x);
```

Listing 5: Matlab function for simulation of javelin flight. The program can be downloaded from: http://www.bme.ccf.org/isb/tgcs/software/bogert.

4.10 REFERENCES

Alexander, R.M. (1989) Sequential Joint Extension in Jumping. *Human Movement Science.* **8,** pp. 339-345.

Alexander, R.M. (1992) Simple Models of Walking and Jumping. *Human Movement Science.* **11,** pp. 3-9.

Alexander, R.M. and Vernon, A. (1975) The Mechanics of Hopping by Kangaroos (Macropodidae). *J. Zool.* **177,** pp. 265-303.

Alexander, R.M., Bennett, M.B., and Ker, R.F. (1986) Mechanical Properties and Functions of the Paw Pads of Some Mammals. *J. Zool.* **209,** pp. 405-419.

Anderson, F.C. and Pandy, M.G. (2001) Dynamic Optimization of Human Walking. *J. Biomech. Eng.* **123,** pp. 381-390.

Andrews, J.G. (1974) Biomechanical Analysis of Human Motion. *Kinesiology IV.* Amer. Assoc. for Health, Phys. Ed., & Rec., Washington, D.C. pp. 32-42.

Andrews, J.G. (1985) A General Method for Determining the Functional Role of a Muscle. *J. Biomech. Eng.* **107,** pp. 348-353.

Anton, M.G. and Nigg, B.M. (1990) An Optimal Control Model for Running. *Proc. 6th Biennial Conf. of the Can. Soc. for Biomechanics.* pp. 61-62.

Arnold, V.I. (1978) *Mathematical Methods of Classical Mechanics.* Springer Verlag, New York. Berlin, Heidelberg.

Asai, T., Kaga, M., and Akatsuka T. (1997) Computer Simulation of the V Style Technique in Ski Jumping using CFD. Proc. 6th Int. Symp. Computer Simulation in Biomechanics. Tokyo, Japan. pp. 48-49.

Audu, M.L. and Davy, D.T. (1985) The Influence of Muscle Model Complexity in Musculo-skeletal Motion Modelling. *J. Biomech. Eng.* **107,** pp. 147-157.

Ballreich, R. and Brüggemann, G. (1986) Biomechanik des Weitsprungs. *Biomechanik der Leichtathletik* (eds. Ballreich, R. and Kuhlow, A.). pp. 28-47.

Blankevoort, L. and Huiskes, R. (1996) Validation of a Three-dimensional Model of the Knee. *J. Biomech.* **29 (7),** pp. 955-961.

Blickhan, R. (1989) The Spring-mass Model for Running and Hopping. *J. Biomech.* **22 (11-12),** pp. 1217-1227.

Bowman, F. (1953) *Introduction to Elliptic Functions with Applications.* Dover, New York.

Brand, R.A., Pedersen, D.R., Davy, D.T., Kotzar, G.M., Heiple, K.G., and Goldberg, V.M. (1994) Comparison of Hip Force Calculations and Measurements in the Same Patient.*J. Arthroplasty.* **9**, pp. 45-51.

Cavanagh, P.R. and Lafortune, M.A. (1980) Ground Reaction Forces in Distance Running. *J. Biomech.* **13,** pp. 397-406.

Clauser, C.E., McConville, J.T., and Young, J.W. (1969) Weight, Volume, and Centre of Mass Segments of the Human Body. *AMRL Technical Report (TR-69-70).* Wright Patterson Air Force Base, Ohio. pp. 69-70.

Cole, G.K., van den Bogert, A.J., Herzog, W., and Gerritsen, K.G.M. (1996) Modelling of Force Production in Skeletal Muscle Undergoing Stretch. *J. Biomech.* **29,** pp. 1091-1104.

Crowninshield, R.D. and Brand, R.A. (1981a) The Prediction of Forces in Joint Structures: Distribution of Intersegmental Resultants. *Exerc. Sport Sci. Reviews* (ed. Miller, D.I.). **9,** pp. 159-181.

Crowninshield, R.D. and Brand, R.A. (1981b) A Physiologically Based Criterion of Muscle Force Prediction in Locomotion. *J. Biomech.* **14 (11),** pp. 793-801.

de Groot, G. (1995) Air Friction and Rolling Resistance During Cycling. *Med. Sci. Sports Exerc.* **27,** pp. 1090-1095.

Delp, S.L. (1990) Surgery Simulation: A Computer Graphics System to Analyze and Design Musculo-skeletal Reconstructions of the Lower Limb. Ph.D. Dissertation, Stanford University, Stanford, CA.

Delp, S.L. and Loan, J.P. (1995) A Graphics-based Software System to Develop and Analyze Models of Musculo-skeletal Structures. *Comput. Biol. Med.* **25,** pp. 21-34.

Dempster, W.T. (1958) Analysis of Two-handed Pulls Using Free Body Diagrams. *J. Appl. Physiol.* **13 (3),** pp. 469-480.

Dempster, W.T. (1961) Free Body Diagrams as an Approach to the Mechanics of Human Posture and Motion. *Biomechanical Studies of the Musculo-skeletal System* (ed. Evans, F.G.). Thomas, Springfield, IL. pp. 81-135.

Denoth, J. (1980) Ein Mechanisches Modell zur Beschreibung von passiven Belastungen. *Sportplatzbeläge* (eds. Nigg, B.M. and Denoth, J.). pp. 45-53.

Denoth, J. (1980) Materialeigenschaften. *Sportplatzbeläge* (eds. Nigg, B.M. and Denoth, J.). pp. 54-67.
Denoth, J. (1985) The Dynamic Behaviour of a Three Link Model of the Human Body During Impact with the Ground. *Biomechanics IX-A* (eds. Winter, D., Norman, R.W., Wells. R.P., Hayes, K.C., and Patta, A.E.). Human Kinetics, Champaign, IL. pp. 102-106.
Denoth, J. (1986) Load on the Locomotor System and Modelling. *Biomechanics of Running Shoes* (ed. Nigg, B.M.). Human Kinetics, Champaign, IL. pp. 63-116.
Denoth, J., Gruber, K., Ruder, H., and Keppler, M. (1984) Forces and Torques During Sports Activities with High Accelerations. *Biomechanics Current Interdisciplinary Research* (eds. Perren, S.M. and Schneider, E.). Martinuis Nijhoff Pub., Dordrecht, Netherlands. pp. 663-668.
Dul, J., Johnson, G.E., Shiavi, R., and Townsend, M.A. (1984) Muscular Synergism - II: A Minimum-fatigue Criterion for Load Sharing Between Synergistic Muscles. *J. Biomech.* **17 (9),** pp. 675-684.
Eisenberg, D. and Crothers, D. (1979) *Physical Chemistry with Applications to the Life Sciences*. The Benjamin/Cummings Publishing Co., Menlo Park.
Elftman, H. (1938) Forces and Energy Changes in the Leg During Walking. *Am. J. Physiol.* **125 (2),** pp. 339-356.
Fischbein, E. (1987) *Intuition in Sciences and Mathematics*. D. Reidel Publishing Company, Dordrecht, Netherlands.
Flash, T. and Hogan, N. (1985) The Coordination of Arm Movements: An Experimentally Confirmed Mathematical Model. *J. Neurosci.* **5 (7),** pp. 1688-1703.
Gasser, H.S. and Hill, A.V. (1924) The Dynamics of Muscular Contraction. *Proc. Roy. Soc.* **96 (B),** pp. 398-437.
Gear, C.W. (1971) Numerical Initial Value Problems in Ordinary Differential Equations. Prentice-Hall, Englewood Cliffs, NJ.
Gerritsen, K.G.M. and van den Bogert, A.J. (1993) Direct Dynamics Simulation of the Impact Phase in Heel-toe Running. *Proc. 4th International Symposium on Computer Simulation in Biomechanics.* BML1-6 - BML1-13.
Gerritsen, K.G.M., van den Bogert, A.J., and Nigg, B.M. (1995) Direct Dynamics Simulation of the Impact Phase in Heel-toe Running. *J. Biomech.* **28,** pp. 661-668.
Gerritsen, K.G.M., Nachbauer, W., and van den Bogert, A.J. (1996) Computer Simulation of Landing Movement in Downhill Skiing: Anterior Cruciate Ligament Injuries. *J. Biomech.* **29,** pp. 845-854.
Gerritsen, K.G.M., van den Bogert, A.J., and Herzog, W. (In press) Force-length Properties of Lower Extremity Muscles Derived from Maximum Isometric Strength Tests. *Eur. J. Appl. Physiol.*
Gordon, A.M., Huxley, A.F., and Julian, F.J. (1966) The Variation in Isometric Tension with Sarcomere Length in Vertebrate Muscle Fibres. *J. Physiol.* **184,** pp. 170-192.
Grieve, D.W., Pheasant, S., and Cavanagh, P.R. (1978) Prediction of Gastrocnemius Length from Knee and Ankle Joint Posture. *Biomechanics VI-A* (eds. Asmussen, E. and Jorgensen, K.). University Park Press, Baltimore. pp. 405-412.
Gruber, K. (1987) Entwicklung eines Modells zur Berechnung der Kräfte im Knie- und Hüftgelenk bei sportlichen Bewegungsabläufen mit hohen Beschleunigungen. Ph.D. Thesis, Universitat Tübingen, Germany.
Gruber, K., Denoth, J., Stüssi, E., and Ruder, H. (1987) The Wobbling Mass Model. *Biomechanics X-B* (ed. Jonsson, B.). pp. 1095-1099.
Gurlanik, D.B. (ed.) (1979) *Webster's New World Dictionary.* William Collins Publishers, Ohio.
Hardt, D.E. (1978) A Minimum Energy Solution for Muscle Force Control During Walking. Ph.D. Thesis, Dept. of Mech. Eng., Massachusetts Institute of Technology.
Hatze, H. (1976) The Complete Optimization of a Human Motion. *Math. Biosci.* **28**, pp. 99-135.
Hatze, H. (1978) Sportbiomechanische Modelle und myokybernetische Bewegungsoptimierung: Gegenwartsprobleme und Zukunftsaussichten. *Sportwissenschaft.* **4,** pp. 1-13.
Hatze, H. (1981) A Comprehensive Model for Human Motion Simulation and its Application to the Take-off Phase of the Long Jump. *J. Biomech.* **14 (3)**, pp. 135-142.
Hatze, H. and Venter, A. (1981) Practical Activation and Retention of Locomotion Constraints in Neuromusculo-skeletal Control System Models. *J. Biomech.* **14,** pp. 873-877.
Haug, E.J. (1989) *Computer-aided Kinematics and Dynamics of Mechanical Systems. Basic Methods*. Allyn & Bacon, Boston.
He, J., Levine, W.S., and Loeb, G.E. (1991) Feedback Gains for Correcting Small Perturbations to Standing Posture. *IEEE Trans. Automatic Control.* **36,** pp. 322-332.
Herzog, W. (1987) Individual Muscle Force Estimations Using a Non-linear Optimal Design. *J. Neurosci. Methods.* **21,** pp. 167-179.

Herzog, W. and Leonard, T.R. (1991) Validation of Optimization Models that Estimate the Forces Exerted by Synergistic Muscles. *J. Biomech.* **24 (S-1)**, pp. 31-39.

Herzog, W. and Binding, P. (1992) Predictions of Antagonistic Muscular Activity Using Nonlinear Optimization. *Math. Biosci.* **111**, pp. 217-229.

Herzog, W. and Binding, P. (1993) Co-contraction of Pairs of Antagonistic Muscles: Analytical Solution for Planar Static Non-linear Optimization Approaches. *Math. Biosci.* **118**, pp. 83-95.

Hill, A.V. (1938) The Heat of Shortening and the Dynamic Constants of Muscle. *Proc. Roy. Soc.* **126 (B),** pp. 136-195.

Hörler, E. (1972a) Mechanisches Modell zur Beschreibung des isometrischen und des dynamischen Muskelkraftverlaufs. *Jugend und Sport, Nov.* pp. 271-273.

Hörler, E. (1972b) Hochsprungmodell. *Jugend und Sport, Aug.* pp. 381-383.

Hörler, E. (1973) Clumsiness and Stature: A Study of Similarity. *Biomechanics III* (eds. Cerquiglini, S., Venerando, A., and Wartenweiler, J.). University Park Press, Baltimore. **8,** pp. 146-150.

Hubbard, M. (1980) Dynamics of the Pole Vault. *J. Biomech.* **13,** pp. 965-976.

Hubbard, M. and Rust, H.J. (1984) Simulation of Javelin Flight using Experimental Aerodynamic Data. *J Biomech.* **17,** pp. 769-776.

Huiskes, R. and Chao, E.Y.S. (1983) A Survey of Finite Element Analysis in Orthopedic Biomechanics: The First Decade. *J. Biomech.* **16 (6),** pp. 385-409.

Jones, J.B. and Hawkins, G.A. (1986) *Engineering Thermodynamics* (2ed). John Wiley & Sons, New York.

Kane, T.R. and Levinson, D.A. (1985) Dynamics: Theory and Applications. *McGraw-Hill Series in Mechanical Engineering.* McGraw-Hill Book Company, New York.

Kemeni, J.G. (1959) *A Philosopher Looks at Science.* D. van Nostrand Pub. Comp., Princeton, NJ.

Kestin, J. (1979) *A Course in Thermodynamics.* Hemisphere Publishing Co., Washington.

Komi, P.V., Salonen, M., Jarvinen, N., and Kokko, O. (1987) In-vivo Registration of Achilles Tendon Forces in Man. *Int. J. Sports Medicine.* **8,** pp. 3-8.

LaFortune, M.A. (1991) Three-dimensional Acceleration of the Tibia During Walking and Running. *J. Biomech.* **24,** pp. 877-886.

Lemm, R. (1978) *Passive Kräfte bei sportlichen Bewegungen.* Unpublished Diplomarbeit, ETH Zürich.

McLean, S.G., Su, A., and van den Bogert, A.J. (2003) Development and Validation of a 3-D Model to Predict Knee Joint Loading during Dynamic Movement. *J. Biomech. Eng.* **125**, pp. 864-874.

McLean, S.G., Huang, X., Su, A., and Van Den Bogert, A.J. (2004) Sagittal Plane Biomechanics Cannot Injure the ACL during Sidestep Cutting. *Clin. Biomech.* **19**, pp. 828-838.

McMahon, T.A. and Greene, P.R. (1979) The Influence of Track Compliance on Running. *J. Biomech.* **12 (12),** pp. 893-904.

Miller, D.I. (1970) *A Computer Simulation Model of the Airborne Phase of Diving.* Ph.D. Dissertation, Pennsylvania State University, University Park, PA.

Misevich, K.W. and Cavanagh, P.R. (1984) Material Aspects of Modelling Shoe/foot Interaction. *Sport Shoes and Playing Surfaces* (ed. Frederick, E.C.). Human Kinetics, Champaign, IL. pp. 47-75.

Morlock, M. and Nigg, B.M. (1991) Theoretical Considerations and Practical Results on the Influence of the Representation of the Foot for the Estimation of Internal Forces with Models. *Clin. Biomech.* **6,** pp. 3-13.

Morrison, J.B. (1968) Bioengineering Analysis of Force Actions Transmitted by the Knee Joint. *Bio-Medical Eng.* **3,** pp. 164-170.

Müller, W., Platzer, D., and Schmolzer, B. (1996) Dynamics of Human Flight on Skis: Improvements in Safety and Fairness in Ski Jumping. *J. Biomech.* **29,** pp. 1061-1068.

Neptune, R.R. and van den Bogert, A.J. (1998) Evaluation of Mechanical Energy Analyses: Application to Steady-state Cycling. *J. Biomech.* **31**, pp. 239-245.

Neptune, R.R. and Hull, M.L. (1998) Evaluation of Performance Criteria for Simulation of Submaximal Steady-state Cycling using a Forward Dynamic Model. *J. Biomech. Eng.* **120 (3),** pp. 334-341.

Neptune, R.R. (1999) Optimization Algorithm Performance in Determining Optimal Controls in Human Movement Analyses. *J. Biomech. Eng.* **121**, pp. 249-252.

Neptune, R.R., Wright, I.C., and van den Bogert, A.J. (2000) The Influence of Orthotic Devices and Vastus Medialis Strength and Timing on Patellofemoral Loads during Running. *Clin. Biomech.* **15**, 611-618.

Neptune, R.R., Kautz, S.A., and Zajac, F.E. (2001) Contributions of the Individual Ankle Plantar Flexors to Support, Forward Progression and Swing Initiation during Walking. *J. Biomech.* **34**, pp. 1387-1398.

Nigg, B.M. (1974) Analysis of Twisting and Turning Movements. *Biomechanics IV* (eds. Nelson, R.C. and Morehouse, C.A.). University Park Press, Baltimore. pp. 279-283.
Nigg, B.M. (1986) Biomechanical Aspects of Running. *Biomechanics of Running Shoes* (ed. Nigg, B.M.). Human Kinetics Pub. Inc., Champaign, IL. pp. 1-26.
Nigg, B.M. and Denoth, J. (1980) *Sportplatzbeläge*. Juris Verlag, Zürich.
Nigg, B.M. and Lüthi, S.M. (1980). Bewegungsanalysen beim Laufschuh. *Sportwissenschaft*. **3,** pp. 309-320.
Nigg, B.M. and Anton, M. (1995) Energy Aspects for Elastic and Viscoelastic Shoe Soles and Surfaces. *Med. Sc. Sports and Exercise*. **27 (1)**, pp. 92-97.
Nikravesh, P.E. (1984) Some Methods for Dynamic Analysis of Constrained Mechanical Systems: A Survey. *Computer-aided Analysis and Optimization of Mechanical System Dynamics (NATO ASI-F9)* (ed. Haug, E.J.). Springer Verlag, Berlin.
Pandy, M.G., Zajac, F.E., Sim, E., and Levine, W.S. (1990) An Optimal Control Model for Maximum-height Human Jumping. *J. Biomech*. **23 (12),** pp. 1185-1198.
Pandy, M.G. and Zajac, F.E. (1991) Optimal Muscular Coordination Strategies for Jumping. *J. Biomech*. **24 (1),** pp. 1-10.
Pandy, M.G., Anderson, F.C., and Hull, D.G. (1992) A Parameter Optimization Approach for the Optimal Control of Large-scale Musculo-skeletal Systems. *J. Biomech. Eng.* **114 (4),** pp. 450-460.
Passerello, C.E. and Huston, R.L. (1971) Human Attitude Control. *J. Biomech*. **4 (2)**, pp. 95-102.
Paul, J.P. (1965) Bioengineering Studies of the Forces Transmitted by Joints - II. *Engineering Analysis, Biomechanics, and Related Bioengineering Topics* (ed. Kennedy, R.M.). Pergamon Press, Oxford. pp. 369-380.
Pedotti, A., Krishnan, V.V., and Starke, L. (1978) Optimization of Muscle-force Sequencing in Human Locomotion. *Math. Biosci*. **38,** pp. 57-76.
Pennycuick, C.J. (1993) Mechanical Limits to Evolution and Diversity. *Proc. 14th Biomechanics Congress*. pp. 20-21.
Pierrynowski, M.R. and Morrison, J.B. (1985) Estimating the Muscle Forces Generated in the Human Lower Extremity when Walking: A Physiological Solution. *Math. Biosci.* **75,** pp. 43-68.
Press, W.H., Teukolsky, S.A., Vetterling, W.T., and Flannery, B.P. (1992) *Numerical Recipes in FORTRAN* (2ed). Cambridge University Press, Cambridge, UK.
Riddle, C. and Kane, T.R. (1968) Reorientation of the Human Body by Means of Arm Motions. *Technical Report 182, 62 (NASA CR-95362)*. Division of Engineering Mechanics, Stanford University, Palo Alto, CA.
Riener, R. and Edrich, T. (1999) Identification of Passive Elastic Joint Moments in the Lower Extremities. *J. Biomech.* **32**, 539-544.
Roth, R. (1989) On Constructing Free Body Diagrams. *Int. J. Appl. Eng.* **5 (5)**, pp. 565-570.
Seireg, A. and Arvikar, R.J. (1973) A Mathematical Model for the Evaluation of Forces in Lower Extremities of the Musculo-skeletal System. *J. Biomech.* **6 (13),** pp. 313-326.
Shampine, L.F. and Gordon, M.K. (1975) *Computer Solution of Ordinary Differential Equations: The Initial Value Problem*. W.H. Freeman, San Francisco, CA.
Spoor, C.W., van Leeuwen, J.L., Meskers, C.G.M., Titulaer, A.F., and Huson, A. (1990) Estimation of Instantaneous Moment Arms of Lower-leg Muscles. *J. Biomech*. **23 (12),** pp. 1247-1259.
Stryer, L. (1988) *Biochemistry* (3ed). W.H. Freeman and Co., New York.
Treloar, L.R.G. (1975) *The Physics of Rubber Elasticity*. Clarendon Press, Oxford.
Truesdell, C.A. (1969) *Rational Thermodynamics*. McGraw-Hill Book Co., New York.
van den Bogert, A.J. and Schamhardt, H.C. (1993) Multi-body Modelling and Simulation of Animal Locomotion. *Acta. Anatomica.* **146,** pp. 95-102.
van den Bogert, A.J., Schamhardt, H.C., and Crowe, A. (1989) Simulation of Quadrupedal Locomotion Using a Dynamic Rigid Body Model. *J. Biomech.* **22,** pp. 33-41.
van den Bogert, A.J. (1994) Optimization of the Human Engine: Application to Sprint Cycling. *Proc. 8th Congress of the Canadian Society for Biomechanics*. Calgary, Alberta. pp. 160-161.
van den Bogert, A.J., Gerritsen, K.G.M., and Cole, G.K. (1998) Human Muscle Modelling From a User's Perspective. *J. Electromyogr. Kinesiol.* **8,** pp. 119-124.
van Gheluwe, B. (1974) A New Three-dimensional Filming Technique Involving Simplified Alignment and Measurement Procedures. *Biomechanics IV* (eds. Nelson, R.C. and Morehouse, C.A.). University Park Press, Baltimore. pp. 476-482.
van Ingen Schenau, G.J. and Bobbert, M.F. (1993) The Global Design of the Hindlimb in Quadrupeds. *Acta. Anat.* **146,** pp. 103-108.

van Soest, A.J., Schwab, A.L., Bobbert, M.F., and van Ingen Schenau, G.J. (1992) SPACAR: A Software Subroutine Package for Simulation of the Behaviour of Biomechanical Systems. *J. Biomech.* **25,** pp. 1241-1246.

van Soest, A.J., Schwab, A.L., Bobbert, M.F., and van Ingen Schenau, G.J. (1993) The Influence of the Biarticularity of the Gastrocnemius Muscle on Vertical-jumping Achievement. *J. Biomech.* **26 (1),** pp. 1-8.

Visser, J.J., Hoogkamer, J.E., Bobbert, M.F., and Huijing, P.A. (1990) Length and Moment Arm of Human Leg Muscles as a Function of Knee and Hip-joint Angles. *Eur. J. Appl. Physiol.* **61 (5-6),** pp. 453-460.

Walker, S.M. and Schrodt, G.R. (1973) I-segment Lengths and Thin Filament Periods in Skeletal Muscle Fibres of the Rhesus Monkey and the Human. *Anat. Rec.* **178,** pp. 63-82.

Weber, W. and Weber, E. (1836) *Mechanik der menschlichen Gehwerkzeuge.* W. Fischer Verlag, Goettingen.

Whittaker, E.T. (1937) *A Treatise on the Analytical Dynamics of Particles and Rigid Bodies.* Cambridge University Press, Cambridge.

Whittaker, E.T. and Watson, G.N. (1962) *A Course of Modern Analysis* (5ed). Cambridge University Press, Cambridge.

Winter, D.A. (1979) *Biomechanics of Human Movement.* Appendix A. John Wiley & Sons, New York.

Winter, D.A. (1983) Moments of Force and Mechanical Power in Jogging. *J. Biomech.* **16 (1),** pp. 91-97.

Winters, J.M. (1990) Hill-based Muscle Models: A Systems Engineering Perspective. *Multiple Muscle Systems: Biomechanics and Movement Organization* (eds. Winters, J.M and Woo, S.L.-Y.). Springer Verlag, New York.

Winters, J.M. and Stark, L. (1987) Muscle Models: What is Gained and What is Lost by Varying Model Complexity. *Biol. Cyber.* **55,** pp. 403-420.

Woledge, R.C., Curtin, N.A., and Homsher, E. (1985) *Energetic Aspects of Muscle Contraction.* Academic Press, London.

Wright, I.C., Neptune, R.R., van den Bogert, A.J., and Nigg, B.M. (In press) Validation of a Three-dimensional Model for the Simulation of Ankle Sprains. *J. Biomech. Eng.*

Wright, I.C., Neptune, R.R., van den Bogert, A.J., and Nigg, B.M. (1998) Passive Regulation of Impact Forces in Heel-toe running. *Clin. Biomech.* **13**, 521-531.

Wright, I.C., Neptune, R.R., van den Bogert, A.J., and Nigg, B.M. (2000) The Influence of Foot Positioning on Ankle Sprains. *J. Biomech.* **33**, pp. 513-519.

Yeadon, M.R. (1984) *The Mechanics of Twisting Somersaults.* Unpublished Ph.D. Thesis, Loughborough University of Technology, UK.

Yeadon, M.R. (1993a) The Biomechanics of Twisting Somersaults, Part 1: Rigid Body Motions. *J. Sports Sci.* **11 (3),** pp. 187-198.

Yeadon, M.R. (1993b) The Biomechanics of Twisting Somersaults, Part 2: Contact Twist. *J. Sports Sci.* **11 (3),** pp. 199-208.

Yeadon, M.R. (1993c) The Biomechanics of Twisting Somersaults, Part 3: Aerial Twist. *J. Sports Sci.* **11 (3),** pp. 209-218.

Yeadon, M.R. (1993d) The Biomechanics of Twisting Somersaults, Parts 4: Partitioning Performances Using the Tilt Angle. *J. Sports Sci.* **11 (3),** pp. 219-225.

Yeadon, M.R. and Morlock, M. (1989) The Appropriate use of Regression Equations for the Estimation of Segmental Inertia Parameters. *J. Biomech.* **22 (6-7),** pp. 683-689.

Yeadon, M.R., Atha, J., and Hales, F.D. (1990) The Simulation of Aerial Movement - IV: A Computer Simulation Model. *J. Biomech.* **23 (1),** pp. 85-89.

Yoon, Y.S. and Mansour, J.M. (1982) The Passive Elastic Moment at the Hip. *J. Biomech.* **15,** pp. 905-910.

Zahalak, G.I. (1986) A Comparison of the Mechanical Behaviour of the Cat Soleus Muscle with a Distribution-moment Model. *J. Biomech. Eng.* **108,** pp. 131-140.

Zajac, F.E. (1989) Muscle and Tendon: Properties, Models, Scaling, and Application to Biomechanics and Motor Control. *Crit. Rev. Biomed. Eng.* **17,** pp. 359-411.

Zajac, F.E. and Gordon, M.E. (1989) Determining Muscle Force and Action in Multi-articular Movement. *Exercise and Sport Science Reviews* (ed. Pandolf, K.). Williams & Wilkins, Baltimore, MD. **17,** pp. 187-230.

INDEX